Mass (10^{23} kg)	Density (g/cm^3)	Escape Velocity (km/s)	Surface	Atmosphere
3.3	5.4	4	silicates	trace Na
48.7	5.2	10	basalt, granite?	90 bar: 97% CO_2
59.7	5.5	11	basalt, granite, water	1 bar: 78% N_2, 21% O_2
0.7	3.3	2	basalt, anorthosites	none
6.4	3.9	5	basalt, clays, ice	0.07 bar: 95% CO_2
18988	1.3	60	none	H_2, He, CH_4, NH_3, etc.
1.1	1.8	2	dirty ice	none
1.5	1.9	3	dirty ice	none
0.5	3.0	2	ice	none
0.9	3.6	3	sulfur, SO_2	trace SO_2
5685	0.7	36	none	H_2, He, CH_4, NH_3, etc.
1.4	1.9	3	ice? hydrocarbons?	1.5 bar: N_2, trace CH_4
866	1.5	21	none	H_2, He, CH_4, NH_3, etc.
1028	1.8	23	none	H_2, He, CH_4, NH_3, etc.
0.2	2.1	1	ices of N_2, CH_4, CO, H_2O, CO_2	trace N_2, CH_4, CO
0.2	2.1	1	ices of N_2, CH_4, CO	trace N_2, CH_4, CO

The Planetary System

In this image of a small part of the Eagle Nebula, columns of cool, dense interstellar gas and dust remain as hot, massive stars form behind them. Within the pillars, Sun-like stars with planetary systems could eventually appear.

SECOND EDITION

The Planetary System

David Morrison
NASA

Tobias Owen
University of Hawaii at Manoa

Addison-Wesley Publishing Company

Reading, Massachusetts ▪ Menlo Park, California ▪ New York
Don Mills, Ontario ▪ Harlow, United Kingdom ▪ Amsterdam ▪ Bonn
Sydney ▪ Singapore ▪ Tokyo ▪ Madrid ▪ San Juan ▪ Milan ▪ Paris

The front cover shows a picture of Jupiter just after the impact of one of the twenty-odd fragments of Comet Shoemaker-Levy 9 on July 19, 1994. This picture was taken with the 2.3-m telescope at the Australian National Observatory at a wavelength of 2.3 micrometers. The methane gas in Jupiter's atmosphere is a strong absorber of sunlight at this wavelength, so normally only the high-altitude hazes in the planet's polar regions are visible in such pictures, and it is usually possible to trace out the dim outline of the rest of the planet. The bright flash at the lower left is caused by the pulse of heat liberated during the impact. To the right of this flash at the same latitude, clouds of dust particles from three earlier impacts are also visible. Observations at radio wavelengths revealed that carbon monoxide, hydrogen cyanide, and several other gases were produced by these explosions as well.

Sponsoring Editor Jennifer Albanese
Production Supervisor Juliet Silveri
Copyeditor Margo Shearman
Proofreader Joyce Grandy
Art Coordinator/Dummier Jim Roberts
Text Designer Sandra Rigney
Illustrator J.A.K. Graphics
Marketing Supervisor Kate Derrick
Manufacturing Manager Hugh Crawford
Cover Design Supervisor Meredith Nightingale

ISBN 0-201-55450-X

Access the latest information about Addison-Wesley books from our World Wide Web page:
`http://www.aw.com`

1 2 3 4 5 6 7 8 9 10-DOC-99989796

Preface

The final third of the twentieth century will be remembered as the golden age of planetary exploration, the period in which we launched the first spacecraft missions to the planets. Within a few short decades, we have come to know our immediate neighbors in space as individual worlds, permitting us to develop a comparative perspective that would have been impossible without the space program. What has emerged from this exciting period is a new scientific discipline, usually termed *planetary science*. This book, *The Planetary System*, is an introduction to this discipline.

In 1988 we published the first edition of *The Planetary System* as a text for a one-semester introductory college course in planetary science. Such courses are taught in many universities, offered by departments of astronomy, physics, and geology. As the only comprehensive introductory text that covers the entire field, giving roughly equal weight to both solid surfaces and atmospheres, the book found a ready market. The first edition is still being used in the mid-1990s, even though new discoveries have made much of it obsolete. In this fast-moving field, a revised edition is needed that incorporates the many spectacular advances of the past decade.

The general subject matter and target audience are unchanged in the second edition. As before, we intend this book to be accessible to the non-science undergraduate with no prior training in either astronomy or geology. At the same time, we realize that in many universities the first course in planetary science is not offered until the upper-level undergraduate or even first-year graduate level, and we have tried to provide material that is suitable for these courses as well. For this more advanced audience, most chapters include optional quantitative supplements that use simple algebra, together with problems that can be solved with a hand-calculator.

As an interdisciplinary subject, planetary science defies easy description. Traditionally, the Earth and planets have been studied within the fields of both astronomy and the Earth sciences (geology, geophysics, meteorology, etc.). Only in the past two decades have components of these standard disciplines come together to generate the new perspectives of comparative planetology—the study of the Earth as a planet and of the other planets as worlds. Such a synthesis would not have been possible without the spectacular successes of the planetary spacecraft launched by the United States and the former Soviet Union, spurred in part by competition between the capitalist and socialist systems. With the collapse of the USSR, the urgency of competition has evaporated, and the pace of spacecraft exploration has slowed accordingly. The European Space Agency (ESA) and Japan have joined the United States and Russia as players, compensating in part for these cutbacks. Also, as the pace of space missions has slackened, we have seen a resurgence of discoveries about the solar system, using more traditional astronomical approaches and carried out with both ground-based and orbiting telescopes. Planetary science is and will continue to be an important beneficiary of the worldwide advances in astronomical capabilities of the 1990s.

The subject of this book is the solar system. There is some difference of opinion among teachers as to whether it is appropriate to include the Earth and Sun, for purposes of an introductory course on the planets, as members of the solar system. We have always felt that the study of the Earth is essential for a proper perspective on the other planets, but we omitted the Sun from the first edition. Now, at the urging of a number of colleagues, we have added a chapter on the Sun. However, we note that the treatment of this nearest star, as well as of our own planet Earth, is less detailed than that found in most astronomy and geology texts, respectively. As indicated by the title, the focus of this book is on the entire solar system of planets, satellites, comets, and asteroids, of which the Sun and Earth are only two members.

The order of presentation of topics is designed to emphasize the processes that formed and continue to influence the members of the planetary system. The first three chapters provide an introduction to the solar system itself, a brief discussion of the Sun, and an explanation of some of the basic physical and chemical concepts used later. Students who have already had a course in astronomy or the Earth sciences may wish to go directly to Chapter 4, which begins the study of the simplest, most primitive solid objects in the system: comets, asteroids, and meteorites. We like to introduce the meteorites at the beginning of our discussion, since these samples of extraterrestrial material provide an essential "ground truth" for remote studies. Chapter 7 moves us one step up the scale of complexity to examine the small cratered worlds, Mercury and the Moon, as an introduction to planetary geology. Chapters 9–12 deal with the Earth-Venus-Mars triad of terrestrial planets, with their active geological evolution and complex atmospheres. The outer solar system is the subject for the last third of the book, Chapters 13–18—which cover the giant planets, their large and small satellites, and their ring systems. Finally, in Chapter 19 we review the structure of the entire system in terms of its origin and evolution.

Throughout the book we emphasize comparative studies and the common processes acting on the members of the system, but we do so in the context of descriptions of individual objects. Generalizations are important, but we feel they need to be derived from understanding of the specific planets and satellites that we have explored.

A great deal has happened in planetary science since the completion of the first edition. Voyager 2 concluded its historic grand tour with close flybys of Neptune and Triton in 1989, and the spectacularly successful Magellan radar mapper transmitted more information from Venus in the early 1990s than the total of all planetary data from all previous spacecraft combined—thus our treatment of both Neptune and Venus is almost entirely new. The Galileo spacecraft, finally launched in 1989, achieved the first encounters with main belt asteroids Gaspra and Ida en route to its probe of the atmosphere of Jupiter in December 1995. Also contributing to spacecraft studies in this period were the Clementine lunar orbiter and the Phobos Mars mission. Tremendous strides in astronomical instrumentation led to improved determinations of the surface compositions of Io, Triton, Pluto, and Charon, provided radar images of near-Earth asteroids Castalia, Toutatis and Geographos, discovered polar caps on Mercury, and brought us a whole new class of distant objects on the fringes of the Kuiper Belt. Also: SNC meteorites were identified as martian samples, Chicxulub Crater was located on the Earth to cement the connection between impacts and extinctions, Comet Shoemaker-Levy 9 was watched by the entire world as it plunged into the atmosphere of Jupiter in the summer of 1994, and the first extra-solar planets were announced at the end of 1995. We have enjoyed including the most important new discoveries and interpretations in this book. We have not reported novel results for their news value alone, however, but rather have made every effort to blend the new results with what was already known in order to present a more complete and balanced picture of the planetary system and its evolution. We have made changes in every section of every chapter in an effort to provide comprehensive, current perspectives.

For the second edition, the book has grown by about 70 pages (15%) with the addition of four new chapters, including one on the Sun, and a doubling of the number of short quantitative supplements for more advanced students. We are also happy to include a major expansion of the color coverage, allowing us to take advantage of the excellent color imaging provided by the Viking and Voyager missions.

Any scientific discipline has a lot of specialized jargon. Planetary science, which draws from several disciplines, may be worse than most. Some of this jargon is useful and even necessary, but much of it is superfluous in an introductory course. We feel that technical vocabulary should be used as a tool, not learned as an end in itself. For this second edition we have added a few of the standard terms that our colleagues wanted to see that were previously omitted, such as albedo, chemical equilibrium, coma, and fractionation. There are also new terms for newly important concepts such as impact erosion of atmospheres, impact frustration of the origin of life, subnebula, the Kuiper Belt, and coronae on Venus. However, we continue to practice "jargon control," introducing only those specialized terms (about 225 in all) that we feel are really necessary for understanding the concepts in this book. These terms are printed in boldface where first explained in the text and are defined again in the Glossary.

Another problem for authors concerns the practice of making reference to the work of individuals. We decided that a book at this level should not include specific references to literature, although expanded lists of additional reading (both technical treatises and trade books aimed at a general audience) are presented at the end of each chapter. We should not neglect the work of individuals entirely, however, since it would give a false impression of the nature of science to imply that it is an impersonal endeavor. Of necessity, however, we have been able to mention by name only a few of the scientists who have made important contributions to this field, and we apologize to the many others for what may seem to be a rather arbitrary selection.

This book would not have been possible without the encouragement and assistance of our colleagues who have shared their new discoveries and interpretations, explained to us what sort of book they need for their undergraduate classes, and provided constructive criticisms of the first edition. We are especially grateful for the advice and assistance we have received from Jeffery Brown, Joe Cain, Dale Cruikshank, Jeff Cuzzi, Frank Drake, Andrew Fraknoi, Woody Harrington, Jim Houck, David Hughes, Bill Kaula, Jack Lissauer, Frank Maloney, Bill McKinnon, Deane Peterson, Lawrence Pinsky, Tricia Reeves, Seth Shostak, and Dale Smith. In addition, Janet Morrison provided welcome editorial advice. We are also grateful for the consistent support we have received from our publisher, and especially from editors Stuart Johnson, Julie Berrisford, Jennifer Albanese, and Juliet Silveri. We thank them all, and we hope that you, the reader, will find the product of these efforts to be both useful and enjoyable.

David Morrison *Saratoga, California*
Tobias Owen *Honolulu, Hawaii*

About the Authors

David Morrison is head of space science research at the NASA Ames Research Center. He participated in the Mariner mission to Mercury, the Voyager mission to the outer planets, and the Galileo mission to Jupiter, in addition to astronomical studies of the planetary system. He received the NASA Outstanding Leadership Medal for his contributions to protection of the Earth from comet and asteroid impacts, and the Klumpke-Roberts award of the Astronomical Society of the Pacific for education and public service in astronomy. In addition to this book, Morrison is the author of seven other texts and popular books on space science.

Tobias Owen is a professor at the Institute for Astronomy of the University of Hawaii. He has participated in NASA's Viking mission to Mars and the Voyager mission to the outer planets, receiving the NASA medal for exceptional scientific achievement for his work on the martian atmosphere. He is presently an Interdisciplinary Scientist and a member of the Probe mass spectrometer teams on the Galileo and Cassini-Huygens missions, and a member of mass spectrometer teams on missions to Mars and comet Wirtanen. In Hawaii, he pursues a program of spectroscopic studies of the members of our planetary system, using the great telescopes on Mauna Kea. This work has included the discovery and evaluation of deuterium on Mars, hydrogen cyanide on Neptune, and ices of nitrogen and carbon monoxide on Pluto.

Contents

Note: (C) means chapter is in color.

1 Finding Our Place in Space **1**

1.1 WATCHING THE SKY: SUN, MOON, AND STARS 2
1.2 WATCHING THE SKY: THE PLANETS 8
1.3 THE SLOW GROWTH OF REASON 11
1.4 NEWTON AND THE LAW OF GRAVITATION 19
1.5 ESCAPING FROM EARTH 22
1.6 THE SYSTEM REVEALED 24
1.7 QUANTITATIVE SUPPLEMENT: *Kepler's Laws in Mathematical Form* 27
Summary 28
Key Terms 29
Review Questions 29
Quantitative Exercises 29
Additional Reading 29

2 The Sun: An Ordinary Star **31**

2.1 THE SUN AS A STAR 32
2.2 BUILDING BLOCKS: ATOMS AND ISOTOPES 34
2.3 COMPOSITION OF THE SUN 36
2.4 THE SUN'S ENERGY 40
2.5 LIFE HISTORY OF THE SUN 43
2.6 SOLAR ACTIVITY 46
2.7 QUANTITATIVE SUPPLEMENT: *Radiation Laws* 50
Summary 51
Key Terms 52
Review Questions 52
Quantitative Exercises 53
Additional Reading 53

3 Getting to Know Our Neighbors **55**

3.1 BASIC PROPERTIES: MASS, SIZE, AND DENSITY 56
3.2 CHEMISTRY OF THE PLANETS 59
3.3 ORIGIN AND CLASSIFICATION OF ROCKS 61
3.4 PLANETARY ATMOSPHERES 64
3.5 STUDYING MATTER FROM A DISTANCE 65
3.6 EXPLORING THE PLANETARY SYSTEM 72
Summary 76
Key Terms 77
Review Questions 77
Quantitative Exercises 78
Additional Reading 78

4 Meteorites: Remnants of Creation **79**

4.1 THE SOLAR NEBULA 80
4.2 CLASSIFICATION OF METEORITES 84
4.3 AGES OF METEORITES AND OTHER ROCKS 91
4.4 PRIMITIVE METEORITES 93
4.5 DIFFERENTIATED METEORITES 96
4.6 METEORITE PARENT BODIES 98
4.7 QUANTITATIVE SUPPLEMENT: *Radioactive Age Dating* 99
Summary 100
Key Terms 101
Review Questions 101
Quantitative Exercises 101
Additional Reading 102

5 Asteroids: Building Blocks of the Inner Planets **103**

5.1 DISCOVERY OF THE ASTEROIDS 104
5.2 MAIN BELT ASTEROIDS 111
5.3 VESTA: A VOLCANIC ASTEROID 115
5.4 ASTEROIDS FAR AND NEAR 117
5.5 ASTEROIDS CLOSE UP 121
5.6 QUANTITATIVE SUPPLEMENT: *Asteroid Collisions* 128

Summary 130
Key Terms 130
Review Questions 130
Quantitative Exercises 131
Additional Reading 131

6 Comets: Messengers from the Cold **133**

6.1 COMETS THROUGH HISTORY 134
6.2 THE COMET'S ATMOSPHERE 139
6.3 THE COMETARY NUCLEUS 143
6.4 COMET DUST 149
6.5 ORIGIN AND EVOLUTION OF COMETS 153
6.6 QUANTITATIVE SUPPLEMENT: *Albedos and Temperatures* 156
Summary 157
Key Terms 158
Review Questions 158
Quantitative Exercises 158
Additional Reading 159

7 The Moon: Our Ancient Neighbor **161**

7.1 THE FACE OF THE MOON 162
7.2 EXPEDITIONS TO THE MOON 167
7.3 IMPACT CRATERING 173
7.4 DATING CRATERED WORLDS 177
7.5 LUNAR CATASTROPHISM 181
7.6 LUNAR VOLCANISM 186
7.7 THE SURFACE OF THE MOON 188
7.8 QUANTITATIVE SUPPLEMENT: *Impact Energies* 191
Summary 192
Key Terms 193
Review Questions 193
Quantitative Exercises 193
Additional Reading 194

8 Moon and Mercury: Strange Relatives **195**

8.1 AN ELUSIVE PLANET 196
8.2 THE ROTATION OF MERCURY 198

8.3 TIDES AND THE SPIN OF PLANETS 202
8.4 THE FACE OF MERCURY 205
8.5 INTERIORS OF THE MOON AND MERCURY 208
8.6 HISTORIES OF THE MOON AND MERCURY 213
8.7 ORIGIN OF THE MOON AND MERCURY 215
8.8 QUANTITATIVE SUPPLEMENT: *Radar and the Doppler Effect* 218
8.9 QUANTITATIVE SUPPLEMENT: *Synodic Period* 218
Summary 219
Key Terms 219
Review Questions 220
Quantitative Exercises 220
Additional Reading 220

9 The Earth: Our Home Planet 221

9.1 EARTH AS A PLANET 222
9.2 JOURNEY TO THE CENTER OF THE EARTH 224
9.3 THE CHANGING FACE OF THE EARTH 227
9.4 PLATE TECTONICS: A UNIFYING HYPOTHESIS 230
9.5 OCEAN AND ATMOSPHERE 235
9.6 UPPER ATMOSPHERE AND MAGNETOSPHERE 240
9.7 CLIMATE AND WEATHER 243
9.8 IMPACTS AND EVOLUTION 248
Summary 253
Key Terms 254
Review Questions 254
Quantitative Exercises 254
Additional Reading 255

10 Venus: Earth's Exotic Twin (C) 257

10.1 UNVEILING THE GODDESS OF BEAUTY 258
10.2 THE ATMOSPHERE AND THE GREENHOUSE EFFECT 261
10.3 WEATHER ON VENUS 267
10.4 THE HIDDEN LANDSCAPE 270
10.5 CRATERS AND TECTONICS 272
10.6 VOLCANOES ON VENUS 277

10.7 ON THE SEARING SURFACE 280
10.8 QUANTITATIVE SUPPLEMENT: *The Greenhouse Effect* 284
Summary 285
Key Terms 286
Review Questions 286
Quantitative Exercises 287
Additional Reading 287

11 Mars: The Planet Most Like Earth (C) 289

11.1 A CENTURY OF CHANGING PERCEPTIONS 290
11.2 GLOBAL PERSPECTIVE 295
11.3 VIEW FROM THE SURFACE 301
11.4 CRATERS AND CHRONOLOGY 306
11.5 VOLCANOES AND TECTONIC FEATURES 310
11.6 MYSTERIES OF THE MARTIAN CHANNELS 316
11.7 THE POLAR REGIONS 320
11.8 THE MARTIAN ATMOSPHERE 324
Summary 327
Key Terms 328
Review Questions 328
Quantitative Exercises 329
Additional Reading 329

12 Life, Planets, and Atmospheres (C) 331

12.1 LIFE ON EARTH 332
12.2 LIFE AND THE EVOLUTION OF THE ATMOSPHERE 334
12.3 EXTRAPOLATING FROM EARTH TO MARS 338
12.4 THE SEARCH FOR LIFE ON MARS 340
12.5 ATMOSPHERES OF EARTH AND MARS 346
12.6 GOLDILOCKS AND THE THREE PLANETS 350
12.7 QUANTITATIVE SUPPLEMENT: *Atmospheric Escape* 352
Summary 353
Key Terms 354
Review Questions 354
Quantitative Exercises 355
Additional Reading 355

13 Jupiter and Saturn: The Biggest Giants (C) **357**

13.1 THE PIONEER AND VOYAGER MISSIONS 358
13.2 INTERNAL STRUCTURE: JOURNEYS TO THE CENTERS OF GIANT PLANETS 361
13.3 ATMOSPHERIC COMPOSITION AND STRUCTURE 364
13.4 WEATHER AND CLIMATE 373
13.5 CLOUDS, COLORS, AND CHEMISTRY 379
13.6 MAGNETOSPHERES AND RADIO BROADCASTS 382
13.7 QUANTITATIVE SUPPLEMENT: *Building a Giant Planet* 389
13.8 THE GALILEO JUPITER PROBE: PRELIMINARY RESULTS 390
Summary 392
Key Terms 392
Review Questions 392
Quantitative Exercises 393
Additional Reading 393

14 In Deep Freeze: Planets We Cannot See (C) **395**

14.1 DISCOVERIES OF THE OUTER PLANETS 396
14.2 THREE DISTANT WORLDS 400
14.3 ATMOSPHERES OF URANUS AND NEPTUNE 403
14.4 CLIMATE, CLOUDS, AND WEATHER 408
14.5 MAGNETOSPHERES 411
14.6 QUANTITATIVE SUPPLEMENT: *Discovering a Planet* 413
Summary 414
Key Terms 414
Review Questions 414
Quantitative Exercises 415
Additional Reading 415

15 Worlds of Fire and Ice: The Large Satellites of Jupiter (C) **417**

15.1 SATELLITE SYSTEMS 418
15.2 IMPACT CRATERS AND SURFACE AGES 424
15.3 CALLISTO AND GANYMEDE: THE LARGE GALILEANS 427
15.4 EUROPA AND IO: THE ACTIVE GALILEANS 433

15.5 VOLCANOES OF IO 437
15.6 COMPARING THE LARGE SATELLITES 441
Summary 443
Key Terms 444
Review Questions 444
Quantitative Exercises 444
Additional Reading 444

16 Titan, Triton, and Pluto: Icy Objects with Atmospheres (C) 445

16.1 TITAN: A RIDDLE WRAPPED IN AN ENIGMA 446
16.2 TRITON: APPROACHING ABSOLUTE ZERO 454
16.3 PLUTO AND ITS MOON: A MOST UNUSUAL SYSTEM 460
16.4 QUANTITATIVE SUPPLEMENT: *Vapor Atmospheres* 464
Summary 465
Key Terms 466
Review Questions 466
Quantitative Exercises 466
Additional Reading 466

17 Small Satellites (C) 467

17.1 THE SATELLITE SYSTEM OF SATURN 468
17.2 ENCELADUS AND IAPETUS: TWO PUZZLING SATELLITES 472
17.3 THE SATELLITE SYSTEM OF URANUS 476
17.4 THE SMALLEST SATELLITES 479
17.5 QUANTITATIVE SUPPLEMENT: *Searching for New Satellites* 485
Summary 486
Key Terms 486
Review Questions 486
Quantitative Exercises 487
Additional Reading 487

18 Planetary Rings 489

18.1 RINGS OF JUPITER 490
18.2 RINGS OF SATURN 492

18.3 RINGS OF URANUS 498
18.4 RINGS OF NEPTUNE 502
18.5 RING DYNAMICS AND SATELLITE-RING INTERACTIONS 503
18.6 ORIGIN OF RING SYSTEMS 507
18.7 QUANTITATIVE SUPPLEMENT: *The Tidal Stability Limit* 509
Summary 510
Key Terms 511
Review Questions 511
Quantitative Exercises 511
Additional Reading 512

19 The Origin of Planets 513

19.1 BASIC PROPERTIES OF THE PLANETARY SYSTEM 514
19.2 THE LIFE OF A STAR 516
19.3 THE PROBLEM OF ANGULAR MOMENTUM 519
19.4 EVOLUTION OF THE DISK: CONDENSATION, AGGREGATION, ACCRETION, AND DISSIPATION 520
19.5 SMALL BODIES AND THE IMPORTANCE OF IMPACTS 525
19.6 GIANT PLANETS AND THEIR SATELLITE AND RING SYSTEMS 528
19.7 THE SEARCH FOR OTHER PLANETARY SYSTEMS 530
19.8 QUANTITATIVE SUPPLEMENT: *Angular Momentum in the Solar Nebula* 538
Summary 538
Key Terms 540
Review Questions 540
Quantitative Exercises 540
Additional Reading 541

Appendix
Units and Exponential Notation 543

Glossary 545

Figure Credits 557

Index 561

1

Finding Our Place in Space

The Great Galaxy in Andromeda. Our familiar Sun is just one of the hundreds of billions of stars that, together with immense clouds of gas and dust, make up the Milky Way Galaxy. The Milky Way itself is simply our insider's view of the Galaxy we inhabit, as we look around ourselves in the plane in which most of the stars are located. Beyond the boundaries of our own Galaxy there are billions of other galaxies, such as this one, which is very similar to ours.

Standing under the stars on a clear, dark night, you are looking out from the surface of our rocky planet into the vast depths of the universe—the grandest assemblage of matter and space that we know. "This majestical roof, fretted with golden fire," has been made familiar to us by the efforts of our ancestors, who gave names to the stars and the patterns they appear to form. Throughout history, humans have struggled to try to understand the place our planet occupies in this starry realm.

The planetary system that is the subject of this book consists of the nine known planets, the 61 satellites discovered so far, and the innumerable asteroids, comets, and meteoroids that orbit the Sun. We first want to know the shape and size of this system and its internal motions. What moves around what, and why? These puzzles, which have fascinated humans since they first recognized the existence of the planets, are the subject of Chapter 1.

1.1 WATCHING THE SKY: SUN, MOON, AND STARS

Most of us live in cities these days. We can see little in the sky, especially at night when city lights obscure all but the brightest stars. We have thus lost contact with an important window on the universe, one we can look through with our own eyes, unaided by any optical or electronic devices. Even so, some celestial events play an important part in our lives, setting the cycle of the seasons and giving us various measures of time. Occasionally, our attention is focused by the news media on some rare event: a solar eclipse or the appearance of a comet. Otherwise, we are left with random views of the Sun and Moon, paying little enough attention to these.

For many of us, even the time-keeping aspect is rather remote, since we have calendars, watches, radios, television sets, newspapers, and magazines to remind us of daily, monthly, and yearly time scales and events. But to ancient civilizations without such assistance, the sky was a very important part of the natural environment. It was a source of pleasure as well as fear, and its changing aspects gave these peoples the ability to predict seasonal changes that were vital to their livelihood. In trying to understand the reasons for what they were seeing, they initiated the subject of astronomy. We are still struggling with some of the problems they could not solve, such as the origins of the Earth, Moon, Sun, and stars, and the question of whether or not there are other worlds like ours.

Many of the phenomena these early observers found mysterious, however, we can understand. Indeed we take these solutions so much for granted it is useful to precede a review of the modern perspective by returning to earlier problems in order to see how they were solved. In the same spirit, we encourage readers to go out and make some of the simple, basic observations themselves—watching seasonal changes in the altitude of the Sun at noon and the point on the horizon where it sets, following the Moon's apparent motion through the sky, learning the constellations, and identifying the planets. All this can be done with no optical aid whatsoever, and these simple activities help establish a personal connection with the planetary system we inhabit.

The Constellations

To become acquainted with the night, you need to find a location that is far from any city, with a really dark sky. This is the kind of environment that was always at hand to the ancients. Such a night sky is filled with stars of different brightness and colors. With its usual quest for order, the human mind found certain patterns among these apparently random points of light. These patterns are called **constellations** and were given names associated with myths, legends, and historical events in various cultures. Astronomers still use some of these names today. Probably most of you at one time in your lives have identified the Big Dipper, part of the Roman "Great Bear" or the Old English "Hay Wain," and Orion, the brilliant constellation of cold winter nights (Fig. 1.1). (All these descrip-

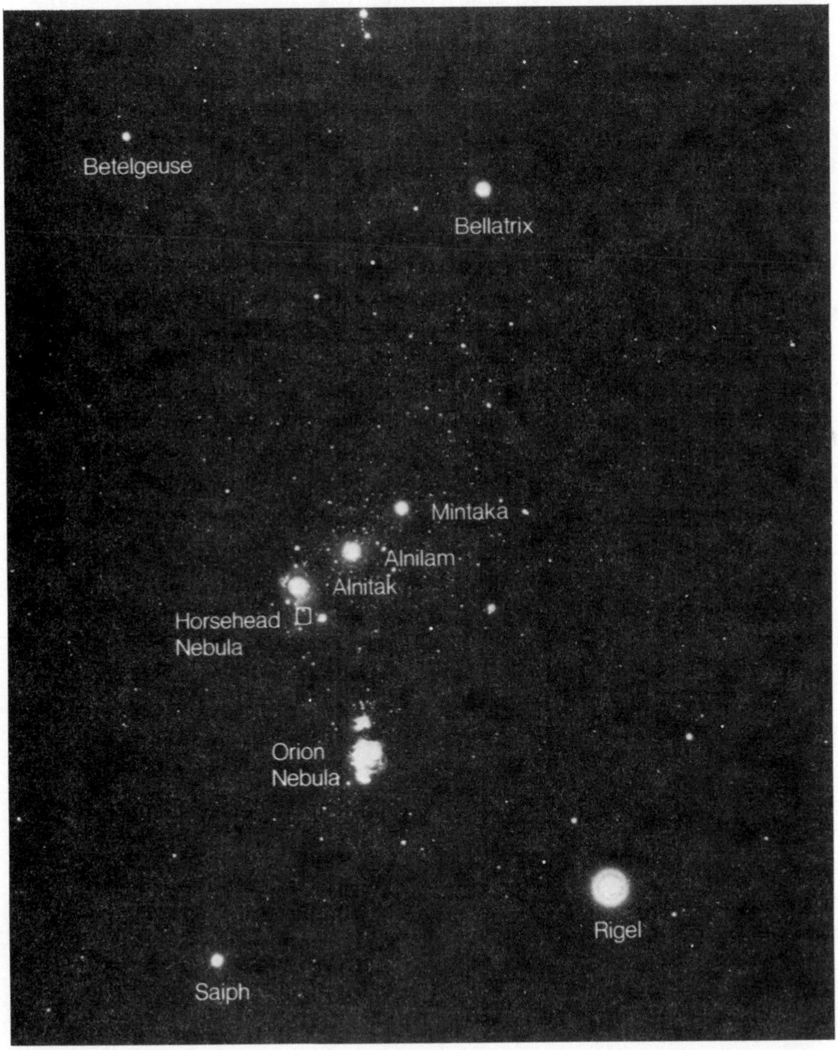

FIGURE 1.1 The constellation Orion, showing the brightest stars and the location of a giant gas cloud, the Orion Nebula, where new stars are forming at the present time.

tions are given for observers living at latitudes from 25° to 50° north of the equator. The sky looks different to people in the Southern Hemisphere.)

In summer, Orion is replaced by other constellations: Lyra and Cygnus nearly overhead, and Scorpius and Sagittarius toward the south. Each time these seasons return, we can find these same constellations in the same places in our skies. While this grand seasonal cycle is taking place, the positions of the stars within each constellation remain fixed. The Big Dipper looked the same to Charlemagne in the year 800 as it does to you. One of the early mysteries the ancients tried to solve was the reason for this daily and annual repetition in the sky. The obvious solution was to suggest that the Earth was in the center of a set of giant spheres that slowly revolved about it, giving rise to day and night and the seasonal change in the visible constellations.

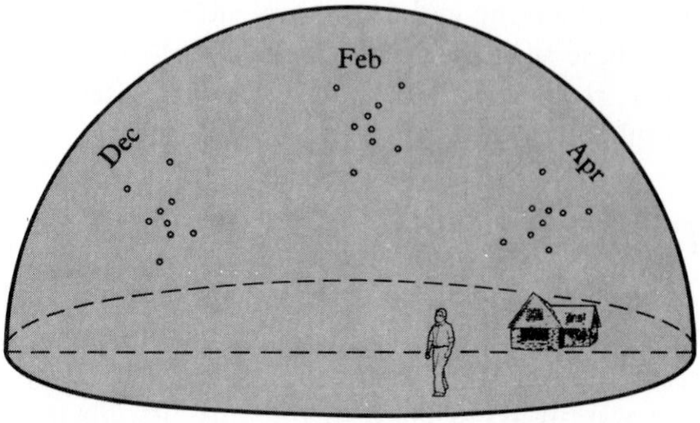

FIGURE 1.2 Appearance of Orion in the sky at 9:00 P.M. on the first of December, February, and April. The seasonal change in the appearance of the night sky is caused by the revolution of the Earth about the Sun.

Apparent Motion of the Sun and Stars

Like the Sun, the stars appear to rise in the East and set in the West. This *diurnal,* or daily, motion can be noticed in an hour or less, and by the end of the night it has brought many new constellations into view to replace the stars present at dusk. A slower *seasonal* cycle is also apparent, however.

If you go out each evening at the same time for several weeks, you will notice that the eastern constellations are farther up in the sky and the western ones farther down than when you first began looking. This seasonal shift ultimately brings back the same constellations to the same place at the same time each year (Fig. 1.2). It is the manifestation in the night sky of the seasonal motion of the Sun through the background of stars.

The Sun's daily motion through the sky is also subject to seasonal changes. In winter the days are shorter than in summer. The Sun is lower in the sky at noon and casts long shadows. Thus a careful observer can detect a seasonal motion of the Sun by plotting its height above the horizon at noon. This motion comes to a stop and reverses at the summer **solstice** (Sun-stationary) when the Sun is highest in the sky, and again six months later at the winter solstice, when the Sun is at its lowest point (Fig. 1.3). Halfway in between we have the spring (or vernal) and autumn **equinoxes,** when days and nights are equal in length.

In repeating this regular cycle, the Sun appears to move through a certain set of constellations. The apparent path of the Sun through the constellations in the course of its seasonal round is an imaginary line in the sky known as the **ecliptic.** It may seem strange that the constellations through which the Sun moves can be determined, since it is impossible to see the stars during the day, but it is simply a matter of watching which constellations are near the Sun in the twilight skies just after it sets and before it rises. This is something you can do yourself, if you have access to clear horizons. When you know the configuration of constellations over the entire sky, you can tell where the Sun is by figuring out the position halfway between the stars seen in morning and evening

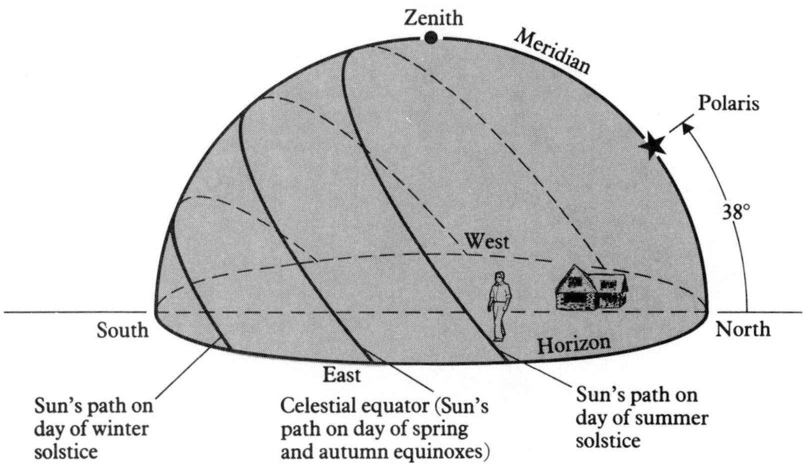

FIGURE 1.3 A person living at north temperate latitudes of the United States will see the Sun rise directly in the East at the spring or fall equinoxes and set directly in the West. At the summer solstice, the Sun rises north of east and sets north of west, reaching its highest noontime elevation in the sky. At the winter solstice, when days are shortest, the Sun rises south of east and sets south of west, reaching its lowest noontime altitude.

twilight. Alternatively, you could note the constellation that is opposite to the Sun, rising when the Sun sets. Either way, you can trace out the annual path of the Sun.

It is easier to see which constellations form the backdrop for the motions of the Moon. It turns out that the Moon moves through the same constellations as the Sun. The Moon nearly follows the ecliptic, only deviating from it by a few degrees, sometimes appearing above it, sometimes below. The set of constellations through which the Sun and Moon appear to move is called the **zodiac,** from the same Greek word that gives us *zoo* and *zoology*. That etymology tells you right away that most of these constellations are named for various real and imaginary animals (Fig. 1.4).

The Sun requires one year or 365.26 days to make its slow, eastward journey through the constellations of the zodiac, returning to its starting point. It is this journey, in fact, that defines our year. The cycle of the Moon through

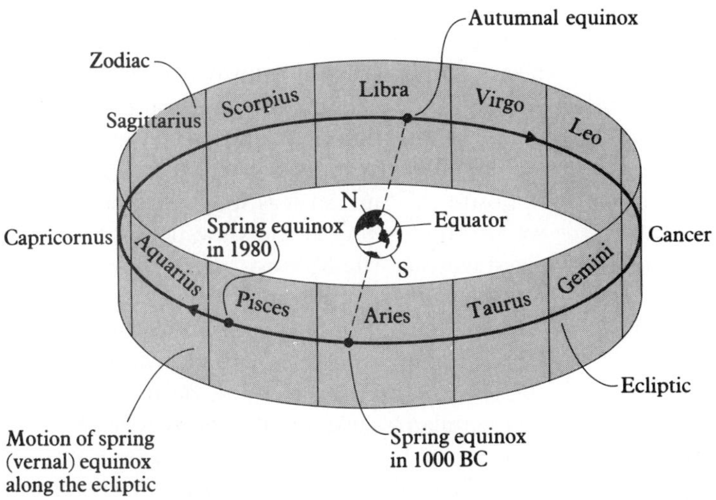

FIGURE 1.4 The planets, the Sun, and the Moon all appear to move through a narrow band of 12 constellations called the zodiac.

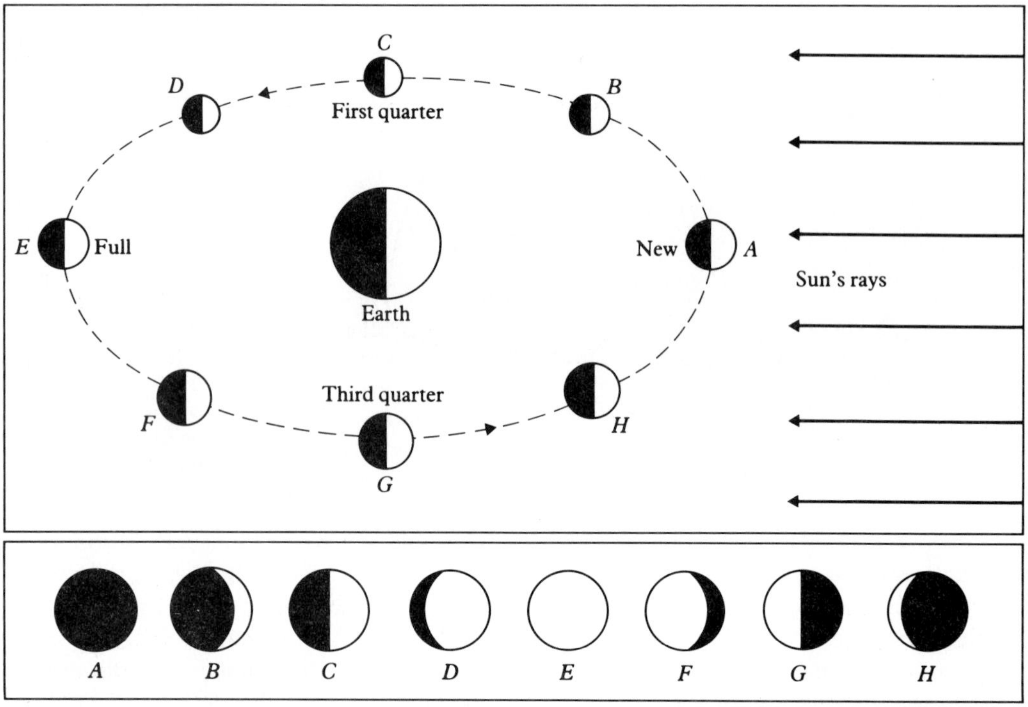

FIGURE 1.5 Phases of the Moon as seen from space (top), and the corresponding appearance of the Moon seen from Earth (bottom). (Figure not drawn to scale.)

the zodiac is much faster, giving us the period of time we call the month.

Phases of the Moon

During the course of its journey through the zodiac, the Moon goes through a complete set of phases, from new to full and back to new (Fig. 1.5). New moon occurs when the Moon is approximately between us and the Sun. In this configuration, the Moon rises and sets with the Sun and is above the horizon only in the daytime sky. The side of the Moon that faces us receives no sunlight and we cannot see it. The night after new moon, we see a thin crescent in the western sky in twilight. The next night the Moon has moved farther east through the stars and the crescent has grown. When the crescent is far enough from the Sun to be seen against a dark sky, we can sometimes see the rest of the Moon faintly illuminated by sunlight reflected from the Earth, a phenomenon called "the old Moon in the new Moon's arms."

The first quarter moon is half-full and rises at noon. At nightfall it is halfway between the eastern and western horizons, and at midnight it sets. The full moon rises at sunset and sets at sunrise. The Moon has now moved to a position opposite the Sun, so it is fully illuminated. But we are still looking at the same hemisphere that faced us when the Moon was new and we could not see it; only the illumination has changed.

Since the full moon is opposite the Sun, in winter full moons occur with the Moon high in the sky, the position the Sun occupies in summer. Similarly, summer full moons are low, in the winter position of the Sun. This opposition of Sun and Moon creates a striking winter effect in places where it snows, since the snow that resists melting because it lies in shadow during the day is brightly illuminated by the full moon at night.

Moving still farther east through the constellations, the waning Moon again reaches half-full phase, rising now at midnight with its curved edge facing the eastern horizon instead of the western one. This is called last quarter moon. And finally we have a thin crescent in the morning sky, rising just before the Sun.

This complete cycle happens every 29.5 days, giving us the basic period of our month. Nature was not kind enough to make the year exactly divisible by the month, but 12 is clearly the closest multiple. Thus we have acquired our present patchwork of 12 stretched and diminished months in one year. And we have 12 constellations in the zodiac, each one corresponding approximately to one of these months (Fig. 1.4). By knowing the location of the Sun with respect to these constellations, the ancients had a good monthly calendar available. Within the months, the phase of the Moon provided a cycle of days.

It should be evident from this discussion that it is easy to account for the apparent motions of both the Moon and the Sun by thinking of them as revolving around the Earth, just like the great sphere holding the stars. This was the popular ancient belief, later codified by Ptolemy of Alexandria and enforced by the medieval Catholic Church. But we know today that the Sun is at the center of our system, and the Earth travels around it in one year, turning daily on its axis, while the Moon makes its monthly circuit of our planet.

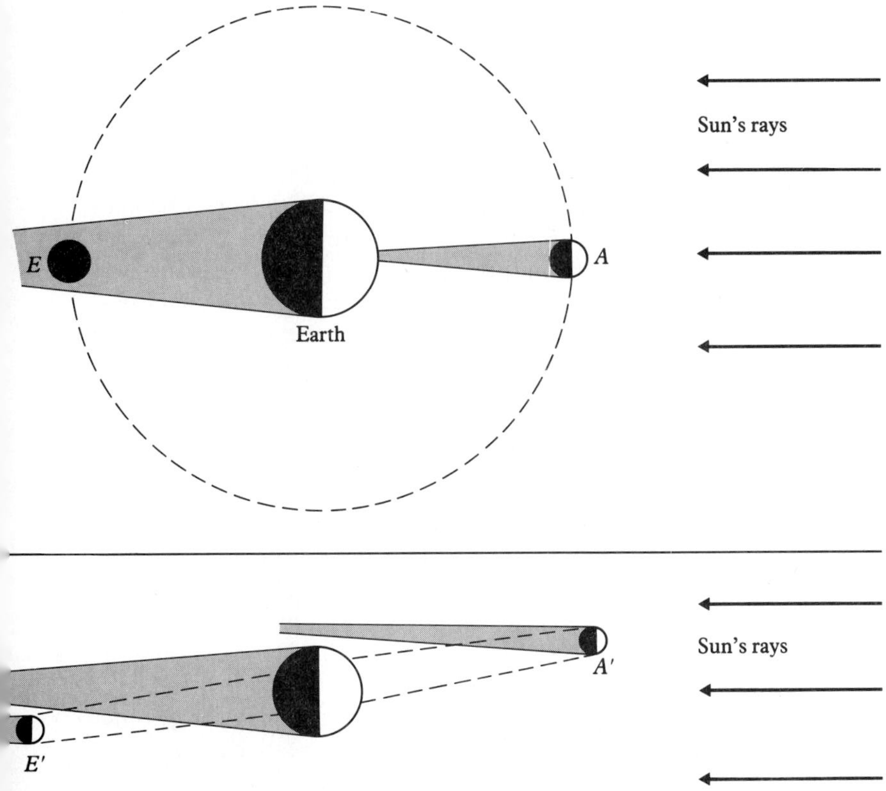

FIGURE 1.6 Looking down at the Moon's orbit from space (top). An eclipse of the Sun can occur at new moon when the Moon is directly between the Earth and the Sun (*A*). An eclipse of the Moon occurs at full moon if the Moon passes into the Earth's shadow (*E*). (This would be experienced as an eclipse of the Sun by an astronaut standing on the Moon.) Side view (bottom). Because the Moon's orbit is slightly inclined relative to the orbit of the Earth about the Sun, eclipses do not occur at each new and full moon. In the configuration shown here, both shadows miss their targets.

Eclipses

The circuit of the Moon around the Earth sometimes produces the spectacular events called **eclipses.** There are two different types of eclipses. The most dramatic, but also most rarely seen, is an eclipse of the Sun. If, at new moon, our satellite is exactly between the Earth and the Sun, its shadow falls on the Earth, blotting the Sun from the view of those in its path (Fig. 1.6). It is a coincidence of nature that at the present time, the Moon's distance from the Earth is such that its apparent size is approximately equal to that of the Sun. The match is so close that the slight deviation of the lunar orbit from an exact circle can carry the Moon far enough from us to appear a little smaller than the Sun. If a solar eclipse occurs when the Moon is at this point in its orbit, a ring of sunlight appears around the Moon and the eclipse is called annular.

The complement to an eclipse of the Sun is an eclipse of the Moon, which occurs when the Moon passes through the Earth's shadow. A lunar eclipse can occur only when the full moon is exactly lined up with the Earth and Sun (Fig. 1.6). But the Earth's shadow is not as dark as the Moon's, since our planet's atmosphere scatters sunlight into the shadow, illuminating the eclipsed Moon with a dull orange glow. Unlike an eclipse of the Sun, which is visible only along the narrow path of the Moon's shadow, a lunar eclipse can be seen from the entire night side of the Earth.

Eclipses also occur on other planets. In Fig. 1.7 we see the shadow of Phobos on Mars. This event, which would be an eclipse of the Sun for any Martians who were present, was actually detected by the U.S. Viking lander that happened to be in the shadow's path. On the giant planet Jupiter, the dark shadow of its satellite Io would produce an eclipse of the Sun for Jovians floating above that planet's colorful clouds. Similarly, the jovian satellites go into eclipse themselves when they pass into their planet's shadow. They can even eclipse each other.

1.2 WATCHING THE SKY: THE PLANETS

While most of the points of light we see in the night sky are fixed stars that form the unchanging patterns of constellations, there are five bright ones that move. "Wild Sheep," the Babylonians called them, while the Greeks used the term *wanderer*, from which we get our word *planet*. These wild sheep do not roam over the entire heavenly meadow. Like the Sun and the Moon, they confine their wandering to the 12 constellations of the zodiac.

Our night skies also contain airplanes, whose flashing lights reveal their motion in a few seconds, and artificial satellites in orbit about the Earth, whose motion is also quickly discernible. So it is important to stress that while planets most definitely move, it usually takes several nights to detect their motion with respect to the fixed stars.

Mercury and Venus

Each ancient civilization had its own names for the planets, often associated with its pantheon of gods. The names we use today are primarily Roman in origin. The fastest moving planet was appropriately named Mercury (Greek Hermes) after the swift messenger of the gods. This planet is visible only when it is close to the sunlit horizon in morning and evening twilight. For this reason, it is not easy to see, and some cultures thought it was actually two objects seen at different times of year: one visible only in the morning and one in the evening.

This same duplicity was sometimes assumed for Venus, which also stays close to the Sun. Venus is the brightest object in the sky after the Sun and Moon, so bright that it is sometimes visible during the day, if you know exactly where to look. Sailing serenely in the silent sky, Venus shines like a brilliant jewel, marking the transition between day and night. The ancient Greeks

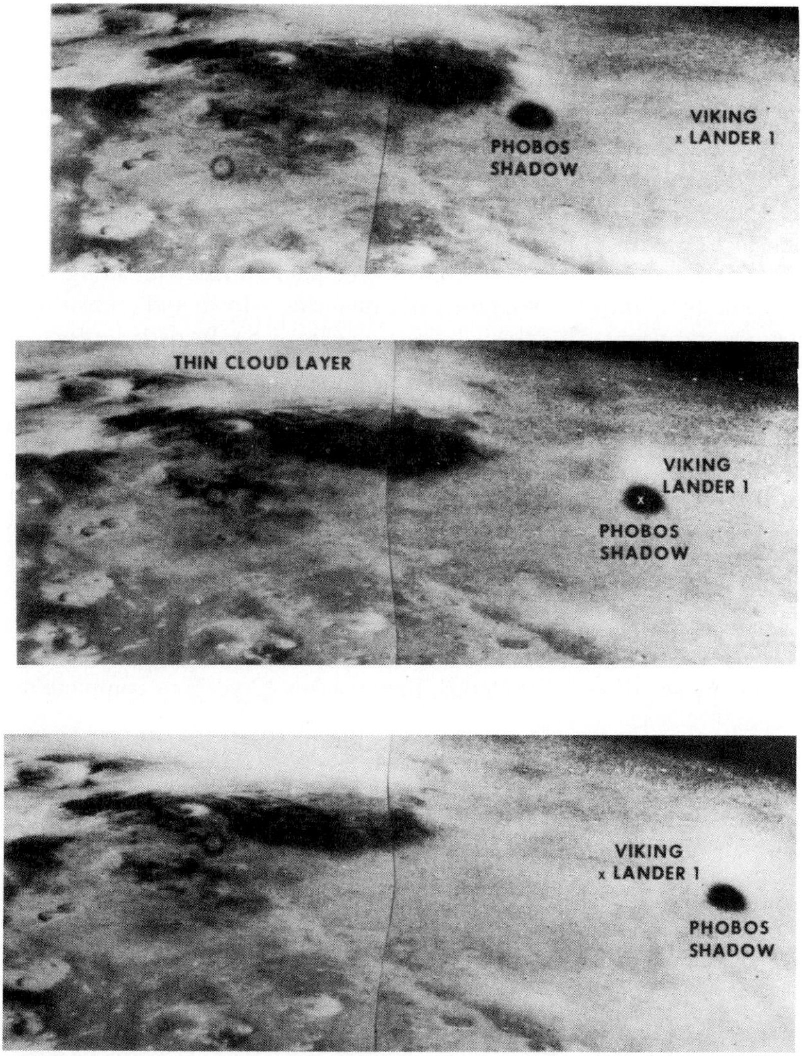

FIGURE 1.7 Eclipse of the Sun for Martians. The shadow of Phobos moving across the surface of Mars as it passes directly over the Viking 1 Lander.

and Romans must have been very impressed, for they named this planet after their goddess of love and beauty (Greek Aphrodite). Venus appears farther from the Sun than Mercury, but never so far that it can be seen in the middle of the night.

Mars, Jupiter, and Saturn

The other three planets easily visible to the unaided eye do not exhibit such a close association with the Sun, and they appear to move through the constellations of the zodiac at a more leisurely pace. There are three unusual characteristics of Mars (Greek Ares) that must have led to its association with the god of war. First, its distinctly reddish hue singles it out from all the other planets, conjuring up images of war. Second, Mars exhibits apparently erratic motions that must have been puzzling to ancient observers. Finally, these erratic motions are associated with large changes in the apparent brightness of the planet.

Retrograde motion is a reversal of the normal direction of movement. Like the Sun and the Moon, the planets usually move eastward through the constellations of the zodiac. Each day they rise and set with the other stars, but from one night to the next, you can notice that they have shifted their position eastward, except that sometimes they reverse themselves and move toward the west! Thus Mercury and Venus will first move eastward until they reach their maximum separation from the setting Sun, then they head west, disappearing in the twilight glow and reappearing in the morning sky, where they proceed to achieve their maximum western separation from the Sun before reversing and repeating the cycle.

Mars, Jupiter, and Saturn behave differently. We might first see Mars in the late night sky, moving slowly eastward through the constellations of the zodiac. As the changing seasons cause the stars to rise earlier and earlier, Mars rises earlier with them, becoming visible shortly after sunset and riding high in the sky in the middle of the night. We now say it is at **opposition,** since it is on the opposite side of the Earth from the Sun.

A few months before Mars reaches opposition, its apparent motion gradually slows and then reverses. The planet speeds up along a retrograde path, then slows down and returns to its normal direction (Fig. 1.8). Its path appears as a loop in the sky, and in traversing this loop, near opposition, Mars becomes much brighter than usual. Although this pattern must have been apparent to ancient observers, its causes remained obscure. Early perceptions of Mars must have included a certain amount of uneasiness at its apparent unpredictability, an appropriate characteristic for the god of war.

The brightest of the planets after Venus, Jupiter (Greek Zeus) was named for the king of the gods. Jupiter takes 12 years to complete one

FIGURE 1.8 Retrograde loops made by Jupiter as it wanders among the stars. The apparent motion of the planet slows down, reverses, slows again, and continues in the original direction.

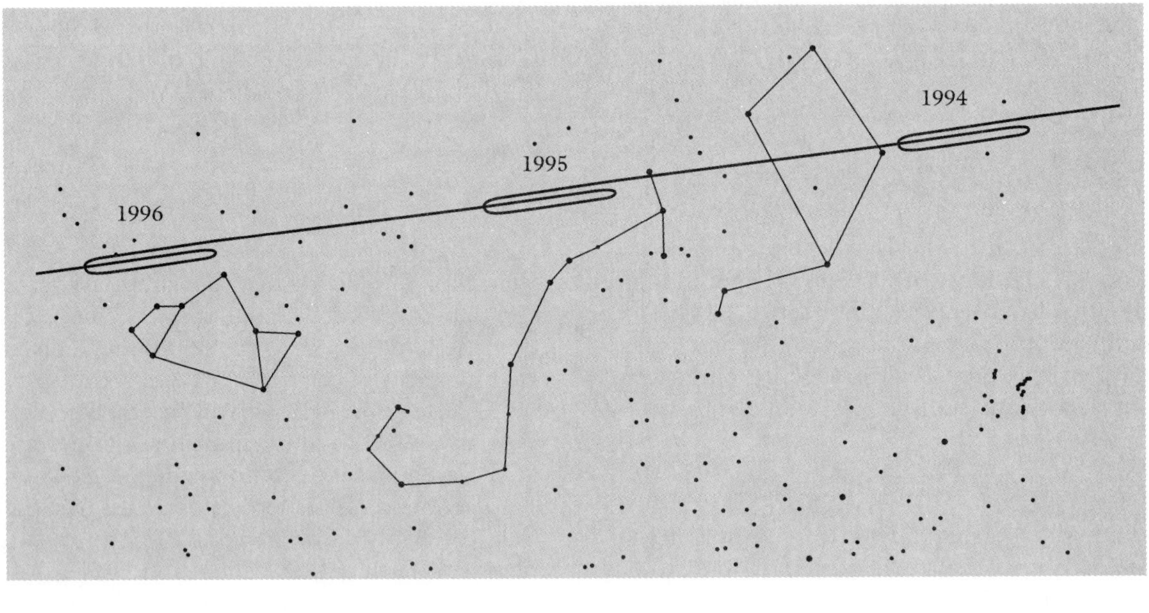

circuit of the zodiac, and exhibits only a small retrograde loop each year.

Saturn (Greek Chronos) is even more dignified. This planet requires a full 30 years to travel through the zodiac. Saturn was Jupiter's father and the ancient god of time. Since 30 years was (and still is) a large fraction of a human lifetime, the name is appropriate.

Ancient Interpretations of Planetary Motion

The various apparent motions of the stars and planets provided the first observations that people could use to understand the Earth's place in the universe. The Sun, Moon, stars, and planets all apparently circle the Earth in some kind of spherical system. The circle was regarded by the ancient Greeks as the most perfect geometric figure, and casual observations of the sky indeed suggested that circular paths were entirely appropriate for the bright objects it contained. The Sun and Moon were clearly circular themselves, giving added support to this approach.

With our space-age perspective, we may regard these ideas as hopelessly naive, but a moment's reflection will indicate that all of your daily needs can be met by this simple model of the universe—a flat Earth around which everything else revolves. It is only if you take a long trip or stand at the seashore watching boats disappear over the horizon (or see a picture of Earth taken from space) that you need to question the flat Earth. And even if you understand (as did the ancient Greeks) that the world is spherical, you can still attribute celestial motions to the revolution of the sky about a stationary Earth.

As observations improved, however, the inadequacy of the simple **geocentric,** or Earth-centered, model of circular motions became apparent. Near the time of the foundation of the Roman Empire 20 centuries ago, astronomers had begun to develop more elaborate models to explain the apparent motions of the planets. They kept the geocentric idea and the perfect circles, but they postulated that instead of moving on a circle with the Earth at the center, each planet moved on a small circle (called an epicycle) that was attached to a larger circle. They also allowed the Earth to be offset from the centers of the planetary circles. By the second century A.D. when the Alexandrian astronomer and geographer Claudius Ptolemy wrote the definitive treatise on classical Greco-Roman astronomy, dozens of cycles and epicycles were employed in the calculation of planetary positions. Cumbersome as this Ptolemaic system may seem to us, it worked very well. It was not until the fifteenth and sixteenth centuries that this approach was abandoned, as a part of the European artistic and intellectual rebirth known as the Renaissance.

1.3 THE SLOW GROWTH OF REASON

Among the basic questions that have been asked in almost all human civilizations are those that concern the size and shape of the Earth, the motions of the stars and planets, and the distances to the Sun, Moon, and planets. These basic elements in the structure of our solar system are now taught in elementary school. Nevertheless, these "facts" were only slowly established, and even today few of us could actually cite the evidence that proves that this modern perspective is correct.

The slow growth of understanding that led to our present view is a fascinating story, but we can mention only a few highlights here. It is useful to review these because they offer some lessons about how we should approach new, unsolved problems today, and they show how the perspective we now take for granted was achieved. Here we shall consider the work of just four people: Copernicus, Tycho, Kepler, and Galileo, moving on to Newton in the next section (Fig. 1.9).

a

b

c

d

FIGURE 1.9 (a) Copernicus, (b) Tycho, (c) Kepler, and (d) Galileo—a gallery of the four Renaissance scientists who changed our perception from an Earth-centered, mysterious universe to a Sun-centered solar system surrounded by distant stars.

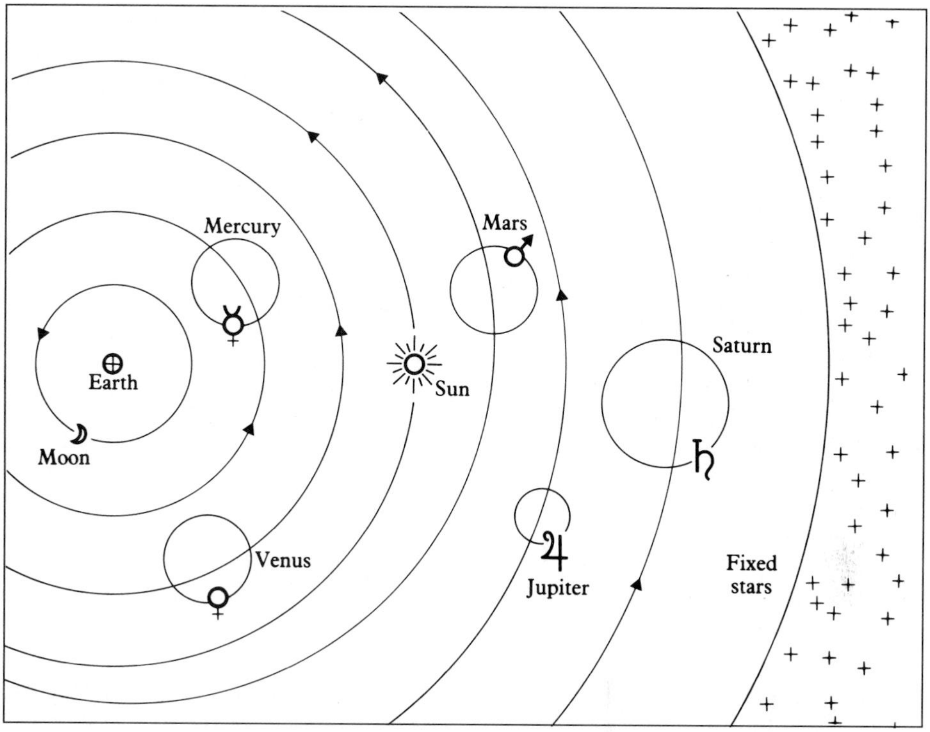

FIGURE 1.10 Earth-centered, or Ptolemaic, system, developed in antiquity to explain the apparent motions of the Sun and planets. Each object except the Earth must move in small epicycles in addition to its circular motion about the Earth.

Copernicus and the Heliocentric System

Nicholaus Copernicus was a Polish astronomer born in 1473. He was a contemporary of some of the great Italian painters of the Renaissance, such as Botticelli and Michelangelo, and he was 19 years old when Columbus made his famous voyage to the New World. Copernicus spent some time as a student in Italy before returning to Poland in 1503, where he was given a position in the Church.

Copernicus was taught the geocentric system (Fig. 1.10) that had been developed by Claudius Ptolemy 1300 years earlier. In the Ptolemaic system, the motionless Earth stood in the center of the universe and everything else circled around it. Copernicus did not like this system because it seemed too complex with its various cycles and epicycles. Aesthetically, it made more sense to him for the radiant, life-giving Sun to be in the center, with the planets revolving around it (Fig. 1.11). This bold hypothesis allowed him to explain the apparent motions of the planets more easily than the geocentric system could. It also allowed the stars to be motionless; this pleased Copernicus since he thought it was more reasonable for the small Earth to move about the Sun than for the great vault of heaven to turn about the Earth. This new perspective is called **heliocentric,** or Sun-centered.

It is hard for us to recapture the challenge to human thought that the Copernican system

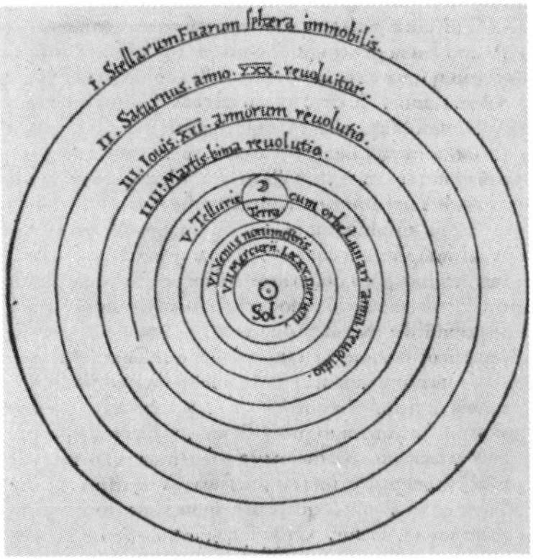

FIGURE 1.11 Heliocentric system: a copy of the original sketch of his plan for the solar system published by Copernicus. Note that only the Moon orbits the Earth, while all the planets (including Earth) move around the Sun.

required. We are so accustomed to the modern heliocentric view that we find the older geocentric system absurd. The geocentric perspective, however, had been supported by most of the philosophers of ancient Greece and Rome as part of a broader framework that argued for a stationary Earth and unchangeable perfection in the heavens. These ideas had become entrenched in the medieval Christian Church, which was then the dominant intellectual and political force in Europe. The new heliocentric theory was seen by many as a direct affront to religion. Moreover, it contained the seeds of a radical alteration in our perception of ourselves. Previously, the Earth was large and the rest of the universe seemed secondary. Humans lived at the center of the cosmos. Copernicus made the Earth small and the universe big, thereby demoting humanity from its central position.

Copernicus had no proof that his heliocentric system was correct. To him it was both more clear and more pleasing than the geocentric system. Others felt differently. Our senses tell us that the Earth is stationary and the sky revolves around it, not an easy point of view to abandon. Nor, in that period of history, were people accustomed to the idea of using experiments or observations of nature to distinguish between different philosophical systems. But consider just one simplifying stroke that Copernicus achieved.

Previously it had been thought that the motions of the planets as seen in the sky were real. Thus when a planet looked as if it were moving backward (retrograde motion) it really was. The retrograde motion was thought to be caused by the planet's motion on small circles called epicycles that were centered on their circular orbits. Thus when a planet needed to go backward, it just traversed the small circle in a direction opposite to that in which the big circle carried it. This complex arrangement may be compared with what happens when the Sun is placed at the center. Now the retrograde motion of the planet is not real, but is simply the effect seen by an observer on Earth as we overtake the planet in our journey around the Sun (Fig. 1.12).

Note that both the geocentric and the heliocentric models for the solar system explain retrograde motion. Copernicus simply does it in a less complex, more elegant way. This is a frequent situation in science, where more than one theory can sometimes explain the same set of observations. However, when the theories are as different as these, they will make different predictions about other phenomena that have not yet been observed. These predictions can then be tested by new observations, which will provide a way of choosing between the two theories.

Tycho and Kepler

In the absence of observations to distinguish the heliocentric from the geocentric system, the ideas of Copernicus remained highly controversial. One of the astronomers who found them interesting but not entirely satisfactory was Tycho Brahe (1546–1601). Born in Denmark just three years after the death of Copernicus, Tycho's early education was in jurisprudence. An eclipse of the Sun he had seen as a boy of 14

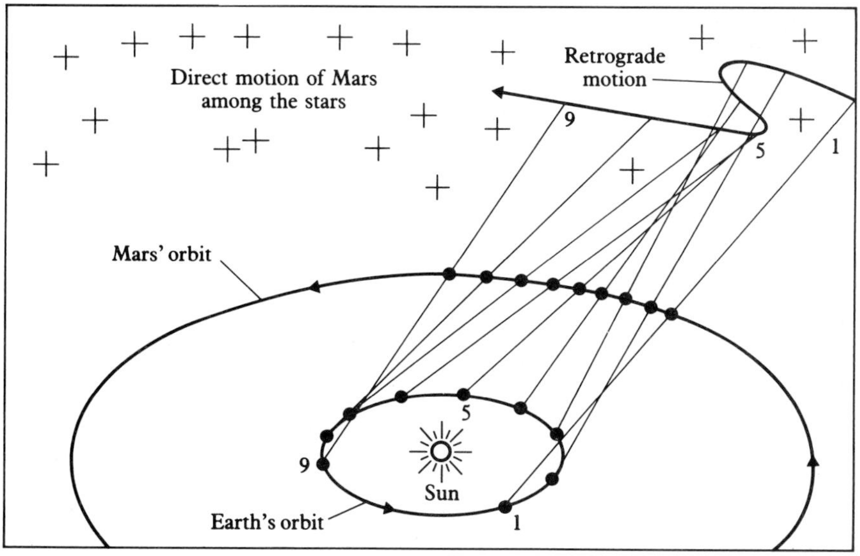

FIGURE 1.12 Retrograde motions as explained by the Copernican system: Outer planets appear to move backward as the faster Earth overtakes them.

made a profound impression on him, however, and he soon abandoned law to become the most accomplished observer of the heavens in the era before the invention of the telescope. Using instruments of a large scale and high angular precision but with no optical parts, Tycho surveyed the heavens, measuring the positions of the stars and planets with an accuracy far greater than that of anyone before him.

In those days, one of the chief tasks of astronomers was the accurate prediction of planetary motions, and the old Ptolemaic tables were no longer adequate for this task. Tycho was determined to improve this situation. He was also a colorful character, rather vain, who wore a silver nose as the result of a dueling injury. Among his own discoveries, his observational proof that a "new star" that appeared in 1572 in the constellation of Cassiopeia was farther away than the Moon was an important step away from ancient ideas. It showed that the heavens were not immutable, that the stars themselves could change.

It was Tycho's successor, Johannes Kepler (1571–1630), who extracted the hidden science from these remarkably precise observations of planetary positions. Kepler was taught the Copernican system at the University of Tübingen, where he was studying to become a Lutheran minister. But the faculty there, recognizing his intellectual gifts, persuaded him to take a post as a mathematician at the University in Graz, Austria, in 1594. (To place the period, note that in England Shakespeare was writing his plays and sonnets at this time.) In addition to exhibiting a real gift for mathematics, Kepler was strongly influenced by a mystical sense of the intrinsic harmony and order in the universe. It was this sense that provided his primary motivation for studying the motions of the planets: he felt that there must be an underlying set of simple principles that governed the structure of the solar system.

Kepler was fortunate in that his early work came to the attention of Tycho Brahe. Tycho recognized the younger man's mathematical abilities and invited Kepler to join him in Prague in 1600. Tycho died a year later, and Kepler took over his post.

Kepler was fascinated by the apparent harmony and order in the Copernican system. Because of his own mystical beliefs, he was cer-

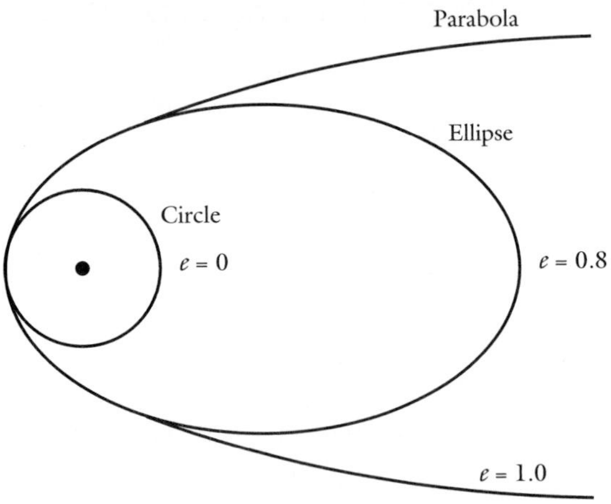

FIGURE 1.13 The circle, ellipse, and parabola represent an increase in eccentricity from 0 (circle) to 1 (parabola). Some comets have elliptical orbits with eccentricities of 0.999, while Earth's orbit has an eccentricity of only 0.02, and that of the orbit of Venus, the most circular of the planets, is only 0.007.

tain that there must be a way of expressing this harmony in simple mathematical relationships. This proved very difficult to do, but nine years after his arrival in Prague, Kepler published a book containing the first two principles of planetary motion that we now refer to as Kepler's Laws.

Kepler's Laws of Planetary Motion

Kepler had been trying to analyze Tycho's observations of Mars, fitting them to the system of orbits described by Copernicus. It simply wouldn't work. The problem lay in the circular orbits that the Copernican system still used. Mars obviously didn't move about the Sun in a circle, not even a circle in which the Sun was not at the center. So Kepler tried a variety of other geometrical shapes, until he finally found the **ellipse.** This worked, not just for Mars but for all the planets.

At first Kepler was unhappy with the ellipse because it doesn't have the perfect symmetry of the circle. The Sun is at one focus, but there is nothing at the center of the ellipse and nothing at the other focus (Fig. 1.13). The ellipse is still elegantly simple, however, and it worked. This then is Kepler's first law:

> *Each planet moves in an elliptical orbit about the Sun, with the Sun at one focus of the ellipse.*

Another problem that had vexed astronomers was the fact that at some times the planets moved faster in their orbits than at others. This seemed random and unexpected, until Kepler found another principle that made the behavior intellectually acceptable. With the planets moving in elliptical orbits, it turns out that the fastest motion occurs when they are nearest to the Sun. Consequently, an imaginary line drawn between the Sun and the planet will sweep out the same area within the ellipse during the same interval of time, regardless of where the planet is in its orbit (Fig. 1.14). This discovery restored a certain

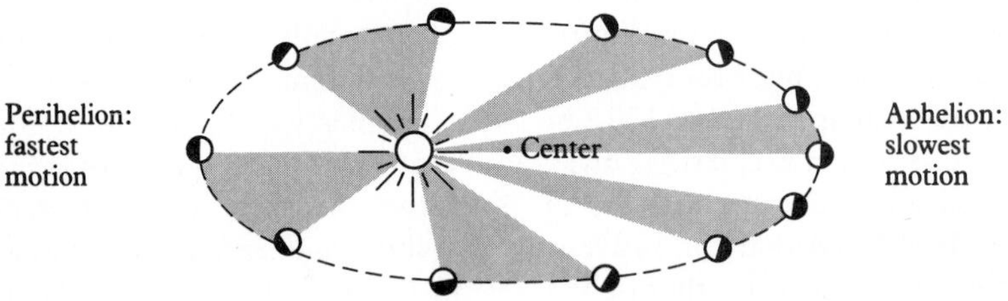

FIGURE 1.14 Kepler's second law. A planet moves fastest when it is closest to the Sun (perihelion). Yet the area created by a line connecting the planet to the Sun is always the same for a given period of time. Thus the areas of the light and dark triangles are all equal.

order to nature and became another simple mathematical description of what was observed. It is Kepler's second law:

An imaginary line connecting the Sun with a planet sweeps out equal areas in equal times as the planet moves about the Sun.

The best was yet to come. Kepler was still searching for some principle that would relate the motions of the planets to one another. He now had an accurate description of how they moved, but he was certain that there must be a harmony in those motions that would explain why the planets moved the way they did. Ten years after publishing his first two laws, he found that harmony. What he had discovered was a simple mathematical relationship between the period of a planet's revolution about the Sun and its distance from the Sun. This is Kepler's third law (Fig. 1.15):

The cube of the distance from the Sun divided by the square of the time required to traverse the orbit is a constant, the same for every planet.

Written as an algebraic formula, the third law is:

$$D^3 = AP^2,$$

where D is the distance of the planet from the Sun and P is the orbital period. The letter A stands for a number whose value depends on the units used; if we measure the period in years and the distance in astronomical units, then $A = 1$. The **astronomical unit (AU)** is the name astronomers give to the average distance of the Earth from the Sun, approximately 150 million km. Kepler didn't know the length of the astronomical unit very precisely, but the third law gave him a way to express all the other planetary orbits in terms of the Earth's orbit.

This third law represented a remarkable achievement. Kepler was overjoyed to have found this underlying harmony of the solar system. His work had greatly strengthened the Copernican system, providing the predictive power it had lacked in its original form when the planets were thought to move in circles. These laws, however, are still descriptive. Kepler's work could be viewed as simply providing a much better way to calculate the positions of planets. The proof that would clearly distinguish between the geocentric and heliocentric systems was still lacking.

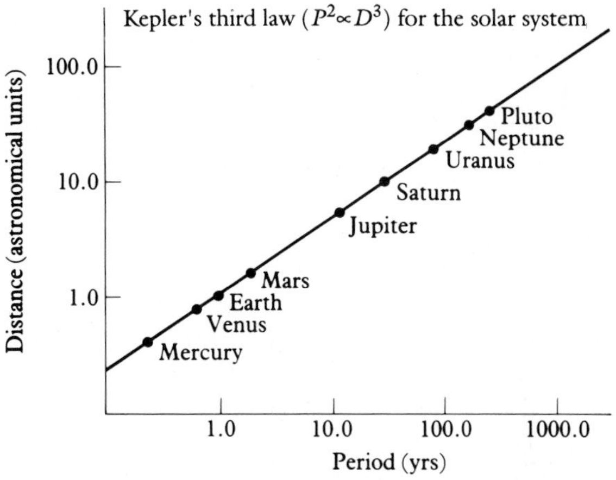

FIGURE 1.15 Graph illustrating Kepler's third law: The square of a planet's period (P^2) is proportional to the cube of its distance from the Sun (D^3). Periods are plotted in years and distances in units of the Earth's distance from the Sun; both scales are logarithmic.

Galileo and the Foundation of Modern Science

Kepler was a contemporary of Galileo Galilei (1564–1642), who was at this time laying the foundations of modern physical science. Galileo's experiments and observations, just as much as Copernicus's and Kepler's theories, were responsible for overturning the geocentric model of the solar system.

Galileo, the son of a musician, began his studies as a candidate for a degree in medicine but was soon attracted to mathematics. In 1609, he was teaching mathematics in Padua when he heard about the invention of the telescope. Unlike many of his contemporary academics, he recognized that this combination of lenses enables one to see better, not just differently (as with a distorting mirror). He quickly made some telescopes of his own and began using them to study the night sky. What he found provided the first experimental demonstration that the geocentric universe and its philosophical underpinnings were not simply complex and inelegant; they were wrong!

One of Galileo's earliest discoveries was the presence of some small "stars" very close to Jupiter. Observing night after night, he found that these stars changed their positions with respect to the planet in an orderly, repetitive way. Evidently, they were satellites moving in four separate orbits around Jupiter. This was a major blow to the geocentric theory, which required

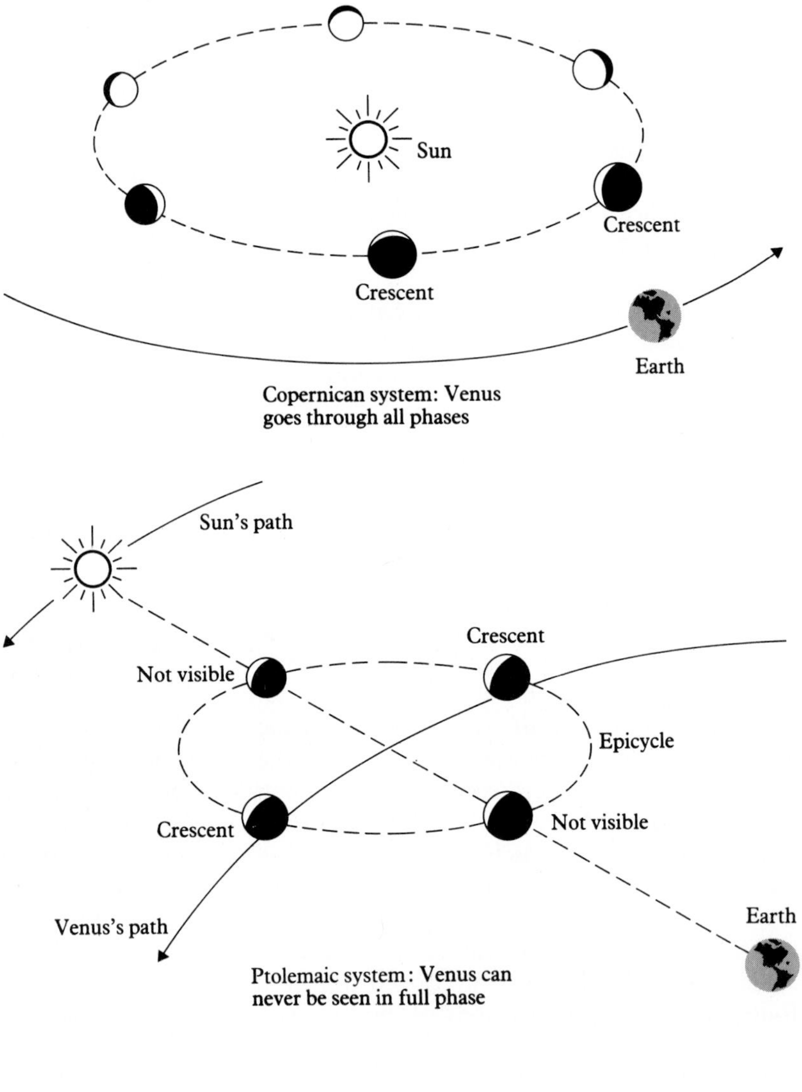

FIGURE 1.16 The Ptolemaic (Earth-centered) view of the solar system could explain crescent phases for Venus (bottom) but would not predict that a full phase would occur (Venus would have to appear in the midnight sky, which it does not do). The Copernican (Sun-centered) view (top) correctly predicts the phases and also the observed fact that the planet at full phase appears much smaller than it does as a crescent—it is simply much farther away.

that everything must orbit the Earth. Galileo saw that what he had found resembled a miniature Copernican solar system itself, in which Jupiter played the role of the Sun and the four satellites that he had found resembled the planets.

The evidence for a heliocentric system was further strengthened when Galileo discovered that the planet Venus exhibited phases like those of our Moon. Only in a heliocentric system could Venus be fully illuminated at some part of its orbit, as seen from the Earth. If this planet orbits the Sun, it also should become larger in apparent size as it nears the Earth, at the same time shrinking to a thinner and thinner crescent. Conversely, as Venus moves around to the other side of the Sun, we can see more of it because of the favorable viewing angle, but it appears much smaller than when it is a crescent since it is now much farther away from us (Fig. 1.16). This behavior, which is exactly what Galileo observed, is impossible in the geocentric system, which would not permit an observer on Earth to see Venus fully illuminated.

The geocentric system was clearly untenable, at least to those who had looked through Galileo's telescope or who believed what he wrote, but the Catholic Church did not accept these proofs and forced Galileo to deny their validity in 1633, 14 years after Kepler (who did not live in a Catholic country) had published his third law. These were perilous times for dissenters: Kepler had to interrupt his work at one point to save his mother from being burned as a witch, and only nine years before Galileo began experimenting with telescopes, Giordano Bruno was burned at the stake for his heretical views of the universe. Galileo was treated much less severely, being sentenced to house arrest for the last eight years of his life. One wonders if he knew that it made no difference how the Church courts ruled, that others would prove him to be correct. In any event, that is exactly what happened, although the Roman Catholic Church did not formally exonerate Galileo until 1992.

1.4 NEWTON AND THE LAW OF GRAVITATION

Isaac Newton was born in 1642, a few months after the death of Galileo. He therefore grew up at a time when the basic discoveries we have just described should have been part of his college curriculum. As an undergraduate at Cambridge from 1661 to 1665, however, Newton found that the university was still teaching the geocentric view of the universe. Nevertheless, he managed to learn about the new ideas, to which he added his own insights. These days undergraduates interested in science and mathematics study calculus. Newton began *inventing* calculus as an undergraduate, while also formulating ideas about optics and mechanics.

Throughout a long career (he died at 84) Newton made many contributions to physics and astronomy (Fig. 1.17). He was one of the first scientists to use *quantitative*, or mathematical, arguments to test scientific ideas. Because

FIGURE 1.17 Isaac Newton, the man who discovered the physical laws governing planetary motions, thereby bringing a sense of order to an apparently random universe.

he was so successful, he fundamentally altered the human perception of the universe and its workings. The impact of his accomplishments was nicely summed up in the following couplet by a contemporary poet, Alexander Pope:

Nature and Nature's laws lay hid in night:
God said, let Newton be! and all was light.

The Laws of Motion

Newton's most important contributions to astronomy were his three laws of motion and the law of gravitation. The first law of motion may be stated as follows:

1. *Every body continues in its state of rest or of uniform motion in a straight line unless it is compelled to change that state by the action of some outside force.*

The first part of this law seems pretty obvious—things don't move unless they are forced to do so—but the second may not be as evident. There are no familiar examples of an object moving uniformly in a straight line. Indeed, the Greeks and Romans had thought that the circle was the natural form of motion of celestial objects. But Newton realized that in the heavens, as on Earth, an object moving without any force on it would go straight. In everyday life, we find that things that are moving eventually slow down and stop. Newton realized that this was because of friction. As friction is reduced, motion continues for a longer time, and one can imagine a situation with no friction at all in which a moving object coasts on forever.

How can we express the relationship between the forces acting on bodies and the motions of these bodies? Newton's second law provided the answer:

2. *The change of motion (acceleration) is proportional to the force acting on the body and inversely proportional to the mass of the body.*

Written algebraically, with A as acceleration, F as force, and M as mass, this law becomes $F = MA$. When expressed in mathematical form, this law provides a way to calculate the motion of a body from the force exerted on it. Or, conversely, by tracking the motion it allows us to calculate the forces. These calculations, carried out today with large computers, are still the way we navigate our spacecraft on their interplanetary journeys.

The larger the force, the greater the acceleration. Also, a given force will have more effect on a body of small mass than on a large one. All of this is well known from everyday experience, where we deal frequently with objects (autos, for example) that change speeds. What may be less obvious is that an object moving at constant speed in a circular or elliptical path is also accelerating. Changing direction is a form of acceleration, which means that a force is acting on the body to produce the curved path.

Newton's third law takes us back to the issue of balancing forces:

3. *To every action there is an equal and opposite reaction.*

This says that when an apple falls toward the Earth, the Earth must also move toward the apple. But since the Earth has about 10^{26} times as much mass as the apple, yet the force acting on the two of them is the same, the Earth accelerates 10^{26} times less. That is a small number indeed! To see this law in action, you might find a partner and try pushing each other on roller skates, or while you are standing on skateboards. This law can also make getting in or out of a canoe rather wet work, while getting in and out of a big boat is easy.

The Law of Gravitation

The framework provided by these three laws of motion seems very close to providing an understanding of why the planets move around the Sun in the way that Kepler described. Obviously, a force is acting on them, or they would be moving in straight lines. But what is that force? And what is the balancing force that allows them to continue this motion instead of slowing to a stop or falling into the Sun?

Newton's brilliant solution to these dilemmas was the law of gravitation. If an apple falls on Earth, it is being attracted to the Earth by a force. Perhaps that same force is attracting the Moon. The Moon doesn't fall all the way to the surface of the Earth because it is moving sideways, approximately at right angles to the force of gravity. The attractive force changes the motion of the Moon from a straight line to a closed curve, an orbit around the Earth. In effect, the Moon is falling around the Earth.

Newton's law of gravitation states that

> *the gravitational force between two objects is proportional to the product of their masses and inversely proportional to the square of the distance between them.*

If the force is F, the masses are M_1 and M_2, and the distance is R, the same law can be written as:

$$F = GM_1M_2/R^2,$$

where the symbol G stands for a number called the gravitational constant.

This is an extremely powerful law (Fig. 1.18). With it, we can determine the mass of a planet or the mass of the entire Galaxy. It also governs the flights of both spacecraft and pole vaulters. And it explains why the planets move the way they do. The planets are "falling" around the Sun in the same way the Moon moves around the Earth. Left to their own devices, they would move through space in straight lines. The acceleration toward the Sun that they experience from gravity is just balanced by the acceleration produced by their orbital motion. Hence they are condemned to move forever in the orbits in which we find them.

One of the important consequences of the law of gravitation is the realization that the gravitational force between two objects depends on the product of their masses. It logically follows that the orbit of one object around another must also depend on their masses. Although the effect of the mass of the planet on its orbit is small, it is important, for it allows us to calculate the masses of the planets.

FIGURE 1.18 A couple holding each other at the intimate distance of 1 cm are exerting a gravitational attraction on one another that is about equal to the attraction of the Moon on either one of them. Fortunately, the gravitational force on both of them exerted by the Earth is about 100,000 times greater, and there are other forces operating between people that are stronger than gravity.

Science Since Newton

"If I saw a little farther than others," Newton wrote in a letter, "it was because I was standing on the shoulders of giants." This modest appraisal of his extraordinary achievements is nevertheless an accurate description of how science proceeds, each step forward building on the work of those who came before. This was particularly the case after Newton, for his carefully reasoned, mathematical approach set the tone for most subsequent scientific inquiry.

Galileo and Newton ushered in a new world of experimental and quantitative thinking. During the following decades, the pace of science quickened, and one discovery followed another. Ever more elegant and powerful mathematical tools were developed, just as scientific instruments become more complex and precise. Even though space travel was still more than two centuries in the future at Newton's death in 1727, the way toward the modern world was clearly indicated.

FIGURE 1.19 Newton's third law applies both to tennis and to rockets. When the girl puts the ball in play with a backhand shot to the left, she and the cart will move to the right.

1.5 ESCAPING FROM EARTH

The laws of motion and of gravitation provide a framework for understanding the orbits of planets and other members of the solar system. They also permit us to understand the operation of rockets and the principles that govern Earth satellites and interplanetary probes.

Rockets

Whether we are launching a space telescope into low Earth orbit or sending a probe to the outer planets, we need to overcome the force of gravity exerted by the Earth. To do that, we must lift a spacecraft above the atmosphere (so friction will not slow it down) and supply it with an amount of energy sufficient to keep it from falling back. The best device to impart this energy to a spacecraft is a rocket.

A rocket accelerates a spacecraft by expelling gases backward at high speed. This is an excellent example of Newton's third law: the exhaust goes in one direction and the reaction drives the spacecraft in the opposite direction (Fig. 1.19). A jet engine works on the same principle, but it draws oxygen from the atmosphere to burn with its petroleum fuel. A rocket carries its own oxidizer as well as its fuel, and therefore it can operate in the near-vacuum of space.

Many kinds of rockets have been developed for spaceflight. For large, brute-force applications there are solid fuel rockets, which look like giant firecrackers and are nearly as explosive. The external boosters of the Space Shuttle are solid fuel rockets. More easily controlled are liquid fuel rockets, such as those of the three Shuttle main engines. Many liquid fuel rockets are designed to be restarted in space, a requirement if we are to make in-flight adjustments on interplanetary trajectories. The most efficient liquid fuel rockets burn hydrogen and oxygen, but these two gases must be kept extremely cold to be used in liquid form, so they are not suitable for engines that must be restarted. Less efficient fuels that remain liquid at less extreme temperatures are needed for flights to deep space. A nitrogen compound called hydrazine is one of these.

Going into Orbit

Suppose that a rocket has lifted a spacecraft above most of the Earth's atmosphere and is ready to inject it into orbit. Enough horizontal speed must be imparted to keep the new artificial satellite from falling back to the Earth. To understand how the trajectory depends on the speed, imagine a mountain that sticks up above the atmosphere, and ask yourself how fast you must throw a ball in the proper direction to keep it in orbit (Fig. 1.20).

The motion of the ball results from the pull of gravity in combination with its forward speed. If the ball is dropped instead of thrown, it simply falls toward the center of the Earth. A moderate forward speed—say a few thousand kilometers per hour—results in a curving fall that takes the rocket partway around the world, on a path similar to that of an intercontinental missile. When the speed reaches a critical value called the orbital velocity, the surface of the Earth curves away as fast as the ball falls, resulting in a low-altitude circular orbit. For Earth, the orbital speed is about 8 km/sec, or about 28,000 km/hr.

FIGURE 1.20 Going into orbit. Throwing a ball horizontally from a mountaintop with greater and greater velocity produces trajectories *D, E, F, G* that will ultimately put the ball into orbit. We assume here that the mountain is on the small martian moon called Phobos, whose gravitational field is so weak that someone with a strong arm could in fact put a baseball into orbit.

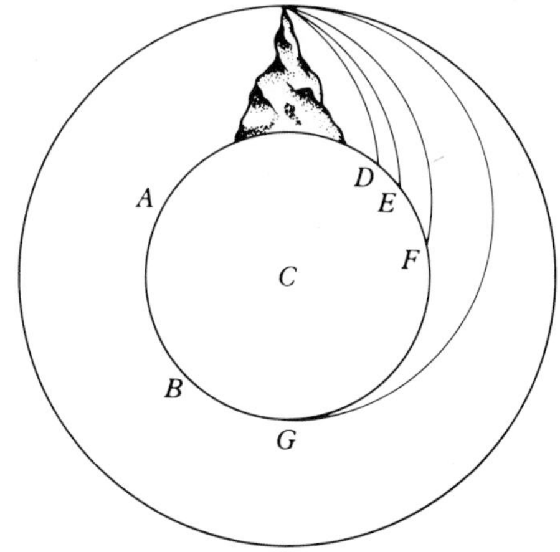

Calculations using Newton's laws show that the square of the orbital velocity is proportional to the mass of the planet. Thus a smaller planet with its weaker gravity requires a lower speed to achieve orbit, just as common sense dictates. In addition, the higher the satellite orbit, the slower its speed. At the distance of the Moon, the orbital speed is only about 1.3 km/sec.

It is easy to calculate the orbital period of an Earth satellite. The period is the distance traveled in one orbit divided by the orbital speed. Since the circumference of the Earth is about 40,000 km, a satellite orbiting just above the atmosphere (at an altitude of about 200 km) will take $P = \text{circumference/speed} = {}^{40{,}000}/_{28{,}000} = 1.4$ hours or about 90 minutes to circle the planet. We can sometimes see these satellites moving against the pattern of fixed stars in our night sky. As it happens, this period for a low orbit is nearly the same for any planet. From a tiny asteroid up to giant Jupiter, the minimum orbital period is between one and two hours.

We know the Moon takes about 29 days to orbit the Earth at a distance of 384,000 km, so there must be a distance between 200 km and 384,000 km where the period of an orbit is just equal to 24 hours, the length of our day. Of course there is, and it is 35,680 km. A spacecraft put into an orbit at this altitude above the equator will appear stationary to an observer on Earth. In fact only observers who are on the right side of the Earth could see such a satellite; it will never become visible to inhabitants of the other hemisphere. This kind of orbit is called **geostationary.** It is very useful for communications, spying, or meteorological satellites on Earth. We shall find when we study Pluto that this planet's natural satellite occupies just such a stationary orbit.

Escape Velocity

Return to Fig. 1.20 and consider what happens when we throw a ball into orbit with greater speed. As the speed increases, the orbit becomes more and more elongated. The part of the ellipse that is closest to the Earth is called the perigee, and the greatest distance (on the opposite side of the Earth) is the apogee. (If the orbit were around the Sun, these points would be called **perihelion** and **aphelion,** respectively.) As the initial speed increases, the perigee moves farther and farther out, until we reach the point where the ball escapes completely.

The **escape velocity** described here is exactly the same as the speed required for a ball projected straight up from the surface to escape entirely from the planet. At this speed, the energy is so large that gravity, pulling backward, can slow but never quite stop the outward motion. Mathematically, the escape velocity is just equal to the orbital velocity multiplied by the square root of two, or about 1.4. For the Earth, the escape velocity is 40,000 km/hr, or just over 11 km/sec.

Escape from a planet becomes easier if the mass of the planet is small. For example, the escape velocity from Phobos, one of the small satellites of Mars, is only about 60 km/hr. This means that you could literally throw a ball into space from Phobos if you have a strong arm, or you could ride into space on a motorcycle, if you could find a smooth highway to serve as your launch pad.

1.6 THE SYSTEM REVEALED

It should be clear from the work of Kepler and Newton why the planets appear to move the way they do. The fact that they confine their motions to a narrow band along the ecliptic is a natural consequence of having their orbits nearly in the same plane. Imagine yourself as a letter in a word on the middle of a single page, taken from this book. Stretching away from you on all sides are the letters that make up the other words you are now reading, so you would be surrounded by words, but only in the plane of the paper. Looking up or down, you would see no writing.

It is the same for us in the solar system. If we stood on the surface of Mars and studied the night sky, we would see the same constellations we see from Earth, and we would find that the planets were still moving through the zodiac. When we look out toward the planets moving in front of these constellations, we are seeing an edge-on view of the solar system. From Mars, the Earth would be a gorgeous blue-white planet in the twilight sky, exhibiting the same kind of apparent motions with respect to the Sun that Venus does from our terrestrial perspective.

The two planets that stray farthest above and below the plane defined by the Earth's orbit are Mercury and Pluto. Their orbits have **inclinations** of 7° and 17°, respectively (Fig. 1.21). All the other planets have orbital inclinations within 4° of Earth's orbit. The orbits of Mercury and Pluto also have the greatest **eccentricity**—that is, the ellipses along which these planets move are more elongated than are the paths of the other planets.

We can also gain an approximate idea of the relative distances and sizes of the planets from simple visual observations, once we have adopted the Copernican point of view. Jupiter appears to move more slowly through the zodiac than does Mars, so it must be farther away. The fact that it is nevertheless the brightest object in the sky after Venus suggests that it must be unusually large. Saturn, still slower, is even farther away and must also be big to be as bright as it is. Mars is much closer than these two giants and must therefore be relatively small.

Directions of Motions

Which way is up? The convention that has come down to us from experience on the Earth is to say that north is up. North in the sky is defined

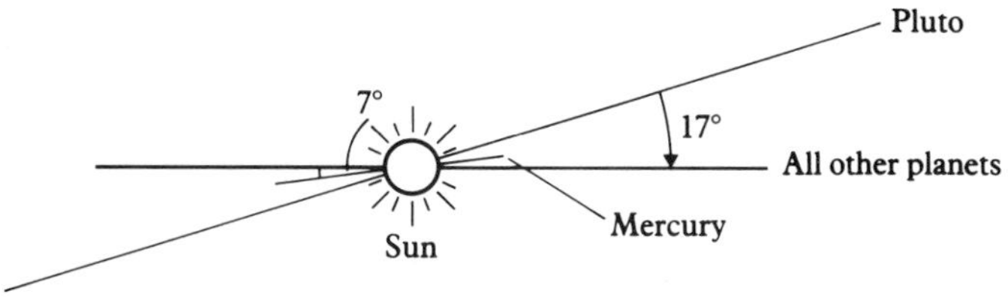

FIGURE 1.21 All the orbits of the planets except for those of Mercury (closest to the Sun) and Pluto (most distant) lie essentially in the same plane.

as the direction in space toward which the Earth's North Pole points. If we imagine ourselves located in space somewhere above the solar system in this sense, we would find when we looked down at the planets that they were all moving about the Sun in a counterclockwise direction. Looking down at the Earth from above its North Pole, we would see that our planet is also rotating on its axis in a counterclockwise direction, in the same direction that the Earth moves around the Sun.

Looking down on Venus from our lofty perch north of the system, we would find that this planet is rotating in the opposite direction, clockwise instead of counterclockwise, even though its revolution about the Sun is in the same direction as all the other planets. Uranus and Pluto would appear tipped over on their sides, with their north poles (defined by rotation) actually a little below their orbital planes. They too are technically rotating backward compared with Earth and the other planets.

The various satellite systems show a different kind of symmetry. Most of the larger satellites stay close to the equatorial planes of their parent planets. The same is true for systems of rings. In the case of Saturn, for example, the ring system and the satellite orbits all share the planet's inclination of 27° to the plane of Saturn's orbit.

Most satellites revolve around their planets in the same counterclockwise direction (as seen from the north) that the planets move around the Sun, but four of Jupiter's moons and one each in the Saturn and Neptune systems revolve in the opposite sense. (Clockwise rotation or revolution is called **retrograde,** the same word used in Section 1.2 for the apparent backward motion of a planet seen in the sky.)

The asteroids are a diverse lot of small bodies with diameters less than 1000 km which orbit the Sun primarily between Mars and Jupiter. There are tens of thousands of them, but their combined mass is less than one-tenth the mass of the Moon. Their orbits show a variety of eccentricities and inclinations, but on average they are inclined about 8° and are more eccentric than most planetary orbits.

The orbits of the comets are even more diverse. In their pristine state, these bodies orbit the Sun at immense distances, some as much as 1000 times the distance of Pluto. Unlike the planets, their orbits show a random distribution of inclinations. When one of these comets appears in our night skies, it is therefore rarely in a constellation of the zodiac. Its orbit may be perpendicular to the plane of the planets, or it might move around the Sun in a retrograde direction. For us to see it, the comet must have sailed in from its distant domain, perhaps perturbed by the gravity of a passing star.

A Scale Model

To gain an appreciation of the size of the solar system and the distances between the various planets, we can make use of a scale model. Let's reduce every dimension in the solar system by a factor of 200 million. On this scale, the Earth is the size of an orange, and our Moon would be a grape, orbiting the orange at a distance of 2 m. The Sun would be a little more than 1 km away, and it would have a diameter of 7 m, the height of a two-story house. It is the great distance of the Sun from the Earth that makes it appear to be about the same size as the Moon when we see both of them in our skies.

Although the Earth and other inner planets in this model are the size only of pieces of fruit, we do have some bigger planets. At this scale, Jupiter would be a large pumpkin, still small compared with the Sun but 11 times larger than Earth. Considering the Saturn system, we find that if we measure it from one edge of the rings to the other, the planet and its rings would just fit between the Earth and the Moon. Saturn's own large satellites would lie far beyond the Moon. Scales in the outer solar system are considerably grander than those we find for the planets close to the Sun.

Moving farther out, we pass Uranus and Neptune, finally encountering Pluto, another grape (but smaller than our Moon) at an average distance of 30 km from the Sun. This is still not the edge of the solar system, however. That lies at a point where the gravitational field of the Sun is challenged by the fields of passing stars. The great spherical cloud of comets extends to a distance of some 35,000 km from our scale model Sun. This means that our scale model solar system is about four times the size of the real Earth.

This exercise illustrates how large the Sun is, how much space exists between the planets, even in the inner solar system, and to what an enormous distance the Sun's gravitational influence extends. The latter emphasizes once again how huge the distances are between the stars. The nearest star (Alpha Centauri) is five times the distance of the comet cloud; to include it, our model would have to extend to a dimension of 200,000 km.

The real solar system is described in Table 1.1. It is useful to try to keep a few dimensions in mind. Jupiter, the closest of the giant outer planets, is about five times as far from the Sun as we are (that is, 5 AU). The next giant planet, Saturn, is about as far from Jupiter as Jupiter is from us. And the next step out, to Uranus, again nearly doubles the size of the system. Neptune and Pluto are closer together, but the average distance between the orbits of these two planets is about equal to the distance of Saturn from the Sun.

TABLE 1.1 Dimensions of the planetary system

Planet	Distance from Sun (million km)	Distance from Sun (AU)	Prediction[a] from Titius-Bode Rule (AU)
Mercury	58	0.39	0.4
Venus	108	0.72	0.7
Earth	150	1.00	1.0
Mars	228	1.52	1.6
Asteroid belt	330–500	2.20–3.30	2.8
Jupiter	778	5.20	5.2
Saturn	1429	9.55	10
Uranus	2875	19.22	20
Neptune	4504	30.12	} 39
Pluto	5900	39.44	

[a]Distance (AU) = 0.4 + 0.3 (2^n), where $n = -\infty$ (Mercury), 0 (Venus), 1 (Earth), 2 (Mars), 3 (Jupiter), and so on for the other planets.

The Titius-Bode Rule

Given Kepler's success in deriving laws governing planetary motion, it was natural for people to try to see whether some law existed that determined why the planets have the spacing we observe. The most famous attempt to unravel this mystery was that of Johann Daniel Titius in 1766. Titius found a relatively simply numerical relationship among the distances of the known planets, as described in Table 1.1. This relationship is sometimes called "Bode's law" or the "Titius-Bode law," since the more famous astronomer Johann Bode played a major role in publicizing it.

In fact, it is not a law at all, since there is no physical reason for this particular numerical progression. It would probably be better to call it the **Titius-Bode rule.** Despite this shortcoming, it played an important part in the history of astronomy. At the time Bode popularized it, there were no known asteroids, and the solar system ended at Saturn. Herschel's discovery of Uranus in 1781 added a planet to the solar system just where the "law" said its orbit should be. This made the gap between Mars and Jupiter all the more evident, and sure enough the first asteroids were discovered at this position at the beginning of the nineteenth century. But Neptune doesn't occur where it should, nor does Pluto. So Bode's "law" seems to be nothing more than one of nature's surprising coincidences, very different from the laws of Kepler and Newton.

1.7 QUANTITATIVE SUPPLEMENT: KEPLER'S LAWS IN MATHEMATICAL FORM

Kepler's laws provide a fundamental description of planetary motion that can be generalized to any system of bodies of small mass in orbit around a larger body (such as the satellite systems of the giant planets). Although Newton and his successors developed many ways of improving the accuracy of orbital calculations by taking into account the gravitational influence of each planet upon the others (called perturbations), the basic utility of Kepler's three laws remains. Kepler's first law states that orbits are ellipses with the Sun at one focus. The formula for an ellipse is

$$r = \frac{D(1 - e^2)}{1 + e\cos\theta}$$

where r is the distance from the Sun, D is the semi-major axis, e is the eccentricity, and θ is the angle that specifies the position of the planet along the ellipse, as seen from the Sun. Perihelion, the smallest value of r, corresponds to $\cos\theta = 1$ and is given by

$$r_{min}/D = (1 - e^2)/(1 + e^2) = 1 - e$$

The maximum value of r (aphelion) is similarly

$$r_{max}/D = 1 + e$$

Kepler's third law can be expressed as

$$D^3 = AP^2$$

when D is in AU and the period, P, is in years. The constant A equals 1 AU^3/yr^2. Newton later showed that the masses of both the Sun and the planet should be taken into account, in which case the third law is written

$$D^3 = A(M_1 + M_2)P^2$$

where M_1 is the mass of the Sun and M_2 is the mass of the planet, both expressed in units of the combined mass of the Sun and Earth. We make this choice of units so that the term in parentheses stays equal to 1 for the case of the Earth orbiting the Sun.

One application of the third law allows us to calculate the mass of a distant planet like Jupiter. We can do this as long as the planet has a satellite (natural or artificial) with a known period and distance from the planet. Ganymede, Jupiter's largest satellite, is observed to orbit the planet with a period of 7.16 days (1.96×10^{-2} yr) with a semi-major axis of 7.15×10^{-3} AU. We thus have

$$M_1 + M_2 = D^3 / P^2 = 9.5 \times 10^{-4}$$

Since the mass of Ganymede is small compared to that of Jupiter, we conclude from this calculation that the mass of Jupiter is about 1/1,000 that of the Sun.

Now consider the orbit of Comet Halley, which has a period of 76 yr and a small mass, negligible in comparison to that of the Sun. From the period, we calculate the semi-major axis of the comet's orbit as follows:

$$D^3 = (76)^2 = 5776$$
$$D = 18 \text{ AU}$$

To find the perihelion distance of the comet from the Sun, we need to know its eccentricity. For Comet Halley, the eccentricity is 0.97. Substituting for D and e in the formula for the perihelion distance, we have

$$r_{min} = 18(1 - 0.97) = 0.54 \text{ AU}$$

This places the comet's perihelion inside the orbit of Venus.

SUMMARY

Before studying the detailed nature of the individual planets as other worlds, it is useful to look up into the sky and try to reconstruct the evolution of ideas about the solar system from the period before spaceflight, or even before the invention of the telescope.

The most basic apparent motions of the lights in the sky (Sun and Moon, stars and planets) are both daily and seasonal. The stars remain in fixed patterns as they rise and set, but the more complicated motions of the Sun, Moon, and planets challenged the ingenuity of ancient peoples to find an adequate interpretation. This was not just an intellectual challenge, but a practical one as well. A working knowledge of the seasons, in particular, was required for the successful development of agricultural society. By 2000 years ago the Greco-Roman world had developed a sophisticated geocentric view of the heavens. People understood time-keeping and seasons, the phases of the Moon, and the causes of eclipses. Their view prevailed in the Western world until about 300 years ago.

The modern worldview of the planetary system was developed between the Renaissance and the eighteenth century, primarily by five extraordinary European scientists. Copernicus (1473–1543) devised a heliocentric theory of the solar system, which he advocated on grounds of simplicity and aesthetic appeal. The heliocentric theory was verified by the first telescopic observations of the planets, carried out in 1610 by Galileo (1564–1642), the founder of modern experimental science. Meanwhile, Kepler (1571–1630), using a remarkable body of pretelescopic measurements of planetary positions made by Tycho (1546–1601), placed the heliocentric theory on a sound mathematical basis by developing his three laws of planetary motion, which still form the foundation for the description of the orbits of planets, satellites, comets, and other solar system bodies.

Kepler's laws are purely descriptive; they tell how the planets move, but not why. The unifying concepts were developed by Newton (1642–1727), who established the physical laws that govern the motion of all bodies, developed the theory of gravitation (a force that acted equally on falling apples and celestial objects), and invented the mathematics required to calculate trajectories and orbits.

Modern observations leave no doubt that the heliocentric model for the solar system is correct. The true distances, motions, and sizes of the planets have been determined, and these properties easily account for the observations so puzzling to our ancestors. Furthermore, we can use this information to understand the operations of rockets and spacecraft that have made possible the exploration of the planetary system.

KEY TERMS

aphelion
astronomical unit (AU)
constellation
eccentricity
eclipse
ecliptic
ellipse
equinox
escape velocity
geocentric
geostationary
heliocentric
inclination
opposition
perihelion
retrograde motion
retrograde rotation
solstice
Titius-Bode rule
zodiac

Review Questions

1. Take a good look at the night sky, and also visit a planetarium if there is one near you. Make sure you understand the apparent motions of the Sun and stars, both daily and seasonal.

2. What is retrograde motion of planets? Explain how this phenomenon was interpreted in both the geocentric and heliocentric systems.

3. What were the main contributions of each of the five scientists discussed in this chapter: Copernicus, Tycho, Kepler, Galileo, and Newton? To what extent was each aware of the accomplishments of the others, and how did they use previous discoveries as a basis for their own contributions?

4. Explain the sequence of phases for both the Moon and Venus, as seen from the Earth. What would be predicted for the phases of Venus according to the geocentric theory? Are they any different from the phases of the Moon, since in the geocentric theory both Venus and Moon orbit the Earth?

5. Describe Kepler's three laws of planetary motion. Write the formula for the period of a planet given its distance in AU. Write the formula for the distance given the period.

6. How is an interplanetary spacecraft launched toward its target? Describe each step: achieving Earth orbit, escaping from the Earth, and going into orbit around the target planet.

7. What evidence could you offer to a friend to support the idea that the Earth revolves around the Sun?

Quantitative Exercises

1. What is the distance from the Sun (in AU) of an asteroid that has a period of revolution of 8 years?

2. What is the revolution period of a hypothetical planet that orbits the Sun at half the distance of Mercury? Or of one that has twice the distance from the Sun of Pluto?

3. What would be the period of revolution for the Earth if the Sun had the same mass but twice its present diameter? What if it had the same diameter but twice its present mass?

4. What is the period of revolution about the Earth for a satellite in circular orbit at an altitude above the surface of 10,000 km?

5. A spacecraft on a trajectory from Earth to Saturn follows an ellipse with perihelion at the Earth's orbit (1 AU) and aphelion at Saturn's orbit (9 AU). If the semi-major axis of this transfer ellipse is halfway from Earth to Saturn (5 AU), what is the time required for the trip from the Earth to Saturn? Using similar reasoning, find the trip time to Mars.

Additional Reading

Beatty, J.K., ed. 1990. *The New Solar System* 3d ed. Cambridge, Mass.: Sky Publishing Corp. An excellent selection of chapters by leading planetary scientists, which can be used as a companion volume to this text.

Boorstin, D.J. 1983. *The Discoverers.* New York: Random House. Historical and philosophical analysis of the human search for knowledge in astronomy, time keeping, and geography, especially during the European Renaissance and Age of Discovery.

Koestler, A. 1959. *The Sleepwalkers.* New York: Macmillan. Famous historical novel that probes the lives and motivations of Tycho, Kepler, and other Renaissance astronomers.

Krupp, E.C. 1983. *Echoes of the Ancient Skies.* New York: Harper and Row. The best popular account of archaeoastronomy, which is the study of astronomy in ancient cultures, especially among the native peoples of the New World.

Morrison, D. 1993. *Exploring Planetary Worlds.* New York: W.H. Freeman. Up-to-date popular overview of the planets.

2

The Sun: An Ordinary Star

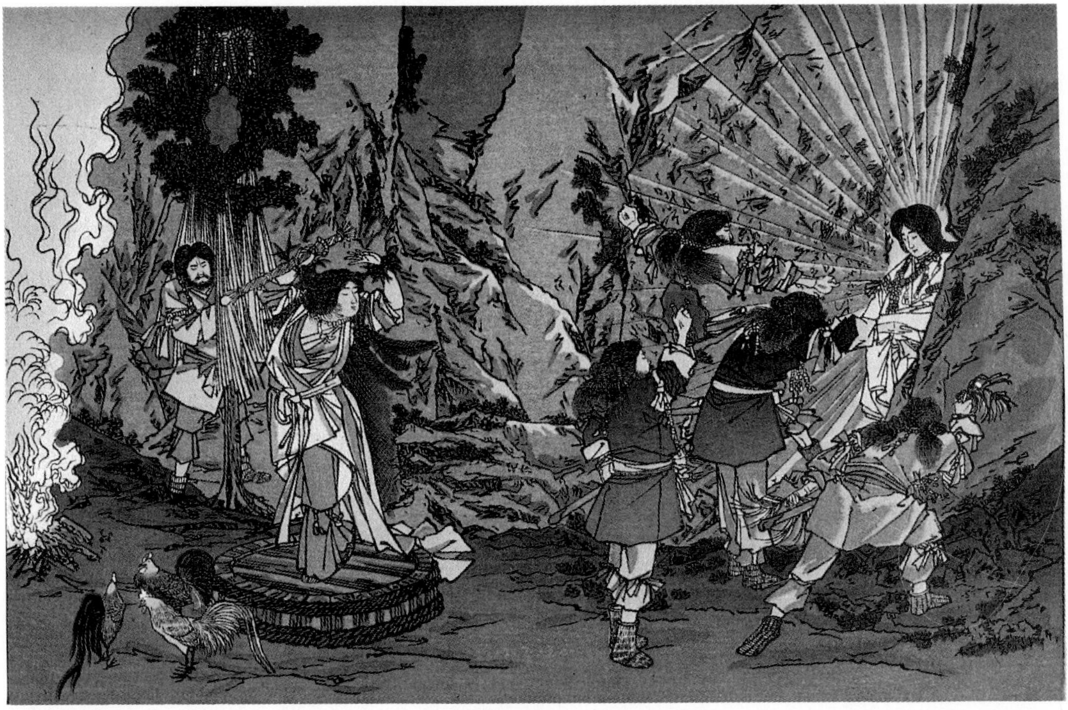

The Japanese sun goddess Amaterasu comes forth from her retreat in a rocky cave in the mountains, symbolizing the return of spring.

The Earth is a planet—this is the fundamental message of Chapter 1. The demonstration by Copernicus and Galileo almost 500 years ago that our own world is one of several planets circling the Sun represented a watershed in intellectual history. This discovery involved more than just a model for planetary motion. It joined together the terrestrial and celestial realms and dethroned humanity from its central place in the universe. Without this perspective, the astronomers could not have developed such concepts as the laws of motion or the universal nature of gravitation. We take the idea of the Earth as a planet for granted today, but we should try to imagine how profound and disturbing this idea was in the time of Copernicus.

The Sun is a star—this is the second great scientific discovery that shaped our image of ourselves and our place in the universe. Galileo, Kepler, and Newton extended the reach of physical law to encompass the planetary system, but they could hardly imagine the vast and varied cosmos that we now recognize, or the insignificant place our own solar system occupies in the great order of things. To them, the Sun was the center of the universe, and the stars were tiny lights of unknown (and unknowable) distance and nature. A few philosophers such as Giordano Bruno speculated on the existence of other suns with other planetary systems, but these ideas could not be evaluated scientifically. Indeed, making such bold and unsupported suggestions led to Bruno's death by fire in 1600.

The Sun is the centerpiece of the solar system. In this chapter we will look at this nearest star, before we begin a detailed examination of the planets and other small bodies that accompany it in space.

2.1 THE SUN AS A STAR

How did scientists first learn that the Sun is a star? Actually, they worked the problem the other way around. They discovered that the stars were suns.

The uniqueness and importance of the Sun for our planetary system stems from two properties: its great mass (more than a hundred times greater than that of all the planets combined) and its prodigious output of energy (a total power of about 4×10^{26} watts). It is this unvarying output of energy that warms the planets and makes life on Earth possible. In contrast, the stars appear to be very faint. Only if they are also very far away can their faintness be understood. The key to recognizing that the stars are suns was to measure their distances, and thus determine their true power output, that is, their **luminosity**.

Distances of the Stars

Early in the nineteenth century, several leading European astronomers undertook the task of measuring the distances to the stars. The only way to measure such distances was by applying the technique of triangulation (as used in surveying) but with extraordinary precision. Their approach was to measure the very slight apparent shift in position of nearby stars as seen from opposite ends of the Earth's orbit (Fig. 2.1). This apparent shift of position from two different viewpoints is called parallax. The stellar parallax the astronomers wanted to measure amounted to a shift of only about 1/10,000°, presenting quite an observational challenge. (The apparent diameter of the Moon in our sky is ½°, and the smallest angle that can be measured without a telescope is about 1/100°.) However, in 1838 success crowned three of these efforts, yielding the distances to the relatively nearby stars Alpha Centauri, 61 Cygni, and Vega. Their distances were extremely large (each greater than 250,000 AU), as they must be if they have luminosities similar to that of the Sun.

Nineteenth-century astronomers also succeeded in measuring the masses of some stars. They found pairs of stars, called double stars or binary stars, in orbit about each other. By measuring the orbital periods and separation of the stars, they could apply Newton's laws to derive their masses. These masses are also similar to that of the Sun. The double stars provided the first evidence that Newton's laws applied outside the solar system, and therefore supported the idea of the universality of physical law.

FIGURE 2.1 As the Earth moves around the Sun from position *A* to position *B*, an observer with a powerful telescope will see the nearby star appear to move from position *a* to position *b* among the background stars. This proves that the Earth moves around the Sun.

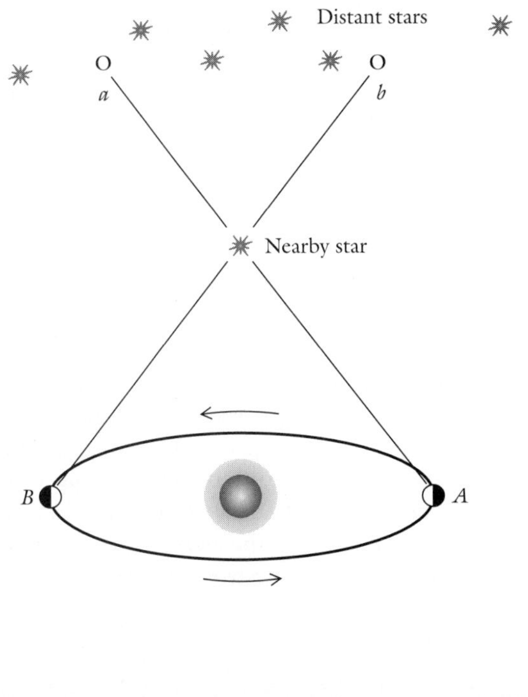

If the stars are suns, and the Sun is a star, then the universe is truly a vast and wonderful place. Thousands of stars can be seen with the unaided eye. Tens of millions of stars can be observed with even a modest telescope. Although they are not exactly alike in such properties as mass, luminosity, and temperature, each of these stars has roughly the same nature as that of our own Sun. Each may therefore be the center of its own planetary system, although we do not have the capability (yet) of distinguishing individual planets in orbit about other stars.

If the universe is filled with stars, then stars must represent one of the most abundant forms of matter. Planets, by comparison, are only small and insignificant fragments of rock, metal, and cool gas.

Although small dark objects of some sort may exist in interstellar space, the planets in our solar system formed together with the Sun, and we believe that planets generally are the result of the same process that forms stars. Understanding the origin of the planets and the Sun is a major theme of planetary research, as we will see throughout this book.

Basic Properties of the Sun

We have already noted that the mass of the Sun is more than a hundred times greater than that of all of the rest of the solar system combined. This mass can be derived by applying Newton's laws to the motion of the Earth and other planets. In metric units, the mass is 2×10^{30} kg, or 2×10^{27} tons. This is a very big number even when expressed in units of the mass of the Earth: 333,000 Earth masses.

The diameter of the Sun is also very much greater than that of even the largest planet. Seen in the sky, the apparent or angular diameter of the Sun is about ½°. Its distance is just equal to the astronomical unit (by definition): 150 million km. To appear so large at such a great distance, the true diameter of the Sun must be greater than a million km, 109 times the diameter of the Earth. The measured value is 1.4×10^6 km. The Sun is thus large enough to hold approximately 1 million Earths within its volume.

In many ways the most important property of the Sun is its luminosity: the quantity of energy that is constantly radiated from its surface in the form of sunlight. This energy fills the solar system, providing light to illuminate the surfaces of the planets and heat to maintain their temperatures. Even at the distance of the Earth from the Sun, this flood of solar energy amounts to 1370 watts per square meter of area. One of the most important questions to be asked concerning the Sun, a question that will be discussed in detail later in this chapter, is the source of this luminosity.

The Sun's energy originates in its deep interior, but the sunlight we see is streaming from the "surface" of the Sun where the temperature is about 5800° on the Kelvin temperature scale (abbreviated as 5800 K). The Kelvin scale uses the same size degrees as the more familiar Celsius scale, but these degrees are measured relative to the absolute zero of temperature, −273°C. Because they are measured with respect to absolute zero, the temperatures in the Kelvin scale are always positive numbers.

At this temperature of 5800 K, the material of the Sun is not solid or liquid but gaseous. When we speak of the solar surface, we refer to the apparent surface, the place where the gas in the solar atmosphere becomes opaque. This apparent surface is called the **photosphere,** meaning the sphere from which the light (photons) originates. Inside the photosphere the Sun must be even hotter, since the energy, which always moves "downhill" from hotter to cooler regions, flows from the interior to the surface. Thus we conclude that the entire Sun is composed of hot, incandescent gas.

TABLE 2.1 Facts about the Sun

Mass	2.0×10^{30} kg 333,000 Earth masses
Diameter	1.4×10^9 m 109 Earth diameters
Luminosity	4×10^{26} watts
Surface temperature	5800 K
Core temperature	15×10^6 K
Average density	1.4 g/cm^3
Distance from Earth	1.5×10^{10} m 1.0 AU
Rotation period at equator	24 days

As the old rhyme goes, "The Sun is a mass/of incandescent gas." The challenge for the astronomer is to learn the composition of this gas, how the Sun's energy is generated, and how long it can continue to shine at its present rate. The past and future of the planetary system, and especially of life on Earth, are intimately bound to the ability of the Sun to provide a continuing, stable source of energy to heat and illuminate the planets. Table 2.1 summarizes some of the most important facts about the Sun for ready reference.

2.2 BUILDING BLOCKS: ATOMS AND ISOTOPES

The matter that makes up both the Sun and the planets consists of pure substances called **elements.** An element is a substance that cannot be decomposed, by chemical means, into any simpler substance; it is the basic building block of matter. The creation of the elements is a natural process in stars, resulting from the nuclear reactions that also generate stellar energy. The details of element formation are the subject for another book. But to understand the composition of the Sun and planets, we need to know something about these fundamental constituents of matter.

The Structure of Matter

The 92 naturally occurring elements are composed of atoms, which consist of nuclei surrounded by orbiting **electrons.** The nucleus of an atom contains almost all of the mass. It is made up of particles called **protons,** which have a positive charge, and **neutrons,** which have no charge at all. The protons and neutrons have about the same mass. The electron carries negative charge and has $1/1836$ the mass of a proton.

Each atom has as many electrons as it has protons, so the atom itself exhibits no net charge in its neutral state. If an electron is removed, however, the atom is said to be an **ion;** it is **ionized** and now exhibits a positive charge. The simplest atom is hydrogen, with a single proton as its nucleus and one electron in orbit about it. If the electron is removed, we have the hydrogen ion, which is just a proton.

The electrons of an atom are capable of both emitting and absorbing light. This ability of each atom to interact with light provides a powerful tool for determining the composition of celestial materials without requiring that actual samples be brought into the laboratory.

Isotopes

Imagine adding a neutron to the nucleus of a hydrogen atom (Fig. 2.2). The mass of the atom is changed by a factor of two, but since the charge of the nucleus is the same, the same single electron is all this new atom needs to remain electrically neutral. This new atom is an **isotope** of hydrogen—that is, a form of an element that differs from others only in the number of neutrons in the nucleus.

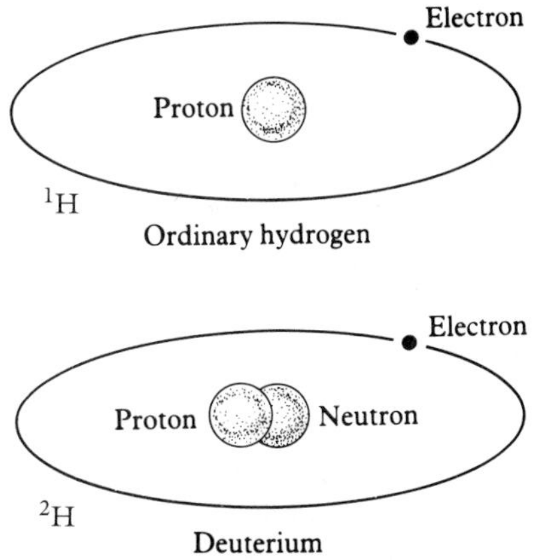

FIGURE 2.2 Ordinary hydrogen atoms consist of one proton as the nucleus and one electron in orbit around it. Adding a neutron to the nucleus changes the mass of the atom but not its electrical charge. The new isotope with one proton and one neutron is called deuterium.

The isotope of hydrogen with one neutron in the nucleus is called deuterium, written ^2H or D. Since deuterium still has just one electron, its chemical properties are virtually identical to those of ordinary hydrogen (^1H), despite its greater mass. Thus deuterium can combine with oxygen to form water, but we now give it the symbol D_2O instead of the familiar H_2O. The surprise comes when an ice cube of D_2O is placed in ordinary H_2O. It sinks! The greater mass of the nucleus of deuterium atoms leads to an increased density for D_2O. More mass is packed into the same volume compared with ordinary H_2O. It is for this reason that D_2O is often called "heavy water."

This is not just an idle thought experiment, since deuterium is found abundantly in nature. Other elements also exhibit more than one isotope. Ordinary carbon, for example, which we find in pencil lead, diamonds, coal, and ourselves, also has two stable isotopes, one of which—with six neutrons and six protons—is 90 times as abundant as the other—which has seven neutrons and six protons. Oxygen has three stable isotopes.

Compounds

Most of the matter we encounter is not in the form of pure elements. Water, carbon dioxide, alcohol, and quartz are all examples of **compounds**, substances that are composed of more than one element. Just as the smallest unit of an element that still preserves the element's chemical identity is called an atom, the smallest unit of a compound is called a molecule. A molecule is formed from the atoms of the various elements making up the compound. Collisions between atoms can form molecules, and ultraviolet light can break molecules apart. Like atoms, molecules can be ionized by losing one or more electrons. The tails of some comets, for example, contain ionized molecules of water, written H_2O^+. The gases around the comet's head contain H and OH, fragments of H_2O molecules that have been broken apart by ultraviolet light.

Both compounds and elements can change their state, becoming gases, liquids, or solids depending on the local temperature and pressure. Of the three possibilities, the liquid state is the most rare, since it requires the most restricted range of temperatures. Water is unusual in remaining liquid over a range of 100° Celsius. Ammonia, for example, is liquid only from −78°C to −33°C, or less than half the range of water.

2.3 COMPOSITION OF THE SUN

The measurement of the composition of objects from a great distance represents one of the triumphs of astronomy. In 1835, the French philosopher August Comte speculated that it would eventually be possible to determine the distances and motions of the stars, but that it would never be possible to determine their chemical composition. Yet within a few decades astronomers were doing exactly that. The key to this accomplishment was the ability to interpret the spectra of the light emitted by the Sun and stars.

Electromagnetic Radiation

The spectrum of sunlight is the collection of its many constituent colors. The colors of the rainbow represent just the visible forms of what is called **electromagnetic radiation**. The light illuminating this page as you read it is an example of electromagnetic radiation. This radiation may be generally defined as the propagation of energy through space by varying electric and magnetic fields. It can be produced and absorbed by interactions of these fields in atoms and molecules.

Electromagnetic radiation can be understood in terms of a wavelike motion of electric and magnetic fields, with each kind of radiation having its own wavelength. But electromagnetic radiation also behaves as if it were made up of particles called **photons**. These particles have no mass, but they do carry energy. The energy of an individual photon is inversely proportional to its wavelength, so long waves correspond to low energies and short waves to high energies.

The Spectrum and Spectroscopy

The **electromagnetic spectrum** (Fig. 2.3) is a way of describing the energy range of electromagnetic radiation. It extends from radiation with very short wavelengths (gamma rays at 10^{-10} cm, x rays at 10^{-8} cm) through the so-called visible region of the spectrum to which our eyes are sensitive (0.35×10^{-4} cm to 0.76×10^{-4} cm), and on through the infrared to the radio region, where wavelengths are measured in meters and kilometers. This can also be thought of as a progression from very high energy (gamma rays) to very low energy (radio waves).

In the part of the electromagnetic spectrum corresponding to visible light, we usually measure wavelengths in nanometers (abbreviated nm). One nanometer is 1 billionth (10^{-9}) of a

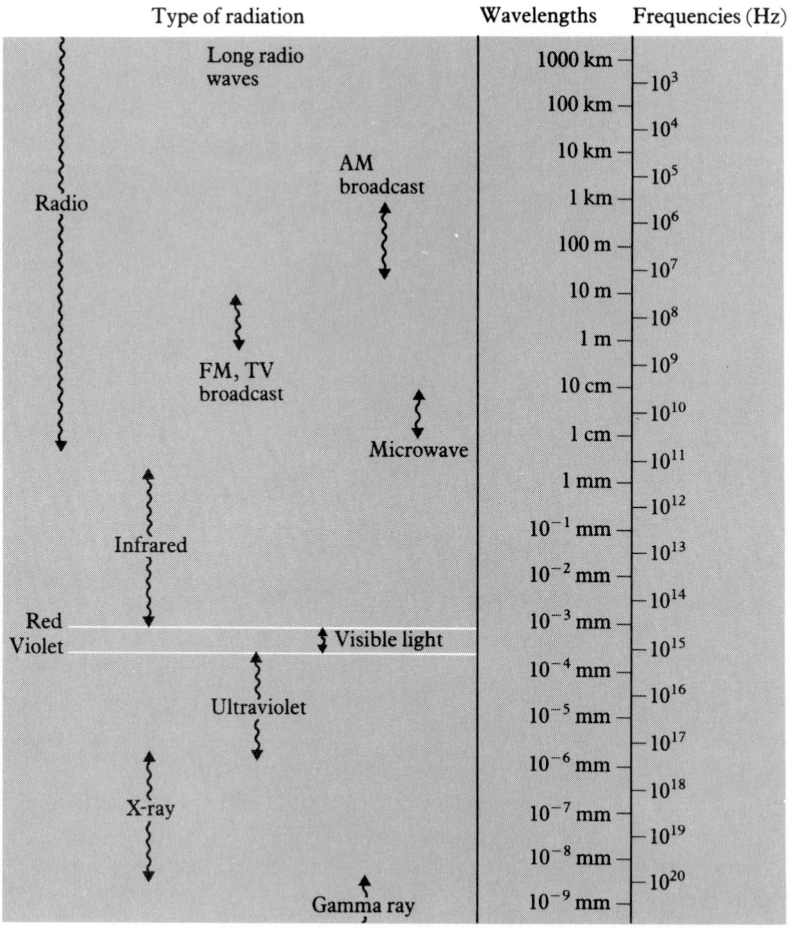

FIGURE 2.3 The electromagnetic spectrum extends from gamma rays with wavelengths shorter than 1 billionth of a centimeter to radio waves with wavelengths measured in millions of centimeters. All of this radiation travels through space at the speed of light. Only certain wavelengths are able to pass through a planetary atmosphere (see Fig. 3.4).

meter, or 10 millionths of a centimeter. In these units, the wavelengths of visible light are from 350 nm (violet) to 760 nm (red). The longer wavelengths of infrared radiation are usually measured in micrometers (μm), where 1 μm is equal to 1000 nm.

The primary tool for the analysis of electromagnetic radiation is spectroscopy. Everyone is familiar with rainbows, the beautiful array of colors created when sunlight passes through a mist of water droplets. The yellow-white light of the Sun is suddenly split into its component colors, which are spread out for us to admire. This is a very simple spectrum, in which radiation is sorted by energy, from violet to red.

We can achieve the same effect in our laboratories by using a prism or a diffraction grating instead of raindrops. Adding a slit to limit the overlapping of the radiation, some lenses to make images, and a detector to record them, we have constructed a **spectrometer** (Fig. 2.4). Now we can spread out sunlight and examine it in great detail.

Formation of Spectral Lines

When we look at the solar spectrum in detail, we find that amidst the beautiful display of colors, which faithfully reproduces the range from violet to red exhibited by rainbows, there are large

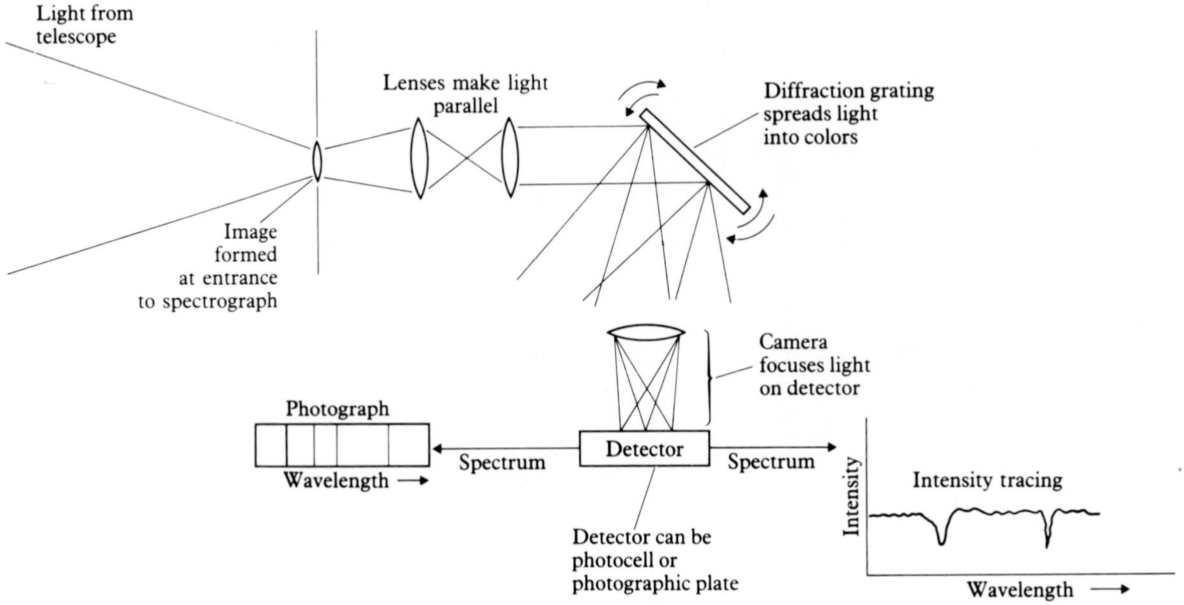

FIGURE 2.4 A spectrometer spreads out the light from a source according to its wavelengths, which we perceive as colors in visible light.

numbers of dark lines called **absorption lines** (Fig. 2.5). These absorption lines represent wavelengths where energy is being removed from sunlight by atoms or molecules in the Sun's atmosphere. We can think of sunlight as radiant energy produced deep in the Sun's interior. This radiation gradually works its way to the Sun's photosphere, from which it escapes into space. But during this last stage it encounters the cooler outer envelope of the Sun, and it is the effects of absorption by atoms and molecules in this solar atmosphere that we see as dark lines in the spectrum.

Why are there discrete spectral lines instead of a general decrease in the intensity of the light, as we experience on a cloudy day on Earth? In

FIGURE 2.5 Bright or dark lines in the spectrum reveal the presence of gases in the source (or between the source and other observer) that absorb or emit specific wavelengths of light. This portion of the visible region of the Sun's spectrum shows dark lines caused by atoms in the outer atmosphere of the Sun. The two prominent absorptions at 589.0 nanometers and 589.6 nanometers are caused by atoms of sodium.

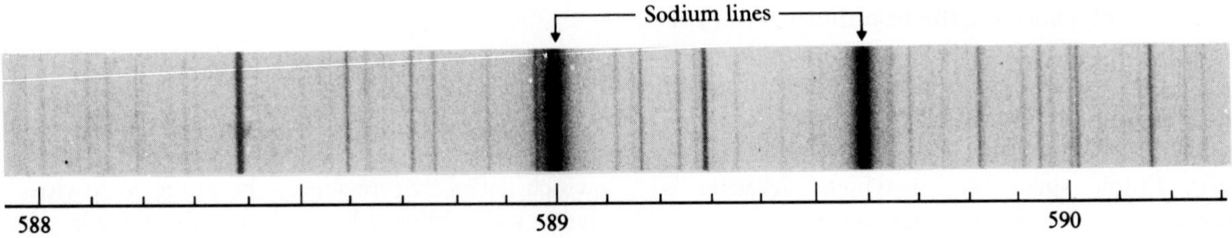

other words, why do atoms and molecules of gas absorb only specific energies or wavelengths, unlike cloud droplets, which absorb all colors? The reason is found in the quantum theory of matter. This powerful theory, which is at the foundation of most modern physics, explains mathematically the discrete nature of the absorption and emission of energy by atoms and molecules.

A violin string produces an approximate analogy to the behavior of an atom of gas. If you pluck the string, you get only one note since the string vibrates at only one frequency. An atom is a bit more complex, but it also operates at only a few specific frequencies or wavelengths, which can be calculated from quantum theory. You can, however, change the violin note (frequency) by changing the length of the string. This would correspond to changing the internal configuration of an atom or molecule, for instance by shifting an electron into a different orbit or removing it to form an ion. In such a changed state, it interacts with a different set of wavelengths or frequencies of electromagnetic radiation to produce a different set of absorption lines.

Atoms and molecules in the solid or liquid state also interact with electromagnetic radiation to produce a characteristic spectral signature. The atomic configuration is generally different for a solid or liquid, however, so the spectrum is distinct from that of the gaseous form. Because the close proximity of atoms to one another in a solid or liquid restricts the vibrations of individual atoms, the solid or liquid is less finely tuned; its spectral features are therefore often broad and fuzzy. To pursue our musical analogy, these forms respond like a bass drum rather than like a violin. Therefore, spectroscopy is a less precise tool for analyzing solids and liquids than it is for gases, and consequently remote sensing has told us less about the nature of solid planetary surfaces than about their gaseous atmospheres.

Since the spectrometer that recorded the spectrum in Fig. 2.5 was located on the Earth's surface, we might expect to find absorption lines produced by gases in our planet's atmosphere also. After all, the sunlight has to pass through the Earth's atmosphere to reach the spectrometer, and along the way the photons are going to encounter a lot of atoms and molecules. It turns out that molecular nitrogen does not absorb visible light, nor does argon. But oxygen does, and some oxygen lines are visible in Fig. 2.5.

Spectral Analysis

It was not until the middle of the nineteenth century that British and German physicists began to use the spectrum as an analytical tool. They found that the solar spectrum contained many absorption lines. Experimenting in the laboratory, they learned that different kinds of gases could absorb specific wavelengths to mimic the solar spectrum. The precise wavelengths that are absorbed depend only on the composition of the gas, not on its density or (over a limited range) its temperature. This is the great strength of spectroscopy as a tool: If we can measure the spectrum of light emitted by the Sun or any other source anywhere in the universe and can match the observed pattern with that of gases measured in the laboratory, we can identify the presence of those gases in the distant source.

The simplest application of spectroscopy occurs when the gas is in atomic form. Fortunately for the nineteenth-century astronomers, the Sun and stars are so hot that most of their material is in the form of individual atoms. In such circumstances, the interpretation of observed spectra is relatively straightforward, once an adequate library of laboratory spectra is available for comparison.

Many of the elements identified spectrally on the Sun are familiar. The prominent lines include the metals iron, magnesium, and sodium. Silicon, an important constituent of terrestrial rock, is clearly present. So are the four elements most essential to life: hydrogen, oxygen, carbon, and nitrogen. But some lines in the solar spectrum do not correspond to common elements on the Earth. Most notable is a prominent set of lines identified in 1899 that seemed to be produced by a simple atom that was not known on Earth. The unknown element was named helium for the Greek word for Sun, *helios*. Shortly thereafter helium was discovered as a trace constituent of the atmosphere of the Earth, where it makes up less than one part in a billion.

Solar Abundances

While it is relatively easy to use spectroscopy to *identify* elements present in the Sun, it is much more difficult to estimate the *relative proportions* of the different elements. Many things in addition to the abundance of the gas influence the strength of the spectral lines. At first scientists assumed that the solar abundances were similar to those on the Earth, perhaps with the exception of a few odd gases like helium. Not until the late 1920s did a Harvard graduate student, Cecilia Payne, succeed in deriving the correct abundances for the major elements in the Sun and stars. She showed that in spite of their differences in size or temperature, the stars are all made primarily of the light gas hydrogen. Subsequently it was found that helium is the second most abundant element, and that fully 99% of the Sun and most other stars is composed of just these two gases. Other more familiar elements, such as oxygen, nitrogen, carbon, silicon, and iron, are just trace constituents of the Sun. Table 2.2 lists the abundances of the major elements in the Sun. Later in this book, when we discuss the origin of the solar system, we will have to explain why the composition of the Sun is so different from that of the Earth and other planets.

2.4 THE SUN'S ENERGY

The composition of the Sun derived from spectroscopy refers to the outermost layers, which is where the sunlight we see originates. Indeed,

TABLE 2.2 Cosmic abundances of the major elements

Element	Symbol	Atomic Number	Number of Atoms per Million Hydrogen Atoms
Hydrogen	H	1	1,000,000
Helium	He	2	80,000
Carbon	C	6	420
Nitrogen	N	7	87
Oxygen	O	8	690
Neon	Ne	10	130
Sodium	Na	11	2
Magnesium	Mg	12	32
Aluminum	Al	13	3
Silicon	Si	14	45
Sulfur	S	16	16
Argon	Ar	18	1
Calcium	Ca	20	2
Iron	Fe	26	32
Nickel	Ni	28	2

almost everything we can determine about the Sun from the analysis of sunlight really applies only to this thin upper layer, consisting of gases in the photosphere and above it. However, the application of a few basic ideas helps us to understand the invisible solar material below the surface.

The Sun is radiating energy from its photosphere, energy that must originate in the interior. In order for this energy to continue to flow out, the interior of the Sun must be hotter than the surface. If the interior were not hotter, the energy would flow from the surface back into the interior, and we would all be in bad trouble. In addition, the upper layers of the Sun must be supported by gas pressure from underneath. If they were not, the Sun would collapse from its own great weight, another bad outcome!

There is evidence that the Sun has been shining for billions of years, and that it has not changed much in either size or luminosity over that period of time. Its present configuration must therefore represent some sort of equilibrium, with a stable source of energy at the center and with the weight of each overlying layer supported by the pressure of the gas below. This assumption of *equilibrium* permits astronomers to calculate the range of pressure and temperature throughout the solar interior. They have found that the central temperature of the Sun is 15 million K (1.5×10^7 K) and that the pressure is 300 billion times greater than the surface pressure of the Earth's atmosphere (3×10^{11} bars). They also verify that most of the solar energy originates in the core, within the innermost 10% of the volume of the Sun. But how is this energy generated?

Energy Source: Early Ideas

The observed luminosity of the Sun is 4×10^{26} watts. This is the energy that the Sun is radiating into space today. If the Sun is in equilibrium, this must also be the rate of energy generation in the core.

The source of energy most familiar to us on Earth is oxidation or burning. Early in the nineteenth century, some scientists suggested that the Sun could obtain its energy by oxidation or other chemical reactions, but calculations quickly showed that this was not possible. Even if the immense mass of the Sun consisted of a burnable material like coal, oxidation could not produce energy at its present rate for more than a few thousand years. With a chemical source excluded, scientists began to look for a physical source for the Sun's energy.

The simplest physical mechanism to produce large quantities of energy is a slow contraction of the Sun. To see how contraction can generate energy, we turn to one of the fundamental laws of nature known to nineteenth-century scientists: the *Law of Conservation of Energy.* This law says that energy cannot be created or destroyed, but it can be transformed from one form to another. The steam engine, for example, relies on the transformation of thermal energy (from the boiler) to the mechanical energy of a piston or rotating wheel. The reverse is also possible, to transform mechanical motion into heat. You do this, for example, when you rub your hands together to warm them.

If an object—a planet, for example—should fall into the Sun, it would be accelerated by the Sun's gravity and would strike with great force, transforming its energy of motion into heat. In a similar way, the German physicist Hermann von Helmholtz and the English physicist William Thomson (Lord Kelvin) proposed that the outer layers of the Sun might "fall" inward and thereby produce heat. Calculations showed that a tremendous amount of energy could be liberated by even a very small contraction. Helmholtz and Kelvin found that the total energy output of 4×10^{26} watts could be provided by an annual contraction of only 40 m. Over the time span of human history (about 10,000 years), such a contraction would amount to about 400 km, or less than 0.03 percent of the Sun's diameter. Such a small contraction would not be measurable.

Helmholtz and Kelvin proposed contraction as the source of the Sun's energy, and they calculated that this energy source could have kept the Sun shining for about 100 million years. At the time, this seemed ample, for the Earth was thought to be only a few million years old.

By the beginning of the twentieth century, the Kelvin-Helmholtz theory began to run into trouble. New observations and analysis showed that the Earth was hundreds of millions of years old—perhaps more than a billion years. Contraction could not provide enough energy to keep the Sun shining for so long. So physicists and astronomers began to search for an alternative energy source.

The Possibility of Nuclear Energy

Albert Einstein (Fig. 2.6) was a young patent clerk in Bern, Switzerland, when he developed the Special Theory of Relativity in 1905. Relativity is a fascinating concept that we cannot discuss in any detail in this text. Here we are concerned with one of the unexpected by-products of Einstein's work: the discovery of the *equivalence of mass and energy.* Nineteenth-century physics had established that while energy was conserved, it could be transformed from one form to another. Einstein discovered that the conservation of energy was not strictly correct: Mass and energy are different forms of the same thing, and what is conserved is a new quantity called mass-energy. In principle, at least, energy can be transformed into mass, and mass into energy.

Einstein's famous equation, $E = mc^2$, expresses the equivalence between energy (E) and mass (m). The other term in the equation is c, the speed of light (about 300,000 km/s). Because c^2 is a very large number, this equation tells us that even a small amount of mass can create a great deal of energy. Here, then, is a possible solution to the dilemma of the Sun's long-term energy source.

If 4 million tons of matter could be transformed each second into energy, 4×10^{26} watts of power would be generated. This sounds like a lot of material, but compared with the total mass of the Sun it is not. The Sun could lose this much mass every second for tens of billions of years without altering its total mass by more than 1%. But can mass actually be transformed into energy? Is there a practical reality associated with this theoretical prediction?

FIGURE 2.6 Young Albert Einstein, the most eminent scientist of the twentieth century.

Thermonuclear Fusion

Several decades of work by physicists were required to identify a reaction in the Sun that might transform mass into energy. From careful laboratory measurements, they found that the nucleus of the helium atom, which can be thought of as four hydrogen atoms combined, has only 99.28% of the mass of the sum of four hydrogen atoms. Thus, if there were a way to combine or fuse four hydrogen nuclei to make one helium nucleus, excess mass would be present. This mass is converted to energy in the fusion process and is liberated primarily in the form of high-energy gamma radiation. As we have seen, the Sun and stars are composed mostly of hydrogen. It also turns out that the high temperatures and pressures in the solar interior are conducive to the fusion of hydrogen into helium. Because these reactions involve atomic nuclei and take place at high temperature, they are called **thermonuclear fusion** reactions.

Physicists have discovered two ways that hydrogen fusion can be accomplished in the core of the Sun and other stars. As we know, they have also found ways to achieve fusion on a much smaller scale, producing the thermonuclear or hydrogen fusion bomb. But that is another story.

The most important set of reactions in the Sun is called the **proton-proton chain** (Fig. 2.7). At temperatures above 10 million K, the hydrogen nuclei (or protons) are moving at speeds of more than 1000 km/s. Even at these speeds, an average proton will rebound from collisions with other protons for about 14 billion years, at a rate of 100 million collisions per second. But very occasionally, two protons will stick together or fuse to form an atom of deuterium, with the release of radiation energy in the form of gamma rays. At this point, the main barrier to the creation of helium is passed. In just six seconds (on average) another proton is absorbed to form light helium, a nucleus consisting of two protons and one neutron, again with the release of gamma radiation. About 1 million years after this the nucleus of light helium collides with another light helium nucleus to form one nucleus of regular helium (two protons and two neutrons) and two protons, which can start this process again.

The mass lost in the proton-proton sequence of reactions is 0.71%. Thus if one kilogram of hydrogen turns into helium, 0.0071 kg of the mass is converted into energy. To maintain the luminosity of the Sun, 600 million tons of hydrogen must be converted to helium each second, with the net conversion of 4 million tons of mass to energy. This energy works its way out through the layers of the Sun until it bursts forth as visible light to illuminate the rest of the planetary system—a process that we discuss in more detail in the next section.

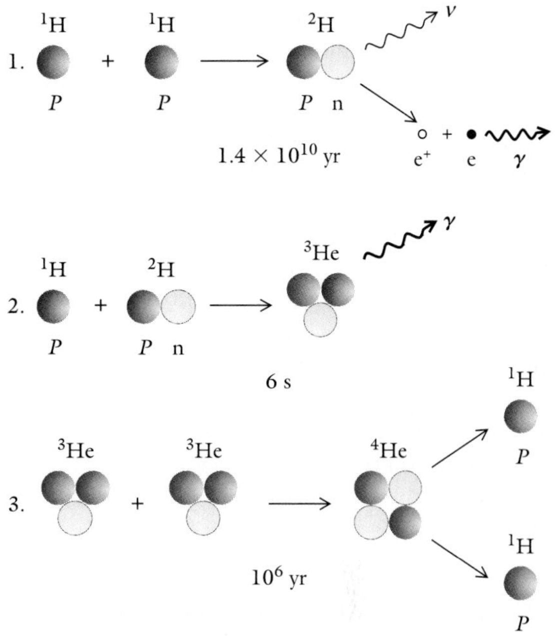

FIGURE 2.7 The Sun generates its energy from the fusion of four hydrogen atoms into one helium atom. The sequence of nuclear reactions illustrated here is called the proton-proton chain, since the first (and most difficult) step is the fusion of two protons (H) to produce one atom of deuterium (D). This step also releases a neutrino (ν) and a positron (e^+), which is a positively charged electron. The positron collides with an electron (e) to produce a gamma ray (γ).

2.5 LIFE HISTORY OF THE SUN

Once scientists recognized the source of the Sun's energy, calculating an approximate limit to the Sun's lifetime was a straightforward matter.

Recall that the Sun must transform 4 million tons of matter each second into energy to satisfy Einstein's equation $E = mc^2$. This quantity of excess mass is produced by the fusion into helium of 600 million tons (or 6×10^{11} kg) of hydrogen per second. Let us suppose that the entire mass of solar hydrogen could be converted to energy in this way. The mass of the Sun is 2×10^{30} kg, of which about 75 percent, or 1.5×10^{30} kg, is hydrogen. The time necessary to exhaust this hydrogen at the current consumption rate is 1.5×10^{30} kg divided by 6×10^{11} kg/s, or about 3×10^{18} s. Since there are 3×10^7 seconds in a year, this lifetime is 10^{11} years—100 billion years. This is a very long time, but is the result realistic? Can the Sun really convert all of its hydrogen into helium?

Energy Generation and Transport

To understand what is happening inside the Sun, we must look more closely at the generation of energy in its core and the flow of energy from the core to the surface. First note that the rate of fusion of hydrogen into helium is sensitive to the temperature. A relatively small increase in temperature causes the protons to move faster and results in a higher fusion rate. More fusion means more energy, and an even hotter gas. If this situation were to continue, the core would explode. However, the Sun has a way of compensating. When the production of energy starts to increase, higher pressures in the core cause the Sun to expand and the core to cool. Conversely, if the energy generation in the core should lag, the Sun would contract, converting gravitational energy to heat and rekindling the nuclear fires in the core. In this way, equilibrium is maintained.

In a steady state, all of the energy produced in the core must find its way outward to the surface, where it is radiated into space. There are three ways for energy to be transported from one place to another. On Earth, the most familiar form of energy transport is **conduction**—the process that causes a pan placed on the stove to become hot. Conduction involves the transport of energy in a solid by molecular motion. Apply heat at one point, and the molecules increase their speed. They bump into their neighbors, and soon all are moving faster—that is, the temperature rises. Conduction plays a major role in the transport of heat inside planets, as we will see later, but inside the Sun it is overwhelmed by more efficient transport mechanisms.

One of these mechanisms is **radiation.** As we have already noted, electromagnetic radiation involves the transmission of energy through a vacuum by electric and magnetic waves. Radiation is the process by which sunlight illuminates and warms the surface of the Earth. It is also the way you are warmed by a fire in a fireplace. Radiation works best in a vacuum, but it can also play an important role in the transfer of energy within the Sun, especially in the core itself. The gamma rays released by thermonuclear fusion work their way outward primarily through the process of radiation.

The other important mechanism for the transport of energy in the Sun is **convection.** Like conduction, convection requires the presence of matter. Now, however, it is not individual molecules that vibrate and collide with each other, but large masses of gas or liquid that move from warmer to cooler regions, carrying energy with them. Convection is a *macroscopic,* not a *microscopic,* process. Where such mass motion is possible, convection becomes the most efficient of these three energy transport processes.

Within the Sun, convection is possible only in the outer layers. Within about the outer third of the Sun, the gas is in a constant state of agitation, with masses of warm gas rising and blobs of cooler gas descending to take the place of the rising columns. Deep in the interior, radiation dominates. In either case, the time required to move energy from its source in the core up to the surface layers is very long—about a million years. The sunlight you see today originated in fusion reactions that took place near the time of the emergence of *homo sapiens* on our planet. This resistance to the outward flow of energy is critical to the stability of the Sun.

Solar Evolution

Astronomers use all of the concepts we have described—the equilibrium of the Sun, hydrogen fusion in its core, and the transport of energy from the core to the surface—to calculate detailed models of the solar interior. They thus develop a quantitative picture of the structure of the Sun as it is today. These same computational tools also permit them to calculate what will happen as the hydrogen in the core is gradually converted to helium. In this way, it is possible to predict the future evolution of the Sun.

The current structure of the Sun is illustrated in Fig. 2.8, which shows the variation of temperature, energy generation, and composition from the center to the surface of the Sun. Note in particular the depletion of hydrogen in the core, where fusion has been at work for billions of years. It is the gradual depletion of hydrogen fuel in the center that causes the Sun to change as it ages.

The response of the Sun to the exhaustion of hydrogen at its center is to readjust its structure so as to increase central temperatures. At the same time, its total energy production also increases. Thus we predict that the Sun will gradually become more luminous with time. After several billion years, the inner core will consist almost entirely of helium and the solar luminosity will be several tens of percent greater than it is today. Then, the calculations predict that the Sun will rapidly expand to enormous size, becoming what astronomers call a red giant. In its bloated state, the Sun will extend at least to the orbit of the Earth. Our planet will be consumed in the solar fires, and the planetary system as we know it will cease to exist.

The same calculations can be extended backward in time to determine what the Sun used to be like. We conclude that the Sun began to convert hydrogen to helium in its core between 4 and 5 billion years ago. Thus the Sun is between 4 and 5 billion years old—a result that is in excellent agreement with the measured ages of the planets, indicating that the Sun and its retinue of planets formed together.

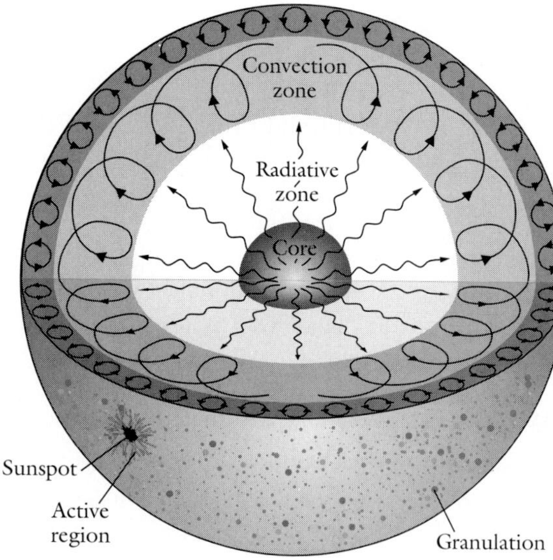

FIGURE 2.8 The internal structure of the Sun has been determined from computer models aided by observations of waves and vibrations in the solar atmosphere. As shown in this illustration, there are three main regions: a hot dense core where nuclear reactions take place; a layer extending more than halfway to the surface in which the energy is transported primarily by radiation; and the outer third of the Sun, where energy is transported mainly by convection.

Early in its history, when hydrogen still made up about 75 percent of its core, the Sun must have been less luminous than it is today. Calculations show that the Sun has increased in luminosity by about 40 percent in the past 4 billion years. Therefore, the Earth and other planets must have been cooler in the past than they are at present. We will return to this constraint on planetary history in Chapter 19, when we discuss the origin and early evolution of the planetary system.

2.6 SOLAR ACTIVITY

We depend on the steady flow of energy from the Sun to the Earth and other planets. If the Sun's luminosity varied substantially, life would be difficult or impossible in the solar system. Fortunately for us, the Sun maintains a stable equilibrium. This equilibrium is only approximate, however, and observations show that our star is capable of changes whose influence is felt throughout the planetary system.

Surface Activity on the Sun

Beneath the visible photosphere of the Sun, great convection currents transport energy upward. These convection currents are characterized by cells of rising warmer gas typically 700–1000 km across, traveling upward at speeds of several kilometers per second. These rising cells are separated by sinking columns of cooler gas. The solar surface seethes and boils like a giant witch's cauldron (Fig. 2.9).

Like breaking waves, the hot convecting gas crashes into the outer layers of the Sun. In rising from the interior, this hot gas carries both thermal and mechanical energy. The rising gas also becomes entangled with the magnetic field of the Sun. One result of these complex interactions is the transport of a great deal of energy to the outer atmosphere of the Sun, called the **corona.** The solar corona consists of very thin gas, heated to over a million degrees, extending for millions of kilometers into space. It is most easily seen in visible light during an eclipse of the Sun, when the Moon blocks the brighter disk of the Sun from view (Fig. 2.10). Such an eclipse is one of the most dramatic sights in nature.

Because it is so hot, the gas of the corona emits most of its radiation in the form of x rays. These x rays cannot be seen from the surface of the Earth, but x-ray images of the Sun are commonly produced from space. Figure 2.11 illustrates the appearance of the Sun in x rays and thus outlines the hottest (brightest) regions of the corona.

FIGURE 2.9 (a) The visible surface of the Sun is a seething mass of gas, with motions driven by convection currents of heat that rise from the interior. It includes dark regions called sunspots as well as many plumes of hot gas called spicules. (b) Close-up of a sunspot, surrounded by the honeycomblike structure of the photosphere. This dark spot on the Sun is about twice the size of the Earth.

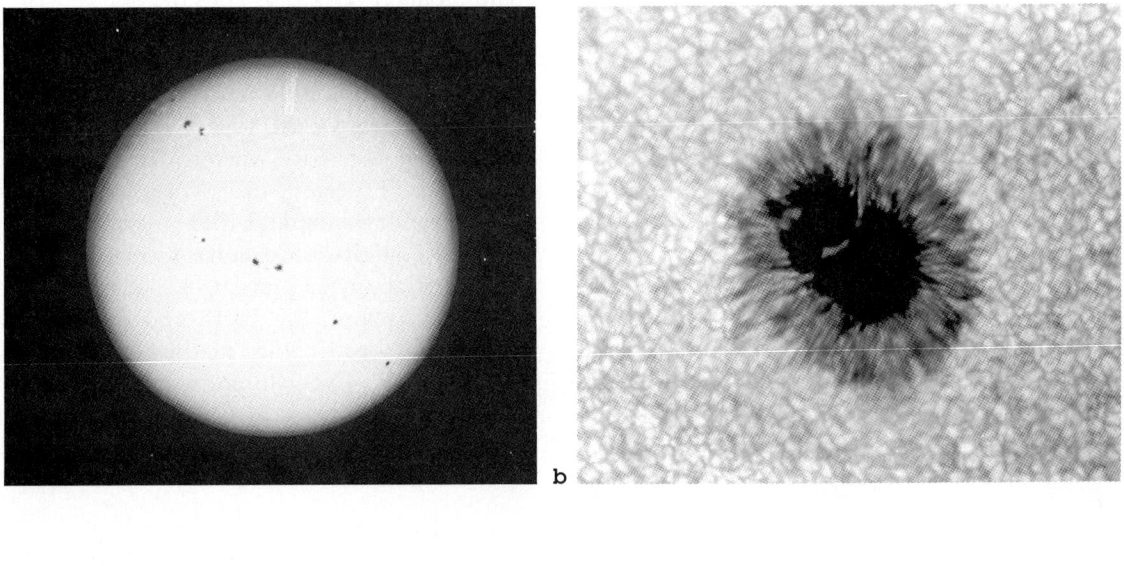

a b

FIGURE 2.10 The tenuous, hot outer atmosphere of the Sun is called the corona; it is most easily seen during an eclipse of the Sun, when the Moon blocks the bright surface from our view.

The Solar Wind

The solar corona does not end suddenly but continues to great distances from the Sun. Because they are so hot, the coronal gases expand rapidly and flow away from the Sun with only minimal resistance from solar gravity. The result is a **solar wind** of hot gas streaming out to fill the solar system.

The gas of the solar wind has essentially the same composition as the atmosphere of the Sun itself. Like the gas of the corona, the atoms in the solar wind are ionized. The great bulk of the solar wind is composed of individual protons (hydrogen nuclei) and electrons. An electrically charged gas of this sort, in which the protons and electrons are separate instead of combined in the form of electrically neutral atoms, is called a **plasma.** This plasma has imbedded within it a part of the magnetic field of the Sun, which is also carried out to the realm of the planets.

At the Earth's distance of 1 AU, the average speed of the solar wind is 450 km/s. The wind particles travel from the Sun to the Earth in about five days. The wind speed declines slowly with increasing distance, but even at the outer planets it is still moving several hundred kilometers per second.

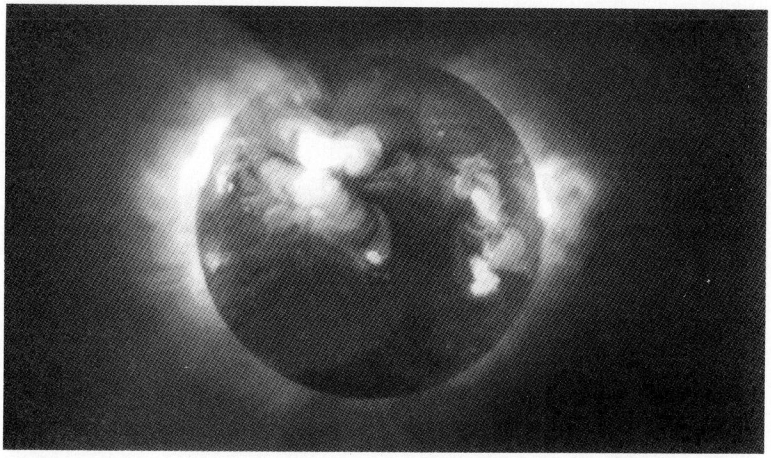

FIGURE 2.11 The hottest regions of the solar corona are strong sources of x-ray emission. Many centers of activity having temperatures of over 2 million K are shown in this image.

a

b

FIGURE 2.12 (a) A photograph of Comet Halley, when it visited the inner solar system in 1986, shows a straight plasma tail with considerable structure. The tail is glowing with light from ionized carbon monoxide gas molecules that are being blown away by the solar wind. (b) A painting of Donati's Comet as it appeared to the naked eye on Oct. 4, 1858, over La Cité in Paris. Note the two straight plasma tails and the curved dust tail. The bright star near the comet's head is Arcturus in the constellation Boötes. (Note that this is an artist's reconstruction, not a drawing from nature.)

The solar wind was discovered before the space age from its effects on the tails of comets. This solar plasma interacts with the gas steaming outward from the center of a comet and carries this gas away from the Sun, in the direction of the motion of the solar wind (Fig. 2.12). By measuring the way individual streamers of gas move outward along the tail of a comet, astronomers can determine the speed of the solar wind and estimate its density. Since about 1960 these indirect astronomical observations have been supplemented by direct measurements of the solar wind made by interplanetary spacecraft.

Sunspots

The Sun has a complex magnetic field that varies with both time and position on the surface. In some parts of the photosphere the magnetic field can become so compressed that it inhibits the flow of hot gas from the interior. The result is a region of reduced temperature called a **sunspot.** If you look at the Sun with a small telescope *equipped with a suitable filter*, you can see the sunspots as small dark blotches against the bright photosphere. Although the sunspots look black, this is mostly a contrast effect. The typical temperature in a sunspot is about a thousand degrees lower than that of the surrounding photosphere, but that is still hot enough that the gas of the spot would radiate brightly if it were seen by itself.

Sunspots vary in size from the smallest we can see with telescopes up to several times larger than the Earth (tens of thousands of kilometers across). They tend to form in groups, often with one large spot accompanied by as many as 100 smaller spots (Fig. 2.13). As the Sun rotates with a period of about 25 days, the spots are carried across the surface. A large sunspot group may have a lifetime of several solar rotations, although individual spots form and coalesce within the group on a timescale of days.

Sunspots are of interest to us primarily because of what they tell us about the changing magnetic field of the Sun. The magnetic field in turn influences a number of energetic outbursts on the Sun that propagate into the corona, into the solar wind, and ultimately to the upper atmospheres of the Earth and planets.

The Solar Cycle

In 1843 Heinrich Schwabe, an amateur astronomer, noted from a long series of his observations that the sunspots come and go in a regular cycle of 11 years. At sunspot maximum there may be dozens of sunspot groups visible

FIGURE 2.13 Sunspots form in groups near regions of enhanced magnetism on the solar surface. This series of photos shows a large group of sunspots as it evolves over a period of nine days.

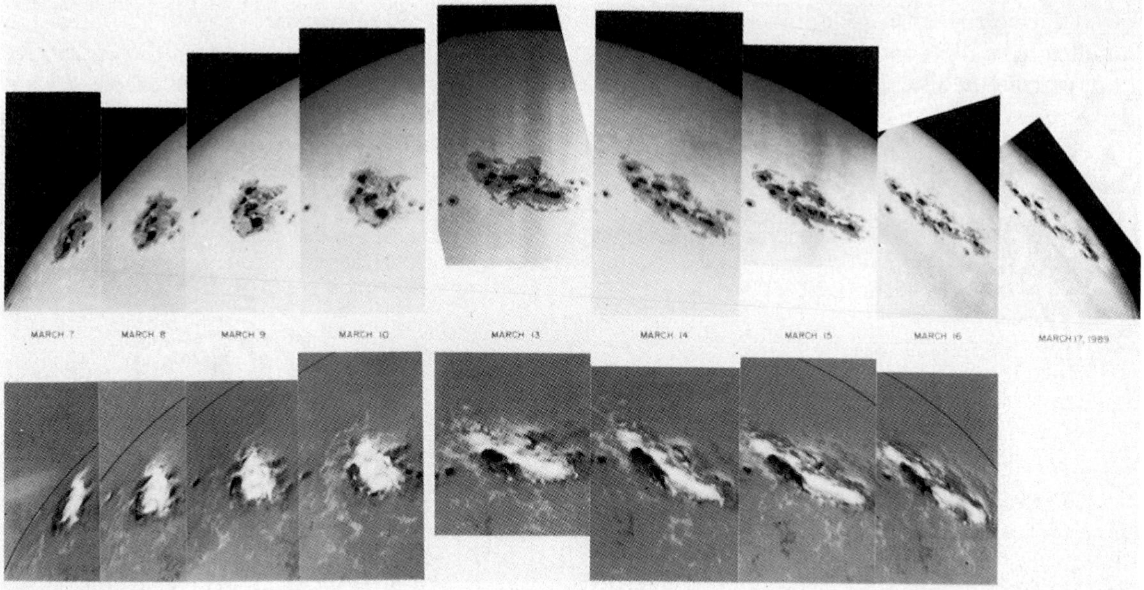

on the solar disk, while at sunspot minimum the number drops nearly to zero (Fig. 2.14). The location of the spots also varies periodically. At the start of a cycle, the sunspots appear at a latitude of about 35° north and south of the equator, but as their numbers increase, the spots are found at lower and lower latitudes. The survivors of one cycle can still be seen near the equator as spots of the next cycle appear at higher latitudes.

When astronomers early in the twentieth century developed techniques for measuring directly the magnetic field of the Sun, they discovered that the 11-year sunspot cycle is closely related to variations in the solar magnetic field. They found that the sunspots in alternate cycles had reversed magnetic polarity, so that the total period for a repeat of spots with the same magnetic orientation was 22 years and not 11. The entire global magnetic field of the Sun also reverses itself in the same 22-year cycle. This 22-year variation is called the **solar activity cycle**.

The changes in solar magnetic fields that take place during the solar cycle influence the total luminosity of the Sun by inhibiting convection of heat from the interior. Near times of sunspot maximum the flow of energy from the interior is strongest, and the Sun is on average about 1 percent brighter than it is near sunspot minimum. The extent and brightness of the corona and the speed and density of the solar wind also share this cyclic variation. It seems reasonable to suppose that there is a corresponding 22-year variation in aspects of the atmospheres of the Earth and planets that are influenced by the solar wind. Although there are no proven indications that these solar effects are coupled to changes in the lower atmosphere of the Earth (weather effects), there are some suggestions that droughts in the U.S. Great Plains, for example, recur at 22-year intervals. Such questions of possible *solar-terrestrial relations* are an active area of contemporary research.

2.7 QUANTITATIVE SUPPLEMENT: RADIATION LAWS

The general nature of the radiation emitted by the Sun can be described by quite simple mathematical formulas. Everything in the universe that has a temperature above absolute zero (–273°C) emits radiation across the entire electromagnetic spectrum. This is called *thermal radiation*, to distinguish it from other more exotic processes that can produce nonthermal

FIGURE 2.14 Sunspots come and go with a period of about 11 years, called the sunspot cycle. This figure shows how the number of sunspots has varied as a function of time over the past 350 years.

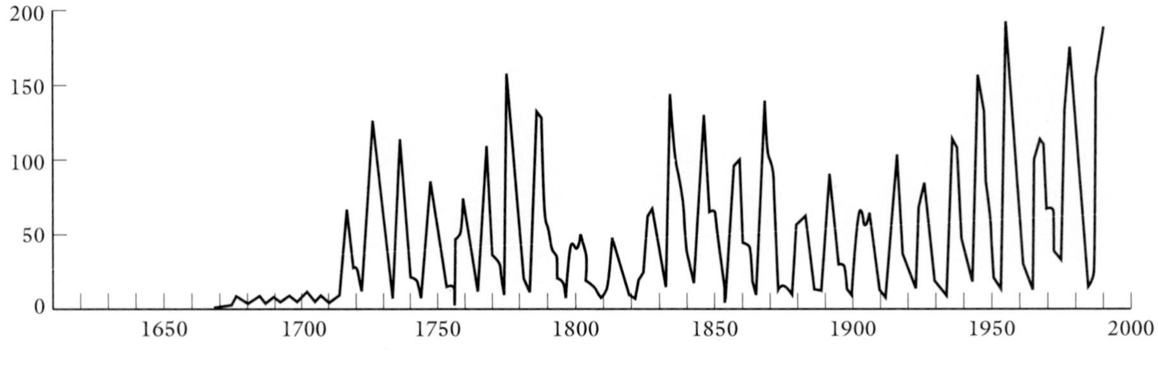

radiation, such as the output of a terrestrial radio transmitter. One of the characteristics of thermal radiation is that its spectrum and intensity depend only on the temperature of the source, and are independent of the chemical composition or other properties of the emitting material. Thus physicists in the nineteenth century were able to deduce formulas, called radiation laws, to describe this ideal thermal radiation, which is sometimes called *blackbody* radiation. Real objects such as the Sun and stars are not perfect radiators, but the basic radiation laws often describe their emission of energy reasonably well.

One radiation law, called the *Stefan-Boltzmann Law* after its discoverers, describes the total emitted power from a perfect thermal radiator. As noted previously, this total energy depends on temperature only. This temperature must be expressed on the absolute, or Kelvin, scale for our calculations. The formula is

$$P = \sigma T^4$$

When P is expressed in units of power (watts), the constant of proportionality, called the Stefan-Boltzmann constant, is given by

$$\sigma = 5.67 \times 10^{-8}\ \mathrm{Wm^{-2}\,deg^{-4}}$$

As an example of the use of the Stefan-Boltzmann Law, we can calculate the relative powers emitted by one square meter on the Sun and by an equal area on the Earth. The solar surface has a temperature of 5800 K, while the so-called effective temperature of the Earth is about 260 K. Thus the ratios of emitted power are

$$\frac{P_{\mathrm{Sun}}}{P_{\mathrm{Earth}}} = \frac{(5800)^4}{(260)^4} = 2.48 \times 10^5$$

The higher the temperature of the source, the shorter the wavelength at which the maximum power is emitted. The *Wien Radiation Law* expresses this as

$$\lambda_{\max} = 0.29 / T$$

when $\lambda_{\max}$ is in centimeters and T is in Kelvins. The solar spectrum has a maximum at $0.29/5800$, or about 5×10^{-5} cm (500 nm). This peak occurs right in the spectral interval to which our eyes are sensitive—visible light—a fact that is surely no coincidence. In contrast, the peak of the terrestrial thermal emission spectrum is at $0.29/260 = 11\ \mu$m, which places this emission mostly in the infrared. The peak for Pluto, with a temperature of about 55 K, is at about 50 μm.

Both of these radiation laws are incorporated in the much more general expression derived in 1899 by Max Planck, which describes the thermal emission spectrum—the power emitted at each wavelength—for a perfect thermal radiator (or blackbody). The *Planck Radiation Law* gives the flux F_λ (in watts/m^2) per centimeter of wavelength:

$$F_\lambda = \frac{(3.74 \times 10^{-8})\,\lambda^{-5}}{e^{c_2 \lambda T} - 1}$$

If the wavelength λ is measured in centimeters, the constant c_2 has a value $c_2 = 1.439$.

Note that all perfect radiators emit at all wavelengths, although of course this emission is strongest near the peak given by the Wien Law. Also, at any given wavelength, a hotter source emits more energy; in other words, the Planck curves corresponding to the emission spectra of sources of different temperatures never cross each other.

SUMMARY

The Sun dominates the solar system: Its gravitation defines the orbits of the planets, and its radiation provides energy to maintain their temperatures. The mass of the Sun is a thousand times greater than that of all the planets combined, and its luminosity is an astounding 4×10^{26} watts.

One of the most important discoveries of the nineteenth century was the great distances of the stars, which proved that the stars were similar to our Sun and might, by analogy with our solar system, possess planets and even life. A second fundamental development was the interpretation of spectra to permit chemical analysis of the Sun and stars from a distance. The analysis of spectra is at the heart of modern astronomy, since the light emitted by astronomical sources is imprinted with the signatures of the elements and compounds that compose the sources of the radiation. These discrete spectral signatures are a consequence of the quantum nature of matter and its interaction with radiation.

The Sun is composed primarily of the two light gases hydrogen and helium, with common terrestrial elements such as silicon and iron mere trace constituents. The tremendous energy output of the Sun is obtained from the fusion of hydrogen nuclei into helium, a process that takes place in the solar core where the temperature is above 10 million K. Hydrogen fusion converts mass to energy (as described by Einstein's equation $E = mc^2$), and this energy, originating as gamma rays, slowly works its way to the solar surface by a combination of radiation and convection. These processes have been at work for nearly 5 billion years, and they will continue for another 5 billion years before the Sun exhausts the hydrogen in its core and expands to consume the inner planets, including the Earth.

The solar interior is in a state of equilibrium, which is fortunate for us, since a stable source of light and heat is required for the survival of life. However, the release of energy at the surface of the Sun shows a 22-year solar activity cycle, which is driven by the constantly changing magnetic field of the Sun. There are cyclic changes in the temperature of the corona and the density of the solar wind, which may in turn have subtle effects on the upper atmosphere of the Earth and even on our weather, although such effects have not been proved to the satisfaction of all scientists. We will continue to watch the Sun carefully, however, for the health of the Sun is a prerequisite for our own well-being.

KEY TERMS

absorption line	luminosity
compound	neutron
conduction	photon
convection	photosphere
corona	plasma
electromagnetic radiation	proton
electromagnetic spectrum	proton-proton chain reaction
electron	radiation
element	solar activity cycle
ion	solar wind
ionize	spectrometer
isotope	sunspot
	thermonuclear fusion

Review Questions

1. Explain how we know that the stars are like the Sun. If some of the stars are markedly different in fundamental properties such as mass, luminosity, or temperature, how could we tell this?

2. What is the electromagnetic spectrum? Consider the spectral analysis of light that originates in the Sun, is reflected from the surface of Mars, and finally reaches our telescope on the Earth's surface. Where might the dark lines seen in this spectrum have originated?

3. Compare the cosmic abundances of the elements given in Table 2.2 with their abundances on the surface of the Earth. Can you see a pattern to the differences that will help us understand how the Earth arrived at such a different composition from the rest of the universe?

4. What is an isotope? How can you distinguish one isotope of the same element from another?

5. What is the energy source of the Sun? Why is energy production confined to the core of the Sun? If the core temperature were higher than 15 million K, would you expect more or less hydrogen fusion to take place?

6. Explain the three ways in which energy is transported from one place to another and give some examples of each from everyday life. Which of these transport mechanisms are most important for the Sun and the Earth?

7. Explain how the concept of equilibrium is used to calculate the interior structure of the Sun. If you could magically reduce the central temperature of the Sun, explain how the structure would readjust itself to restore equilibrium.

8. What is the 22-year activity cycle of the Sun? How might you go about determining whether these changes on the Sun have any direct influence on the Earth?

Quantitative Exercises

1. What would be the period of revolution for a hypothetical planet just above the surface of the Sun, at a distance from the solar center of 1 million km?

2. Estimate the lifetime of the Sun by calculating how long it will require, at the present rate of thermonuclear fusion, to consume all of the hydrogen originally in the core of the Sun. Assume that 20% of the total hydrogen is in the core and thus available to be fused into helium.

3. In the calculation of the solar lifetime carried out in question 2, what is the total percentage loss in mass of the Sun?

4. Use the Stefan-Boltzmann Law to calculate the relative emission of energy per square meter for the Sun (5800 K), for a cool star (2400 K), and for the planet Mercury (600 K).

5. Use the radiation laws to calculate the wavelengths of peak emission and the total emitted radiation from each square meter of the surfaces of Mars (250 K) and Pluto (40 K).

6. Use the Planck Law to calculate the emitted power from the surfaces of Mars and Pluto (question 5) at wavelengths of 10 μm, 100 μm, and 1 mm. Do your results support the statement made in the text that the emission curves for objects of different temperatures never cross?

7. Harnessing thermonuclear fusion on Earth would give us a huge supply of energy. Calculate the mass of the oceans, which are equivalent to a layer of water (H_2O) 3000 m thick over the entire globe. If water consists of 10% hydrogen, how much energy could be produced by converting this hydrogen to helium?

Additional Reading

Frazier, K. 1982. *Our Turbulent Sun.* Englewood Cliffs, N.J.: Prentice-Hall. Good popular discussion of the Sun, emphasizing solar-terrestrial relationships.

Friedman, H. 1985. *Sun and Earth.* New York: Freeman. Discussion of the Sun by a leading solar scientist, especially good in describing space experiments.

Giovanelli, R.G. 1984. *Secrets of the Sun.* Cambridge: Cambridge University Press. Detailed description of solar activity, including the role of magnetic fields.

Phillips, K.J.H. 1992. *Guide to the Sun.* Cambridge: Cambridge University Press. Comprehensive, up-to-date review of solar physics and the importance of the Sun in the solar system.

Washburn, M. 1981. *In the Light of the Sun.* New York: Harcourt Brace Jovanovich. Well-written popular book on the Sun.

Wentzel, D.G. 1989. *The Restless Sun.* Washington, D.C.: Smithsonian Institution Press. A broad popular discussion of many aspects of the Sun, solar activity, and solar-terrestrial relationships.

3

Getting to Know Our Neighbors

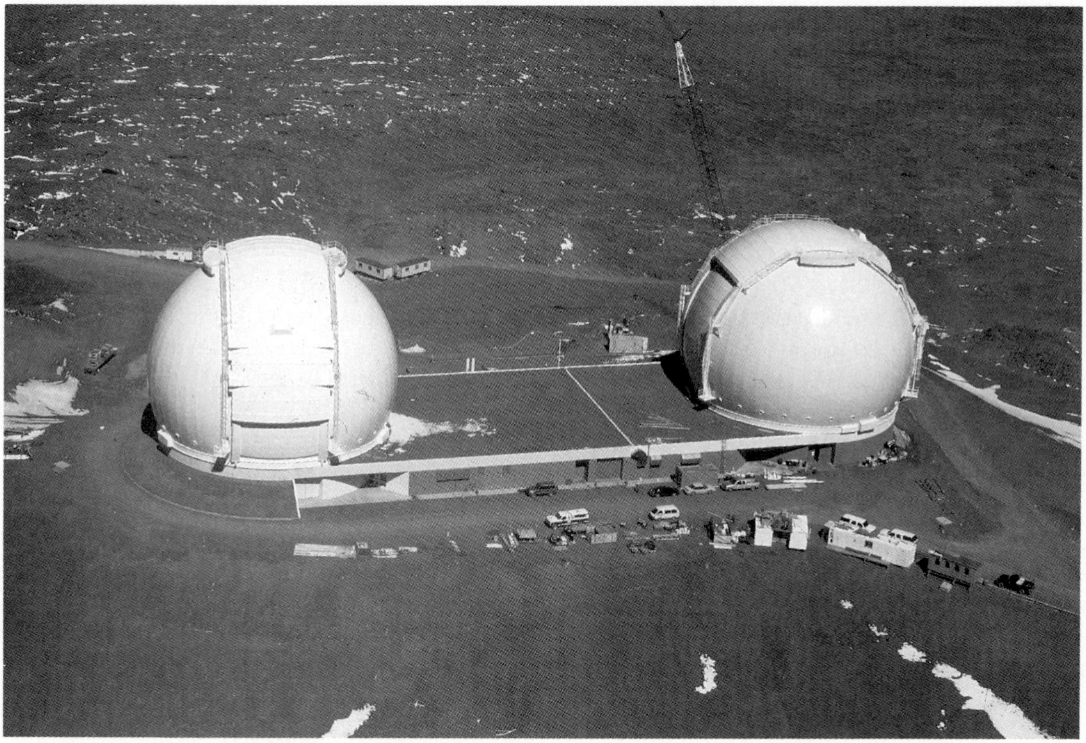

The summit of Mauna Kea, an extinct volcano on the island of Hawaii, is probably the best location on Earth for observational astronomy. Here we see the twin domes of the two Keck 10-m reflectors, the largest optical telescopes that have been built so far. There are six other major observatories on this mountain.

Planets are places, made of rock and metal, ice, and various gases. Some are similar to the Earth; others are very different. We begin Chapter 3 with a discussion of the materials that form the planets. We want to know how the composition varies from one object to another, and how closely individual planets represent the mixture from which the solar system was originally made. Here and throughout the rest of the book we will use a comparative approach, concentrating on the fundamental processes that determine the similarities and differences between one world and another. We also will be asking questions concerning the origin of the objects we see, and the ways they have evolved over the 4.5-billion-year history of the planetary system.

The Earth has a composition very different from that of the Sun, consisting mainly of silicon, oxygen, magnesium, and iron instead of hydrogen and helium. We are living on a kind of cosmic cinder that is greatly depleted in the abundant light elements. Or, to put it more attractively, the Earth is a chocolate chip in the great galactic cookie. One of our tasks in this book is to see how this peculiar concentration of heavy elements came about and how likely it is that there may be other chocolate chips out there among the stars.

3.1 BASIC PROPERTIES: MASS, SIZE, AND DENSITY

What are the members of the planetary system really like? Before we begin the detailed answer to this question (starting in Chapter 4), let's take a quick look from a very basic point of view. We shall describe the *sizes* and *masses* of these bodies, and from these two characteristics, we can then derive their *densities*. Even before we study the planets with spectrographs and spacecraft, a knowledge of the densities gives us an important clue to their composition.

Density as a Guide to Composition

Density is a measure of the amount of mass contained in a given volume. In the metric system we express the units of density as grams per cubic centimeter, or, equivalently, as tons per cubic meter. In these units, water has a density of 1.0 g/cm^3. Ice, since it floats in water, must have a lower density. In fact, it is only a little lower: 0.92 g/cm^3. That's why only the tip of an iceberg shows above the surface of the ocean in which it floats. On the other hand, a piece of pine wood has a density of 0.5 g/cm^3, and a piece of the porous volcanic rock called pumice may have a density of 0.7 g/cm^3. Both the wood and this unusual rock float better than ice. So would the planet Saturn, if we could build a big enough bathtub, since its density is only 0.7 g/cm^3. Metals, in contrast, have high densities; lead has a density of 11, indicating 11 times as much mass in the same volume as a gram of water (Fig. 3.1).

Intuition tells us that we could expect a planet like the Earth to be at least as dense as the rocks we find on its surface. They typically have densities between 2.5 and 3.5 g/cm^3. In fact, the

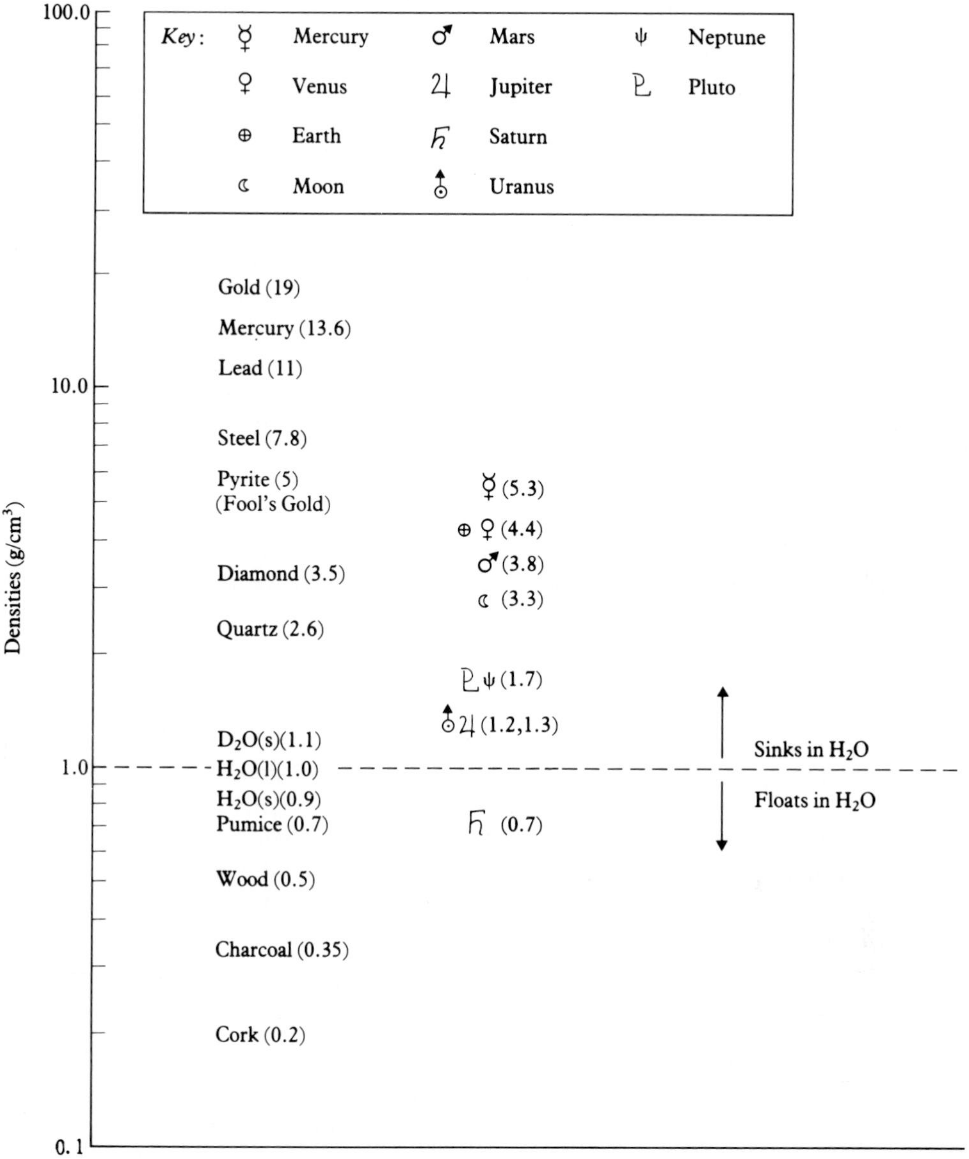

FIGURE 3.1 The densities of various substances compared with planetary densities. The densities of the inner planets are corrected for the effect of compression. Note that just as pumice and Saturn could float in water, steel and Mercury could float in mercury (pumice and Saturn could also).

Earth's overall density is nearly twice as great. There are two reasons: The central core of our planet is composed of iron and nickel, obviously more dense than rock, and these metals are compressed to a greater than normal density by the weight of the overlying material. The result is a density for the Earth as a whole of 5.5 g/cm^3. For more massive planets, we expect higher densities, since the central compression will be greater, and conversely for the less massive ones, if they are made of the same rocky and metallic materials as the Earth. Any deviations from this expectation must indicate differences in composition.

Densities of Planets

The density of a planet can be calculated once we know the mass (which can be determined from its gravitational influence on its satellites or other bodies in the solar system) and its size. The technique requires Kepler's laws, Newton's laws, some simple trigonometry, and a telescope. We need no fancy astrophysical equipment, and the picture of the solar system we can now unfold was already available in the eighteenth century (except, of course, for the new planets that were discovered later). Let's see what this picture looks like.

A summary of the masses, sizes, and densities of the planets is given in Table 3.1. It is evident that the planets can be divided into three categories according to these characteristics, which turn out to be correlated with distance from the Sun: inner planets, outer planets, and Pluto. Our Moon may also be thought of as an inner planet in terms of its composition, but it is distinctly different from Earth, as we'll see in Chapter 7.

TABLE 3.1 The planets: size and density

Planet	Diameter*	Mass*	Density
Mercury	0.38	0.06	5.4
Venus	0.95	0.85	5.3
Earth	1.00	1.00	5.5
Mars	0.53	0.11	4.0
Jupiter	10.8	318	1.33
Saturn	8.9	95	0.68
Uranus	4.1	15	1.2
Neptune	3.8	17	1.7
Pluto	0.2	0.002	2.1

*Both relative to the Earth = 1

Considering only the densities, the inner planets seem very similar. But a closer look reveals that even with this simple approach it is possible to find some distinguishing characteristics. Thus we expect bodies smaller than Earth to have smaller average densities, since their centers will be less compressed. This is certainly true for Mars, but not for Mercury, whose density of 5.4 g/cm^3 is nearly equal to that of our planet. As we shall see in Chapter 8, the solution to this apparent paradox is to assume that Mercury has a disproportionately large core of iron and nickel.

We confront the opposite problem when we consider the planets in the outer solar system. Here we are dealing with bodies far larger than the Earth, and we would expect them to have greater densities. Instead we find that these objects are less dense than any of the inner planets. The conclusion is inescapable: They cannot be made of rock and metal.

The only way to construct massive planets with such low densities is to make them predominantly of the two lightest elements—hydrogen and helium. Thus the outer planets are fundamentally different from the inner ones. In terms of their composition, they are more closely related to the Sun and stars than to the rocky bodies in our immediate neighborhood.

Pluto is an exception. This planet is about two-thirds the size of our Moon, but its lower density (2.1 vs. 3.3 g/cm^3) suggests that Pluto is probably made partly of some material with a lower density than rock or metal. Water ice is a good bet. Pluto therefore resembles one of the large icy satellites of the giant planets more closely than it resembles any of the inner or outer planets.

Satellites, Asteroids, and Comets

Among the satellites of the outer planets, we find an interesting parallel to the solar system itself when we look at the densities of the four moons of Jupiter discovered by Galileo. In order of increasing distance from the planet, the names and densities of these objects are Io (3.3), Europa (3.0), Ganymede (1.9), and Callisto (1.9). Just as in the case of the planets, the satellites exhibit decreasing densities with increasing distance from the center of their system. Evidently proximity to Jupiter has had a comparable effect to proximity to the Sun. You might also ask yourself what such low densities for Ganymede and Callisto imply about the composition of these objects. They are far too small to be made of hydrogen and helium.

Our information about the densities of asteroids and comets is much less secure. Nevertheless, it is clear from the evidence we do have that the asteroids are composed predominantly of rock, while the comet nuclei are made mainly of ice.

3.2 CHEMISTRY OF THE PLANETS

Based on their observed densities and some hints about their surface and atmospheric composition obtained from spectroscopy, we can begin to see the outlines of planetary composition. The most fundamental distinction is between the oxygen-dominated chemistry of the inner solar system and the hydrogen-dominated chemistry of the outer planets.

Hydrogen and Oxygen

Hydrogen and oxygen are two of the most abundant and chemically reactive elements in the universe. Each is capable of forming compounds with many other elements. For example, carbon can combine with hydrogen to form methane (CH_4) or with oxygen to form carbon dioxide (CO_2). Hydrogen and oxygen can also combine with each other to form H_2O. If hydrogen atoms outnumber oxygen atoms, we find abundant hydrogen compounds and conditions are said to be chemically **reducing.** If oxygen predominates, we find compounds containing oxygen and conditions are said to be **oxidizing.**

Hydrogen is the major constituent of the gas and dust that gave birth to the Sun and planets. In the parts of the solar system that retain abundant hydrogen, like the atmospheres of the giant planets, reducing conditions prevail, and we find such reduced compounds as methane and ammonia (NH_3), as well as more complex hydrocarbons (compounds containing atoms of hydrogen and carbon). In general, regions of low temperature have been able to retain hydrogen and are therefore reducing.

Closer to the Sun the temperatures are too high for much hydrogen to remain, and in the absence of hydrogen oxygen takes control. All of the inner or terrestrial planets are at least partially oxidized, and their atmospheres are also oxidizing. The most extreme example of oxidized chemistry is provided by our own atmosphere, which contains oxygen gas (O_2)—an exceedingly rare material in the universe, because it so easily combines with almost any other element. It is only because of the presence of life on Earth that this gas persists in our atmosphere.

Four Types of Matter

A school of ancient Greek philosophers thought matter could be divided into just four categories: Fire, Air, Water, and Earth. As we saw in Chapter 2, matter is actually composed of atoms of pure substances called elements, of which there are many more than four. But this simple Greek view is a useful tool for organizing the objects in the solar system.

For our purposes, it is helpful to change the categories somewhat. We shall consider the following four forms: gas, ice, rock, and metal. Using these four components, we can classify the various members of the solar system according to the relative proportions of these forms of matter that they contain.

Gas We all know what gas is, and we know that planetary atmospheres are composed of gas. The atmospheres of the giant planets Jupiter and Saturn are composed predominantly of hydrogen and helium gas, but at the great pressures of these planets' interiors the hydrogen and helium are in liquid form. Jupiter and Saturn preserve the material that has been least modified chemically since the formation of the solar system.

No other planets resemble Jupiter and Saturn in having a composition similar to that of the Sun. The next closest are Uranus and Neptune. These two giants still contain a great deal of hydrogen and helium, but they have a higher proportion of heavy elements. We see this both in the composition of their atmospheres and from their densities, which are substantially greater than the densities of Jupiter and Saturn, even though Uranus and Neptune are smaller.

The solar system began as a cloud of neutral gas and dust, commonly called the **solar nebula.** The abundances of the elements we find in the atmosphere of the Sun are approximately the same as those that existed in the solar nebula (Table 2.2). This follows directly from the fact that when the solar nebula collapsed to form the solar system, most of the material went to form the Sun. Deep in the solar interior, nuclear reactions are changing this primordial composition, but in the outer atmosphere that we observe, the Sun can serve as a standard for the composition of the matter from which the entire solar system formed.

Ice Ices are molecules that are liquid or gaseous at moderate temperatures but form solid crystals at low temperatures. The most familiar examples on Earth are water (H_2O) and carbon dioxide (CO_2). Other ices found in the planetary system include carbon monoxide (CO), ammonia (NH_3), and methane (CH_4). These molecules are also frequently called **volatiles,** meaning that they melt or sublime at moderate temperatures. The comets and most planetary satellites are composed primarily of ices.

Ices formed from the original solar nebula in regions of low temperature. If it was cold enough to form ice, it was also cold enough for rock and perhaps some metal to be present, since we are considering a solar mixture from which we have simply stripped away two of the lightest gases, hydrogen and helium. Since there are more molecules of water than of any other compound, the total mass of water or ice is about equal to the mass of the (much denser) rocky material. We think that the dense cores of the outer planets consist of a mixture of rock and ice, although at their high pressures and temperatures they do not resemble the kinds of rock and ice we normally encounter.

The next step away from the primordial composition is to eliminate the hydrogen and helium entirely. This would leave us with ice, rock, and metal. If such bodies are large enough and not too warm, they have the potential to retain atmospheres of heavier gases. In our solar system, Pluto and the large icy satellites of Jupiter, Saturn, and Neptune seem to satisfy this description, although only Titan has a significant atmosphere.

Rock If we began with objects composed of mixtures of ice and rock and subsequently heated them, we could imagine the ice evaporating to leave predominantly rocky objects. The Moon is an example of a body composed almost entirely of rock. The most common rocks are **silicates,** which are oxides of silicon, aluminum, and magnesium. We will discuss the rocks and their constituent minerals in Section 3.3.

Metal At still higher temperatures, the rock itself undergoes chemical and structural transformations. In some circumstances, the iron, nickel, and magnesium that are common constituents of rock can separate into metallic form. Small metallic grains were a common constituent of the original solar nebula, but today most of the metal in the solar system is found in the cores of planets. Mercury is an example of a planet that is about 75% composed of metal, and some asteroids are nearly pure iron and nickel.

Density and Composition

We have described a progression from gas to metal, which corresponds to an increase in the amount of heating or other processing that the original material from the solar nebula has undergone. This progression is also correlated with the composition of the gases we find around those bodies capable of holding atmospheres. In the warm inner solar system, we find oxygen compounds in planetary atmospheres, while hydrogen dominates the atmospheres of the outer planets.

If you recall the discussion of density in Section 3.1, you will expect a general progression from low to high density as we move to more evolved objects. This is a natural consequence of losing the abundant light elements, hydrogen and helium, and the most abundant compound, water ice. Thus a good determination of an object's density can tell us a great deal, before we begin the much more difficult work required to make a detailed analysis of chemical composition.

These various trends are illustrated in Table 3.2, which provides a useful one-page introduction to solar system chemistry and planetary evolution.

3.3 ORIGIN AND CLASSIFICATION OF ROCKS

Some of the most important solids we will encounter in our exploration of the solar system are the rocks. These are compounds that include the elements silicon and oxygen. Living on the Earth, we tend to take rock for granted. But as we have seen, hydrogen and helium are far more abundant than silicon in the universe at large, so rock is actually rather rare.

TABLE 3.2 Planets and satellites: overview of composition

Object	Distance from Sun (AU)	Density (g/cm^3)	Bulk Composition
Mercury	0.4	5.4	iron, nickel, silicates
Venus	0.7	5.4	silicates, iron, nickel
Earth	1.0	5.5	silicates, iron, nickel
Moon	1.0	3.3	silicates
Mars	1.4	3.9	silicates, iron, sulfur
Jupiter	5.2	1.3	hydrogen, helium
Callisto	5.2	1.8	water ice, silicates
Ganymede	5.2	1.9	water ice, silicates
Europa	5.2	3.0	silicates, water ice
Io	5.2	3.4	silicates
Saturn	9.6	0.7	hydrogen, helium
Titan	9.6	1.8	water ice, silicates
Uranus	19.2	1.2	ices, hydrogen, helium
Neptune	30.1	1.6	ices, hydrogen, helium
Triton	30.1	2.1	silicates, water, other ices
Pluto	39.4	2.1	silicates, water, other ices

Rocks and Minerals

Even if we restrict ourselves to solid compounds, we can see that water ice must be far more common than rock, since oxygen is 11 times more abundant than silicon in the universe. What this means is that oxygen could combine with every silicon atom in the universe to make quartz (SiO_2), and there would still be plenty of oxygen left over to make an even larger quantity of water ice. When we explore the outer solar system where ice is stable, we will find that ice is indeed the dominant solid. Meanwhile, let us consider the rocks.

The rock we see in the familiar landscapes of our planet is composed of assemblages of compounds or elements called **minerals.** The principal difference between a rock and mineral is in homogeneity. A mineral is composed of a single substance, whereas a rock may be made up of several different minerals. Only a few elements occur in nature in a pure state, so most minerals are compounds. Gold and silver are probably the most famous examples of single-element minerals in the United States. Elemental sulfur, copper, and carbon in the form of graphite and diamonds are also economically important. Minerals composed of single compounds are much more common. Quartz (SiO_2), hematite (Fe_2O_3), iron pyrite or "fool's gold" (FeS_2), and calcite ($CaCO_3$) are all relatively familiar. These four minerals are examples of silicates, oxides, sulfides, and carbonates, respectively, the four most common types of rock-forming minerals.

Igneous, Sedimentary, and Metamorphic Rock

Rocks can be classified in terms of the minerals they contain, as we have just seen. However, it is also convenient to divide them into three large categories on the basis of their origins: igneous, sedimentary, and metamorphic.

Igneous rocks are those that have formed directly by cooling from a molten state. A rock picked up from the slope of a volcano in Hawaii is an obvious example (Fig. 3.2). This is a representative of the family of igneous rock called **basalts,** which we shall encounter repeatedly during our studies of the inner planets. Igneous rocks make up roughly two-thirds of the Earth's crust.

Sedimentary rocks are composed of fragments of other rocks that are cemented together. Limestone and sandstone are examples. On Earth, the fragments are produced by various weathering (erosional) processes that break up the parent rocks. The most effective weathering processes on our planet involve liquid water. Fragmentation can also be produced when one rock bangs into another, creating a different type of sedimentary rock on the lunar surface, for example.

Metamorphic rocks are produced from either igneous or sedimentary rocks that have been buried far below the Earth's surface, modified by the high pressures and temperatures they encountered there, and then returned to the surface. This process of burial and return is part of the great cyclical movement of the Earth's continental plates which we shall discuss in Chapter 9. At the present time, we don't know whether any other planet in our solar system exhibits this phenomenon, so Earth may be the only place where metamorphic rock exists. Marble is probably the best known metamorphic rock.

Primitive Rock and the Origin of the Earth

In this review of rocks and minerals, we have actually been talking about reprocessed material, since all the rock on Earth was melted after the planet formed. Are there more ancient rocks in the solar system? There certainly are. We can establish a fourth category called **primitive** rock, which has never melted and which has been affected only moderately by chemical and physical processes since the solar system began. We expect such rock to have the same abundances of the nonvolatile elements (elements that are solid rather than gaseous at low temperatures) that exist in the Sun or the giant planets. Examples of this kind of rock are found among the meteorites, as we shall see in Chapter 4. We expect that there is primitive rock lodged in the ices of the comets. Some asteroids and many of the smaller planetary satellites must also be composed of this material (Fig. 3.2b).

The reason we don't find primitive rocks on Earth has to do with the way in which a planet forms and evolves with time. Current theories for the formation of the inner planets start with solid material from the solar nebula. This material is indeed primitive in the sense described above, and the meteorites are remnants of it. But as this dust and rocky debris come together to form a planet, the solid body that develops is heated. This heating results in part from the energy released by the impacting material that is bombarding the forming planet and causing it to grow, and partly from radioactivity deep inside the body.

A large planet can retain heat better than a small one. The generation of heat usually depends on the volume of the planet, which is proportional to the cube of the radius (remember, volume = $\frac{4}{3}\pi R^3$). But the planet can lose heat only through its surface, and the surface area is proportional to the square of the radius (area = $4\pi R^2$). Thus, since the generation of heat increases faster than its loss, a large planet will tend to grow warmer than a small one. If the internal temperatures rise enough, the rocks melt and the central part of the planet becomes a liquid. At this stage, the denser materials are free to migrate to the center, and the lighter materials rise to the top.

FIGURE 3.2 (a) This ropy pahoehoe lava on the floor of Hawaii's Kilauea volcano was molten rock just a few weeks before the photo was taken. The crack in the left foreground is about 2 cm wide. (b) Astronomer Dale Cruikshank holding a 2-year-old rock (Hawaii lava) in his left hand, and a 4.5-billion-year-old rock (a primitive meteorite) in his right hand.

a

b

The process of separation according to density is known as **differentiation,** and it leads to the development of *differences* in internal composition. All the inner planets have undergone differentiation, as we would expect. An interesting puzzle, however, is posed by the presence of asteroids that are also differentiated, in spite of their small sizes. What processes have heated these tiny bodies? We shall return to this problem in Chapters 4 and 5.

We can see why our own planet does not have any primitive rocks. In fact, we even have trouble finding *old* rocks on Earth, since an active geology and erosion have effectively eliminated the first billion years of our planet's geological record. To probe this ancient history, we must go to other places—the Moon, Mars, the asteroids—where less alteration of primitive conditions has taken place.

3.4 PLANETARY ATMOSPHERES

An important characteristic that distinguishes one solar system object from another is the nature of its atmosphere. Earth is 11 times smaller than Jupiter and over four times as dense, with a totally different composition. Yet both bodies have atmospheres, although the composition of these atmospheres is very different. On the other hand, Jupiter's satellite Ganymede has no atmosphere while Saturn's slightly smaller moon Titan has an atmosphere denser than Earth's. What accounts for these differences?

Getting and Holding an Atmosphere

There are two basic ways in which a planet can obtain an atmosphere: it can form with one (a primordial or captured atmosphere), or it can produce one from the material of which it is made (secondary or outgassed atmosphere). Often when there are two explanations, a third can be created by combining the first two. In fact, capture plus outgassing appears to be the most likely process for the formation of the atmospheres of the giant planets. They are a blend of captured solar nebula gases and gases produced from the planetary cores.

In order for a planet to hold on to an atmosphere over the 4.5-billion-year life span of the solar system, the molecules in that atmosphere must not move fast enough to escape from the planet's gravitational field. In other words, their speed must be less than what we call the escape velocity. Otherwise they will soar off into space, eventually leaving a denuded planet behind.

The escape of gas molecules occurs from a layer that is so tenuous that a molecule moving in an upward direction will not encounter any other molecules. It is free to leave the planet if it is moving fast enough. Scientists have called this layer of the atmosphere the **exosphere,** since the gases here can exit from the Earth if they have escape velocity.

Physics tells us that the velocities of molecules in a gas are determined by the temperature of the gas and its composition. If the temperature is high enough and/or the gas is sufficiently light, individual atoms and molecules in the exosphere may reach escape velocity and be lost. For a planet to keep an atmosphere, it should have a large mass and thus a high escape velocity. It should also be cold, so that velocities of the gas molecules will be low (Fig. 3.3). Gases with large molecular or atomic weights (for example, N_2, with weight 28) are more easily retained than lighter ones (e.g., H_2, with weight 2).

These two conditions are well met in the outer solar system where we find the giant planets. These bodies have such high escape velocities (typically tens of kilometers per second) that they can retain thick atmospheres of even the light gases hydrogen and helium.

Saturn's satellite Titan is not massive enough to retain hydrogen, but it is sufficiently massive and cold to keep an atmosphere of heavier gases, such as the nitrogen and methane that we find there. Similar conditions apply to Ganymede, so the reason this massive satellite of Jupiter does

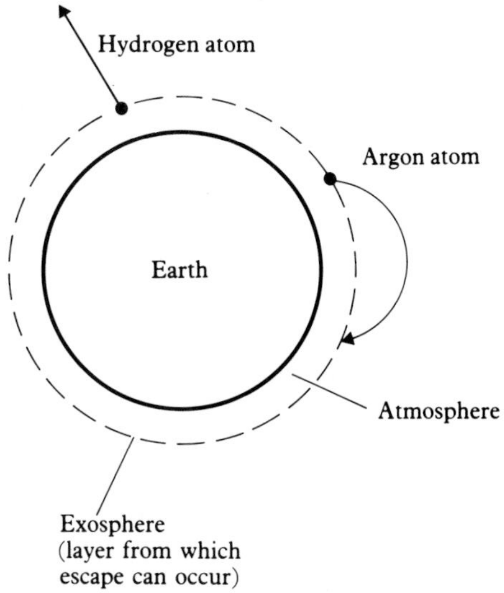

FIGURE 3.3 The light hydrogen atoms (atomic mass = 1) move much faster than the heavier argon atoms (atomic mass = 40) at the same temperature. Thus hydrogen can escape from the Earth, while argon cannot.

not have an atmosphere must be found elsewhere. (We will find it in Chapter 15.) Our Moon and the planet Mercury are both too warm and insufficiently massive to maintain atmospheres.

Why Atmospheres Are Different

There are thus several reasons for the diversity of atmospheres that now exist around various bodies in the solar system (when they exist at all). We will come back to this discussion as we consider each planet individually, but let's briefly consider the implications of what we have learned.

Large planets form with sufficient mass to capture hydrogen and helium (and everything else) from the solar nebula, which is why they exhibit hydrogen-rich atmospheres today. Small planets must produce their own atmospheres, so the composition of these atmospheres will depend on the materials that make up the planet and on the planet's geological and chemical evolution. Thus we find atmospheres on Mars and Venus that are dominated by carbon dioxide. Earth should exhibit the same thing. The fact that it doesn't is a result of the presence of liquid water and life on our planet. Why is methane much more abundant in Titan's atmosphere than carbon dioxide? We will return to this question when we discuss this fascinating satellite in Chapter 16.

3.5 STUDYING MATTER FROM A DISTANCE

We have reviewed several basic characteristics of the planets and some of the properties of the matter that composes them. We now want to tie these two themes together by asking what we can learn about these distant worlds by studying the light, heat, and radio emissions that they send us across the vast emptiness of space.

The process of investigating distant objects by the analysis of their radiation is often referred to as **remote sensing,** to distinguish it from in situ (or "on site") studies. Astronomy is almost entirely a remote sensing science, although astronomers rarely use the term. Today, we study the members of the planetary system by a combination of astronomical studies from Earth, remote sensing carried out from flyby or orbiting spacecraft, and direct studies from entry probes and landers.

Spectroscopy

The most powerful technique of planetary remote sensing is spectroscopy. In Chapter 2, we discussed the formation of spectral lines and their use in determining the composition of a gaseous object like the Sun. The same techniques are used to study the spectra of the planets. However, in planetary remote spectroscopy, some of the most important clues are derived from the invisible parts of the electromagnetic spectrum.

Leaving the visible spectrum and considering shorter wavelengths, we find that we cannot detect sunlight at the Earth's surface in the ultraviolet region. This is not because the Sun does not radiate here, but because of a layer of **ozone** in the Earth's atmosphere. Ozone has the chemical formula O_3. It is formed from oxygen molecules in the Earth's upper atmosphere and shields the Earth's surface from high-energy ultraviolet photons. This is beneficial for life on Earth, since this high-energy radiation is capable of destroying many of the molecules that make up living organisms.

Moving toward longer wavelengths past the red end of the visible region of the spectrum, we encounter the infrared, first detected by English astronomer William Herschel (1738–1822), who is best known for his discovery of Uranus. Here again we encounter absorption from the Earth's atmosphere, primarily from water vapor and carbon dioxide. Despite their relatively low abundances in the Earth's atmosphere, these two gases play a dominant role in absorbing infrared radiation. Unlike the short wavelength end of the spectrum, however, some infrared light can penetrate our planet's atmosphere

FIGURE 3.4 The Earth's atmosphere absorbs radiation at most wavelengths at various altitudes above our planet's surface. Only visible light, and some infrared and radio waves, can penetrate the atmosphere and reach the surface. These regions of the spectrum in which radiation can pass through the atmosphere are called atmospheric windows. The same situation (with variations depending on atmospheric mass and composition) will exist on all other planets with atmospheres.

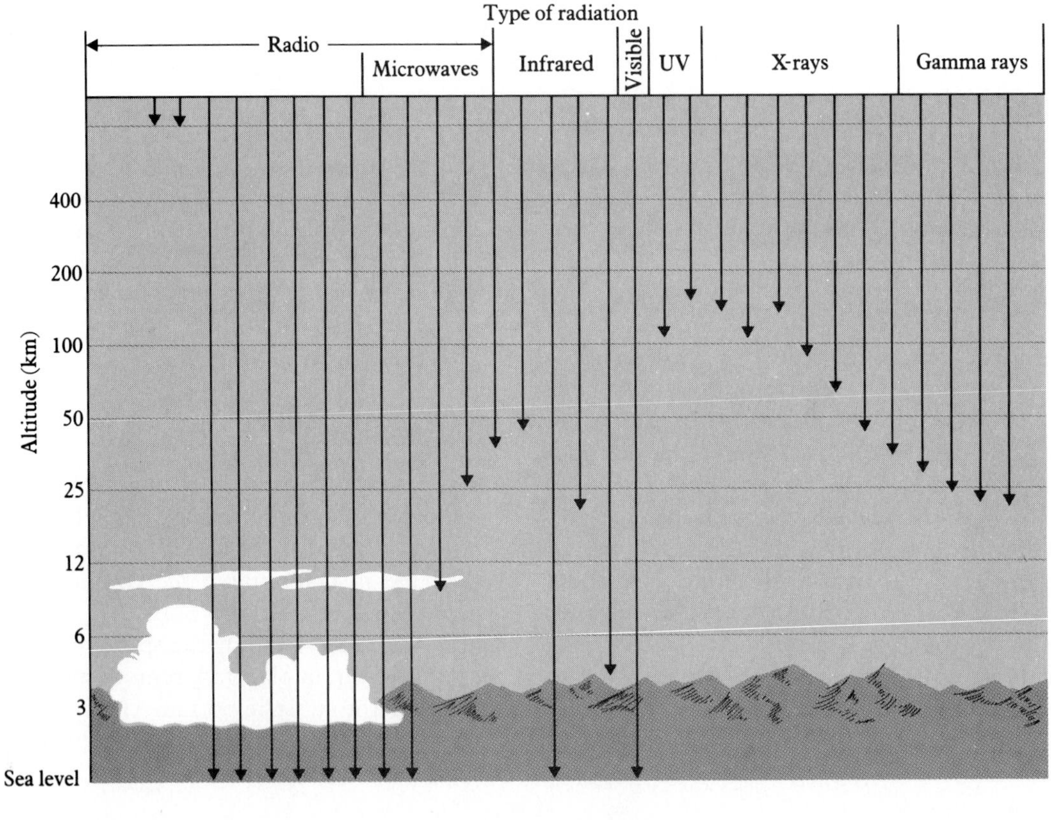

and reach the Earth's surface. This occurs in regions of the spectrum that lie between the strong absorption bands of water vapor and carbon dioxide (Fig. 3.4). Such regions of transparency are called **atmospheric windows.** The atmosphere acts as a kind of color filter, letting some wavelengths (colors) through and absorbing others.

This pattern of windows and absorptions continues out to longer and longer wavelengths. The atmospheric absorption grows progressively stronger until it ceases at a wavelength of about 1 millimeter. The atmosphere is transparent to radiation of longer wavelengths, providing another clear window on the universe. The long-wave limit on this window is at about 30 cm. This window includes the microwaves, radar, television, and FM broadcasts, which are often lumped together as "radio." AM broadcasts occur at still longer wavelengths and are reflected back to the surface by our planet's upper atmosphere.

Spectra of Other Planets

Orbiting the Sun, the planets shine by reflected light. Just as sunlight passing through the Earth's atmosphere is absorbed at certain wavelengths by our planet's atmospheric gases, we can expect the same thing to occur in the atmospheres of the other planets. By examining the reflected sunlight, we can then hope to discover new absorption lines that will tell us what gases are present on that planet (Fig. 3.5).

Not all the sunlight that strikes the planets is reflected, however. Astronomers call the fraction

FIGURE 3.5 Sunlight, whose spectrum (*A*) contains absorptions (*S*) from gases in the Sun's atmosphere, penetrates the atmosphere of another planet, where some of it is absorbed by planetary gases, producing new lines (*P*) in the spectrum of the planet (*B*). Reflected by clouds or the planet's surface, the light continues its journey to Earth, where some of it may be absorbed by gases in our planet's atmosphere, producing new lines (*E*) before it reaches a telescope and spectrograph to form the observed spectrum (*C*).

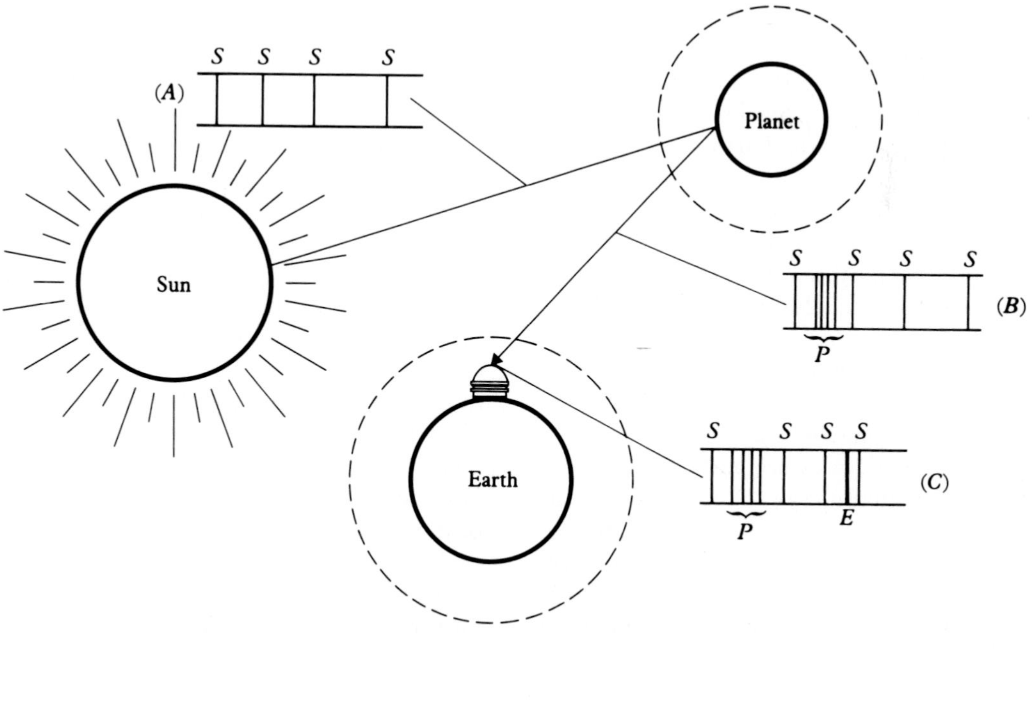

of sunlight that is reflected the **albedo** of the object. Our Moon, for example, reflects only 11% of the sunshine that illuminates it, despite the fact that it seems so bright in our night skies. This low albedo is characteristic of an exposed, rocky surface. In contrast, cloud-covered Venus sends back 75% of the incident sunlight, while icy Enceladus (a satellite of Saturn) has an albedo of nearly 100%. What happens to the radiation that is not reflected? It is absorbed by the planet, heating its surface or atmosphere. Thus each planet, satellite, asteroid, or other body assumes a temperature that will depend on its distance from the Sun, its albedo, and what kind of an atmosphere it has, if any. If there is a large internal source of heat, that will also be important, as we'll see in the case of some of the giant planets.

When the surface of a planet absorbs sunlight and becomes warm, it begins to radiate energy at infrared and radio wavelengths. Imagine you are studying the spectrum of such a planet. After looking at the visible region, where you would be analyzing reflected sunlight, you might move on to the infrared. Here you detect thermal radiation from the planet itself. Since there are no solar lines in the spectrum to confuse you, the absorptions you find in the planet's infrared spectrum are caused by gases in its atmosphere or in the atmosphere of the Earth.

Spectroscopy provides a powerful tool for studying planets from a distance. By analyzing the light they reflect as well as the radiation they emit, we can learn about the composition of their atmospheres and the nature of their surfaces. In fact we can do even more than this. The strengths of absorption lines can tell us the amount of the absorbing gas. By observing many lines of the same gas, we can often derive an average atmospheric pressure and temperature. Comparing the intensities of thermal emission at several different wavelengths, it is possible to determine physical and chemical properties of the emitting surface. In short, there is a huge bag of tricks that can be played with planetary radiation, once we have enough of it to work with. To achieve that, we need telescopes.

Telescopes and Observatories

Astronomers would like to collect as much radiation from the planet as possible, to view the planet from all angles, and to study small regions as well as the entire globe. There are two ways to do this: (1) use a large collecting area that focuses all the radiation falling on it into an image that can then be examined (but this doesn't broaden the angle of view), or (2) go closer to the object. The first approach employs a telescope; for the second a spacecraft is essential. In practice, the optical instruments on spacecraft also use small telescopes, thereby further improving their performance.

The reasons scientists need to make this effort to study the planets are quite simple. First, even the brightest planets don't send us very much light. Second, when a spectrograph spreads out the planetary radiation to form a spectrum, it is essentially sorting photons by wavelength or energy. Instead of just detecting the planet by adding up all the photons in the visible spectrum, which is what our eyes do, the detector in the spectrograph is only sampling a few discrete energies. Unless a lot of radiation is collected before it is spread out in this way, there won't be enough photons to stimulate the detector.

An optical telescope performs two critical functions for the astronomer: It collects light, and it forms a magnified image of the object being studied. The collection of light is accomplished by the mirror or lens that forms the principal optical element of the telescope. This lens or mirror is the heart of the telescope, and its diameter, or **aperture,** is the measure of telescope power. Thus, for example, we speak of the 200-inch Hale Telescope on Palomar Mountain, or the 10-meter Keck Telescope on Mauna Kea.

All the light from a single source such as a planet which falls on the entire area of the lens or mirror is brought to a focus at a single point.

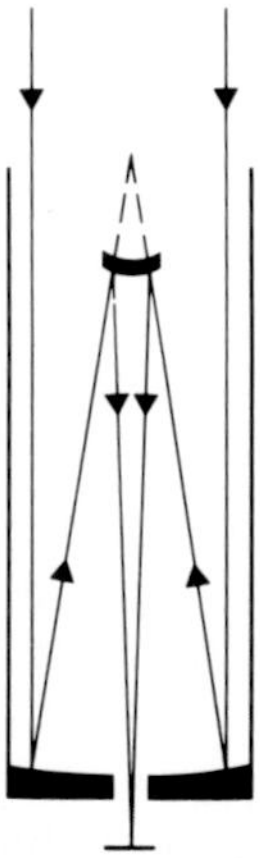

FIGURE 3.6 The Cassegrain focus of a modern reflecting telescope. Light from a planet or star is reflected by the large primary mirror to a smaller secondary mirror that sends it back through a hole in the primary. Instruments to examine the light may then be mounted on the back of the telescope.

When you look at such a planet without a telescope, your retina is detecting only the amount of energy that passes through the few square millimeters of area corresponding to the size of the lens in your eye. But gazing through the eyepiece of a 2-meter reflecting telescope, you have access to all of the light falling on the 3 million square millimeters of the primary mirror (Fig. 3.6). The eyepiece, in turn, is constructed to magnify further the image of the planet formed by the telescope. Using eyepieces composed of different lenses, different magnifications can be achieved.

A good pair of binoculars is a small but sophisticated pair of telescopes. Typically each has a main lens 35 mm in diameter, collecting several hundred times more light than the eye alone. The magnification might be 6, in which case we would call this a pair of 6 × 35 binoculars. An amateur astronomer's telescope, such as are sold widely in camera stores, might typically have a main mirror diameter of 10–20 cm and magnifying powers from 20 to 100.

Binoculars can be held in the hand, but large astronomical telescopes require permanent mountings (Fig. 3.7). The motion of the telescope is computer controlled to permit precise pointing and tracking. The telescope, its computer drive system, and all the auxiliary instrumentation are housed in an observatory. This is a building with a dome that has a slot in it through which the telescope can be pointed to view the heavens. The dome rotates to give the telescope complete access to the sky.

With appropriate modifications, these same considerations apply to radio telescopes. Once again a large collecting area is used to gather the incident radiation. In this case the collectors may be an array of antennas or a parabolic dish,

FIGURE 3.7 The 4-m Mayall telescope of Kitt Peak National Observatory. The telescope tube is mounted in a yoke and is in a vertical position in this picture. Note the figure on the platform at lower right for scale.

shaped to bring the radiation to a focus. The long wavelengths of radio photons lead to a requirement for very large aperture antennas or even collections of antennas linked by computer to allow studies of just a small area in the sky (Fig. 3.8).

Seeking Better Observatory Sites

To take full advantage of a telescope's radiation-collecting ability, it is necessary to observe from a favorable location. If you have ever looked at a distant scene near the surface of the Earth on a hot day, you will have noticed that motions in the air through which you looked distorted the appearance of what you were trying to see. In extreme cases, layers of air at different temperatures can produce mirages. A good observatory site should minimize these distortions, as well as provide as many cloudless nights as possible.

When you look at a star or a planet in the night sky, you are looking through the Earth's entire atmosphere, so the light reaching your eyes has passed through many different layers of air, at various temperatures and moving at various speeds in various directions. The effect of all this is to make the stars appear to twinkle, when actually they are shining with a very steady light.

A telescope magnifies these effects, since it is collecting light over a much greater area than the eye, thus allowing even more random properties of the atmosphere to affect the image. That is why even the best pictures of planets taken from telescopes on Earth appear blurry.

This situation can be improved by placing the telescope at a high-altitude location, where the air is thinner and steadier. Such a mountaintop location has several additional advantages. Mountains are usually far from cities so the sky is darker and clearer, free from dust, smog, and the

FIGURE 3.8 The Very Large Array (VLA) of radio telescopes of the National Radio Astronomy Observatory near Socorro, New Mexico. Each of the 27 telescope antennas has a diameter of 25 m (82 ft). They can be moved on railroad tracks laid out in the shape of *Y* to produce an effective aperture of nearly 35 km. Even though this aperture is not filled by the antennas, they are linked by a computer so the whole array functions as a single instrument.

light from street lamps, cars, and buildings. Furthermore, a sufficiently high altitude also takes the observer above most of the water vapor in the Earth's atmosphere, thereby opening wider the infrared atmospheric windows. Some sites, particularly isolated mountains surrounded by ocean, are better than others because the local topography and wind patterns reduce turbulence and yield sharper images.

The best sites for astronomical observations in the world today are both high and far from population centers. Examples are Mauna Kea in Hawaii, the Pic du Midi in France, and Cerro Tololo in Chile. Mauna Kea, with an altitude of 4.2 km above sea level, is the premier site on Earth for astronomical observations, and most of the very large telescopes under construction are located there (Fig. 3.9). These include 8-m national telescopes of Japan and the United States, as well as twin 10-m telescopes of the Keck Observatory, jointly operated by the University of California and the California Institute of Technology.

Even a mountaintop is not ideal, however. The next step is to use a telescope in an airplane, in order to reach still higher altitudes. This has proved particularly effective for infrared studies, since the absorptions by Earth's water vapor are reduced dramatically. For the past two decades, NASA has operated the Kuiper Airborne Observatory, an infrared telescope that flies above 13 km in altitude. To do still better, one must use balloons and rockets, and ultimately send a telescope into orbit about the Earth, totally outside our planet's atmosphere. Then the entire electromagnetic spectrum becomes accessible.

FIGURE 3.9 The finest observatory site on Earth is located at the summit of Mauna Kea, a 4.2-km high extinct volcano in Hawaii. Among the large telescopes recently completed or under construction are the two 10-m Keck telescopes and two 8-m instruments, one (called Subaru, or Pleiades) built by Japan and the other built by the U.S. National Optical Astonomy Observatories.

Orbiting Observatories

In the discussion of good sites for observatories, we have concentrated on observations made in the visible and infrared regions of the spectrum. What about the ultraviolet? The ozone layer is so high that even an airplane doesn't get above it, and only instruments in rockets or satellites will do. Nor is this the only advantage of placing a telescope above the Earth. Without the distorting effects of the atmosphere, a space telescope can see fainter sources and distinguish finer detail than its counterpart on Earth, as well as have access to the entire electromagnetic spectrum.

While this approach has obvious advantages, putting it into practice is not so easy. Launching a large optical instrument into orbit with the ability to find and track faint sources, to detect and analyze their radiation, and then to send the information to Earth—all these are major challenges. Nevertheless, these challenges have been met, although the resulting space observatories are many times more expensive than those built on the ground.

As we write this, the International Ultraviolet Explorer (IUE) is well into its second decade of service. This satellite contains a 70-cm telescope and two spectrographs for recording ultraviolet radiation. The Infrared Astronomy Satellite (IRAS) carried out a survey of the entire sky during 1983. Among many other discoveries, it found a fast-moving comet that passed within 5 million km of the Earth.

In 1990 a larger 2.4-m telescope was placed into orbit (Fig. 3.10). Called the Hubble space telescope (HST), after the astronomer Edwin Hubble, who discovered the expansion of the universe, this instrument was equipped with four instruments initially, working in the ultraviolet and visible parts of the spectrum. The primary objectives of the HST concentrated upon achieving very high spatial resolution—better than is possible with any ground-based telescope. Unfortunately, the primary mirror of the telescope was incorrectly shaped, so that its resolution was no better than that of the best ground-based telescopes. This tragic error was corrected when astronauts installed new instruments in December 1993, and since that time HST has performed flawlessly, generating superb pictures of the planets as well as probing distant reaches of the universe.

FIGURE 3.10 Orbiting the Earth at an altitude of 500 km, the Hubble space telescope (HST) was captured for repairs by the Space Shuttle Endeavor, Dec. 9, 1993. Perched atop a foot restraint on the shuttle's remote manipulator system arm, astronauts F. Story Musgrave (top) and Jeffrey A. Hoffmann wrap up the last of five space walks required to refurbish the telescope. The west coast of Australia is in the background.

3.6 EXPLORING THE PLANETARY SYSTEM

Even the largest telescopes on the ground or in orbit are limited by the immense scale of the

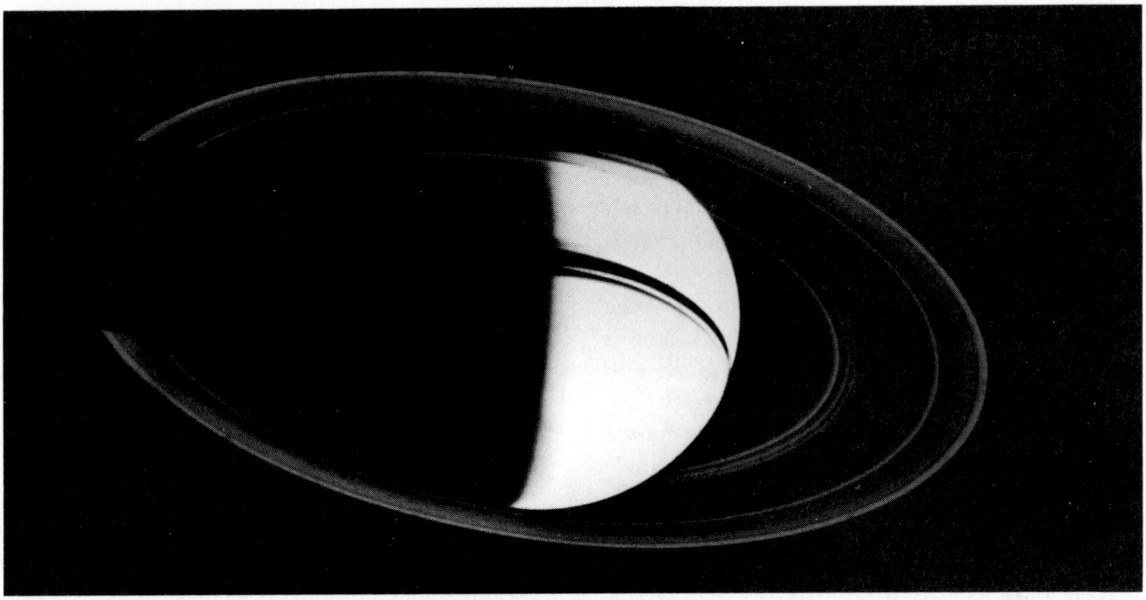

FIGURE 3.11 Looking back at Saturn from Voyager 2; a view from behind and below the planet, showing the unlit side of the rings, as the spacecraft set forth on its journey to Uranus.

solar system. To see the craters on the moons of Jupiter, to explore the rings of Saturn, to search for hydrocarbon seas on Titan, to sift the sands of Mars for signs of life—all these things are beyond our reach at any terrestrial observatory. To accomplish objectives like these, we have no choice: We have to go there. Direct planetary exploration, begun in 1959, has by now resulted in spacecraft visits to every planet except Pluto. Humans have walked on the Moon and brought nearly half a ton of lunar rocks and soil back to Earth. Instrumented probes have landed on the surfaces of Mars and Venus, and probed Jupiter's atmosphere. Other spacecraft have flown past comets and asteroids. In the 1990s, we have come to know every planet in our solar system except Pluto as a familiar, three-dimensional world.

The results of these missions form most of the substance of this book. A list of the successful spacecraft is given inside the back cover. We will also discuss some of the individual missions when we consider the discoveries they made. Here we simply want to indicate some of the advantages of studying the planets close-up.

We have already mentioned the most obvious one. Getting closer allows a much more detailed inspection of the target object. Equipped with telephoto lenses, the cameras on the spacecraft can record details that even the Hubble Telescope will not discern. Spectrographs can study small areas of a planet, to see whether the composition of one lava flow or type of cloud resembles that of another. But spacecraft can also view planets from angles we can never achieve from Earth. The far side of the Moon was the first such realm to be explored in this way, in 1959 by the Soviet spacecraft Luna 3. A more recent example is shown in Fig. 3.11, where we see a view of Saturn we can never achieve from Earth. This is a farewell image recorded by Voyager 2 in 1981.

Interplanetary Flight

All the spacecraft that leave our planet completely are moving like bullets shot from a gun. They are given a terrific boost to escape from the Earth's gravitational field, but then they coast most of the way to their targets. Small thruster rockets are fired occasionally to provide midcourse corrections, nudging the spaceships into more accurate trajectories. If one of these robot vehicles is to go into orbit around the Moon or another planet, it must reduce its speed and change its direction as it gets close. Otherwise it will fly past in a trajectory that becomes curved in response to the pull of gravity, or it will simply crash into the target.

An important technique that is used to change the velocity of a spacecraft without using up precious fuel is to perform a close flyby of one planet in order to reach another one. This kind of cosmic billiard shot (without the impact!) helped Mariner 10 to travel from Venus to Mercury, sent Pioneer 11 completely across the solar system from Jupiter to Saturn, and permitted Voyager 2 to accomplish a grand tour of all four giant planets (Fig. 3.12). These gravity-assist trajectories work by using the target planet's moving gravitational field (as the planet itself moves in its orbit about the Sun) to accelerate the spacecraft. This acceleration can be a change both in direction and in speed. Gravity-assist has proved to be a very effective technique.

Both of the Voyager spacecraft as well as Pioneer 10 and 11 have achieved escape velocity with respect to the Sun, not just with respect to Earth. Thus they are on trajectories that will take them out of the solar system, into the vast realms of space among the stars. But while we may feel proud to be starting on this adventure, we should understand just how inadequate our current technology is for the task. The nearest star, Alpha Centauri, is 4.3 light years away. This is

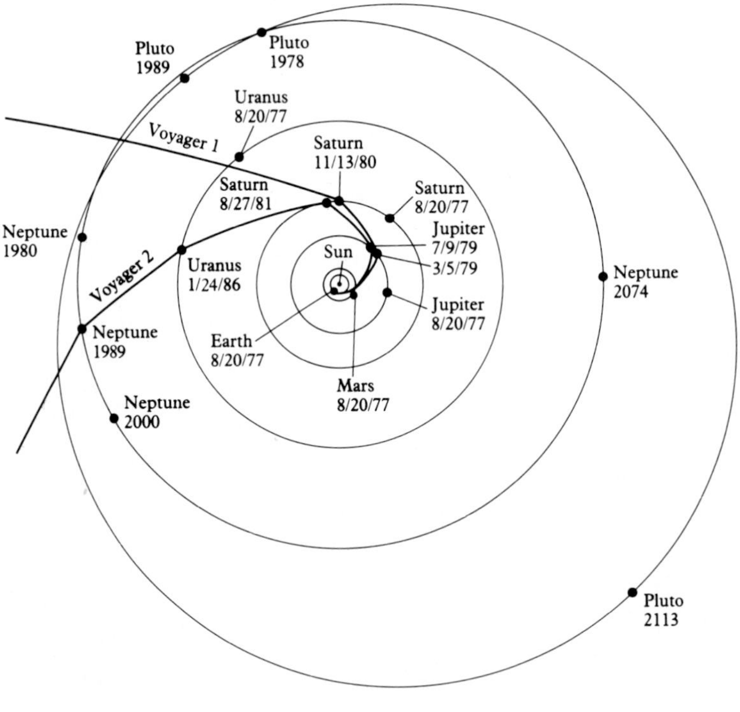

FIGURE 3.12 The two Voyager spacecraft were both launched in 1977 but followed divergent paths in their journeys past the planets. Both spacecraft are now exiting the solar system.

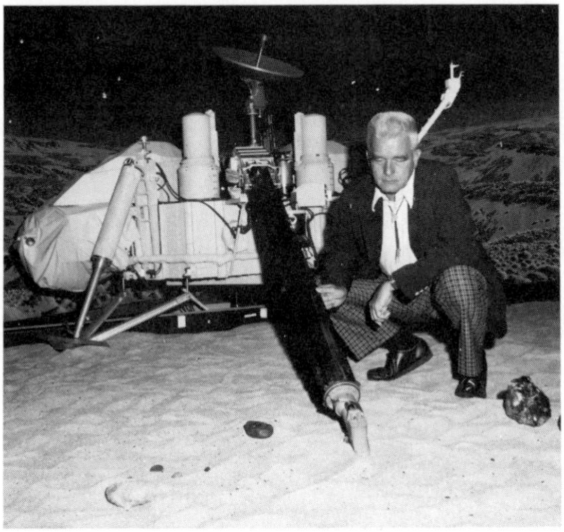

FIGURE 3.13 The science test model of the Viking landers that reached the surface of Mars in 1976. Viking Project Manager Jim Martin provides useful scale. This lander is currently on display at the Smithsonian Air and Space Museum in Washington, D.C.

the distance light can travel in 4.3 years, which is about 25 trillion miles. If these speedy spacecraft were directed toward that star (and none of them is), it would take them about 100,000 years to get there. In fact, we may lose communication with them before the end of this century.

How a Spacecraft Works

The trip to a planet can require just a few months, if the target is Venus, or as much as a decade if our objective is in the outer solar system. The spacecraft come in all different shapes and sizes, depending on the functions they are to perform, how far they are traveling, the capability of the launch vehicle, and such mundane things as how much money is available to build them. The Viking landers (Fig. 3.13) that are resting on the surface of Mars are about the size of a subcompact car and weigh about a ton. The Venus Pioneer spacecraft was smaller, Galileo and the Voyagers somewhat larger (Fig. 3.14). Recently there has been progress in miniaturization, and future spacecraft may be much smaller.

These spacecraft are crammed with sophisticated instrumentation, the electronics and computers required to make them run, antennas for communication with Earth, and panels of solar cells or tiny nuclear generators to provide the power to make everything work. The amount of power required is amazingly small, since everything has been carefully miniaturized. A 100-watt lightbulb consumes more energy than some of these spacecraft.

What happens when one of these remarkable devices reaches its intended target depends on its sophistication. Usually it measures the intensity of the magnetic field and the density of charged electrons and protons in the interplanetary medium all along its trajectory. When it reaches

FIGURE 3.14 Linda Morabito with a picture of Io in front of the science test model of the Voyager spacecraft. It was with this picture that she discovered the plumes indicating the intense volcanism that wracks this satellite of Jupiter.

its destination, these measurements are continued and the magnetosphere of the planet is mapped out in detail. These kinds of instruments require no particular pointing. Indeed, it is helpful for some of them to be mounted on a spacecraft that spins slowly around one axis so they can sample the full 360° of angle around the trajectory.

To record images or to make spectroscopic observations requires a movable platform that is attached to a spacecraft that does not spin. Such a spacecraft uses star sensors and small rockets (or thrusters) to maintain a fixed orientation as it proceeds along its track. The large radio antenna that it carries to communicate with Earth remains pointed toward our planet, while the scan platform carrying the cameras and spectrographs is turned toward the target planets and satellites in accord with a sequence of commands stored in the on-board computer.

Orbiters, Probes, and Landers

If orbit around another planet is desired, an additional rocket engine must be carried to slow down the spacecraft just enough to permit the planet's gravitational field to capture it. In this case, the spacecraft is called an orbiter. Examples include the U.S. Magellan and the Soviet Veneras 15 and 16. These three Venus orbiters were all equipped with radar antennas and transmitters, in order to penetrate the cloudy veil of Venus and gradually produce topographic maps of the planet's surface after many orbits. The U.S. Mariner 9 and Viking orbiters and the Soviet orbiters Mars 3 and Phobos performed the same function for Mars, using cameras instead of radars.

An atmospheric probe or a lander might be carried on the orbiter, from which they are released to coast down to the planet itself. The former USSR mounted a very successful series of Venus probe missions in the 1970s and early 1980s, bringing us much new knowledge about that planet's atmosphere and surface. Many of these probes could properly be called landers, since they survived on the surface, sending back pictures and compositional analyses. The United States has also sent probes to Venus and two highly successful Viking landers to Mars. Galileo deployed a probe into the atmosphere of Jupiter in 1995, and the Cassini mission under development as a cooperative venture between NASA and the European Space Agency is expected to probe the atmosphere of Saturn's satellite Titan in 2004.

SUMMARY

As a first approximation to studying what planets are composed of, we can look at four types of matter: gas, ice, rock, and metal. For the planets, a simple measurement of density already gives important clues concerning composition, allowing us to distinguish a rocky from an icy world or to deduce the overall composition of the giant planets.

In the outer solar system hydrogen dominates the composition, resulting in a reducing chemistry; closer to the Sun oxygen dominates, and conditions are oxidizing. The inner planets are relatively small, dense objects composed predominantly of rock and metal. Any primitive material they once contained has been strongly modified, producing the familiar igneous, sedimentary, and metamorphic rocks. In contrast the outer giants have low densities and are made primarily of hydrogen and helium, the two lightest and most abundant elements in the universe. These giant planets captured most of their hydrogen and helium from the primordial solar nebula. These planets are surrounded by satellites that form systems reminiscent of the solar system itself.

The distinction in composition between inner and outer solar system reflects a difference in the degree to which the matter in the planets has been processed. The primitive outer planets (and their satellites) are rich in gases and ices, reflecting more closely the original hydrogen-rich composition of the nebula. The oxidized inner planets are differentiated bodies and are much more evolved. Some have outgassed thin atmospheres that are oxidizing in chemistry.

The ability of a planet to retain an atmosphere depends on its mass and temperature. Large, cold planets do better than small warm ones, and heavier gases are easier to retain than light ones. The gases in planetary atmospheres were captured from the original nebula or outgassed from the material making up the planets themselves.

To learn about the rocks and gases of another planet, we must study the radiation the planet sends us—either reflected sunlight or thermal emission. Using a spectrograph to analyze the radiation, we can determine the nature of the reflecting or emitting surface and the composition, density, and temperature of any surrounding atmosphere.

We help ourselves enormously in such studies by using powerful telescopes to collect the radiation we wish to investigate. These telescopes must be located in observatories on remote mountaintops if they are to realize their full potential. To do still better, we must leave the Earth and its obscuring atmosphere, and launch our telescopes into space.

Most of what we are learning about the planets today is the result of visits by instrumented spacecraft. With spacecraft cameras we can see details we could never hope to glimpse from Earth, even using a telescope in orbit. Spacecraft also offer the opportunity to send probes and landers to a planet to sample its environment directly. Ultimately, such missions can even bring samples back from some distant world, allowing detailed studies in our laboratories.

KEY TERMS

albedo
aperture
atmospheric window
basalt
differentiation
exosphere
igneous
metamorphic
mineral
oxidizing
ozone
primitive
reducing
remote sensing
sedimentary
silicate
solar nebula
volatiles

Review Questions

1. Why is density such an important quantity to measure for a planet? List the densities of some common materials, including gold and lead. Determining whether an item is made of pure gold has been of interest for many centuries; look up how the ancient Greek scientist Archimedes solved this problem based on a determination of density, and relate this to the problem of determining the composition of distant planets.

2. Distinguish between rocks and minerals. What are some common examples of each? On a cold planet (such as the moons of the outer planets) ices of various kinds are present; are these rocks or minerals?

3. Why are there no primitive rocks on the Earth? Where in the solar system are primitive rocks likely to be found? Why is the search for them important?

4. What is differentiation? Explain why a large planet is more likely to differentiate than a small one. What other properties, in addition to size, are likely to be important in determining whether a planet differentiates?

5. Distinguish among gas, ice, rock, and metal, giving examples of each. Which of these dominate for the various members of the planetary system?

6. Explain the origin and loss of planetary atmospheres. Is it clear why some objects have atmospheres while others do not?

7. What are the differences between direct measurements and remote sensing? Explain how each might be used to study a planetary surface, a planetary atmosphere, or the plasma that makes up a planetary magnetosphere.

8. What are the various advantages and disadvantages of studying the planets by ground-based telescopes, by space telescopes, by flyby spacecraft, by orbiters, or by probes/landers?

Quantitative Exercises

1. Suppose you wanted to send a spacecraft to Pluto by the minimum energy trajectory, with perihelion at the Earth and aphelion at the orbit of Pluto. Calculate the one-way flight times for the case (a) of Pluto near its perihelion, which it reached in 1989, and (b) near its aphelion, which it will reach early in the twenty-second century.

2. What would be the mass of the Earth if it retained its present diameter but was composed entirely of (a) water ice, (b) silicate rock, or (c) iron metal?

Additional Reading

Burrows, W.E. 1990. *Exploring Space: Voyages in the Solar System and Beyond.* New York: Random House. History of the exploration of the solar system by spacecraft, with emphasis on the often complex politics behind the missions.

Field, G.B., and E.J. Chaisson. 1985. *The Invisible Universe.* Boston: Birkhäuser. Description of how astronomers use radiation from all parts of the electromagnetic spectrum, including justifications for observatories in space.

Morrison, D., S.C. Wolff, and A. Fraknoi. 1995. *Abell's Exploration of the Universe.* 7th ed. Philadelphia: Saunders College Publishing. Comprehensive introductory textbook in astronomy, with detailed discussion of observatories and the operation of telescopes and astronomical instruments.

Poynter, M., and A.L. Lane. 1981. *Voyager: The Story of a Space Mission.* New York: Atheneum. Highly readable history of one of the most successful of all space missions, written for a high school audience.

4

Meteorites: Remnants of Creation

Admiral Robert Peary, the first man to reach the North Pole, brought the Anighito ("tent") meteorite from Greenland to the American Museum of Natural History in 1897. At 34 tons, the Anighito is the largest meteorite on display in a museum. Here it is being loaded aboard Peary's ship, the *Hope*, for the voyage to New York City.

Some of the most fundamental questions about the planets concern their origin. When was the solar system created, and how? Are the processes that formed the planets unique, or might they be a common feature of the birth and evolution of stars, implying that multitudes of planetary systems might exist throughout the Galaxy?

Unfortunately, the planets themselves tend to be mute on questions of origin, because they (and their larger satellites) are geologically and chemically active bodies that have evolved considerably since their formation. Just as a study of an adult population of humans provides limited and ambiguous data on birth and childhood development, so mature planets retain only limited and ambiguous memories of the conditions of their formation and early history. Much more helpful would be examples of smaller, unmodified objects—planetary building blocks—that have changed little since the formation of the solar system. An important development in modern planetary science has been the recognition that some of these remnants of creation have survived as the comets and asteroids. Even more important, fragments of material from comets and asteroids are available for laboratory study on Earth in the form of the meteorites.

A small piece of debris in space, before it encounters the Earth, is called a **meteoroid.** Most impacting meteoroids are fragmented and consumed by atmospheric friction before reaching the surface. Plunging into the atmosphere at speeds of tens of kilometers per second, they burn away at high altitudes in fiery trails of glowing gas. This death-flash of a small meteoroid is called a **meteor,** or, more familiarly, a shooting star or falling star. If a meteoroid survives its brief passage through the air and lands on the ground, we call it a **meteorite.** The story revealed by studies of meteorites is the subject of this chapter.

4.1 THE SOLAR NEBULA

The planetary system began with the primordial *solar nebula,* the cloud of gas and dust out of which the solar system formed about 4.5 billion years ago. Astronomers can identify many similar condensing clouds of interstellar material in our Galaxy today, at locations where other stars and perhaps other planetary systems are now being born. Before we begin to study the individual objects in our planetary system, we need to sketch some current ideas on the origin of our solar system.

Fundamental Properties of the Solar System

It is easy to give a name to the solar nebula. Characterizing it is harder. In order to deduce the properties of this disk of dust and gas, we must work our way backward from conditions in the solar system today. As we noted in Chapter 2, more than 99% of the material in the solar system is in the Sun, which is composed almost entirely of the two lightest gases, hydrogen and helium. In this respect the Sun is like the other stars, and indeed like the interstellar material between the stars. Thus we can conclude that the solar nebula also had approximately this same cosmic composition.

The largest planets, Jupiter and Saturn, have almost the same composition as the Sun. The smaller bodies, however, are depleted in hydrogen, helium, and other light gases. In addition, the inner planets were probably formed without ices or other volatile materials. Thus it appears that some processes involved in planetary formation selectively concentrated the less volatile compounds, permitting lighter materials to escape. Such chemical sorting is often called **fractionation.** While the Sun may have the same composition as the original solar nebula, the processes that gave rise to smaller solid bodies were highly selective in the building materials they employed.

Another basic property of the solar system is its rotation. All the planets have orbits that are approximately circular and lie roughly in the same plane, and all revolve in the same direction around the Sun. Further, the Sun itself rotates in this same direction, and its equator lies essentially in the same plane as the orbits of the planets. All these bodies apparently formed from a solar nebula that was rotating. Moreover, the material that formed the planets must have been confined to a disk in order for their orbits to have settled so closely to the same plane.

Two fundamental properties of the solar system must therefore be related to the solar nebula: (1) its rotation and disk shape; and (2) its chemical composition, with the Sun composed of unmodified cosmic material, while the solid planets and smaller members of the system are made up primarily of rarer rocky, icy, and metallic substances.

Formation and Condensation of the Nebula

In order for the Sun to form at all, the thin clouds of interstellar gas and dust must have become concentrated enough to collapse under their own weight. Astronomers see many similar gravitationally contracting clouds at sites of active star formation today (Fig. 4.1), but we still do not understand just what triggers the collapse. Anyway, it happens, or we would not be here to discuss the process. Let us start our story of the solar system at the stage of a contracting solar nebula with a mass at most twice that of the Sun today and a composition representative of our corner of the Galaxy.

As the solar nebula collapsed, its central parts were heated by the infalling material. As we described in Chapter 2, gravitational contraction generates heat, as the energy of falling material is transformed into heat energy. At the same time, the shrinking nebula began to spin faster, and its outer parts flattened into a disk.

The formation of a rapidly spinning disk of material from a larger, more spherical cloud is an example of the principle called *conservation of angular momentum.* This familiar physical law is based on the fact that the **angular momentum** of a body depends on three quantities: mass, rotation, and size. In order for angular momentum to be conserved, or held constant, if one of these quantities decreases, another must increase proportionately. In our example, the size of the nebula decreases while the mass remains the same, so the law tells us that the rotation rate must increase. Just as a turning ballet dancer on point increases her spin by drawing in her arms, the nebula speeds up as it contracts. Thus the contracting solar nebula develops a hot central core surrounded by a rapidly spinning disk. The core will become the Sun when its internal temperature is hot enough and dense enough to sustain thermonuclear reactions, while the disk will give rise to the planets. The outer parts of the

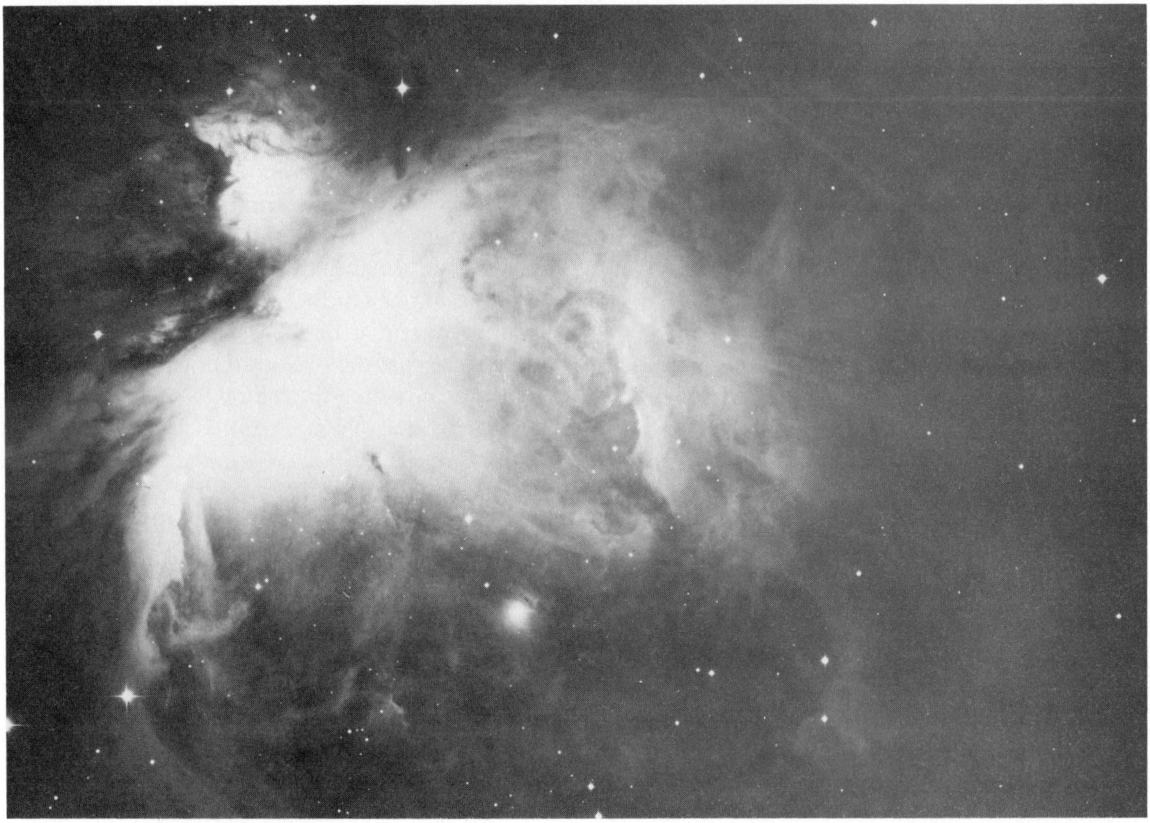

FIGURE 4.1 The Great Nebula in Orion is a giant cloud of gas and dust 1500 light years away. Deep within this huge complex—partially illuminated by light from embedded and nearby stars—new stars, planets, comets, asteroids, and meteoroids are forming.

nebula continue to fall into the disk, which increases in both mass and temperature.

Let us focus our attention on the disk (Fig. 4.2). As the initial rapid infall of nebular material subsides, temperature differences develop in the disk. The outer parts cool while the inner parts are heated by the proto-Sun, just as the temperatures of the inner planets are higher than those in the outer solar system today. At each point, solid grains begin to condense like hailstones from a storm cloud, but they are composed of only those chemical substances that have freezing temperatures below the local temperature of the nebula. Table 4.1 indicates the composition of the condensate expected in the outer regions of the solar nebula, relative

TABLE 4.1 Expected abundances in the outer solar nebula

Material	Percent (by mass)
Hydrogen (gas)	77
Helium (gas)	22
Water (ice grains)	0.6
Methane (ice grains)	0.4
Ammonia (ice grains)	0.1
Rock and metal (solid grains)	0.3

Cloud rotates
more rapidly as
it contracts.
Cloud flattens to
pancake-like
configuration.
Proto-Sun
Planets accrete
at their present
distances from Sun.

FIGURE 4.2 As the cloud of gas and dust that formed the solar system contracted, it rotated more rapidly, leading to the formation of a flattened disk, which in turn broke up into condensations that became the planets and their satellites. The Sun, accumulating most of the mass of the system, formed at the center.

to the hydrogen and helium gas. These are the main materials available to form a giant planet such as Jupiter or Saturn.

In warmer parts of the solar nebula not all of these materials were able to form solid grains. For instance, water ice could condense at a distance of 5 AU from the proto-Sun, but not at 2 AU. Similarly, many common silicate rocks could form at 1 AU, but not at the higher temperatures closer to the center. Inside of about 0.2 AU no solids formed and the solar nebula remained gaseous. In this way, a sequence of chemically distinct grains formed at different distances from the center of the nebula. This temperature-related *chemical condensation sequence* provides an explanation for some of the major differences we see today in the compositions of the planets and other members of the planetary system.

Accretion and Fragmentation

We now have a solar nebula in which solid grains have formed, chemically sorted according to distance from the proto-Sun. The next step in formation of the planetary system requires that these grains come together to form larger aggregates. Initially, much of this aggregation results from the sticking together of the grains as they bump into one another, quickly building up to objects a few tens of kilometers in diameter, which astronomers call **planetesimals** (little planets). As they grew, the planetesimals began to attract each other gravitationally, beginning the process of **accretion,** in which the particles of the solar nebula grew by their mutual gravitational attraction. Ultimately the innumerable tiny grains were swept up into bodies that reached hundreds and then thousands of kilometers in diameter—the **protoplanets.**

The two largest protoplanets eventually grew so massive that they were able to attract and hold the uncondensed gas in their part of the nebula; they thus grew to giant proportions, becoming the Jupiter and Saturn we see today. Some of the loose gas was also attracted to Uranus and Neptune. Elsewhere in the nebula, however, only the condensed solids were available as building materials for accretion.

As the planetary objects became fewer in number and their individual sizes larger, they exerted stronger gravitational forces upon each other and their smaller neighbors, which were jostled about like floating Ping-Pong balls stirred with a stick. As a result, the speeds at which one body impacted another grew higher, and the violence of the impacts increased sharply. When a small body struck a large one, it gouged out a crater, or, if the body were not solid, splashed liquid fragments in all directions. When two bodies of similar size crashed, the results were even more catastrophic, ending in mutual disruption.

These processes, by which impacts break down objects rather than building them up, are called **fragmentation.** Thus we have two competing processes: accretion, which makes planets grow, and fragmentation, which breaks them down. As the system evolved, fragmentation became more important, ultimately dominating over accretion for all but the largest bodies. Thus the big got bigger, but most of the smaller protoplanets were reduced to fragments—some of which survive today in the form of asteroids and comets.

Final Stages

In the end, three processes eliminated the gas and dust of the solar nebula. One was condensation and accretion, which led to the formation of protoplanets. A second process, which accounted for a much greater loss of the original material, was the increasing activity of the young Sun at the center. Early in their lifetimes, stars go through a stage of *mass loss* in which they expel material at high speed from their surfaces. These streams of outflowing solar material—a kind of super solar wind—effectively swept away most of the gas in the solar nebula, as well as any grains that had not yet accreted into the larger bodies. Third, gravitational interactions with the newly formed planets eliminated most of the remaining solid fragments. Either these fragments impacted the solid surfaces of the planets or they came sufficiently close to be expelled gravitationally.

We have provided this sketch of solar system formation as an introduction to the *meteorites.* Much of what we know about the processes of condensation, accretion, and fragmentation in the solar nebula has been learned from these remnants from that early epoch of our system's history. The meteorites are fragments from comets and asteroids, which are themselves the rare survivors of the vast numbers of building blocks that formed from the solar nebula. When you contemplate a meteorite, don't think of it as just a piece of ugly dark rock that fell from the sky. Think of it as a survivor from the time when the planets formed, providing us a unique perspective on our origins.

4.2 CLASSIFICATION OF METEORITES

Meteorites are defined as those extraterrestrial fragments that collide with the Earth and survive to reach the surface. Until the last century, however, the idea that extraterrestrial materials were reaching the surface of the Earth was scoffed at by educated persons, who placed stories of falling stones in the same category with tales of fairies and dragons. U.S. President Thomas Jefferson, himself a distinguished amateur scientist, is reported to have reacted to information about an 1807 meteorite fall in Connecticut by commenting that he could more easily believe that Yankee professors would lie than that stones would fall from the sky. Such events were so infrequent and unpredictable, and so rarely observed by "reliable" witnesses, that it was easy to dismiss them (Fig. 4.3).

By the end of the eighteenth century, however, the special compositions of some meteorites were becoming recognized, and a case was made that they came from beyond the Earth. For most scientists of the time, the extraterrestrial origin was proved in April 1803 when a fall of stones from the sky was reported in the village of l'Aigle, France. The French Academy of

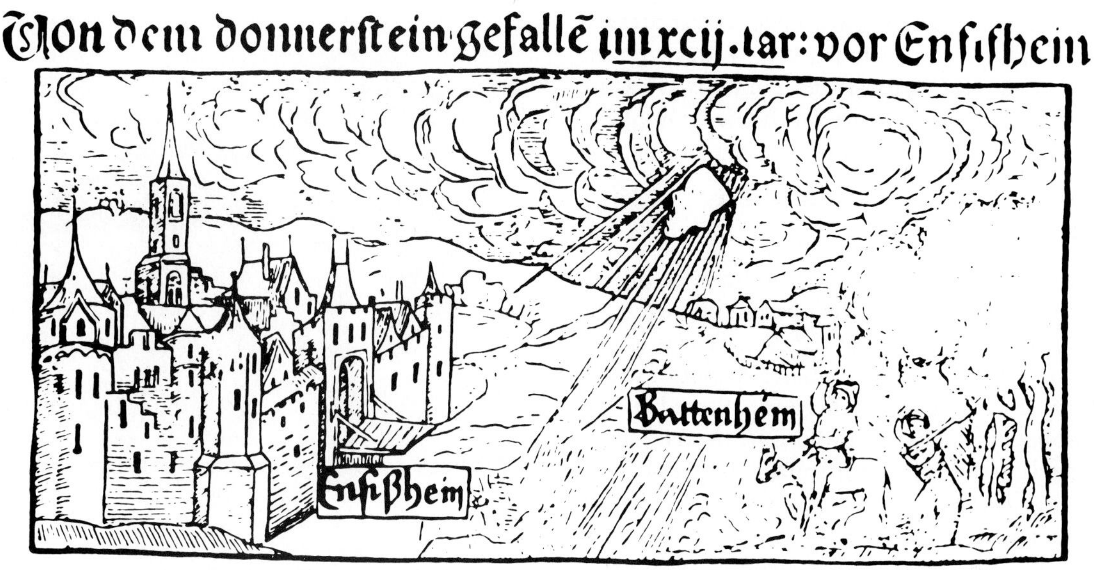

FIGURE 4.3 An old woodcut showing the fall of the meteorite near Ensisheim, France, in 1492. The German caption reads: "of the thunderstone (that) fell in xcii (92) year outside of Ensisheim."

Science sent a team of reputable scientists to investigate, interview the witnesses, and collect the fallen stones; further investigation confirmed that these stones were unlike any ordinary rocks. Thus the authenticity of meteorites was established.

Irons, Stones, and Stony-Irons

Meteoritic materials are constantly reaching the Earth, and several falls are observed and recovered every year. The fragments that reach the ground are usually stones or metallic masses of only a kilogram or two, small enough to be held comfortably in your hand. Larger falls are produced when a mass of hundreds or thousands of kilograms strikes the atmosphere, often breaking up to scatter meteorites over many kilometers. Even larger projectiles can also strike at intervals of thousands of years, some of them exploding in the atmosphere, while others produce impact craters when they crash into the surface. We will return to the process of impact cratering in Chapter 7, when we consider the heavily scarred surface of the Moon.

A surprising variety of rocks is in our meteorite collections, suggesting the existence of many different **parent bodies**—the term we use for the asteroids or other objects from which the meteorites originated through cratering or breakup of the parent. The traditional descriptions of these meteorites are based on their appearance and bulk composition. Three classes are generally used: iron, stony, and the rarer stony-iron types.

The **irons** are nearly pure metallic nickel-iron and are readily recognized from their high density of more than 7 g/cm^3. Their extraterrestrial origin is obvious when we recall that iron and most other metals normally occur on Earth in the form of oxides rather than in the pure metallic state. In fact, iron meteorites were one of the first sources of iron, and their use helped to stimulate the transition from Bronze Age to Iron Age cultures. The second group, the **stones,** more closely resembles terrestrial rocks; these are not generally recognized as being of extraterrestrial origin unless their fall was witnessed. The third group, the **stony-irons**, contains a mixture of stone and metallic iron, as the name implies.

Primitive and Differentiated Meteorites

A more useful categorization of the meteorites is based on the history of their parent bodies. If the chemistry of a meteorite indicates that it is representative of the original materials out of which the solar system was made, little altered by the subsequent chemical evolution of its parent body, we refer to it as a **primitive meteorite.** All primitive meteorites are stones. Since many of the meteorites that reach the Earth are primitive, we can be confident that material still exists in the solar system which has remained relatively unchanged since before the planets were formed.

Although the surface of a meteorite is heated to incandescence during its brief plunge through the atmosphere, the heat pulse penetrates no more than a few centimeters into the interior, most of which remains cool and undisturbed. Even the outer layers have normally cooled by the time the meteorite strikes the ground, as shown by objects that have fallen on ice or snow without melting it. Thus primitive meteoritic material remains largely unaltered by its violent arrival at Earth.

Meteorites that have experienced major chemical change since their formation are called **differentiated (or igneous) meteorites.** Like the igneous terrestrial rocks, they solidified out of a molten state. Differentiated meteorites appear to be fragments of differentiated parent bodies that experienced major episodes of heating, along with loss of volatile materials. All of the iron and stony-iron meteorites, and many of the stones as well, are examples of differentiated meteorites.

While many meteorites are lumps of material of fairly uniform composition, others of both the primitive and differentiated types show evidence of having been repeatedly broken, mixed, and welded by impact processes on their parent bodies. These fragmented and recemented rocks are called **breccias** (Fig. 4.4). Brecciated rocks, whether meteorites or lunar samples, have experienced impact cratering in the crustal soil of their parent bodies. Some meteorite breccias are especially interesting, because each contains fragments from a variety of regions on the crust of the parent body, thus providing a sampling over a wide area.

FIGURE 4.4 A meteorite sawed in half to show the appearance of its interior. This is a breccia, consisting of light and dark rocky components. The dark material contains chondrules.

Falls and Finds

Meteorites can also be distinguished in terms of the way they are identified on Earth. The most obvious meteorites are ones that are seen to fall. A flaming plunge through the atmosphere may terminate in an aerial explosion that scatters fragments over many square kilometers. Meteorites located in this way are termed **falls.**

A second group of meteorites is called **finds.** These are objects whose falls are not witnessed but which are later recognized to be of extraterrestrial origin. Because stony meteorites look superficially like ordinary rocks, they are rarely recognized unless their fall is seen or unless they land in an unusual location, such as a snowfield. Most finds therefore consist of the much more obvious irons and stony-irons.

Most of the meteorites exhibited in museums are irons, in part because they are more easily identified. Table 4.2 illustrates the frequency of the different types of meteorites in three population groups: finds, falls, and the Antarctic meteorites discussed below. It is clear from this table that the primitive stones are the most common type of meteorite.

FIGURE 4.5 The fifth meteorite found at Allan Hills (Antarctica) in 1981 (ALHA 81005). This basaltic breccia is a sample of the lunar surface delivered to Earth without human intervention.

TABLE 4.2 Abundances of major meteorite types (as percent of total)

	Finds	Falls	Antarctic
Primitive stones	52%	87%	85%
Differentiated stones	1%	9%	12%
Irons	42%	3%	2%
Stony-irons	5%	1%	1%

Antarctic Meteorites

The Antarctic is an important source of meteorites that does not really conform to the traditional distinction between falls and finds. In some parts of the Antarctic continent the slow movement and subsequent evaporation of ice transports and concentrates any meteorites that fall over areas of tens of thousands of square kilometers (Fig. 4.5). In these "blue ice" regions thousands of meteorites have been identified and collected, representing the accumulation of hundreds of thousands of years.

Although they are technically finds, not falls, the Antarctic meteorites do not suffer from the selection effects that cause stony meteorites, including nearly all of the most primitive samples, to be overlooked among meteorite finds. The fact that essentially every meteorite that has fallen within the area of ice flow can be spotted and collected is one reason why the Antarctic meteorites constitute such a valuable addition to our inventory of extraterrestrial materials.

The first Antarctic meteorites were discovered in 1969 by a team of Japanese scientists. Since then expeditions have been sent out nearly every southern summer to search by helicopter and on the ground for the stones, many of which are little bigger than pebbles. Because they have remained frozen in the ice since their fall, these meteorites have suffered relatively little weathering and atmospheric contamination. In the United States, many of the Antarctic meteorites are being stored under carefully controlled conditions at the NASA Johnson Spaceflight Center in Houston, along with the lunar samples brought back by the Apollo and Luna missions. From there, these samples are distributed to scientists for analysis in their laboratories. As of the end of the 1995 season, more than 15 thousand meteorites had been collected in the Antarctic. The total number of meteorites in our collections prior to the Antarctic discoveries was under a thousand (where fragments from the same fall are not counted separately).

Nomenclature

Each individual meteorite is given a name, usually for a town or other geographic feature near its point of recovery. Thus, for instance, we have the Allende meteorite, a large collection of stones that fell in 1969 near Pueblito de Allende in northern Mexico. In the case of the Antarctic meteorites, where thousands have been found in just a few locations such as the Yamato Mountains and the Allan Hills, a number is also used for identification, such as ALHA 81005, a first-discovered lunar meteorite. The first two digits give the year of the find (in this case 1981), while the last three represent a running index. Decisions on meteorite nomenclature are made by a committee of the Meteoritical Society, an international organization of meteorite researchers.

Table 4.3 summarizes the types of meteorites, proceeding from primitive to modified.

TABLE 4.3 Characteristics of the main meteorite types

Type	Composition
Primitive meteorites (chondrites)	
Carbonaceous	silicates, carbon compounds, water
Other primitive stones	silicates, iron
Differentiated meteorites (achondrites)	
Differentiated stones	igneous silicates
Stony-irons	igneous silicates, iron, nickel
Irons	iron, nickel

Orbits and Origins

In order to determine where the meteorites are coming from, we would very much like to know their orbits before they struck the Earth. Of course, once a meteorite has fallen, it is too late to go back and reconstruct its path before its orbit intersected that of the Earth. Since the meteorites we deal with are all far too small to have been sighted telescopically before impact, the only solution is to make accurate enough measurements of the path of the meteorite through the atmosphere during its fall to reconstruct its pre-impact orbit.

Four major efforts have been made to acquire such data, in the United States, Canada, England, and Czechoslovakia. Of course, we cannot know when a meteorite is coming, so we must set up our cameras, photograph the sky continuously every clear night, and hope for luck. During about 20 years of such searches, three meteorites (all of them primitive) have been photographed in flight and subsequently recovered: Pribram, in Czechoslovakia (1959); Lost City, in the United States (1970); and Innisfree, in Canada (1977). These meteorites proved to be on eccentric orbits that carried them in from the main asteroid belt between the orbits of Mars and Jupiter to cross the orbit of the Earth (Fig. 4.6). While not identifying any specific parent bodies, these three observations do suggest that the asteroids might have been the source of these meteorites. With the advent of video cameras, amateurs can also contribute to the determination of pre-impact orbits; such video records of the entry fireball of the 1993 Peekskill meteorite permitted scientists to reconstruct an approximate orbit for this object as well.

Hits and Misses

At this point, a reader may wonder whether any of these interplanetary wanderers ever impact human lives, literally or figuratively. The answer is yes, in both categories. There is only one confirmed report of a meteorite striking a human

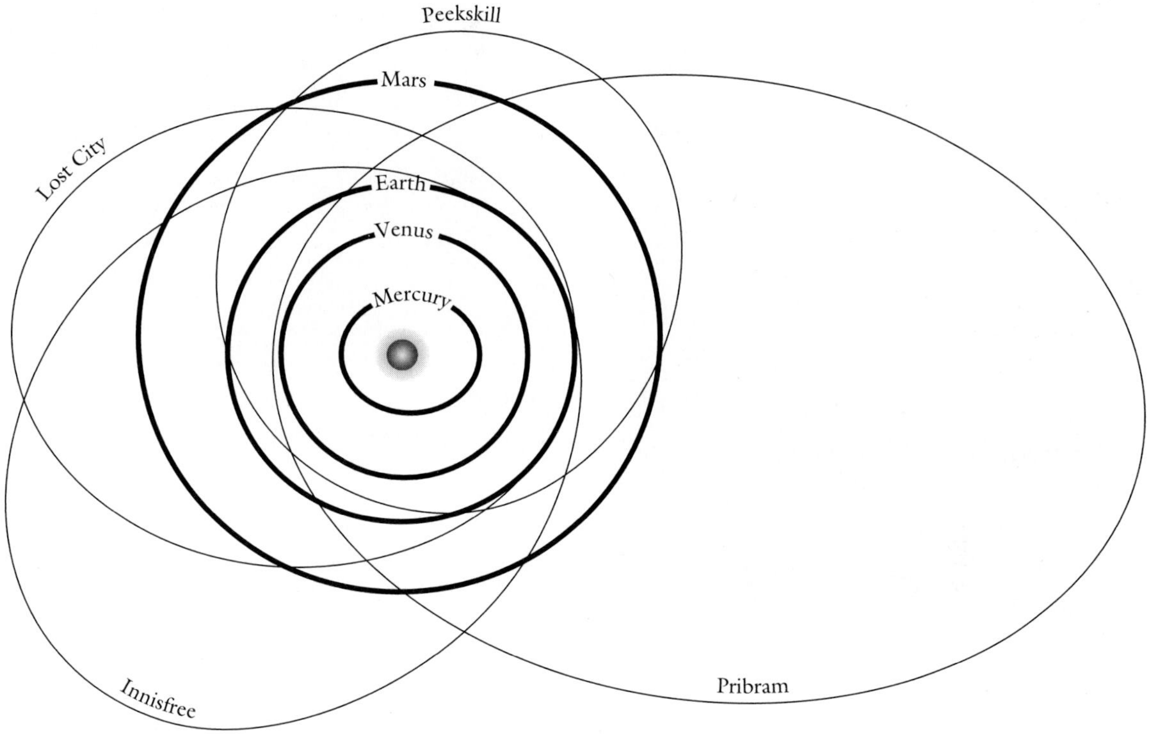

FIGURE 4.6 The reconstructed orbits of four meteorites, Peekskill, Lost City, Innisfree, and Pribram, whose tracks through our atmosphere were photographed. All four orbits extend into the asteroid belt.

being (Annie Hodges of Sylacauga, Alabama, bruised by a ricochet on November 30, 1954) but there are several reports of animals being killed by direct hits. For example, one of the meteorites that is now known to have originated on the planet Mars killed a dog when it fell to Earth in Nakhla, Egypt, on June 24, 1911.

Impacts on human structures are much more common. On November 12, 1982, a six-pound meteorite crashed through the roof of the Donahue home in Wethersfield, Connecticut, continued on through the living room ceiling, bounced off a carpeted wooden floor (cracking a supporting beam in the process) and eventually came to rest under the dining room table (Fig. 4.7a). Defying all odds, this was the second meteorite to hit a home in Wethersfield in 12 years!

Another recent meteorite was seen first as a brilliant meteor, lighting up the evening sky over several eastern states before it smashed into the back end of a Chevrolet in Peekskill, New York (Fig. 4.7b, c). Scientists were able to analyze video tapes fortuitously recorded by people under the meteor's path who had aimed their cameras at this totally unexpected phenomenon. As a result, it was possible to add a fourth orbit to the three deduced from the specific research described in the previous section (Fig. 4.6). As with the other three, it is clear that the Peekskill meteorite originated in the asteroid belt, this time near the inner edge of the main belt.

a

a (inset)

b

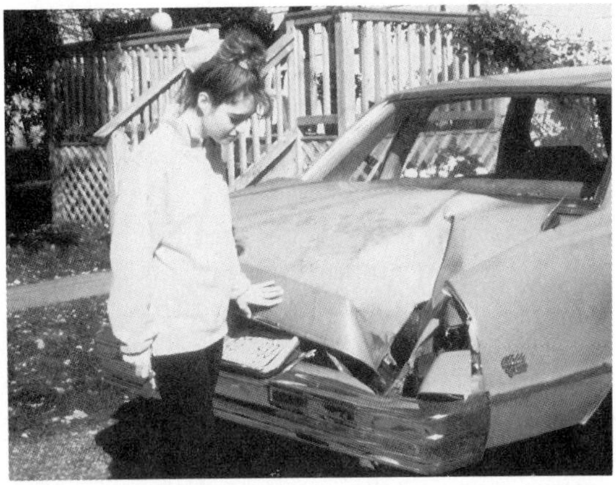

c

c (inset)

FIGURE 4.7 Examples of meteorites that have affected human lives: (a) Wethersfield Fire Department Assistant Chief John McCauliffe surveys the damage caused by a 6-pound meteorite (inset) that hit the home of Robert and Wanda Donahue shortly after 9:00 P.M. on Nov. 8, 1982. (b) The path of the Peekskill meteor, projected on the ground, as it flashed downward through the Earth's atmosphere at 7:45 P.M. on Oct. 9, 1992. (c) A Chevrolet that suffered a most unusual collision! Inset: the meteorite that did the damage.

4.3 AGES OF METEORITES AND OTHER ROCKS

We have asserted several times in this book that the solar system, including the Earth, was formed 4.5 billion years ago. How was this figure determined? The ability to date the formation of rocks and planets is one of the triumphs of modern science, with implications for many fields ranging from astrophysics to paleontology. Rock dating, for example, provides the absolute time scale for biological evolution on Earth, permitting us to determine the time intervals between different eras originally assigned only relative ages on the basis of geological context and fossil remains.

Natural Radioactivity

The dating of rocks depends on the properties of radioactive decay, discovered early in the twentieth century. Radioactivity is the natural process whereby some isotopes of certain elements spontaneously change into other isotopes. This is a one-way process, accompanied by the release of energy in the form of gamma rays, electrons, or alpha particles, which are the nuclei of helium atoms. It is these emitted radiations that make high levels of radioactivity damaging to living tissue.

Today much of our concern about atomic radiation involves radioactivity from nuclear reactors or nuclear bombs. Such artificially created radioactive materials can pollute the environment and, in the aftermath of a nuclear war, could result in the death of much of the life on our planet. In contrast, the radioactivity that we will be discussing here is natural, the product of minute quantities of uranium, thorium, potassium, and other radioactive elements that are distributed throughout most rocks and minerals.

As a radioactive atom decays to produce a new, nonradioactive isotope, the concentration of the original, or parent, material decreases and that of the new, or daughter, material increases. The relative concentrations of parent and daughter are the basic information from which an age can be derived.

The Rate of Radioactive Decay

In order to interpret measurements of the parent/daughter ratios, we must know the rate at which radioactive decay is taking place. Fortunately, this rate is a fixed property of the parent material and is independent of external conditions, such as temperature, pressure, or even the chemistry of the minerals in which the radioactive material is located. Although there is no way to predict when any particular individual atom will undergo this change, we know that in a fixed interval of time some specific fraction of a collection of similar atoms will decay.

Usually the decay rate is expressed in terms of a **radioactive half-life,** defined as the time for one half of the atoms to decay into their daughter product. For instance, if a rock initially contained 64 micrograms of a radioactive isotope with a half-life of 1 million years, we would find that only 32 micrograms remained at the end of the first million years, 16 at the end of 2 million, 8 at the end of 3 million, 1 at the end of 6 million years, and so forth. The daughter product would increase as the parent declined.

To determine the ages of meteorites and other ancient rocks we require parent elements with half-lives in the range of hundreds of millions to tens of billions of years, commensurate with the ages of the rocks we are studying. The most useful reactions, together with their half-lives, are given in Table 4.4.

Different Rock Ages

Suppose that you make a careful measurement of the concentrations of one of the parent/daughter pairs in Table 4.4 and calculate an age from the given half-life. Just what does such an age mean? This is an important question, especially since different parent/daughter pairs may not

TABLE 4.4 Radioactive decay reactions used to date meteorites and planets

Parent	Daughter	Half-life (billions of years)
Samarium (Sm-147)	neodymium (Nd-143)	106
Rubidium (Rb-87)	strontium (Sr-87)	48.8
Thorium (Th-232)	lead (Pb-208)	14.0
Uranium (U-238)	lead (Pb-206)	4.47
Potassium (K-40)	argon (Ar-40)	1.31

yield the same calculated age. Essentially, the age measured is the time interval over which the daughter product has been able to accumulate undisturbed from the decaying parent. Since in a liquid the parent and daughter tend to become mixed with other materials, the measured age is the time since the material became solid, and it is referred to as the **solidification age.**

In the case of the potassium-argon pair, the daughter product is a gas, and the age measured is the **gas retention age**. Thus, if we calculated a meteorite age as 18 million years by the K/Ar method and as 4.3 billion years by the Rb/Sr method, we would probably conclude that the rock solidified from the liquid 4.3 billion years ago but that a subsequent shock (perhaps the breakup of the parent body) allowed the argon gas to escape 18 million years ago, in effect resetting the clock that measures the gas retention age.

In order to realize the full power of the radioactive age measurement techniques, we must have a way to determine the original ratio of parent to daughter isotopes in the rock at the time of solidification. This ratio is calculated using other nonradioactive isotopes of the same elements. These sister isotopes behave in exactly the same way as the radioactive isotopes in any chemical processes that alter the composition of the rock melt, thereby removing this source of uncertainty. Additional checks can be applied by making independent age measurements for different mineral grains from a single sample.

Our confidence in the measured ages of meteorites and lunar rocks is the result not only of the high precision achieved in the measurement of the concentrations of parent and daughter products, but also of the number of checks that can be carried out to establish the consistency of the results. Usually it is possible to apply three or more separate techniques, based on different radioactive elements with very different half-lives, and also to apply the measurements to a variety of mineral grains of differing chemistry. Generally, these results agree to within a few percent in the derived age. Where differences do arise, they can usually be attributed to subsequent events, such as the example cited earlier where the gas retention age was shorter than the solidification age as a consequence of impact resetting of the gas retention clock.

Table 4.5 illustrates the solidification ages derived for a variety of meteorites. Almost all of them are near 4.5 billion years. These values, which cluster together so well in time, provide the best measure of the age of the solar system, defined as the time interval since its most primitive constituents condensed out of the cooling gas and dust of the solar nebula.

TABLE 4.5 Measured solidification ages of primitive meteorite groups*

H group chondrites	4.50 ± 0.04 by
L group chondrites	4.43 ± 0.05 by
LL group chondrites	4.51 ± 0.03 by
E group chondrites	4.45 ± 0.03 by

*Rubidium-strontium ages, adapted from data given in Robert T. Dodd, *The Meteorites: A Petrologic-Chemical Synthesis* (Cambridge: Cambridge University Press, 1981).

4.4 PRIMITIVE METEORITES

Primitive meteorites have chemical compositions that have remained relatively unchanged since they formed in the cooling solar nebula about 4.5 billion years ago. Except for a shortage of gaseous and other volatile constituents such as hydrogen, helium, argon, carbon, and oxygen, the composition of the primitive meteorites is thought to be the same as that of the Sun. In fact, these meteorites are used to define the presumed solar abundances for some rare elements that cannot be observed directly in the Sun.

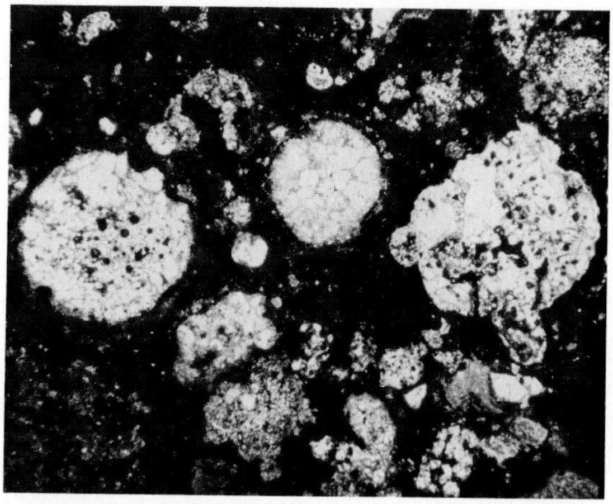

FIGURE 4.8 Chondrules are small congealed drops of rock; this image is an enlargement of a thin section of the Allende meteorite showing chondrules (the largest a few millimeters across) embedded in a dark matrix.

Composition and History

The primitive meteorites are also called **chondrites**, named for the small round **chondrules** that they contain. These chondrules (typically about 1 mm in diameter) appear to be frozen droplets that condensed from molten material (Fig. 4.8). Because not all chondrites contain chondrules, we prefer the more general term *primitive meteorite.* In appearance, the primitive meteorites are light to dark gray rocks, often with a darker crust produced during their fiery descent through the atmosphere. Many are breccias. Their densities are about 3 g/cm^3, similar to that of many crustal rocks on the Earth.

Chondrites often show evidence of limited alteration by heating or the effects of liquid water that was present during some early epoch of their evolution. Some scientists call these metamorphic meteorites. Such altered meteorites are less primitive, and scientists must make due allowance for their past history in order to relate them properly to the solar nebula.

Many of the primitive meteorites contain grains of metallic iron as a major constituent, 10%–30% by weight. The quantity of iron leads to a useful classification of the primitive meteorites by their metal content. Thus scientists speak of H chondrites for those with high iron content, L chondrites for those with low iron, and so forth. The most abundant elements after iron are silicon, oxygen, magnesium, and sulfur.

It is interesting to note the uniform age of the primitive meteorites (Table 4.5). Apparently they all formed together, and none has intruded from beyond the solar system. This fact indicates just how isolated the planetary system is within the Milky Way Galaxy, even though our system has completed some 20 revolutions around the center of the Galaxy in the 4.5 billion years of its existence.

Carbonaceous Meteorites

The most primitive meteorites are a special group called the **carbonaceous meteorites,** some of which have no chondrules at all. These carbonaceous meteorites are relatively rich in carbon (a few percent by weight), as their name implies, and also in volatile compounds such as water, which combines chemically with other minerals to form clays. They contain very little metallic iron. Carbonaceous meteorites (Fig. 4.9) are dark gray to black in color and are phys-

FIGURE 4.9 One of the meteorites from the Allende fall. The part of the stone facing us has been broken off and shows unaltered meteoritic material.

ically very weak; a fragment can generally be crushed between the fingers. Because of their relatively high proportion of water, they are less dense than other meteorites, typically about 2.5 g/cm^3. From their composition we conclude that they were formed in a cooler region of the solar nebula than were the other primitive meteorites.

One of the special properties of the carbonaceous meteorites is the evidence that some of them have been altered by liquid water, which has dissolved certain minerals and redeposited them in veins. Apparently there existed parent bodies within which water could assume a liquid form, at least for some stage of early solar system history. Many also contain up to 3% of complex carbon compounds, as discussed next.

Organic Matter in Meteorites

Carbon compounds in which this element is joined with hydrogen and other elements are called **organic compounds,** since scientists once thought that only living organisms could produce them. On Earth, natural carbon minerals are rare except for calcium carbonate, and these complex compounds (which make up coal and petroleum, among other things) are indeed produced as a part of the life process. The organic carbon compounds in meteorites, however, are not an indication of life in the primordial solar nebula.

Many chemical reactions could have taken place in the hydrogen-rich environment of the solar nebula to produce organic compounds. This kind of matter is actually rather common in the solar system, both among the asteroids and in the satellite families of the outer planets. Let us look at the organic materials in carbonaceous meteorites and see what they can tell us about the origin of the planetary system.

Most of the carbon compounds in the primitive meteorites are complex, tar-like minerals that defy exact characterization. They also include substances that we recognize as being of fundamental importance for life: components of proteins and nucleic acids. These were first identified with certainty in the Murchison meteorite, a carbonaceous meteorite that fell in Australia in 1969, a few months after the Allende meteorite fell in Mexico. Murchison and Allende are probably the two best-studied meteorites.

Murchison yielded 74 separate amino acids, including 55 that are rare on Earth. The most remarkable thing about these and other amino acids in meteorites is that they include equal numbers of left-handed and right-handed forms. Amino acids can have either of these kinds of symmetry, but in practice life on Earth has evolved using only the left-handed versions of these compounds to make its proteins. The fact that both symmetries are present in the meteorites demonstrates that no contamination has taken place since they arrived on Earth, and indicates that these organic compounds formed in space without the intervention of living things. This conclusion is supported by the fact that the ratio of the two stable isotopes of carbon,

C-12/C-13, is different in the meteorite from its value in terrestrial organic materials.

Peculiar Inclusions

A few carbonaceous meteorites contain evidence on the very earliest periods of planetary history, stretching back perhaps even to before the solar nebula formed. Much of this evidence is to be found in small grains of foreign matter called inclusions. These inclusions are light-colored irregular grains, not to be confused with the round chondrules. The most revealing results on the light-colored inclusions have been derived from studies of the Allende meteorite, but a number of other carbonaceous meteorites also contribute to this field of research.

The Allende inclusions have yielded one of the most precise solidification ages ever measured: 4.566 ± 0.002 billion years. Some scientists argue that this value provides our best estimate of "the age of the solar system," but others are concerned that these inclusions may slightly predate the formation of the solar nebula. In this book, we will continue to use the round number 4.5 billion years as the nominal age of the planetary system.

The most important way in which these meteorites differ from terrestrial and lunar rocks is in their isotopic compositions. In all previous measurements for materials from the Earth, the Moon, or other meteorites, the ratios of the three common oxygen isotopes (O-16, O-17, and O-18) have followed the same pattern, independent of the chemical context or the place of origin. Inclusions in some of the carbonaceous meteorites, however, show different isotopic ratios, suggesting that a mysterious source of nearly pure O-16 has been mixed with the "normal" type of oxygen. Apparently this anomalous oxygen was injected into the solar nebula at the time these meteorites were forming, a process associated with the collapse of the solar nebula.

Interplanetary Dust

Perhaps the most primitive material available to us arrives at Earth in the form of interplanetary dust particles (IDPs), which are collected at high altitudes by specially instrumented airplanes

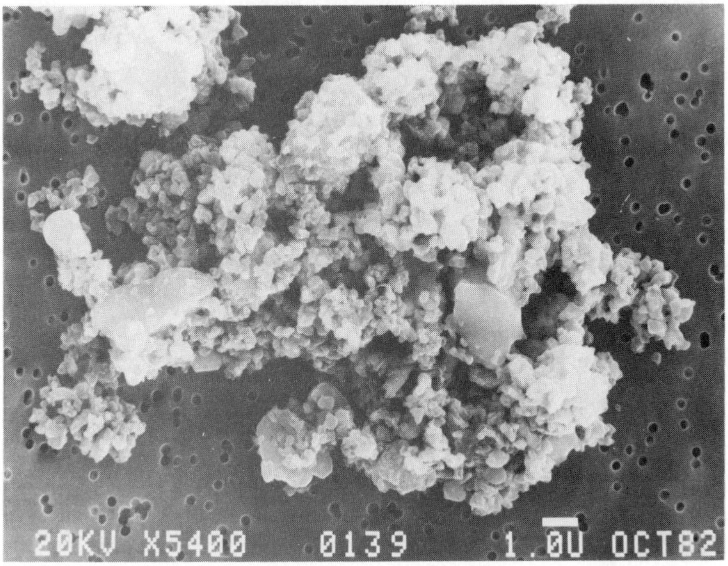

FIGURE 4.10 These meteoric particles (often called "Brownlee particles" after their discoverer, Donald Brownlee of the University of Washington) were captured in the Earth's stratosphere by a U-2 aircraft equipped with a special dust sampling device. A steady rain of particles like these (here magnified ×10,000) is constantly falling on the Earth. We are thus continuously breathing in "comet dust" along with everything else our air contains.

(Fig. 4.10). These dust particles, typically no larger than the width of a human hair, are not properly meteorites at all. Most of them have a cometary origin, and we will discuss them further in Chapter 6. Laboratory analysis shows that they are rich in organic carbon, and many of them also contain unusual amounts of deuterium (heavy hydrogen), which is interpreted as a signature of interstellar rather than interplanetary material. Perhaps some of the interplanetary dust particles are survivors of the original interstellar dust from the solar nebula. Unfortunately, the quantities of material are too small to permit us to date the formation of these strange particles.

Because the carbonaceous meteorites and interplanetary dust particles are the most primitive samples of solar system material available for laboratory study, they reveal the most about the earliest stages of the planetary system. In the search for a key to unlock the secrets of the past, these cosmic fossils are proving to be our most valuable artifacts.

4.5 DIFFERENTIATED METEORITES

The differentiated or igneous meteorites are composed of materials that have been melted subsequent to their condensation from the solar nebula, indicating that their parent bodies underwent heating and differentiation. Often they are also referred to as the achondrites, a term that signifies the absence of the chondrules that are characteristic of most primitive meteorites.

Irons and Stony-Irons

The most obviously differentiated meteorites are the irons, which consist of almost pure metallic nickel-iron, with trace quantities of sulfur, carbon, and metals such as platinum (Fig. 4.11). The nickel content is usually about 10% by weight. Iron meteorites make up only 2% of the falls and of the Antarctic meteorites, which probably represent an unbiased sample of meteoritic material. These meteorites can be quite beautiful when cut and polished, and subsequent etching further reveals a unique crystalline pattern, created by slow cooling of the nickel-iron melt over millions of years (Fig. 4.12).

Presumably the iron meteorites are fragments of the metal cores of their differentiated parent bodies. The cores of the Earth and other intact planets will remain forever beyond our direct investigation, but these meteorites provide a unique glimpse into the very heart of a former differentiated planetary body. Detailed chemical analysis indicates that there were a number of different parent bodies, at least several dozen, for the known iron meteorites.

Related to the iron meteorites are the stony-irons, composed of a mixture of nickel-iron and

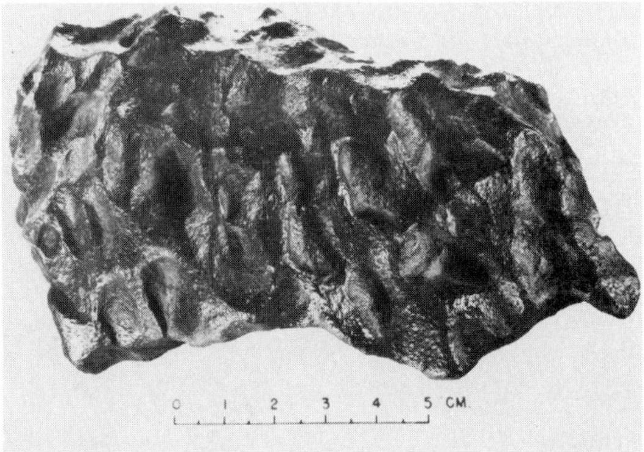

FIGURE 4.11 The Henbury iron meteorite, showing a pattern of "thumb prints." This pattern is caused when the thin skin of melted iron produced by the friction of the meteorite's passage through the atmosphere is pushed back away from the direction of motion.

FIGURE 4.12 A slice of an iron meteorite that has been polished and then etched with dilute nitric acid to show a crisscross crystallization pattern, called a "Widmanstättn" pattern.

silicate minerals. The most beautiful of these meteorites, called pallasites, consist of large crystals of olivine, a green semiprecious stone, in a setting of shiny meteoritic iron. The pallasites are thought to be fragments from the interface between the core and mantle of their parent bodies. Fewer than 1% of the meteorites reaching the Earth are stony-irons.

The age of the iron cannot be measured by radioactive technique, but age measurements can be carried out for the silicate fraction in stony-irons or for small silicate inclusions found in some iron meteorites. These ages lie between 4.4 and 4.5 billion years, suggesting that the processes of heating, differentiation, and cooling of their parent bodies all took place very early in solar system history.

Basaltic Meteorites

The primary group of differentiated *stony* meteorites appears to be derived from the crusts of their parent bodies. These rocks were formed by crystallization from a cooling body of basaltic lava, which in turn represents the lower-density material that rose to the surface during differentiation. Basalt is the most common form of lava on the Earth, and we will see in Chapter 7 that basalt is also the material that makes up the dark lunar "seas." Most of the basaltic meteorites are breccias, indicating their long residence near the surface of their parent bodies, where they were fragmented and mixed together by impact cratering.

The best-studied group of basaltic meteorites are the **eucrites** (Fig. 4.13), of which more than 30 examples are known, having identical oxygen isotopic ratios and closely related compositions. It is generally believed that the eucrites are derived from a single parent body, the asteroid Vesta (cf. Section 5.3), and that their compositions are representative of a series of different lava flows.

Another group of about a dozen unusual basaltic meteorites are called the **SNC meteorites,** short for the names of the prototype subgroups: shergottites, nakhlites, and chassignites. Like the eucrites, these are chemically related lavas, apparently derived in this case from a

FIGURE 4.13 An example of a eucrite showing the glossy black crust caused by melting during atmosphere entry and the underlying gray rocky material, which is free of chondrules. The eucrites, which are composed of basalt, are fragments from the asteroid Vesta.

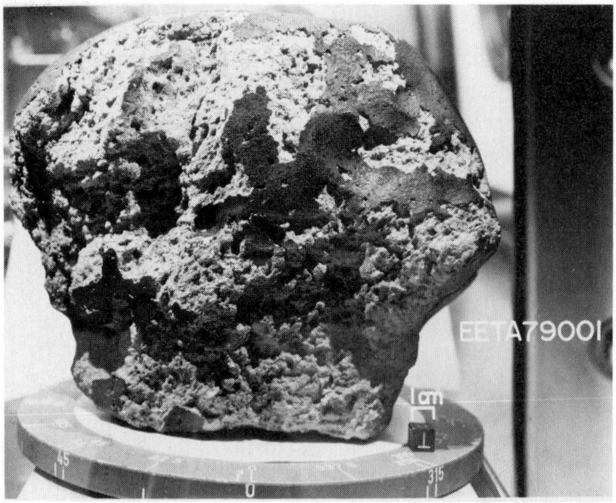

FIGURE 4.14 An example of one of the SNC meteorites. Here is a rock from Mars, only slightly altered by the impact that sent it to us and by the heating it experienced upon entry into the Earth's atmosphere.

source region in the mantle of their parent body (Fig. 4.14). All except the nakhlites show evidence of violent shocks, perhaps associated with the impact that liberated them. Most remarkably, however, these are the youngest meteorites, with relatively recent solidification ages of about 1.4 billion years. The parent body of the SNC meteorites remained volcanically active long after the asteroids, and even the Moon, had subsided into geological old age. The parent body for the SNC meteorites is almost surely Mars, and we will discuss them further in Chapter 11.

Finally there are the rare lunar meteorites, which are similar to the Moon rocks collected in the Apollo program; these are discussed in Chapter 7.

4.6 METEORITE PARENT BODIES

The lunar and martian meteorites are exceedingly rare. For the more common meteorite types, both comets and asteroids have been suggested as parent bodies. An exploded planet was also once hypothesized. Before pursuing these ideas further, it would be useful to summarize what we know about the parent bodies from our study of the meteorites.

Sources of Primitive Meteorites

Primitive meteorites must have originated in or on relatively small bodies that formed directly from dust condensing out of the cooling solar nebula. Large parent bodies are not possible because they would have retained too much internal heat generated by natural radioactivity; the thicker layers of material act as a blanket to keep the radioactive heat confined instead of allowing it to escape to space. Substantial heating, of course, would have altered the material from its primitive state.

Calculations show that in order to have escaped unacceptable heating levels, the parent bodies must have been no more than a few hundred kilometers in diameter. Some of these must have included organic chemicals and liquid water. The chemical and isotopic variety among the primitive meteorites indicate that many different parent bodies are represented in our meteorite collections.

Sources of Differentiated Meteorites

The differentiated meteorites are fragments of differentiated parent bodies. From detailed chemical analysis, it is clear that several dozen distinct parent bodies were involved. We can set some limits on their size from the iron meteorites, which retain in their crystal patterns (Figs. 4.11 and 4.12) indications of the rate at which they cooled. These cooling times of the order of a million years indicate objects no more than about a hundred kilometers in diameter. Thus the differentiated parent bodies were at least as small as the parent bodies of the primitive meteorites, and some characteristic other than size was responsible for the fact that one group of objects differentiated while the other remained in a primitive state. No one knows why one set of parent bodies differentiated and the other did not.

From the existence of basaltic meteorites we conclude that some parent bodies experienced surface volcanism, and in the case of the eucrite parent body the chemistry of these lava flows is well defined. On the SNC parent, volcanism persisted at least 3 billion years after formation, pointing toward a relatively large object such as Mars. A number of lunar meteorites have been found, with ages typical of the time of extensive lunar volcanism, between 3.3 and 3.9 billion years ago.

Asteroids or an Exploded Planet?

As we will see in Chapter 5, most of the meteoritic evidence points toward the asteroids as parent bodies, although comets are possible as parents for some of the primitive meteorites and interplanetary dust. What is clearly excluded, however, is the old idea of an exploded planet as the source of both the asteroids and the meteorites.

It is easy to see why the exploded planet theory fails. More than 90% of the meteorites show evidence of parent bodies no more than 200 km in size, and several lines of evidence suggest dozens of distinct parent objects. In addition, the prevalence of breccias points to the impact-stirred surfaces of *airless* precursors as a common environment for many meteorites. None of these observations would be true if the meteorites were derived from a single planetary parent body as has sometimes been proposed. In addition, the breakup or explosion of such a planet seems to be an intrinsically unlikely (and perhaps impossible) event.

4.7 QUANTITATIVE SUPPLEMENT: RADIOACTIVE AGE DATING

Rocks are dated—that is, their solidification ages are determined—by measuring the amount of a radioactive parent isotope that has changed into a daughter isotope since a particular crystal or grain of material formed. This process is relatively straightforward if the only source of the daughter isotope is decay of a single parent. In this case, we can assume that the rock solidified with little of this daughter present, and the current measured ratio of daughter to parent in a mineral grain will establish the age. In practice, however, the situation is a little more complicated.

One common method of dating the formation of the Earth involves the decay of uranium and thorium to produce the lead isotopes Pb-206 and Pb-208, respectively (Table 4.4). But not all of the Pb-206 and Pb-208 present on the Earth today resulted from decay of thorium and uranium. To determine the original quantity of these two isotopes, we look at a third lead isotope, Pb-204, which is not produced by radioactive decay. One can calculate what the primordial ratio of Pb-206 and Pb-208 must have been to Pb-204, and use the measured amount of Pb-204 to subtract the original quantity of the other two isotopes before calculating the age.

One of the simplest ways to use the presence of other, nonradiogenic isotopes to ensure that correct ages are being calculated is to graph the results, as we now describe. Consider the decay of Rb-87 to Sr-87, which is one of the most useful reactions for determining the ages of meteorites. After a time t the amount of Sr-87 is the sum of the initial quantity present (Sr-87*) plus that produced by decay of Rb-87:

$$\text{Sr-87} = \text{Sr-87}^* + \text{Rb-87}^* \, (e^{kt} - 1),$$

where k is a constant related to the half-life of the reaction. How do we determine Sr-87*? We note that since all isotopes of the same element have the same chemical properties, the amount of Sr-87 initially bound into the minerals of the rock should be exactly proportional to the amount of other strontium isotopes, such as Sr-86.

For each individual crystal or grain in a rock, make a plot of the measured ratio of the two strontium isotopes Sr-87/Sr-86 against the ratio Rb-87/Sr-86. Such a plot is illustrated in Fig. 4.15, which shows data from the Guareña meteorite. Grains with a great deal of initial Rb-87 will be toward the lower left. Now consider what happens as the rock ages. Rb-87 is converted to

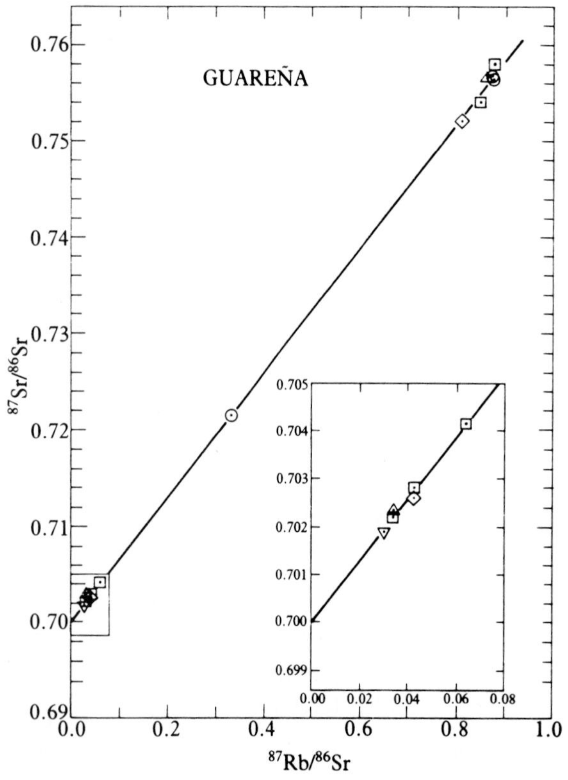

FIGURE 4.15 The Rb-Sr isochron for the Guareña H6 chondrite. The derived age is 4.46 ± 0.08 billion years.

Sr-87, causing the Rb-rich grains to shift down and to the right on the plot. The older the rock, the more the line drawn through the data points rotates in a clockwise direction, until eventually it will be nearly horizontal. Such a line is called an *isochron,* and its slope is a measure of the age of the rock, independent af any assumptions about the original quantity of Sr-87 present. In the example illustrated in Fig. 4.15, the Rb-Sr isochron indicates an age of $(4.46 \pm 0.08) \times 10^9$ years.

SUMMARY

The meteorites are rocky and metallic fragments that have reached the Earth and survived their plunge through the atmosphere. They provide information on conditions at the birth of the solar system, when solid material condensed out of a contracting and cooling solar nebula of interstellar gas and dust. Generally they are fragments of parent bodies that formed very early in the history of the planetary system.

Most of the meteorites that fall on Earth are primitive in composition, meaning that they and their parent bodies have not suffered significant changes since their birth. In this respect they differ from the rocks of Earth and Moon, both of which are the product of a history of planetary heating and geological activity. The most important of the primitive meteorites for understanding the solar nebula are the carbonaceous meteorites, some of which contain small inclusions that appear to have survived from the period before the solar nebula.

Differentiated meteorites are samples from differentiated parent bodies which, like the Earth and other planets, have experienced chemical changes due to melting. The evidence indicates, however, that the parent bodies of these meteorites were not full-fledged planets, but small bodies that were heated and differentiated very early in solar system history and subsequently broken up by impacts. The iron meteorites appear to be fragments from the metallic cores of such bodies, and the stony-irons come from the boundary between core and silicate mantle.

A few differentiated meteorites have come to us from the crusts of objects that experienced surface volcanic activity. These basaltic meteorites include samples from the Moon, the eucrites, which are widely thought to originate on the asteroid Vesta, and the SNC meteorites, which come from Mars.

Today there are thousands of individual meteorites in scientific collections around the world, and hundreds of scientists are working on problems of their analysis. In some laboratories, instruments of tremendous sophistication, many developed for study of the Apollo lunar samples, can determine the detailed chemical and isotopic composition of even a single crystal or grain almost too small to be seen with the naked eye. In the exotic surroundings of these superclean laboratories, some of the most important results of modern planetary science are produced.

KEY TERMS

accretion
angular momentum
breccia
carbonaceous meteorite
chondrites
chondrule
differentiated meteorite
eucrite
fall
find
fractionation
fragmentation
gas retention age
iron meteorite
meteor
meteorite
meteoroid
organic compound
parent body
planetesimals
primitive meteorite
protoplanets
radioactive half-life
SNC meteorite
solidification age
stony meteorite
stony-iron meteorite

Review Questions

1. Distinguish among meteoroids, meteorites, and meteors. What do any of these objects have to do with the weather? (The word *meteor* is closely related to *meteorology*.)

2. What is meant by primitive bodies? Which objects are the most primitive? What kinds of information do they provide on the origin of the solar system?

3. Describe the competing processes of accretion and fragmentation in the solar nebula.

4. What are meteorite falls and finds, and why are the relative numbers of irons and stones so different for falls and finds?

5. How do we measure the age of a rock, and what exactly does the measured age mean?

6. Why do you think there are differentiated meteorite parent bodies? What might these bodies have been like? Can you explain how they might have been heated enough to cause differentiation?

7. Could the meteorites have originated in an explosion of a planet between the orbits of Mars and Jupiter? Give the arguments pro and con.

Quantitative Exercises

1. Consider the relative proportions of meteorites in Table 4.2. If the observed falls and the Antarctic meteorites represent the true distribution of meteorite classes in space, and if the iron meteorites are the remnants of the cores of differentiated parent bodies, what percentage of the total would you expect to be from the mantles of these same parent bodies? What can you conclude about the relative fraction of primitive and differentiated parent bodies?

2. Suppose that the eucrite parent body was a differentiated asteroid with diameter 500 km which is now completely disrupted and reduced to fragments. Further suppose that the basaltic crust of this parent body had a thickness of 10 km and the mantle below it had a thickness of 200 km. What is the relative amount of basaltic and mantle material that would be expected when this body fragmented?

3. Consider the radioactive decay of Rb-87, Th-232, and U-238 (Table 4.4). After 5 billion years, what fraction of each of these isotopes remains?

4. Look up some real data for meteorites or lunar samples as measured in modern laboratories to see how well the isochron ages agree for different dating methods on the same samples. Possible sources of data include Dodd's *Meteorites* and Wasson's *Meteorites.*

Additional Reading

Burke, J.G. 1986. *Cosmic Debris: Meteorites in History.* San Francisco: University of California Press. Interesting discussion of the role of meteorites in history and our changing attitudes toward these remarkable rocks.

Dodd, R.T. 1986. *Thunderstones and Shooting Stars: The Meaning of Meteorites.* Cambridge: Harvard University Press. Excellent popular-level introduction to meteorites by a leading researcher in the field.

*Dodd, R.T. 1981. *The Meteorites: A Petrologic-Chemical Synthesis.* Cambridge: Cambridge University Press. Detailed, technical treatment; an excellent reference source.

Hutchison, R. 1983. *The Search for Our Beginnings.* New York: Oxford University Press. Well-written popular account, with emphasis on the message of the meteorites concerning the origin and early evolution of the solar system.

*Kerridge, J.F., and M.S. Matthews, eds. 1988. *Meteorites and the Early Solar System.* Tucson: University of Arizona Press. Definitive technical volume with chapters by more than 50 authors; part of the Arizona Space Science Series of graduate-level textbooks.

Wasson, J.T. 1985. *Meteorites: Their Record of Early Solar System History.* New York: Freeman. Semipopular introduction to the meteorites by a leading researcher.

*Indicates the more technical readings.

5

Asteroids: Building Blocks of the Inner Planets

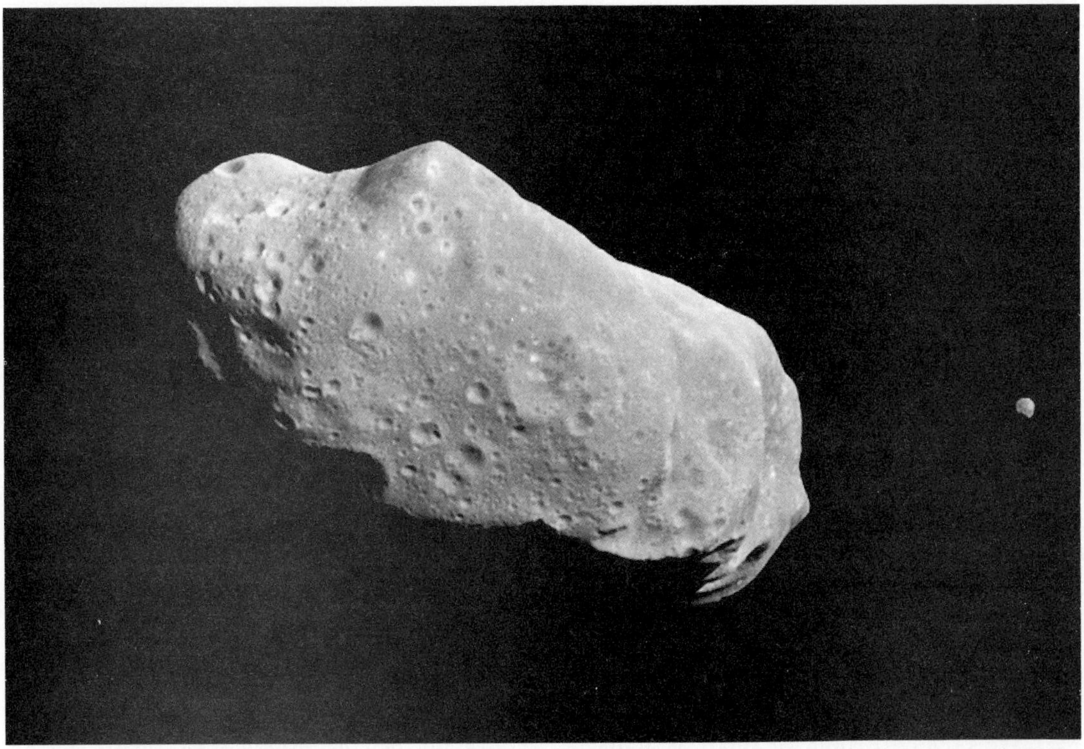

This is the asteroid Ida with its satellite Dactyl, as imaged by the camera on the Galileo spacecraft in August 1993.

In order to interpret the scientific message of the meteorites, we need to have some idea of their origin. These interplanetary wanderers appear to be fragments of larger parent bodies that have been broken up and eroded by impacts in space. The two classes of objects in space that are the likely parent bodies of most meteoroids are the comets and the asteroids, both of which are remnants of the population of small bodies that date back to the formative stages of the planetary system.

The primary difference between comets and asteroids is one of composition. Most asteroids are rocky objects, composed of the same sorts of materials as the inner planets. Comets, in contrast, contain a substantial quantity of water ice and other frozen volatiles in addition to rock dust and organic compounds. Comets that make frequent passes around the Sun may lose their volatiles and become indistinguishable from rocky asteroids. We now believe that most meteorites come from asteroids, and therefore we concentrate in this chapter on the asteroids, deferring discussion of the comets until Chapter 6.

Many astronomy textbooks refer to the comets, asteroids, meteors, and meteorites as *debris*. We prefer to think of them as artifacts from which we may be able to deduce events in the distant past. We study them in much the way an archaeologist sifts through the ruins of past civilizations, in the hope that we too can locate a Rosetta stone that will unlock some of the secrets of the birth of the planetary system.

5.1 DISCOVERY OF THE ASTEROIDS

There is a pattern to the orbits of the planets. As we discussed in Section 3.5, the distances from the Sun of most of the planets can be represented by a numerical progression called the Titius-Bode rule. However, this rule creates a problem: It predicts a planet at 2.8 AU where none exists.

In earlier times, before scientists had developed a sense of the workings of physical laws, numerical schemes such as the Titius-Bode rule were taken very seriously. Thus late in the eighteenth century many astronomers became disturbed by this problem, and a group of six German observers set out on a systematic search for this missing planet. Just as their effort was beginning, however, they were scooped by an Italian, Giuseppe Piazzi, who was measuring the positions of stars from his observatory in Palermo, Sicily.

The First Asteroids

On New Year's Day, 1801, Piazzi discovered the first asteroid, which he named Ceres for the Roman patron goddess of Sicily. This faint object, invisible without a telescope, had just the expected orbital distance of 2.8 AU, and it was at first hailed as the missing planet.

The German astronomers did not give up, however, and soon they discovered three more minor planets—Pallas, Juno, and Vesta—also

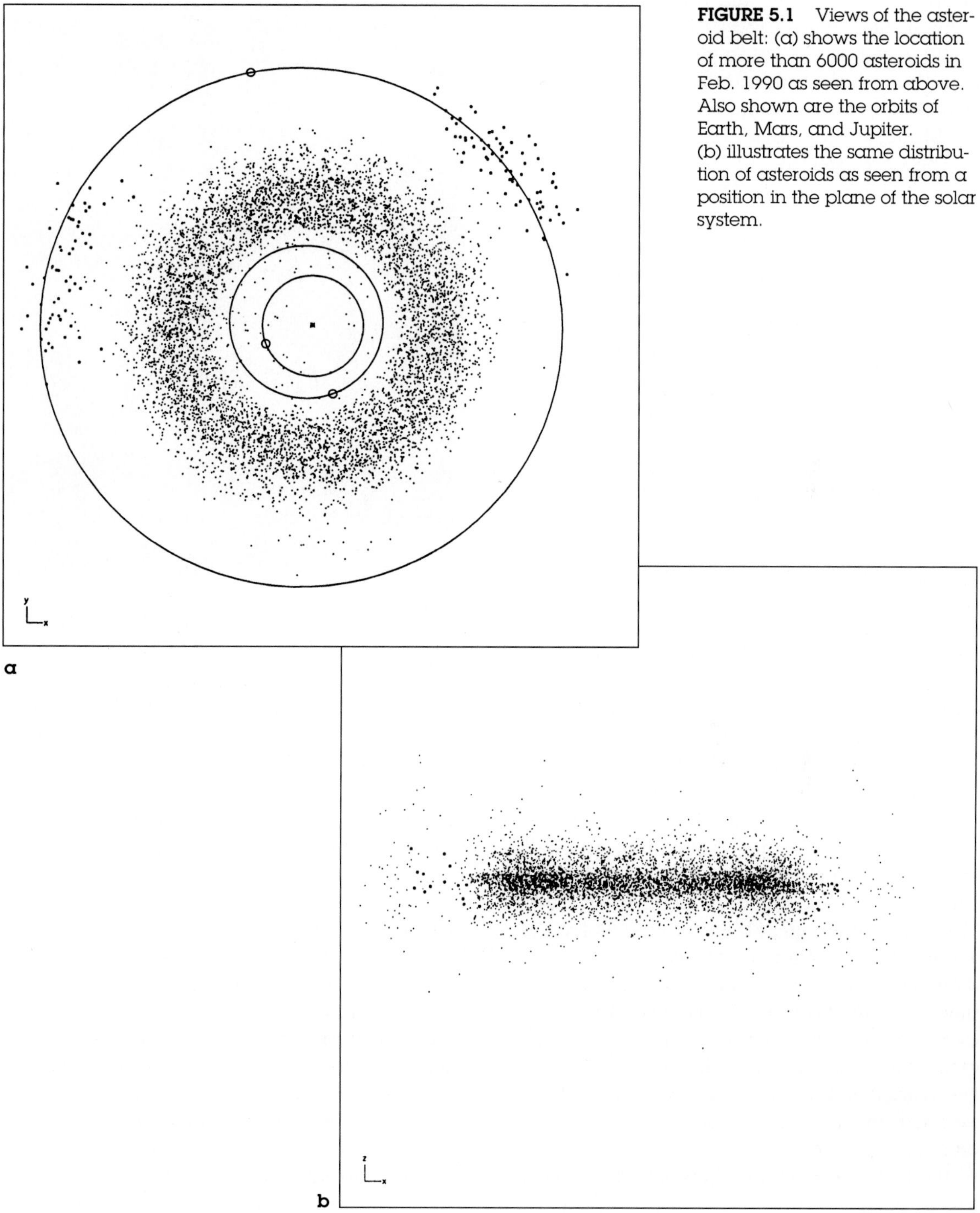

FIGURE 5.1 Views of the asteroid belt: (a) shows the location of more than 6000 asteroids in Feb. 1990 as seen from above. Also shown are the orbits of Earth, Mars, and Jupiter. (b) illustrates the same distribution of asteroids as seen from a position in the plane of the solar system.

orbiting between Mars and Jupiter. These were even smaller than Ceres, although Vesta is slightly brighter due to its more reflective surface. Even when combined, however, the masses of these four objects came nowhere near that of the Moon, let alone that of a real planet.

The next asteroid was not discovered until 1845, but from then on they were found regularly by visual observers who scanned the sky looking for them. By 1890 the total number had risen to 300. At that time photographic patrols began, and the number of known objects rapidly increased, reaching 1000 (named Piazzia) in 1923, 3000 in 1984, and 5000 in 1990. While not confined entirely to the space between Mars and Jupiter, the majority of asteroids do occupy this part of the solar system. We define the **main asteroid belt** as consisting of asteroids with average distance from the Sun between 2.2 and 3.3 AU (Fig. 5.1).

Today new asteroids are relatively easy to find, and accidental discoveries often annoy stellar astronomers by cluttering their long-exposure images of stars and galaxies (Fig. 5.2). To be entered on the official list of asteroids, however, an object must be well enough observed to establish its orbit and permit its motion to be accurately calculated many years into the future. Thus fewer than 8000 asteroids have been catalogued so far, although many thousands more could be added to this list.

The responsibility for cataloguing asteroids and approving new discoveries is assigned to the International Astronomical Union Minor Planet Center in Cambridge, Massachusetts. As soon as a preliminary orbit is calculated, a temporary designation is given to the object and predictions of future positions are circulated to interested astronomers. Only after observations extend to a second observing season (usually two years after discovery) is a permanent number assigned and the object added to the official catalogue.

In addition to their numbers, which are given in order of discovery, most asteroids have names, usually selected by the discoverer. Initially these were the names of Greek and Roman goddesses, such as Ceres and Vesta, later expanded to include female names of any kind. When masculine names were applied, they were given the feminine Latin ending, as in the example of 1000 Piazzia mentioned above. More recently the requirement of female gender has been relaxed, and asteroids today are named for a bewildering variety of persons and places, famous or obscure. For instance, asteroid 2410 is named Morrison, for one of the authors of this book, and asteroid 2309 is named Mr. Spock.

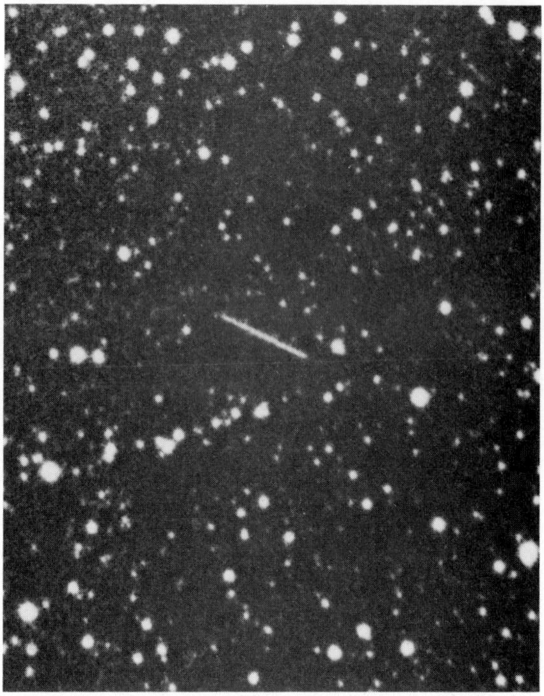

FIGURE 5.2 A photograph of the night sky taken with a telescopic camera that was slowly moved to compensate for the Earth's rotation. The stars appear as dots in this time exposure, while the rapidly moving asteroid named Aten appears as a streak.

Basic Asteroid Statistics

While many thousands of small asteroids remain undiscovered, we can draw some general conclusions from present data. Ceres is the largest, with a diameter of just under 1000 km. The next largest asteroids are about half this size (Fig. 5.3). The total mass of all the asteroids amounts to only 1/2000 of the mass of the Earth (less than 1/20 the mass of the Moon), indicating that the gap in the planetary system between Mars and Jupiter is real, a topic we return to in Chapter 19.

Our census of the larger asteroids is by now fairly complete. Probably we have discovered all main-belt asteroids 50 km or more in diameter, and discovery should be at least 50% complete for diameters down to 10 km. Our knowledge is better for the closer asteroids in the inner part of the asteroid belt, and most of the larger undiscovered bodies are probably beyond 3 AU from the Sun.

A few attempts have been made to sample the fainter asteroids. In the early 1950s sky photographs were taken for this purpose at McDonald Observatory in Texas, and another photographic survey was made a decade later in a joint project between Palomar Observatory in California and the University of Leiden in the Netherlands. The most comprehensive effort, however, did not involve ground-based telescopes at all. In 1983, the first infrared astronomical satellite (IRAS, a joint U.S.-Netherlands-U.K. project) detected thousands of new asteroids from their thermal radiation. All these efforts have produced valuable statistical data on the numbers and distribution of the smaller objects.

FIGURE 5.3 This scale drawing shows the relative sizes of some of the larger asteroids compared with that of the planet Mars. The numbers next to the names give the rotation periods in hours. The horizontal scale gives the mean distance from the Sun in astronomical units (AU).

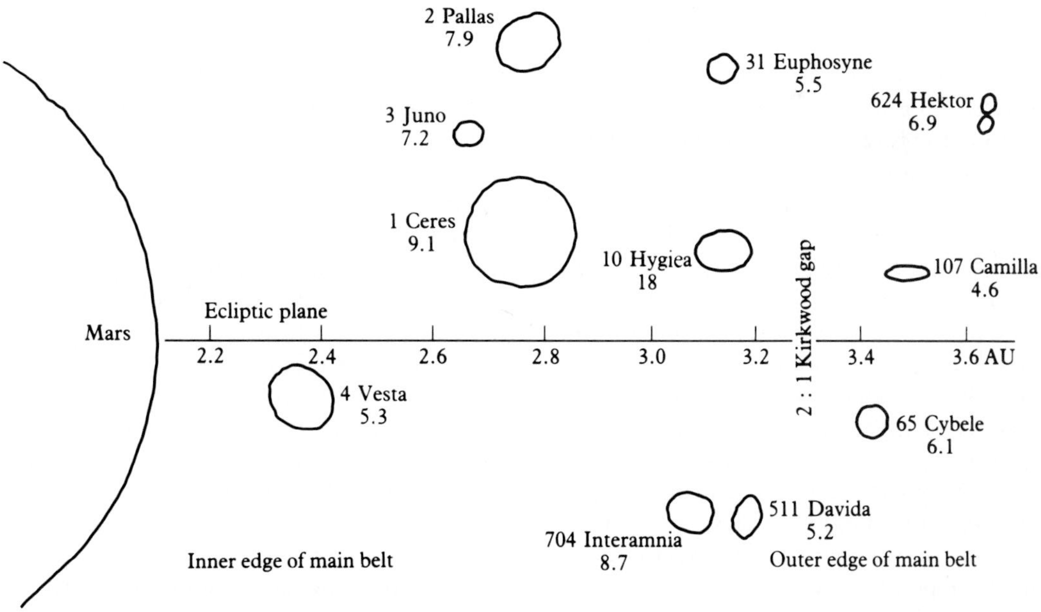

Size-Frequency Distribution

There are many more small asteroids than large ones. An estimate of the relative numbers of objects of each size is interesting as a characterization of the asteroid population, and it will also prove important when we look at the size distribution of lunar craters produced by asteroidal and cometary impacts.

As a rule, many processes in nature, including those of fragmentation, result in approximately equal masses of material in each size range. Consider asteroids with diameters of 100 km, 10 km, and 1 km, respectively. Each 100-km asteroid has 1000 ($10 \times 10 \times 10$) times the mass of a 10-km asteroid, which in turn has 1000 times the mass of a 1-km object. Therefore, if the total mass is to be the same in each size range, there must be 1000 times more 10-km objects than there are 100-km ones, and a million (1000×1000) more at 1 km than 100 km. In other words, the number of objects of a given diameter (D) is inversely proportional to the cube of their diameter ($1/D^3$).

Actual measurements of the asteroids indicate that the numbers do not rise quite this fast with declining size. Instead of the number increasing as the inverse cube of the diameter, it increases more nearly as the inverse square ($1/D^2$), resulting in a distribution with most of the mass in the larger objects (Fig. 5.4). This is why we are relatively certain of the total mass of the asteroids, even without having counted all of the small ones.

Think a moment about a size distribution in which number is proportional to $1/D^2$. What would we say, for example, is the *average size* of the objects? The great majority are very small, near the lower limit of size, but most of the mass is in the larger objects. The concept of average size is not very meaningful, at least not in the way we usually think about the term. As we shall see later, this type of size distribution applies to lunar craters and to the particles in planetary rings as well as to asteroids.

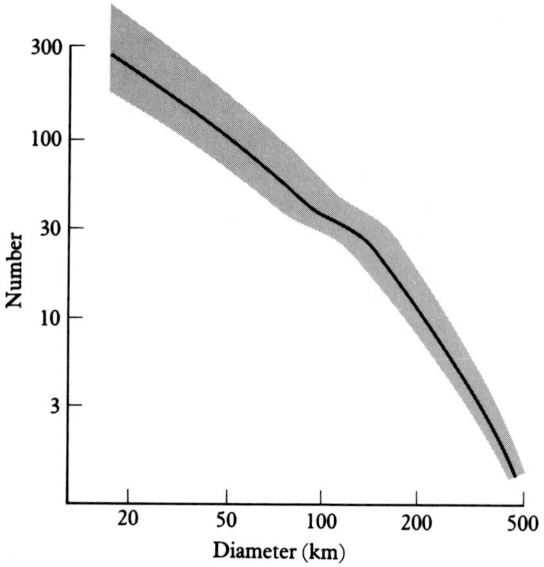

FIGURE 5.4 The number of asteroids increases rapidly with decreasing size.

Physical Studies: Size and Reflectivity

As seen through a telescope, an individual asteroid is an unresolved starlike point. (The word *asteroid* means starlike.) Before about 1970 almost nothing was known about the physical nature of asteroids, and research was confined to discovery and charting of orbits and the determination of rotation rates from observations of periodic brightness variations. In the past 25 years, however, the application of new observing techniques and larger telescopes has revealed a great deal about the physical and chemical nature of the asteroids.

One of the first challenges was to measure the size and reflectivity of an object too small to appear as a disk in the telescope. How can we make such measurements? The most accurate technique involves timing the passage of an asteroid in front of a star. Since we know exactly how fast it is moving against the stellar background, a measurement of how long the star is obscured yields an accurate size for the asteroid.

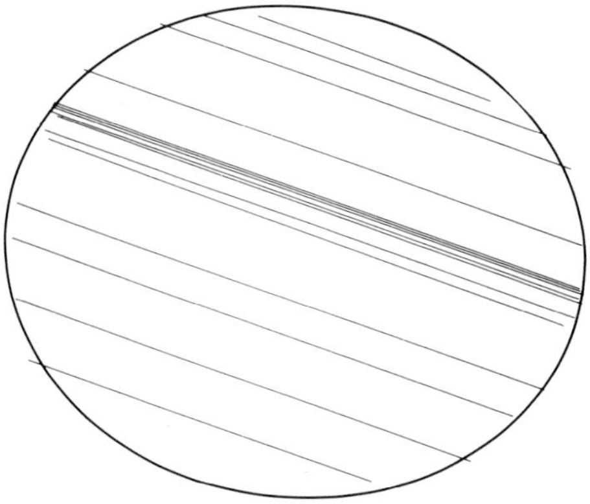

FIGURE 5.5 The approximate shape of the asteroid Juno as determined from its passage in front of a star. This occultation was observed from several different locations on Earth, hence from several different viewing angles. Each straight line in the figure represents the apparent path of the star as viewed from each observing location on Earth as the asteroid passed in front of it.

If timings of the same event made from different locations on Earth are combined, the profile of the asteroid can also be derived. An example of the results for Juno is shown in Fig. 5.5. Unfortunately, however, such events are rare, and only a dozen asteroids have been measured successfully by this method.

More generally, we would like to have a way to determine whether an asteroid of a given brightness is large and dark or small but highly reflective. The most useful technique that can be applied to large numbers of asteroids involves measurement of the heat radiated as well as the solar light reflected. A dark object absorbs most of the incident sunlight and re-emits it in the infrared as heat; it will appear relatively bright in the infrared and faint in the visible. A highly reflective object will be bright in the visible but will be a weaker source of heat (Fig. 5.6). Application of techniques based on this principle has resulted in the determination of diameters and reflectivities for more than a thousand asteroids.

Physical Studies: Spectrometry

A second important type of information about asteroids is obtained from measuring the spectra of sunlight reflected from their surfaces. Different minerals have different colors. Variations with wavelength in the reflectance of the surface material can indicate the composition of the

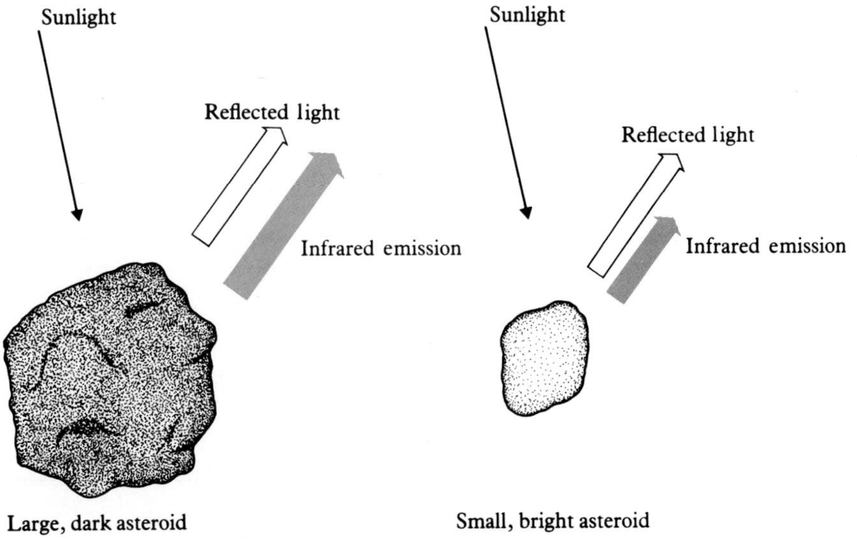

FIGURE 5.6 Comparison of a large, dark asteroid with a small, bright (highly reflective) asteroid. In this example, both reflect the same amount of sunlight, so both appear equally bright in visible light. The large, dark asteroid emits much more infrared thermal radiation, owing to its larger size and its hotter surface. Thus measurement of reflected and emitted radiation allows us to determine both size and reflectivity.

asteroid. While not as rigorously diagnostic as the sharp spectral lines produced when light passes through a gas, the broad absorption features in the spectrum of an asteroid are often sufficient to identify the major minerals present. This identification is aided by the similarities between the spectra of asteroids and of many meteorites, permitting a laboratory comparison between individual meteorites and asteroids (Fig. 5.7).

Much of the pioneering work in remote sensing of asteroids was carried out by Thomas B. McCord and his colleagues and students. This work began at MIT in 1970 when McCord's group obtained visible and near-infrared spectra of Vesta and found features diagnostic of basaltic lavas (Section 4.5). During the next few years, this team, led by McCord's former student Clark R. Chapman (Planetary Science Institute, Tucson), measured about 200 asteroids, finding a surprising variety of spectral types. By 1975 the database was large enough to begin the classification and interpretation of asteroid mineralogy.

The development of asteroid studies during the 1970s illustrates how rapidly a new field can blossom even without spacecraft missions. McCord, Chapman, and their co-workers (Fig. 5.8) required only five years from their first measurements to development of the statistical database and classification system discussed in the following pages. By 1979 a 1181-page book on asteroids would be published, with 69 scientists as chapter authors. Contrast this history with the slow pace of scientific progress in previous centuries, as described in Chapter 1.

The use of spectral and reflectance data to characterize asteroids has yielded preliminary determinations of composition for more than a thousand objects. In most cases the minerals inferred to be present on asteroid surfaces are

FIGURE 5.7 A comparison of reflection spectra from four asteroids (circles) with laboratory-determined spectra of meteorites (solid lines).

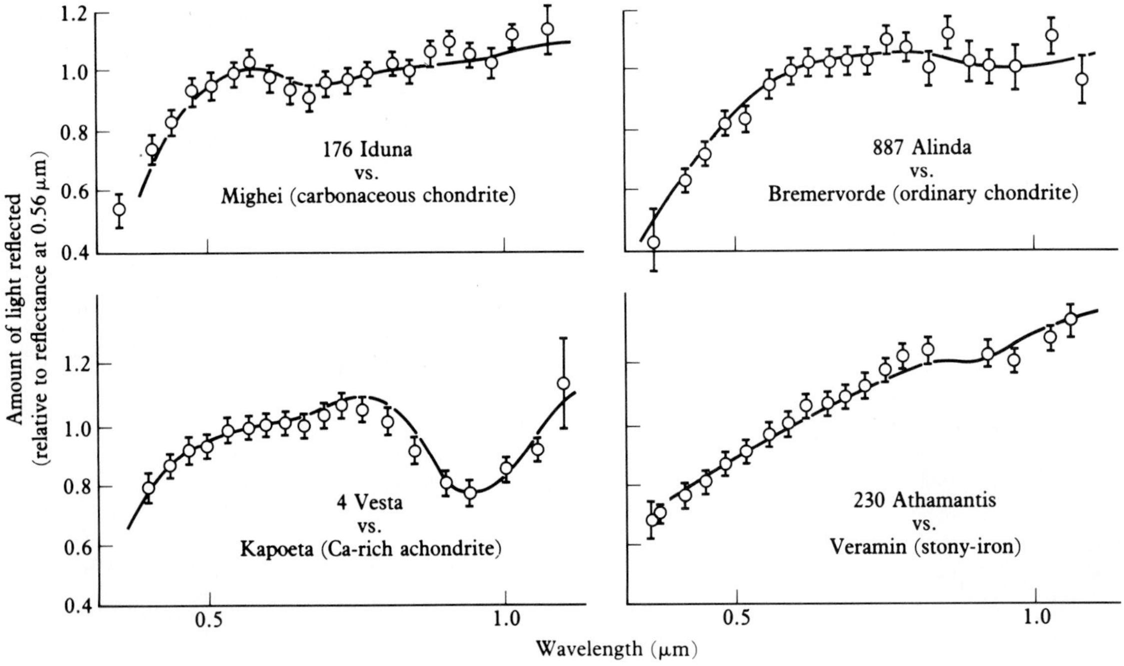

FIGURE 5.8 Three pioneers in the study of asteroid sizes and compositions, from right to left: Clark Chapman, Benjamin Zellner, and David Morrison.

similar to those in the stony meteorites. Exact identifications are difficult, however. Our current state of knowledge is more nearly equivalent to the verbal description of an unknown criminal suspect. We can categorize in terms analogous to height, weight, age, sex, and hair color, but we cannot specify the unique properties that identify an individual. With some notable exceptions (cf. Section 5.5), contemporary asteroid research therefore tends toward broad statistical studies rather than detailed investigation of particular objects.

5.2 MAIN BELT ASTEROIDS

The great majority of the asteroids are located in the asteroid main belt, at average distances from the Sun between 2.2 and 3.3 AU, corresponding to orbital periods between 3.3 and 6.0 years. From the observed distribution of sizes we can estimate that there are more than 100,000 down to a diameter of 1 km.

Although 100,000 sounds like a lot of objects, space in the asteroid belt is still empty. The belt asteroids occupy a very large volume, roughly doughnut shaped, about 100 million km thick and nearly 200 million km across. Typically they are separated from each other by millions of kilometers. They pose no danger to passing spacecraft; in fact, it was difficult to locate asteroids near enough to the trajectory of the Jupiter-bound Galileo spacecraft to allow close asteroid flybys.

Orbits and Resonances

The orbits of the belt asteroids are for the most part stable, with eccentricities less than 0.3 and inclinations below 20° (Fig. 5.9). In the past,

FIGURE 5.9 (a) The number of asteroid orbits having various inclinations. (b) The number of asteroid orbits having various eccentricities.

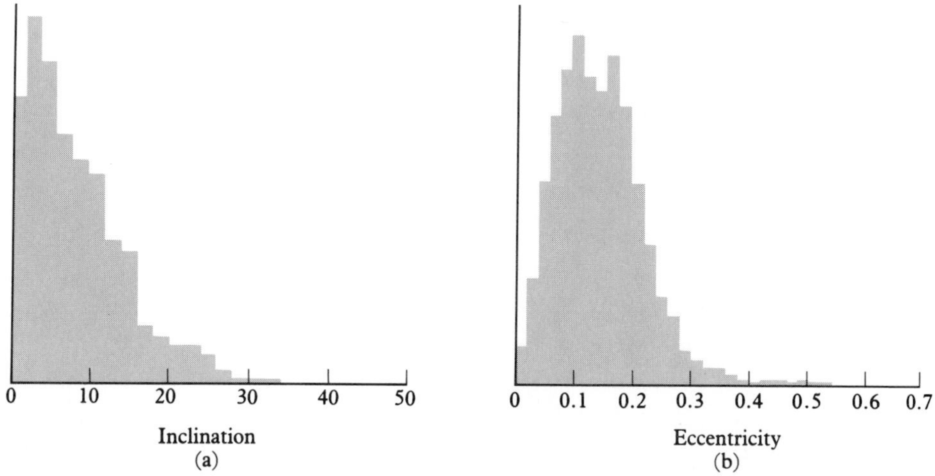

when presumably there were more asteroids in this region of space, collisions may have been common, but by now the population has thinned to the point where each individual asteroid can expect to survive for billions of years between collisions. Still, with 100,000 objects, a major collision somewhere in the belt is expected every few tens of thousands of years. Such collisions, as well as lesser cratering events, presumably yield some of the fragments that eventually reach the Earth as meteorites.

The orbits of asteroids within the main belt are not evenly distributed. As shown in Fig. 5.10, some orbital periods seem to be preferred, while others are nearly unpopulated. These unpopular sections of the belt are **resonance gaps,** also known as the Kirkwood gaps for the nineteenth-century American astronomer Daniel Kirkwood who discovered them.

The gaps in the asteroid belt occur at orbital periods that correspond to **resonances** between these periods and the orbital period of Jupiter. A resonance is defined as follows. If an asteroid should find itself in a resonant orbit, repeated small gravitational effects of Jupiter would add together to alter the orbit, in much the same way that repeated small pushes on a swing, applied at the proper time, lead to a large oscillation. A resonance takes place when the orbital period of one body is an exact fraction of the period of another. In this case, the underpopulated asteroid orbits correspond to periods that are ½, ⅓, ¼, etc., that of the period of Jupiter.

Consider an asteroid with a 6-year period; it would revolve about the Sun exactly twice for each revolution of Jupiter. Its distance from the Sun would be 3.3 AU. It is closest to the giant planet once each 12 years, at exactly the same place in its orbit. The cumulative effects of Jupiter's gravity, always applied at the same place, will alter the orbit of the asteroid; in contrast, an object on a nearby orbit that is not resonant with Jupiter experiences gravitational nudges all around its orbit with no net effect. In this way Jupiter will eliminate asteroids in resonant orbits. Indeed, the period corresponding to 3.3 AU is one of the observed resonance gaps, the one that marks the outer edge of the main asteroid belt. Similar processes involving the satellites of Saturn play an important role in creating resonance gaps in Saturn's rings (Chapter 18).

The resonance gaps represent orbital *periods* that are missing, rather than physical gaps in the belt. Since most asteroids have orbits of significant eccentricity, their in-and-out motions as they orbit the Sun effectively fill the space within the belt. In contrast, the resonance gaps in the

FIGURE 5.10 The number of asteroid orbits with various average distances from the Sun. The positions of orbital resonances with Jupiter are shown (2:1 = ½ Jupiter's orbital period, etc.).

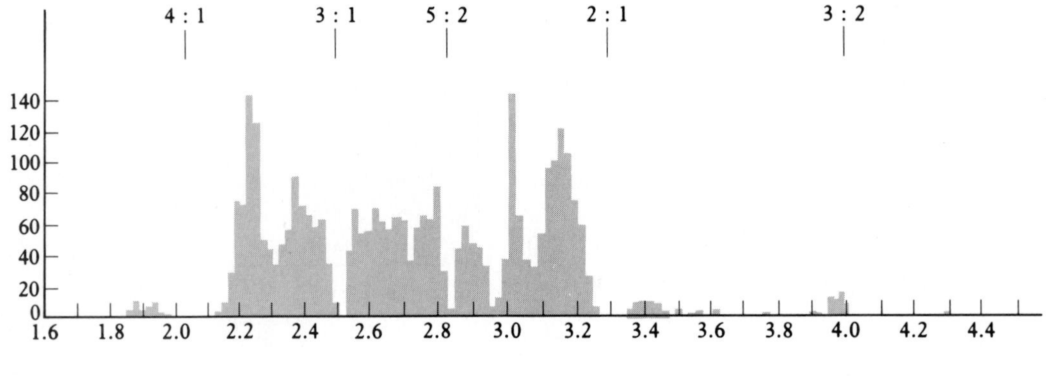

rings of Saturn show up as *physical* gaps also, since the Saturn ring particles are in essentially co-planar circular orbits.

Asteroid Families

Asteroid orbits display other patterns in addition to the resonance gaps. An **asteroid family** is defined as a group of objects with similar orbits, suggesting a common origin. About half of the known belt asteroids are members of families, with nearly 10% belonging to just three: the Koronos, Eos, and Themis families. Although not clustered together in space at the present, the members of an asteroid family were all at the same place at some undetermined time in the past.

Members of the same family tend to have similar reflectivities and spectra. Apparently the family members are fragments of broken asteroids, shattered in some ancient collision and still following similar orbital paths. According to some estimates, almost all the asteroids smaller than about 200 km in diameter were probably disrupted in earlier times, when the population of asteroids was larger. The families we see today may be remnants of the most recent of these interasteroid collisions.

The Largest Asteroids

The largest asteroid in the main belt is Ceres, the first to be discovered. Its diameter is just under 1000 km, and its mass of 1/5000 that of the Earth amounts to nearly half of the total mass of all the asteroids. Two other asteroids are about 500 km in diameter: Pallas and Vesta. The 20 largest asteroids (all larger than 220 km) are listed in Table 5.1. Note that several of the larger asteroids, such as Interamnia, Davida, and Patientia, were not among the early discoveries. These objects are in the outer part of the belt and have low reflectivities, making them faint in spite of their large sizes.

Compositional Classes

In order to characterize the asteroids further, we use the classifications based on measurements of reflectivity and spectrum discussed in the previous section. As soon as the data on a substantial sample of asteroids were examined, it became apparent that most of them fell into one of two classes based on their reflectivity: They were either very dark (reflecting only 3%–5% of incident sunlight) or moderately bright (15%–25% reflectivity). A similar distinction exists in their spectra; the dark asteroids are fairly neutral reflectors with no major absorption bands in the visible range to reveal their composition, although some of them show spectral evidence for H_2O in the infrared. Most of the lighter asteroids are reddish and show the spectral signatures of common silicate minerals such as olivine and pyroxene. Since the dark gray asteroids had spectra similar to the *carbonaceous* meteorites, they were called **C-type** asteroids. The lighter class was named the **S-type**, indicating *silicate* or *stony* composition. A third major group appears to be *metallic* (like the iron meteorites) and is called the **M-type**.

Table 5.1 includes the assigned classes for the asteroids larger than 220 km. Among the largest asteroids, Ceres and Pallas are in the C class, but Vesta fits into none of the three classes we have just defined. The special case of Vesta is discussed in Section 5.3.

Given the uncertainties inherent in deducing composition from such limited data, few scientists are willing to state positively that C-type asteroids are made of exactly the same thing as carbonaceous meteorites, but the connection seems probable. In any case, the extreme darkness of their surfaces and the presence of H_2O indicate that C-type asteroids are primitive and carbon-rich. The interpretation of the rarer M type as metallic is even clearer, thanks to radar studies of the large M asteroid Psyche; the high radar reflectivity of this object is a clear signature of the presence of metal on its surface. Least certain is the identification of the S-type asteroids with the primitive (chondritic) stony meteorites, but that is the best hypothesis available today. As we will discuss later, this is the interpretation of observations made from the Galileo spacecraft in 1993, when it flew past the S-type asteroid Ida. However, some scientists argue that other S-type asteroids may be igneous stony-iron objects, and careful comparisons between the spectra of S-type asteroids and of various types of meteorites are not able to resolve this controversy.

Using the current classification of about a thousand asteroids, we can look at the distribution in space of the C, S, and M types. At the inner edge of the belt, near 2.2 AU, the S asteroids predominate. Moving outward, the fraction of C-type objects increases steadily, and in the belt as a whole these dark, carbonaceous objects make up 75% of the population, compared to 15% S and 10% other types. Beyond the main belt, all the asteroids appear to be carbonaceous. The M types are located near the middle of the belt, but they constitute only a few percent of known asteroids.

Clearly, most asteroids are composed of relatively primitive material. If the asteroids are still near the locations where they formed, we can use the distribution of asteroid types to map out the composition of the solar nebula. Since carbonaceous meteorites formed at lower temperatures than the other primitive stones, the concentration of C-type asteroids in the outer belt is consistent with their formation farther from the Sun, where the nebular temperatures

TABLE 5.1 The 20 largest asteroids

Name	Discovery	Semimajor Axis (AU)	Diameter (km)	Class
1 Ceres	1801	2.77	940	C
2 Pallas	1802	2.77	540	C
4 Vesta	1807	2.36	510	*
10 Hygeia	1849	3.14	410	C
704 Interamnia	1910	3.06	310	C
511 Davida	1903	3.18	310	C
65 Cybele	1861	3.43	280	C
52 Europa	1868	3.10	280	C
87 Sylvia	1866	3.48	275	C
3 Juno	1804	2.67	265	S
16 Psyche	1852	2.92	265	M
451 Patientia	1899	3.07	260	C
31 Euphrosyne	1854	3.15	250	C
15 Eunomia	1851	2.64	245	S
324 Bamberga	1892	2.68	235	C
107 Camilla	1868	3.49	230	C
532 Herculina	1904	2.77	230	S
48 Doris	1857	3.11	225	C
29 Amphitrite	1854	2.55	225	S
19 Fortuna	1852	2.44	220	C

*Vesta has a very unusual (once thought unique) basaltic surface.

were lower. It is also possible, however, that the asteroids formed elsewhere and were herded into their present positions by the gravity of Jupiter and the other planets. In that case, the C-type asteroids could have formed far beyond Jupiter and subsequently diffused inward to their present positions in the outer part of the asteroid belt. Similarly, the S-type asteroids near the inner edge of the belt could either have formed where we see them today or have been gravitationally scattered to their present locations from still closer to the Sun. The solar nebula temperatures that we would deduce from the application of these two alternative models are quite different. So far, however, we have not been able to settle on which model for the origin of the asteroids is to be preferred.

If the C and S asteroids both represent the *primitive* meteorites, where did the *differentiated* meteorites come from? Are there enough differentiated parent bodies among the asteroids to account for these meteorites? We still aren't sure. Presumably the M asteroids are the cores of differentiated parent bodies as well as the probable sources of the iron meteorites. Several other minor classes of asteroids might also be evolved objects. But the best case for a differentiated asteroid is Vesta, which is discussed in detail in the next section.

5.3 VESTA: A VOLCANIC ASTEROID

If planetary scientists were asked to pick a favorite asteroid, many would surely name Vesta, the brightest, third largest, and the fourth discovered of the minor planets. Vesta is unique, and that is why we have learned so much about it. It is the only asteroid identified as the probable parent body of a specific group of meteorites (the eucrites), and it is the one differentiated asteroid known to have experienced volcanic activity on its surface.

Similarities to the Eucrites

The uniqueness of Vesta derives not from its position, which is in the midst of the main belt, but from its surface composition. It is one of the most reflective (30%) of the asteroids, and its spectrum (Fig. 5.7) exhibits two deep and well-defined absorptions at wavelengths near 1 and 2 μm. These absorption bands indicate the presence of minerals associated with igneous rocks, basalt in particular. A careful search of the spectra of more than a thousand other asteroids has failed to identify any other large object with this same basaltic surface composition.

Recall (Section 4.5) that we introduced the eucrites as a related group of about 30 basaltic meteorites. When the spectra of the eucrites were measured in the laboratory, it was found that they matched the spectrum of Vesta closely, although not exactly. Subsequent more precise telescopic measurements of Vesta then showed that the shapes and wavelengths of its prominent spectral features changed slightly as the asteroid rotated, indicating that the surface composition of Vesta is not uniform.

The eucrites are not all exactly alike either. A more careful comparison now makes it appear that there are regions on the surface of Vesta that have the compositions of the main subgroups of the eucrites. As the asteroid rotates, these regions—essentially large lava flows of slightly different composition—are viewed successively. The new data therefore provide an even more convincing association between the eucrites and Vesta.

The Eucrite Parent Body

Is this spectral similarity perhaps fortuitous? Might the eucrites be fragments from the breakup of another large, differentiated asteroid similar to Vesta, rather than samples of Vesta itself? This seems like a reasonable hypothesis, and to evaluate it we must apply a careful chain of logic.

The lava rock of the eucrites represents the crust of the parent body. From the chemistry of the eucrites it is possible to predict the composition of the much thicker mantle of the parent. Yet an examination of our meteorite collections reveals no meteorites with this predicted mantle composition. Since there would be more mantle debris than crust debris produced by the total breakup of this hypothetical parent, it follows that no such breakup took place. Therefore, the "eucrite parent body" is still largely intact, and the eucrites themselves were chipped from its crust by impact cratering.

If we accept this argument for an intact eucrite parent body and combine it with the fact that no other large asteroid exists with the same basaltic surface as Vesta, we are drawn to the conclusion that Vesta really is the parent body of the eucrites. This conclusion is reinforced when we consider that only a rather large asteroid could have been heated sufficiently to generate a basalt surface.

If the connection between Vesta and the eucrites is correct, we can apply the meteoritic evidence to this asteroid. Vesta then becomes the fourth solar system object for which we have samples (the others are Earth, Moon, and Mars). Thus we can date the Vesta lava flows at 4.5 billion years ago, the solidification age of the samples, and fix the impact that released the meteorites at about 3 billion years ago from the gas retention age. Whatever triggered the volcanic activity of this asteroid was short-lived, as we might expect for such a small object.

Is Vesta Really the Eucrite Source?

As a check on the line of reasoning presented here, astronomers would like to locate the intermediate objects that link eucrites with Vesta—large fragments from Vesta that are the immediate source of the meteorites. The first step along this path was the discovery about 1990 of three small asteroids, between 0.5 and 1.5 km in diameter, with spectra similar to that of Vesta and the eucrites. Each of these is in a Mars-crossing orbit and is capable of coming close to the Earth.

While these three small Vesta-like asteroids may be the immediate sources of the eucrites, the link with Vesta is still tenuous. Is it possible for fragments this large to have been blasted from the surface of Vesta without destroying it? If not, then these must be fragments of a second large differentiated asteroid, now destroyed, and the chain of logic that uniquely connects Vesta with the eucrites falls apart.

The missing link was apparently discovered in 1992 by Richard Binzel of MIT, who found about a dozen additional small basaltic asteroids in the main belt. These objects have orbits generally similar to that of Vesta but are more widely dispersed, as we might expect if they were expelled from Vesta in one or more collisions. Some of them are in locations where a modest additional gravitational shove from Jupiter could carry them into the inner solar system, like the three basaltic Mars-crossing asteroids already observed. Thus we now have a chain of evidence stretching from the eucrites back to the vicinity of Vesta itself.

In solving one problem, however, we have created a new difficulty. It is not easy to understand how a collision could have created so many relatively large (up to 10 km) fragments from the surface of Vesta without disrupting the asteroid. Apparently our understanding of the impact process requires additional work.

Although a great deal has been learned about Vesta from its link with the eucrites, this knowledge raises more questions than it answers. Why did Vesta become volcanically active while its larger neighbors, Ceres and Pallas, did not? Why is Vesta unique? Were there once similar differentiated asteroids long since destroyed by collisions? What were the parent bodies of the iron and stony-iron meterorites? These are questions that trouble scientists trying to understand the processes of heating and differentiation of planets and asteroids.

5.4 ASTEROIDS FAR AND NEAR

A few asteroids—perhaps 1% of the number in the main belt—have orbits that cross the orbits of Mars or the Earth. Most of these wanderers are fragments of main belt asteroids spun into more eccentric orbits as a result of collisions in the belt. There are also asteroids outside the main belt. Beyond 3.3 AU, the distribution of asteroids changes character, with most of the objects concentrated at just a few orbital periods with wide gaps in between. Many more asteroids are concentrated near the orbit of Jupiter, and a few have even been found beyond Jupiter.

Trojan Asteroids

A particularly interesting group of dark, distant asteroids is orbitally associated with Jupiter. Although the gravitational attraction of this giant planet generally has the effect of making nearby asteroid orbits unstable, exceptions exist for objects with the same orbital period as Jupiter, but leading or trailing it by 60° (Fig. 5.11). These two stable regions are called the leading and trailing Lagrangian points, named for the French mathematician (Joseph Louis, Comte Lagrange) who demonstrated their existence in 1772. While he was examining mathematically the possible motions of three mutually gravitating bodies, Lagrange found two regions where a small object could occupy a stable orbit within the gravitational fields of two larger objects. If the larger objects are Jupiter and the Sun, a small object in one of the Lagrangian points occupies one corner of an equilateral triangle, with the Sun and Jupiter at the other two points.

The regions of stability around the two Lagrangian points are quite large, and each contains several hundred known asteroids. The first of these Lagrangian asteroids was named Hektor when it was discovered in 1907. All of them are named for the heroes of the *Iliad* who fought in the Trojan War, and collectively they are known as the **Trojan asteroids.** Their spectra are distinctive, suggesting that they represent a group of special, primitive objects that have been trapped in this region of space since the birth of Jupiter. If we could detect the fainter members of these Trojan clouds, we might find that the Trojan asteroids are nearly as numerous as those in the main asteroid belt.

As the most massive planet, Jupiter has the

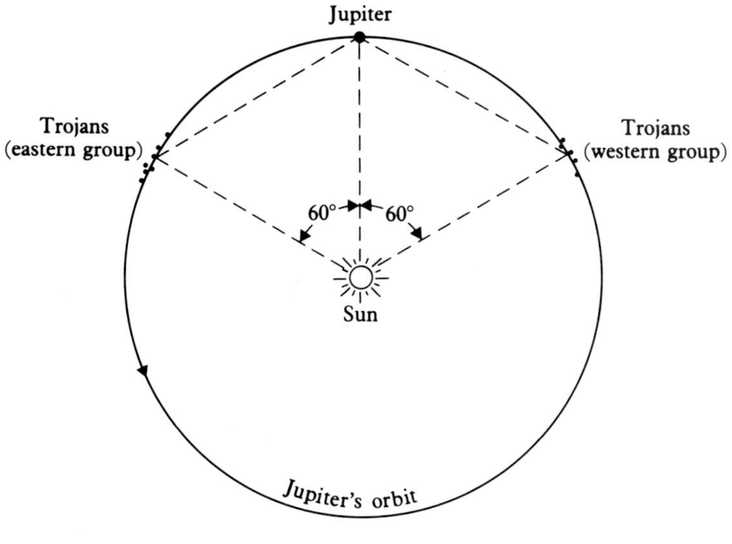

FIGURE 5.11 The Trojan asteroids are located 60° ahead of, and behind, Jupiter in that planet's orbit. Mars and Neptune have also captured asteroids in their Trojan points, and we will also find small satellites in the Saturn system occupying similar positions with respect to Saturn (in the role of the Sun) and a larger satellite (in the role of Jupiter).

most stable Lagrangian orbits. Once located in one of the Trojan clouds, an asteroid should remain there for the lifetime of the solar system. Other planets can support their own Lagrangian orbits, but they tend to be less stable and would not be expected to contain so many objects as the jovian Trojans. Several small asteroids have been discovered in the Lagrangian regions for Mars and Neptune, but none for Earth or the other outer planets. Within the Saturn satellite system there are also tiny Lagrangian satellites, as we will see in Chapter 17.

Although they are dark and carbonaceous, the detailed compositions of these more distant asteroids appear to be different from those of the C-type asteroids of the main belt. Their colors are redder, and they do not look the same as any known carbonaceous meteorite. Because these objects do not seem to be represented in our meteorite collections, scientists hesitate to commit themselves concerning their composition. It is generally believed, however, that they are primitive objects, and that a fragment from one of them would be classed as a carbonaceous meteorite, although of a different kind from those already encountered.

Asteroids Beyond Jupiter

A few asteroids are known to venture beyond Jupiter. The first to be discovered was Hidalgo, a dark object occupying a cometlike orbit of high inclination and eccentricity that carries it from the inner edge of the main belt out almost as far as Saturn. Second is Chiron, discovered in 1977; its orbit carries it from 8.5 AU, near Saturn, out to 19 AU, near Uranus. A still more distant object discovered in 1992 is called Pholus, with an orbit that takes it as far as 33 AU from the Sun, beyond the orbit of Neptune, and more of these faint objects are continuing to be found. The names *Chiron* and *Pholus* are taken from the two good centaurs of Greek mythology. Pholus is the reddest object in the solar system, indicating a very strange (but yet unknown) surface composition. Figure 5.12 illustrates the orbits of Chiron and Pholus, together with those of Comet Halley and of the first object discovered in the Kuiper comet belt (both discussed in Chapter 6).

Should distant objects with highly eccentric orbits be classed as asteroids or comets? Their orbits suggest that they could be comets, but the true distinction between a comet and an asteroid is based on the nature of the volatile compounds the body contains. For objects in the inner solar system, this distinction in volatility translates into a difference in telescopic appearance: If the object develops a visible atmosphere, it is a comet; if it appears starlike, it is an asteroid. However, at the great distances of these objects from the Sun, no cometary atmosphere would be expected, even if the objects were composed in large part of water ice. If they contain frozen nitrogen or carbon monoxide, however, enough material might evaporate from their surfaces to be detected even at the orbit of Neptune. We suspect that if these three objects came closer to the Sun they would develop atmospheres and be reclassified as comets.

This speculation became a reality in 1988, when astronomers discovered that Chiron had brightened by more than a factor of two from the previous year, presumably as the result of the formation of a tenuous atmosphere of gas and dust. The following year this atmosphere was photographed (Fig. 5.13), confirming the transition of Chiron from asteroid to comet. By 1988 Chiron's orbital motion had brought it within 10 AU of the Sun, and its higher temperature was apparently sufficient to begin the release of volatiles from the interior.

Our knowledge of asteroids (or comets) beyond Jupiter is very incomplete, and it is difficult to guess how many other, smaller objects of the centaur type might be present. Chiron and other trans-jovian objects are in unstable orbits, and ultimately they will either impact a planet or be gravitationally ejected from the solar system.

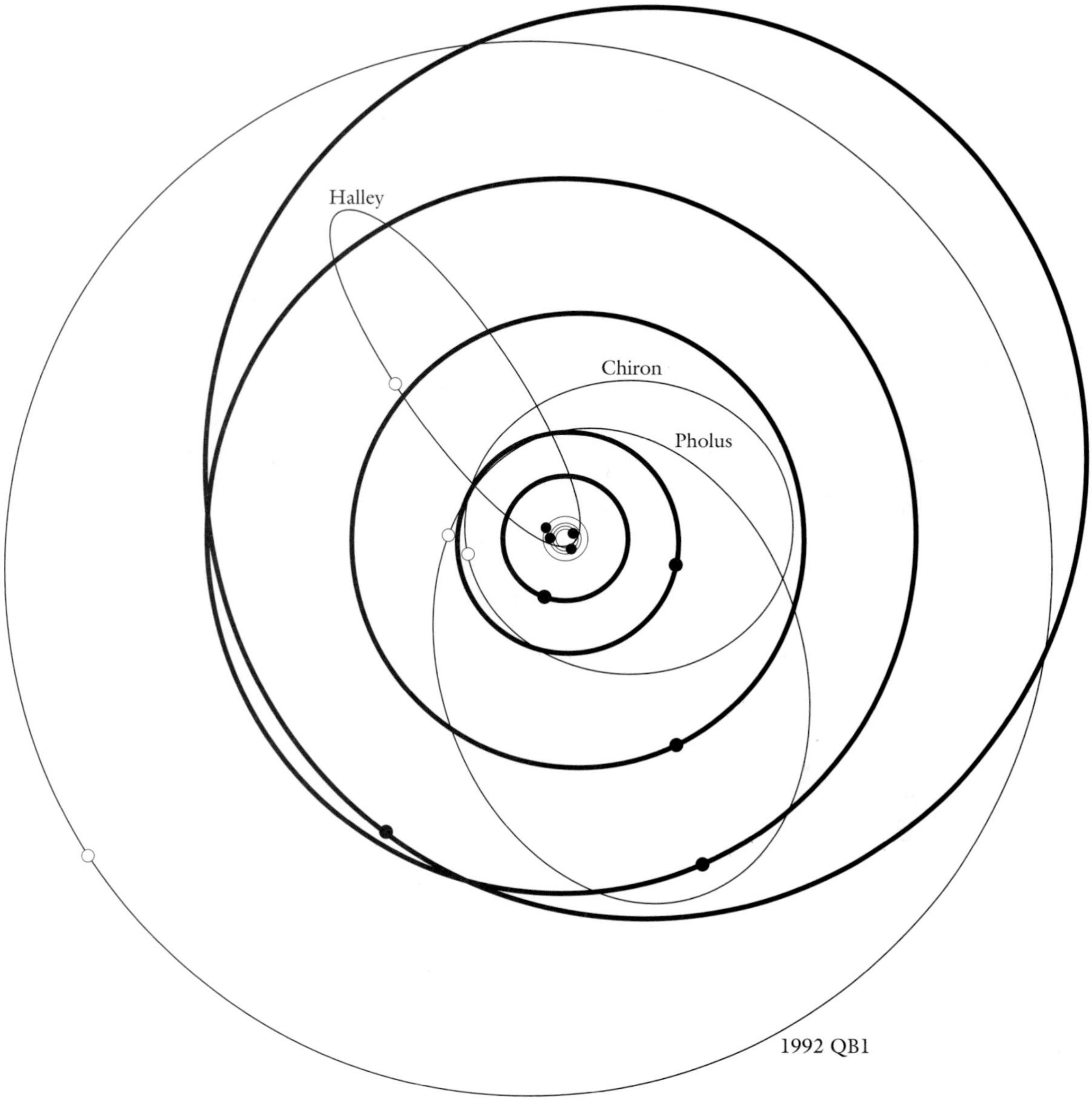

FIGURE 5.12 View of the solar system from above, showing the orbit of Comet Halley and of three of the newly discovered objects in the outer solar system: Chiron (initially designated an asteroid but subsequently seen to display cometary activity); Pholus, the reddest known object in the solar system; and the first object found in the Kuiper Belt (discussed in Chapter 6).

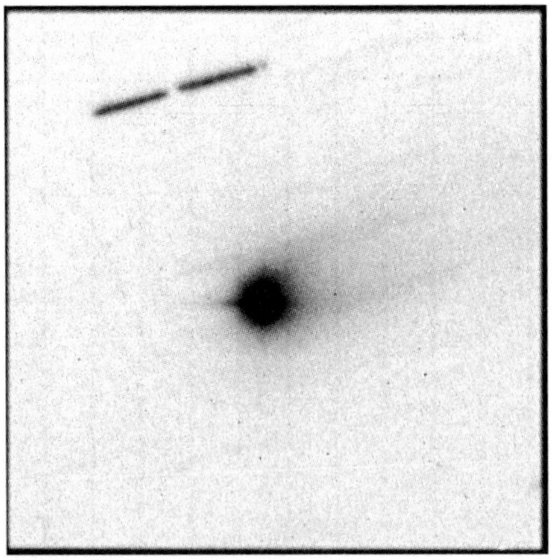

FIGURE 5.13 The tenuous atmosphere of Chiron, photographed in 1989 when the object was 12 AU from the Sun (negative image).

Near-Earth Asteroids

While only about 1% of the asteroids stray inside the asteroid belt and approach the Earth, we have a special interest in those that do. The Earth-approaching asteroids are potentially useful to humanity as sources of raw materials in space, and they also pose a threat to us if they should collide with the Earth (as many eventually will; see Fig. 5.14).

The first asteroid found on an orbit that crossed the Earth's orbit (coming within less than 1 AU of the Sun) was Apollo, discovered in 1948, and thus the Earth-crossing objects are often called the Apollo asteroids. Here we use the somewhat broader term **near-Earth asteroid** (NEA) to include potential Earth-crossers, objects whose orbits do not now intersect that of the Earth but which could do so under the shifting gravitational nudges of the other planets.

Continuing searches for near-Earth asteroids carried out primarily at Palomar Observatory and at the University of Arizona yield several new objects each month, most of them less than 1 km in diameter. Today more than 200 are known. The largest is Eros, a highly elongated asteroid about 30 km in length, but most of the

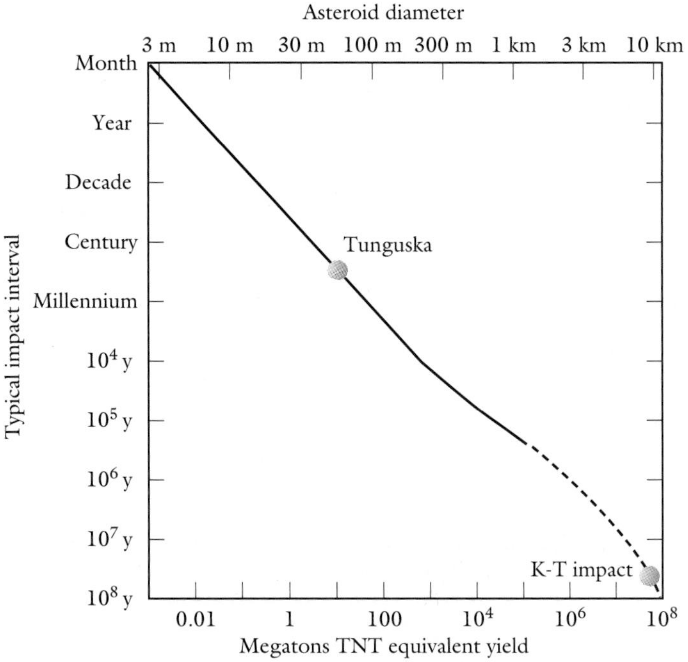

FIGURE 5.14 The frequency of asteroid impacts on the Earth, showing how the frequency increases as the size of the objects declines. Also indicated is the impact energy (in megatons of TNT) for asteroids of various diameters. Objects smaller than about 50 m diameter (10 MT energy) disintegrate in the upper atmosphere and do no damage. Shown for reference is the 1908 Tunguska event (15 MT airburst, average interval of several hundred years) and the K-T impact of 65 million years ago (100 million MT; average interval about 100 million years).

known Earth-approachers are much smaller. The asteroid that has the smallest perihelion distance is Icarus, which comes inside the orbit of Mercury. It is estimated that nearly 2000 Earth-approaching asteroids exist down to 1 km in diameter, and about a third of a million down to 100 m in diameter.

The roughly 50 near-Earth asteroids that have measured reflectivities and spectra are a varied lot. Most are S-type objects, but there are some of C type and others with spectra that are uncommon among main belt asteroids. Some of these may be dead comets, as we discuss later. Since so few of existing Earth-approachers have been measured, it is probably premature to try to compare them in detail with the meteorites.

Origin and Fate of the Near-Earth Asteroids

The orbits of near-Earth asteroids are unstable, as a result of the constantly varying gravitational influence of Earth, Mars, and Venus. Any object that finds itself caught in such an orbit will experience one of two fates, each about equally probable. The object may be gravitationally ejected as the result of nearly missing a planet, or it may terminate its existence dramatically in a crater-forming impact. Most of the craters on the Moon and on the terrestrial planets are caused by impacts with near-Earth asteroids.

Calculations indicate that about one-third of the asteroids whose orbits currently cross that of the Earth will eventually hit our planet. An Earth-approaching asteroid will either impact a planet or be ejected after an average interval of about a hundred million (10^8) years. A hundred million years may seem like a long time to us, but it is short in comparison with the 4.5-billion-year history of the planetary system. Any asteroids that began as Earth-approachers at the time the solar system was formed would long since have been ejected or impacted a planet. Therefore the near-Earth asteroids present today must have come from somewhere else, and there must be a continuing source for these objects.

One source for the Earth-approachers is collisions among main belt asteroids, several of which take place every million years. Another possible source is the comets. If a solid core remains after a comet has exhausted its volatiles, this core might be indistinguishable from an asteroid. This latter possibility is supported by the existence of several near-Earth asteroids with orbits similar to those of the Jupiter-family comets, which we discuss in the next chapter. The IRAS orbiting observatory also discovered an Earth-approaching asteroid, called Phaethon, in association with a meteor stream (the Geminids), another property normally associated with comets, and subsequently several asteroid-meteor connections have been made.

The occasional impact on our planet of near-Earth asteroids or comets is an important aspect of our solar system environment. Although these impacts have negligible effect on the planet as a whole and are far too small to influence the rotation or orbit of the Earth, they can have severe environmental effects. Both the initial ejection of molten rock from such an impact and the subsequent presence of fine dust in the Earth's atmosphere can first heat and then cool the surface, with disastrous consequences. Only recently have scientists recognized that impacts have had a major influence on the extinction and evolution of life, a topic we will return to in Chapter 9.

5.5 ASTEROIDS CLOSE UP

On its long flight path to Jupiter, the Galileo spacecraft made close flybys of two main belt asteroids, Gaspra and Ida; its observations provide our best information on the appearance of these celestial fragments. Radar images have also been obtained of several near-Earth asteroids. We can also learn about the asteroids from spacecraft measurements of the small satellites of Mars, Phobos, and Deimos. Because these two small moons are probably captured asteroids, acquired by Mars shortly after the formation of the planet, they should be able to tell us something about their asteroid cousins.

Gaspra

Gaspra is a small asteroid from the inner part of the main belt, with an average distance from the Sun of 2.2 AU and an orbital period of 3.3 years. It is one of many members of the Hungaria asteroid family and is classed as S type on the basis of its spectrum. Telescopic observations of its light variations showed long before the spacecraft encounter that Gaspra is highly elongated in shape, about twice as long as it is wide (Fig. 5.15). Yet none of this information influenced the selection of Gaspra as the first asteroid to be visited by spacecraft. This reason was simple: Gaspra was the easiest known asteroid for Galileo to reach on its predetermined path to Jupiter. Only a minuscule tweak of the spacecraft's flight path was required to bring it within 1600 km of this particular asteroid.

A spacecraft image of Gaspra is shown in Fig. 5.16, representing a nearly end-on view as the irregular object rotates with a period of 7 hr. The measured dimensions are $16 \times 11 \times 10$ km, and the surface reflectivity is 22%. The highly irregular shape demonstrates that this object is a fragment, produced in some earlier collision that broke up a larger preexisting asteroid. This may have been the catastrophic collision that shattered the Hungaria parent body and created the Hungaria asteroid family, or it might represent a subsequent collision involving one or more of the Hungaria family members.

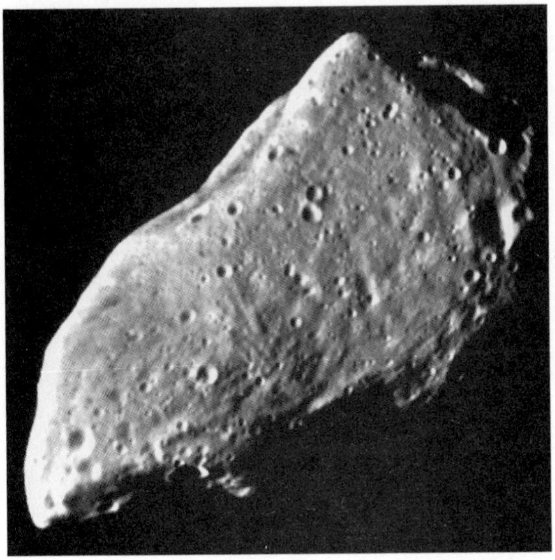

FIGURE 5.16 Image of Gaspra obtained during the 1992 Galileo flyby. The dimensions of this main belt asteroid are $16 \times 11 \times 10$ km.

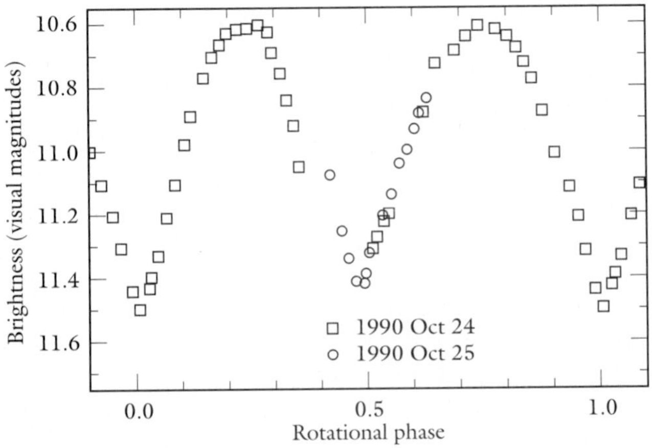

FIGURE 5.15 Variation in the brightness of asteroid Gaspra as it rotates with a period of 7 hours. Such large-amplitude lightcurves, which are common among small asteroids, indicate highly elongated and irregular shapes.

From the accumulation of small impact craters on its surface, we can estimate how much time has passed since the catastrophic birth of Gaspra. (We will see in Chapter 7 that such crater counting is an extremely useful way to estimate the ages of many objects in the planetary system.) Apparently this collision took place less than 500 million years ago, making Gaspra one of the younger members of the solar system. Based on the numbers of objects in the inner asteroid belt where it orbits, Gaspra is likely to experience another catastrophic collision sometime within the next billion years, at which time it will break up into many still smaller fragments.

Ida

The second Galileo flyby was of Ida, a larger (53 km long) S-type asteroid and a member of the Koronos family. Even better images were obtained (Fig. 5.17), and like Gaspra Ida was found to be elongated, irregular, and heavily cratered. In this case, however, the degree of cratering was a surprise, because the age of the Koronos family was thought from dynamical considerations to be less than that of the Hungaria family. Some scientists began to think that some special event had generated most of the craters on Ida.

The real excitement of the Ida flyby was the unexpected discovery that this asteroid has a small satellite, which was named Dactyl (Fig. 5.18). The origin of this little object (just 1500 m across) is something of a mystery, and may be related to the anomalous number of craters on Ida. One suggestion is that Dactyl itself is a large boulder ejected from the surface of Ida in a recent cratering event.

The greatest value of Dactyl is that it allows us to measure the mass of Ida, using Kepler's laws.

FIGURE 5.17 Image of asteroid Ida obtained by Galileo in 1993. Resolution of this image is about 30 m. This asteroid is more heavily cratered than Gaspra, suggesting a greater age since the catastrophic event that created it.

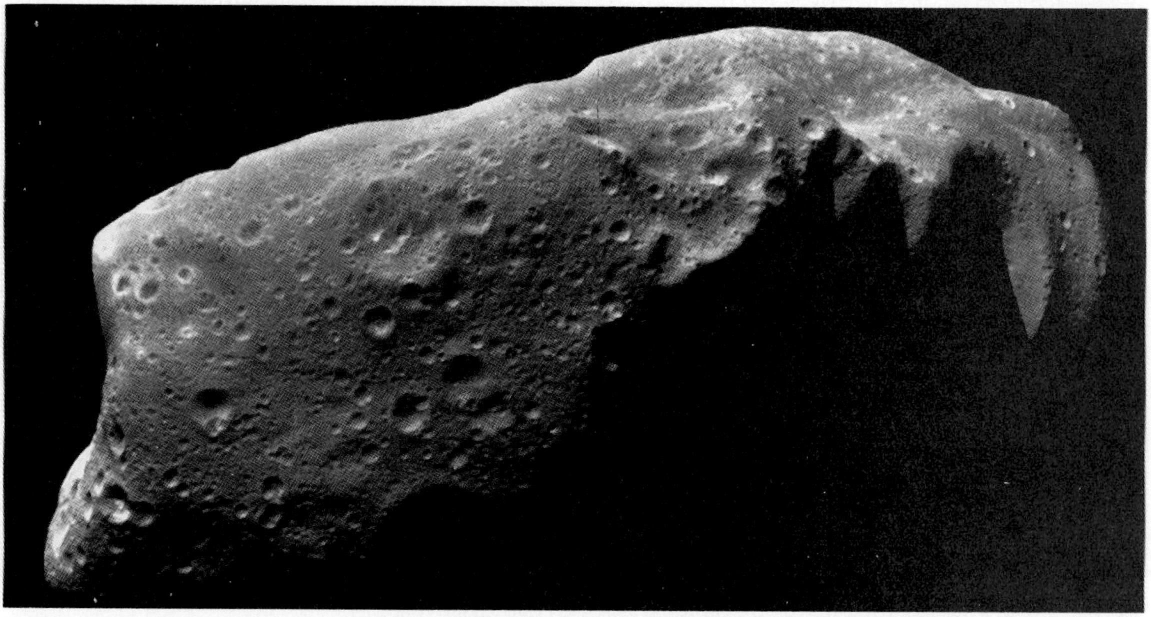

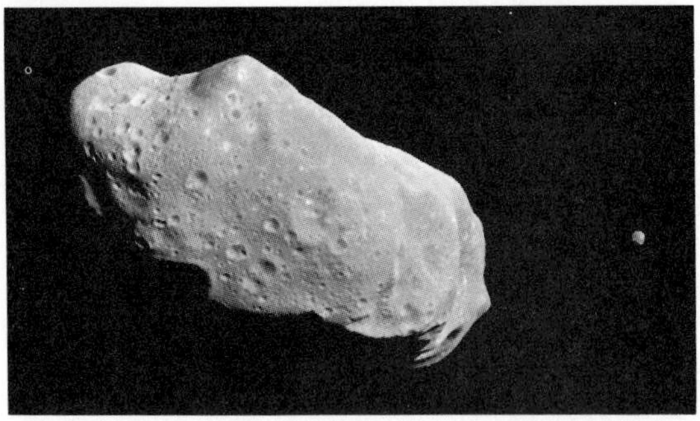

FIGURE 5.18 Ida and its moon, Dactyl (to the right), imaged by the Galileo spacecraft on Aug. 28, 1993. Dactyl is about 1.5 km in diameter and is at a distance of 100 km from Ida.

The little satellite is in an elliptical orbit with a 23-hr period at a distance of about 100 km from Ida. There is only a limited region in which such an orbit can be stable; if it were much closer to the asteroid it would eventually crash into the surface, and if it were much farther away it would escape altogether. From the mass we can calculate the density of Ida, which is between about 2.2 and 2.8 g/cm^3. This low value effectively excludes the presence of much metal, indicating that Ida is not composed of stony-iron meteoritic material. The only alternative for an object with the spectrum of Ida is a primitive composition, similar to that of the ordinary chondritic meteorites. By inference, the other members of the Koronos family, and indeed most of the S-type asteroids in the inner part of the asteroid belt, are also primitive objects.

Radar Images

Toutatis is a near-Earth asteroid discovered by French astronomers in 1988 and named for a popular French cartoon character. In December 1992 Toutatis passed within 3 million km of the Earth—less than ten times the distance to the Moon. Radar has been used for several years to track the orbits of asteroids that come near the Earth, but in this instance it was possible to go much further and use radar data to construct a crude image of the object. Steven Ostro of the Jet Propulsion Laboratory is one of only a tiny handful of astronomers who use radar to probe objects in the solar system, and he led simultaneous efforts to obtain images of Toutatis at both the NASA radar site in California and the 300-m Arecibo radar facility in Puerto Rico—succeeding in highly demanding experiments carried out in two places at once.

The appearance of Toutatis as revealed by radar is shown in Fig. 5.19. The asteroid (which is 4.6 km long) turns out to be highly elongated and lumpy, suggesting the possibility that two or more preexisting fragments may have collided at low speed and stuck together. These observations also reveal that Toutatis has a remarkable spin state, consisting of two apparent simultaneous rotations around different axes, with periods of 5.4 and 7.3 days. Radar images of a yet smaller asteroid called Castalia (Fig. 5.20) are even stranger. Castalia consists of two objects of similar size, barely in contact as they spin about each other. If other near-Earth asteroids turn out to be as interesting as these two, we can expect a bonanza of exciting results from additional radar imaging.

Phobos and Deimos: Discovery and Orbits

After Galileo's discovery of the four large satellites of Jupiter, scientists began to speculate about moons orbiting other planets. Since Venus had no satellite, Earth had one, and Jupiter had four, some concluded that Mars, the planet between Earth and Jupiter, should have two (another example of numerology). Kepler was one of those who suggested the existence of two martian satellites, and these hypothetical objects were widely publicized when Jonathan Swift included them in his popular satire *Gulliver's Travels,* published in 1726. Swift wrote of the astronomers of his mythical Laputia that they had discovered

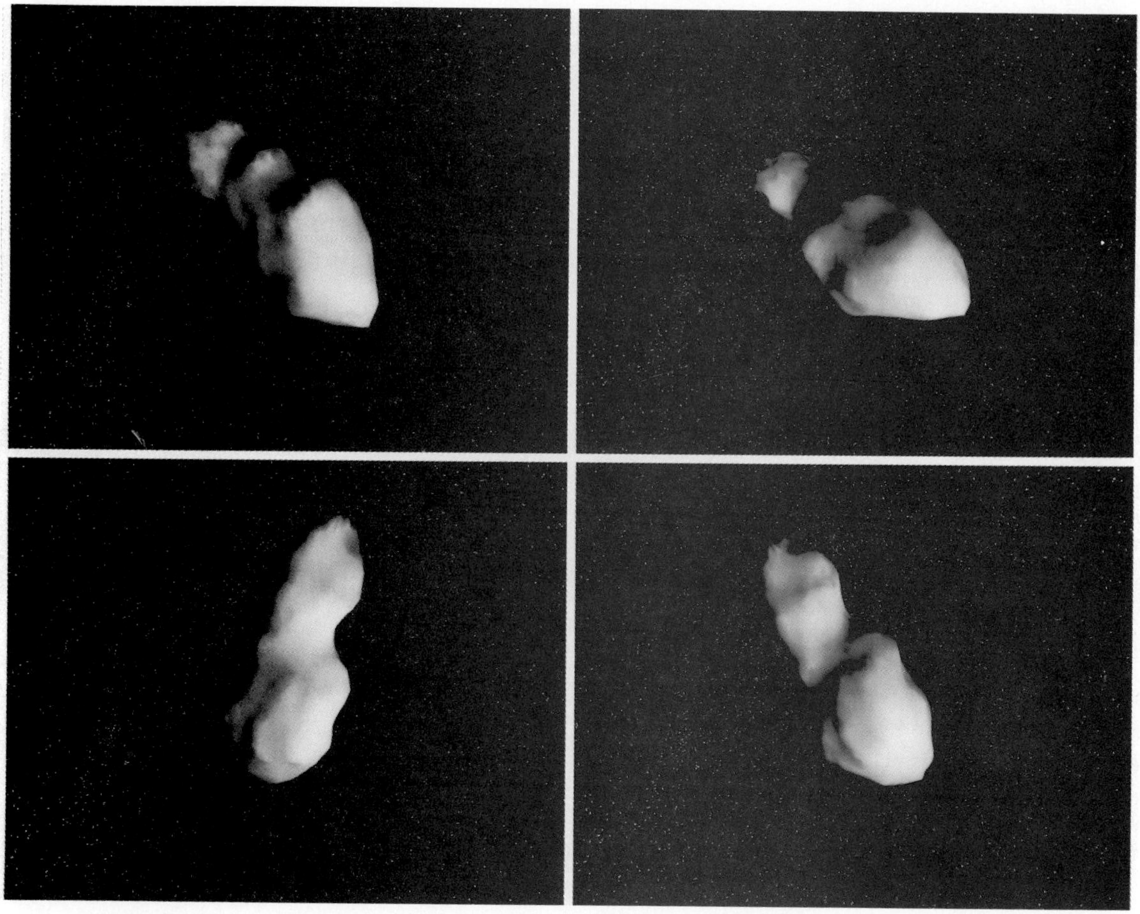

FIGURE 5.19 Four views of the near-Earth asteroid Toutatis reconstructed from radar images by Steven Ostro and Scott Hudson. The asteroid is 4.6 km long and is shown at a resolution of about 100 m. The narrow neck suggests that two originally separate asteroids collided at low velocity to produce this composite object.

two satellites, which revolve about Mars, whereof the innermost is distant from the center of the primary planet exactly three of the diameters, and the outermost five; the former revolves in the space of ten hours, and the latter in twenty-one and a half; so that the squares of their periodical times are very near in the same proportion with the cubes of their distance from the center of Mars . . .

Note that Swift is explaining Kepler's third law to his readers! It was to be another 150 years before these fictional moons were actually discovered, however.

This was among the last important planetary discoveries made by the human eye looking through a telescope. After many nights of careful searching, the American Asaph Hall spotted both satellites during the favorable martian opposition of 1877, the same year that Schiaparelli first reported canals on Mars (see Chapter 11). The satellites were named Phobos (fear) and Deimos (panic), for the horses that pulled the chariot of Ares (Mars) in Greek mythology.

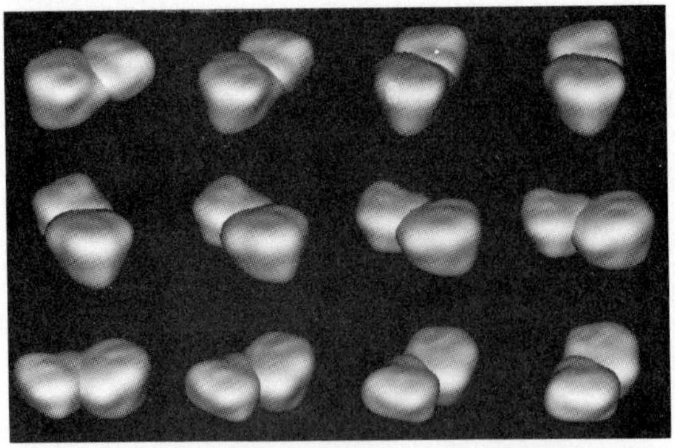

FIGURE 5.20 Looking a lot like a double dinner roll, the asteroid 4769 Castalia is shown here, as modeled by a computer from radar observations of the object during a close (5.6 million km) approach to the Earth. The asteroid is 1.8 km across at the widest point and appears to be a "contact binary."

Phobos, the larger of the two satellites (about 21 km in diameter), is closest to Mars; its distance (from the center of the planet) is 9378 km and its orbital period is 7 hr 39 min. Because Phobos revolves faster than Mars rotates, the satellite is seen (and was photographed by the Viking landers) to rise in the west and set in the east. Deimos, which is about half the size of Phobos, is 23,459 km from Mars and has an orbital period of 30 hr 18 min. Both satellites always keep the same side turned toward Mars, just as the same hemisphere of our Moon always faces the Earth.

Because Phobos and Deimos seem to have so little in common with Mars, and because scientists would not expect the processes that formed Mars to generate two small satellites as well, these objects are assumed to have been captured from elsewhere. To be captured, a passing asteroid must be slowed near the planet, losing energy and falling into orbit—the energetic equivalent of firing a retrorocket on a spacecraft. Probably the capture of Phobos and Deimos took place while Mars still had an extended early atmosphere, and friction with this envelope of gas provided the capture mechanism. We can imagine many objects approaching Mars while the atmosphere was denser and being slowed so much that they crashed into the surface. Later, the atmosphere was too thin to affect passing bodies. Thus Phobos and Deimos represent just

FIGURE 5.21 Views of Phobos (a) and Deimos (b) from the Viking orbiters. Deimos is seen in full phase, so craters are visible only as bright spots.

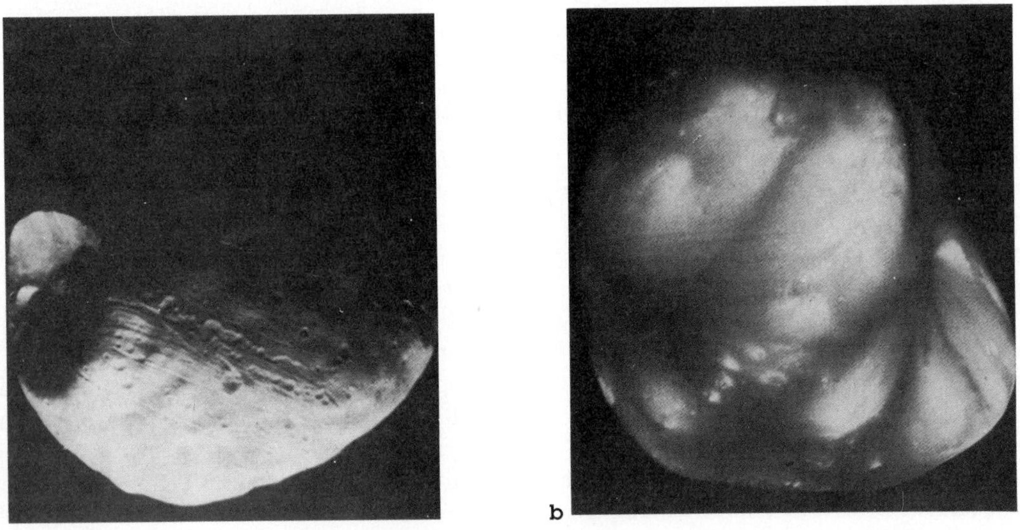

a

b

two of many potential captured bodies, the two that happened to come along at the right moment in the evolution of the early atmosphere of Mars.

Phobos and Deimos: Composition and Appearance

Both martian satellites were photographed by the U.S. Viking spacecraft in 1977. They are dark, cratered, irregular objects (Fig. 5.21). Phobos has dimensions of 28 × 23 × 20 km, and like Gaspra it appears to be a collisional fragment. Deimos is smaller and less irregular in shape.

With reflectivities of about 6%, Phobos and Deimos are almost as dark as the C-type asteroids. Their colors are similar to those of C-type asteroids and of carbonaceous meteorites. Perhaps most diagnostic, the Viking orbiters were able to measure the densities of both satellites, which came out to be approximately 2.0 g/cm^3. Table 5.2 compares these densities with those derived for several main belt asteroids. The density of the martian moons is below that of silicate rock and is lower than that of the other measured asteroids. Probably the composition of Phobos and Deimos is similar to that of the most volatile-rich carbonaceous meteorites, possibly with the addition of some water ice deep in their interiors.

TABLE 5.2 Masses and densities of asteroids

Name	Mass (Moon = 1)	Density (g/cm^3)
Ceres	1.4×10^{-2}	2.4
Pallas	2.9×10^{-3}	2.6
Vesta	3.7×10^{-3}	3.8
Hygeia	1.3×10^{-3}	2.6
Ida	5.6×10^{-5}	2.5
Phobos	1.5×10^{-6}	2.0
Deimos	2.4×10^{-7}	2.0

Surface Processes on Small Bodies

Late in the Viking mission to Mars, special efforts were made to obtain very high resolution photographs of Phobos and Deimos. The spacecraft orbits were adjusted to be resonant with the satellite periods (3/1 for Phobos and 5/4 for Deimos), and with careful navigation a flyby of Phobos at just 88 km distance was achieved in February 1977, followed in May by a 28-km flyby of Deimos. This Deimos encounter, just 23 km above the surface, is the closest any planetary spacecraft has approached an object without actually hitting it.

The highest-resolution picture of Deimos is seen in Fig. 5.22, showing an area about 2 km on a side. The many craters are subdued, as if the surface were blanketed with dust. Individual boulders the size of small houses can be seen.

The blanketing of the surface of Deimos, and to a smaller extent also of Phobos, is a result of the proximity of these objects to Mars. Because

FIGURE 5.22 The highest-resolution Viking picture of Deimos, showing details only 5 m across.

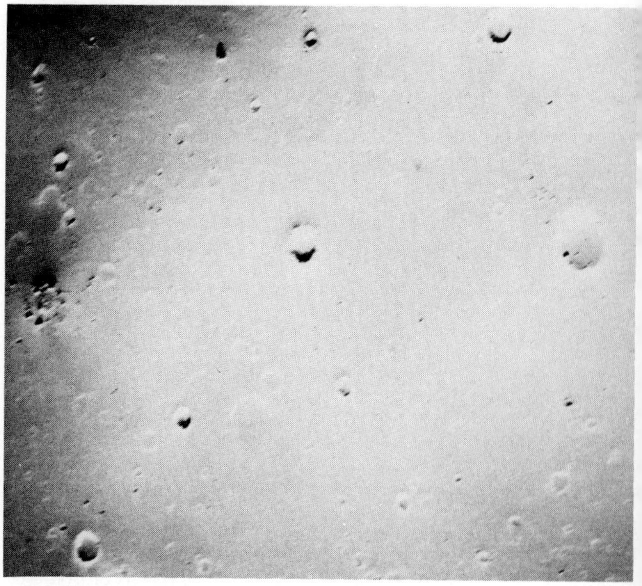

both satellites are small, dust and debris ejected from their surfaces by impact will tend to escape. The martian gravity continues to hold this ejected material, forming tenuous dust rings around the planet in approximately the satellite orbits. Eventually the ejecta returns to the satellite, where it is deposited as a porous dust layer perhaps many tens of meters thick.

Close-up views of Phobos (Fig. 5.23) are even more spectacular than those of Deimos. In addition to its heavy cratering, Phobos displays a unique set of parallel grooves or valleys, typically several hundred meters across and several tens of meters deep. Apparently the grooves are fractures, related to the impact that formed Stickney, the largest crater on Phobos (10 km in diameter). Indeed, this crater is so large relative to the size of the satellite that its formation must very nearly have broken Phobos apart.

Although the best photos of Ida and Gaspra have a resolution of only 30–50 m, poorer than the best Viking pictures of the martian satellites, we can see evidence of some of the same processes. The subdued appearance of the craters suggests the presence of a dust layer, although perhaps thinner than the dust on Deimos, and there is evidence for long parallel grooves similar to those on Phobos. The significance of these similarities is not clear, however, and we may never get higher-resolution data on these two objects. We can expect more spacecraft and radar images of other asteroids, however, from which we will try to derive some general conclusions about asteroid surfaces and histories.

FIGURE 5.23 Seen in this close-up view, the surface of Phobos is marked with a series of linear grooves that appear to be surface fractures caused by the impact that formed the largest crater on this satellite.

5.6 QUANTITATIVE SUPPLEMENT: ASTEROID COLLISIONS

Although there are about 10^5 asteroids larger than 1 km in diameter in the main belt, they are widely spaced and collisions are rare. It is quite easy to make some approximate calculations to estimate collision frequencies.

First let us estimate the average spacing between asteroids. The asteroid belt can be considered a volume stretching around the Sun from 2.2 to 3.3 AU and about 0.2 AU thick. Its volume (circumference × width × height) is thus about 4 AU^3, or about 10^{25} km^3. Dividing this total volume by the number of asteroids gives 10^{20} km^3 per asteroid. The average spacing is approximately the cube root of this, or 5×10^6 km. Thus it should come as no surprise that spacecraft crossing the asteroid belt rarely come within a million kilometers of any asteroids unless they are specifically targeted to do so.

We can now calculate the frequency of collisions. Observations have shown that the typical asteroid has a speed relative to its neighbors of about 4 km/s, resulting from their different orbital eccentricities and inclinations. Take an asteroid with diameter 10 km as average. Its cross section (given by πR^2) is about 100 km^2, and it thus sweeps out a volume in space relative to its neighbors = to its cross section × its speed, or 400 km/s. Since there are 3×10^7 seconds in a year, the volume swept out each year is equal to about 10^{10} km^3.

To estimate the frequency of collisions, we compare the volume swept out by each asteroid in a year with the volume of 10^{20} km^3 associated with each asteroid, which we calculated earlier. The ratio is 10 billion years, older than the age of the solar system. This tells us that most asteroids do not collide with others, even over billions of years, if there are 10^5 asteroids larger than 1 km in diameter. However, there is some current evidence (based in part on the heavy cratering observed on asteroid Gaspra) that there are as many as 10^6 asteroids of this size. In that case, you can see that collisions would be 10 times more frequent and nearly every kilometer-size asteroid would have suffered several such catastrophes.

Finally, we can estimate how often a collision of the sort just described takes place in the asteroid belt today. This is simply the probability of an impact for one asteroid in one year (10^{-10}) times the number of main belt asteroids (10^5). The answer is one collision every 100,000 years. This frequency is sufficient to generate a great deal of dust and meteoritic fragments, and such collisions are probably responsible for the prominent asteroid dust bands discovered by the IRAS satellite.

SUMMARY

Asteroids are interesting in their own right, as small members of the planetary system, and also as possible parent bodies for the meteorites. Because they are much smaller than the planets, many (but not all) of the asteroids have remained relatively unmodified since the origin of the solar system.

Most of the asteroids are found in the main asteroid belt, between the orbits of Mars and Jupiter. In this location, and in the Trojan Lagrangian clouds that lead and trail Jupiter in its orbit, are the only stable orbits where small objects could have survived for the lifetime of the solar system. Even though their orbits are stable, however, there have been many collisions among asteroids over the history of the solar system, and the asteroids we see today are a highly fragmented remnant of the original population of small bodies.

The largest asteroid (Ceres) is only 1000 km in diameter, and the total mass of the asteroids is only about 1/20 the mass of the Moon. Asteroids are classified by analogy with the meteorites as C, S, and M type. Indeed, the asteroids appear to be similar in composition to many of the meteorites, and most scientists believe that they are the parent bodies of most meteorites. The asteroids, like the meteorites, include both primitive bodies (which are in the majority) and differentiated objects. The best-known differentiated asteroid is Vesta, which has a basaltic surface and is probably the parent body of the eucrites.

Four asteroidal objects have been observed at close range from spacecraft: the main belt asteroids Gaspra and Ida, and the two small satellites of Mars, Phobos and Deimos. All are irregular, cratered, and apparently the product of catastrophic fragmentation of their parent objects. Radar imaging of the near-Earth asteroids Toutatis and Castalia shows that the latter is a double object, with the two fragments rotating together in contact with each other.

Near-Earth asteroids are a separate class, sharing many orbital properties with the comets while physically resembling small main belt asteroids. Most are probably derived from the main belt, but some may also be extinct comets that have lost most of their volatiles. They have relatively short lifetimes (100 million years), and about one-third of those with Earth-crossing orbits will eventually impact our planet. The largest of these impacts are capable of causing global environmental catastrophe and thus influencing the evolution of life on Earth.

KEY TERMS

C-type asteroid	resonance
families of asteroids	resonance gap
M-type asteroid	S-type asteroid
main belt asteroid	Trojan asteroid
near-Earth asteroid	

Review Questions

1. Describe the geography of the main asteroid belt and the kinds of objects that populate it.

2. Summarize the kinds of basic information that can be obtained for an asteroid from telescopic studies. Compare this information with that available for various planets and satellites in the planetary system.

3. What is meant by a size-frequency distribution in which the number of objects is proportional to $1/D^2$? Compare this with the size-frequency distribution of people in a shopping mall crowd, or of trees in a forest.

4. Distinguish between main belt asteroids and near-Earth asteroids in terms of their orbits and their physical and chemical properties.

5. Summarize the evidence that meteorites come from asteroids, and indicate which kinds of asteroids are the probable parent bodies of the different kinds of meteorites.

6. Explain the phenomenon of resonance gaps in the asteroid belt. Where else in the planetary system would you expect similar resonance effects to be seen?

7. Compare Gaspra and Toutatis in size and shape. One is a main belt asteroid and one is a near-Earth asteroid; compare their probable origin and fate.

8. In what ways are the martian satellites, Phobos and Deimos, like the asteroids? And in what ways have they been influenced by their environment in orbit about Mars?

Quantitative Exercises

1. Use Kepler's laws to find the asteroid semimajor axes that correspond to resonances with Jupiter—that is, to periods that are ½, ⅖, ⅓, and ¼ that of Jupiter.

2. Ceres has a diameter of 940 km. How does its volume compare with that of the Earth and Moon? How many objects the size of Ceres would you expect to find if the asteroids had originated from the breakup of a planet the size of the Earth, if all the results of this explosion were the size of Ceres?

3. Use reasoning similar to that in Section 5.6 to estimate the probability that a spacecraft crossing the asteroid belt will collide with an asteroid of 1 km diameter or larger. Use the fact that there are about 10^{20} km^3 per asteroid, and assume that the spacecraft has an area (cross section) of 10 m^2 and that it spends 6 months crossing the belt at an average speed of 8 km/s.

4. Use the same reasoning as in the preceding question to estimate how close the spacecraft will pass by an asteroid of 1 km diameter during its 6-month trip through the belt.

Additional Reading

*Binzel, R., T. Gehrels, and M.S. Matthews, eds. 1989. *Asteroids II.* Tucson: University of Arizona Press. The definitive volume on asteroids, part of the Arizona Space Science Series of graduate-level textbooks on planetary science.

Chapman, C.R. 1982. *Planets of Rock and Ice* (Chapter 3). New York: Scribner's. Well-written and authoritative discussion by a leading asteroid researcher.

Chapman, C.R., and D. Morrison. 1988. *Cosmic Catastrophes.* (Chapter 5). New York: Plenum Press. Popular discussion of a variety of cosmic disasters, with emphasis on the collision of comets and asteroids with the Earth.

Kowal, C.T. 1988. *Asteroids: Their Nature and Utilization.* New York: John Wiley and Sons. Up-to-date treatment that includes the potential of asteroids as resources for future development in space.

*Indicates the more technical readings.

6

Comets: Messengers from the Cold

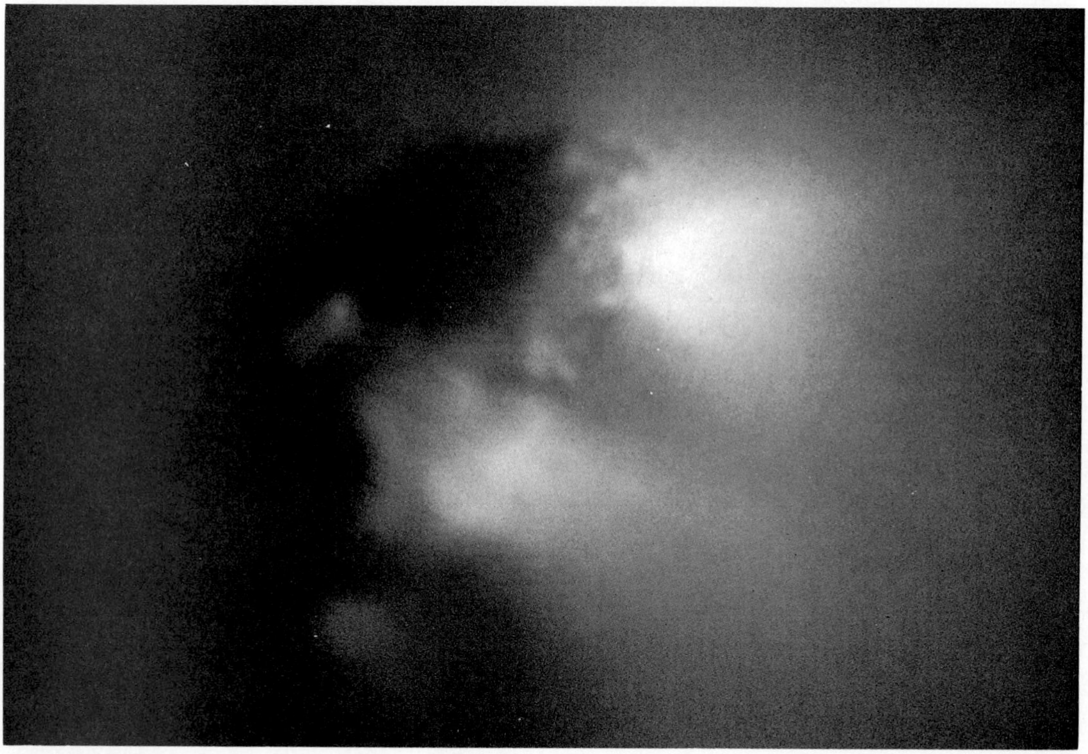

The nucleus of Comet Halley, showing the jets of gas and dust that produce the comet's spectacular tails.

Comets are small primitive objects like the asteroids, but their composition is different and so also must be their place of origin. Comets contain a substantial quantity of water ice and other frozen volatiles. When a comet approaches the Sun, these ices evaporate to form a tenuous, transient atmosphere. It is this extensive but short-lived cloud of gas and dust that makes the comet readily visible and forms its characteristic long tail. (The name *comet* is derived from the Latin word for hair, suggested by the appearance of comet tails.) In order to have preserved these volatiles, the comets must have formed at low temperature and must have spent most of their existence in the far outer reaches of the solar system.

The comets and asteroids provide complementary information about the formation of the solar system. The asteroids are remnants of material that formed in the inner part of the solar nebula and became the building blocks of the inner planets. In contrast, the comets represent a much larger reservoir of material from cooler parts of the solar nebula, more closely related to the outer planets and their systems of satellites and rings.

6.1 COMETS THROUGH HISTORY

Comets have been known since antiquity. Approximately one comet visible to the naked eye appears each year, and a moderately bright comet comes along an average of once per decade. Written records of these long-tailed wanderers go back to at least 1140 B.C. in the Middle East, and nearly as far in China.

Perhaps because the times of their appearances were unpredictable and their forms were constantly variable, comets have frequently been regarded with apprehension (Fig. 6.1). Often the apparition of a comet was believed to herald some remarkable event on Earth, usually unfavorable. Thus Shakespeare wrote in *Julius Caesar*, "When beggars die, there are no comets seen/The heavens themselves blaze forth the death of princes"; while Milton characterized Satan as a comet that "from its horrid hair/ Shakes pestilence and war." No one knows how comets acquired such a bad reputation, but par-

FIGURE 6.1 Representation of Halley's Comet on the Bayeux Tapestry commemorating the conquest of England in 1066. For King Harold of England (shown seated here) the comet was a bad omen, but presumably William the Conqueror felt differently about it. The Latin words say, "They marvel at the star."

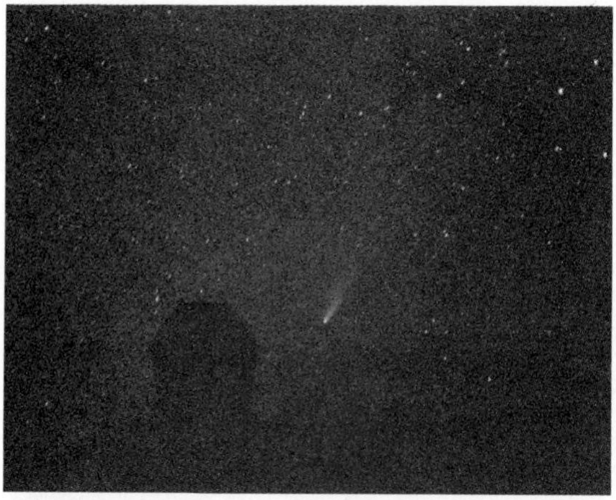

FIGURE 6.2 Comet Halley as seen from Mauna Kea in the spring of 1986. Like most bright comets, it appeared as a small nebulous path of light with a tail. Note the silhouette of one of the observatories. This photograph is a good approximation of a naked-eye view.

ticularly during the medieval period in Europe it seems that no good could be associated with them. In Asia, however, comets were treated more like other celestial phenomena, capable of association with either good or evil events.

The scientific study of comets may be traced back 2000 years to the Greeks and Romans. Because of their belief that the celestial realm was unchanging, most of the classical philosophers thought that comets were located in the atmosphere, and efforts were made to use them to predict the weather. As late as the sixteenth century, the scientists of Renaissance Europe were arguing over whether comets were astronomical or meteorological phenomena.

Appearance of a Comet

Modern students who read of the great attention afforded comets throughout history are often disappointed when they actually see a comet. Perhaps they expect a spectacular light show, something akin to a fireworks display. Instead, they see a faint nebulous patch of light smaller than the apparent size of the Moon and a great deal fainter (Fig. 6.2). Nor is the comet distinguished by its motion, which is imperceptible unless you watch for many hours. To see it at all, you must escape the smog and light pollution of our cities, but even under a dark sky you may require binoculars for a good view.

When we see a comet, we are observing primarily the extended atmosphere that is its trademark. The small solid body from which this atmosphere is released is called the **nucleus.** This is the real heart of the comet, even though it is generally too small to be seen. The atmosphere that surrounds the nucleus is called the head or **coma** of the comet. The long streamers of less substantial gas and dust sweeping away from the Sun are called the **tail.** By definition, all comets have a nucleus and coma, and all except the faintest have tails.

There are at least three reasons why comets seem less important to us than they did to our ancestors. First is the problem of light pollution, which obscures the sky and overwhelms the celestial pageant unfolding above us (Fig. 6.3).

FIGURE 6.3 Satellite photograph of the United States at night, showing the strong artificial illumination of populated areas. Every year the level of light pollution increases, making it more and more difficult to see the stars at night.

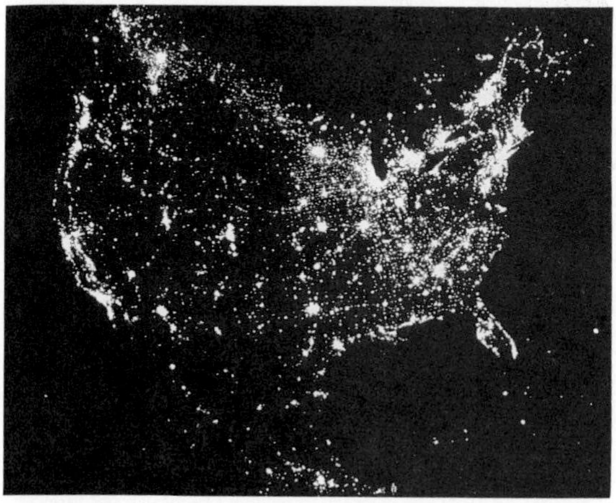

Second, we are simply not as familiar with the sky as were people in the past, nor do we think of the heavens as influencing our daily lives. And finally, the fact is that there has been no comet since 1910 that was really bright. Perhaps Comet Hale-Bopp, which reaches perihelion in the spring of 1997, will make up for this long dry spell. It was more than 200 times brighter than Comet Halley at the distance of Jupiter, but this is no guarantee that it will put on a spectacular show nearer the Sun.

Tycho and Halley

A turning point in cometary studies came with the Great Comet of 1577, which was extensively investigated by Tycho Brahe, greatest of the pretelescopic astronomical observers. He carefully measured the position of the comet against the background of stars, finding that it did not shift back and forth with the change in viewing angle afforded by the Earth's rotation, as would a nearby object. (This is an example of parallax, discussed in Section 2.1.) Thus Tycho concluded that this comet was not in the atmosphere of the Earth, but well beyond even the Moon, whose parallax he could measure.

At about the same time that Tycho was studying the comet of 1577, others demonstrated that comet tails always point away from the Sun, rather than back along the direction of motion. After Johannes Kepler's elaboration of the heliocentric motion of the planets, it was quickly realized that comets also were bound to the Sun, although their orbits were so eccentric that they were essentially parabolic rather than nearly circular ellipses.

One astronomer whose name has been immortalized for his work on comets was Edmund Halley (Fig. 6.4), a contemporary of Newton in late seventeenth-century England. It was Halley who first realized that cometary orbits could be closed ellipses, with a given comet reappearing at regular intervals. He reached this conclusion from a study of the Great Comets of 1531, 1607, and 1682, all of which had similar orbits. Halley concluded that these were successive appearances of the same comet, which he predicted would return in 1758.

FIGURE 6.4 British astronomer Edmund Halley successfully predicted the return in 1758 of the comet that now bears his name.

Although Halley did not live to see his calculations verified, the comet returned just as he had expected, and it was quickly hailed as Halley's Comet, a name it retains today. Comet Halley, with its 76-year orbit, is the brightest of the short-period comets. It appeared twice during the twentieth century, in 1910 and in 1986. Not until 1820 was a second periodic comet identified: the much fainter Comet Encke, with a period of only 3.3 years.

Understanding the physical nature of comets has proved even more difficult than pinning down their elusive motions. Until Tycho's work on the comet of 1577, comets were believed to be "exhalations" in the atmosphere of the Earth. Tycho's determination that this comet was beyond the Moon also implied that the visible head or coma of the comet was very large, comparable in size to the Earth. Soon astronomers were measuring cometary tails millions of kilometers in length, and public fear of plagues from comets was augmented by fear of collision with them.

Comets in the Nineteenth and Twentieth Centuries

The fact that stars could easily be seen through the cometary coma and tail shows that in spite of their large size, comets are basically rather tenuous objects. Isaac Newton was among the first to suggest that most of what was seen was thin gas expelled from a solid nucleus as it was heated by the Sun. Nineteenth-century astronomers also correctly attributed the streaming of the tail

FIGURE 6.5 A series of photographs of Comet Halley in 1910 showing the changes in the comet's appearance as it approached and then receded from the Sun.

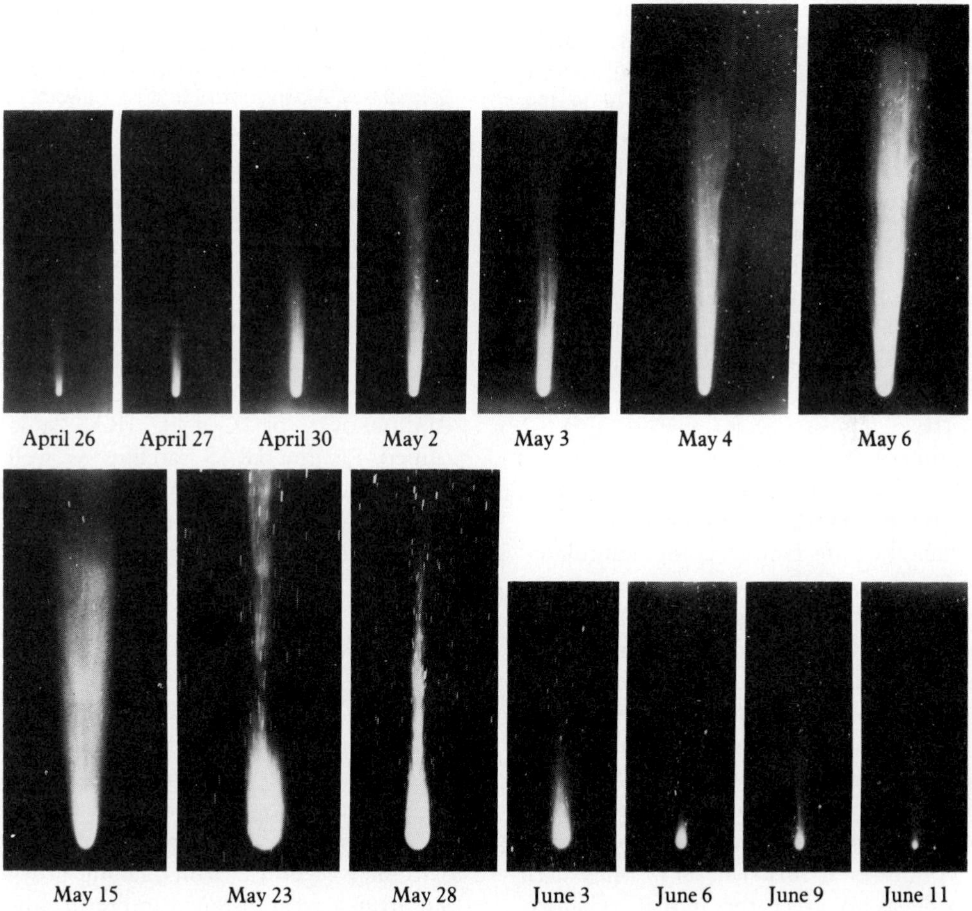

away from the Sun to the force of solar radiation on the insubstantial atmosphere of the comet.

During the second half of the nineteenth century, public interest in comets throughout the United States and parts of Europe provided a substantial boost to astronomy, leading to the endowment of a number of observatories. In 1910 Comet Halley was particularly well placed for viewing, and the Earth even passed through the tail of the comet on May 20 of that year (Fig. 6.5). A few months earlier, an even brighter new comet had appeared, making 1910 a memorable year for comet watchers.

Since the early part of this century, the gradual but steady increase in smog and light pollution has blotted the night sky from the view of city dwellers, and several fine comets (Bennett in 1969, West in 1976) have gone practically unnoticed by the public. The 1986 return of Halley was also a disappointment in comparison with 1910. The comet was very poorly placed; when it was closest to the Sun and therefore brightest, it was located in its orbit on the opposite side of the Sun from the Earth. Although there was a great deal of public interest and media attention, many persons who tried to see the comet, especially from Canada, Europe, and the northern parts of the United States, were frustrated in their efforts.

The primary interest in the most recent appearance of Comet Halley did not depend, however, on ground-based observations. The pending arrival of this famous comet stimulated the USSR, the European Space Agency, and Japan to send instrumented spacecraft to fly into the comet a few weeks after perihelion. Of the major space-faring nations, only the United States failed to contribute to this Halley "armada." Later in this chapter we will discuss the results from these missions (Fig. 6.6).

In spite of the difficulty of seeing comets in our modern world, a devoted group of comet hunters continues to discover new ones each year, as well as to greet the return of short-period comets. This interest is encouraged by the International Astronomical Union policy of naming comets for their discoverers, a custom that began about a century ago. Comet Hale-Bopp, for example, is named for two American amateur astronomers. Up to three independent discoverers can be recognized, leading to some awkward names, such as Comet Honda-Mrkos-Pajdusakova, or Comet IRAS-Araki-Alcock, named for the IRAS satellite as well as two astronomers. Discovering a comet is the only sure way of getting your name assigned to an astronomical object. In recent years, about a dozen new comets have been discovered annually.

FIGURE 6.6 Part of the European team that developed the Giotto mission to Comet Halley, celebrating a successful launch in French Guiana: in back, F. D'Allest; middle row, left to right, J. Lions, G. Delouzy, A. Ammar, and R. Lüst (director of ESA); front row, R. Bonnet, A. Brahic, and G. Vedrenne.

Orbits of Comets

Comets are classified on the basis of their orbits as long-period comets or short-period comets, with the division placed rather arbitrarily at 200 years. Most **long-period comets** are on extremely eccentric orbits, falling in toward the Sun from a great distance and returning again to the depths of space. Their ellipses are so elongated that they can be approximated very well by a parabola with the Sun at one focus, with corresponding orbital periods of a million years or more.

Of the 855 comets discovered through 1992, 681 are long-period comets and 21 are of intermediate period (20–200 years). Most of the 153 comets with periods of less that 20 years have aphelia near the orbit of Jupiter; they are called Jupiter family comets. Among the short-period comets, Halley, with its aphelion near Neptune, is exceptional.

Because the paths of comets cut across those of the planets, their orbits are inherently unstable. Like the Earth-approaching asteroids, they must be coming from elsewhere. As we will discuss later under the heading of cometary evolution, scientists believe there are two sources for comets. One is a very large diffuse cloud of comets at the outer fringe of the solar system, containing perhaps a trillion (10^{12}) objects; the second is a smaller disk of comets beyond the orbit of Neptune. The comets we see represent the tiny fraction that leak into the inner solar system from these two immense reservoirs. Table 6.1 lists some famous comets from history.

6.2 THE COMET'S ATMOSPHERE

The visible part of a comet is its atmosphere, a term we use to include both the coma and the tail. Since most astronomical studies of comets deal with the atmosphere, we discuss it first before turning to the tiny nucleus that is the source of the atmospheric gas and dust.

The Coma

The brightest part of a comet is the inner coma, consisting of gas and dust recently ejected from the nucleus (Fig. 6.7). Sometimes there appears to be a starlike condensation of bright material at the center, fooling observers into thinking they are seeing the nucleus itself, when actually it is only the brightest inner part of the atmosphere, a few hundred to a few thousand kilometers in diameter. Sometimes this inner atmosphere is symmetric about the nucleus, but more often it is brightest in the direction of the Sun, and frequently it displays structure in the form of fans of denser material apparently streaming away from gas jets on the nuclear surface.

TABLE 6.1 Some famous comets

Name	Period	Significance
Great Comet of 1577	long	Found by Tycho to be farther away than the Moon
Great Comet of 1811	long	Largest coma (2 million km across)
Great Comet of 1843	long	Brightest ever (visible in daylight)
Daylight Comet of 1910	long	Brightest of twentieth century
West (1976)	long	Best recent comet; nucleus split
Hale-Bopp	long	Unusually bright at distance of Jupiter (1995)
Swift-Tuttle	133 yrs	Parent comet of Perseid meteors
Halley	76 yrs	Studied by multiple spacecraft in 1986
Chiron	51 yrs	Largest nucleus (formerly designated as an asteroid)
Biela	6.8 yrs	Broke apart and never returned
Giacobini-Zinner	6.5 yrs	First spacecraft encounter (1985)
Encke	3.3 yrs	Shortest known period
Shoemaker-Levy 9	—	Broke up (1992) and crashed into Jupiter (1994)

The gases released from the nucleus are quickly broken down by ultraviolet sunlight to create such molecular fragments as OH, CH, and NH. Most of these reactions are straightforward and easy to calculate, but others, such as the source(s) of the carbon molecules C_2 and C_3, remain mysterious. There is a great deal of complex chemistry that takes place in the inner atmosphere of a comet, within minutes of the release of gas from the surface, and not all of it is understood.

Near the surface of an active comet, the gas density is only about a millionth as great as the density of the Earth's atmosphere, and this density falls off rapidly with distance as the gas streams away from its source. At distances as far as 10,000 km the gas flows smoothly; beyond this range, the individual molecules cease interacting with each other and with the embedded dust. Surrounding the rest of the atmosphere is a cloud of glowing hydrogen atoms that can extend more than a million kilometers from the nucleus. Beyond a hundred thousand kilometers, however, most of the gas molecules become ionized, and this plasma is swept away by the blast of the solar wind.

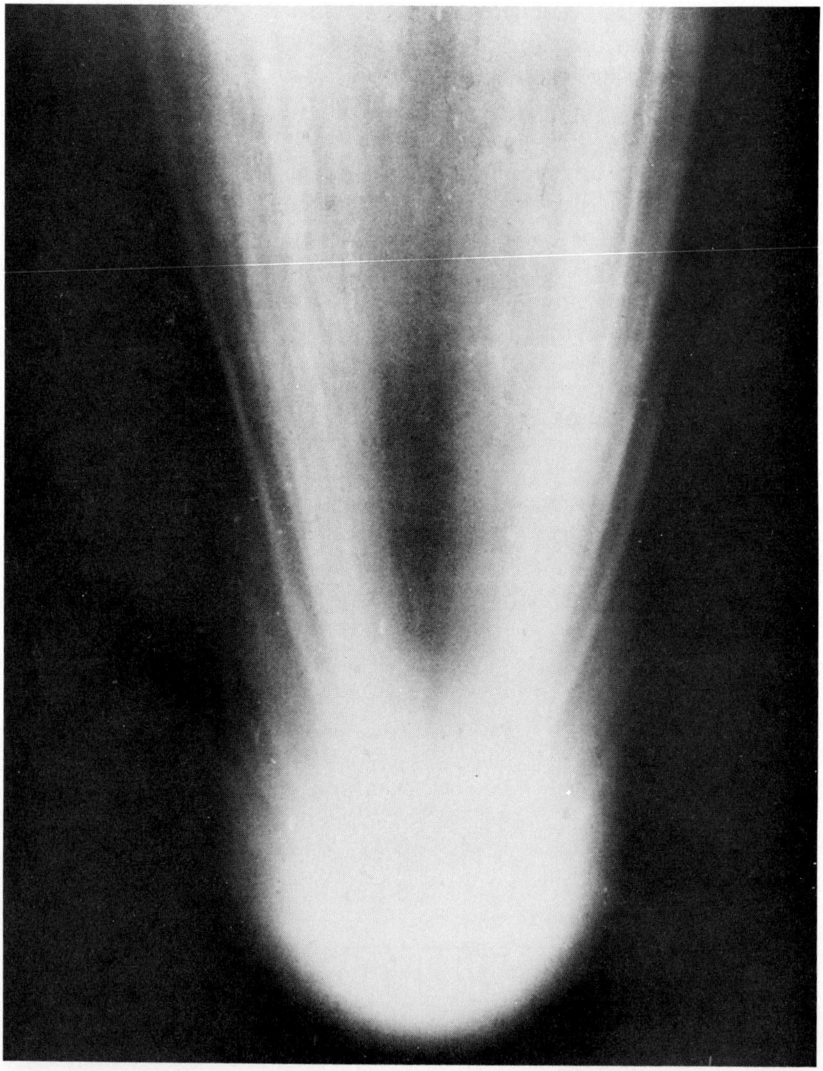

FIGURE 6.7 The head of Comet Halley as photographed on May 8, 1910.

Composition of the Gas

Spectroscopic studies of comets generally refer to the inner coma, which is the source of most of the light from a comet. Astronomers measure the composition of this gas and then try to work backward to identify the parent materials that were on the nucleus. The central role of water ice in comet chemistry has been confirmed by observations of the spectra of various comets. When water vapor is released from the nucleus, it is quickly ionized to become H_2O^+, and it is then broken down by sunlight into its constituent parts: hydrogen (H) and hydroxyl (OH). Emission by OH molecules is a common feature in spectra of bright comets. In recent years both H and OH have been detected from observatories in space, while emission from H_2O^+ plasma has been detected optically and measured by radio astronomers. Moreover, in 1986 neutral water itself was detected in the atmosphere of Comet Halley using the telescope of the NASA Kuiper Airborne Observatory.

In addition to water, emissions have been seen from carbon gases: C_2, C_3, CO, CO_2, and CH_3OH (methanol). These indicate the presence of additional ices, consisting of both hydrocarbons (compounds containing hydrogen and carbon) and oxides of carbon. It is unusual to find both reduced and oxidized compounds of the same element on the same object, since simple chemical reactions will tend to convert from one to the other. The preservation of these compounds on comets confirms the low-temperature history of these objects.

Nitrogen compounds, including N_2, HCN, NH_3, and CH_3CN have also been detected in cometary spectra, as well as several sulfur compounds. HCN is hydrogen cyanide, a gas that is lethal to humans; the deduction that HCN was in the coma of Comet Halley frightened people in 1910 when they were told that the Earth would pass through the comet's tail. A summary of constituents detected in cometary atmospheres (as of 1993) is given in Table 6.2.

Comet Tails

The tenuous gases composing the comet's coma are not gravitationally bound to the nucleus. As they expand, these gases are ionized and caught up by the solar wind to assume the characteristic wispy form of the cometary tail. This tail is called a **plasma tail.** We see plasma tails primarily by the blue glow of ionized carbon monoxide (CO^+) stimulated to fluorescence, although

TABLE 6.2 Atoms and compounds observed in comets*

Metals	Other Atoms	Diatomics	Triatomics	Polyatomics
Na	H	CH	H_2O	H_2CO
K	C	CO	CO_2	CH_3OH
Ca	N	CN	NH_2	NH_3
V	O	C_2	HCN	CH_3CN
Cr	S	CS	C_3	
Mn		OH	HCO	
Fe		NH		
Co		S_2		
Ni		CN		
Cu				

*Adapted from a summary by M. A'Hearn.

molecular emissions due to water, carbon dioxide, and other ionized gases have also been detected. Plasma tails are straight, point away from the Sun, and usually are made up of individual streamers or rays only a few thousand kilometers across. An active comet such as Halley can develop a plasma tail 100 million kilometers in length, almost long enough to stretch from the Sun to the Earth.

Within a plasma tail the actual density of material is incredibly low, typically only a few hundred molecules per cubic centimeter. Thus a giant tail tens of millions of kilometers long contains no more molecules than a modern supertanker—perhaps half a million tons of mass. For most comets, the mass of material in the tail is far smaller.

Since the plasma tail is driven by the solar wind, it provides a kind of celestial weather vane to record the speed and direction of flow. By tracking individual kinks and twists as they move downstream, astronomers can measure properties of the solar wind. Typical solar wind velocities determined from comets are about 400 km/s, sufficient to traverse the entire length of the comet tail in less than a day. It is not surprising, therefore, that plasma tails can alter their form from hour to hour and that the overall appearance of a comet even as seen by the naked eye can change dramatically from one night to the next.

Most comets have two tails: the plasma tail, which we have been discussing, and which consists of charged molecules caught in the solar wind, and a second tail consisting of dust grains, appropriately called the **dust tail.** Although generally shorter than plasma tails (usually under 10 million km long), dust tails can be as bright or even brighter. They are readily distinguished by their color (yellow-white, from reflected sunlight) and from the fact that they are curved rather than straight (Fig. 6.8).

Dust tails consist primarily of small grains only a few micrometers in size, indicating that much of the solid material in the nucleus must be of the same consistency, similar to the finest

FIGURE 6.8 Comet Mrkos photographed during six nights in 1957. Note the large changes in the plasma tail that occur from night to night, while the curved dust tail remains relatively unchanged.

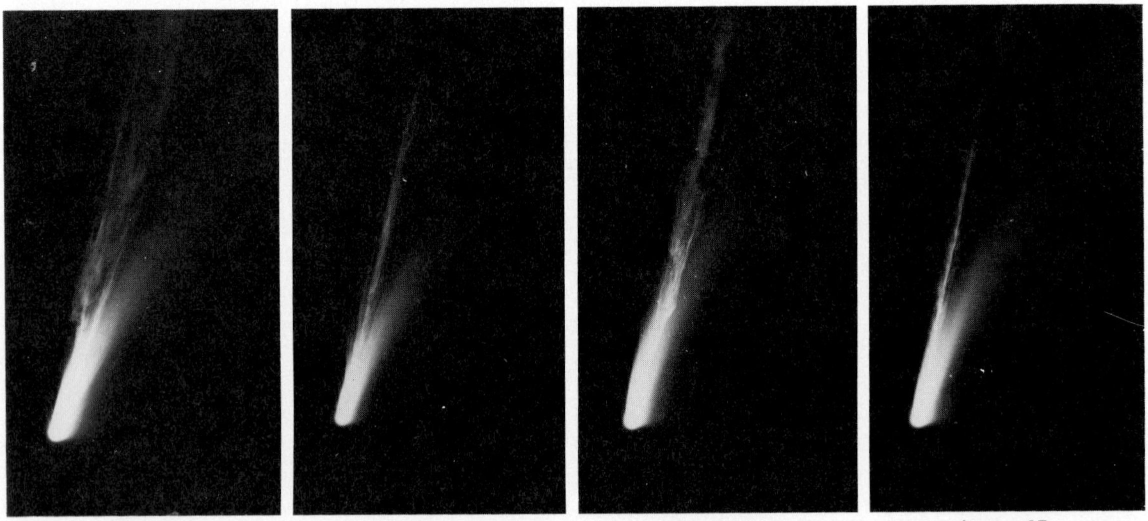

August 22 August 24 August 26 August 27

cake flour. Once decoupled from the expanding and thinning gas, the dust grains follow their own orbits, moving under the joint influence of solar gravity and of pressure generated by solar radiation. Because they move relatively slowly, the dust grains tend to mark the location of the comet at the time of their release. Thus the dust tail traces the path of the comet across the sky. Measurements of the material in dust tails suggest a mass that is roughly comparable to the gas in the plasma tail, indicating that similar quantities of ice and dust are present in the nucleus.

6.3 THE COMETARY NUCLEUS

The most important part of a comet is its *nucleus;* the rest is insubstantial show. We know a nucleus must exist, since the transient gases of the atmosphere come from somewhere. Telescopic observers cannot see the nucleus directly, however, since it is lost in the glare of the head of an active comet. Only recently has radar penetrated the atmospheres of several comets to provide unambiguous detection of their nuclei, while the spacecraft that flew through Comet Halley in 1986 were able to photograph the nucleus of that comet at close range.

The Dirty-Snowball Model

Before we recognized that comets contain a great deal of water ice, which evaporates under solar heating to generate the atmosphere, some scientists thought that the nucleus consisted of a loose agglomeration of grains and boulders similar in composition to the primitive meteorites. Because this meteoritic material contains limited quantities of trapped gas, it was necessary to hypothesize a great number of particles to account for the observed gas production near the Sun.

In 1950, however, Fred L. Whipple of Harvard University (Fig. 6.9) made a fundamental step forward when he developed a model for the nucleus as a single, relatively small "dirty snowball," made up of roughly equal quantities of silicates and ice. Whipple suggested that the rock and ice were intimately mixed, as might be expected for a primitive body that formed directly from the solar nebula or even from dust and gas clouds in interstellar space.

FIGURE 6.9 Fred Whipple (left), the originator of the icy-snowball model for comet nuclei, listening to Roald Sagdeyev (right), the director of the VEGA mission and of the USSR Institute for Space Research, as the data from the 1986 comet flyby were received at Earth. The VEGA pictures provided the first visual evidence that Whipple's model—conceived 36 years earlier—was indeed correct.

Composition of the Nucleus

At first very little was known about the composition of the nucleus, although Whipple concluded that water (H_2O) ice was likely to be the major constituent. This hypothesis was supported by observations that most cometary activity turns on as the approaching comet reaches about 3 AU from the Sun. At this distance we can calculate that the surface temperature, produced by the steadily increasing intensity of sunlight, should reach the point (about 210 K) at which water ice begins rapid evaporation. Other ices, with different evaporation rates, would begin activity at different temperatures and hence at different distances from the Sun. A comet like Hale-Bopp, which develops a large coma at more than 5 AU from the Sun, must contain other more volatile materials than H_2O, such as CO and N_2.

We have some direct measurements of the gases evaporating from the nucleus of Comet Halley, obtained by the VEGA and Giotto probes within a few thousand kilometers of the nucleus itself. In addition there are spectroscopic measurements of the atmospheres of many comets (Table 6.2). It is clear that water ice is the predominant volatile component, but that a few percent of the ice also consists of frozen carbon monoxide (CO), carbon dioxide (CO_2), and methyl alcohol (CH_3OH), plus traces of hydrocarbons and other compounds. One possible model for the ice chemistry of the nucleus is illustrated in Table 6.3.

In addition to the volatiles listed in Table 6.3, the cometary nucleus also contains substantial quantities of dark carbonaceous and silicate dust. The 1986 spacecraft measurements of Comet Halley showed that, for this comet at least, carbon and hydrocarbon dust predominates over silicate dust. As the ices evaporate, they release the dust particles, which stream into space carried along by the flow of gas. This dust contributes to the cometary tail and is the primary constituent of the meteors. Larger rocky masses may also be present in the comet, but whether these are represented in our meteorite collections remains an open question.

TABLE 6.3 Volatile abundances in comets*

Molecule	Abundance (percent by mass)
H_2O	65–80
CO	5–20
CO_2	2–10
CH_3OH	2–10
CH_4	<1
NH_3	<1
HCN	<1
H_2S	<1
Hydrocarbons	<1

*Adapted from a 1993 summary by M. Mumma, P. Weissman, and A. Stern.

It is possible that some of the carbonaceous and silicate material in comets predates the origin of the solar system, having originated in the interstellar material before the solar nebula formed. Some of the ices might also represent direct condensation from interstellar clouds. Many of the known or suspected volatiles in comets have also been identified by radio astronomers in interstellar "molecular clouds" throughout our Galaxy. It is possible that some of the cometary volatiles are materials frozen out directly from such interstellar sources. If this idea is correct, it implies that comets may be vehicles for carrying molecules created out among the stars to the surfaces of satellites and planets throughout the solar system. This includes the possibility of bringing organic material to the early Earth.

Physical Nature

How large is the nucleus of a comet? For most comets only a rough estimate can be made. Once the atmosphere of a comet forms, the tiny nucleus is lost in the midst of the bright dust and gas cloud that forms the coma, and its size cannot be measured.

One way to penetrate through the obscuring gas and dust is to use radar. The first radar signal was bounced from the nucleus of Comet Encke in 1980, and in 1983 radar data were used to derive a diameter of between 5 and 10 km for the faint comet IRAS-Araki-Alcock. Radar observations not only can determine the diameter of the nucleus, but also have been used to detect clouds of large particles (boulders?) surrounding the nucleus.

The only cometary nucleus that has been photographed in detail is that of Halley. During

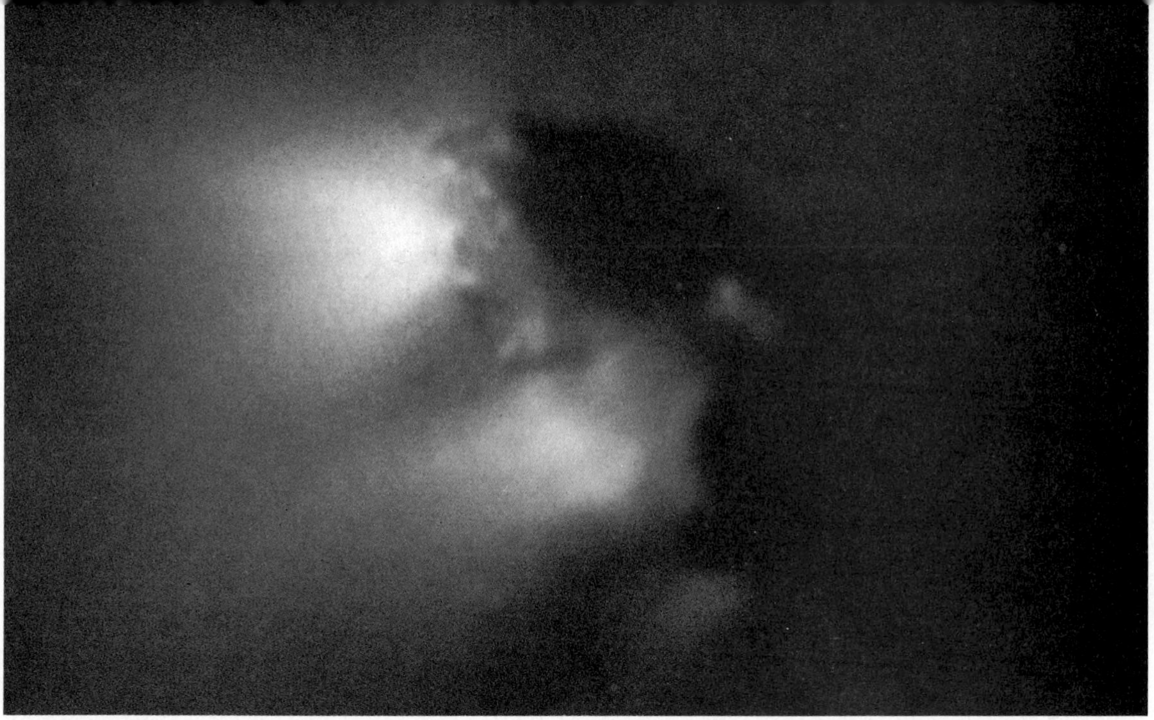

a

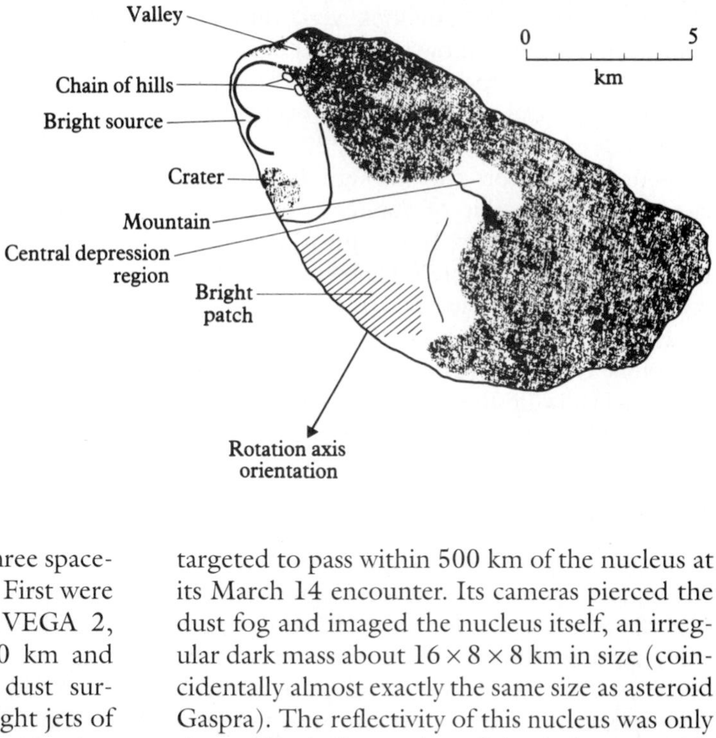

b

FIGURE 6.10 (a) The nucleus of Comet Halley is about twice as large as the island of Manhattan. It is composed primarily of water ice, but frozen CO_2, CO, NH_3, and other gases are present, all covered by a layer of dark carbon-rich material like a giant ice cream bar. This view was produced from 60 different images obtained by the Giotto spacecraft in March 1986. The smallest details are only 60 m across. Jets of dust and gas emanate from at least three bright regions along the left (sunward) side of the nucleus. (b) Schematic diagram illustrating the main features visible on the Halley nucleus.

the week beginning March 6, 1986, three spacecraft flew into the heart of this comet. First were the Soviet spacecraft VEGA 1 and VEGA 2, which each came within about 8000 km and photographed a thick envelope of dust surrounding the nucleus, with several bright jets of gas and dust erupting from it. Using navigation data from the VEGAs, the European Giotto craft targeted to pass within 500 km of the nucleus at its March 14 encounter. Its cameras pierced the dust fog and imaged the nucleus itself, an irregular dark mass about 16 × 8 × 8 km in size (coincidentally almost exactly the same size as asteroid Gaspra). The reflectivity of this nucleus was only about 3%, similar to that of the darker asteroids (Fig. 6.10).

The very low reflectivity of Halley, indicative of primitive, carbonaceous material, apparently confirms the suggestions made by NASA astronomer Dale P. Cruikshank and a few other observers that most comets are covered by dark material similar to that found on some distant asteroids. Cruikshank originally encountered great skepticism when he reported that these icy objects were as black as coal, but now we see that the nucleus is a very dirty snowball indeed. Presumably the loss of ice and other volatiles from the upper layers of the nucleus allows a crust of dark dust to accumulate over the entire surface. A similar effect darkens the toe of a melting glacier on Earth by concentrating nonvolatile material as the ice evaporates.

Both Halley and IRAS-Araki-Alcock have diameters near 10 km, in spite of a difference in activity and hence brightness of a factor of a hundred, indicating that comets with nuclei of about the same size can look very different, depending on the outflow of gas and dust from the nucleus. A few spectacular comets from past centuries are estimated to have been much larger. For example, a comet in 1727 was visible without a telescope even though it came no closer than the orbit of Jupiter, where most comets are exceedingly faint. It must have had a diameter on the order of 100 km. Chiron, which now seems firmly established as a comet, has a measured diameter of about 200 km. The corresponding range in masses is from about 10^{12} tons for Halley up to greater than 10^{15} tons for Chiron and the monster comets of the past.

Cometary Activity

As a typical comet nucleus approaches the Sun, the rapid evaporation of its ices begins at a surface temperature of a little above 200 K, out in the main asteroid belt. By the time it crosses the orbit of Mars, the comet develops a full-scale atmosphere and tail. Nearer the Sun, the solar energy evaporates more and more ice as well as heating the surface. Eventually, most of the energy goes into evaporation, although the darker parts of the cometary crust can become quite warm.

Measurements of the total brightness of comets have yielded estimates of the mass of gas and dust lost during one trip through the inner solar system. Typically this is 10–100 million tons for an active comet, corresponding to roughly 0.1% of the total cometary mass. Clearly, at this rate of loss the comet will not last forever, but it will exhaust its store of ices after a few thousand passes through the inner solar system. If the solid or dirty component is carried away with the evaporating ices, the comet will simply shrink down to nothing. On the other hand, if a residual core of solid material remains after the ices are gone, this core would be indistinguishable from a dark near-Earth asteroid. We do not know which of these scenarios for the death of a comet is correct. Nor are they mutually exclusive; some comets may end one way, some the other.

Nongravitational Forces

While a comet remains active, the escaping gases can have an effect on its orbit. Recall the way a rocket engine operates, generating thrust from the ejection of mass according to Newton's third law of motion. The same thing occurs with a comet. Evaporation of ice takes place primarily in jets on the warmest part of the comet, which corresponds to the "afternoon" side facing the Sun. As dust and gas stream away from the nucleus at a speed of hundreds of meters per second, the reaction creates a force in the opposite direction, away from the Sun (Fig. 6.11). Acting continuously over a period of weeks, such a "nongravitational" force can have an easily measurable effect on the orbit, and the influence of such forces must be taken into account when predicting the future paths of comets.

One of the motivations for Whipple's icy-snowball model for the nucleus was its ability to explain nongravitational forces by the rocket

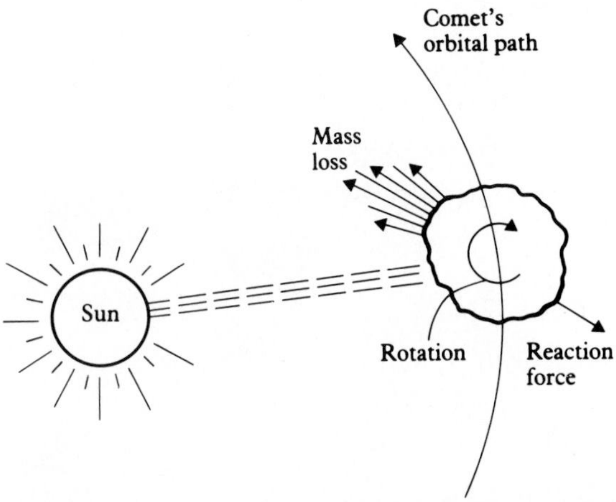

FIGURE 6.11 A schematic illustration of the rocket effect exerted by gases escaping in jets from a comet nucleus.

effect. A nucleus consisting of a loose collection of gravel and boulders would not behave in this way. In his original 1950 paper on the cometary nucleus, Whipple showed how the rocket effect could explain the observed changes in the orbit of Comet Encke, which had been tracked by astronomers for more than a century. Subsequent analysis has also yielded the spin rates of comets, which are typically a few hours. Comet Halley, as measured in 1986, has an unusually long rotation period of three days.

While the dirty-snowball model provides a satisfactory understanding of many cometary phenomena, it is incomplete, leaving unanswered important questions concerning details of the activity. What would it really be like to ride along on a comet as it approaches the Sun? The Halley probes seemed to indicate a highly dynamic environment, with huge jets of material like geysers dominating the activity near perihelion. Perhaps eruptions burst out from different spots on the nucleus, as the dark surface absorbs sunlight and heats the ices below. Gases trapped in the ice could cause the activity of distant comets. Mysteriously, Halley brightened suddenly in 1991, when it had already retreated 14 AU into the cold; we do not know what triggered this event. Finally, we do not know the ultimate fate of comets: Do they become asteroids or just fade away?

The Comet Crash of 1994

The fate of one comet is not in question: In 1994 a multiple comet called Shoemaker-Levy 9 crashed into Jupiter, ending its life in a spectacular display of celestial fireworks. The presence of craters on the surfaces of planets and satellites throughout the solar system is evidence that comets sometimes impact into planets, but never before had such a cosmic collision been witnessed by astronomers.

This comet, called S-L 9 for short, was discovered in 1993 by Gene and Carolyn Shoemaker and their amateur colleague David Levy as part of a regular telescopic patrol carried out with a small telescope at Palomar Observatory. At the time of its discovery, S-L 9 was no ordinary comet, for it had about 20 separate nuclei spread out like pearls on a string (Fig. 6.12). A backward calculation of the orbit indicated that the comet had passed just 35,000 km above the clouds of Jupiter in the summer of 1992, close enough for its nucleus to have been disrupted by the gravity of the giant planet. Detailed calculations of this disruption process indicated that the original nucleus was between 2 and 3 km in diameter and consisted of many loosely bound fragments—not the orbiting gravel bank of pre-1950 models, but lacking any significant structural strength. These calculations also provided the first actual estimate of the density of a comet, yielding a value for the density near 1.0 g/cm^3.

When the positions of S-L 9 were calculated into the future, even more surprises were in store. It turned out that the comet was not technically in orbit about the Sun at all, but that it had been captured into orbit about Jupiter. In effect, Jupiter had acquired about 20 new small satellites. However, this jovian orbit was not sta-

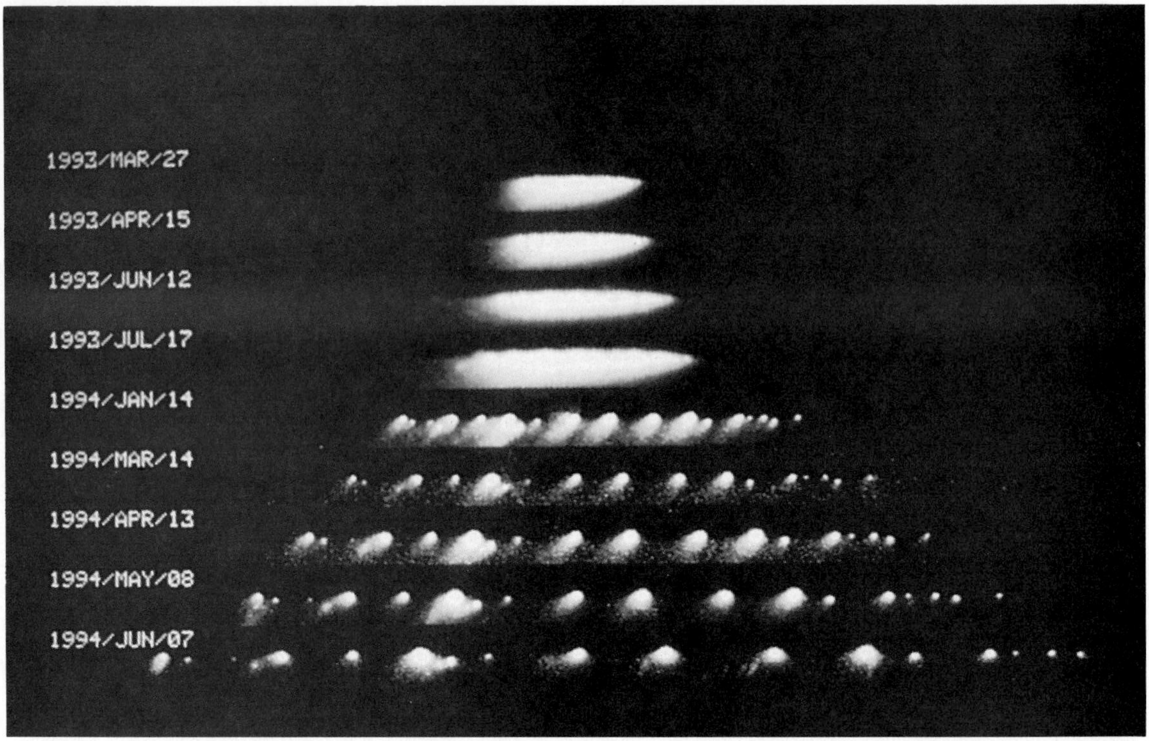

FIGURE 6.12 Comet Shoemaker-Levy 9 (S-L 9) as it appeared during the 16 months prior to its collision with Jupiter in July 1994. These images are all the same scale; they show the fragments increasing in number and moving apart with time.

ble, and all of the fragments would smash into the planet on their next close approach, in July 1994.

The impacts all took place on the back side of Jupiter, just over the horizon as seen from the Earth. The only direct view of the impact sites was obtained from the Galileo spacecraft, then about 200 million km from Jupiter. However, telescopes all over the Earth (and in orbit above the Earth) were trained toward Jupiter during the week when these unprecedented impacts took place, and the results were not disappointing. The Hubble space telescope and other instruments were able to see the impact explosions as they rose above the edge of Jupiter, and each of the events left huge dark smudges in the planet's atmosphere that were visible even with small amateur telescopes.

Each comet nucleus smashed into the jovian atmosphere with a speed of 60 km/s, creating a brilliant meteor and violently disintegrating as it plunged into the deeper layers of the atmosphere. The largest nucleus was probably about 1 km in diameter, and some were not more than a few hundred meters across, yet they exploded with energies of millions of megatons—each one hundreds of times greater than all the nuclear weapons on Earth put together. Each explosion generated a hot fireball that erupted upward to an elevation of about 3500 km, easily

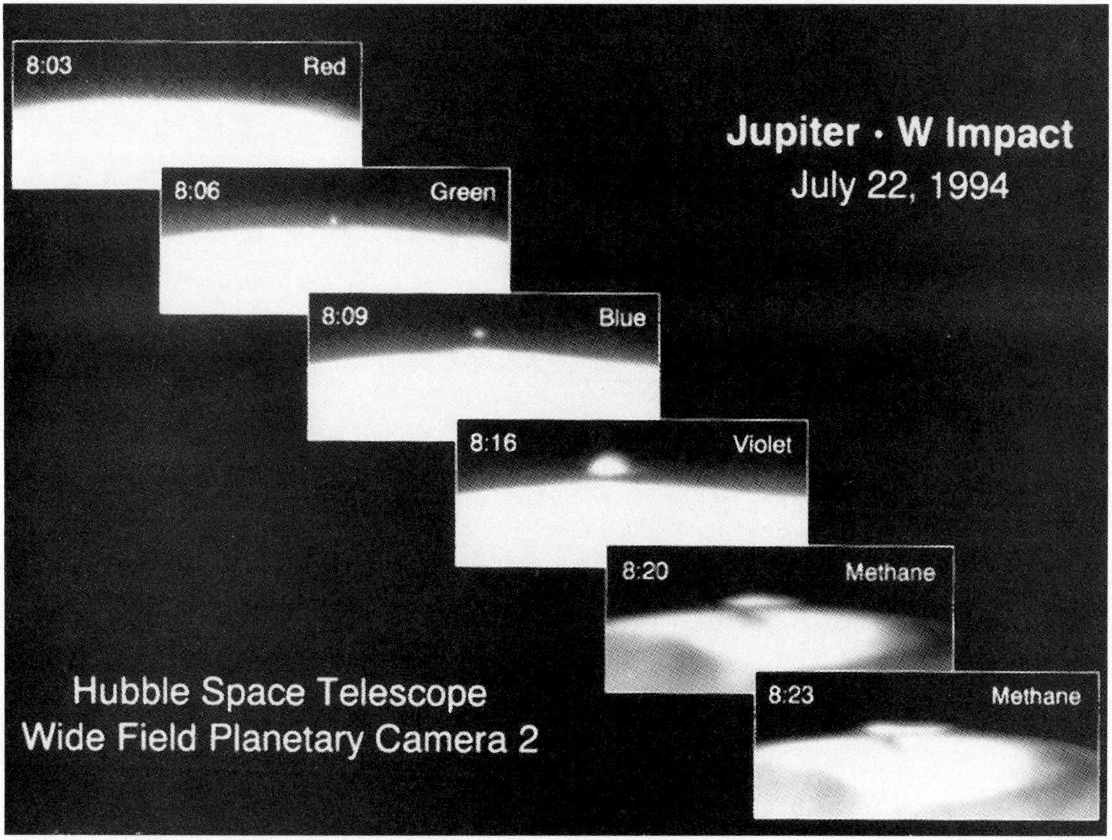

FIGURE 6.13(a) Successive views of a single fragment (W) of S-L 9 as it struck Jupiter, obtained with the Hubble space telescope.

seen protruding over the horizon (Fig. 6.13). The material from the fireball spread out to distances of more than 10,000 km from the point of impact as it collapsed and heated the jovian upper atmosphere. As the rotation of Jupiter carried each impact site into view, astronomers saw a huge dark cloud of cometary material, which remained visible in the jovian atmosphere for weeks (Fig. 6.14). In retrospect, astronomers realized that the impact geometry was nearly ideal, permitting them to view the erupting fireballs in profile against dark space.

In addition to the information revealed about the comet and the jovian atmosphere, these events raised an obvious concern about similar impacts on Earth. Many people asked themselves, What if the same thing happened to our planet? We will return to this interesting question in Chapter 9.

6.4 COMET DUST

The dirty-snowball model for the comet nucleus predicts that comets will release large quantities

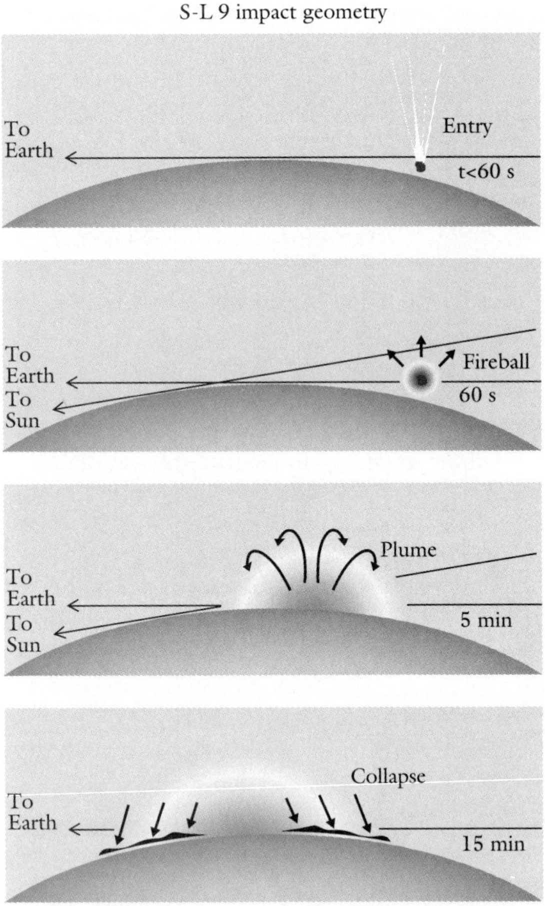

◀ **FIGURE 6.13(b)** A diagram indicating what was happening in Jupiter's atmosphere during the events recorded in (a).

FIGURE 6.14 A Hubble space telescope image of Jupiter showing the effects of the impacts of two of the fragments of S-L 9 on Jupiter's atmosphere. The large circular feature with a dark spot inside it and a crescent-shaped feature to the right were produced by fragment "G." The image was recorded less than two hours after impact. The dark spot to the left was produced a day earlier by fragment "D." ▼

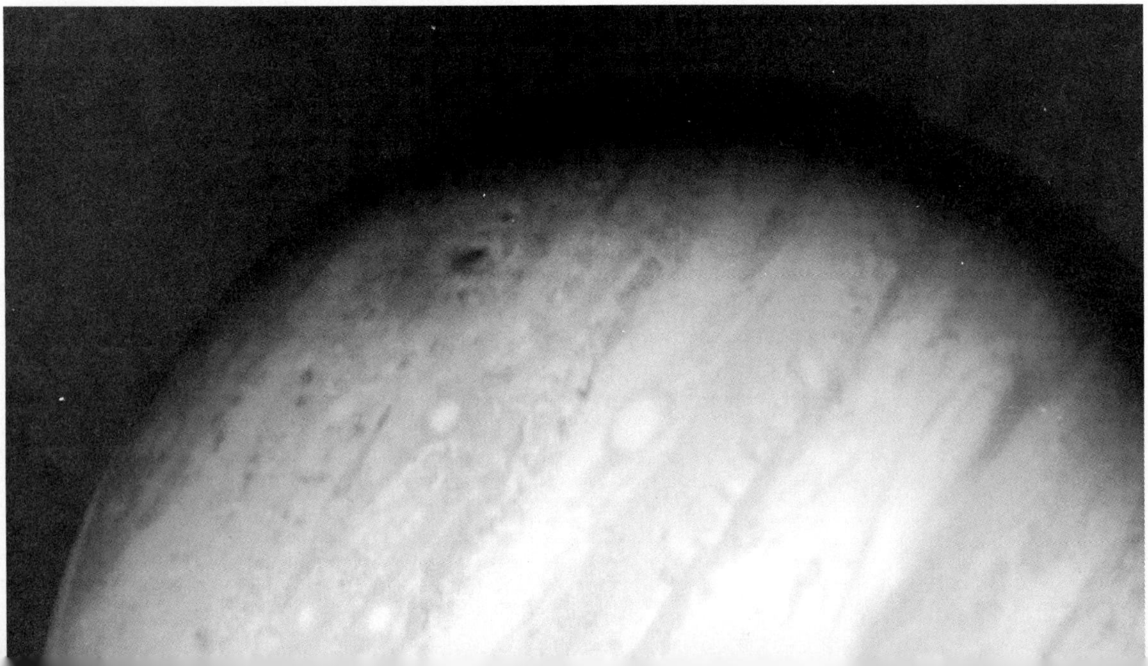

of dust or larger solid material into the inner solar system. The production of solid material is further supported by the curving dust tails of comets. What is the ultimate fate of this dust?

Meteors and Meteor Showers

Comet dust fills the inner solar system. Most of it either falls into the Sun or is swept outward by the solar wind, but a tiny fraction of it strikes the Earth, burning up in the atmosphere to produce the meteors.

Meteors or shooting stars can be seen on any clear night, usually at a rate of several per hour. The typical meteor is no larger than a pea, but this is sufficient to generate a much larger cloud of glowing gas high in our planet's atmosphere which can be seen as far away as 200 km. Perhaps as many as 25 million meteors bright enough to be seen strike the Earth every day, amounting to hundreds of tons of cosmic material added to our planet's atmosphere every 24 hours.

Many meteors are produced by bits of material in random orbits. These can come from any direction at any time, and they are called sporadic meteors. Sometimes, however, the Earth encounters a stream of particles moving together along similar orbits around the Sun, and we see a **meteor shower.** It is the shower meteors that are most directly linked to the comets.

To illustrate the connection between cometary dust and meteors more specifically, we can look to one of the famous comets of the last century called Comet Biela, discovered in 1826 on a 6.8-year orbit. In 1846 this comet split in two, and upon its next return in 1852 both components were again present, separated by 2 million kilometers. Neither part of Comet Biela was ever seen again; between 1852 and 1866, it simply ceased to exist. Nevertheless, astronomers watched with interest when the Earth passed through the orbit of Comet Biela in 1872, and they were not disappointed. Instead of the comet they saw a wonderful meteor shower, with thousands of meteors visible from any spot on Earth during the night the Earth crossed the comet's orbit. The comet had transformed itself into a stream of meteoric particles. In 1983 the IRAS satellite succeeded in imaging some of these meteor streams in their thermal infrared emissions (Fig. 6.15).

The most dependable annual meteor showers are listed in Table 6.4. In addition, however, an unpredicted shower can occur any time the Earth encounters a clump of meteoritic dust in space.

TABLE 6.4 Meteor showers

Name	Date of Maximum	Associated Comet	Comet's Period (yr)
Quadrantid	Jan. 3	unknown	
Lyrid	April 21	Thatcher	415
Eta Aquarid	May 4	Halley	76
Delta Aquarid	July 30	unknown	
Perseid	Aug. 11	Swift-Tuttle	133
Draconid	Oct. 9	Giacobini-Zinner	7
Orionid	Oct. 20	Halley	76
Taurid	Oct. 31	Encke	3
Andromedid	Nov. 14	Biela	7
Leonid	Nov. 16	Tempel-Tuttle	33
Geminid	Dec. 13	Phaethon	1.4

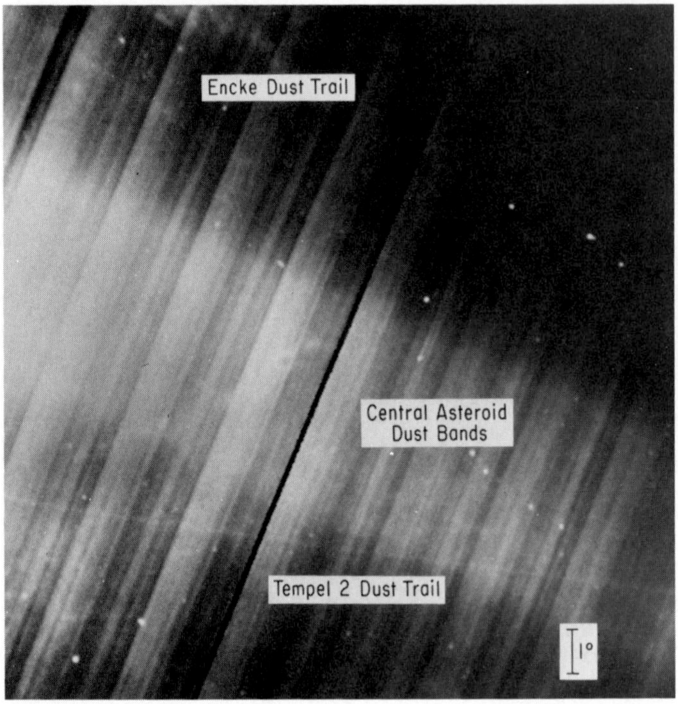

FIGURE 6.15 The IRAS image in which comet dust trails were first detected directly. The thick, bright band cutting diagonally across the image consists of debris from asteroid collisions, scattered throughout the main asteroid belt. This band is centered on the ecliptic. The dust trail from Comet Encke is seen to the north of the ecliptic. The Tempel 2 dust trail crosses the ecliptic from below, at a distance of 1.3 AU. Both comets are located several degrees off the left of the image.

The Nature of Meteoric Material

Most of the meteors that strike our atmosphere are associated with showers, and therefore with meteor streams in space. Many of the nonshower, or sporadic, meteors are probably also the remnants of dispersed meteor streams. And most known meteor streams in turn are associated with comets.

If most *meteors* are really cometary dust, we naturally ask if *meteorites* might not also be from comets. However, meteorites are not associated with meteor showers, and even on the rare occasions when the sky is filled with "falling stars" no actual meteorites fall to Earth. The shower meteors are not simply small meteorites that do not reach the surface. They are fundamentally different material.

Something of the nature of the cometary meteors can be inferred from observations made during their flights through the atmosphere. Their densities, calculated from the rate at which they slow due to atmospheric friction, are less than 1 g/cm^3, in contrast to the densities of meteorites, which are between 3 and 7 g/cm^3.

Efforts have also been made, as noted in Chapter 4, to collect tiny meteoric particles from the upper atmosphere. These fragments of cosmic dust are strange, fluffy bits of chemically primitive matter. With such low densities and fragile construction, it is no surprise that none of the cometary particles survives its fiery plunge through the atmosphere. Still, just because most meteors are made of stuff different from that of the meteorites, we cannot be sure that some meteorites might not also be of cometary origin.

The Zodiacal Dust Cloud

Cometary dust also contributes to another phenomenon of the planetary system called the **zodiacal dust cloud.** In reflected sunlight, this cloud appears as a faint, diffuse band of light stretching around the ecliptic, called the zodiacal light. You can see the zodiacal light just after twilight, if you have a clear, dark horizon well away from city lights. In the northern hemisphere, the visibility of the zodiacal light is best in March after sunset and in September before sunrise, since the ecliptic is nearly perpendicular to the horizon at these times. In the thermal infrared part of the spectrum, the glow of the zodiacal cloud is one of the brightest features of the entire sky. Both the reflected sunlight and the thermal glow originate in small dust particles that fill the inner solar system, near the plane in which the planets orbit the Sun.

Maintenance of the zodiacal dust cloud requires the release of about 10 tons of dust per second into the inner solar system. Much of this dust comes from comets, but part may also originate in the asteroid belt. In 1983 the IRAS infrared observatory discovered several distinct dust bands within the asteroid belt (Fig. 6.15). As with the meteorites, we find that we cannot make a definitive choice between comets and asteroids as the source of the dust in the solar system.

6.5 ORIGIN AND EVOLUTION OF COMETS

The comets we see are temporary residents of the inner solar system. Many are traveling on nearly parabolic orbits and will not return for millions of years. Even the **short-period comets** have limited lifetimes. After a few thousand orbits—less than 10,000 years for many of them—their volatile ices will be exhausted and they will no longer be comets, although the dead nucleus might remain as an Earth-approaching asteroid. They are also dynamically unstable, however. Like all objects with planet-crossing orbits, they run the risk of either impacting a planet or being gravitationally expelled from the solar system. Their average residency in the inner solar system is no more than a few million years. Therefore, it is important to determine the source of the comets.

We would not have to look for a continuous source of new comets if the comets were young, having all formed in some catastrophe that took place within the last few million years. Although such a hypothesis cannot be absolutely disproved, there is no real evidence to support it, and it contradicts our understanding of other solar system phenomena such as asteroids and meteorites. Much more satisfactory is an explanation that does not require a unique event, but rather can be understood in terms of a general model for the origin and evolution of the solar system.

The Oort Comet Cloud

The first satisfactory theory for the origin of the comets was proposed by the Dutch astronomer Jan Oort in 1950, the same year that Whipple published his model for the nucleus. Oort noted that in all cases where the orbits of long-period, nearly parabolic comets had been carefully determined, the orbits indicated an aphelion at a distance of about 50,000 AU, a thousand times more distant than Pluto. Very few comets seemed to come from greater distances, and none showed evidence of originating outside the solar system in interstellar space. He therefore suggested the existence of a comet cloud associated with the Sun but very far beyond the known planets. Since the orbits of new comets are not confined to the ecliptic but show a full range of inclinations, Oort concluded that this cloud was roughly spherical. This distant reservoir of comet nuclei is now called the **Oort comet cloud.**

At any given time virtually all the comets in the Oort cloud are too far away for detection, and their ices will be preserved indefinitely in the cold of space. Even if the orbits of these comets bring them in as close as Neptune, they remain frozen and are unlikely to be discovered. However, occasionally some of these comets can be perturbed by the slight gravitational tugs of nearby stars to bring their perihelia into the inner solar system. Only then will they develop the coma and tail we associate with comets.

In order to account for the several new comets discovered each year, Oort calculated that the comet cloud contained about 100 billion (10^{11}) comets. More recent calculations have suggested that the true number is ten times greater. If there are a trillion comets in the Oort cloud, and ten of these are lost each year by falling too near the Sun, the total number of comets lost to the cloud is still only about 5% of the original population over the history of the solar system.

FIGURE 6.16 Gerard P. Kuiper was the leading planetary astronomer of the midtwentieth century. A pioneer in infrared astronomy and the founder of several observatories, Kuiper was influential in planning the early NASA space missions.

The Kuiper Belt

The idea of a large, spherical Oort cloud accounts for most of the new comets, which can descend upon us from any direction at any time. However, most short-period comets share the general sense of revolution of the planets and do not have particularly large orbital inclinations. Detailed calculations show that most of these must originate in a region that is shaped like a flattened disk and not like a large sphere centered on the Sun. This disk-shaped source region is called the **Kuiper Belt** and is thought to extend outward from about the orbit of Pluto. It is named for another Dutch astronomer, Gerard P. Kuiper, who immigrated to the United States in the 1930s and became the country's leading planetary astronomer (Fig. 6.16).

The first direct evidence for the existence of the Kuiper Belt was the discovery in 1992 by David Jewett and Jane Luu of an exceedingly faint object (designated 1992QB) orbiting beyond Pluto with a period of 296 years (Fig. 5.12). Its inclination is only 2°. About two dozen such objects (all near 100 km in diameter) were discovered in the next three years, and in 1995 Anita Cochran and her colleagues used the Hubble space telescope to find evidence for many more smaller objects in similar orbits.

Thus we have identified two distant sources for the comets. In addition to the outer, spherical halo of icy objects identified by Oort, there is also a smaller, flattened disk of cometary objects. Presumably additional clouds of icy objects exist between the observed Oort cloud and Kuiper Belt, and these various regions may contain as much or more mass than all of the planets combined.

The Fate of Comets

Although a comet may rest peacefully in deep space for billions of years, once it is diverted into the inner solar system its life expectancy is limited. There is some chance that it will not survive even its first plunge toward the Sun; several comets have been seen to impact the Sun or to pass so near its surface that they are destroyed by the heat in a single pass (Fig. 6.17). Other comets die even before reaching the Sun; in 1906 and 1913, for example, new comets that had been expected to put on a good show simply faded to invisibility as they neared perihelion.

FIGURE 6.17 The collision of Comet 1979XI with the Sun on Aug. 30, 1979, photographed with an orbiting coronagraph telescope. The white disk shows the size of the Sun, which is actually hidden behind a disk in the coronagraph to mask the Sun's glare. The comet approaches the Sun from the lower right in the sequence of frames (a). The bright spot in the upper left is the planet Venus. The series of frames (b) shows the dissipation of material from the comet after impact with the Sun.

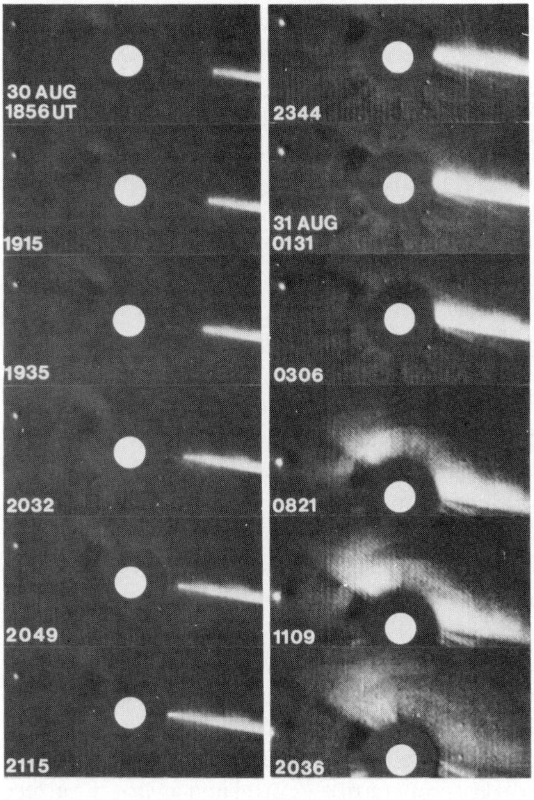

Comets may also break apart, for reasons not always understood. Shoemaker-Levy 9 was torn apart by the gravity of Jupiter. But what happened to Comet West, which apparently split three times within a few days early in March 1976? The four components drifted apart, extending more than 10,000 km by the end of the month. The smallest fragment survived only a few days, but the other three components of the nucleus still retained their individual identities as the activity level declined with increasing distance from the Sun. We do not know why this breakup occurred.

More likely than hitting the Sun or breaking apart, a new comet may pass near enough to some planet—usually Jupiter—to suffer a change in orbit. This change can either increase its energy, ejecting it completely from the solar system, or cause it to lose energy, making it a short-period comet. The reason most short-period comets have aphelia near the orbit of Jupiter is that they were captured by the gravitation of this giant planet. But the comet has now gone out of the frying pan and into the fire, for it still runs a significant risk of impact with one of the terrestrial planets. Meanwhile the heat of the Sun is also consuming its icy substance at a rapid rate.

There is considerable debate among astronomers concerning the aging of comets. As a comet suffers repeated heating and loss of gas and dust each time it travels around the Sun, does it change or simply grow smaller? If it is of uniform, homogeneous composition, and if the dust is stripped away together with the evaporating ices, then we can expect the nucleus to shrink without otherwise changing much over time. Alternatively, if comets have a layered structure, the successive stripping away of the surface (typically about a meter each orbit) will expose differing materials as the comet ages. To date there is no evidence for such inhomogeneities, since the spectra of old comets are not different from those of young ones. A third possibility is that the nucleus is homogeneous, but the nonvolatile debris that builds up on the surface ultimately insulates the interior under a thick crust of dark material. So far the jury is out on this question.

Origin of the Comet Cloud

The idea of the Oort cloud and the Kuiper Belt provides a framework for understanding the continuing supply of comets to the inner solar system, but it begs the question of the ultimate origin of the comets. How did the comets get into the Oort cloud or Kuiper Belt in the first place? Might the comets have been formed in place, tens of thousands of AU from the Sun? We have already mentioned this possibility, but at such distances in the solar nebula it seems difficult to imagine solids condensing, except possibly when the solar system was passing through a dense molecular cloud. Most astronomers believe instead that the comets were formed in the realm of the planets, at the same time as the other members of the planetary system.

In this scenario, the comets represent bodies that condensed in cooler parts of the solar nebula. The presence of a variety of ices in comets suggests formation temperatures in the range of 30–100 K, corresponding to the region of space now occupied by Uranus and Neptune as well as the inner edge of the Kuiper Belt. Perhaps all the Kuiper Belt comets formed in approximately the locations where we find them today. However, many of the protocomets must have been ejected into the Oort cloud from regions where they interacted gravitationally with the giant planets, while others were gravitationally dispersed throughout the solar system or expelled from the system entirely.

The comets are representative of the original condensates from the outer part of the solar nebula, just as the primitive asteroids are remnants of the materials that condensed in the inner part of the nebula. The question of whether comets can be separated into different compositional classes like the asteroids is unanswered. There may be substantial variations among comets—for example, between those derived from the Oort cloud and the Kuiper Belt—but current observations are inadequate to demonstrate such differences.

6.6 QUANTITATIVE SUPPLEMENT: ALBEDOS AND TEMPERATURES

In this text we use the simple term *reflectivity* to indicate the fraction of incident sunlight that is reflected from a planetary surface. Generally, what we mean is the *normal reflectivity*—that is, the fraction of the light reflected straight back from a beam incident normal (perpendicular) to the surface. But there are several other ways in which reflectivity can be defined, depending on the application.

The normal reflectivity is a property of a particular area of the surface. Unless we have close-up photos from a spacecraft, however, we may not be able to distinguish individual surface elements. Suppose, instead, that we are interested in the global properties of a planet or asteroid. We define the *albedo* (more properly, the *Bond albedo*) as the ratio of total reflected light to total incident light. The Bond albedo, which is generally numerically smaller than the normal reflectivity of the surface, is a measure of the global reflection and absorption of sunlight.

Since the Bond albedo (called A) tells us how much solar energy is absorbed, we can use it to calculate an effective temperature for a planet. Remember that the Stefan-Boltzmann law expresses the radiation emitted by a perfect radiator at temperature T. A planet is not a perfect radiator, nor is its surface all at the same temperature, but we can define effective temperature as the value of T that satisfies the energy balance when the absorbed sunlight is equated to the emitted thermal power:

$$(1 - A)\,(P_{\mathrm{Earth}}/D^2)\,(\pi R^2) = \sigma T^4 \times 4\pi R^2$$

Here P_{Earth} is the sunlight per unit area at the distance of the Earth from the Sun (the solar constant), D is the distance of the planet from the Sun (in AU), and R is the radius of the planet. Note that the area (πR^2) on the left side is the cross section of the planet, while that on the right side ($4\pi R^2$) is the total area of a sphere; we assume that the planet radiates from its entire surface, not just the part facing the Sun. Simplifying this equation, we have

$$T^4 = (1 - A)\,(P_{\mathrm{Earth}}/4\sigma D^2)$$

Another kind of albedo is defined to describe the apparent brightness of an object. This is the *geometric albedo* (called p), and it is the ratio of the global brightness, viewed from the direction of the Sun, to that of a hypothetical, white, diffusely reflecting sphere of the same size at the same distance. Numerically, the geometric albedo is approximately the same as the normal reflectivity. Note that some objects can be more reflective than the hypothetical white reference sphere, so that geometric albedos greater than 1.0 are possible; the geometric albedo of Saturn's satellite Enceladus is about 1.1. The Bond albedo can never be greater than 1.0, however.

Finally, there is a relationship between the two albedos for any object, called the *phase integral*, q. The phase integral, which depends on the scattering properties of the surface, is defined by the relation

$$q = A/p$$

For the Moon and many other objects in the solar system, the value of the phase integral has been measured to be near ⅔; in other words, for typical surfaces of planets, the Bond albedo is equal to about two-thirds of the geometric albedo.

SUMMARY

The comets are the most primitive members of the planetary system. Composed in large part of water ice and other volatiles, they were formed at temperatures below 100 K and have spent most of the past 4 billion years in the deep freeze beyond Pluto. Only when they are diverted into the inner solar system do we see these small icy bodies and identify them as comets.

As it is heated by sunlight, a comet develops a thin but extensive atmosphere. Spectral analysis shows that the cometary atmosphere contains a complex mixture of water, carbon monoxide, carbon dioxide, and methanol, as well as many complex hydrocarbons. Once ionized, this atmosphere streams away from the Sun to form the comet's tail, which can extend for millions of kilometers. Copious dust is also released, consisting of both silicate and carbonaceous particles.

The comets can best be understood in terms of Whipple's dirty-snowball model, in which the small solid nucleus is composed of approximately equal quantities of ice (primarily water ice) and silicate and carbonaceous material (largely in the form of small dust grains). Rapid evaporation of the ice under the influence of solar heating leads to the formation of the atmosphere. Preferential evaporation from the Sun-facing hemisphere of the nucleus also gives rise to jets, which produce the nongravitational forces that alter the orbits of many comets. The gas released from the nucleus dissipates into space, while the dust contributes to meteor showers when the Earth intersects the orbit of a comet.

Comets are on unstable orbits and must originate elsewhere. Two source regions have been identified: the Oort cloud, a roughly spherical halo of icy bodies that extends to about 50,000 AU from the Sun; and an inner, flattened Kuiper Belt beyond the orbit of Pluto. The bodies in these regions (as many as a trillion of them) are thought to be leftover building blocks from the outer solar system, perhaps the same materials that made up the cores of the giant planets as well as smaller bodies such as Pluto and the outer-planet satellites.

KEY TERMS

coma	meteor shower
comet nucleus	Oort comet cloud
dust tail	plasma tail
Kuiper belt	short-period comet
long-period comet	zodiacal dust cloud

Review Questions

1. Distinguish among comets, main belt asteroids, and near-Earth asteroids, in terms of their orbits and their physical and chemical properties. What relationships exist among these three classes of small bodies?

2. Describe the parts of a comet and their appearance from the Earth. Can you think of any reason why comets have traditionally been associated with disasters, such as wars and plagues?

3. Describe the dirty-snowball model for the nucleus of a comet. Compare this model with the observations. Explain in particular how it accounts for the nongravitational forces that disturb the orbits of comets.

4. Imagine yourself riding on a comet. Describe what you would see as the comet approached the Sun. Do you think that an instrumented lander on a comet nucleus would survive for long? What kind of measurements might such a lander make?

5. Describe how the atmosphere and tail of a comet are formed. How do the phenomena we see in the head and tail relate to the properties of the underlying nucleus?

6. What are the relationships between comets and meteors? If the shower meteors come from comets, isn't it likely that the meteorites come from comets too? Explain.

7. Describe the Oort comet cloud. While a comet is in this cloud, how might it change over time? What is the orbit of a comet in the cloud like? Under what circumstances will a comet in the Oort cloud be able to enter the inner solar system?

8. What is the difference between the Oort cloud and the Kuiper belt? Which one supplies the short-period comets? Do you think the comets in these two reservoirs have different compositions? Explain.

Quantitative Exercises

1. What is the eccentricity of a comet with perihelion at the Earth's orbit (1 AU) and aphelion at Jupiter's orbit (5 AU)? What are its semimajor axis and period of revolution?

2. If the comet in exercise 1 has a Bond albedo of 0.03, what is its effective temperature at perihelion and aphelion? (For comparison, the effective temperature of a black sphere [$A = 0$] at 1 AU is 270 K.)

3. Saturn's satellite Enceladus has a geometric albedo of 1.1 and a Bond albedo of 0.8. Find its phase integral and effective temperature.

4. The equilibrium temperature of a black surface facing the Sun at 2 AU from the Sun is 270 K. What is the effective temperature for a spherical object, also black ($A = 0$), at 2 AU? What is its effective temperature at 2 AU if its Bond albedo is 0.5? What are the effective temperatures of both objects at 1 AU?

Additional Reading

Brandt, J.C., and R.D. Chapman. 1992. *Rendezvous in Space: The Science of Comets.* New York: Freeman. A semitechnical work by two comet experts, this is one of the few books that includes analysis of the observations of Comet Halley in 1986, including the spacecraft missions to the comet.

*Grewing, M., F. Praderie, and R. Reinhard, eds. 1987. *Exploration of Halley's Comet.* Berlin: Springer-Verlag. Final technical reports by dozens of scientists on results from the spacecraft missions to Comet Halley.

*Nature: encounters with Comet Halley, 1986. *Nature* 321: 6067, Supplement. Preliminary technical reports on the Giotto and VEGA missions to Comet Halley.

Sagan C., and A. Druyan. 1985. *Comet.* New York: Simon and Schuster. Beautifully written popular summary of our knowledge of comets on the eve of the 1986 apparition of Comet Halley, with discussions of cultural and historical comet lore.

Whipple, F.L. 1985. *The Mystery of Comets.* Washington: Smithsonian Institution Press. Popular pre-Halley summary of our knowledge of comets by the leading cometary scientist of our time.

Yeomans, D.K. 1991. *Comets: A Chronological History of Observation, Science, Myth, and Folklore.* New York: John Wiley and Sons. Fascinating historical information on bright comets through history and changing cultural attitudes toward these space visitors.

*Indicates the more technical readings.

7

The Moon: Our Ancient Neighbor

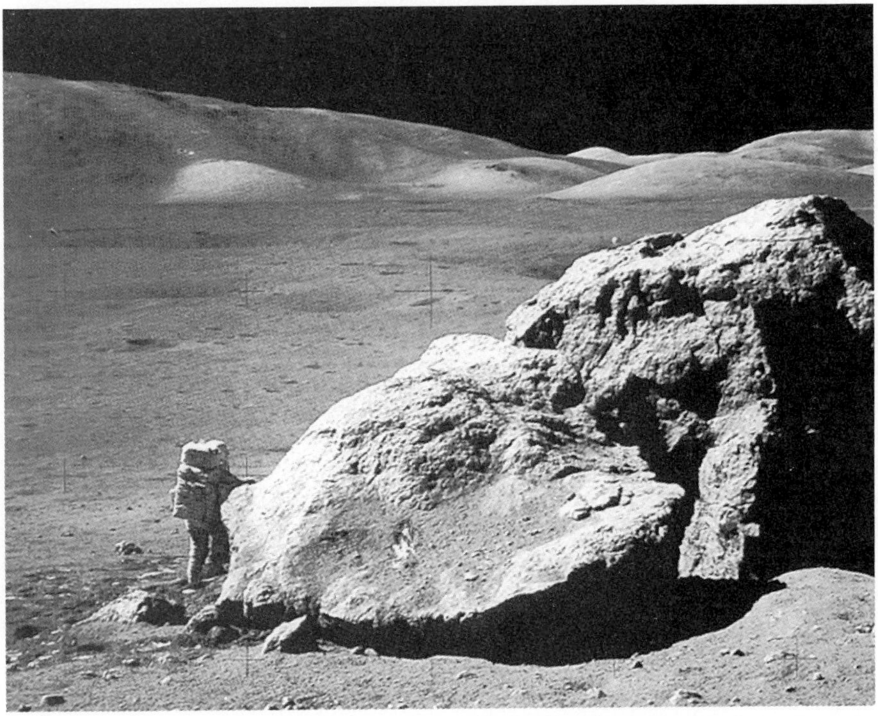

Astronaut Harrison "Jack" Schmitt examining a huge rock on the surface of the Moon. The gray lunar landscape stands out in sharp relief against the black, airless sky.

The objects in our planetary system are remarkably varied. As a general rule among planets, larger size leads to geological complexity, since internal heat can be retained by larger objects to power an active surface geology. It is therefore reasonable to expect that smaller planets will be less active, and hence probably simpler to understand, than larger ones. This is why we began in the previous two chapters with the smallest bodies, and it is why we now take up the study of the Moon before we tackle the larger and more complex Earth.

A relatively simple geologic history is not the only reason for beginning a detailed study of individual planets with the Moon. The lack of a lunar atmosphere also eases our task considerably. There are no obscuring clouds or vapor, no wind or precipitation to erode the surface, and no oxygen or water to weather the rocks or alter their chemistry. Change on the Moon is a slow process; a million years from now, the footprints of Apollo astronauts in the lunar soil will still be fresh. Unlike the Earth, where the surface geology is often hidden under vegetation, and rock and soil chemistry are altered by local conditions, on the Moon what you see is what you get.

The Moon also provides an appropriate introduction to the other planets and satellites because of the dominant influence of impacts in shaping its surface, producing mountains and abundant craters. Such impacts provide a common thread weaving together the histories of different worlds. By first studying impacts on objects like the Moon where such processes are dominant, we will gain the understanding needed to interpret more complex planets where both impacts and internally driven processes are important.

7.1 THE FACE OF THE MOON

Although large in comparison to the comets and asteroids discussed in Chapters 5 and 6, the Moon is a small planet. With a diameter of 3476 km, its total surface area is about the same as that of North America. A flight from Los Angeles to New York traverses a greater distance than the lunar diameter. The mass of the Moon is less than 2% that of the Earth, although still substantially greater than that of all of the asteroids combined.

Chemically the Moon differs from both the Earth and the primitive objects such as comets and C-type asteroids. The Moon is not primitive; it has been heated and differentiated, and it lacks water, ice, carbon, and organic compounds. From its relatively low density of 3.3 g/cm^3, we can be sure that the Moon also lacks a metallic core like that of the Earth. The Moon has its own life history, one very different from that of our planet.

Resolution of the Surface

The Moon is the only planetary body that can be distinguished with the naked eye as a globe, and even without a telescope we can see that its surface is not uniform. Many cultures have associated names and myths with the markings familiarly known as the "Man in the Moon," but it was not until Galileo Galilei turned his first small telescopes on the Moon in 1610 that it became clear that the surface of our satellite was rugged and mountainous like that of the Earth. Galileo's observations provided the foundation for considering the planets as other worlds and thus indirectly gave birth to the science of planetary geology.

A sharp-eyed observer can see features on the Moon as small as ⅟15 of its apparent diameter, or approximately 200 km across. This is the **resolution** of the human eye, the dimensions of the smallest features of a planetary surface that can be seen or imaged. At 200 km resolution, the larger light and dark markings can be distinguished, but topographic features, such as mountains and craters, are undetectable (Fig. 7.1). You can demonstrate this conclusion to your own satisfaction by sketching the Moon at different phases and using these sketches to produce a naked-eye map. This exercise is especially instructive when you note that the best Earth-based *telescopic* views of other planets, such as Venus and Mars, have about the same resolution as *naked-eye* maps of the Moon.

A telescope, with its higher resolution, is able to reveal lunar surface topography. Galileo's early telescope was barely sufficient to distinguish the largest craters and mountains. You can do as well today with a good pair of 7 power or 8 power binoculars, which actually perform better than Galileo's best 30 power telescope. Today's astronomical telescopes are far more powerful. As we noted in Section 3.5, telescopes on the surface of the Earth are limited in resolution by the atmosphere. For the Moon, this limiting resolution is about 1 km.

The angle at which sunlight illuminates a surface plays an important part in determining our ability to distinguish topographic detail. When the Sun is low, features cast shadows, while at moderate angles of illumination slopes and contours are revealed by shadings that vary as the surface is tilted toward or away from the Sun. When the sunlight streams down from directly

FIGURE 7.1 Seen at full phase, the Moon shows little topographic detail even through a telescope. The dominant features visible under these lighting conditions are the dark volcanic maria and the bright rays associated with young impact craters.

FIGURE 7.2 The nature of the lunar surface is better revealed at quarter phase, when the sunlight strikes obliquely, highlighting topographic features such as craters and mountains. ▶

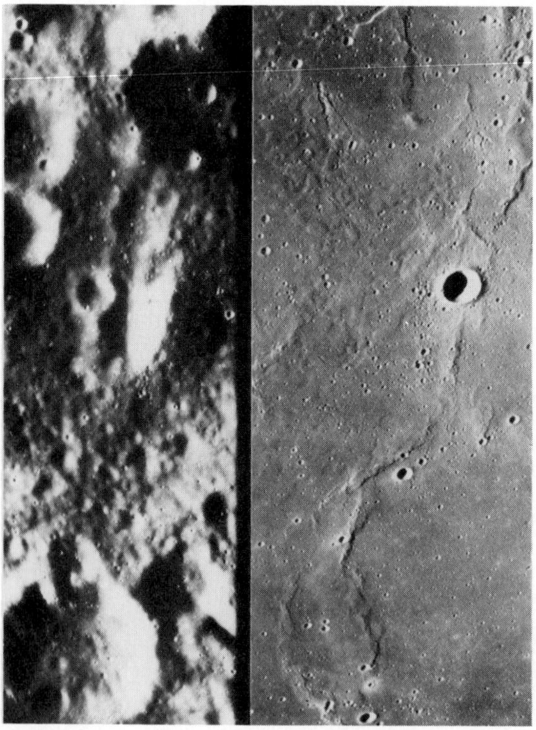

FIGURE 7.3 The contrasting nature of the lunar highlands and maria is clearly shown in these Apollo orbital photos of the two types of terrain. The crater density is 10–20 times greater on the highlands (left) than on the mare (right). ▼

behind the observer, details of topography become indistinguishable. Thus an image of the full moon emphasizes differences in reflectivity between light and dark areas but suppresses topography. Craters and mountains are best studied near first and last quarter, where we see the border between night and day. Here the early morning or late afternoon Sun is low in the lunar sky, and every topographic detail is sharply etched (Fig. 7.2).

Lunar Highlands and Maria

When we look at the Moon through binoculars or a small telescope, we are immediately aware of two different kinds of surface terrains. The predominant surface type is relatively light (reflectivity about 15%) and extremely rugged, with craters of all sizes piled one upon the other. Since these lighter, heavily cratered regions also generally lie at higher elevations, they are called the lunar **highlands.** The second surface type is darker (reflectivity about 8%) and smoother, with relatively few large craters. These regions, which make up the features of the Man in the Moon, are called **maria** (singular: **mare**). *Mare* is the Latin word for sea, and when the term was first applied to the Moon in the seventeenth century these darker regions were thought to be water oceans. Figure 7.3 contrasts the appearance of these two terrains.

Most of the maria are found on the side of the Moon that faces the Earth. The opposite, or far side, had never been seen until 1959 when the Soviet Luna 3 radioed back to Earth the first rudimentary photos (Fig. 7.4). To the surprise of almost everyone, no major maria appeared on the lunar farside, which was almost all highland terrain. Any explanation for the existence of highlands and maria must address this basic asymmetry in mare distribution. Subsequent mapping of the entire Moon revealed that only 17% of the surface area consists of mare material.

Maria and highlands differ in many ways. Most obvious is the distinction in color and reflectivity, implying that their chemical compositions differ. Second, a difference in cratering suggests different geologic histories. Finally, there is a difference in elevation, related to the largest scale forces that have molded the lunar surface. The maria and highlands are representative of different chemical and geological regimes, providing windows into two different periods of solar system history.

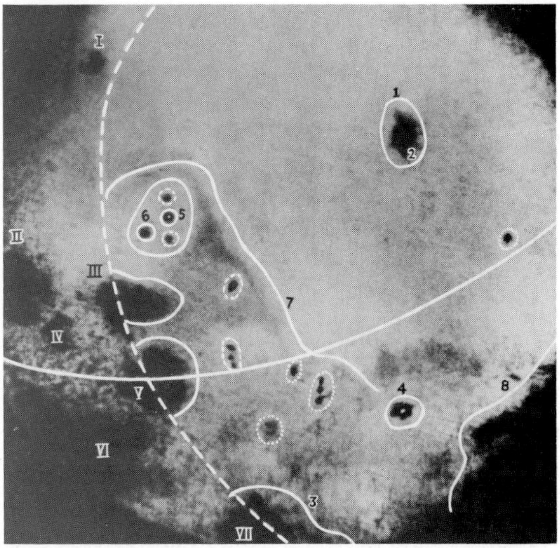

FIGURE 7.4 The first view of the side of the Moon that is always invisible from the Earth was obtained by the USSR's Luna 3 spacecraft in 1959. The continuous line across the chart is the lunar equator; the broken line indicates the border between the hemispheres visible and invisible from Earth. The feature marked II is Mare Crisium, which we can see, while 4 is the crater Tsiolkovsky, which we cannot. Note the absence of maria on the Moon's farside.

Lunar Craters

The prevalence of the characteristic circular features called **craters** is one of the most striking features of the lunar surface. The word *crater*, derived from the Greek for cup or bowl, refers to the shape of the feature. Note that a crater is a *depression;* the popular application of the term to a volcanic mountain, such as Vesuvius in Italy or Haleakala in Hawaii, is incorrect. A volcano frequently has a crater at its summit, but the mountain itself is not a crater.

Even binoculars or a small telescope can reveal the larger lunar craters, which are hundreds of kilometers in diameter. At the best Earth-based resolution of 1 km, 30,000 craters can be identified on the hemisphere of the Moon that faces the Earth. These craters range from sharp, new-looking depressions of nearly perfect bowl shape to old, battered craters. In the highlands, the craters are packed shoulder to shoulder, and it seems obvious that any new additions must occur at the expense of those craters already present. In some highland areas, the surface is so battered that individual craters almost become lost in the accumulation of jumbled, mountainous debris. In contrast, the craters on the maria are rather widely spread, indicative of a younger surface on which there has been less time for craters to accumulate.

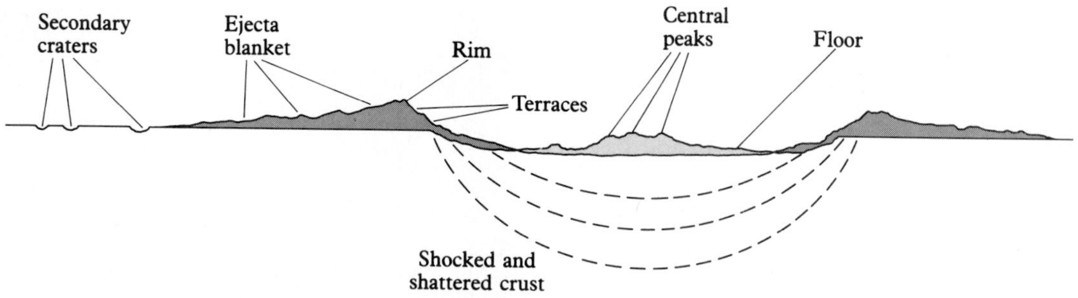

FIGURE 7.5 A typical large lunar impact crater is illustrated here in cross section.

Although the interiors of lunar craters are depressions, crater rims are raised above the surrounding surface. Figure 7.5 illustrates a cross section through a typical fresh lunar crater. For a mare crater several kilometers in diameter, the rim typically rises hundreds of meters above the surrounding plain, while the interior drops more than a kilometer below the mare surface. On the Moon, craters in this size range usually have a depth, measured from the rim crest down to the lowest part, that is between 1/10 and 1/5 of their diameter. Larger craters are shallower and more complex, as we will describe in Section 7.3.

At the largest scale, the face of the Moon is dominated by a few **impact basins,** which are roughly circular in outline. On the side of the Moon facing the Earth, these basins tend to be the sites of the darker mare materials. Examples readily seen with binoculars or on a photograph of the full moon are Mare Crisium, Mare Serenitatis, and Mare Imbrium (Fig. 7.1). Even more striking would be the bull's-eye feature Mare Orientale (Fig. 7.6), nearly a thousand kilometers in diameter, but Orientale is just barely visible from the Earth. If the Moon were rotated 90° so that this feature faced us, we can imagine the myths that might have developed to explain this "eye" staring down on us from space.

For a long time, the origin of the lunar craters and basins was a major issue of debate, as important an issue as that of the distinction between maria and highlands. Largely by analogy with the Earth, most geologists in the nineteenth and early twentieth centuries believed the lunar craters had a volcanic origin. Now we recognize that the great majority, including all the larger ones, were produced by impacts.

When the Moon nears full phase and the sunlight strikes it directly, a few craters brighten to become the most prominent features on the

FIGURE 7.6 The Orientale Basin is the youngest of the great lunar impact basins. Because it has escaped flooding by mare volcanism, Orientale Basin shows most clearly the double-ringed, or "bull's eye," form of large impact structures.

Moon. In contrast with the maria or even the lighter highlands, these craters and their surroundings appear brilliant white. Among the brightest of the light ray-craters is the highland crater Tycho, which is named for Tycho Brahe; like most lunar craters, it commemorates a scientist or ancient philosopher. Tycho is 84 km in diameter and appears to be among the youngest major features of the Moon, only 100 million years old.

Lunar Stratigraphy

The identification of bright craters like Tycho as younger than other lunar features is an example of **stratigraphy,** a geological technique for establishing the relative sequence of geologic events. Stratigraphy is based on the principle that younger surface features will generally lie on top of older ones. Thus a crater that overlaps the rim of another is determined to be younger, or a colored deposit that lies on another is concluded to have been laid down more recently. Although no *absolute* ages can be determined by this technique alone, careful stratigraphic mapping of the visible face of the Moon carried out in the 1960s established the main outlines of *relative* lunar ages.

The most important stratigraphic benchmark for the near side of the Moon is the formation of the Imbrium Basin. The great impact that excavated this 1200-km feature blanketed much of the Moon with a recognizable debris layer called the Fra Mauro formation. Features that lie below this layer are pre-Imbrium; those on top of it are post-Imbrium in age. After the Apollo missions brought back rock samples to be dated, we learned that the Imbrium event took place 3.8 billion years ago.

Examination of Mare Imbrium itself further illustrates the sequence of events that followed the formation of the basin. There are craters inside the basin that have been flooded by mare lavas. Obviously, these craters were produced by impacts that postdated the great Imbrium event itself. Equally clear from stratigraphic principles is the fact that the eruption of mare lavas took place after these craters formed. Therefore we conclude that the emplacement of the lavas followed the basin-forming impact by a considerable time, and that these lavas were not produced as a direct result of the impact. This is an example of the kind of lunar studies that were carried out using telescopic images with resolutions of a few kilometers in the years before the Apollo program. But the greatest increases in our knowledge have come as a result of the lunar expeditions and from subsequent laboratory studies of the returned lunar samples.

7.2 EXPEDITIONS TO THE MOON

On July 20, 1969, half a billion inhabitants of Earth watched on live TV as two Americans stepped onto the surface of the Moon and began the first human exploration of another world. By any standard, the Apollo Moon program was one of the supreme achievements of history. That we landed on the Moon at all was remarkable; that we did so in so short a time, and without losing a single astronaut in space, is little short of miraculous.

Apollo was the culmination of a decade of lunar research, designed with two goals, that of paving the way for a safe landing by humans and that of providing scientific information regarding the Moon. No one pretends that Apollo or its precursor missions were primarily scientific programs. Nevertheless, the Moon program of the 1960s and early 1970s yielded a level of knowledge about our satellite far surpassing what we have learned about any other world beyond the Earth. Even though the last Apollo mission was completed in 1972, scientific results continued to accrue from automatic instruments that operated on the lunar surface until 1978, and analysis of the priceless collection of samples brought back by the Apollo astronauts continues today.

Robot Exploration of the Moon

The USSR dominated initial efforts to explore the Moon by spacecraft. Luna 1, the first spacecraft to escape the Earth, passed the Moon on January 4, 1959, and Luna 2, which crash-landed in September 1959, was the first craft actually to reach the lunar surface. A month later, Luna 3 returned the first photographs of the far side of the Moon (Fig. 7.4). Finally, in February 1966, Luna 9 successfully landed and radioed back to Earth the first pictures and other scientific data from the lunar surface.

Meanwhile, the United States had begun a three-part effort at unmanned exploration of the Moon in anticipation of the Apollo landings. The first of these, aiming only at crash landings and oriented primarily toward acquisition of close-up photos during final approach to the Moon, was called Ranger. Between 1961 and 1965, nine Rangers were launched, but only the last three of this problem-plagued program were successful. Subsequently five successful Lunar Orbiter spacecraft mapped the surface of the Moon, with special emphasis on the identification of potential Apollo landing sites. Finally, the Surveyor program was designed to achieve the first controlled soft landings on another planet, testing the detailed physical and chemical properties of the lunar surface. Surveyor 1 touched down successfully on June 2, 1966, just two months after its Russian counterpart (Fig. 7.7). After four successful landings were made at potential Apollo landing sites, the final space-

FIGURE 7.7 Apollo 12 astronauts visited the Surveyor 3 spacecraft, which had helped pave the way for the first human landings on the Moon. The Apollo 12 landing module is visible on the horizon in this photo.

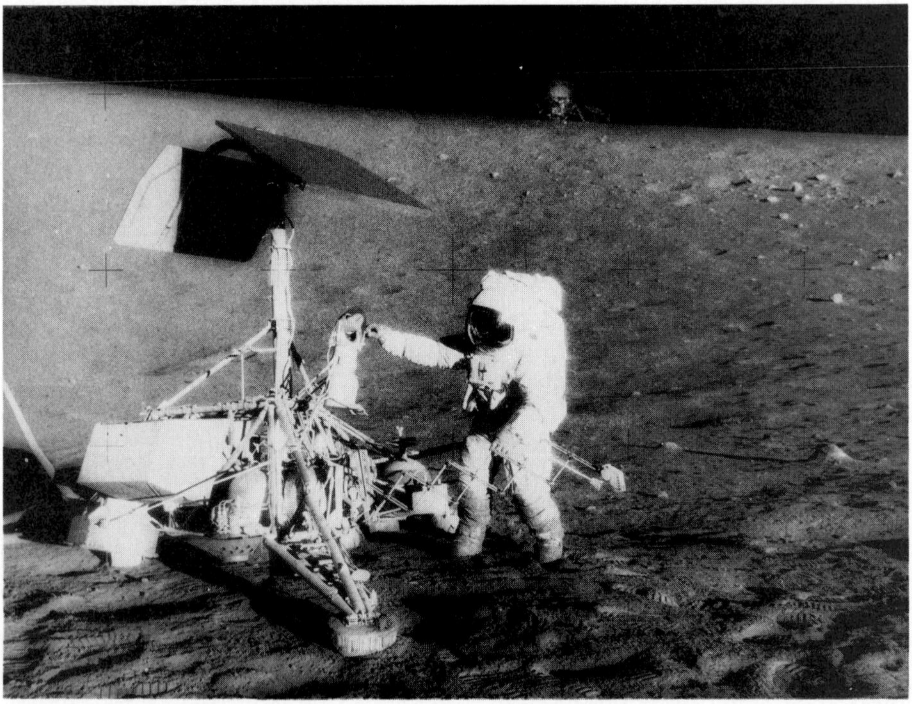

craft was sent to a much more exciting and less accessible highlands area near the prominent ray-crater Tycho. Tycho is therefore the only large, fresh lunar crater for which we have lander data.

The Apollo Program

The dramatic history of the Apollo program (Fig. 7.8) is too long and complex to tell here. Following a decade of development work, Apollo 11 astronaut Neil A. Armstrong took his historic first step on the Moon in July 1969. Eleven others followed him in six successful lunar landings, each more ambitious and productive than its predecessors. For a short period, it seemed that humanity had broken the bonds of Earth and truly begun a new space age. But at the peak of its success, Apollo lost its political appeal, and the program was abruptly terminated. The last human footprints were planted in the lunar soil by Harrison H. Schmitt, the only scientist-astronaut to reach the Moon, before he reentered the Apollo 17 spacecraft on December 14, 1972. The human exploration of the Moon had lasted only 30 months.

Several aspects of the Apollo exploration deserve mention in even a brief overview of the scientific results of the program. Perhaps most important is the role of the lunar samples returned to the Earth. Sample collection was a primary scientific objective of every Apollo landing. The first thing each astronaut did upon alighting on the surface was to scoop up a contingency sample of soil and seal it in a pocket of his spacesuit, thus assuring the return of some material even if the moonwalk had to be abandoned. Special tools were designed to pick up and store samples of different kinds. In the later missions, rocks were carefully documented: They were measured and photographed in place before being picked up, and kept in individual labeled bags for the return trip to Earth. Core samples more than 2 meters long were obtained by forcing specially designed hollow tubes into

FIGURE 7.8 Launch of one of the Apollo missions to the Moon.

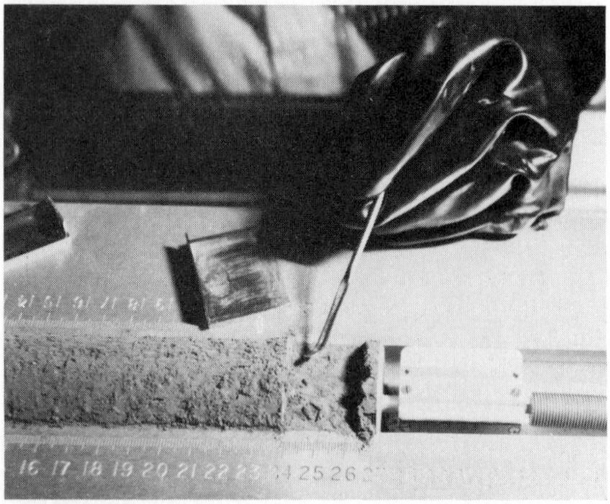

FIGURE 7.9 Among the samples of lunar material returned to Earth in the Apollo program were cores, obtained by drilling into the lunar soil. This Apollo core (lower left) is being inspected in the laboratory at NASA's Johnson Space Center, Houston.

the surface (Fig. 7.9). All samples were transported back to the Houston Lunar Receiving Laboratory in sealed containers, where they were inspected and catalogued under sterile conditions, exposed only to filtered dry nitrogen and thus protected from contamination or corrosion by oxygen or water vapor in the Earth's atmosphere. Most of the 382 kg of returned samples are still in Houston (Fig. 7.10) and are made available to scientists from all over the world.

On the lunar surface, each Apollo landing after the first left behind an automated surface laboratory called ALSEP (Apollo Lunar Surface Experiments Package; see Fig. 7.11). Powered by its own nuclear generator, the device contained instruments to measure the solar wind, analyze any trace of thin lunar atmosphere, and measure heat flow from the deep interior. Perhaps most important, each ALSEP included a seismometer to measure moonquakes. Because the lunar environment is much quieter than that of Earth, with no winds, waves, truck traffic, footsteps, and so forth, these seismometers were much more sensitive than their terrestrial counterparts. With the installation of the third ALSEP by Apollo 15, a network of stations became available that could identify the locations of moonquakes through triangulation. In the later missions, the spent Saturn upper-stage rockets and the lunar landing modules themselves were deliberately sent crashing into the Moon to generate artificial moonquakes to be tracked by the ALSEP seismic instruments. The ALSEPs continued to collect data until 1978,

FIGURE 7.10 NASA stores lunar samples (and some meteorites) at its Extraterrestrial Materials Curatorial Facility in Houston. The samples are studied in this laboratory, and small quantities are also made available to qualified scientists throughout the world.

FIGURE 7.11 Automated instruments called ALSEPs (Apollo Lunar Surface Experiment Packages) were deployed on the Moon by the crews of Apollos 12 through 17. Shown here is the Apollo 16 ALSEP, the only instrument package in the lunar highlands.

when they were turned off by NASA as a cost-savings measure.

A third important aspect of Apollo involved the scientific study of the Moon from the orbiting command modules. These spacecraft carried a variety of instruments, including the highly sophisticated mapping cameras flown on the last three missions. Film from these cameras could be returned directly to Earth, yielding much better images than had previously been transmitted by radio from the lunar orbiter robot craft. Orbital maps were also made of magnetic fields, chemical composition, and surface radioactivity as the command modules passed over the Moon. Unfortunately, since all the Apollos flew on nearly equatorial orbits, this intensive mapping did not extend to higher latitudes.

Apollo Flights to the Moon

Figure 7.12 shows the locations of the six Apollo landing sites. A brief summary follows of each of the Apollo missions that went to the Moon; other numbers in the series represented Earth-orbital tests.

Apollo 8: December 1968. First human circumlunar flight and lunar orbit. Excellent photography from hand-held cameras as the spacecraft orbited the Moon ten times.

Apollo 10: May 1969. First rendezvous in lunar orbit, in a dress rehearsal of the Apollo 11 landing. LM (landing module) maneuvered to within 14 km of the surface.

Apollo 11: July 1969. First lunar landing, in Mare Tranquillitatus (Tranquillity Base), a site chosen for its smoothness and freedom from hazards. Astronauts spent 23 hours on the Moon. First lunar samples (22 kg) returned.

Apollo 12: November 1969. Landing at another mare site, in Oceanus Procellarum. First ALSEP deployed. Landed 200 meters from Surveyor 3 spacecraft, retrieved parts of it for return to Earth. Thirty-four kg of samples collected.

Apollo 13: April 1970. Mission aborted after explosion of oxygen tank in command module. Astronauts returned safely after harrowing flight around the Moon in their disabled spacecraft.

Apollo 14: January 1971. Landing on the Fra Mauro ejecta from the Imbrium Basin. First use of lunar rickshaw to haul equipment and

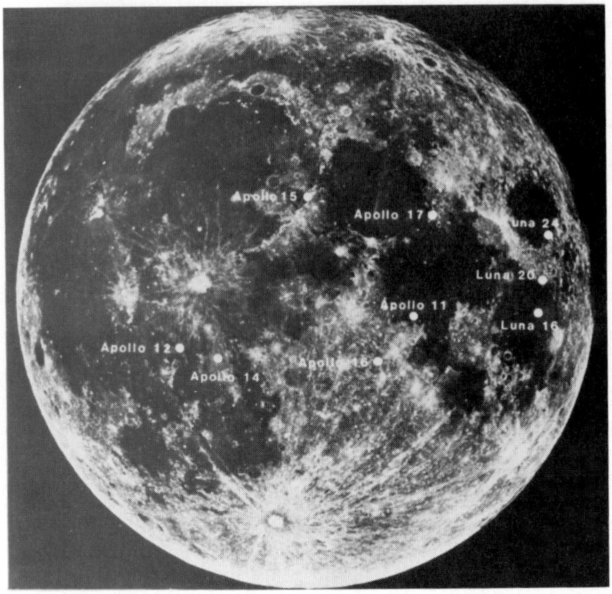

FIGURE 7.12 The landing sites of the six Apollo and three Soviet Luna sample return missions are shown on the NASA map of the nearside of the Moon.

samples; astronauts traversed 5 km on foot. Forty-three kg of samples collected.

Apollo 15: July 1971. Landing at edge of the Imbrium Basin at foot of Apennine Mountains. First in new phase of missions with extended scientific capability. Carried improved command module instruments, launched subsatellite into lunar orbit. First use of lunar rover vehicle permitted 24-km traverse, including visit to Hadley Rille, an ancient lava channel. First measurement of lunar heat flow. Seventy-seven kg of samples collected.

Apollo 16: April 1972. Only landing in a highland site, near crater Descartes. Additional detailed science from the surface and orbit. Ninety-five kg of samples collected.

Apollo 17: December 1972. Final Apollo landing, in Taurus-Littrow Valley on margin of Mare Serenitatis. Included only scientist-astronaut to visit Moon, geologist Harrison Schmitt. One hundred eleven kg of samples collected.

During the development of the Apollo program the United States competed with the USSR to be the first to land humans on the Moon. The Soviet effort finally faltered when their large rocket, comparable to the U.S. Saturn 5, failed several crucial tests in 1968 and 1969. However, the Soviets pursued a vigorous robotic exploration program during the early 1970s. Three robot sample return missions, Lunas 16, 20, and 24, brought back 300 g of material, including one core sample, and in 1970 and 1973, two mobile vehicles called Lunakhods were successfully operated on the lunar surface.

It is ironic that a quarter century after Apollo neither the United States nor any other nation has a capability for human lunar exploration. Tourists gawk at the giant Saturn rockets, rusting on the grass at Cape Canaveral and Houston (Fig. 7.13). Leftover Apollo spacecraft, built at costs of hundreds of millions of dollars, take the place of honor in museums instead of resting where they were intended to, on the surface of the Moon. No forecaster of the future or writer of science fiction had ever predicted that humans, having once attained the Moon, should so quickly have abandoned it. The scientific legacy of Apollo continues, but what

FIGURE 7.13 The giant Saturn 5 rockets of the Apollo program were the largest rockets ever built. Unfortunately, Apollo was terminated before its completion, and several Saturn 5s, such as this one on public display in Houston, were left to rust on Earth rather than being sent to the Moon.

of the exploration potential? Where has that legacy gone?

In the 1990s interest has revived in scientific study of the Moon, but on a much more modest scale. Small orbiter spacecraft have been successfully sent to the Moon by Japan and the U.S. Department of Defense, and NASA plans to launch a low-cost orbiter called Lunar Prospector in 1997. In effect, we are returning to the pre-Apollo approach of the 1960s.

7.3 IMPACT CRATERING

Craters are the dominant geological features on the Moon, readily visible to generations of telescopic observers. Yet their origin, in the impacts of meteoroids, was not widely recognized until about 50 years ago. It is interesting to examine why this fundamental principle of planetary science remained hidden for so long, and to see what finally provided convincing evidence in favor of an impact origin for the lunar craters.

Volcanic or Impact Origin?

Throughout the nineteenth century, geologists thought that the lunar craters were volcanic in origin. The argument was really fairly simple. No impact craters had been recognized on Earth, and the largest projectiles known to strike our planet were meteorites. In those days, before the discovery of near-Earth asteroids, there was no evidence for large meteoroids in the inner solar system that could impact either Earth or Moon. Volcanoes, on the other hand, were well known to terrestrial geologists, and volcanoes do have craters. Therefore, the craters of the Moon, by analogy with the Earth, must also be volcanic.

The first detailed arguments against the volcanic crater hypothesis were presented in the 1890s by G.K. Gilbert, then director of the U.S. Geological Survey. Gilbert was among the few geologists who were interested in the Moon, and he was a strong proponent of quantitative measurements rather than subjective descriptions of what the Moon looked like through a telescope.

Assembling data on the sizes, shapes, and distribution of lunar craters and of terrestrial volcanoes, Gilbert pointed out the many differences between the two. Most significant was the fact that lunar craters do not appear at the summits of mountains. Gilbert was among the first to emphasize that the floors of lunar craters actually lie well below the level of the surrounding plains (Fig. 7.14). Using these and similar arguments, he developed a strong case for the dissimilarity between lunar and terrestrial craters. With the apparent similarity demolished, the argument by analogy crumbled.

Through elimination of alternatives, meteoroidal impacts emerged as the most probable cause of lunar craters, and Gilbert presented as strong a case for this hypothesis as was possible in the 1890s. He did, however, perpetuate a fundamental misconception about the impact cratering process. In his mind, the formation of an impact crater on the Moon was similar to the way a small crater is formed when one throws

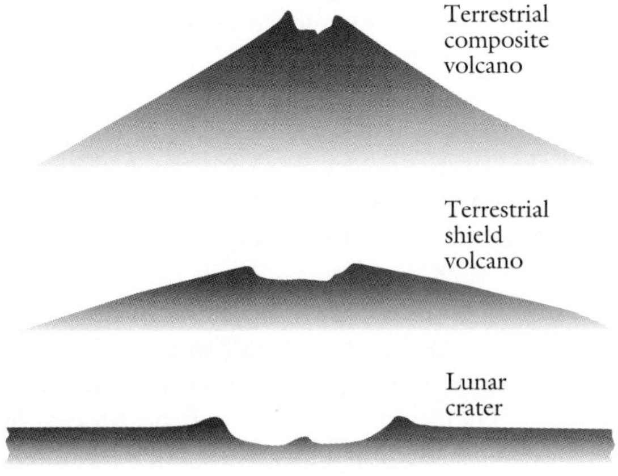

FIGURE 7.14 Comparison of the profiles of typical lunar impact craters and terrestrial volcanic craters.

stones into mud or sand. Such sandbox craters are round only if the stone strikes the surface from above; an oblique impact produces an elongated crater. Since virtually all lunar craters are circular, Gilbert had to hypothesize peculiar circumstances in which the meteoroids would fall on the Moon from nearly overhead. These efforts were not very convincing, undermining his arguments for an impact origin of lunar craters.

The Process of Impact Cratering

Gilbert's fundamental problem lay in his failure to understand that the high-speed impact of a meteoroid on the lunar surface generates an explosion similar in many ways to that of a bomb. (Actually, there are significant differences between impact craters and explosion craters which are apparent to the trained geologist, but we will not probe that deeply into the crater-forming process.) Once the explosion is under way, it obliterates any evidence of the direction from which the projectile struck the surface. To cite a modern analogy, shell and bomb craters are always circular, independent of the direction from which the bombardment occurred.

Why does the impact of a meteoroid produce an explosion? Not because fragments of asteroids or comets are inherently explosive. Rather, it is because of the tremendous energy acquired by the meteoroid as it falls to the surface. For the Moon, the minimum speed of impact (which is equal to the escape velocity) is 2.4 km/s, and for Earth, with its stronger gravitational pull, the minimum impact speed is 11 km/s. To these values must be added the original orbital speed of the projectile, relative to the target. Such speeds endow the meteoroid with energy greater than that of an equivalent mass of TNT, so that it explodes upon impact, regardless of its own composition or the nature of the target. For more information on these impacts, see Section 7.8.

Imagine a large meteoroid striking the Moon at a speed of several kilometers per second. Its energy is so great that it penetrates two or three times its own diameter below the surface before it stops. The force of the blow shatters the surface and generates seismic waves—moonquakes—that rapidly spread throughout the Moon. Meanwhile, most of the energy goes into heating the projectile and its immediate surroundings. The material forms a pocket of superheated gas, and the expansion of this hot, high-pressure gas contributes to excavation of the crater.

Figure 7.15 illustrates the stages of crater formation. At the speeds associated with lunar impacts, the energy released is sufficient to form a crater with a diameter about ten times that of the projectile and a depth, even after much of the ejecta falls back into the crater, of about ⅕ the crater diameter. The material removed from the crater, which consists of the original projectile plus several hundred times its own mass in excavated rock, is thrown upward and outward. Part of it falls back into the hole, partially filling it, while the rest spreads over the surrounding area.

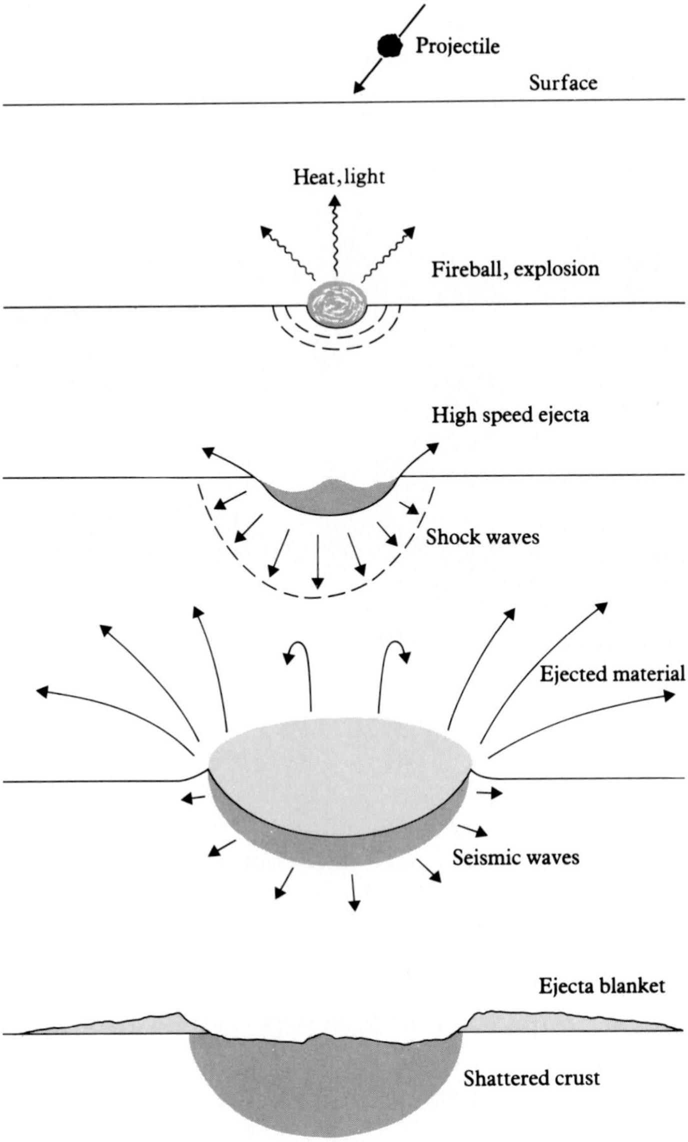

FIGURE 7.15 This diagram illustrates the formation of a large lunar crater from the explosive impact of an asteroid or comet.

The Ejected Material

The ejected material from a cratering explosion consists primarily of broken and shattered rock fragments mixed with gas and liquid droplets generated by the heat of the explosion. The bulk of it falls within one crater diameter of the rim, where it produces a rough, hilly deposit known as an **ejecta blanket.** There may also be a surge of back-falling debris that flows outward from the explosion center. On the Earth, similar hot, fluid surges are a characteristic (and very dangerous) aspect of some volcanic explosions, such as that of Mount Saint Helens (Washington, 1980). Impact craters formed on Mars, where

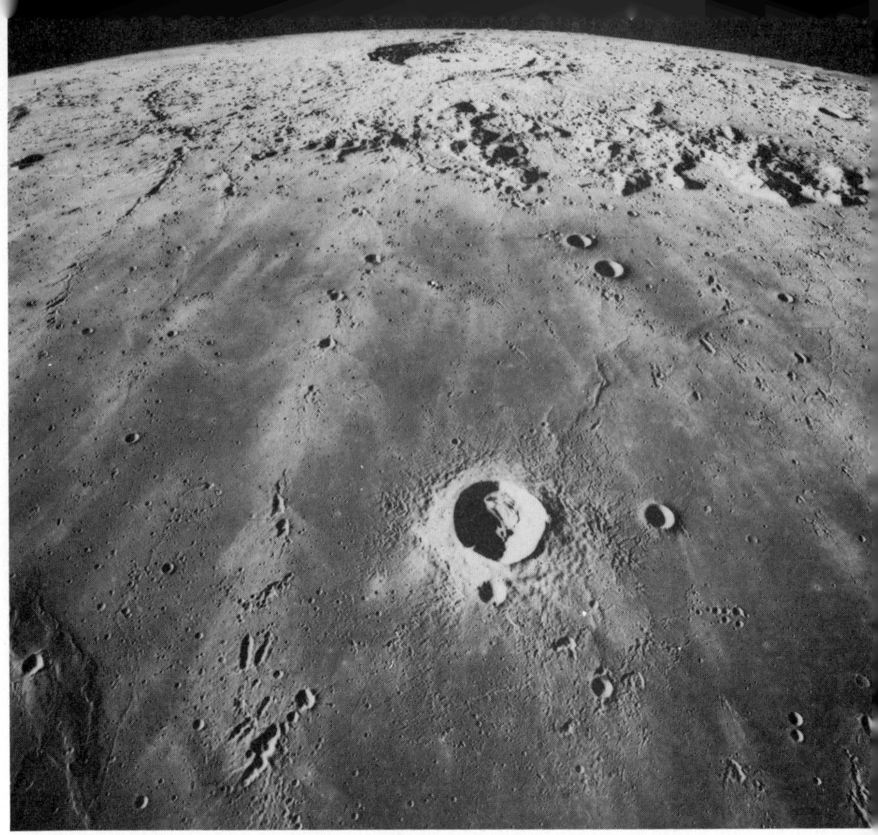

FIGURE 7.16 This spectacular view of the Imbrium Basin with the crater Copernicus on the horizon illustrates the ejecta patterns from a large lunar impact. Extending toward the viewer are many rays and patterns of secondary craters formed by the Copernicus impact about a billion years ago.

the subsurface is (or was) saturated with ice, have a unique form of ejecta blanket, as we will see in Chapter 11.

Large fragments thrown out by an explosion rain down on the surrounding lunar surface, where they generate their own small craters, called **secondary craters.** Sometimes a group of these fragments strikes the surface together to produce a chain of secondary craters.

A third type of crater ejecta consists of high-speed streams of material that are spewed out of large impacts like the splash created when a careless diver plunges into a pool. Such streamers can arc for hundreds or even thousands of kilometers across the lunar surface. Where they strike the ground, they produce many small secondary craters and generally stir up the surface, lightening its color. These long streamers or rays are well illustrated in Fig. 7.16, which shows an oblique Apollo view of the Imbrium Basin with the 91-km crater Copernicus in the distance. Several rays stretch toward the observer across the dark mare surface. Near the foreground, we can see the association between the light ray material and clusters of irregular secondary craters. Since the impact speed of secondaries is rather low (less than 1 km/s), they do not generate true explosions. Therefore these craters can be irregular or elongated, just like the pits produced by throwing stones into a sandbox.

Evolution of the Crater

Meanwhile, back at the crater, the lunar crust is trying to adjust in the aftermath of the impact. For a small crater, up to perhaps 10 km in diameter, little adjustment is necessary, and the crater retains approximately its original bowl shape. Larger craters, however, cannot be supported for long once the explosion is past. Under the force of gravity, sections of the crater wall collapse and slide downward, partially filling the center and creating a series of steplike terraces along the walls, reducing the depth/diameter ratio below the original ⅕. Near the center of the crater, where a great weight of overlying material has been removed, the crust may rebound to create a central peak or group of peaks.

Large craters tend to have flat floors, often flooded by later lava flows. Let us return to Tycho, where you may imagine yourself landing in the interior. Standing on the floor, you would not have the sensation of being in a bowl-shaped depression. Instead, you would find yourself on a rather level rocky plain, with the distant crater rim visible as a low line of mountains serrating the horizon. The central peak would appear as a huge pile of rubble.

7.4 DATING CRATERED WORLDS

The number of impact craters on a planetary surface is a measure of the age of that surface. On an active planet like the Earth, erosional and other geologic processes rapidly degrade and destroy craters. However, there is little degradation on the Moon, and the numbers of craters accumulate with time.

Crater Densities and Surface Ages

We define the **crater retention age** as the time over which the surface has been sufficiently stable to preserve craters once they are formed. On the Moon, practically the only events that can destroy craters are later impacts or lava flooding during periods of large-scale volcanism. In the lunar maria, the crater retention ages generally represent the time elapsed since the most recent major eruptions.

When we see differently cratered regions (for instance, in Fig. 7.3), we naturally interpret the differences as the result of different crater retention ages. After all, we know the impacting meteoroids came from outside, and there is no reason to think that they preferentially struck in some regions, the highlands, for instance, while sparing others. The situation resembles that of a city street in the midst of a long, windless snowstorm. As you walk along, you will find the sidewalk in front of some houses deeply covered with snow, while in other places the depth is less, with a few areas of sidewalk nearly clear. Do you conclude that different amounts of snow have fallen in front of the Joneses' house than at the Smiths' next door? No; you attribute the differing depth to the time that has passed since that section of walk was shoveled. The less snow, the shorter the "snow retention age." It is just the same with craters.

To determine a crater retention age, we must first count the number of craters that can be seen on a particular terrain being studied. The number of craters on a given area is called the **crater density.** This density has nothing to do with the material density used throughout this book to characterize the bulk composition of an object. It is simply the expression of the number of craters of a given size on a well-defined area of the surface, usually taken as 1 million square kilometers, about the size of the state of Texas on Earth or of Mare Imbrium on the Moon.

Crater Size Distributions

Look at the photographs of the Moon throughout this chapter. You will note that there are always many more small craters than large ones. The reason is obvious: As we saw in Chapter 5, small meteoroids are much more abundant than large ones. Fragments in space, from asteroids down to pebbles, tend toward a size distribution in which there are more than a hundred 10-km meteoroids for each one that is 100 km in diameter, more than a hundred 1-km objects for each 10-km one, and so forth. Since each lunar crater has a diameter approximately ten times that of the impacting meteoroid, the craters will display the same sort of size distribution, scaled up by a factor of ten. Indeed, scientists have turned the argument around. By counting lunar craters in

FIGURE 7.17 (a) The relative number of objects that could impact the Moon, according to a law in which the number of objects increases in inverse proportion to the square of the object's radius. (b) A random distribution of craters made by the population of objects shown in (a).

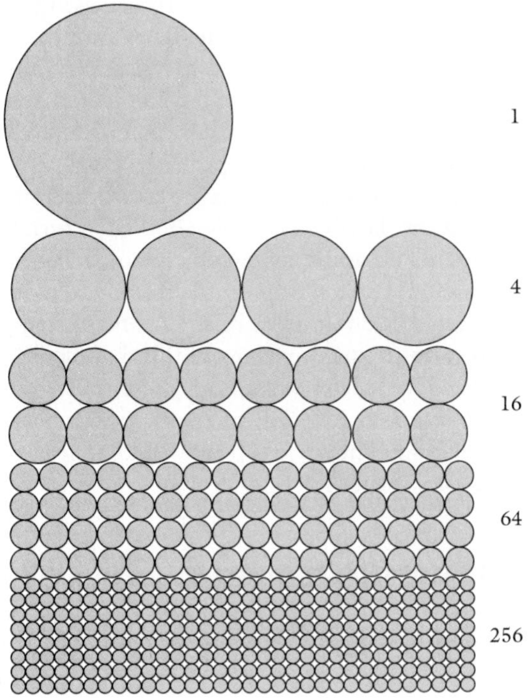

a

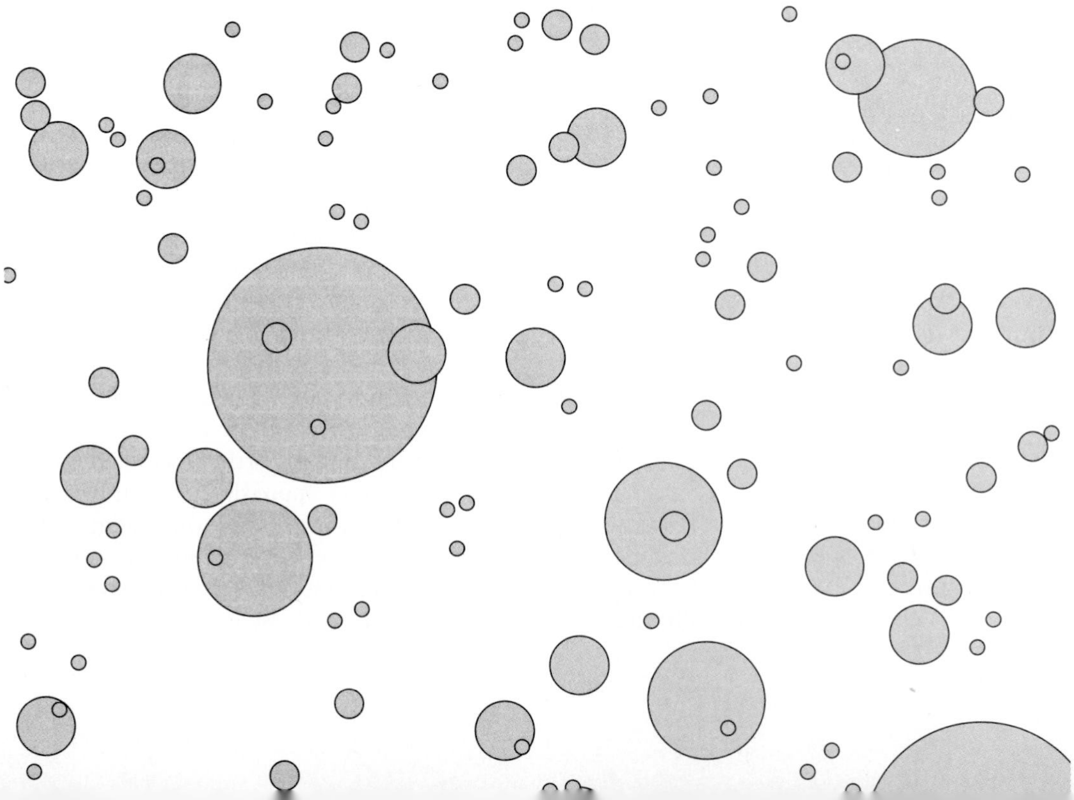

b

different size ranges, they have determined the size distribution of the meteoroids.

Figure 7.17(a) illustrates schematically the crater size distribution observed on the Moon for craters larger than 3 km in diameter. There are many more small craters than large ones, as we expect. In Fig. 7.17(b), exactly these same craters are randomly distributed over a surface, illustrating the typical appearance of a planetary surface where there is no erosion and all craters are retained once they form.

The measure of crater density that is commonly used is a cumulative one. That is, we count all craters larger than a given minimum size, frequently 10 km. For example, the 10-km cumulative crater density on the average mare is 50 per million square kilometers, while that on the lunar highlands is 1000 per million square kilometers. Cumulative crater densities for a number of areas of the Moon are listed in Table 7.1.

TABLE 7.1 Cumulative crater densities on the lunar surface

Lunar Region	Crater Density (10-km craters per million km^2)	Age (by)
Highlands	1000	> 4.0
Fra Mauro (Apollo 14)	130	3.9
Apennine (Apollo 15)	95	3.9
Tranquillitatis (Apollo 11)	50	3.7
Fecunditatis (Luna 16)	20	3.4
Putredinis (Apollo 15)	25	3.3
Procellarum (Apollo 12)	20	3.3
Copernicus Crater	10	0.9
Tycho Crater	2	0.2

How much of the lunar surface is actually covered by craters? Following is an approximate calculation. For the lunar crater size distribution, the average size crater is about twice as large as the lower limit counted. In our case, where we are considering craters down to 10 km diameter, the average diameter is about 20 km, so the area of each crater is about 400 km^2. A quick multiplication shows that on the maria, where the crater density is 50 per million square km, only about 2% ($50 \times {}^{400}\!/_{1,000,000}$) of the surface is cratered. You can verify these calculations by examining the photos of highland and mare terrain in this chapter, and counting the craters yourself.

Absolute and Relative Ages

The greater the crater density on a planet, the older the surface. But how much older? Only for the Moon, where scientists have measured both the crater densities and the associated radioactive ages from returned samples, can this question be answered with any precision. On Earth, there are too few craters, while on other planets there is no absolute age scale defined by returned samples.

All of the lunar maria have roughly similar crater densities, and all therefore appear to have comparable ages. This fact was recognized before the Apollo missions, but without returned samples there was a lively debate among lunar scientists concerning the absolute age of the maria and, hence, the period since the cessation of major lunar volcanism. If one assumes that the rate of impacts has remained constant throughout solar system history, then the fact that the mare crater densities are about $\frac{1}{20}$ of the highland densities suggests that the maria are only $\frac{1}{20}$ as old as the highlands. If the highlands are as old as the Earth, 4.5 billion years, the maria would be only about 200 million years old.

Is this conclusion consistent with what we know about the current impact rates from comets, near-Earth asteroids, and other meteoroids? In the 1960s many scientists thought not. Starting from the known numbers of objects in near-Earth space, they calculated that it would require closer to 4 billion years, rather than 200 million, to accumulate the observed mare crater density. In other words, the flooding of the maria by lava flows was restricted to the early period of lunar history. They therefore argued for an inactive Moon, on which the volcanic fires cooled billions of years ago.

This explanation poses a problem, however. If the maria and the highlands are both billions of years old, how can the crater density in the highlands be 20 times greater than on the maria? In order for these parts of the Moon to have accumulated so many craters, a much higher early impact rate would have been required, between the formation of the highland crust and the period of lunar volcanism represented by the maria.

When the first lunar samples were returned to Earth in 1969, the most eagerly awaited scientific result was the radioactive age determination for mare materials. The solidification age, for a variety of Mare Tranquillitatis rocks and soil samples, was 3.7 billion years. Subsequent measurements on samples from other mare sites confirm that all the mare solidification ages are between 3.2 and 3.9 billion years. Highland rocks, in contrast, have ages greater than 4 billion years, while the estimate for the age of the entire Moon (since differentiation) is about 4.5 billion years.

A Unifying Concept of Cratering on the Inner Planets

The combination of lunar crater counts, dated lunar samples, and measurements of the numbers of meteoroids in near-Earth space today, has given rise to the current concept of early planetary history. Judging from the lunar evidence, the impact rates on all of the inner planets have not changed greatly during the past 3.8 billion years. Presumably the impacting bodies have been the comets and near-Earth asteroids that we recognize today. If similar numbers of meteoroids have struck Mercury, Venus, or Mars, then the lunar chronology can be applied to crater densities on these planets as well.

According to this theory, however, quite different conditions prevailed during the first 700 million years of solar system history. Before about 3.8 billion years ago, the rate of impacts must have been much higher; otherwise the high crater densities on the highlands cannot be understood. From lunar research, it appears that the impact rate was a thousand times greater at 4.0 billion years ago than at 3.8 billion years ago (Fig. 7.18). Earlier it may have been much higher yet, but the evidence concerning lunar history before 4.0 billion years is limited.

This period around 3.9 billion years ago is variously called the late heavy bombardment or the terminal bombardment of the Moon. It is presumed to have involved a different source of impacting bodies from those present today.

FIGURE 7.18 Schematic illustration of the variation over history in the flux of projectiles on the Moon. The rate of formation of craters was much higher during the first few hundred million years of lunar history, but has remained nearly constant for the past 4.8 billion years.

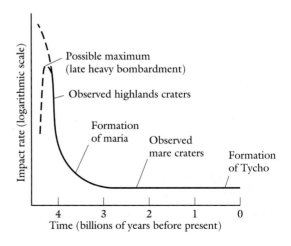

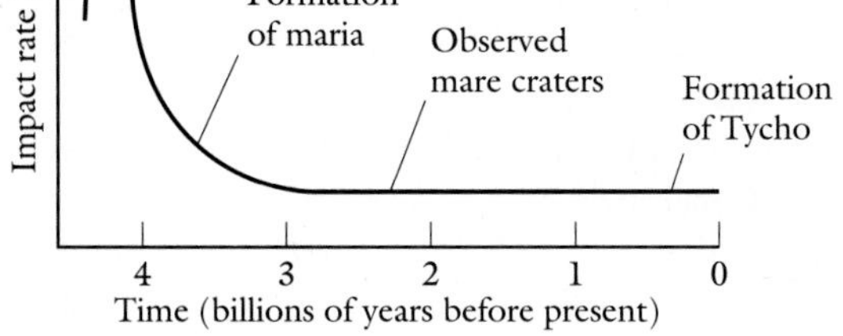

Because the impact rates were very high at this time, many scientists suspect that they were the result of a unique event, such as the disruption of a large asteroid, which scattered fragments through the inner solar system. Alternatively, the terminal bombardment may represent the final stages of planetary accretion. Either way, this period of heavy cratering has largely obliterated direct evidence of the first half-billion years of lunar history.

The idea that the inner solar system experienced a late heavy bombardment 3.9 billion years ago, followed by nearly constant impact rates extending to the present, constitutes one of the *paradigms,* or generally accepted theories, of modern planetary science. One of the many scientists contributing to this paradigm is Eugene Shoemaker of Lowell Observatory and the U.S. Geological Survey branch of astrogeology (Fig. 7.19). Shoemaker has dealt with impact craters all his life, writing his doctoral dissertation on the geology of Arizona's Meteor Crater. In the 1960s he was one of the Apollo geologists who guessed the wrong lunar chronology, predicting relatively young mare ages. After Apollo, Shoemaker began a major observational and theoretical program to understand the orbits and size distributions of meteoroids in Earth-Moon space today, and it is this program that has yielded the recent discovery of many near-Earth asteroids (Section 5.4). Thanks to this work, we now recognize that the number of Earth-crossing meteoroids is consistent with the lunar impact rates deduced from mare crater counts. Shoemaker has extended this cratering paradigm to the satellites in the outer solar system, a subject we will return to in Chapter 15.

FIGURE 7.19 Eugene Shoemaker of the U.S. Geological Survey and Lowell Observatory has played a leading role in developing a theory of the role of impacts from comets and asteroids over the history of the solar system. He is shown here with his wife, Carolyn, who has discovered more comets than anyone else in history.

Application of the cratering paradigm to other planets, particularly Mercury and Mars, has yielded important insights concerning solar system history. But all such unproven theories must be treated with some caution. We know the lunar chronology, but its application elsewhere carries a certain risk, at least so long as we do not know the origin of the late heavy bombardment. We will try to keep this reservation in mind when we come to interpret crater counts on other planets.

7.5 LUNAR CATASTROPHISM

Catastrophism is a term used in geology to refer to sudden or violent events that substantially modify the landscape. Often it is contrasted with **uniformitarianism,** which seeks explanations of the world around us in terms of very slow processes acting over vast time spans. Large impacts and massive volcanic eruptions are examples of catastrophic events, while the gradual accumulation of ocean sediments is the product of uniformitarian processes. Although proponents of the two concepts have been in conflict in the past, we now recognize that both types of geological processes are important.

The Lunar Highlands

On the Moon, most of the catastrophic geological events took place early in its history. The lunar highlands (Fig. 7.20) are the oldest surviving part of the lunar crust. The degree of highland cratering far exceeds that on the maria. Since the highland craters are packed virtually shoulder to shoulder, we have no way of telling how many impacts took place in the past. In these areas, the crater density has apparently reached a state of *saturation,* in which new impacts do not create additional craters; they simply replace old craters with new ones.

An indication of the magnitude of highland cratering is provided by the behavior of moonquakes studied by the Apollo ALSEP instruments. The reverberation of seismic waves in the highlands indicates that the crust has been shattered to a depth of 25 km, and that the fragmented surface layer of rubble has an average depth of hundreds of meters. In contrast, the fragmented surface layer on the maria is only about 10 meters deep.

As might be expected from this history, the rocks of the lunar highlands are breccias, composed of recemented pieces broken and scattered by previous impacts (Fig. 7.21). Many

FIGURE 7.20 This view of the heavily cratered lunar highlands was taken from the Apollo 17 command module.

FIGURE 7.21 Many lunar rocks are breccias, like this sample from the highlands. These rocks, consisting of individual fragments cemented together, display a long history of impacts. Most lunar breccias formed during the high bombardment period of the first half-billion years of lunar history.

highland breccias are extremely complex, indicating three and even four generations of shattering and reforming of the rocks. Often the highland breccias contain frozen droplets of impact-melted material. By comparison, the meteoritic breccias discussed in Chapter 4 are relatively simple, consistent with their formation on smaller bodies.

Highland Samples

The oldest dated highland samples are not whole rocks but tiny individual fragments within breccias. Most of these show solidification ages near 4.2 billion years, although a few go back as far as 4.4 billion years. The apparent age cutoff of 4.2 billion years does not mean that the surface remained molten up to that time. More likely, the radioactive clock was repeatedly reset for highland samples by heating and intense shock from multiple impact events.

Dating of Apollo samples indicates that the period of lunar catastrophism extended up to nearly 3.8 billion years ago, when the rain of impacts subsided to more nearly its present low rate. As lunar scientist Stuart Ross Taylor has observed, "The meteoritic bombardment has drawn a curtain across the landscape through which we peer dimly to discern the earlier history. The [original] structure of the lunar crust is obliterated, and there is no vestige of a beginning."

One way to try to pull this curtain aside is to look at the bulk composition of the highland lunar crust. This material, like all lunar samples, is severely depleted in volatile elements—that is, in those elements that can be evaporated or vaporized at relatively low temperatures. Ices are extremely volatile, but geologists also consider any element that evaporates at the temperature of lava, about 1000 C, to be volatile. On the Moon water is absent; the lunar minerals include no clays or other familiar terrestrial forms that incorporate chemically bound water. Other common volatiles that are depleted include nitrogen, carbon, sulfur, chlorine, and potassium. The lack of both metals and volatiles on the Moon is an important clue to its origin, as we will discuss in the next chapter.

The primary material of the highland crust is a class of igneous silicate rocks called **anorthosites.** The lunar anorthosites consist primarily of mineral oxides of silicon, aluminum, calcium, and magnesium. Anorthosites are rocks that might form out of an originally molten Moon, and their presence argues for an early differentiation, before the lunar crust solidified. Battered though these fragments of the crust may be, they still provide some information on this earliest stage of lunar history.

During the hundreds of millions of years of heavy bombardment that followed the formation of the original crust, many episodes of remelting triggered by impacts probably occurred. We can imagine large projectiles breaking through the crust to release floods of still liquid rock from the interior. At some point after the crust had stabilized and thickened, more conventional volcanic activity also began. Although now nearly obscured by subsequent events, indications exist of the formation of the first basalts as early as 4.2 billion years ago.

Lunar Basins

The surviving lunar impact basins were formed during the final stages of the heavy bombardment. We use the term *basin* to refer to any impact feature more than about 300 km in diameter. The impacting meteoroids that blasted out these features must have been from 30 to 100 km in diameter. Possibly many basins were formed from a cluster of asteroidal impacts that occurred during the terminal bombardment, in the relatively brief interval from about 4.1 to 3.9 billion years ago. Today lunar scientists have identified about 30 ancient basins, including many on the lunar farside. The largest known basin, which was confirmed in 1994 by data from the U.S. Department of Defense Clementine orbiter, is located near the south pole and has a diameter of roughly 2200 km.

The youngest of the great basins are Imbrium and Orientale (Figs. 7.16 and 7.6), each about the size of the state of Texas. The Imbrium event can be precisely dated using samples from Apollo 14 and 15. The time of the impact was just over 3.9 billion years ago, and the extensive ejecta, which overlie older features on much of the nearside of the Moon, provide a reference marker for lunar stratigraphy. The ejecta from Orientale are found on top of Imbrium features, indicating that the Orientale event occurred

FIGURE 7.22 The Fra Mauro formation, shown in this view of the Apollo 14 landing site, is the hummocky material extending through the center of the frame. This material was ejected from the impact that formed the Imbrium Basin, located about 600 km to the north of the location shown here.

later. Although no Apollo landing took place on Orientale ejecta, indirect evidence places this last great impact at between 3.8 and 3.9 billion years before the present.

Both Imbrium and Orientale are mountain-ringed circular features; today they are different in appearance primarily because the Imbrium Basin was later flooded by lavas to produce Mare Imbrium, while the Orientale Basin escaped major volcanic modification. The outer mountain ring of Imbrium, consisting of the lunar Carpathians, Apennines, and Caucasus, rises as high as 9 km above the present mare material, similar to the height of the Hawaiian volcano Mauna Loa above the ocean floor. The diameter of this outer ring is 1200 km. The Fra Mauro formation, which is the ejecta blanket from Imbrium, extends outward nearly a thousand kilometers and varies in thickness from more than a kilometer down to a fraction of a meter (Fig. 7.22). The Orientale Basin is defined by a 900-km outer mountain ring, the Cordilleras, and a 600-km inner ring, the Rook Mountains. The floor of this unflooded basin consists of rough and hilly contours and patches of impact-melted rock.

The process of formation of basin-rim mountains is not entirely understood, but it apparently involved a combination of uplift, produced by the blast itself, and later subsidence along concentric cracks as the lunar surface adjusted following the impact. A cross section of the Apennine front is shown in Fig. 7.23, which reveals layers in the uplifted basin rim. Other, isolated mountains are probably rootless piles of ejecta.

A Perspective

Compared with conditions today, the era of heavy bombardment was a time of great violence, with impact rates tens of thousands of times greater than those experienced now. Thinking back to that time, we tend to picture a sky filled with projectiles or a surface pockmarked with fresh craters. A bit of arithmetic will quickly convince us otherwise, however. Today a

FIGURE 7.23 One of the best views of mountains on the Moon was obtained by the Apollo 15 crew, who landed near the base of the Apennine Mountains. Mount Hadley, shown here, was formed by uplift associated with the impact that created the Imbrium Basin.

1-km crater is formed over a million square kilometer area about once per million years. When the impact rate was 10,000 times greater, such a crater was still formed only once per century. An individual witness is aware of events only to a radius of about 100 km, or about 4% of a million square kilometers. Thus a hypothetical observer standing on the Moon or Earth during the period of heavy bombardment would have witnessed formation of a 1-km crater only once in 2500 years! Events that seem catastrophic on a planetary time scale are still rare from our human perspective.

7.6 LUNAR VOLCANISM

Even before the period of heavy bombardment had ended, volcanic vents were erupting on the lunar surface. However, the major period of lunar volcanism apparently did not begin until shortly after the formation of the Imbrium and Orientale Basins, about 3.8 billion years ago. During the following half-billion years, repeated outpourings gradually filled the nearside basins with dark basalt to create the familiar pattern of lunar maria.

The Lunar Maria

The story of lunar volcanism is essentially the story of the maria. Although there may have been isolated examples of volcanic activity elsewhere, they were of negligible significance compared with the great outpourings of mare lava. Fortunately, several landing sites on the flat mare surfaces were selected by the safety-conscious Apollo planners, thereby providing us with many observations and large quantities of returned mare rocks and soil.

The mare basalts resemble their terrestrial counterparts and are believed to have originated in a similar way, from subsurface melting and eruption of lava. However, lunar basalts have a higher iron content, including metallic iron, and are free of water and its effects. Since all lunar rocks are igneous and formed in an environment lacking both water and atmospheric oxygen, they have simpler compositions than terrestrial rocks. Only about a hundred separate minerals have been identified on the Moon, as against more than 2000 on the Earth. There is no clay or other water-bearing rock, and very little carbonaceous material.

The oldest lunar basalts, obtained from the Apollo 14 and Apollo 17 sites, have solidification ages of 3.8–3.9 billion years. The flows on the margin of Mare Imbrium sampled by Apollo 15 are dated at 3.3 billion years, which presumably fixes the late stages of volcanism in this basin. Apollo 12, which landed on Oceanus Procellarum, yielded the youngest mare basalt, with a solidification age of just under 3.2 billion years.

The Maria-Forming Eruptions

Unlike most of the volcanic activity with which we are familiar on Earth, the lunar eruptions did not normally create volcanic mountains. Instead, the eruption of large volumes of fluid lava from long fissures yielded thin, flat flows of enormous extent. On Earth, similar *flood basalt* eruptions are responsible for the lava plains of eastern Washington State and western India.

Most of the lunar lava flows have been buried by subsequent eruptions, but the final large outpourings can still be identified. On Mare Imbrium, the last major eruptions originated from a fissure about 20 km long and spread more than a thousand kilometers from the vent, covering a total area of 0.2 million square kilometers, or about the size of the state of Utah. Along the flow fronts the thickness is typically 30–50 m. A hundred flows of this thickness are required to account for the total depth of the mare, which is estimated at about 5 km. As much as a million years might have elapsed between individual flows.

After the eruption of the mare material, some settling and subsidence took place, perhaps as a

FIGURE 7.24 The Aristarchus Plateau, seen in this Apollo 15 view, shows the many wrinkle ridges that characterize the lava plains that make up the lunar maria.

result of shrinkage as the lava cooled. In places, "high-water marks" show that lava once reached up to 100 m higher than the current level. Many cracks have formed around the margins of the maria, often concentric with the center of the basin. In other areas the lavas seem to have been compressed to form wrinkle ridges resembling the wrinkles formed in a rug by pushing it against a wall (Fig. 7.24).

Volcanic Valleys and Mountains

Among the most remarkable geologic features of the maria are the curving valleys called sinuous rilles. Before Apollo, many scientists thought these broad, meandering channels, which resemble terrestrial rivers, had been carved by running water. We now know that the Moon is, and always has been, dry. Since the sinuous rilles are found in the maria, it seems likely that they too represent some kind of volcanic phenomenon. Subsequently the Magellan spacecraft revealed even larger lava channels on Venus, a subject we will discuss in Chapter 10.

Apollo 15 was targeted to visit one of the largest sinuous rilles, Hadley Rille, near the edge of Mare Imbrium (Fig. 7.25). When the astronauts drove their rover to the edge of the channel, they found a curving valley 1.2 km wide and 370 m deep, littered with fallen boulders, some as large as a small house. On the far wall they could make out an exposed 50-m-thick layer of solidified lava. Hadley Rille seems to be a channel scoured by an ancient lava river.

Although volcanic mountains are extremely rare on the Moon, they may not be completely absent. A group of domes in Oceanus Procellarum, the Marius Hills, are thought by some geologists to be true volcanoes. The Marius Hills were to have been visited by a later Apollo flight, but when the program was canceled in 1972 this mission was among the casualties.

Sources of Lunar Eruptions

Why are the maria confined primarily to the side of the Moon facing the Earth? The main reason seems to be that the average surface elevation is lower on the Earth-facing hemisphere. The subterranean pressures that forced lava to the surface appear not to have been great enough to produce major eruptions in the highlands, including most of the farside. In this respect the maria do indeed resemble seas, filling low-lying areas just as the oceans of Earth occupy the basins between the continents.

Substantial information on the source regions of the lunar eruptions has been derived from a detailed study of the composition of the lunar

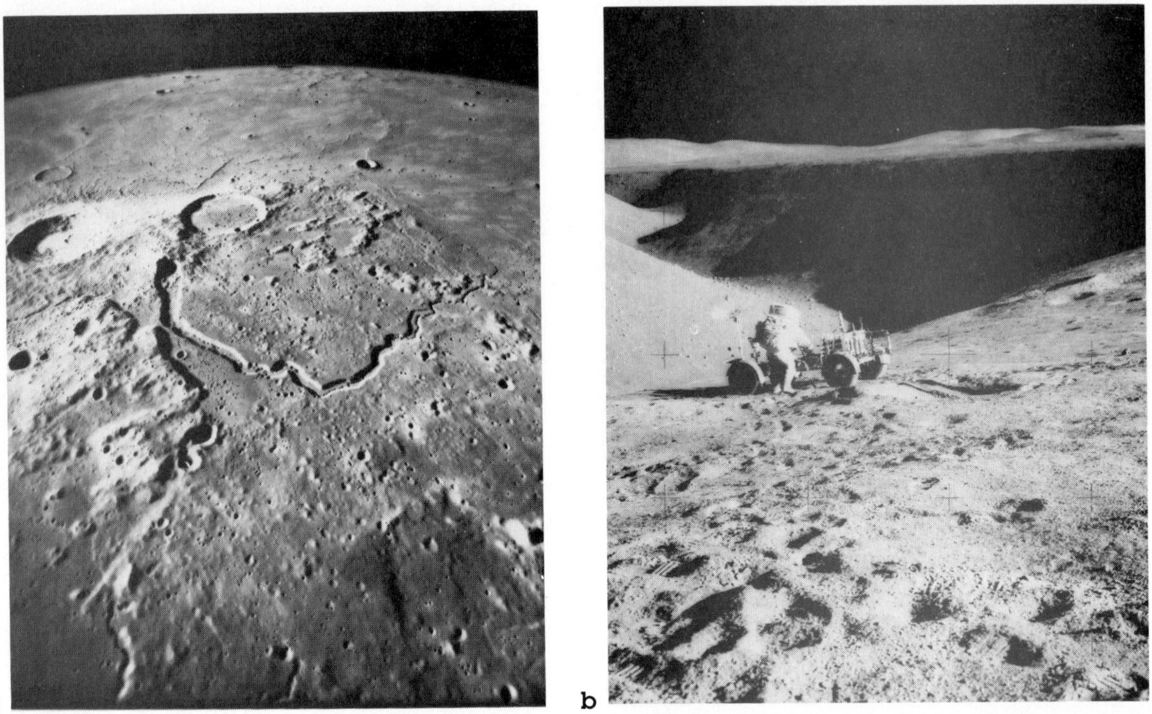

a b

FIGURE 7.25 One of the largest of the ancient lava rivers on the Moon is Hadley Rille, visited by the Apollo 15 astronauts; (a) the curving valley, about a mile in width, is seen from orbit; (b) the view from the surface.

basalts. Liquified rock in the interior of a planet is called **magma.** The mare lava samples are derived from magma that has been three times chemically separated or *fractionated* from the original material of the solar nebula. First there was the loss of water and other volatiles that is characteristic of the entire Moon. Second was differentiation, in which the lunar interior separated into core, mantle, and crust. Finally, there was a process of "partial melting," in which the more easily melted minerals of the mantle were liquified to create large magma reservoirs hundreds of kilometers below the surface.

The major lunar volcanic activity ceased a little more than 3 billion years ago. This means that if we saw the Moon 3 billion years ago it would look much as it does today, whereas the continents of the Earth's crust would have been unrecognizable even a few hundred million years ago.

7.7 THE SURFACE OF THE MOON

When the first Apollo astronauts stepped onto the lunar surface they found themselves in a stark but beautiful world. The airless sky was deep black, the surrounding plains a dark brownish gray. The force of gravity was just one-sixth that at the surface of the Earth, making their bulky spacesuits seem relatively light. With no haze to obscure the view, distant details stood out as sharply as those in the foreground. The mare plain itself was flat and covered with scattered irregular rocks of all sizes and shapes. Some distance away the low profiles of small craters could just be identified, but at first sight there was little evidence to establish that they stood on a heavily cratered planet.

The Lunar Regolith

Once they began to move about, the astronauts quickly became aware that the surface of the Moon was layered with a fine dark dust. Every step raised clouds of this material, which soon coated their spacesuits and eventually found its way into every item of their equipment. In spite of this omnipresent dry dust, they found the lunar soil to be firm underfoot. Their boots sank only a few centimeters into the soil, and the bootprints they made were crisp and sharp-edged, as if they had been made in damp dirt or crunchy snow (Fig. 7.26).

FIGURE 7.26 This astronaut's bootprint in the lunar soil illustrates the strength and cohesiveness of the fine upper layer of the lunar regolith.

Although later Apollo flights landed in rougher and more scenic locales than that first site in Mare Tranquillitatis, the basic character of the immediate surroundings was the same everywhere. The entire Moon appears to be covered with fragmented rock and dust, which represents the ejecta from impact craters, far and near. This continuous debris blanket has been named the lunar **regolith.** The finer component is generally referred to as lunar *soil*, while the term *regolith* includes the full range from fine dust to house-sized blocks.

The depth of the regolith varies with the age of the surface. On the maria, it is typically 10 m thick. In the older Littrow Valley area visited by Apollo 17, the measured thickness ranged from 6 to 40 m, while in the highlands the regolith thickness may be as great as hundreds of meters. As shown by core samples, the regolith is composed of overlapping layers of ejecta, each typically a few centimeters thick. One Apollo 17 core, which penetrated nearly 3 m of regolith, revealed 42 distinct layers, each representing ejecta from a single impact.

The lunar regolith accumulates at an average rate of about 2 mm per million years, or 2 m per billion years. In comparison with terrestrial erosion and deposition processes, this is slow change indeed. Along with the buildup by addition of ejecta layers, the upper part of the regolith is frequently disturbed by the impacts of very small meteoroids, which stir or "garden" the top few millimeters of material and help to maintain the loose fine dust.

Studies of the processes that form the regolith indicate that most of the ejecta at any one site is derived from relatively nearby impacts. If this were not the case, the sharp boundaries between mare and highland materials would have been blurred by horizontal mixing, and we would not see the Man in the Moon so clearly. Only about 5% of the material comes from as far away as 100 km, and only 0.5% from 1000 km. These calculations apply to the normal cratering process, not to ejecta from the major basins, which have blanketed substantial fractions of the lunar surface to depths of many meters.

In addition to original lunar material, the regolith includes meteoritic fragments from billions of years of cosmic bombardment. Typically, between 1% and 2% is meteoritic, apparently of about the same composition as the primitive meteorites collected on Earth. Most of the small amount of carbon in the lunar soil is derived from impacting carbonaceous meteorites.

Lunar Soil

When examined under a microscope, all lunar soils are found to contain large quantities of glass in the form of spherules up to a millimeter in size (Fig. 7.27). *Glass* is simply silicates that have been melted and then rapidly cooled so that no crystals can form. These glasses, which are characteristic of the lunar soil, are the product of melting of lunar rock during meteorite impact. Since its atmosphere protects the Earth from these micrometeorites, our soil does not contain similar glass spherules, except at a few sites associated with terrestrial impact craters. Most of the lunar glass is dark, and the presence of these tiny spherules contributes to the dark color of the lunar surface. The tiny spherules can themselves be pitted by even smaller micrometeorites when they are exposed on the lunar surface (Fig. 7.28).

FIGURE 7.27 Viewed under a microscope, the lunar soil is composed of a variety of small rock fragments mixed with glass spheres produced by impact melting.

In addition to the micrometeorites, the lunar surface is also exposed directly to ions of the solar wind and to **cosmic rays,** which are not rays or radiation at all but high-speed atomic and subatomic particles, leaving tracks that can be studied on Earth. The existence of cosmic ray tracks, and of the similar implantation of solar wind particles in lunar materials, has given rise to a new kind of astronomy. In effect, the lunar surface has served as detector and recorder of solar and cosmic particles over geological time scales. For example, studies have shown that the flux of cosmic rays from the Galaxy has remained constant over the past billion years, that the rate of major solar eruptions has not changed greatly over that past 100,000 years, but that the isotopic abundances in solar wind nitrogen have changed substantially over the past 2 billion years (for reasons we don't understand). Thus study of returned lunar samples provides information never before available on long-term variations in solar and galactic energetic charged particles.

Surface Conditions and Erosion

In the absence of moderating air or oceans, the surface of the Moon experiences much greater temperature ranges than does the Earth. The contrast between day and night is further magnified by the fact that day and night are each two weeks in duration. At the near-equatorial Apollo sites, the maximum surface temperature is about 110 C, higher than the boiling point of water. During the long lunar night, the surface temperature drops to −170 C, only about 100 K. It is a tribute to the design of the Surveyor spacecraft and the ALSEP instruments that they were able to operate over this huge temperature range. The astronauts themselves were never subjected to such extremes, since they always landed in the lunar morning when the temperatures were similar to those on the Earth.

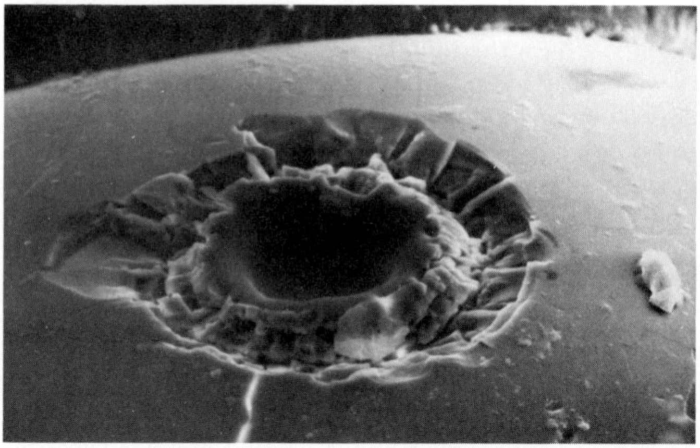

FIGURE 7.28 This photograph of a droplet of lunar glass, made with a scanning electron microscope, shows a beautiful microcrater produced by the high-speed impact of a tiny fragment of interplanetary dust.

The last three Apollo flights visited mountainous areas: the Apennines, the Descartes highlands, and the Taurus Mountains. Notable in each of these landscapes were the gentle, rounded contours of the lunar mountains and hills. The illustrations in science fiction stories had depicted the mountains of the Moon as extremely steep and spiky, but the reality was otherwise. A partial explanation for the soft contours of lunar mountains can be found in the ejecta that blankets all lunar features, but the primary reason why there are no sharp peaks or steep cliffs on the Moon is that there is no water or ice erosion to sculpt them. Water and ice erosion on Earth do not simply wear down mountains; they also cut the valleys and shape the peaks. In the absence of such natural forces, the mountains on any planet will be as gentle as those of the Moon.

7.8 QUANTITATIVE SUPPLEMENT: IMPACT ENERGIES

In order to determine the size of a crater formed from the impact of an asteroid or comet, we need to calculate the energy released by such an impact. This energy depends on the mass (m) and speed (v) of the projectile, as expressed by the standard equation for kinetic energy:

$$E = \frac{1}{2} m v^2$$

The speed is made up of two components: the speed of the projectile in space before it approaches the planet, and the additional speed it gains as it falls to the surface accelerated by the planet's gravitation. The minimum impact speed is the escape velocity of the planet, to which the initial relative speed must be added. For large planets, the speed gained during the fall generally exceeds the initial speed.

The escape velocity of the Earth is 11 km/s or 1.1×10^4 m/s. Thus we calculate the minimum impact energy per kilogram of the projectile to be 6×10^7 joules. This is about ten times larger than the chemical energy of TNT, which is 4×10^6 joules/kg, which is why we have emphasized that the crater-forming event is in many ways similar to a bomb explosion.

Since the escape velocity of the Moon is only 2.4 km/s, impacts of much lower energy are possible. Typically, a projectile strikes the Moon with only about ½ the speed with which it would strike the Earth, resulting in an energy only about ¼ as great. The mass (and volume) of material ejected in the impact is roughly proportional to the energy, and the linear dimensions (diameter and depth) of the crater are proportional to the cube root of the volume. Thus a projectile striking the Moon would produce a crater with diameter only $\sqrt[3]{(1/4)} = 0.63$ times the diameter of the crater it would make on the Earth. On the basis of this calculation we expect that a given impacting body makes a lunar crater only ⅔ as large as a terrestrial crater (about 10 and 15 times the projectile diameter, respectively).

Just how large are these impacting energies? A 10-km-diameter asteroid with a density of 2.5 g/cm^3 has a mass of about 1.3×10^{15} kg. Striking the Earth with a speed of 20 km/s, it would have an energy of about 2.6×10^{23} joules. We have already noted that one kilogram of TNT is equivalent to 4×10^6 joules, so the energy of impact can also be expressed as 65×10^{15} kg of TNT = 65,000,000 megatons of TNT = 65 gigatons of TNT. For comparison, the fission bombs dropped on Hiroshima and Nagasaki had a yield of 20 kilotons, and the largest nuclear device ever tested was a Soviet bomb with a yield of 53 megatons.

SUMMARY

The Moon is the best-studied planetary body after the Earth, as the result primarily of the Apollo expeditions and their treasure of returned lunar samples. Unlike the Earth, however, our satellite is no longer geologically active, nor does it possess an atmosphere. Therefore, the Moon is a much easier place to understand, although still complex enough to leave us with a number of fundamental questions unanswered.

The Moon is a differentiated planet, with a highland crust of anorthosite that solidified at least 4.4 billion years ago, much older than any rocks that still are present on the Earth. During its first half-billion years, the Moon (and probably the other inner planets as well) was subjected to an intense meteoroidal bombardment, which saturated its surface with craters and created a deep regolith of broken rock. The major basins, including Orientale and Imbrium, were formed near the end of this period. About 4 billion years ago a final burst of impacts occurred, after which the bombardment dropped to approximately its current rate. The highland crust preserves a record of this early period, while the much less cratered lunar maria represent younger surfaces.

The maria, which cover 17% of the surface, are composed of very fluid basaltic lava flows that erupted between 3.9 and 3.2 billion years ago from magma reservoirs deep beneath the lunar crust. They generally occupy low-lying areas on the nearside of the Moon, primarily within impact basins. There is no evidence of any major activity during the past 3 billion years, a period when the lunar surface has been modified only by impact cratering, including the creation of a fine, glass-rich soil from the eroding effects of many small impacts.

The dominant landform of the lunar surface is the impact crater, ranging in size from basins with diameters of 1000 km down to microscopic pits. We discuss the craters in some detail, since they are a common phenomenon on other planets; their characteristic shapes (including central peaks, ejecta blankets, and secondary craters) are the result of violent explosions caused by the sudden impacts of fast-moving meteoroids. The density of craters is determined by the rate at which meteoroids strike the surface and by the age of the surface. Crater counting provides a good measure of relative surface ages, but it must be used with caution when trying to estimate absolute ages.

The Moon differs greatly from the Earth. In spite of their proximity in space, the Moon's bulk chemical composition is depleted in both metals and volatiles relative to the Earth. This is a major mystery, which we return to in Chapter 8. In addition, its small mass, less than 2% that of the Earth, has led to much lower levels of geological activity. With its relatively cool, solid interior, the Moon has been unable to form the continents, mountain ranges, volcanoes, and other characteristic geological features of our own planet. Also because of its low mass, the Moon has no atmosphere, creating a very different surface environment of extreme temperatures, with constant bombardment by micrometeorites and cosmic rays. Although unlike the Earth, the Moon is rather similar to one other planet, Mercury. In the next chapter we compare the Moon and Mercury, and probe further into the origin and evolution of both.

KEY TERMS

anorthosite	magma
catastrophism	mare
cosmic rays	maria
crater	regolith
crater density	resolution
crater retention age	secondary crater
ejecta blanket	stratigraphy
highlands	uniformitarianism
impact basin	

Review Questions

1. Describe the appearance of the Moon, noting what kinds of features can be seen at different levels of resolution.

2. Impact cratering is an important process in planetary geology. Describe how craters are formed and note what features are characteristic of impact craters. How can you tell an impact crater from a volcanic crater? Why did it take so long for scientists to recognize that the lunar craters are of impact origin?

3. Determining the age of a planet's surface is another basic part of planetary geology. Discuss what is meant by stratigraphy, with illustrations from the history of formation and flooding in the Imbrium Basin. Discuss the concept of crater density and crater retention age. How are these relative ages translated into absolute ages for the lunar surface?

4. Describe the standard paradigm for the cratering history of the inner solar system. Note how our detailed knowledge of the Moon is used to define the stages in this history. What role is played by the asteroids and comets, and is this theory consistent with what we learned about asteroids and comets in Chapters 5 and 6?

5. Discuss volcanism on the Moon. When and where did the magmas form? Why do the lava flows occupy the nearside basins almost exclusively? Why are there so few volcanic mountains? Why did the volcanic activity cease billions of years ago?

6. Discuss the soil on the Moon. What erosional processes are at work? How do large impacts create and modify the surface layers? What are the roles of micrometeorites and cosmic rays? What forms the lunar glass?

Quantitative Exercises

1. Compare the impact energy of a 1000-ton primary impact on the Moon (at 15 km/s) with the energy of a secondary impact of the same mass but at a speed of 1.5 km/s.

2. According to the crater scaling law discussed in this chapter, the diameter of a stony asteroid impacting the Moon is about 1⁄10 the diameter of the resulting crater. Calculate the diameter and mass of the projectile that formed Tycho. What was the energy of this impact in megatons if the impacting speed was 15 km/s?

3. Consider a meteoroid impacting the surface at exactly the escape velocity. If it makes a 1-km crater on the Earth, what size crater would it make on asteroid Ceres, which has an escape velocity given by $\sqrt{2GM/R}$, where G is the gravitational constant (6.67×10^{-11} newtons m^2 kg^{-2}) and M and R are the mass and radius, respectively?

4. It can be estimated that the gravitational binding energy on the Moon is given approximately by GM^2/R, where G is the gravitational constant (6.67×10^{-11} newtons m^2 kg^{-2}) and M and R are the Moon's mass and radius. An impact that releases more than this amount of energy is capable of breaking the

Moon apart. Calculate the diameter of an impacting asteroid that would generate this amount of energy. Also calculate the size of the resulting crater, using the simple arguments for crater size scaling given in this chapter, and compare this size with the diameter of the Moon itself. This will give you an idea of the largest possible crater or basin that could be made on the Moon without disrupting the object entirely.

Additional Reading

Chapman, C.R. 1982. *Planets of Rock and Ice: From Mercury to the Moons of Saturn* (Chapters 2, 3, 5, and 8). New York: Scribner's. Popular introduction to planetary geology by a leading planetary scientist.

Cooper, H.S.F. 1970. *Moon Rocks.* New York: Dial Press. Journalist's account of the acquisition and study of lunar samples, originally published in *The New Yorker.*

Cortwright, E.M., ed. 1975. *Apollo Expeditions to the Moon* (NASA SP-350). Washington, D.C.: U.S. Government Printing Office. Beautifully illustrated NASA publication describing the Apollo missions and their scientific harvest of information.

French, B.M. 1977. *The Moon Book.* New York: Penguin Books. Highly readable introduction to the Moon, now somewhat dated.

*Guest, J.E., and R. Greeley. 1977. *Geology on the Moon.* London: Wykeham. Textbook and reference by two lunar geologists; a good introduction to planetary geology.

Masursky, H., *et al.*, eds. 1978. *Apollo over the Moon: A View from Orbit* (NASA SP-362). Washington, D.C.: U.S. Government Printing Office. Beautifully illustrated book of the best Apollo orbital photos, with informative text on lunar geology.

*Taylor, S.R. 1975. *Lunar Science: A Post-Apollo View.* New York: Pergamon Press. Authoritative and well-written monograph on lunar geology and geochemistry, aimed at the science student (now somewhat dated).

*Indicates the more technical readings.

8

The Moon and Mercury: Strange Relatives

Mercury was traditionally known as the messenger of the gods. Fleet of foot, he is usually depicted with wings on his heels and a winged helmet, as in this bronze sculpture by Francesco Righetti (1749–1819).

The next step toward larger-size planets takes us from the Moon to Mercury. In this chapter, we discuss Mercury and compare this planet with our satellite. The five terrestrial bodies—Mercury, Venus, Earth, Moon, and Mars—are an appropriate group for comparison. The larger three, Venus, Earth, and Mars, all have substantial atmospheres and long histories of geological activity. In contrast, Mercury and the Moon invite comparison to one another as smaller, airless, less active worlds.

Although the Moon and Mercury are similar in many ways, they have one outstanding difference: their bulk composition. The Moon is a rocky planet, depleted in metals. In contrast, Mercury is a planet made mostly of metal, as is evident from its high density. Even though the two objects look alike, we must remember that the resemblance is only skin deep: In their interiors they are very different.

Although the Moon and Mercury are simpler planets to understand than Venus, Earth, or Mars, we should not delude ourselves into thinking we have solved all of their mysteries. Both are complex worlds, and even for the Moon our available information is infinitesimal in comparison with what we know about the Earth. In the case of Mercury, we have never even seen half the surface! It is sobering to remind ourselves that, with all the missions that have studied the Moon and hundreds of kilograms of returned samples, we are still not certain how and where our satellite originated. How much more uncertain may be the conclusions drawn about other, more distant worlds?

8.1 AN ELUSIVE PLANET

Because of its proximity to the Sun, Mercury is a difficult planet for astronomers to study. There is a story, perhaps apocryphal, that the great theorist Copernicus said he had never seen the planet. Yet it was well known to observers in antiquity, although not all of them recognized that this twilight "star" was the same object when seen in the morning before sunrise or in the evening after sunset.

Appearance of Mercury in the Sky

If you observe the motions of Mercury or Venus for several months, you can see that these two planets are closely tied to the Sun, never moving very far from it in the sky. Because Mercury and Venus have smaller orbits than that of the Earth, they are sometimes called *inferior* planets. The maximum angular distance of Venus from the Sun is about 45°, compared with 30° for Mercury. In the following discussion we will focus on Mercury, but with due allowance for its larger orbit, the same descriptions apply to Venus.

Figure 8.1 shows the apparent position of Mercury as evening star over a two-month period as you might draw it each evening about half an hour after sunset. Mercury is actually quite bright, about as bright as the brightest stars, so it is one of the first objects to become visible after sunset. Mercury never strays far from the twilight sky, but Venus, with its larger orbit, can set as long as three hours after the Sun, allowing it to be seen against the dark sky.

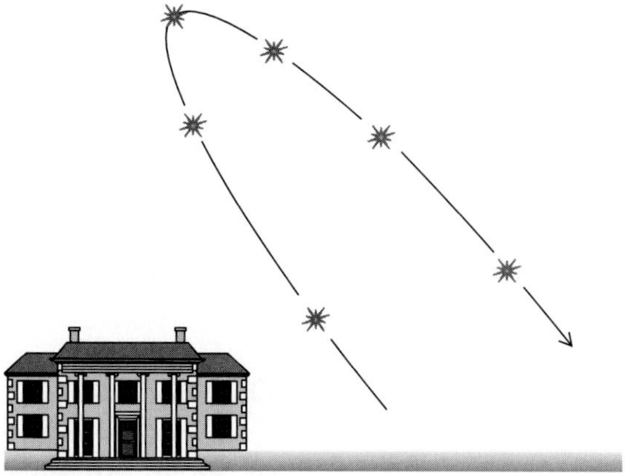

FIGURE 8.1 Appearance of Mercury as an "evening star," showing its apparent position at weekly intervals, as seen in the western sky a few minutes after sunset.

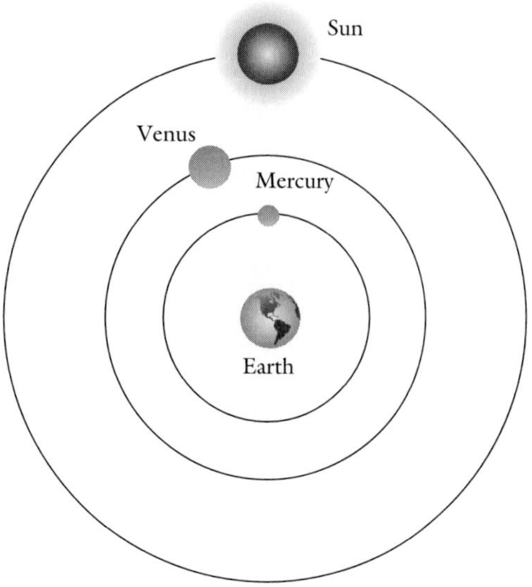

Geocentric and Heliocentric Perspectives

Do the planets circle the Earth or the Sun? This fundamental question was debated in the Western world for two millenia and not settled until the time of Kepler and Galileo. During most of that time, the geocentric (Earth-centered) ideas prevailed. Yet the apparent motions of Mercury and Venus draw us naturally to the idea of a Sun-centered system. Figure 8.1 shows how Mercury appears to move away from the Sun until it reaches a maximum distance (eastern elongation). It then appears to reverse its motion and gradually swings back to disappear in the Sun's brilliance. A few weeks later the planet reappears in the morning sky, moves out to its greatest western elongation, and then reverses. The entire pattern repeats every four months. Venus does the same thing, but with a period of about 20 months.

In the geocentric cosmology, Mercury and Venus each circled the Earth, but they were attached to an imaginary line that connected the Earth with the Sun (Fig. 8.2a). In this theory their apparent motion was not due to their path

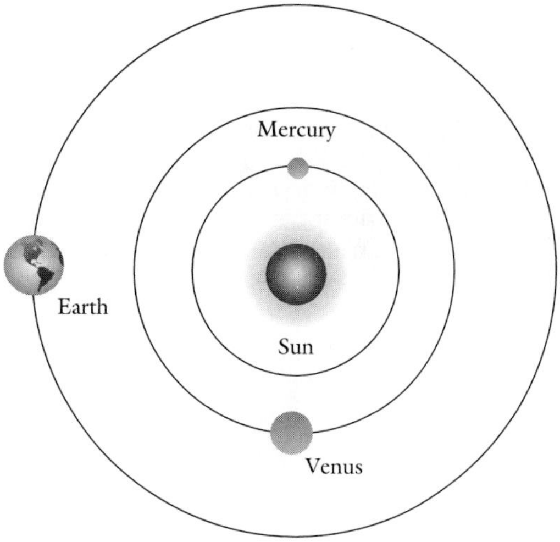

FIGURE 8.2 Motion of Mercury as described in the geocentric and heliocentric systems. (a) Geocentric view: Mercury is constrained to move along only a part of its supposed orbit around the Earth, staying always near the Sun in the sky. (b) Heliocentric view: Mercury orbits the Sun just like all the other planets. (Compare Fig. 1.16.)

around the Earth but was the result of motion on a second circle, called an epicycle, that rotated about a hub on the Sun-Earth line. According to this theory, these planets were always illuminated from behind by the Sun, and they should therefore always display a crescent phase.

Figure 8.2(b) illustrates the heliocentric theory for the motion of Mercury. The description is now simplified; if Mercury really circles the Sun, then its close attachment to the Sun is immediately understood without a requirement for epicycles or other artificial contrivances. But note also that the phases of the planet are different. It is now crescent in phase only during the half of its orbit when it lies between the Sun and Earth. For the other half of its orbit, Mercury is on the far side of the Sun as seen from the Earth and is therefore more fully illuminated. The same sequence of phases applies to Venus as well. Galileo's telescopic observations of the phases of Venus provided strong early support for the Copernican theory, as we discussed in Chapter 1.

8.2 THE ROTATION OF MERCURY

Mercury is a disappointment to any telescopic viewer, with its small size and the difficulties of observing it near the Sun. It is a lucky (and per-

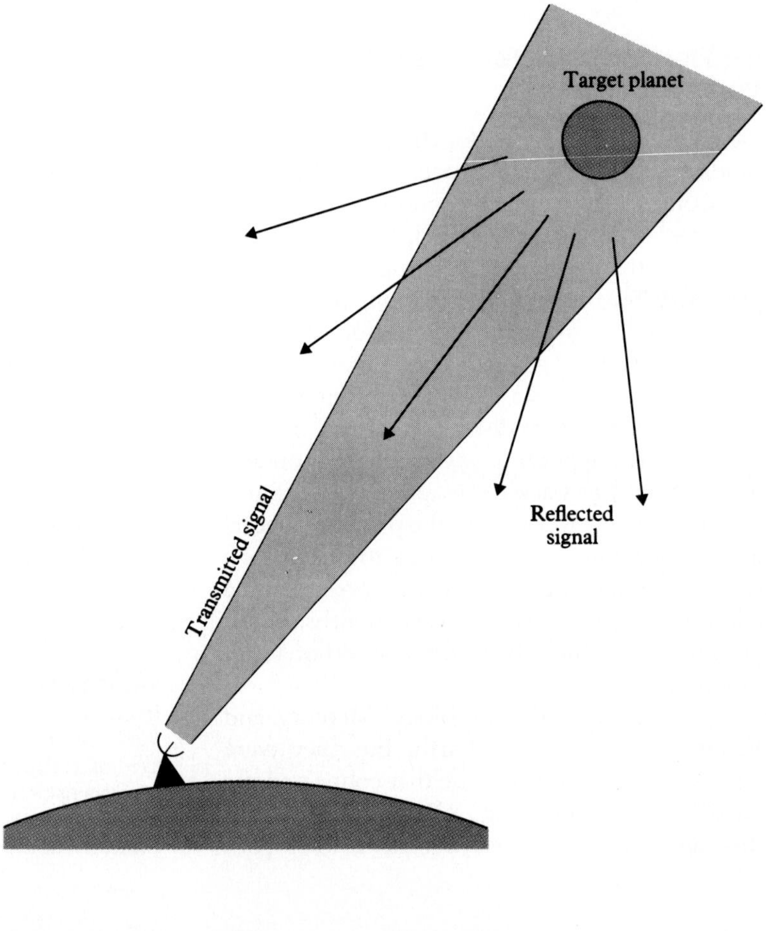

FIGURE 8.3 A radar transmitter sends a powerful signal into the Earth-based antenna (lower left). This antenna then focuses this radiation into a parallel beam and directs it toward a planet (upper right). The planet in turn reflects the beam, and a small fraction of this reflected radiation reaches the radio telescope or antenna, which is equipped with a receiver in order to detect this returning signal.

sistent) astronomer who has succeeded in seeing any markings at all on its surface. For this reason, Mercury was a neglected object until the development of radar astronomy in the early 1960s.

Radar Astronomy

Radar is a powerful tool of the modern planetary scientist. The principles of this technique are essentially the same whether applied to a distant planet or a nearby airplane on its approach to a foggy airport. An intense pulse of radio radiation is emitted by a transmitter in the direction of the target, and a very, very much weaker echo reflected from the target is picked up by a sensitive radio receiver (Fig. 8.3).

The most powerful planetary radar system in the world uses the 1000-foot antenna suspended in a bowl-shaped valley near the town of Arecibo in Puerto Rico (Fig. 8.4) as both a transmitter and a receiver. The radar echo from Mercury is typically only 1 millionth of a trillionth (10^{-18}) as strong as the transmitted pulse, and the round-trip time (at the speed of light) is more than 10 minutes.

By transmitting a radio pulse of carefully controlled duration and frequency, a radar system can reveal much more about a target than would be achieved by simply illuminating it with ordinary light. The time required for the radio pulse to cover the two-way path to the target and back provides a measure of the distance to the target. In addition, the motion of the target toward or away from the observer can be measured using the Doppler effect.

The **Doppler effect** describes the dependence of radio wavelength or frequency on the relative motion of the source and receiver. Actually, we experience the Doppler effect most commonly in the realm of sound, which is propagated, like light and radio, in the form of waves. A sound source moving toward a listener has a higher pitch (shorter wavelength), while motion away results in lower pitch (longer wavelength) (Fig. 8.5). We hear these effects daily in the sounds of police and fire truck sirens or in the familiar "whoosh" as trains or cars pass by us, their pitch suddenly dropping as the relative motion of the vehicles changes from approaching us to moving away.

FIGURE 8.4 The 300-m (1000-foot) Arecibo radio telescope in Puerto Rico is the most powerful radar instrument on our planet. It is this telescope that first determined the true rotation period of Mercury.

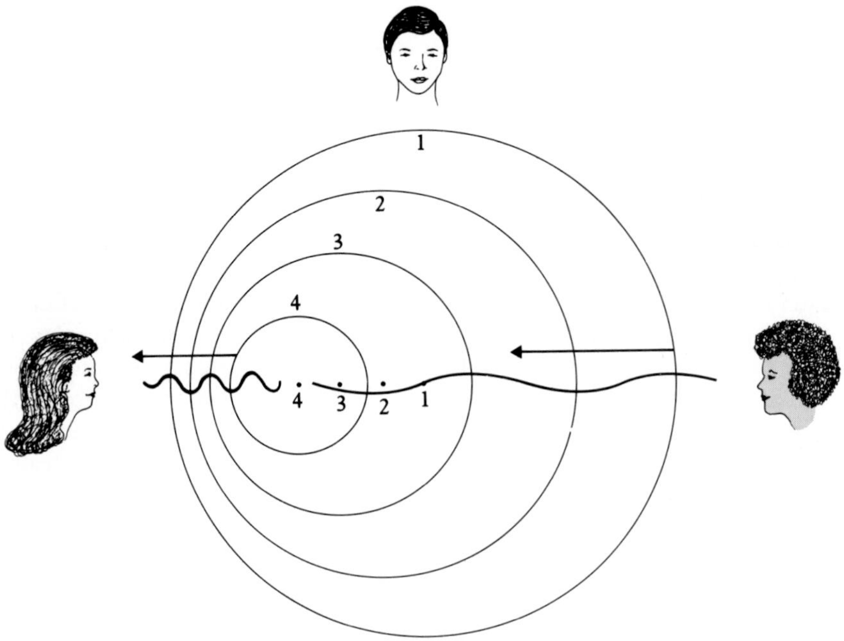

FIGURE 8.5 The Doppler effect. Suppose a source of sound or light is moving from position 1 to position 4. Motion toward the observer at left crowds the waves of sound or light together, so the wavelength is shorter and the frequency is higher. Meanwhile, the motion away from the observer at the right spreads the waves, increasing the wavelength and decreasing frequency. The observer at the top sees no change in wavelength or frequency, since the source moves neither toward nor away from him.

When either sound waves or radar waves are reflected from a moving target, the Doppler shift in frequency is proportional to the target's relative speed toward or away from the transmitter/receiver system. In the case of planetary radar systems, the frequency of the transmitted pulse is controlled precisely, so that even tiny changes in the frequency of the echo can be detected and analyzed.

The first results from planetary radar were the determination of the radar reflectivities of the Moon, Venus, and Mercury, and precise measurements of their distances. Careful tracking of the motion of Venus, in particular, was critical in defining the distance scale of the solar system and thereby making possible the accurate navigation of spacecraft on complex interplanetary trajectories. The planetary radar observations of the 1960s were an essential prelude to the Mariners, Pioneers, and Voyagers of the 1970s. They also provided new data on the planets.

Rotation of Mercury

Radar provided our first measurements of the rotation rates of Venus and Mercury. The determination of a planet's rate of rotation by radar uses the Doppler effect in a clever way. Since the incident radar beam illuminates an entire planetary hemisphere, the rotation of the planet as well as its orbital motion alters the frequency of the reflected signal (Fig. 8.6). As seen from the Earth, the effect of rotation is to make one side of the planet approach us while the other recedes, relative to the overall motion. The echo from the approaching side of the planet is shifted toward higher frequency, and that from the receding side is lowered in frequency. As a result, the single frequency of the transmitted pulse is *broadened* in the returned echo, which includes reflections from a full hemisphere. The faster the rotation, the broader the frequency range in the echo.

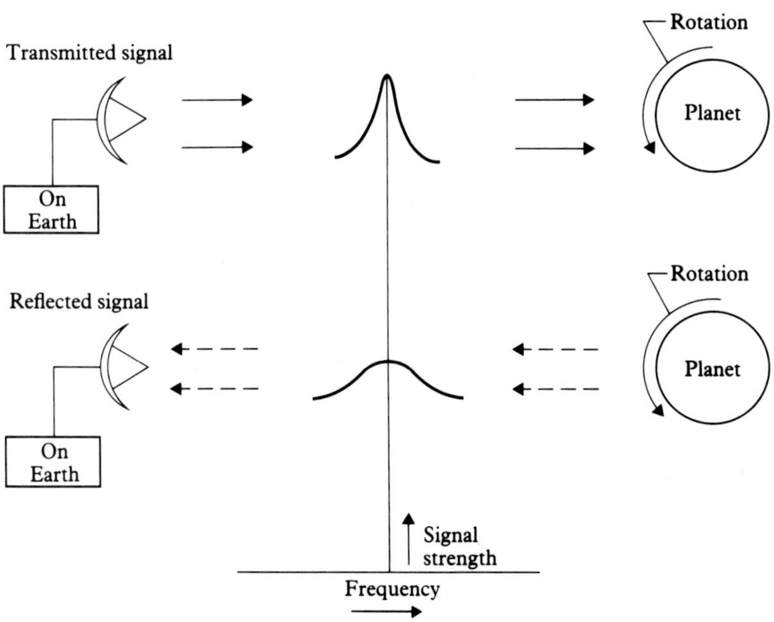

FIGURE 8.6 The broadening of a radar signal by planetary rotation. The side of the planet approaching the radar shifts the signal toward higher frequencies. The net effect is to broaden the frequency of the transmitted signal.

When the Doppler radar technique was first applied to Mercury in 1965, the frequency width of the echoes did not conform to astronomers' expectations. Ever since nineteenth-century maps had been made showing faint markings on Mercury, it was believed that Mercury always kept the same face toward the Sun, just as the Moon does with respect to the Earth. Astronomers thought that Mercury, unlike the Moon, really had a dark side, a hemisphere never illuminated by the Sun. Mercury was therefore thought to be simultaneously the hottest and coldest place in the planetary system. (We now know it is neither.) All maps of the visible markings on the planet were of the supposed Sun-facing side only.

The radar results did not yield the expected 88-Earth-day rotation period, equal to the period of revolution. Instead, the indicated rotation period was approximately 59 days. Subsequent data have demonstrated that the rotation period of Mercury is exactly two-thirds of its 88-day period of revolution, or 58.65 days.

It is no coincidence that the rotation period of Mercury is an exact fraction of the orbital period. Section 8.3 further examines the probable reason for this curious spin-orbit coupling, but first we examine the consequences it has for Mercury itself. In the process, we must consider just what we mean by the terms *day* and *year* as applied to another planet. The year on a planet is its period of revolution about the Sun, the time necessary to complete one orbit. The term *day* is more ambiguous, at times referring to the rotation period—that is, the time for one complete spin relative to the fixed stars. Here we shall adopt the other, more familiar, meaning of day as the time interval between successive sunrises. This is also called the **solar day.**

A year on Mercury lasts approximately 88 Earth days, during which time the planet rotates one and a half times on its axis. Since the orbital

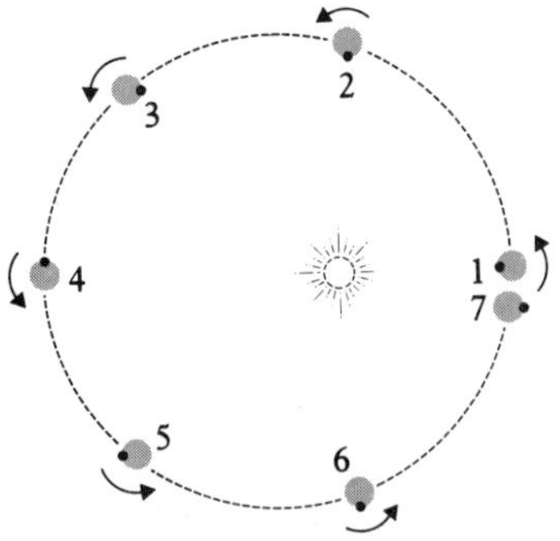

FIGURE 8.7 Mercury rotates on its axis once every 58 ⅔ days and orbits the Sun every 88 days. Hence the planet rotates three times (3 × 58 ⅔ = 176 days) for every two revolutions about the Sun (2 × 88 = 176 days). Starting at position 1, the planet completes one full rotation when it reaches position 5. At 7, the point on the surface that originally faced the Sun now points directly away from it. After the second orbit, this point will again face the Sun.

motion and the spin are in the same direction, however, one partially cancels the other, and the Sun appears to move only halfway through its diurnal cycle in this span of time (Fig. 8.7). The solar day on Mercury is equal to two orbits and three complete rotations, or about 176 Earth days. Thus on Mercury the solar day is longer than the year.

The situation for Mercury is further complicated by the large eccentricity of its orbit, which takes it from 0.308 AU away from the Sun at perihelion to an aphelion distance of 0.467 AU. The corresponding range in apparent diameter of the Sun is from 1.6° to 1.1°, compared with an apparent size of the Sun as seen from the Earth of about 0.5°. Each solar day on this planet includes two perihelion and two aphelion passages.

Strange Days and Seasons

Let us imagine the appearance of the Sun to a race of hypothetical Mercurians, beginning with a group living at a longitude where the Sun is overhead at perihelion. There are two such hot longitudes, on opposite sides of the planet. At sunrise and sunset the Sun will be at its most distant. The Sun, therefore, rises small and increases in size as it approaches the zenith. At the same time, the combination of rotation and orbital motion causes the apparent motion of the Sun to slow, until it hangs nearly motionless overhead while the planet races through perihelion. Speeding up as it shrinks, the Sun then dips toward its setting point. The stars will meanwhile be moving through the sky about three times as fast as the Sun. A star that rises with the Sun will set before noon and will rise again before sunset.

Now consider an observer situated at a longitude 90° away. For her, the Sun will be small and rapidly moving at noon but large and nearly stationary at sunrise and sunset. Will such a Mercurian wonder about the favored lands 4000 km to the east or west where the Sun stands still at noon, or will she be grateful that the Sun lingers at rising and setting when its light is most desired? What interesting conversations two Mercurians separated by 90° in longitude would have were they to compare their cosmologies!

In addition to their different cosmological perspectives, the two longitudes that face the Sun at perihelion clearly have a very different thermal history from the longitudes that face the Sun at aphelion. The perihelion longitudes are called the hot longitudes, because they receive extra heating as a result of both the proximity of the Sun at noon and the way it lingers overhead for many Earth days. In contrast, the maximum surface temperatures at the cool longitudes 90° away are more than 100° C lower, although still a sizzling 550 K.

8.3 TIDES AND THE SPIN OF PLANETS

Both the Moon and Mercury have rotation periods closely linked to their orbital motion. For the Moon, the two periods are equal and the same side always faces the Earth. Mercury has a unique 2:3 resonance between the two periods. Many outer planet satellites also experience spin-orbit coupling. How were these linkages forged, and what are the consequences for the evolution of planets and satellites?

The most effective agents for altering the spins of planets are **tidal forces**—the forces that generate tides. A tide is a distortion in the shape of one body induced by the gravitational pull of another nearby object. The tides with which we are most familiar, of course, are those generated in the Earth's oceans by the attraction of the Moon and Sun.

The Nature of Tides

If a planet were completely rigid, the gravitational attractions of other bodies would act on it as if its mass were all concentrated in a point at its center, the so-called *center of mass*. However, there must also be a *differential* gravitational force that results from the fact that planets are not point masses, and that the part of one body facing toward another is more strongly attracted than the part that is turned away. This tendency to pull harder on nearer parts and thereby to distort the shape of one planet when it is near another is a *tidal force*. It is easy to see that the part of the Earth that faces the Moon is more strongly attracted than is the center of mass. Note, however, that there is an equally effective differential force acting on the opposite side of the Earth and directed away from the Moon since this part of our planet's surface experiences less attraction than does the center of mass.

The tidal force of the Moon on the Earth sets up horizontal forces that cause the water of the oceans to move toward two areas, one approximately facing the Moon and one on the opposite hemisphere. The result is two tidal bulges in the ocean (Fig. 8.8), oriented toward and away from the Moon. As the Earth rotates, each point on the surface passes through these bulges, experiencing two high tides and two low tides each day. Similar but smaller tidal bulges result from the pull of the Sun, and the highest tides occur twice a month when the two bulges are aligned (Fig. 8.9).

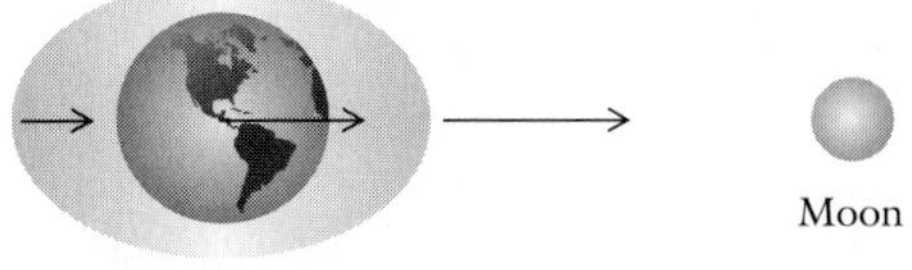

FIGURE 8.8 Tidal forces: The gravitational pull of the Moon is strongest on the side of the Earth facing the Moon and weakest on the opposite side. The arrows (exaggerated) illustrate these tidal forces.

Tidal forces are strongly dependent on the separation of the objects; mathematically they are proportional to the mass of the perturbing body and to the inverse cube of its distance (M/D^3), as opposed to an inverse square law for gravity (Section 1.4). Thus tides affect most strongly satellites in close orbits around their primaries, and tides raised by the Sun are much more important for Mercury than for any more distant planet.

Tidal Friction and Orbital Evolution

If the Earth did not rotate with respect to the Moon, the lunar-induced tidal bulges would align directly with the Moon. On a rotating Earth without friction between the oceans and the land, the bulges would maintain their lunar alignment while the solid Earth rotated beneath. This is approximately the case, and an observer on the shore watches the tide rise and fall twice a day. Friction occurs, however, as the restless seas wash against the shores of the continents, inducing a lag in the tidal bulges. This lag, which varies from one part of Earth to another, explains why high tide does not take place when the Moon is overhead (or underfoot) but usually follows by several hours.

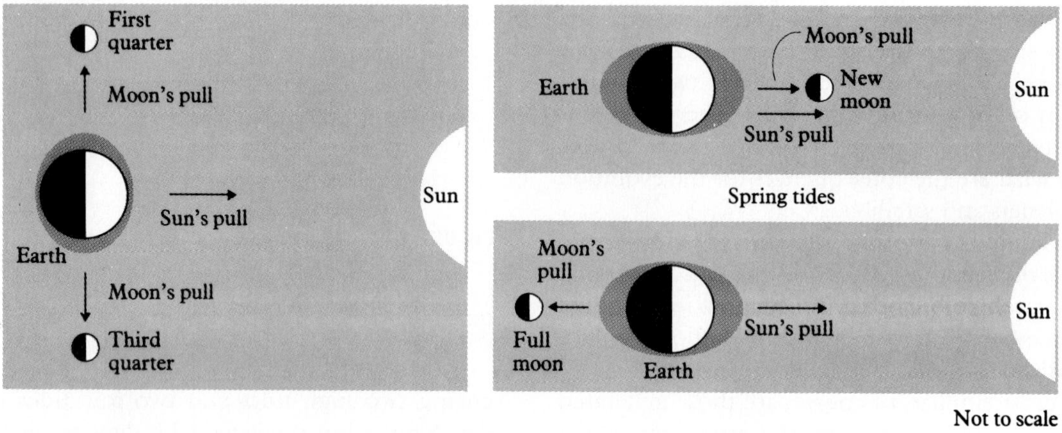

FIGURE 8.9 When the Sun and the Moon pull on the Earth at right angles to one another, the tides are smaller than when they pull together or opposite one another.

The constant friction between the ocean tides and the land has several important consequences. First, it slows the rotation of the Earth. About 2 billion horsepower is dissipated by ocean tides, lengthening the day by about 500 millionths (5×10^{-8}) of a second per day. Part of the energy lost by the spinning Earth results in a very small frictional heating of the ocean and our planet's surface. Part also goes to alter the orbit of the Moon.

As the Earth slows its spin, the Moon is gradually forced away into a larger orbit, an example of the conservation of angular momentum. The total angular momentum of the Earth-Moon system includes the spin of both objects and the revolution of the Moon around the Earth. If the Earth's rotation is slowed, losing angular momentum, some other part of the system must gain equivalent momentum. The Moon absorbs this additional momentum by moving outward.

Long ago, the Moon was much closer to the Earth and the length of the month much shorter. Calculations suggest that at one time the Moon may have circled the Earth in as little as a week, while at the same time the solar day on the Earth would have been as short as six hours. Barring unforeseen events, the coupled evolution of the lunar orbit and the rotation of the Earth will continue until the Earth has slowed down to the point where it keeps the same side always facing the Moon.

We noted in Section 1.1 that total eclipses of the Sun are possible on Earth because the Moon is just large enough in apparent size to cover the face of the Sun. We now see that this situation applies only briefly on a cosmic time scale. In the past, the Moon was nearer and solar eclipses were more common, while in the future the Moon will be too far away to block the Sun completely, and there will be no total eclipses of the Sun.

Everything we have said about the effects of the lunar tides on the Earth is equally applicable to the tidal effects of the Earth on the Moon. To be sure, the Moon has no oceans, but a significant tidal bulge can be formed even in a solid planet by gravitational distortion of the rock. Such body tides can also dissipate energy. As the Moon rotates with respect to the bulge, it is subjected to forces that compress and bend its rocky crust and mantle. Some energy is released through tidally induced moonquakes, some by heating of the rocks. The seismometers deployed as part of the ALSEP experiments were able to record these moonquakes. Just as in the case of ocean tides, this release of energy slows the

Moon's, or any planet's, spin until the same side faces its companion.

In the distant past the effects of energy dissipation on the Moon from Earth-induced tides slowed the rotation of our satellite until it reached a rotation period precisely equal to its period of revolution. This **synchronous rotation** is a highly stable situation since it minimizes frictional energy loss. Once synchronous rotation is reached, it will be maintained; if the orbital period should change slightly, the rotational period adjusts itself accordingly. Most of the satellites in the solar system are in synchronous rotation about their planets.

Mercury and the Effects of Solar Tides

If synchronous rotation is favored and Mercury is the planet most strongly affected by solar tides, why does it not keep the same face toward the Sun? How did it get into its present peculiar rotational state? The answer lies in its large orbital eccentricity, which trapped this planet into a 2:3 synchronous rotation.

Imagine that Mercury started its existence with a relatively rapid rotation and that early in its history it was slowed by solar tides. At the same time its eccentric orbit carried it first closer to the Sun, then farther away. Since tidal forces are so sensitive to distance (varying as the inverse cube), most of the tidal dissipation took place near perihelion. As Kepler's second law (Section 1.3) tells us, however, the orbital velocity of the planet near perihelion is much higher than the average, and in Mercury's case the orbital speed at perihelion matches a spin period near ⅔ of the orbital period. That is, near perihelion, when the tidal forces are strongest, Mercury finds itself turning on its axis in apparent synchronism with its orbit for a rotation period near 60 days. This is the same effect that causes the Sun to appear to stand still in the mercurian sky near perihelion.

Once Mercury's tidal evolution carried it into the 2:3 synchronous rotation period, it was trapped. Given the orbital eccentricity, this spin minimizes tidal dissipation near perihelion, where it is most effective. Therefore, the planet remains in a stable rotational state, with two fixed tidal bulges that face the Sun at alternate perihelion passages.

What would happen if we magically intervened and placed Mercury into a 1:1 synchronous rotational state? It could not speed up to return to the 2:3 state, since friction can only slow a body's spin and not increase it. In spite of keeping the same face toward the Sun on the average, however, Mercury would still experience significant **tidal heating.** Each time it neared perihelion, the increased force of solar gravity would swell its tidal bulge, while the changing direction to the Sun would twist and distort it. Jupiter's satellite Io finds itself in precisely this situation today, where the resulting tidal heating causes a high level of volcanic activity (Chapter 15). Past tidal heating has also been implicated in the thermal histories of many bodies in the planetary system, as we will see in later chapters.

8.4 THE FACE OF MERCURY

Almost everything we know about the geology of Mercury was learned from a single spacecraft, Mariner 10, which made three flybys of the planet in 1973 and 1974. For a few months this small planet held center stage in the planetary exploration program, before settling back again into relative obscurity.

The View from Mariner 10

Mariner 10 provided resolution and coverage of Mercury very similar to that of the Moon seen by telescopes from the Earth: one hemisphere was photographed at a resolution of approximately 1 km, while the other hemisphere was not seen at all. The spacecraft cameras revealed a planet that looked remarkably similar to the

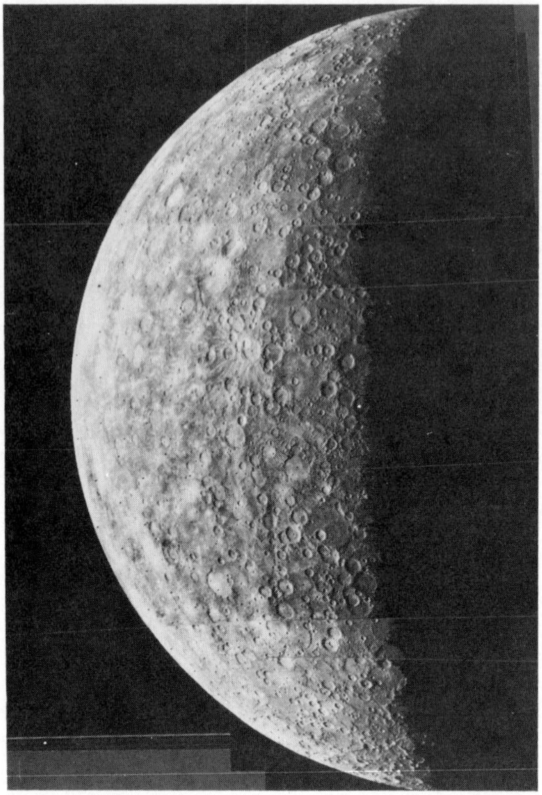

a

b

FIGURE 8.10 Two mosaics produced by the Mariner 10 spacecraft in 1974 show the visible hemisphere of Mercury as the spacecraft approached (a) and as it receded (b) from the planet. Mariner 10 was able to examine about 45% of Mercury's surface. The remaining 55% remains unexplored.

Moon (Fig. 8.10). Although an abundance of impact craters had been expected for this small, airless world, the degree of similarity of the mercurian surface to the lunar highlands was surprising. Most geologists had anticipated greater evidence of internal geological activity, since Mercury is larger than the Moon, but the Mariner photographs revealed only one lava-flooded impact basin, suggesting the mercurian volcanism was less than on the Moon, or it had been expressed in different ways.

While the composition of the surface material on Mercury has not been measured directly, the colors and reflectivity of this planet are similar to those of the lunar highlands; the surface therefore likely consists of igneous silicate rock. Basaltic lavas such as those of the lunar maria are possible, but the generally higher reflectivity of the surface is suggestive of compositional differences, perhaps associated with a higher metal content in the mercurian soil. There is no way to know for certain whether the crust of Mercury is depleted in volatiles like that of the Moon, or whether it shares any of the other chemical peculiarities of our satellite. Nor is there any way to measure the age of the surface, beyond noting that the heavy cratering probably implies great antiquity. In the absence of returned samples, interpreting the composition and history of another world is very difficult!

On a small scale, we know that Mercury has a regolith and that its surface soil has a texture resembling that of the Moon. These similarities can be deduced from the similar way the two objects reflect sunlight and emit thermal radio energy. With our post-Apollo knowledge of the Moon as a benchmark, we can interpret these astronomical studies of Mercury with greater confidence. Surface temperatures have also been measured, ranging from a high of about 675 K at noon at the hot longitudes (high enough to melt zinc), to a low of about 90 K over most of the night hemisphere (low enough to freeze carbon dioxide). The lowest temperatures on Mercury are nearly equal to those on the Moon, since the higher daytime temperatures are compensated by a longer cooling period during the 88-Earth-day mercurian night.

One of the most surprising discoveries about Mercury in the years since Mariner 10 is the presence of polar caps, as revealed by radar. A **polar cap** is a permanent deposit of water or other frozen volatiles in the cool polar regions of a planet. The Earth and Mars have polar caps, but no one expected such a phenomenon on Mercury, which is much closer to the Sun. Yet both poles of Mercury show enhanced radar reflectivity of the sort expected from surface or near-subsurface deposits of water ice.

Since the spin axis of Mercury is almost exactly perpendicular to the plane of its orbit, very little sunlight falls on the poles and they are therefore cool. Strictly in terms of temperature, ice caps at the poles are possible. But where did the ice come from? Without an atmosphere, the water cannot have been transported to the poles from lower latitudes. Is this primordial water dating from the formation of Mercury? And if Mercury has polar caps, why not the Moon? These questions do not yet have satisfactory answers, although a radar experiment carried out with the Clementine lunar orbiter *suggested* the *possibility* that there might be a similar ice deposit near the north pole of the Moon.

Nomenclature for Surface Features

Most of the craters on the Moon are named for scientists or philosophers of the ancient world. Mars, which was the next planet to be extensively mapped, follows a similar scheme, although many of the scientists commemorated on Mars are from the nineteenth and early twentieth centuries. When Mariner 10 arrived at Mercury, however, astronomers decided to adopt a different scheme, naming the newly discovered craters for great artists, writers, composers, architects, and other heroes of world culture.

Among the most prominent craters on Mercury are Bach, Shakespeare, Tolstoj, Mozart, and Göthe. Other craters bear the names of Al-Hamadhani, Balzac, Bashō, Bramante, Imhotep, Ibsen, Juda Ha-Levi, Mark Twain, Phidias, Raphael, Rublev, Valmiki, Verdi, Vyasa, and Wang Meng (to name a few).

Geology of Mercury

The ubiquitous impact craters on Mercury bear many resemblances to their lunar counterparts. Where differences occur, they are probably due to the higher surface gravity on Mercury (0.38 of the terrestrial value, as opposed to 0.21 for the Moon). As a result, ejecta blankets are smaller, rarely extending more than one crater radius beyond the rim, and the secondary craters are similarly more confined in their distribution.

The density of craters on Mercury varies from values near that of the lunar highlands down to about one-tenth of this value (from approximately 1000 down to 100 10-km craters per million square kilometers). There are no large regions with crater densities as low as those found on the lunar maria. If Mercury experienced a period of volcanism that produced a counterpart to the lunar maria, this period must

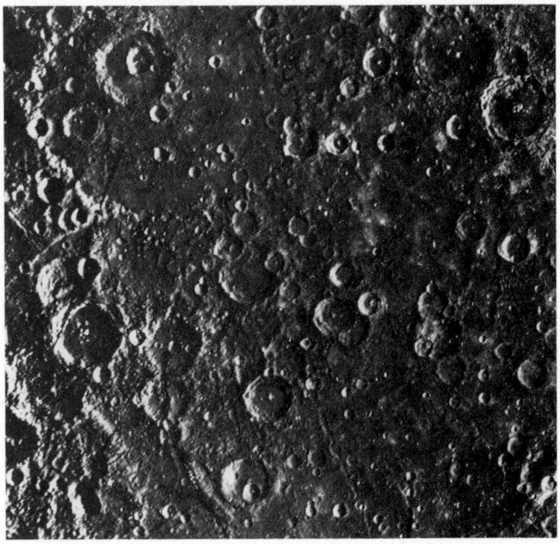

FIGURE 8.11 The cratered plains of Mercury are less heavily cratered than the lunar highlands but are more heavily cratered than the lunar maria. They may be lava plains similar to the maria but somewhat older, or they may have some other origin.

have been contemporary with the late heavy bombardment rather than occurring after the impact flux had dropped, as was the case for the Moon.

The less cratered regions of Mercury do not look like lunar maria (Fig. 8.11). They do not differ from their surroundings in brightness or color, nor are they as flat as maria. No flow fronts, sinuous rilles, or other distinctively volcanic features have been identified in the Mariner pictures. Thus geologists are left with a fundamental question: What destroyed some older craters to produce these moderately cratered plains? Was it volcanism that took place during the later stages of heavy meteoritic bombardment but did not leave the characteristic marks of the lunar maria? Or was some other process responsible, such as blanketing of the surface by impact-produced ejecta? No one knows, although a vote among geologists would probably favor some kind of volcanic process.

The largest impact basin in the half of Mercury seen by Mariner is located near one of the two hot longitudes and is appropriately named the Caloris (Latin for heat) Basin. With a diameter of 1300 km, the Caloris Basin (Fig. 8.12) resembles Orientale or Imbrium on the Moon. The most important difference, however, arises in the distinctive patterned interior floor of Caloris. If this material is basalt, we ask, why does it lack the smooth surface and dark color of Mare Imbrium and other ancient lunar basalt flows? If it is not basalt, what is it? One hypothesis proposes that the floor of Caloris is made of material that melted as the direct result of the impact, rather than being filled in later as were the lunar maria.

Geologically, the most remarkable features on Mercury are compressional scarps or cliffs that have no lunar counterpart (Fig. 8.13). These scarps were produced when the planet's crust was compressed by shrinking of the interior. The total extent of such features indicates that the crust has lost approximately 4% of its area, probably as a result of a 2% decrease in the radius of the planet. Since they formed after most of the craters (as you can see in Fig. 8.13), these scarps apparently represent an internal event on Mercury that took place many hundreds of millions of years after the solidification of the crust. Interior changes resulting from the slowing of the planet's spin to achieve the present 2:3 resonance have been suggested as a possible cause.

8.5 INTERIORS OF THE MOON AND MERCURY

Planetary interiors are mysterious places, forever inaccessible to direct measurement. What little is known about them must be learned indirectly and is inevitably subject to misinterpretation. Yet the effort must be made, for without some understanding of the bulk of a planet which lies beneath the visible surface, an intelligent discussion of planetary evolution is impossible.

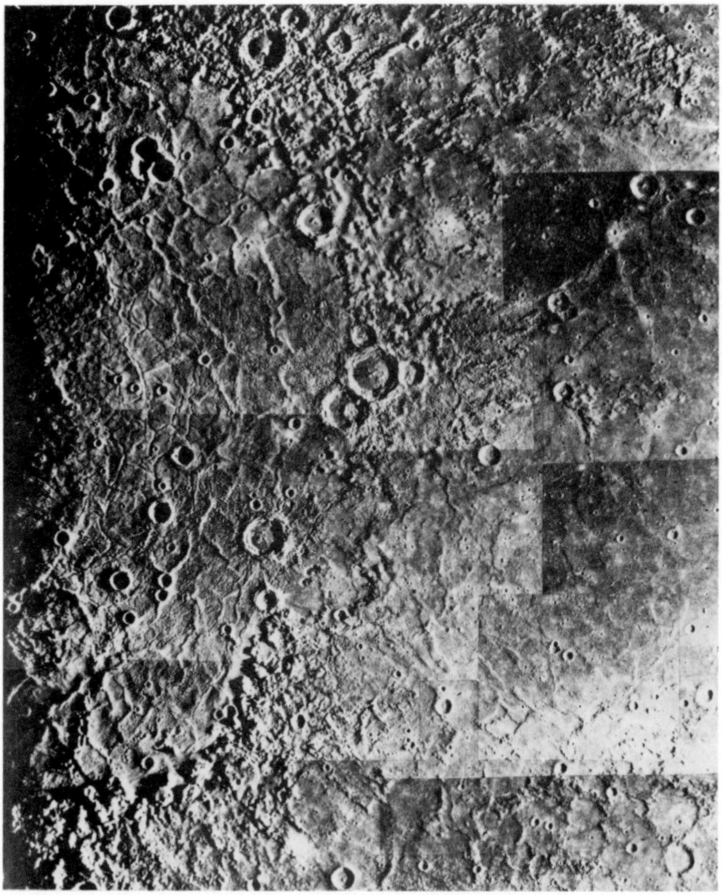

FIGURE 8.12 This mosaic of images obtained by Mariner 10 shows the half of the Caloris Basin that the spacecraft was able to see. It is interesting to compare this feature with the Orientale Basin on the Moon (see Fig. 7.6).

The Significance of Density

The most important clue to the bulk composition of a planet is provided by its density (Section 1.6). The Earth has a density of 5.5 g/cm^3; the Moon, 3.3 g/cm^3; and Mercury, 5.4 g/cm^3. The fact that these densities differ suggests differences in composition, but we must be careful; the average density is also affected by the size of a planet. Big planets with stronger gravity are able to compress their interior materials and increase their densities. For example, the iron in the Earth's core has a density of about 15 g/cm^3, while ordinary iron on the surface of the Earth has a density of only 9 g/cm^3. Thus a better starting point for a chemical interpretation would be an **uncompressed density**—that is, the density corrected for the effects of self-compression in a planet.

The uncompressed densities of Earth, Moon, and Mercury are, respectively, 4.5 g/cm^3, 3.3 g/cm^3, and 5.2 g/cm^3. The Moon, because of its relatively small size, has experienced little internal compression. Mercury has the highest uncompressed density of any planet, with the Earth second. Therefore, Mercury must have the largest proportion of high-density constituents, which are the metals, and the Earth has the second-highest allotment.

FIGURE 8.13 The scarp named Discovery stretches more than 500 km across Mercury's surface. In places it is 3 km high, about the same height as Mount Olympus in Greece. This scarp is younger than the craters it crosses. It probably formed when Mercury's crust contracted as the planet cooled early in its history.

The uncompressed density of the Moon is only slightly greater than that of ordinary rocks, including the returned Apollo samples. For any planet, the density must increase with depth, since it is hard to imagine the alternative in which heavier materials could have "floated" above lighter ones when the Moon was young and molten. Therefore the materials near the center of the Moon can be at most only slightly denser than the surface rocks, and there simply is no place to hide a substantial metallic core. One of the fundamental properties of the Moon is its lack of metals. If Earth and its satellite formed together from the solar nebula, this basic difference in composition is hard to understand.

In contrast, the only plausible model for a planet as dense as Mercury requires a lot of metal, roughly 60% by mass. If 60% of Mercury consists of an iron-nickel core similar to that of the Earth, this core must have a diameter of 3500 km and extend to within about 700 km of the surface. Mercury can then be thought of as a metal ball the size of the Moon covered with a 700-km-thick crust of dirt.

The proper way to look at this situation is to realize that Mercury is *deficient* in silicate rocks, rather than *enriched* in metals. Either temperatures in the solar nebula were so high that only metals and some high-temperature silicates could condense and form the solid grains that ultimately became the building blocks of Mercury, or else the planet formed initially with a composition more like that of the Earth and Venus but subsequently lost part of its share of silicates. We will discuss these two possibilities later in this chapter and try to choose between them.

Moonquakes and Heat Flow

In the case of the Moon, considerable insight into interior conditions has been derived from the Apollo ALSEP experiments, as described in Chapter 7. The lunar seismometers measured the response of the whole Moon to impacts, such as those of spent Apollo rockets (Fig. 8.14), while an additional experiment on Apollos 15 and 17 directly measured the rate of heat flow from the interior.

Relative to the Earth, the Moon is a very quiet place. No moonquake measured by any of the Apollo instruments was large enough to have been felt by an astronaut standing on the Moon, and the total energy released each year by moonquakes is a hundred billion times less than that of earthquakes. Internally generated moonquakes arise primarily at a depth of around 1000 km. Below this depth, temperatures may be high enough to produce some melting of rocks (Fig. 8.15a).

The heat flow from the interior of the Moon also suggests that temperatures are higher at

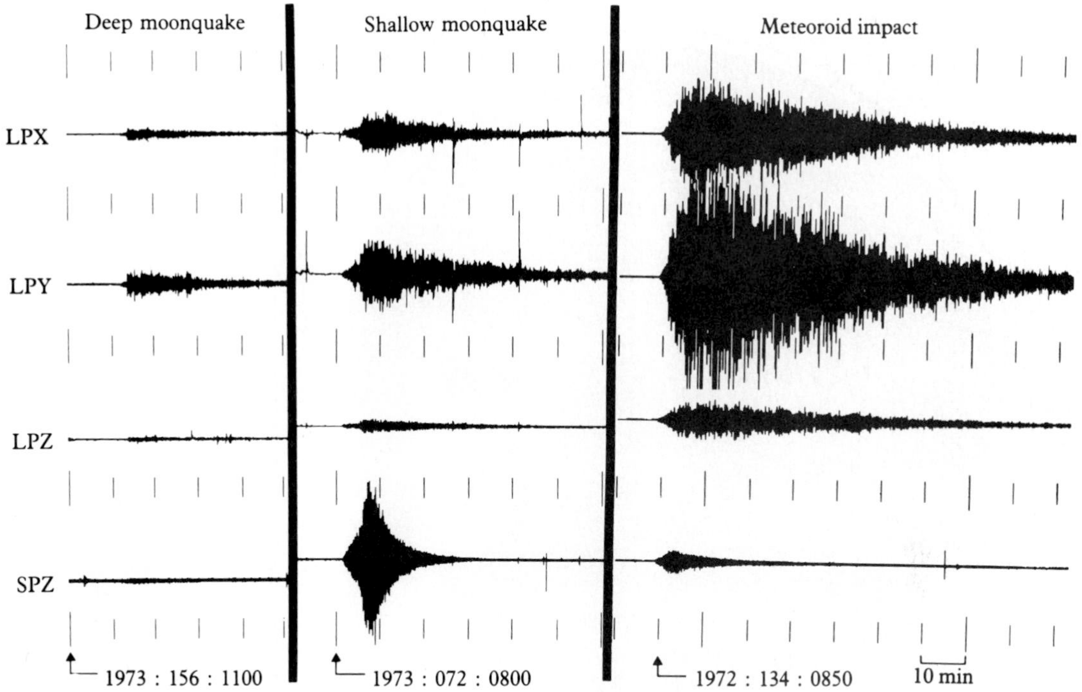

FIGURE 8.14 A comparison of seismograms obtained from various types of seismic events. The long "ringing" of the Moon in response to meteoroid impacts is especially noteworthy.

great depths. Most of this heat is being generated today by radioactive elements in the Moon. If a similar heat flow measurement were made on the surface of Mercury, it would be possible to determine whether the metal core is liquid, but in the absence of a lander no such data exist.

Another line of evidence on interior structure comes from the measurement of a planet's magnetic field. Although the exact mechanisms for generating such a field are not understood, it is thought that a strong field requires motions in a liquid, electrically conducting core. Since the Moon has no metallic core, it was no surprise when early space missions showed no global magnetic field. But no one knew what to expect from Mercury, which has a metal core, but not necessarily a *liquid* metal core.

The Magnetic Field of Mercury

One of the discoveries made by Mariner 10 was the presence at Mercury of just such a planetary magnetic field after none had been detected for the Moon, Mars, or Venus. Although only about 1% as strong as the field of the Earth, the magnetic field of Mercury is similar in character and seems to indicate the existence of a liquid core. Two decades after the Mariner flybys the issue of a liquid core remains unclear, since we still have no data on the state of the planet's interior. But the presence of a magnetic field on Mercury is undeniable, and most scientists conclude that the metal core must therefore be at least partially molten (Fig. 8.15b).

Mercury presents a paradox to challenge planetary scientists. Its interior is in many ways

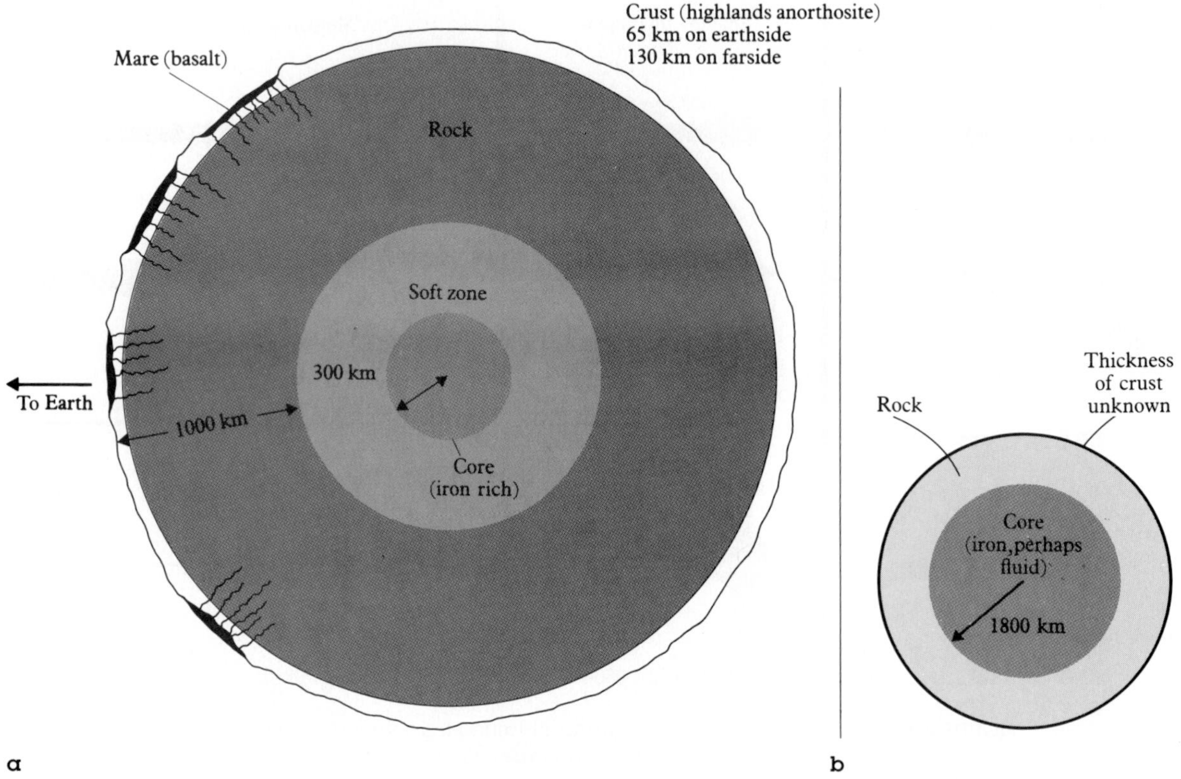

FIGURE 8.15 (a) A model of the interior of the Moon derived from seismic data. It is still uncertain whether there is a small liquid core at the Moon's center. (b) A model of the interior of Mercury based on the planet's high density and its requirement for a large, iron-rich core. (Note different scales in (a) and (b).)

like that of the Earth, with a large metal core and a planetary magnetic field. Yet its surface is like that of the Moon, with little evidence of geological activity for at least several billion years. If the interior is hot enough to generate a magnetic field, why has it not also reworked more of the surface? Or have we incorrectly estimated the age of the surface, and the cratering on Mercury was produced more recently than that on the Moon? Or, perhaps, the magnetic field does not indicate a molten core? These are important questions, but without additional missions to Mercury they may remain unanswered for a long time.

Atmospheres of Mercury and the Moon

We have spoken at length about the surfaces and interiors of these two similar, battered worlds, but what about their atmospheres? Because both objects are small, their gravity is not strong enough to hold any substantial atmosphere over the lifetime of the solar system. The escape of atmospheric gas is further accelerated by their high surface temperatures, up to 400 K on the Moon and 675 K on Mercury. For most purposes, therefore, we may consider both of these to be airless worlds.

Nevertheless, both Mercury and the Moon have extremely tenuous, transient atmospheres produced by the slight buildup in density of the solar wind as it flows around these bodies. These flickering envelopes of gas consist mostly of hydrogen and helium, the primarily atoms in the solar wind. Since they are partly ionized, the main effect of these thin atmospheres is to provide a degree of electrical conductivity around the object, thus influencing the interaction between the planet and the solar wind.

The atmospheres of both objects also contain atoms that are being either released (outgassed) from the interior or dislodged from their surfaces by the impact of ions in the solar wind. The most easily observed of these atoms are sodium and potassium, both metals. Observations of these elements and others that may be detected in the atmosphere of Mercury are especially useful because they could provide additional clues to the mercurian surface composition. So far, no clear picture of surface chemistry has emerged from such studies, however.

8.6 HISTORIES OF THE MOON AND MERCURY

The geological history of a planet is written in its rocks and surface topography. This record can be read using the techniques of stratigraphy supplemented by detailed study of selected rock samples. Because both Mercury and the Moon are relatively inactive internally and lack atmospheric weathering and erosion, the span of solar system history available for study covers a great expanse of time. Approximately 4 billion years of geologic evidence is exposed on the Moon, and the surface of Mercury probably preserves a similar record, although in the absence of dated samples we cannot be sure. Yet there is a limit beyond which we see dimly if at all. As noted before, heavy meteoroidal bombardment during the first half billion years of solar system history has destroyed most of the evidence from this epoch.

Models of Planetary Evolution

Direct geological evidence from the surface can be augmented by theoretical calculations of the probable *thermal evolution* of a planet—the changes in its internal temperature over time. To trace this life history, we must begin with what we know today and calculate backward and forward in time. Such calculations require the powers of the largest computers if they are to represent realistically the laws of physics and chemistry that govern planetary evolution.

Models also require knowledge of the behavior of materials at the unfamiliar pressures and temperatures of planetary interiors. Near the center of the Earth, for instance, the pressure is about 4×10^6 bars (that is, 4 million times the sea-level pressure of the Earth's atmosphere), and the temperatures are measured in thousands of degrees. These calculations also require certain assumptions concerning the initial state of a planet, especially assumptions about its chemical composition. If these assumptions are reasonable and the computer program works properly, an outline or **scientific model** of planetary history is produced.

The output of a planetary model calculation is many columns of numbers, each representing the pressure, temperature, and chemical and physical properties at various points inside a planet. All of these values are calculated for a series of time steps, separated by perhaps a few million years. From these numbers, it is possible to trace the changing conditions at each interior point over a span of billions of years.

Lunar history is comparatively well defined by the wealth of evidence obtained by Apollo, while Mercury's past remains more speculative. One way to approach Mercury is to adopt the lunar-based paradigm for the cratering history of the inner solar system (Section 7.3). If Mercury experienced impact rates similar to those of the Moon, including a terminal heavy bombardment about 3.9 billion years ago, a consistent picture emerges. Of course, consistency does not guarantee truth, but it does suggest that planetary scientists may be moving in the right direction.

Early History

We will discuss the actual origin of the Moon and Mercury in the next section. For the present discussion, we begin about 4.4 billion years ago, when we think both objects were largely molten and fully differentiated. By this time the lunar highland crust had begun to form by freezing of the surface of a magma ocean. The oldest anorthositic highland rocks provide us samples from this period of planetary history. During this period the heavy rain of debris still impacting the lunar surface repeatedly shattered and remelted the crust, destroying any detailed evidence of its early structure. The entire surface of the Moon came to consist of overlapping ejecta from innumerable impacts. By 4.2 billion years ago the magma ocean of the Moon had cooled and solidified, while at the same time the terminal bombardment was occurring, perhaps as the last stages of an accretion process that had continued since the formation of the Moon.

Mercury's earliest history differed from that of the Moon primarily because of its different composition. Mercury was depleted in silicate rocks rather than in metals, giving it the highest metal concentration of any planet. Differentiation must have taken place early, resulting in a magma ocean (shallower than that of the Moon) and a rocky crust. Early heating and expansion of the planet may have triggered widespread volcanism near the end of the period of heavy bombardment.

From the Terminal Bombardment to the Present

On the Moon, the first 700 million years are called the pre-Imbrium era, ending 3.8 billion years ago with the formation of the Imbrium Basin and widespread blanketing of the nearside of the Moon with ejecta from this impact. Shortly thereafter the formation of the Orientale Basin marked the last of the large asteroidal impacts. As the meteoritic impacts declined, large-scale volcanism began to fill the deep impact basins on the lunar nearside. The magma sources were located hundreds of kilometers below the surface, in regions being heated by internal radioactivity. Slight expansion triggered by changes in internal temperature may have opened fractures, permitting basaltic lava to rise and flood the basins. Imbrium was flooded in this way between 3.4 and 3.3 billion years ago.

On Mercury at approximately the same time, the presence of a huge iron-nickel core apparently led to a different evolutionary path. Instead of expanding, calculations show that Mercury began a slow contraction, amounting to perhaps 2% of its radius. Cracks in the crust were squeezed shut, inhibiting large-scale volcanism, and compressional forces became great enough to form the scarps we see today on the surface. Shrinkage rather than expansion of Mercury may explain the apparent paradox that this planet seems to have had less surface volcanic activity than the Moon in spite of higher interior temperatures.

Large-scale volcanic activity on the Moon ceased about 3 billion years ago, and slow cooling of the interior continued until the entire object solidified. Nothing much has happened on the surface for the last 3 billion years except for rare impacts by comets or near-Earth asteroids and continuing slow erosion by smaller impacts. Mercury is similarly winding down. Although it may still have a partially liquid iron core, the outer parts of the planet have cooled, and major geological activity ended long ago.

It is interesting to contrast these histories with that of the Earth. Except for very rare rocks in the most ancient parts of the continents, the oldest terrestrial rocks have all formed since the end of the period of lunar volcanism. Even the young (billion-year-old) lunar crater Copernicus formed before the appearance of hard-shelled lifeforms in our oceans. The youngest major lunar crater, Tycho, formed only 100 million years ago, about the time the outlines of the major terrestrial continents were established. In many ways the lunar and mercurian records are the perfect complement to that of Earth, with all the action on the smaller bodies ending at just about the time our own geologic story begins to unfold.

8.7 ORIGIN OF THE MOON AND MERCURY

All explanations for the origin of the Moon are improbable.

—H.C. Urey

In some ways the Moon is like the Earth, and in other ways it is different. Reconciling these facts in a self-consistent theory of the origin of the Moon has proved difficult. Before lunar exploration began, "What is the origin of the Moon?" was a major question for planetary scientists. Thirty years later, it remains partly unanswered, but current thinking attributes the origin of the Moon to catastrophic impacts in the inner solar system and links its early evolution (conceptually) to that of Mercury.

Moon and Earth: Similarities and Differences

The evidence for the origin of the Moon contains some serious contradictions. Study of the Apollo lunar samples revealed a fundamental similarity between the Earth and the Moon in the detailed isotopic composition of their rocks. Not only are many of the lunar minerals similar to those found on our own planet, but the relative proportions of different isotopes of oxygen and other elements are also the same. Recall from Chapter 4 that these isotopic ratios differ among the meteorites and that these differences are used to infer incomplete mixing of material in the early solar nebula. The close similarity of the Earth and Moon in this respect suggests that these two bodies formed together out of the same mix of materials condensing from the solar nebula.

In spite of the isotopic similarities, the Moon is fundamentally different from the Earth in its bulk composition. In many ways the composition of the Moon resembles that of the terrestrial mantle, which also has a density of about 3.3 g/cm^3, but significant differences remain between their bulk compositions. Water and other volatiles are severely depleted on the Moon. There is five times less potassium, while the uranium has about the same abundance. Other elements, such as calcium, aluminum, and titanium are more abundant on the Moon. If the Moon and Earth formed together, we would not expect such differences in composition.

Three Theories of Lunar Origin

There are three traditional theories regarding the origin of the Moon. While none of them is now believed to be wholly correct, they illustrate the range of alternatives considered. These theories are sometimes called the daughter theory, the sister theory, and the capture theory. Let us look at each and confront it with the evidence cited above concerning the similarities and differences between Earth and Moon.

The daughter, or fission, theory proposes that the Moon formed from the Earth. It was first proposed in 1880 by astronomer and mathematician George Darwin, the son of the famous biologist Charles Darwin, who calculated that a rapidly spinning molten Earth could split (fission) to form a double planet. If the alleged split took place after the differentiation of the Earth, the smaller Moon might be formed purely from mantle material and thus be depleted in metals. Furthermore, this idea seemed consistent with the tidal evolution of the Moon, which showed that our satellite is slowly moving away from the Earth. There are serious flaws in the daughter theory, however, many of them associated with the differences in composition between the Earth's mantle and the Moon. The deficiency of volatile elements on the Moon, for example, is not explained by this theory. There are additional mechanical problems in the theory as originally proposed by Darwin, which has caused the idea of planetary fission to be rejected by modern scientists.

The sister theory suggests that the Earth and Moon formed close together from a spinning cloud of dust, in much the same way that most scientists believe the large satellite systems of Jupiter, Saturn, and Uranus originated. The problem here is simple: The sister theory provides no explanation for the compositional differences between the Earth and Moon. At least the daughter theory suggests how the Moon might have formed without much metal, but sisterhood does not seem to be compatible with the chemical evidence.

The third possibility is the capture theory. The idea here is that the Moon formed elsewhere in the solar system as an independent body and was subsequently captured into orbit around the Earth. The compositional differences are understood as representing different condensation conditions in separate locales in the solar nebula, but again there are two problems. First, the isotopic ratios seem to indicate that the Earth and Moon formed from the same pool of material. Second, the Earth could not have captured the Moon into a stable orbit unless a third body intervened, and no helpful go-between has been identified. Capture by atmospheric drag, the method proposed in Section 5.5 for the capture of the martian satellites, does not work for an object as large as our Moon. Today, the capture theory has few if any defenders.

A Possible Synthesis

If the Moon is neither daughter, sister, nor interloper, what is it? Apparently the three basic theories just described are too simple, and we must seek a more complex scenario. What is required is a mechanism that permits the Moon or its precursor materials to form initially in the same part of the solar nebula as the Earth and then to undergo some fractionation process or processes that can remove most of the metals and volatile elements before solidification. The daughter theory comes closest, but we need a better way to get a piece of the terrestrial mantle into orbit where it might be able to evolve into the Moon.

The most likely answer to this dilemma is that one or more giant impacts during the period of planetary accretion ejected huge quantities of mantle material from the Earth, at the same time heating this material sufficiently for most of the volatiles to escape. This event must have taken place after the Earth differentiated, but as we have seen differentiation occurred quickly, probably as the Earth was forming. Impact heating and fractionation would change the chemical composition of the ejected material without altering the isotopic ratios for individual elements, which would remain the same in both the ejected material and the rest of the Earth.

Subsequently, some fraction of the ejecta could aggregate in orbit to form the Moon. Initially, the Moon was in a close orbit, but tidal forces would drive it away to its present position. Figure 8.16 outlines this scenario as currently envisioned.

The idea of a giant impact appears to match what we know of the composition of the Moon, but does it make sense otherwise? How large would this impact have been? Is it reasonable that an impact could eject material into orbit rather than blasting it away from the Earth entirely? In the late 1980s, a number of scientists investigated these questions. From their work, it appears that the impactor must have had a mass about one-tenth that of the Earth (about the size of Mars), and it must have struck at an oblique angle. If the mass had been much larger, the Earth would have split apart; much less and insufficient material would have been ejected. An impact of this magnitude could have ejected about 10% of the mass of the Earth into space,

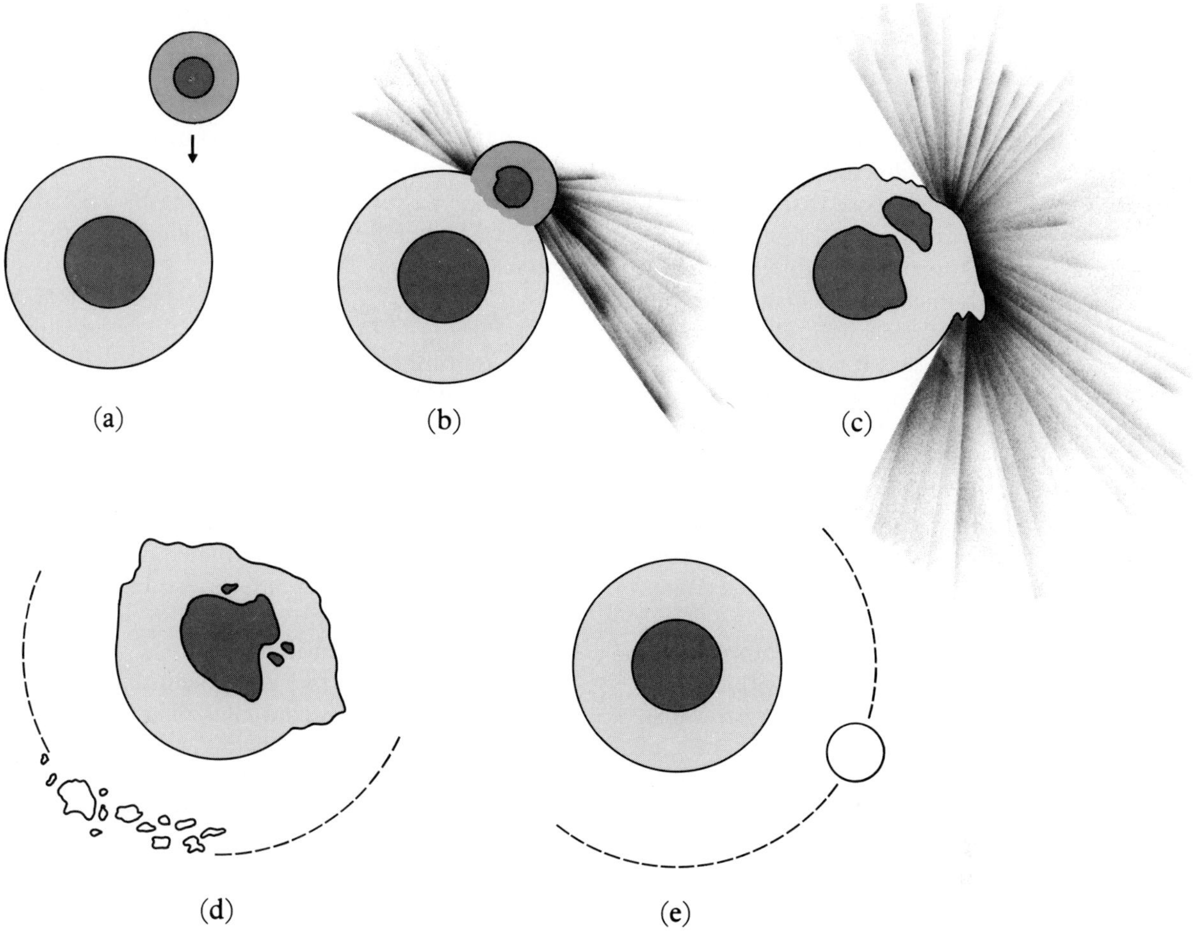

FIGURE 8.16 The formation of the Moon apparently resulted from the impact of a Mars-sized body with the early Earth. Part of the ejecta from this oblique collision went into orbit about the Earth and subsequently aggregated to form the Moon.

with perhaps 10% of the ejecta (1% of the mass of the Earth) remaining in orbit. Since the mass of the Moon is only about 1% the current mass of the Earth, the giant impact idea seems to pass this plausibility test.

Origin of Mercury

If the Earth was struck by a Mars-size projectile during its formation, we may ask if other planets were similarly disrupted by impacts. George Wetherill of the Carnegie Institution of Washington has carried out extensive calculations on the early evolution of the terrestrial planets. His models indicate that as many as a hundred protoplanets of lunar to martian size were probably formed in the inner solar system, many of them on highly eccentric orbits. In such chaotic circumstances, large impacts would have been a natural part of the final accumulation of the planets. Indeed, Wetherill's models not only predict the existence of giant impacts, they also suggest that several would-be planets were probably shattered beyond repair in this cosmic demolition derby. The four remaining terrestrial planets are the survivors.

If Wetherill is correct, then giant impacts also provide an explanation for the anomalous composition of Mercury. Suppose that Mercury was originally larger, with about half the mass of the Earth or Venus. It would have differentiated to form a core and a thick silicate mantle with the same proportions of iron and silicates as the Earth. Suppose then that it was hit by a Mars-size projectile. In this case a large fraction of its mantle mass might have been ejected, leaving a thin silicate layer above a now disproportionately large core. Thus the same theory of giant impacts early in solar system history provides us a plausible explanation for both the low metal content of the Moon and the high metal content of Mercury.

If Wetherill is correct in his hypothesis that a multitude of protoplanets formed with eccentric orbits, then each of the surviving terrestrial planets would be expected to have about the same composition, except for the fractionation effects of giant impacts. We will return to these and other ideas of planetary formation in the final chapter of this book.

8.8 QUANTITATIVE SUPPLEMENT: RADAR AND THE DOPPLER EFFECT

Radar provides one way to determine the rotation rate of a planet, using the Doppler effect. As an illustration, let us calculate the effect of the rotation of Mercury on a radar signal transmitted at a wavelength of 10 cm. The equivalent frequency of this radar wave is given by

$$f = c/\lambda = 3 \times 10^9 \text{ Hz} = 3 \text{ GHz}$$

where c is the velocity of light (3×10^8 m/s).

Mercury has a diameter of 4878 km. Its circumference is 1.5×10^4 km = 1.5×10^7 m. If it rotates in 59 days (5×10^6 s), the equatorial rotational speed is a little more than 3 m/s. Thus one edge will be moving away from us at 3 m/s, and the other edge will approach at 3 m/s, relative to the center of mass of the planet. The total range in apparent velocity is 6 m/s, corresponding to a frequency (or wavelength) spread ($\Delta\lambda/\lambda$) in the reflected radar signal given by the Doppler formula:

$$\Delta\lambda/\lambda = \Delta f/f = v/c = 2 \times 10^{-8}$$

If the transmitted wavelength is exactly 10 cm and there is no motion of the center of mass of Mercury relative to the Earth, the wavelengths of the returned signals would lie between 9.999999 and 10.000001 cm. By measuring this bandwidth in the returned signal, the rotation rate of the planet can be determined.

Note that one would rarely measure the particular apparent rotation calculated in the above example. At any given time, the apparent rotation of Mercury has two components, one due to its actual rotation, and the other resulting from our changing perspective as Mercury passes by the Earth. Thus we must know the relative orbits of Mercury and the Earth before we can use such radar observations to derive the true rotation rate.

8.9 QUANTITATIVE SUPPLEMENT: SYNODIC PERIOD

The apparent motion of a planet in the sky does not represent its true period of revolution about the Sun. We look out from a moving platform, the Earth, and the motion we see results from a combination of the true motion of a planet with that of the Earth. Consider, for example, the case of a distant planet like Saturn, which has an orbital period of 30 years, making it nearly fixed among the stars. Yet the Earth swings around in its orbit every year. The result is that the apparent period of Saturn as it shifts from morning to evening sky follows a cycle of nearly one year which is dominated by the motion of the Earth. This apparent period is called the *synodic period* of a planet; for Saturn the synodic period is 378 days.

The case of nearby planets is a little more complicated. Mercury has a synodic period of 115 days; that of Venus is 596 days; and that of Mars is 780 days. However, there is a simple formula for calculating the synodic period of any object from its true period of revolution, as given by Kepler's laws. You take the reciprocal of the true period (in years), subtract 1, and take the reciprocal again. This is easy with a hand calculator; try it! More generally, we can find the synodic period for any two planets with the formula

$$1/P_{\text{inner}} - 1/P_{\text{outer}} = 1/P_{\text{synodic}}$$

where P_{inner} is the period of the inner of the two planets and P_{outer} corresponds to the outer of the two planets.

SUMMARY

Because Mercury is small and near the Sun and has been visited by only one spacecraft, Mariner 10, our database is limited. One important source of information, however, has been planetary radar. We saw how radar astronomy worked, and in particular noted the power of the technique of using Doppler shifts in the frequency of the reflected radio pulse to determine planetary rotation. Radar has revealed the unusual rotation period of Mercury, which is exactly two-thirds of its 88-day period of revolution around the Sun. We examined how tides raised by one body on another can slow planetary rotation and modify orbits, providing an explanation for the fact that the Moon always keeps the same face toward the Earth, as well as for the more unusual 2:3 synchronous rotation of Mercury. Tidal forces are important for many planets, satellites, and rings, as we will see later in this book.

Mercury is a small airless world heavily scarred by impact craters. Its surface looks very much like that of the Moon, although the indications of past volcanic activity are less obvious. However, Mercury is very different in its bulk properties. Its great density (the highest of any planet) indicates that much of its interior is made of iron, like the core of the Earth, and this hypothesis is further supported by the presence of a small planetary magnetic field. Roughly speaking, Mercury has an interior like that of the Earth and a surface like that of the Moon.

The lunar interior is unique. Our satellite has lost not only volatile elements, but also most of its expected allotment of iron and other metals. As a result, the Moon has at most a very small iron core and no global magnetic field. The compositional differences between Earth and the Moon are great enough to frustrate simple theories in which the two objects formed together, out of the same raw materials in the solar nebula.

We looked at three simplified theories of the origin of the Moon, and found difficulties with each. Apparently the correct answer is more complex, involving the ejection of terrestrial mantle material in a giant impact, and subsequent aggregation of the iron- and volatile-depleted ejecta into the Moon. Paradoxically, the same giant impact theory can explain Mercury's loss of silicate mantle material. Apparently giant impacts were an important element of the early evolution of the terrestrial planets.

KEY TERMS

Doppler effect
polar cap
scientific model
solar day
synchronous rotation
tidal force
tidal heating
uncompressed density

Review Questions

1. Describe how radar is used in the study of the planets. In particular, make sure you understand how the rotation of a planet broadens the returned signal by the Doppler effect, even though the signal is reflected from an entire hemisphere.

2. Discuss how tides are formed. Why are there two high tides every day on Earth? Why does high tide not generally correspond to the time when the Moon is overhead? What sort of tides are raised on the Moon by the Earth?

3. Compare the tidal and rotational histories of the Earth, the Moon, and Mercury. Can tidal theory account for their different rotational periods today? For each planet, is the present rotational state stable, or is evolution continuing? What will the final rotation period be?

4. Compare the geology of the Moon and Mercury. What do they have in common? Consider in particular the craters, impact basins, and plains (or mare) areas.

5. Describe and contrast what an astronaut would see and feel on the surfaces of the Moon and Mercury. Could you easily tell whether you were standing on Mercury or the Moon?

6. Use what you now know about Mercury to critique the standard paradigm for solar system history. How would you expect the meteoroidal bombardment history of Mercury to compare with that of the Earth and Moon? Is there any evidence for a late heavy bombardment on Mercury? What are the likely ages for the craters, basins, and plains on Mercury, according to the standard model?

7. Compare the interiors and histories of the Moon and Mercury. What does the geology of each tell us about the interiors? How useful are interior models? How would you go about distinguishing among several models to see which is most likely to represent correctly the thermal history of a planet?

8. Describe the three traditional scenarios for the origin of the Moon. How does each compare with the data we have on the Moon? Show how an appropriate combination of elements from these theories may be consistent with what we know about the Moon. How does the origin of Mercury relate to this theory for the Moon?

Quantitative Exercises

1. Calculate the synodic period for each of the planets from Mercury through Saturn.

2. Find the synodic periods of Mercury and Earth as viewed from Venus. How does the synodic period of Venus as seen from the Earth compare with the synodic period of Earth as seen from Venus?

3. Calculate the energy of an impact with the Earth by a projectile the mass of Mars striking at a speed of 20 km/s. Compare this energy with the gravitational binding energy of the Earth, given by GM^2/R, where G is the gravitational constant (6.67×10^{-11} newtons $m^2\ kg^{-2}$) and M and R are the Earth's mass and radius.

4. The Arecibo radar instrument is sometimes used to study Earth-approaching comets at a wavelength of 12 cm. Suppose the nucleus of such a comet has a diameter of 5 km and its period of rotation is 10 hours. What are the relative speeds of its leading and trailing edges as seen from Arecibo? What is the total broadening of the 12-cm transmitted signal due to this rotation?

Additional Reading

Dunne, J.A., and E. Burgess. 1978. *The Voyage of Mariner 10: Mission to Venus and Mercury* (NASA SP-424). Washington, D.C.: U.S. Government Printing Office. Interesting account of the Mariner 10 mission and its results.

*Murray, B.C., M.C. Malin, and R. Greeley. 1981. *Earthlike Planets: The Surfaces of Mercury, Venus, Earth, and Mars.* San Francisco: Freeman. Excellent introduction to planetary geology, with good discussions of Mercury and the Moon.

Strom, R. 1987. *Mercury, the Elusive Planet.* Washington, D.C.: Smithsonian Institution Press. Comprehensive semipopular book on Mercury, combining results from Mariner 10 with other information about this planet.

*Vilas, F., C.R. Chapman, and M.S. Matthews, eds. 1988. *Mercury.* Tucson: University of Arizona Press. The definitive technical compendium, from the Arizona Space Science Series.

*Indicates the more technical readings.

9

The Earth: Our Home Planet

Looking across the battered, airless surface of the Moon, we see the brilliant blue and white globe of the Earth as it rises to greet the Apollo 11 astronauts, who took this picture from lunar orbit. The contrast between the dry, barren Moon and the only planet we know with abundant liquid water and life is enhanced by the black background of space.

It is hard to think of the Earth as a planet. We humans are so intimately involved with our world that we have difficulty establishing a planetary perspective. Yet the Earth is a planet, quite an ordinary one in terms of location and size, and it is subject to many of the same processes that shape the other worlds of the solar system. It is one of the achievements of the space age that humans have begun to appreciate the close relationships among the planets, and to use information about one to gain insight into another. The other planets have much to teach about the Earth. But even more, since we know it so well, the Earth can teach us a great deal about other planets.

In order to study the Earth the way we do other planets, we must step back from the wealth of detail that surrounds us and try to address the same kinds of questions that motivate our study of other worlds. Where and how did the Earth form? What is its composition and internal structure? What sources of energy maintain its geologic activity, and how have these energy sources varied over time? Why does the Earth, alone of all the planets, have oceans of liquid water and an atmosphere rich in oxygen? What maintains the circulation of the oceans and atmosphere, and how has each evolved with time? And perhaps most intriguing of all, why is the Earth home to living things? All of these questions are addressed in this chapter.

FIGURE 9.1 Earth is the only planet in the solar system with oceans of water, blue skies, and abundant life. On average, more than half of the planet is shrouded by water clouds at any time.

9.1 EARTH AS A PLANET

Let us begin our study of the Earth by imagining ourselves visitors from another solar system, making our preliminary reconnaissance of the planets. Undoubtedly our first attention would have been directed to the giant planets with their spectacular systems of rings and satellites. Later, perhaps, we would check up on the smaller worlds of the inner solar system. Among these the largest would be Earth, the only inner planet with a major satellite.

The View from Space

Even from a distance, Earth would appear unique (Fig. 9.1). It is the only planet with liquid water, and seen from the correct angle sunlight glints brightly off its oceans. Ice permanently blankets its south polar continent as well as Greenland and other isolated land areas in the northern regions. The polar caps show large seasonal changes; in winter, sea ice covers the polar seas and a transient snow cap invades the temperate land masses, extending halfway to the equator. These surface changes could be seen only dimly through a shifting canopy of white clouds that reflects nearly half the sunlight striking the Earth. A few tawny brown land areas would normally be clear of clouds, while other regions, primarily near the seas, are eternally clouded. Enigmatic hazes of dust sometimes obscure even the cloud-free areas, and hints of changes might appear in surface color or reflectivity with the passage of the seasons.

More remarkable than its surface appearance would be the chemistry of the terrestrial atmosphere. Nitrogen, the primary gas, makes up 78% of the atmosphere, but the real surprise is oxygen, the second most abundant gas at 21%. On Mars and Venus, oxygen, a highly reactive gas, is trapped in minerals in the surface rocks. Perhaps equally surprising is the absence of more than a trace of carbon dioxide, which is the primary constituent of the atmospheres of Venus and Mars. The atmosphere and oceans of Earth represent a triple enigma: too much oxygen, too much water, and not enough carbon dioxide. All of these chemical anomalies are related to the presence on Earth of abundant and varied life, a subject we will return to in Chapter 12.

Terrestrial Geology

The Earth's oceans contain enough water to cover the entire globe to a depth of 3 km, yet nearly a third of the surface of the planet rises above this watery layer. Most of this land area is concentrated in the six massive continents, much larger than can be explained by individual volcanoes or upthrust mountain ranges. Near the edges of some continents, mountains rise nearly 9 km above sea level, while in places the sea floor, also near the continental margins, can drop as low as 11 km below sea level.

A closer examination would reveal to the space visitor many additional indications of geological activity: erupting hot springs and volcanoes; frequent earthquakes; youthful mountain ranges with sharp ridges and peaks sculpted by the forces of water and ice; and subtle variations in gravity, revealing that parts of the crust are not in equilibrium but are currently undergoing uplift or being drawn down by internal forces.

To learn something about the interior of this planet, the visitor might look at space surrounding the Earth. There, electrons and ions from the solar wind are trapped by a magnetic field produced in the planet's core. Noting the high density of the Earth, the space visitor might conclude that the Earth had differentiated and that its core consists primarily of iron, a cosmically abundant metal.

Cyclic Processes and Planetary Evolution

The next step in the initial investigation of the Earth might be to examine the *processes* taking place. Earth is a remarkably active planet in many ways, in dramatic contrast to the Moon and Mercury. How should we look at this activity? One possibility is that we are seeing evidence of rapid evolution and change: The Earth is in transition from one state to another in much the same way we viewed the history of the Moon and Mercury, although most of their transitions took place a long time ago. But a little thought quickly shows the fallacy of this perspective when applied to Earth. Consider the rivers flowing into the sea. Do they indicate that the sea is filling up, and that soon the land will be submerged? Certainly not, because the water evaporates from the ocean as fast as it is added from the land. What we have here is a *cycle,* called the water cycle, in which input and outflow balance.

Many processes on Earth are cyclic. Volcanoes deposit lava on the surface, but the new rock is eventually drawn down again to the interior to be recycled through new volcanic eruptions. The land is eroded into the sea, but the sea does not silt up; instead the sediment is used to create new continental crust. Marine animals trap carbon dioxide to form vast deposits of carbonate rocks, but eventually some of these rocks also are reprocessed, and their carbon dioxide is released again into the atmosphere. The Earth behaves like a giant machine, with many interlocked cycles turning at different rates.

The idea of Earth as a machine is useful for understanding many processes, but not all. True evolution does take place. Life, for instance, has clearly evolved, and along with it the atmosphere has changed greatly. There are long-term trends in geology that can be separated from the cyclic events. One objective in looking at Earth as a planet is to distinguish the *long-term evolution* from *shorter-term cyclic processes*, in order to see the many ways in which our planet's history resembles that of other planets.

9.2 JOURNEY TO THE CENTER OF THE EARTH

Few places in the planetary system are less accessible to direct study than the Earth beneath our feet. The very deepest bore holes have penetrated less than 10 km, barely one-tenth of 1% of the distance (6378 km) to the center. Determining the structure and composition of the interior of the Earth is a difficult scientific puzzle.

Probing the Interior

Our first clue to the composition of the interior of the Earth is found in its relatively high density of 5.5 g/cm^3. Even when corrected for the compression in the interior, the uncompressed density of our planet is 4.3 g/cm^3, a much higher value than that of the Moon. Evidently the Earth has an invisible metallic core in addition to its obvious rocky upper layers.

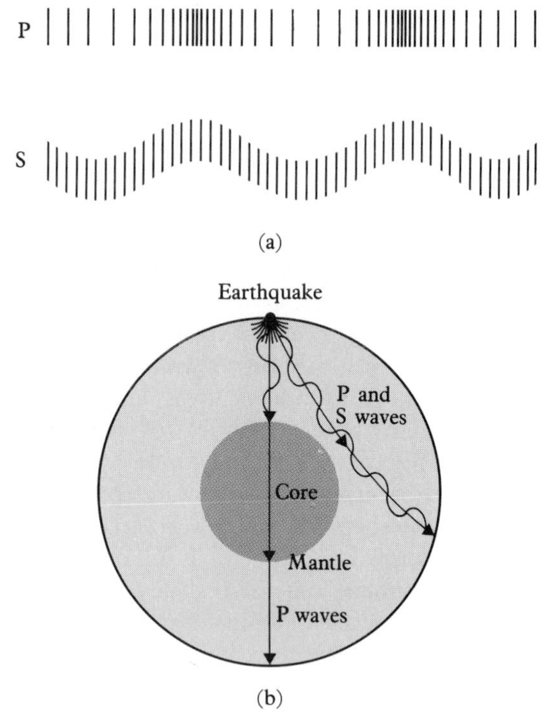

FIGURE 9.2 (a) Schematic diagram of P (pressure or compression) waves and S (shear) waves. (b) Paths of such waves through the Earth. S waves cannot propagate through the core, indicating that it must include a liquid component.

A powerful technique for exploring the interior structure of the Earth is provided by earthquakes. Whenever movement of the brittle crust produces an earthquake, low-frequency vibrations are generated. These vibrations (Fig. 9.2) are analogous to sound, but with much longer wavelengths, typically several kilometers. The shaken Earth responds like a giant bell, and its interior structure determines the tones that will be detected at different places on the surface. By measuring the vibrations of the Earth from many locations, we can reconstruct the temperature and pressure of the deep interior. The waves generated by earthquakes are called **seismic waves**, and instruments that measure them are seismometers.

Additional information on the interior of the Earth is provided by measurements of the chemical and physical properties of rocks. Most of the crustal rocks of the Earth have been derived, like the lunar basalts, by partial melting of subsurface layers, and they retain some information on the composition and nature of their source regions. In addition, there are places on Earth where magma from as deep as 200 km has reached the surface carrying rocks called kimberlite. Kimberlite, named for Kimberly in South Africa, is best known as the source of most of the world's diamonds. Diamond, formed from carbon at great pressures, is an artifact from the mysterious world a few hundred kilometers beneath us.

Even the deepest rocks sample only the upper few percent of the planet, but some useful information on the deep interior is provided by the magnetic field of the Earth. Although no one understands all the details of the mechanisms that generate a planetary magnetic field, it is clear that an electrically conducting, turbulent liquid is required. The presence of a strong magnetic field therefore demonstrates that at least some part of the interior is both conducting and liquid. Combining this information with that obtained from rock chemistry and from earthquakes allows us to undertake a scientific journey into the interior of our planet.

There are six major and several minor landmarks along the path to the interior. We will list them here and discuss each one in more detail later. The six major divisions (Fig. 9.3) into which it is convenient to divide the Earth are, starting from the outside:

1. The magnetosphere, the region of charged atomic particles that extends from the upper atmosphere, about 200 km high, out to the boundary with the solar wind and true interplanetary space, at an altitude of about 100,000 km.
2. The atmosphere, the layer of gas that extends from about 200 km altitude down to the surface.
3. The ocean or hydrosphere, the layer of liquid and frozen water that covers about three-fourths of the surface of the Earth.
4. The crust, the solid surface of the Earth bounded on the top by the oceans and atmosphere. Its thickness is typically about 10 km.
5. The mantle, consisting of solid but plastic rock extending from the bottom of the crust down to a depth of 2900 km. The mantle makes up two-thirds of the total mass of the Earth.
6. The core, composed of dense metal, divided into an outer liquid core and an inner solid core.

We begin our journey at the surface and work our way down; discussion of the ocean and atmosphere will come later.

The Crust

The crust of the Earth is defined as the uppermost solid layer, derived from the mantel by partial melting. Under the ocean basins, the crust is thin, averaging only 6 km in thickness, and composed of basaltic rocks, generally similar to the lunar and meteoritic basalts. This crust, which covers 55% of the surface, is all relatively young, having solidification ages of less than 200 million years. Although apparently analogous to the basaltic maria of the Moon, the origin of the terrestrial basalts is rather different. The ocean crust is formed from magma rising from the interior along well-defined rifts or spreading centers (discussed later), whose position is marked by several midocean ridges. Formed at these rifts, the ocean crust persists until it is destroyed by sinking into the mantle to be reheated and recycled. This process of formation and destruction of the oceanic crust is detailed in Section 9.4.

The continental crust is thicker, older, and of lower density than the oceanic crust. It covers 45% of the surface and constitutes 0.3% of the mass of the Earth. Its primary constituents are rocks called **granites**. Granitic rocks, like basalts, are igneous, but their composition is different, and they solidified below the surface under great

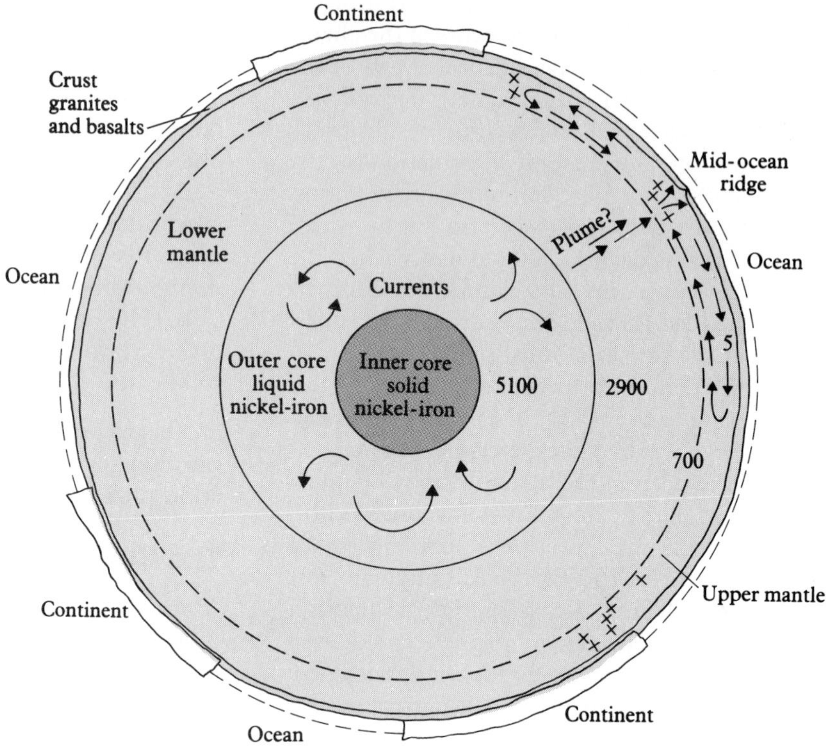

FIGURE 9.3 The interior of the Earth, according to current thinking. A small solid core is surrounded by a liquid region, which in turn gives way to the mantle. Rapid convection in the liquid core generates the planet's magnetic field. Slow convection in the mantle causes the motions of crustal plates, moving the continents, forming mountains, and subducting eroded material. (Highly schematic.)

pressure. The thickness of the continental crust varies from about 20 km to more than 70 km. The continental crust can be thought of as floating on top of the denser mantle. In addition to the igneous granites, the continental crust includes a great deal of sedimentary and metamorphic rock accumulated over hundreds of millions, and even billions, of years. All of the oldest rocks on the Earth are found in the continental crust.

The Mantle

Below the crust is the mantle, composed of higher-density minerals than the crust. Heated by its own radioactive elements and by heat leaking from the hotter core below, the mantle is more plastic than brittle. Within the mantle, three distinct layers can be identified.

First is the rigid upper part of the mantle, down to a depth of about 100 km. This upper mantle and the crust to which it is attached are together called the **lithosphere**. The boundary between the crust and upper mantle represents a change in composition; the boundary between the lithosphere and the mantle underneath represents a change in mechanical properties. The lithosphere is not strongly attached to the mantle beneath, and it can slide over it like an ice cube on a smooth tabletop.

Next, below the lithosphere and extending to a depth of about 700 km is the part of the mantle that *convects*, transporting heat upward by a slow mechanical turnover of material. These sluggish convection currents provide the motive force that moves the plates of the lithosphere, creates mountains and continents, and generates the heat to power volcanoes.

Finally, the massive lower mantle extends from a depth of 700 km down to the core boundary at 2900 km. Within it, the increasing pressure causes the minerals to alter to a denser phase in which convection cannot take place.

The Core

The boundary between mantle and core marks a major discontinuity in the interior composition of the Earth. In the core the primary constituent is iron, probably containing nickel, sulfur, and other cosmically abundant elements. At the temperature and pressure corresponding to this boundary—about 4500 K and 1.3 million bars, respectively (where one **bar** is defined as the atmospheric pressure at the surface of the Earth)—iron is liquid, so the core-mantle boundary is also a region of transition from solid to liquid phase. At these pressures, the iron has a density of at least 10 g/cm^3. The radius of the core is almost 3500 km, substantially larger than the planet Mercury, and it accounts for one-third of the total mass of the Earth.

Still deeper, at a radius of 1200 km from the center, the increase of pressure to 3.2 million bars triggers another phase change, with the iron reverting to solid form. This solid inner core continues all the way to the center of the Earth, where the pressure is nearly 4 million bars and the temperature is about 5000 K.

Within the liquid outer core the iron can move rather easily, driven either by heat (causing convection currents) or by chemical changes. Because the Earth is spinning, these motions lead to turbulent flows, which in turn generate the magnetic field that extends through the surface and into the magnetosphere. Thus the force that aligns a compass needle is the result of a special combination of temperatures, pressures, composition, and rotation deep in the interior of the planet.

Periodic magnetic field reversals demonstrate just how turbulent the motions are within the core. We are accustomed to having our compass needles point approximately northward, but there have been many periods in the relatively recent geologic past when our present compasses would have pointed south. As a matter of fact, the strength of the Earth's field has been steadily dropping for the past few decades, and if this trend should continue we might experience a field reversal within the next 200 years. More likely, however, we are seeing only a short-term fluctuation.

9.3 THE CHANGING FACE OF THE EARTH

All around us we see the processes of erosion. Look at a road cut, and you will see the strata, often tilted and twisted, of sedimentary rocks deposited by erosion in the past (Fig. 9.4). The mountains and hills are worn down, yet there are still high plateaus and jagged peaks; therefore invisible processes must be acting to raise new mountains. Volcanoes erupt, and earthquakes testify to movements in the Earth's crust. There is an almost total absence of impact craters, signifying destruction of craters as fast as they are formed.

The Geologic Timescale

The geologic history of the Earth is read in its rocks, which record the events of the past 4 billion years. The strata deposited in successive layers are like the pages of a history book. Unfortunately, however, the pages have frequently been shuffled or even lost in subsequent upheavals. To read the record, therefore, the geologist must find a way to number the pages, independently of their present sequence.

There are two ways to order the record: by stratigraphy, to deduce a *relative* sequence, often aided by noting the fossils embedded in the rock; or by radioactive dating, to produce an *absolute* or chronometric age scale. The standard **geologic timescale** was first determined by

FIGURE 9.4 Examples of layers of rock folded and lifted by the slow but inexorable motions of the Earth's crust. Thin-bedded limestone in Texas (left). Sandstone in Maryland (right).

stratigraphy, and specific dates were established by radioactive dating only in the second half of the twentieth century.

The simplest major division of the geologic record is into two time periods, the **Precambrian**, before fossils of multicelled organisms became widespread (about 590 million years ago), and the **Phanerozoic**, which extends from the Precambrian to the present. The next level of detail involves the division of geologic time into eras: the Archean (from the origin of the Earth to 2500 million years ago); the Proterozoic (the era of one-celled creatures, extending to 590 million years ago); the Paleozoic (an era that saw the development of fish and the occupation of the land by plants and early reptiles, ending 248 million years ago); the Mesozoic (dinosaurs, early mammals, flowering plants, ending 65 million years ago); and finally the Cenozoic (from the extinction of the dinosaurs 65 million years

TABLE 9.1 Comparison of geologic timescales for Moon and Earth

Time before Present (billion yrs)	Earth Era or Eon	Events on Earth	Events on Moon
0.03	Cenozoic	Mammals	
0.2	Mesozoic	Dinosaurs, K/T Impact	Crater Tycho
0.4	Paleozoic	Land plants, fish	
1.0	Proterozoic	Early fossils, O_2	Crater Copernicus
2.0	Proterozoic	One-celled life forms	
3.2	Archaean	Reducing atmosphere	End of mare volcanism
3.6	Archaean	Earliest record of life	Peak of mare volcanism
3.8	Archaean	Oldest rocks	Last basin impacts
4.0	Archaean	<no record>	Late heavy bombardment
4.4	Archaean	<no record>	Oldest rocks

FIGURE 9.5 The Alps, like most other high mountains on Earth, represent a balance between uplift from pressures in the crust and rapid erosion from ice and water. It is only because of ice and water that terrestrial mountains have sharp peaks.

ago to the present). In Table 9.1 this timescale for Earth is compared with that of the Moon, the only other planetary body for which precise radioactive dates can be assigned to geologic events.

Much of the geologic structure that we see around us is comparatively young. The jagged peaks of the Rocky Mountains or the Alps (Fig. 9.5) have been sculpted by ice within the past hundred thousand years; the volcanic islands of the Hawaiian chain are at most a few million years old; and even so large a feature as the Grand Canyon in Arizona has been cut by the Colorado River within the past 10 million years. Thus many of the major landforms around us have ages comparable to the short span of human existence on this planet.

Volcanoes

Volcanism is one of the most spectacular forms of geologic activity. On our planet, heat from within the mantle drives several types of volcanic activity. The largest but rarest types of eruptions lay down vast plains of fluid lavas very similar to the lunar maria; examples (each about 60 million years old) are the Columbia River flood basalts that cover eastern Washington State (400,000 km^3 in volume) and the even larger Deccan lavas of the Indian subcontinent. More familiar, however, are the eruptions that build the distinctive mountains we call *volcanoes.*

Although geologists distinguish among many types of volcanoes, we will consider only two basic classes, both of which have been found on other planets as well as our own. First are the **shield volcanoes**, which are dome-shaped mountains built up by the repeated eruption of relatively fluid lavas from a single vent or fissure. Although the individual lava layers are typically only a few meters thick, the mountains, such as Mauna Loa in Hawaii (Fig. 9.6), can reach

FIGURE 9.6 Mauna Loa in Hawaii is a shield volcano, with gradual slopes formed from successive flows of fluid lava. This photo shows the summit eruption of 1982. Note the constellation of the Southern Cross above the mountain.

heights as great as 9 km. Typical shield volcanoes have slopes of about 10° and are topped by large, shallow, flat-floored craters called **calderas**, produced by subsidence rather than explosions.

At the opposite extreme are a variety of explosive eruptions that build (and often then destroy) steep-sided **composite volcanoes**. Their distinctive cone shapes are formed by the fallback of plumes of viscous lava fragments ejected to great heights by the force of the eruption. The slopes of composite volcanoes are typically about 40°. Perhaps the best known example is Mount Fuji in Japan (Fig. 9.7). The most violent kinds of eruptions, such as the eruption of Vesuvius in A.D. 79 that destroyed Pompeii, can eject many cubic kilometers of rock dust into the atmosphere as well as devastate the surrounding countryside.

Other Geologic Activity

Another easily recognized form of geologic activity is erosion by ice. In glacial regions, the movement of ice sheets can gouge deep, U-shaped valleys in times as short as a few thousand years. It is primarily glacial action that is responsible for the sharp peaks that characterize the best-known mountains of the Earth. As we have seen on the Moon, mountains tend to be low and smooth in profile on a world without ice erosion.

FIGURE 9.7 Painting of Mount Fuji by the Japanese artist Hokusai. The slopes of this volcano are not as steep as the artist has painted them, but they are much steeper than those of Mauna Loa. The volcanism here is of a different type than that in Hawaii.

Finally, **sedimentation** is a process that can alter the landscape rapidly. Shallow lakes can silt up in a few decades, and the recession of the sea due to sedimentation has left the docks of once-great ports like Ephesus, Miletus, and Pisa dry and deserted. Most sediment is deposited in the sea. The fact that the seas have not filled up is one of the properties of the Earth that requires an explanation.

By mapping the distribution of rock types and landforms, and determining ages from the fossil record, radioactivity, and rock magnetism, scientists have begun to read the book of geological history, although the records of events earlier than the last few hundred million years are blurred or the pages lost entirely. Beyond 2 billion years ago, entire chapters are missing, and of the earliest periods of terrestrial history not a trace remains.

9.4 PLATE TECTONICS: A UNIFYING HYPOTHESIS

In the 1960s the science of geology underwent a dramatic change that altered the basic assumptions of the field. Until then, almost all geological science had been based on the idea that the major features of the Earth—the continents and ocean basins—were fixed. This was not an unreasonable hypothesis, given that most geologists thought the mantle was solid and unyielding. Within this framework, geology was interpreted in terms of periods of slow uplift to form mountains, in competition with the forces of erosion gradually tearing them down again. This outlook is still reflected in older books and museum displays that explain the landscape in terms of alternating uplift and subsidence, as though the crust were part of a seesaw, always moving up and down, never sideways.

Wegener and Continental Drift

The first serious challenge to the orthodox geological viewpoint came from Alfred Wegener, a German meteorologist and amateur geologist. Early in this century Wegener became convinced that the strong similarities between rocks on either side of the Atlantic could best be understood if the east coasts of North and South America had once fitted, like the pieces of a jigsaw puzzle, against the west coasts of Europe and Africa. Even casual inspection of a world map suggests that this fit is rather good. Wegener found other places where contintents, now separated by thousands of kilometers, appeared once to have been joined together, and in Iceland he studied the active volcanoes, concluding that this mid-Atlantic island was located on the rift where the New World and the Old were still separating. Based on this evidence Wegener developed a theory of **continental drift**, in which the continents slowly moved while the lower crust and ocean basins remained relatively fixed.

The theory of coninental drift was rejected by most geologists, however, because its proponents could suggest no plausible force to explain the migration of continents across the solid oceanic crust, to which they seemed firmly attached. In spite of the empirical evidence that supported this theory, the motion of the continents proposed by Wegener was deemed physically impossible.

Plate Tectonics

The idea of continental drift was finally accepted when, in the 1960s, observations clearly showed that the Atlantic Ocean was widening, as a line of volcanoes, approximately halfway between the two continental masses, injected new lava along the ocean floor. These data showed for the first time that the ocean basins themselves were part of the movement, and that the continents do not drift over an unyielding crust. Instead, the crust itself is being forced apart by the injection of magma. Soon many other pieces of the puzzle fell into place. The result is a unifying geological concept called **plate tectonics**.

The term **tectonic** refers to the forces that stress a planet and the way that the crust responds to such forces. On planets with active geology, such as the Earth, Venus, and Mars, many of the mountains, valleys, and other landforms are of tectonic origin. What makes terrestrial geology unique is the organization of the crust into six major and about ten smaller plates that float like sheets of ice upon the plastic mantle beneath (Fig. 9.8). The plates constitute the lithosphere, about 100 km thick and made up of both the crust and the upper mantle. When tectonic forces are applied in the Earth's crust, these plates are capable of lateral motion—plate tectonics.

The motive power for continental drift comes from convection in the mantle. Driven by this convection, the lithospheric plates move at the rate of a few centimeters per year. Since there is no free space on the surface of the Earth, the plates bump into each other as they move.

Four kinds of boundaries between plates are possible. First there is the rift zone or spreading center, such as the mid-Atlantic ridge, where the plates are separating. Second is the subduction zone, such as the deep trenches off the east coast of Asia, where plates collide and one is forced under the other. Third is a region of collision where neither plate slides under the other so that folded mountains are produced, such as the Himalayas. Finally, there are faults where one plate scrapes along parallel to another, such as the San Andreas fault in California. We will discuss each of these cases in turn.

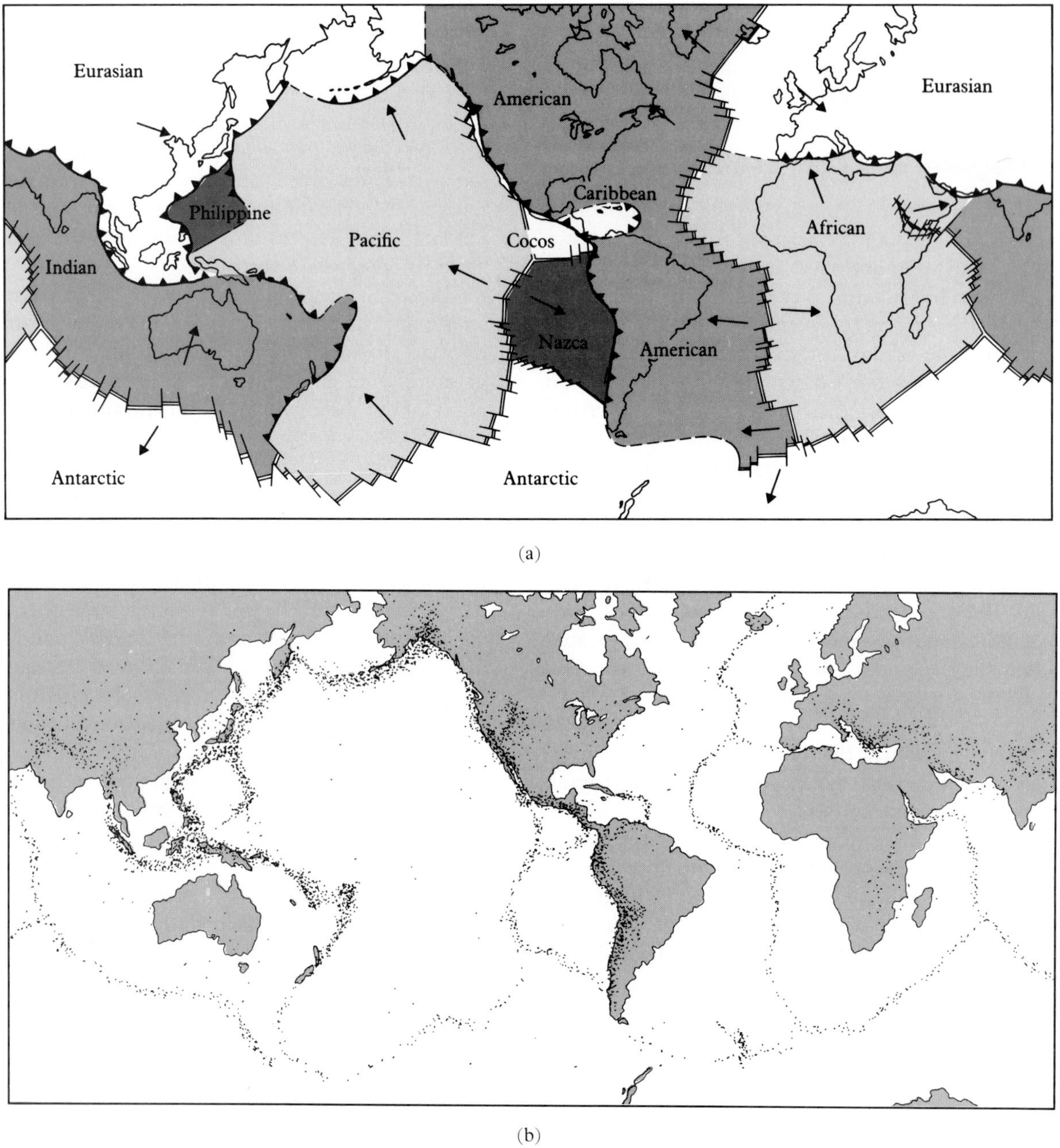

FIGURE 9.8 (a) The main crustal plates on the Earth. (b) The boundaries of these plates are delineated by the source points (epicenters) of earthquakes observed over a seven-year period.

Formation and Destruction of Ocean Crust

Most of the surface of the Earth is ocean floor. Sediment washed down from the land is deposited first on the continental shelves, but gradually it moves out and down into the oceans proper, where it typically accumulates to a depth of two or more kilometers. Much of the ocean floor consists of vast, gently rolling plains deeply buried in sediment.

There are 60,000 km of active **rifts** in the ocean where new crust is formed by basaltic magma rising from below. The few rifts on land, such as the great rift valley that extends through central Africa, probably represent places where a continent is being torn apart to be replaced by new ocean. The average spreading rate is a few centimeters per year, corresponding to the creation of about 2 km^2 of new oceanic crust each year. Since the total area of oceanic crust is 260 million km^2, we can see that the lifetime of a given piece of oceanic basalt is only a little more than 100 million years (260 million km^2 divided by 2 km^2). Otherwise the entire planet would be covered by oceanic crust.

Oceanic crust is destroyed by **subduction** when convection pulls a plate into the mantle. As two oceanic plates are forced together, one generally slides under the other (Fig. 9.9) to produce a deep ocean trench. The sinking slab grinds down into the mantle, generating deep earthquakes, which permit its descent to be traced. At a depth of 200–300 km it succumbs to increasing heat and melts. Some of the melted rock can rise to the surface in the form of volcanoes, which also mark the position of the subduction zone, but most is recycled back into the upper mantle to await eventual release at another rift zone.

Volcanoes are common along both rift zones and subduction zones. In addition, volcanoes can erupt in the middle of lithospheric plates above *mantle hot spots*, which represent rising plumes that heat the crust from below. The Hawaiian Islands are an example of volcanoes produced by a mantle plume. Since the Pacific plate is moving with respect to the plume, the volcanic center has traced out an island chain over a period of tens of millions of years (Fig. 9.10). It is not the plume that moves, however, but the plate.

In Chapter 2 we discussed the three ways that heat can be transported: conduction, convection, and radiation. At the temperatures of the Earth, radiation is not important, but both conduction and convection play a role in the movement of energy to the surface. First there is heat conduction through the crust, which is thought to be the dominant process on most other plan-

FIGURE 9.9 Here a slab of oceanic crust is being subducted. As it slides down under the granitic continental plate, earthquakes are generated and volcanoes form at the plate boundary.

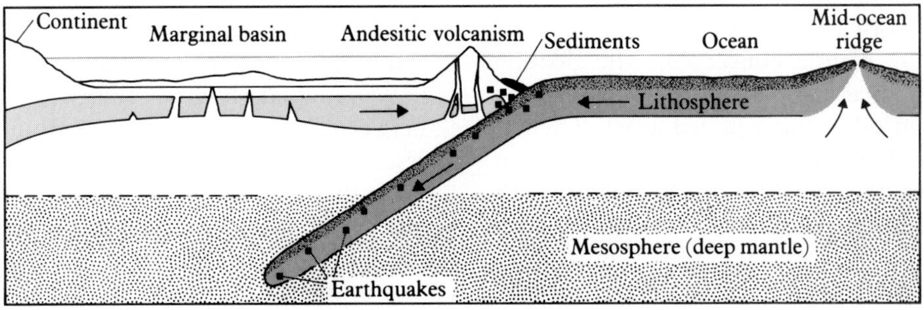

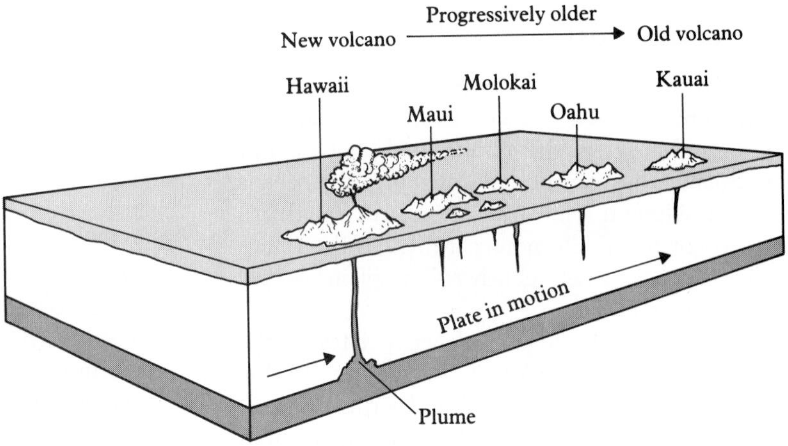

FIGURE 9.10 The gradual motion of the Pacific plate past a long-lasting hot spot (representing the top of a mantle plume) accounts for the formation of the chain of Hawaiian islands.

ets. Second is the volcanic activity associated with hot spots. But the most important way heat is released from our planet is at the rifts and subduction zones produced by tectonic activity. Without plate tectonics, the heat balance of the Earth's interior would be very different.

The Continental Crust

The continental crust, composed mostly of granite, has a different history from that of the oceans. Like a frothy slag floating on the top of the plates, it is rarely subducted. Continental granite is produced much more slowly than oceanic basalt, but it also lasts longer. The creation and destruction of the ocean floor is a *cyclic* process, but the continents *evolve*, slowly increasing in total area as more new granite is added by tectonism than is destroyed by subduction or worn away by erosion.

Jostled by mantle convection, the floating continental masses are sometimes pushed together and other times pulled apart. The last time nearly all of the land mass of the Earth was jumbled together was about 200 million years ago. This supercontinent has been named Pangaea. As illustrated in Fig. 9.11, Pangaea has gradually been pulled apart and rearranged by continental drift. Not everything is coming apart, however. Some of the land masses are colliding with each other, most notably the Indian subcontinent with Asia. At such points mountain ranges are folded and uplifted by the pressure of the moving plates. The Himalayan Mountains and Tibetan Plateau produced by this collision have the highest elevations on our planet.

In many places one plate is forced to move parallel to another along a **fault** or crack in the crust. California's San Andreas fault, for instance, marks the break where the Pacific plate (which includes the southern coastal regions of California) slides northward at a rate of a few centimeters per year with respect to the main part of the North American continent. At this rate, Los Angeles will be next to San Francisco in about 20 million years. Instead of sliding smoothly, however, the jagged edges of the plates tend to stick until the strain becomes so large that the restraints suddenly give way. The resulting motion is sudden and violent, generating earthquakes. In California, the interval between major plate motions is about a century, resulting in typical movements of several meters when the strain is finally released. The longer the interval between earthquakes, the larger they are likely to be. The last major earthquake in the part of the San Andreas near San Francisco took place in 1906, while the southern part of California has not moved significantly since the Fort Tejon earthquake of 1857. The 135-year buildup of tension along the southern part of this fault is obviously a matter of concern for the residents of southern California.

The geological processes outlined here are highly dynamic, with much of the landscape of our planet produced by forces deep in its interior. Because the Earth is much larger than either the Moon or Mercury, our planet has cooled less and these forces continue to act today. Most of the processes discussed here are cyclic, involving repeated acts of creation and destruction. However, the production of continental rock is a continuous process, and the continents thus evolve with time.

9.5 OCEAN AND ATMOSPHERE

Floating on top of the solid Earth are two layers of great importance to us: the hydrosphere and the atmosphere. The liquid water that covers 70% of the surface, and the oxygen-rich atmosphere that shields us from solar ultraviolet light and replenishes the land with life-giving moisture, are unique to our planet. Without them there would be no life. In return, both atmosphere and ocean are shaped by the evolution of the very life they make possible. In this and the following section we examine the intricately interwoven threads of ocean, atmosphere, and life on Earth.

FIGURE 9.11 Motions of the continents with respect to the South Pole (cross) through geologic time. The dashed circle represents the position of the Earth's equator.

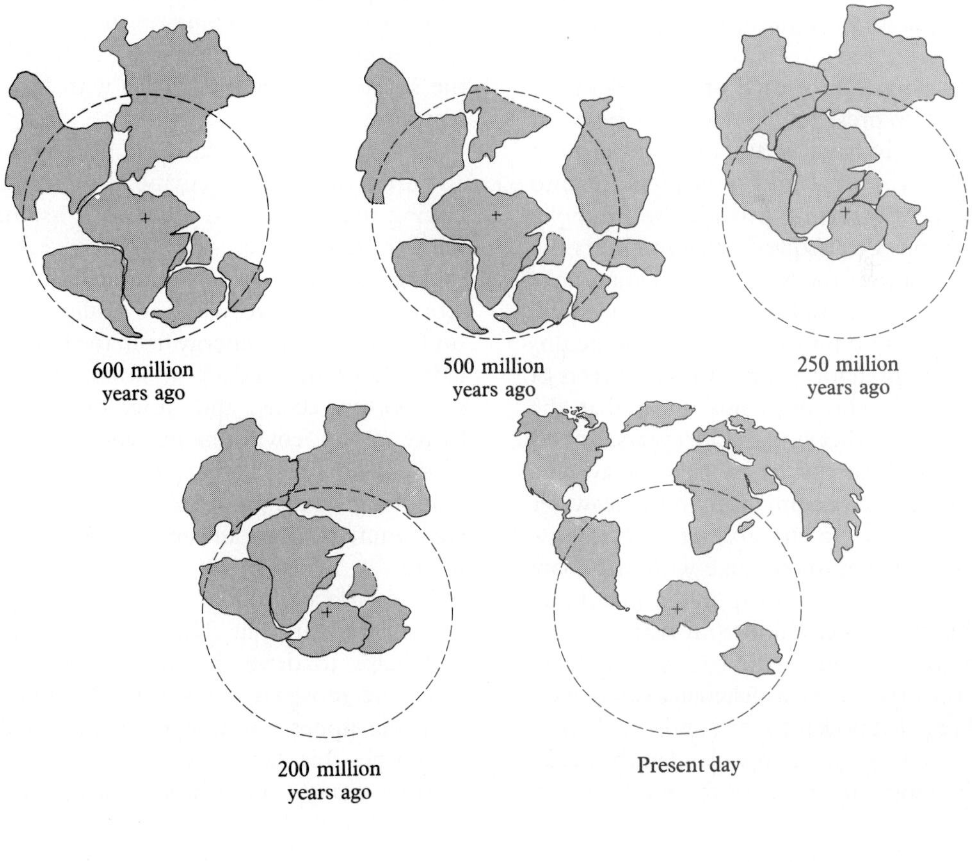

Composition of the Oceans

The hydrosphere of the Earth is defined to include the freshwater lakes and streams, the ice caps that blanket the polar land masses and high mountains, and the sea ice that floats on much of the arctic and antarctic seas. Its main component, however, is the oceans. The oceans, with a volume of 1.3 billion km^3 and a mass of more than 10^{18} tons, are the great reservoir of water on our planet.

Ocean water is not pure H_2O. As we all know, seawater is salty. In addition to common NaCl, the ocean contains a variety of other water-soluble salts, which together make up about 3.5% of the mass of the oceans. Water is an excellent solvent (which is why we wash with it), breaking down many compounds into their individual atoms. Six elements constitute more than 90% of the salts dissolved in the ocean: chlorine, sodium, magnesium, sulfur, calcium, and potassium.

The concentration of each of these elements in seawater represents a balance between inflow—primarily from weathering and erosion of continental rocks—and precipitation into ocean sediment. Calcium provides an example. About 10^8 tons of dissolved calcium enters the ocean each year, yet the total calcium content of the ocean is only a million times larger—10^{14} tons. Thus the amount of calcium in the oceans could be accounted for by only 1 million years of input. However, this does not mean that the oceans have existed only a million years, as was once thought. Rather, it means that the average residence time of a calcium atom in the seawater is a million years. On the average, this is how long it takes calcium to combine with carbonate molecules and become incorporated into limestone. This cycle of calcium from surface rocks to ocean to limestone sediment is completed when the ocean sediment is melted in a subduction zone and recycled back into the continental crust.

Gases as well as salts are dissolved in the ocean water. The most important of these are oxygen and carbon dioxide. The oxygen in the water makes possible the existence of advanced forms of marine animal life, such as the fish, which use gills to extract the oxygen they require from the water. However, the total oxygen content of the ocean is less than that of the atmosphere. In contrast, about 60 times as much carbon dioxide is dissolved in the oceans as is present in the air. As a result, the role of the oceans is critical in determining the amount of carbon dioxide in the atmosphere, and therefore in establishing the climate of the Earth.

Ocean Temperatures

Because of its huge mass and thermal capacity, the ocean serves as a great heat reservoir. The average temperature of the ocean is 4 C, only a little above the freezing point. Although the surface temperature varies from about 30 C near the equator to below freezing on the polar ice caps, the temperature a kilometer or more below the surface changes very little with location or season.

In a number of areas near plate boundaries and areas of volcanic activity, heated water containing many dissolved minerals is issuing from massive underwater vents. These remarkable thermal springs support unique forms of life. Creatures live here without sunlight, deriving both nutrients and energy from the hot, mineral-laden water. In this dark, mysterious world, giant sea worms, clams, and other exotic creatures thrive in an environment where no life was thought possible. Perhaps there is a lesson here in looking for life on other worlds: The variety of environments that can support life-forms is very broad.

The surface of the ocean is heated by the Sun and stirred by the air. The wind generates waves and helps to drive the major ocean currents. Even the most violent storms, however, affect only the upper few tens of meters of the sea. Although this fact is of no comfort to the mariner caught in a wild storm, we should remember that the most fearsome hurricane or typhoon is a superficial event, of little consequence for the placid reservoir of water beneath the surface.

Composition of the Atmosphere

Above the ocean is the atmosphere, which consists primarily of molecular nitrogen (N_2) and molecular oxygen (O_2), which make up 78% and 21%, respectively. The remaining 1% of dry air is primarily argon (Ar), with carbon dioxide (CO_2) the fourth most abundant gas at 0.036% (or 360 parts per million). In addition, the atmosphere contains considerable water vapor, ranging from 3% in wet, warm regions, down to much less than 1% in cold, dry locations. The compositional measurements are summarized in Table 9.2.

The atmosphere exerts a *pressure* on the surface, which is just the *weight* of the gas. On the Earth, the force of gravity and the amount of gas generate a sea-level pressure of 1 bar, the same pressure that would be exerted by a 10-meter-thick layer of water. Note, however, that the pressure is not a measure of the total amount of gas in an atmosphere. Pressure is the product of the amount of gas multiplied by the force of gravity. On Saturn's satellite Titan, for instance, about ten times as much gas is required to generate a pressure of one bar as would be the case on Earth, since the surface gravity of Titan is only about a tenth that of our own planet.

TABLE 9.2 Composition of Earth's atmosphere

Gas	Fraction (percent by number)
N_2	78.1
O_2	20.9
H_2O	3.0–0.1 (variable)
Ar	0.93
CO_2	0.036 (increasing)
Ne	0.0018
He	0.0005

Nitrogen is a chemically lazy gas, little inclined to react with other substances. Even more inert is argon, one of the noble gases discussed in more detail in Chapter 12. Nitrogen and argon are also heavy enough to resist escape from the upper atmosphere. The present abundance of argon therefore reflects the total accumulation released into the atmosphere over the history of the Earth. In contrast, both oxygen and carbon dioxide are highly reactive chemically, so the abundances of both must represent an equilibrium between competing reactions that produce and destroy them. Nitrogen is also removed from the atmosphere, primarily by organisms. As we will see later, the primary sources and sinks of both oxygen and carbon dioxide also involve life, and their atmospheric abundances have evolved along with life itself.

Both water and carbon dioxide are underrepresented in the current atmosphere. Both are in the atmosphere only as trace gases, but consider the difference that a slightly higher temperature would make. If the oceans warmed only a little, both water vapor and dissolved CO_2 would be released into the atmosphere. If the warming continued up to 100 C, the boiling point of water, the entire ocean would vaporize and become part of the atmosphere.

How much water vapor would be produced if the ocean boiled away? We have noted that the average depth of the ocean is about 3 km, at which depth the weight of the water is 300 bars. (Remember that a 10-meter thickness of water exerts a pressure of 1 bar.) The weight of this water would be the same if it were in vapor form rather than liquid; therefore the pressure of the atmospheric water, if the oceans boiled away, would also be 300 bars. Water vapor would thus be the dominant constituent of the atmosphere. The CO_2 content would also rise, particularly if the carbonate minerals in the ocean sediment were broken down and their CO_2 released into the atmosphere. We will return to the Earth's carbon reservoirs later.

If the Earth were hotter, it might have a massive atmosphere consisting primarily of H_2O and CO_2. If it were just a little cooler, the oceans would freeze and there would be no liquid layer on the surface. Only within a narrow temperature range can the Earth maintain liquid water oceans and a nitrogen-dominated atmosphere.

Life and Atmospheric Chemistry

Liquid water is essential to the survival of life on Earth. As far as we know, life originated in the seas of Earth, and the presence of life today testifies to the continuous presence of liquid water on Earth over the past 4 billion years. Throughout this vast expanse of time, life has played an important role in the evolution of the Earth's atmosphere.

In the absence of life we would expect the Earth to have maintained a modestly oxidized atmosphere composed primarily of CO_2, N_2, and Ar, like the contemporary atmospheres of Venus and Mars. The predominant gas should be CO_2, with a surface pressure of tens of bars. Such an atmosphere can persist in chemical equilibrium with the ocean and land. Instead, we find a relatively thin and much more oxidized atmosphere that is deficient in CO_2 and contains molecular oxygen, O_2.

The transformation of the atmosphere from its initial state to its present composition is primarily the result of the development of photosynthetic life. Atmospheric oxygen is manufactured from H_2O and CO_2 by green plants, and in the absence of such plants it would quickly recombine with surface minerals and be lost to the atmosphere. Studies of ancient rocks show that the transformation to an oxidizing atmosphere took place gradually between 2 and 1 billion years ago, as a consequence of the proliferation of life and the burial of carbon in sediments. The removal of carbon inhibited the recombination of oxygen and carbon to form CO_2 and left us the present unique atmospheric chemistry.

The chemical interaction of life with the atmosphere is a complex saga. As far as we can tell, life could never have originated in a strongly oxidizing environment. Yet life is responsible for the gradual oxidation of the atmosphere. As the atmosphere changed, life evolved ways to protect itself from the toxic oxygen and ultimately to utilize that oxygen for more efficient metabolism than could ever have been possible in the environment of early Earth. It is this evolutionary path that led to the development of the animal kingdom and eventually to us.

We will discuss the interaction of life and the atmosphere further in Chapter 12, in a comparative study with the evolution of the atmospheres of Venus and Mars.

Structure of the Atmosphere

The density of air at sea level on Earth is about 0.001 g/cm^3, and the pressure is, by definition, 1 bar. This pressure drops off rapidly with elevation, reaching one-half its sea-level value at 5.5 km (Fig. 9.12). Water vapor is even more strongly concentrated near the surface, and it declines to half its sea-level value at about 2 km; this is why mountain air is dry and skiers and mountain hikers suffer so much from chapped skin.

The lower 10–15 km of the atmosphere is called the **troposphere**. This region, which contains 90% of the mass, is the location of nearly all of the phenomena we call *weather*. The main characteristic of the troposphere is the convection of air within it. Heating by sunlight causes warmer air near the surface to expand and rise under its own buoyancy, to be replaced by downdrafts of cooler air from above. Convection in the troposphere maintains a steady decrease in temperature with altitude of 6 C per kilometer. The top of the troposphere, called the tropopause, is at a temperature of about −60 C or 213 K.

One of the most important cycles linking the ocean and the lower atmosphere is the *hydrologic*

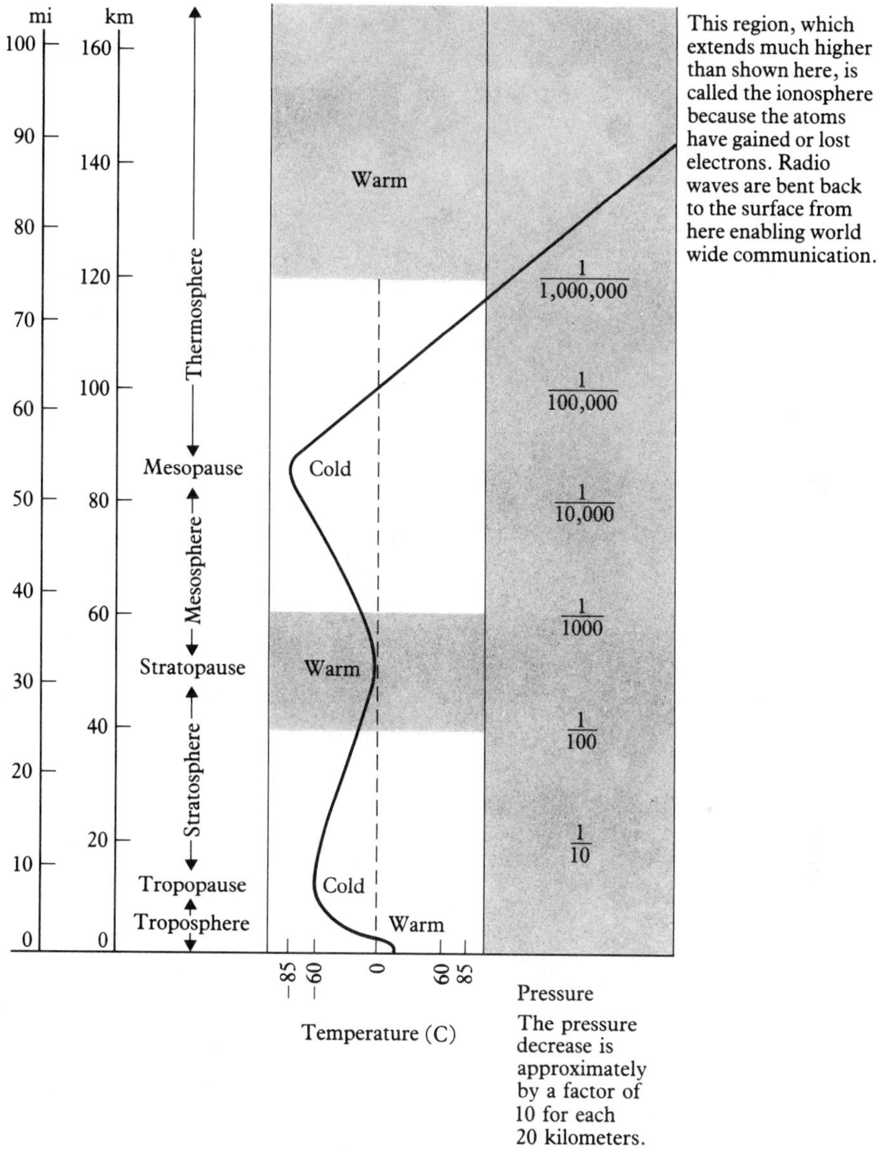

FIGURE 9.12 The vertical structure of the Earth's atmosphere showing the approximate altitudes of the principal regions and the change in pressure and temperature with altitude.

or *water cycle*. Water in the atmosphere forms clouds and precipitates as rain and snow on both the land and the sea. Some of the precipitation is stored temporarily in the ground or in lakes, but most finds its way to the greatest reservoir, the ocean. Evaporation back into the atmosphere completes the cycle. In the process of evaporation and condensation, moreover, the water can store and release large amounts of solar energy. This energy as well as direct heating by sunlight drives tropospheric winds and generates weather.

9.6 UPPER ATMOSPHERE AND MAGNETOSPHERE

Above the troposphere the atmosphere is too thin to sustain life. However, chemical reactions that occur at high altitudes are critical to the maintenance of the environment near the surface. Still higher, the tenuous outer fringes of the atmosphere are influenced by the Sun and the solar wind, linking our planet directly with the plasma that fills interplanetary space.

The Stratosphere

Above the troposphere is the **stratosphere**, in which very little vertical mixing takes place and the temperature is nearly uniform at about −60 C. In the stratosphere water is almost entirely frozen out and the atmosphere is very dry. In the upper part of the stratosphere the absorption of solar ultraviolet light creates a layer of **ozone** (O_3), an unusual form of oxygen with three atoms per molecule instead of two. The ozone layer, sometimes called the ozonosphere, extends from about 30 to 50 km. Without it, solar ultraviolet light would penetrate to the surface, doing irreparable harm to nonmarine living organisms, both plant and animal.

Because ozone is so important to the health and even the survival of life on the Earth, we are especially concerned about the damage to the ozonosphere caused by the industrial chemicals called **CFCs** (chlorofluorocarbons). These inert, nontoxic gases have been manufactured in huge quantities for use in air conditioning and refrigeration systems and for cleaning electronic components. Unfortunately, CFCs have the unintended property of interacting with sunlight at high altitudes to destroy O_3. Beginning in the early 1980s, an increasing quantity of the stratospheric ozone over the Antarctic continent has disappeared every spring to generate the Antarctic "ozone hole." Each year the hole has expanded, and by the 1990s significant ozone depletion was observed in temperate latitudes as well, in both hemispheres. This ozone depletion is especially scary because the lifetime of CFCs in the atmosphere is more than a century; thus even if we discontinued all manufacture of CFCs immediately, their effect on the chemistry of the stratosphere would not be reversed until the twenty-second century.

The Ionosphere

A convenient if somewhat arbitrary altitude to mark the beginning of the tenuous upper atmosphere is 100 km. This altitude represents the highest level at which meteors are seen and the lowest level at which an Earth satellite can complete an orbit without being brought down by atmospheric friction. It is also about the level at which solar ultraviolet light breaks apart molecules into atoms more rapidly than they can recombine. Finally, and perhaps most important, 100 km represents approximately the lower boundary of the ionosphere of the Earth.

The **ionosphere** is the region of the upper atmosphere in which many of the atoms are ionized—that is, broken apart into positively charged ions and negatively charged electrons. The atmosphere in this region is therefore a *plasma*, albeit a weak one, since only a small fraction of the atoms are ionized at any given time. The ionosphere was discovered early in the twentieth century by its ability to reflect radio waves transmitted from the ground. Before the advent of communications satellites, the ionosphere provided the primary means for long-distance communications, and it is still critical for the propagation of AM and short-wave radio.

The degree of ionization of the atmosphere is measured by the density of electrons. Below 80 km ions and electrons recombine as quickly as they form, and there is no ionization. The electron density rapidly increases with altitude, forming a first maximum at 100 km of a little more than 10^5 electrons/cm^3. A stronger ionization peak occurs at about 140 km, where the electron density reaches somewhat more than 10^6 electrons/cm^3. Above 140 km, the electron density falls off gradually with altitude. At each altitude, the electron density represents a balance between ionizing solar ultraviolet and x-ray radiation, recombination of the ions and electrons, and the total amount of gas available. Above 500 km most of the gas is ionized, and the declining electron density with altitude simply reflects the thinning of the atmospheric gas.

Within the ionosphere, between altitudes of about 150 and 400 km, occurs one of the most beautiful of natural phenomena: the **aurora** or polar lights. Named for the Greek goddess of the dawn, the auroras are regularly seen at polar latitudes but only rarely grace the night sky in temperate zones. The displays can take on a variety of forms, with the most common being pale rays or curtainlike sheets of green or red color that silently sway and dance across the night sky. Viewed from space (Fig. 9.13), the auroras are seen to be concentrated in rings centered on the north and south magnetic poles. Auroras are the result of electric currents flowing through the ionosphere, stimulating the gas to glow much as an electric current produces light from a fluorescent lamp. The green color comes from the fluorescing oxygen, while red is contributed by hydrogen atoms. The currents themselves originate at higher altitudes, in the magnetosphere, to which we now turn our attention.

The Magnetosphere

The Earth's **magnetosphere** was discovered in 1958 by instruments on board the first U.S. Earth satellite, Explorer 1 (Fig. 9.14). The scientist who built the high-energy charged-particle detectors on Explorer 1 was University of Iowa professor James Van Allen, and his name has been given to the primary features of the

FIGURE 9.13 Auroral arcs in the upper atmosphere, photographed from space.

FIGURE 9.14 From right to left: Werner von Braun, James Van Allen, and William Pickering holding aloft a model of the Explorer 1 satellite that discovered the Earth's radiation belts in February of 1958. Von Braun organized the development of the launch vehicle, Van Allen designed and built the instruments, and Pickering was the director of the Jet Propulsion Laboratory, which built and operated the satellite.

inner magnetosphere, the Van Allen belts. These are regions of space containing large numbers of protons and electrons, trapped in the magnetic field of the Earth.

Unlike the ions and electrons of the ionosphere, the magnetospheric charged particles are highly energetic, spiraling back and forth within the magnetic field at speeds of thousands of kilometers per second. When they strike solid material, such as the skin of a spacecraft, they produce additional subatomic particles and gamma- and x-radiation, leading to the misnomer "radiation belts." But it is not radiation that is trapped in the Van Allen belts, it is energetic electrons, protons, and other ions.

The configuration of the magnetosphere is the result of interactions between the solar wind and the magnetic field of the Earth. From the direction of the Sun, the charged particles of the solar wind stream toward the Earth at a speed of about 400 km/s. As they near our planet, however, these particles are deflected by the Earth's magnetic field. This deflection distance, which represents the point at which the magnetic field strength is just sufficient to balance the pressure of the solar wind, is at about ten Earth radii, or 60,000 km. The cavity within which the planetary magnetic field dominates over the solar wind is the magnetosphere. It is shaped rather like a wind sock or a stubby comet tail, pointing away from the Sun. Downstream, it extends just about to the distance of the Moon. The outer boundary of the magnetosphere is called the magnetopause (Fig. 9.15).

Charged particles from three different sources contribute to the magnetosphere of the Earth. First there are the solar wind particles, a few of which leak across the magnetopause and

FIGURE 9.15 Diagram of Earth's magnetosphere. The planet's magnetic field forms a shield that staves off the solar wind. The tilt of the field lines close to the planet results from the inclination of Earth's rotational axis. The belts of electrons and protons (discovered by James Van Allen) surround the Earth within the magnetosphere.

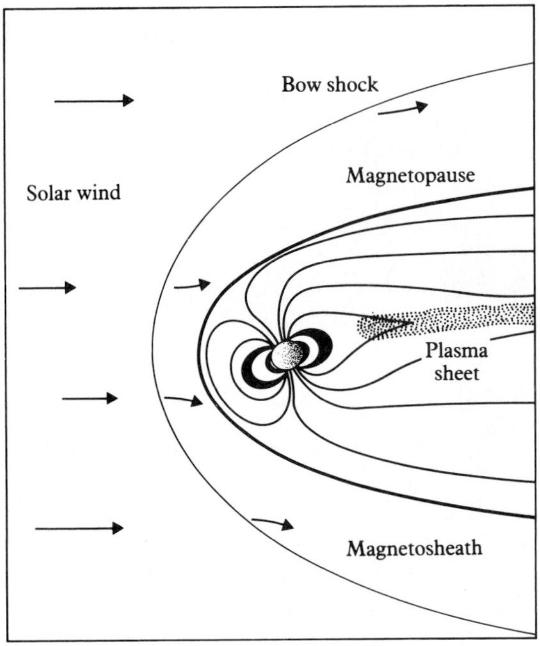

become trapped inside. Second are escaping atoms from the atmosphere, which contribute most to the population of the inner magnetosphere. Third are particles generated from impacts on the atmosphere of the very high energy galactic cosmic rays, which contribute primarily to the middle belt a few thousand kilometers up. To balance the sources of charged particles, there must be corresponding sinks: the escape of particles from the downstream side of the magnetosphere back into the solar wind, and the impact of charged particles on the atmosphere of the Earth. Both sinks are triggered when the boundary between the solar wind and the magnetosphere becomes unstable as the result of fluctuations in solar wind pressure. These fluctuations, generated by outbursts on the Sun, produce the bright auroral displays and ionospheric disruptions associated with periods of intense solar activity.

9.7 CLIMATE AND WEATHER

Weather and climate are both aspects of the closely coupled systems of air, land, and water on our planet. *Weather* refers to the state of this system at a given place and time and its short-term variations, while *climate* is concerned with average conditions and their long-term trends. While there is no sharp line of demarcation between the two, climate is generally thought of as referring to changes that take place over a period of ten years or more. More rapid changes are considered a part of the phenomena of weather.

Heating by the Sun

The source of energy to drive motions in both the atmosphere and the water is the Sun. Of the energy incident on the planet, about 30% is reflected back to space from the atmosphere or surface. The principal reflectors are water and ice clouds, atmospheric dust, surface snow and ice, and unvegetated deserts. Thus the basic energy balance of the planet is affected, through changes in reflectivity, by the amounts of dust in the atmosphere, the degree of forestation of the land, and the size of the polar deposits of snow and ice.

The 70% of incident sunlight absorbed by the planet corresponds to an average power of 240 watts for each square meter of surface, or a total input of more than a hundred billion megawatts (1.2×10^{17} W). (Imagine the consequences of more efficient ways of harnessing this solar energy!) If the climate is to be stable, an exactly equal amount of energy must be radiated from the Earth back into space, primarily at infrared wavelengths. Part of this infrared radiation escapes directly from the surface, but most is absorbed and reradiated within the troposphere.

The average surface temperature of the Earth is about 10 C, a value about 25 C higher than would be the case without an atmosphere. This increase in surface temperature is the direct result of the blanketing effect of the atmosphere, primarily caused by the carbon dioxide greenhouse effect.

The Greenhouse Effect

The **greenhouse effect** is named after the gardener's greenhouse, which provides a simple means of passively heating a room. The glass in the greenhouse roof allows visible sunlight to enter and be absorbed by the plants and soil within. These objects then heat up and radiate at infrared wavelengths, just like the Earth itself. But glass is largely opaque to infrared radiation. It acts like a color filter, letting short wavelengths through, but limiting passage of longer-wave thermal radiation. Since most of the heat can't get out, the interior of the greenhouse warms up, until enough infrared radiation escapes through the glass to balance the energy coming in. A similar effect occurs in a car left out in the sunlight on a hot day. Once again, sunlight passes easily through the glass in the windows, heating the upholstery and metal of the car's interior. Infrared radiation is trapped inside, and soon the temperature inside the car is much higher than that of the surrounding air.

The same thing happens in a planetary atmosphere, with the atmospheric gas playing the role of the glass in retaining infrared energy radiated by the surface. The magnitude of the greenhouse effect is determined largely by the infrared opacity of the gas. Both water vapor and carbon dioxide are effective at blocking thermal emission, and they are thus sometimes called *greenhouse gases.* Other trace constituents of the Earth's atmosphere, including CFCs and some hydrocarbons, are also effective greenhouse gases.

On the Earth, CO_2 dominates the greenhouse effect. Even at only 0.0036% of the atmosphere, the CO_2 is sufficient to raise the surface temperature by about 25 C. As we will see in the next chapter, Venus has a massive atmosphere composed primarily of CO_2, and its greenhouse effect is much greater than that on our own planet.

Circulation of the Atmosphere

The Earth is heated more near the equator than in polar regions, and therefore the average surface temperature decreases with increasing latitude. However, this temperature decrease is less dramatic than it would be if there were no oceans or atmosphere to redistribute the solar energy. Air that is warmed near the equator tends to rise and move toward the poles, carrying much of this energy with it (Fig. 9.16). Transport of energy from warm regions to cooler ones generates the most important forces that drive the circulation of the atmosphere.

FIGURE 9.16 The global circulation of Earth's atmosphere in simplified form.

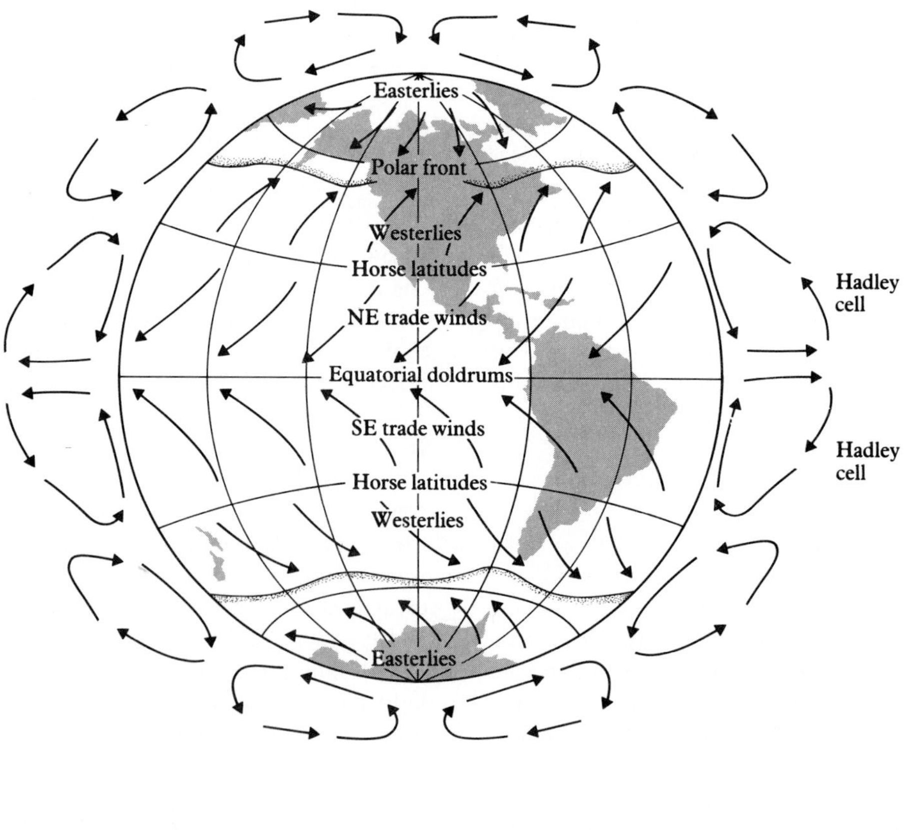

Because the rotation axis of the Earth is tilted by about 23°, we have seasons on our planet. Each year the familiar cycle is repeated, as the latitudes of greatest solar heating move north and south, bringing with them the succession of spring, summer, autumn, and winter. If the axis were not so tilted, the atmosphere might be able to sustain a single pattern of atmospheric circulation. The constantly shifting deposition of energy encourages the circulation pattern to break up into the cyclones and storm fronts and other phenomena we call weather.

Near the equator the weather is simplest. Between 35° N and 35° S latitude the large-scale winds blow toward the equator near the surface and away at higher altitudes, carrying warm moist air upward near the equator and recirculating cooler, dryer air back to complete the cycle. The rising air generates a region of high precipitation within 10° of the equator, producing the tropical rainforests of the Earth. In contrast, most of the deserts of the Earth are at latitudes between 15° and 35°, where the dry air sinks and flows back toward the equator.

Poleward of 35°, in the temperate and polar regions of the Earth, the basic character of the circulation changes. Here the predominant winds are east-west, rather than north-south, in direction. Instead of a smooth flow of warmer air toward higher latitudes and cooler air back toward the tropics, a series of large wavelike patterns develops, with the primarily east-west winds veering north and south. As the waves develop, masses of cool air are carried toward the tropics, and masses of warm air penetrate toward the poles. At any one time, about a dozen of these waves, each a couple of thousand kilometers across, are likely to exist in the temperate regions of each hemisphere.

Planetary Rotation and the Coriolis Effect

The rotation of a planet, as well as thermally driven atmospheric motions, is important in determining its weather. If it were not for the relatively rapid spin of the Earth, we would not have the rotating weather systems that are variously called cyclones, hurricanes, and typhoons. All of these phenomena are generated when the motion of air is deflected by the **Coriolis effect**.

To understand the Coriolis effect, we must imagine ourselves riding along on a north- or south-moving wind, perhaps as passengers suspended from a hot air balloon. Suppose we start our trip at 30° N latitude, and that the wind is initially blowing northward. The air, of course, shares the rotation of the surface, which at 30° N corresponds to an eastward speed of 1200 km/hr. As we move north, we continue going east at this speed, but soon notice that the landscape below us is moving east more slowly, since it has less distance to go to spin once around in 24 hours. Thus our extra eastward momentum causes us to turn toward the right. In a similar way, a balloonist moving south from the same starting point is deflected to her right because her initial eastward speed is lower than that of the ground nearer the equator. If both are deflected toward the right, the result is a counterclockwise circular motion (Fig. 9.17).

The same line of reasoning will convince you that in the Southern Hemisphere the direction of turning is reversed, and that air diverging from a point is deflected toward the left. A similar mental exercise shows that air blowing inward toward a center is deflected in the opposite way from that blowing outward. You may even have experienced the Coriolis effect directly, if you have ever tried to walk quickly from the outside edge to the center of a carousel.

As a result of the Coriolis effect, air converging toward a center sets up a left-handed spin in the Northern Hemisphere and a right-handed spin in the Southern. This sense of motion, corresponding to a low-pressure region, is called **cyclonic**. A cyclone is a circular storm system moving around a low-pressure region. Diverging air spins the opposite way and is called **anticyclonic**. Because high-pressure regions do not

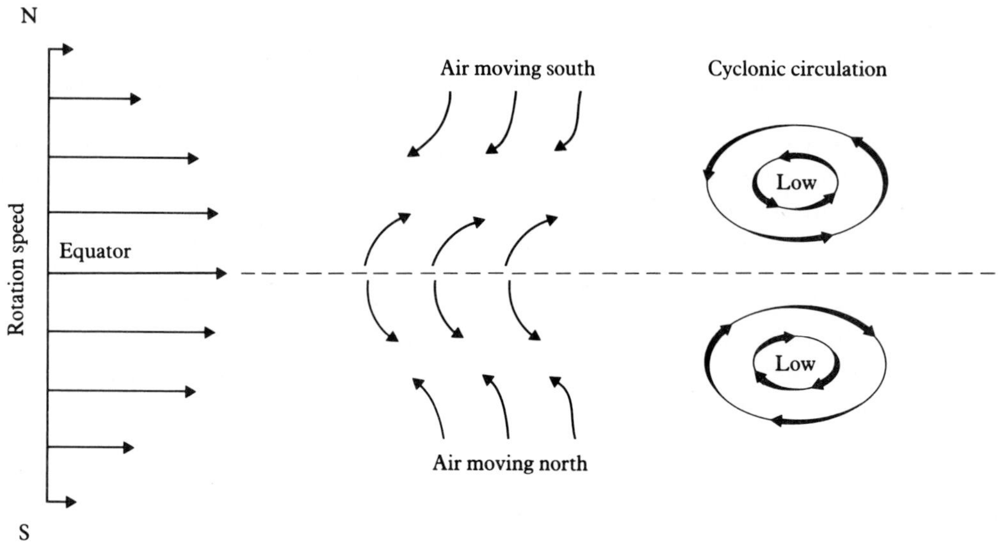

FIGURE 9.17 The Coriolis effect on air moving north or south on a rotating planet. You can experience this same force by trying to walk toward or away from the center of a spinning carousel.

generate clouds and rain, however, we rarely refer to an "anticyclone" when discussing weather.

A low-pressure cyclone has the potential to become a major storm. When the inward-moving air is moist, it may cool as it nears the center of the system and is carried to higher altitudes. The resulting condensation of water vapor not only produces rain, it also releases energy to increase the speed of the wind. (Since it requires addition of energy to water to produce steam, it follows that the opposite phase change, when water vapor condenses back to liquid, must release energy.) A situation can then develop in which the growing storm feeds upon itself, drawing in more and more moist air and releasing more and more energy. The result is a tropical storm, more commonly known as a hurricane or typhoon. Only on Earth, where water is abundant, can self-sustaining storms of this type be formed. When such storms move over land, they immediately begin losing their strength since their source of water (and hence energy) is cut off.

Changes in Climate: The Ice Ages

Weather on Earth is an extremely complex phenomenon, difficult to understand and almost impossible to predict with any high degree of accuracy. Climate presents fewer problems, in that changes are small. However, change in the climate of the Earth has much more far-reaching consequences than any storm.

We know that the climate on Earth has changed, even during the relatively brief period that human records have been kept. At the time civilization was developing in the fertile crescent of the Middle East, rainfall in that region was higher than it is now, and vast forests supported lions and other animals now confined to sub-Saharan Africa. About a thousand years ago, when Norse seafarers founded colonies in Greenland and present-day Canada, the climate in these regions was warmer than it is today, and when a global cooling took place around 1400 A.D. these colonies did not survive. Even more recently, there is evidence of recurring droughts in the plains of North America at approximately 22-year intervals.

The most dramatic climatic changes for our planet are those associated with the great ice ages of the past million years. At intervals of about 100,000 years the average temperature of the Earth has dropped by about 3 C, sufficient to produce vast ice sheets up to 3 km thick over much of the Northern Hemisphere land masses. During an ice age, the sea level drops and atmospheric circulation patterns alter significantly. The last such glacial period ended only about 10,000 years ago, and the Earth today is in an unusually warm period. It is so warm, in fact, that the sea ice that still covers the Arctic Ocean and the thick ice cap on Greenland are in some danger of melting. If this should happen, the sea level would rise and submerge parts of our coastal cities.

The primary cause of the great ice ages is now believed to be changes in the orbit of the Earth and the tilt of its rotation axis. Astronomers have calculated the changes expected from the gravitational influence of other planets over the past million years, and the pattern of ice ages follows the orbital changes closely. Even though the resulting variations in solar heating are small, they seem to be sufficient to shift the Earth from an interglacial equilibrium, such as we have at present, to an ice age condition, with about equal intervals of time spent in each climatic state.

Global Warming

One of the consequences of the Industrial Revolution of the nineteenth and twentieth centuries has been the release of immense quantities of CO_2 into the atmosphere from the burning of fossil carbon (coal, oil, and natural gas). Increasing population has also led to massive deforestation of the land, with associated burning of biological carbon. As a result, the CO_2 content of the atmosphere and oceans has risen and continues to do so. It is inevitable that this change in atmospheric composition will have some effect on global climate.

Figure 9.18 illustrates the CO_2 content of the atmosphere as measured from a special atmospheric observatory high on the slopes of Mauna Loa in the middle of the Pacific, far from any local sources of pollution. Calculations show

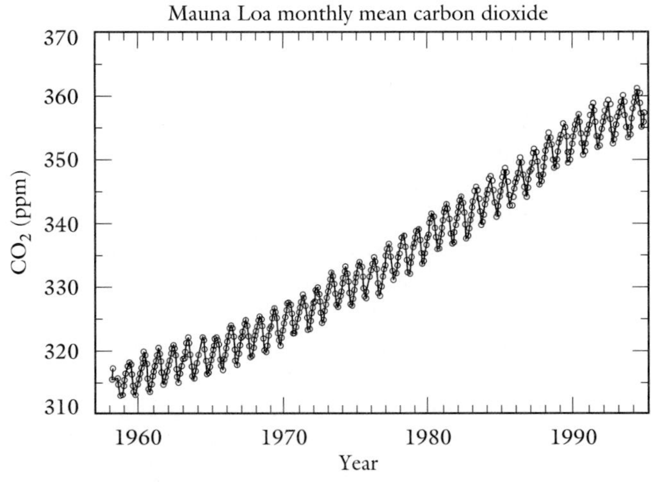

FIGURE 9.18 Variations over time in the carbon-dioxide content of the Earth's atmosphere, as measured from an atmospheric observatory on the upper slopes of Mauna Loa. The content has been rising by about 1% per year, mostly as a consequence of consumption of fossil fuels in the advanced nations and destruction of tropical forests in the developing nations.

that such increases in CO_2 will lead to an enhanced greenhouse effect and an average increase of the Earth's temperature, relative to pre-industrial values, of about 2 C. These calculations appear to be confirmed by direct measurements of temperatures, which reached all-time highs in most land areas of the Earth during the 1980s and 1990s. However, such effects are difficult to distinguish from shorter-term variations in temperature, such as the global cooling that followed the volcanic eruptions of El Chicón (Mexico) in 1982 and Pinatubo (Philippines) in 1991.

In order to assess the consequences of continued oxidation of carbon, especially of the fossil fuels that have lain in the ground for more than a hundred million years, we need to predict the changes in the greenhouse effect that will follow the release of CO_2 into the atmosphere. Most models suggest that global temperatures can be expected to rise another several degrees by the middle of the twenty-first century, but there are substantial uncertainties in these numbers. One of the most important contributions to society that can be made by planetary scientists is to improve these greenhouse models. The future of our industrial society depends in part on our ability to predict the impact of our activities on global climate.

Note that in the debates concerning global warming some commentators seem to question the reality of the greenhouse effect. This is foolish. The greenhouse effect is very real, and without it the Earth would be frozen. The questions concern not the reality of the greenhouse effect, but rather the degree of greenhouse warming to be expected from production of additional CO_2. That there will be some global warming is inevitable, but the question of just how much is crucial.

There is a great deal that scientists do not yet understand about both climate and weather. Billions of dollars are spent each year in collecting data and attempting to make accurate predictions. The economic impact of even a single major storm can reach hundreds of millions of dollars, and the consequences of a shift in climate are almost incalculable. One way of helping to understand our own planet, of course, is to compare it with other planets. We will frequently emphasize such comparisons as we look at the atmospheres of the other planets.

9.8 IMPACTS AND EVOLUTION

In this chapter, we have discussed many features of terrestrial geology, but conspicuously absent have been the craters that dominate the surfaces of the Moon and Mercury. Why does the Earth have so few impact craters, relative to the Moon? Both objects occupy the same region of space, and Earth, with its greater surface area and stronger gravity, should receive many more impacts than does the Moon. The presence of an atmosphere does not explain the absence of craters, for while the air effectively filters out the smaller debris, any projectile capable of digging a large crater will punch through the atmosphere as easily as a fist through a cobweb.

Where Have All the Craters Gone?

How did the Earth avoid being heavily cratered? The fact is, it did not avoid impact cratering. The difference between Earth and the Moon lies not in the rate at which craters are formed, but in the rate at which they are destroyed. We are deluded into thinking we are immune to large impacts because the evidence of past catastrophes has largely been removed. Remember, in this connection, that most of the craters on the Moon were formed more than 3.8 billion years ago.

The primary agents that obliterate impact scars are tectonic and erosional. New crust is formed along rift zones, and old crust is destroyed in subduction zones. Elsewhere, collisions between plates squeeze the crust into mountain ranges, while volcanism periodically floods sections of the continental blocks. In addition, the processes of erosion and sedimentation fill in and wear away the craters generated by impacts. Only recently, largely with the aid of surveys from Earth orbit, have geologists learned to recognize the faded scars of ancient impacts on the continental crust of our planet.

Meteorites penetrate the atmosphere daily, but only about once a decade does one strike that is large enough to form a small crater. In terms of significant effects on the surface of our planet, however, we must consider the much rarer impacts of objects that weigh from hundreds of thousands up to billions of tons. Calculations show that the atmosphere effectively shields us from most impacting bodies with mass less than 100,000 tons, corresponding to an energy of about 10 million tons (10 megatons) of TNT. Above this threshold, incoming projectiles will either explode in the lower atmosphere or reach the surface to form a crater.

FIGURE 9.19 A small part of the forest destroyed by the 1908 explosion of a small (60-m diameter) asteriod in the Earth's atmosphere near Tunguska, Siberia. The energy of the airburst was about 15 megatons of TNT.

Recent Impacts

A remarkable impact event took place above the wilderness of Russian Siberia on June 30, 1908. Witnessed by only a few of the native population, an explosion of unprecedented magnitude occurred, registering its shock wave on atmospheric instruments around the world. A brilliant fireball in the sky briefly rivaled the Sun in brightness, followed by explosion and fire. Over a region of a thousand square kilometers the forest was flattened, the trees stripped of leaves and branches and toppled in parallel rows radiating from the blast center (Fig. 9.19). Thousands of caribou died, and a man in a trading post 60 km away was thrown from his seat and knocked unconscious. Yet, in spite of all this violence, no crater was formed, and by now the Siberian forests are returning to erase the last evidence of this explosion.

From the damage caused by the Tunguska blast we conclude that its energy was equivalent to that of a large nuclear weapon, probably between 10 and 20 megatons. The diameter of the stony projectile was thus about 60 m, the size of a large city office building. It is estimated that such an impact should occur every few centuries somewhere on Earth. Fortunately, most of the Earth's surface is covered by ocean and much of the land area is still relatively unpopulated, so on statistical grounds we can hope that the next such impact does not take place near a populated area.

For an example of a much rarer impact from an iron meteoroid, we turn to Meteor Crater in northern Arizona, near the volcanic region of the San Francisco Peaks. The time is 50,000 years ago, before there were humans living in the region to witness the impact of 100,000 tons of iron on the Earth. There was no chance that this solid mass of metal would break up in the atmosphere. Rather, the force of its impact carried it several hundred meters below the surface before it finally came to a halt, transferring its energy into an explosion that destroyed the impacting body and blasted a hundred million tons of pul-

a b

FIGURE 9.20 (a) Meteor Crater, near Winslow, Arizona, is 50,000 years old, only yesterday on the geological timescale. A similar impact could occur anywhere on Earth at any time. Meteor Crater is about 1.3 km across and 200 m deep. A similar crater would be a tiny pit on the lunar landscape as viewed from Earth, but quite a nuisance if it were formed today in downtown Chicago. (b) The Manicougan Lakes Crater in northern Quebec. This circular feature, 70 km in diameter, was caused by an impact in the late Triassic period, about 200 million years ago.

verized rock into the atmosphere, leaving a crater a kilometer in diameter. Around the edges of the crater the thick layers of sediment were twisted up and folded back (Fig. 9.20). Fragments of metal from the impacting object were spread over hundreds of square kilometers around the crater.

Today, Meteor Crater is the best known and most obvious impact crater on Earth. Tens of thousands of tourists take the short detour from Interstate 20 each year to visit it. Although erosion has partly filled the interior with sediment and has washed away much of the ejecta, the characteristic uplifted and folded-over strata around the edges clearly testify to the explosive origin of this feature, and tiny metal fragments can still be picked up in the surrounding desert. Yet even here, the recognition that this is the scar of an extraterrestrial object has come only during the past 50 years; previously, most geologists thought it was an unusual volcanic crater associated with the nearby San Francisco Peak volcanic field.

No equally large impact crater has been located on Earth that is younger than Meteor Crater, but recent searches have found three craters with diameters of 7–18 km—two in Russia and one in Ghana—which formed within the last 3 million years. Small craters much older than this become very difficult to identify, especially if they are less than a few kilometers in diameter.

Death of the Dinosaurs

An important discovery of the 1980s is the critical role played by impacts in the evolution of terrestrial life. One of the most important catastrophes of the last few hundred million years was caused by the impact in the Yucatán state of

Mexico of a meteoroid 10–15 km in diameter having a mass of more than a trillion (10^{12}) tons. This impact released energy of about 10^8 megatons (100 gigatons) of TNT and blasted out a crater (Chicxulub) almost 200 km in diameter, the largest so far positively identified on Earth.

This impact took place 65 million years ago precisely at the boundary between the Mesozoic and Cenozoic eras. This is better known as the boundary between the Cretaceous and Tertiary periods, and the event is known as the **K/T event**. (If this seems a strange choice of letters, note that "Cretaceous" is spelled with a *K* in German.) The K/T boundary has long been recognized by geologists as one of the major breaks in the history of life on Earth. It corresponds to what is called a **mass extinction**, in which a great many species suddenly became extinct, to be succeeded by new species in subsequent strata. In the K/T mass extinction, more than half of the major marine species were destroyed. At about the same time the dinosaurs also became extinct, presumably as a consequence of the same global environmental catastrophe.

The key to understanding the nature of the K/T event was the discovery in 1980 of a thin sedimentary layer of remarkable composition exactly at the boundary where the fossil record indicated the mass extinctions (Fig. 9.21). In this boundary layer, which has subsequently been recognized all over the globe, the quantities of a number of rare elements, notably the metal iridium, are dramatically enhanced.

Iridium is very rare on the surface of the Earth because it dissolves readily in iron, and most of our planet's allotment of iridium is thought to be locked up in the core. The relative concentrations of isotopes of the element osmium are also peculiar, being typical of meteoritic materials rather than terrestrial. Both of

FIGURE 9.21 A layer of clay just a few inches wide marks the boundary between the Cretaceous and Tertiary eras on Earth. Here we see it exposed on a cliff wall near Trinidad, Colorado. The clay is strongly enriched in iridium and other metals, suggesting an extraterrestrial origin for part of this material.

these anomalies can be understood by the sudden addition of about a trillion tons of typical stony meteorite material to the Earth's atmosphere, which would include 200,000 tons of iridium. The impact of a 10-km stony asteroid (or a 15-km comet) would do just that.

The energy of the explosion as the asteroid struck the Earth is almost beyond imagining: the equivalent of 5 billion bombs of the size that obliterated Hiroshima and Nagasaki. The direct blast and associated tsunami (tidal wave) would have killed almost any living thing within a thousand kilometers, and the spray of molten rock fragments that fell back into the atmosphere ignited forest and range fires over much of the planet. Even more devastating for the biota of Earth was the dust cloud raised by the explosion, the same dust that is responsible for the iridium-enriched layers that mark the K/T boundary. With its mass of a hundred trillion tons, this cloud blotted out the sunlight from the surface of the Earth for a period variously calculated at several weeks to several months. Either way, the temperatures would have dropped drastically, plants would have died from lack of sunlight, and the climate of the planet would have been seriously disrupted, perhaps triggering the onslaught of an ice age. Probably 99% of all living things were killed over the whole Earth by a combination of conflagration followed by darkness and cold.

Horrible though this event may sound, it had profound effects for us. The new species that rapidly evolved to fill the ecological vacuum of this mass extinction gave rise to most of the life on Earth today. The tiny mammals that succeeded the dinosaurs were our ancestors. Had it not been for the K/T impact, who knows what course evolution would have taken on the Earth?

There are half a dozen mass extinctions recognized in the Phanerozoic, and several of them may be impact-associated. We can be sure that the K/T impact was not unique. The crater densities on the lunar maria indicate that several impacts of objects 10 km or more in diameter have taken place on Earth during the Phanerozoic. The large crater Tycho, which was caused by an asteroid about the same size as the one that triggered the K/T extinction, is the most recent example from the nearside of the Moon. Surely these other Earth impacts have also placed their mark upon the planet and its life-forms.

If impact-induced extinctions have played an important role in the evolution of life, then we must look at natural selection from a new perspective. It may be that the most important traits for long-term survival of a species have not been size or speed or even intelligence, but rather the ability to survive random global catastrophes of fire and ice. Natural selection works the way Darwin suggested, but the terms of survival have changed.

Contemporary Hazard of Impacts

If it happened to the dinosaurs, could we also succumb to an impact catastrophe? This seems entirely possible, although the near-term odds of such an event are extremely small. We know from the study of comets and asteroids, and from the graphic example of the collision of Comet Shoemaker-Levy 9 with Jupiter in 1994 (see Chapter 6), that such collisions still take place in the solar system. Probably the average interval between collisions of the magnitude of the K/T event is 100 million years, but that is only an average, and as far as we know such events are random—meaning that one could happen any time with a probability of about $^1/_{100,000,000}$ per year. This is a very low probability, but the consequences of such an impact are so great that we must consider it seriously.

From the perspective of contemporary hazards, we are more at risk from smaller but much more frequent cosmic impacts. Calculations suggest that the greatest danger is from objects 1–2 km in diameter, which are too small to cause a mass extinction but are large enough to precipitate an ecological catastrophe that might lead to worldwide crop failures and mass starvation. The chance of such an impact occurring within any given century is about one in a thousand. For you as an individual, the risk of death as a result of such an impact is probably at least as great as the risk of death from such more common natural disasters as earthquakes and tornadoes. That is why governments are beginning to think about ways of detecting threatening asteroids or comets and even of developing defense systems to deflect or destroy them with nuclear bombs before they can strike the Earth.

SUMMARY

The Earth is the planet we know best, and it therefore is an appropriate object with which to compare the other worlds studied in this book. We did not begin with the Earth only because of its complexity, which led us to start with the smaller, simpler planets Mercury and the Moon. Now, with our own planet, we must look at a much wider range of phenomena related to its active geology, its oceans and atmosphere, and the unique influences of life on the evolution of the planet.

The various layers of the Earth, above and below its surface, have given their names to similar divisions on other planets. Thus we must understand how the interior is divided into crust, lithosphere, mantle, and the liquid and solid cores. Similarly, we saw the division of the atmosphere into the troposphere, stratosphere, and ionosphere, with the magnetosphere extending still farther out to the edge of interplanetary space.

The complex geology of the Earth can be understood in terms of plate tectonics, the process by which heat released in the interior causes convection currents in the mantle, in turn exerting forces on the lithospheric plates floating on the top. These plate motions result in the formation of new oceanic crust at rifts, the destruction of the crust at subduction zones, and the generation of earthquakes and volcanoes along faults where one plate scrapes against another. In this process the ocean crust is recycled on a timescale of about 100 million years, while the floating granitic continents are moved about and occasionally compressed and raised up to form great folded mountain ranges.

Above the crust are the oceans and the atmosphere. The short-term variations called weather and the longer-term conditions that constitute climate both result from varying solar heating coupled with the large-scale motions of ocean and atmosphere. We looked at atmospheric circulation and saw the way Coriolis forces give rise to cyclones and anticyclones in the temperate regions of the Earth. Also considered were the processes that can affect the climate, such as changes in the inclination of the Earth's axis of rotation and the shape of the orbit, or the injection of dust from volcanoes or asteroidal impacts.

The role of impacts on the Earth is different from that on the Moon or Mercury because of our oceans and atmosphere. A large impact not only produces a crater; it can also profoundly modify the climate, with disastrous effects on life, leading in some cases to mass extinctions. The best-documented mass extinction arising from an impact is the K/T event, 65 million years ago.

Atmospheric chemistry is another complex topic. Our present atmosphere is the result of outgassing from the interior, primarily through volcanic eruptions, combined with changes due to interaction of the gas with the crust, the oceans, and especially with life. It is life that has removed almost all of the carbon dioxide (now mostly in the form of carbonate deposits) and has generated free oxygen through photosynthesis.

The Earth is a remarkable planet, like the others in many ways, but also uniquely influenced by its liquid water oceans, its oxygen-rich and carbon dioxide–poor atmosphere, and its abundant and varied life-forms. As we will see in the next chapter, Venus is the other planet most like the Earth, yet it has evolved to a strikingly different condition.

KEY TERMS

anticyclonic
aurora
bar
caldera
CFC
composite volcano
continental drift
Coriolis effect
cyclonic
fault
geologic timescale
granite
greenhouse effect
ionosphere
K/T event
lithosphere
magnetosphere
mass extinction
ozone
Phanerozoic period
plate tectonics
Precambrian period
rift
sedimentation
seismic waves
shield volcano
stratosphere
subduction
tectonic
troposphere

Review Questions

1. In simple terms, how does the Earth rate as a planet? Compare its size, density, composition, and other basic properties with those of its neighbors. How would you expect study of other planets to help us understand our own?

2. Distinguish between evolutionary processes and cyclic processes. What are examples of each that we have encountered so far in this book? Why are cyclic processes more common on the Earth than on the Moon?

3. Make sure you understand how plate tectonics works. Can you describe how this process gives rise to mountains, volcanoes, deep-sea trenches, and earthquakes? What does plate tectonics tell us about the fate of deep-sea sediments, including the carbonates that trap most of the carbon dioxide that would otherwise be in our atmosphere?

4. The changes in geological thinking that resulted from the acceptance of plate tectonics and continental drift constituted one of the major scientific revolutions of the twentieth century. Why do you think these ideas were so slow in being widely adopted? You may wish to compare this scientific revolution with others of the past century: Darwin's discovery of the role of natural selection in biological evolution; Pasteur's proof of the role of germs in causing disease; the insights into human psychology provided by Freud; and the revolution in physics represented by relativity, quantum mechanics, and the discovery of the nature of the atom.

5. Compare the ocean crust and the continental crust, in terms of composition, origin, and evolution. Are these two divisions of the terrestrial crust at all analogous to the division of the lunar crust into highlands and maria? Explain.

6. Discuss the chemical balance that exists between the land, the ocean, and the atmosphere. Imagine what would happen to the other two if any one of these were dramatically changed.

7. Consider the role of impacts in influencing the history of the Earth. Why don't we see more impact craters on the Earth today? What would conditions have been like during the period of terminal heavy bombardment? Could life have existed then? Explain how a mass extinction works. Consider the effect if a 10-km asteroid struck the Earth today, either on land or in the ocean. Would humans survive, or would we suffer the fate of the dinosaurs?

Quantitative Exercises

1. Typical motions of plates in the Earth's crust amount to about 5 m per century. At this rate, how long will it take Los Angeles to move up next to San Francisco? How much older is the Hawaiian island of Kauai than the currently active island of Hawaii, about 300 miles away?

2. The impacts that formed Meteor Crater in Arizona and the 1908 Tunguska event in Siberia are both estimated to have released about 15 megatons of energy. The Meteor Crater projectile was an iron object with a density of 7 g/cm^3, while the Tunguska projectile was a rocky asteroid with an estimated density of 2.5 g/cm^3. If both impacts took place at 20 km/s, calculate the diameters of each projectile.

3. Suppose a mass extinction is caused by a large impact on the Earth once per 30 million years on average. If such an event takes place during your lifetime you will die from this cause. What is the probability that this will happen, and thus that an asteroid or comet will be the cause of your death?

4. Suppose that an impact of the sort that takes place once per million years has a 50% chance of killing you (due to global crop failures and starvation) if it happens during your lifetime; find the probability that this will be the cause of your death. Compare the results of these calculations with your probability of death from other causes, such as auto accidents, homicide, earthquakes, or hurricanes.

Additional Reading

*Gehrels, T., ed. 1994. *Hazards Due to Comets and Asteroids.* Tucson: University of Arizona Press. Technical papers on impacts, hazards, and planetary defense in a volume of the *Arizona Space Science* series.

Goldsmith, D. 1985. *Nemesis: The Death Star and Other Theories of Mass Extinction.* New York: Walker. Fascinating account of impacts and mass extinctions, especially the possibility that impacts are periodic and due to some astronomical influence (the "death star").

Lewis, J.S. 1996. *Rain of Fire and Ice: The Very Real Threat of Comet and Asteroid Bombardment.* New York: Addison-Wesley. Balanced, popular overview of the history of impacts and the threat they pose today.

Miller, R. 1983. *Continents in Collision.* Alexandria, Va.: Time-Life Books. Beautifully illustrated popular account of the fundamentals of plate tectonics.

Raup, D.M. 1991. *Extinction: Bad Genes or Bad Luck?* New York: Norton. Examination of the role of impacts in evolution, including the idea that most extinctions may have an astronomical origin.

Scientific American, special Earth Science issue (1983), vol. 249, no. 3. Excellent series of articles by leading experts on various topics in Earth science.

Steel, D. 1995. *Rogue Asteroids and Doomsday Comets.* New York: Wiley. Contemporary account of the impact hazard, with a slight tendency toward the overdramatic.

Sullivan, W. 1985. *Landprints.* New York: Times Books. Popular book by a leading science journalist on the role of plate tectonics in shaping the geology and landscape of the Earth.

*Indicates the more technical readings.

10

Venus: Earth's Exotic Twin

The Birth of Venus, by Sandro Botticelli (1445–1510). Aphrodite ("born of the foam") is being gently blown ashore at the island of Cythera, in Greece. Botticelli was born some 30 years before Copernicus, but their lives did overlap in time. One could imagine the two of them discussing the origin of Venus over a glass of wine, from rather different points of view!

Earth and Venus are more nearly twins than are any other pair of planets. They have essentially the same diameters (12,756 and 12,104 km, respectively), nearly the same densities (5.5 and 5.3 g/cm^3), and presumably very similar bulk compositions. As we might expect from their large size relative to Mars or Mercury, these two planets also have the most active geology in the inner solar system. Both planets also have major atmospheres.

The twins also differ in important ways. Earth has a relatively large Moon; Venus has none. Earth rotates directly in 24 hours; Venus rotates retrograde in 243 days. Earth experiences active plate tectonics; Venus does not. Earth has extensive oceans of liquid water; Venus is dry. And perhaps most dramatically, Earth has a climate that can support abundant life, while Venus has developed an oppressive atmosphere, sulfurous clouds, and blistering surface temperatures. Twins these two planets may be, but certainly not identical twins. Understanding the reasons for their divergent atmospheric and surface evolution is one of the outstanding problems of planetary science, with important implications for the future of the Earth.

10.1 UNVEILING THE GODDESS OF BEAUTY

Venus is a fascinating planet in its own right, in addition to the insights it may provide concerning the Earth. With its surface hidden under a perpetual blanket of cloud, it remained largely mysterious while other planets were yielding their secrets to telescopic observers and the reconnaissance of spacecraft. To understand the geology of Venus and its history, it is necessary to penetrate the clouds with radar. Only recently has Venus been mapped with radar, achieving the resolution and coverage necessary to reveal its geology.

Basic Properties

Venus is the nearest planet to the Earth. When we see it glowing brightly in the twilight skies as the morning or evening star, we can understand

FIGURE 10.1 A picture of Venus taken in visible light with the Palomar 5-m (200-in) telescope when the planet was relatively close to us. It shows no detail, indicating the presence of an atmosphere filled with featureless clouds.

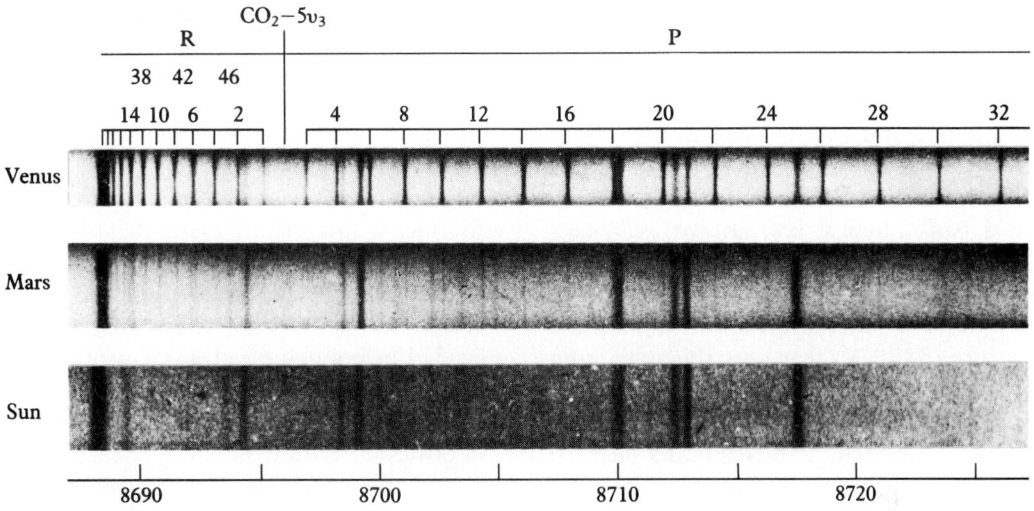

FIGURE 10.2 A comparison of portions of the spectrum of Venus (recorded photographically at the McDonald Observatory in Texas), of Mars, and of the Sun. Carbon dioxide absorptions appear strongly in the Venus spectrum, weakly in that of Mars, and not at all in the spectrum of the Sun. We conclude that even Mars has more CO_2 in its atmosphere than our planet does, while Venus must have an enormous amount of this gas.

why this planet was named for the goddess of love and beauty. Venus owes its brilliance both to its proximity to the Earth and Sun and to its layer of clouds, which serves as an excellent reflector of sunlight. The Moon, with a cloudless surface of gray rock, reflects only 12% of the sunlight that strikes it, whereas Venus has a reflectivity of about 75%.

Both Venus and Earth have substantial atmospheres and brilliant clouds. However, Venus is completely covered by its clouds, unlike the ragged canopy of Earth. The clouds prevent astronomers from seeing the planet's surface, and they frustrated early attempts to measure even so basic a property as the rotation rate (Fig. 10.1). The temperature of those clouds, first measured in the 1930s, is −35 C, similar to the temperature that would be measured in our own stratosphere, and astronomers of that time generally assumed that the clouds of Venus were composed of water, like those of Earth.

Less than 30 years ago, the only gas that had been detected spectroscopically in the atmosphere of Venus was carbon dioxide (Fig. 10.2). Spectroscopic observations, however, did not preclude the existence of other gases, since observational difficulties could not rule out undetected amounts of nitrogen, oxygen, and water vapor. Therefore it seemed possible, near midcentury, that the atmosphere of Venus was primarily nitrogen, like our own, and that carbon dioxide might be a relatively minor component. All of these observations and inferences suggested to those who wished to believe it that Venus might have oceans of water beneath its brilliant clouds and a climate conducive to the existence of life.

Surface Temperature

The optimistic picture of Earthlike surface conditions changed dramatically in the late 1950s when radio telescopes were first used to measure the thermal radiation from Venus. Unlike visible light, radio waves easily penetrate clouds. (You know this from your experience in listening to radio broadcasts or watching television on cloudy and rainy days.) Furthermore, any object having a temperature above absolute zero (0 K or −273 C) radiates some energy at all wavelengths, including radio wavelengths.

It is a basic physical law, known as Wien's Law, that the hotter an object is, the more the maximum energy output shifts toward shorter wavelengths. The outer layers of the Sun are at a temperature (5800 K) that puts this maximum in the range of visible light. A planet with a temperature like Earth's is radiating most of its energy in the infrared, at wavelengths near 10 micrometers (0.001 cm). But some of the thermal energy from both Sun and Earth is radiated at much longer wavelengths in the centimeter to meter range, the region of the spectrum where radio telescopes operate.

Since Venus is closer to the Sun than we are, astronomers expected its temperature to be higher than the Earth's. On the assumption that both planets rely on solar radiation alone to warm them, it is possible to calculate what temperature Venus should have. Its greater proximity to the Sun is countered to some extent by its high albedo; its bright clouds reflect most of the incident sunlight back into space. Such calculations showed that the equilibrium temperature of Venus should be roughly 15 C warmer than that of our planet, or about 280 K.

In 1958, however, radio astronomers found that the amount of thermal energy emitted by Venus at radio wavelengths implied a surface temperature greater than 600 K. Venus was radiating more than twice as much energy at radio wavelengths as had been expected. This result was so surprising that at first it was not accepted, and scientists sought alternative explanations for the high intensity of the radio radiation. One proposal was that emission from a dense ionosphere on Venus gave the appearance of a high surface temperature.

Early Radar Observations

One of the primary scientific objectives of the first interplanetary spacecraft was to test whether the radio emission from Venus was thermal radiation, arising from the surface, or whether it might instead be ionospheric in origin. In 1962 Mariner 2 successfully made the 14-week flight to Venus and radioed back the critical result: The radiation really did come from the surface. At about the same time, radar astronomers succeeded in bouncing radar waves off the surface of Venus and found that not only was the ionosphere transparent but the planet was rotating backward incredibly slowly. By defining the location of the planet's surface relative to the top of the cloud layer, the radar data also indicated that the atmosphere was massive, with a surface pressure at least 50 times higher than the 1-bar pressure on the Earth.

Subsequent radar observations have determined the exact rotation period of 243.08 days in a retrograde direction. In other words, Venus rotates on its axis in the opposite direction from the course of its motion around the Sun. On Venus, if you could see it through the clouds, the Sun would appear to rise in the west and set in the east.

The length of a day on Venus requires further definition. While not quite so peculiar as Mercury, Venus also has a day that is longer than its year. The rotation period, measured with respect to the stars, is 243.08 days, about 19 days longer than the period of revolution around the Sun, 224.7 days. This "day" is the length of time required for Venus to make one rotation about its axis. But as we saw for Mercury, the other way to determine the length of a day is to measure the time between two successive "noons" (or sunsets, sunrises, and so forth). The time between successive noons—the solar day—is much shorter, amounting to 116.67 Earth days, with the Sun moving across the sky in the wrong direction, from west to east (Fig. 10.3).

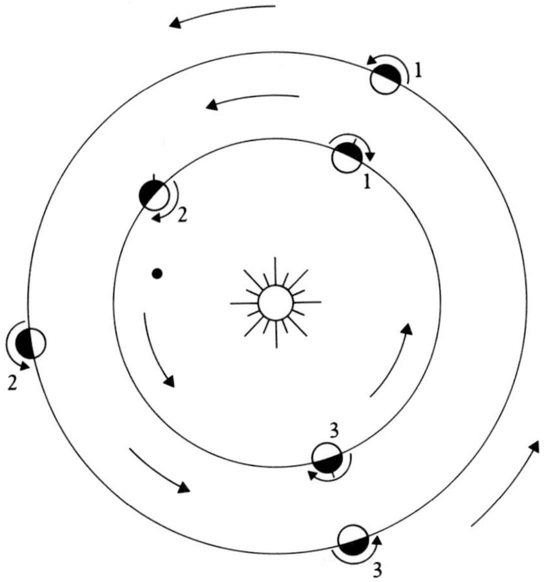

FIGURE 10.3 Venus rotating as it revolves around the Sun. At position (1) Venus is at inferior conjunction, with a feature on its surface pointing toward the Earth; 486 days later the two planets are at the positions labeled (2). Venus has undergone two complete retrograde rotations and slightly over two revolutions around the Sun. Earth has undergone 486 rotations and 1⅓ revolutions around the Sun. Ninety-eight days later the two planets have moved to position (3). Venus is again at inferior conjunction 584 days after position (1), and the surface feature has almost rotated into position to face the Earth.

The Venera Missions

The next major step in the exploration of Venus was taken by the USSR in 1967 with the deployment of a probe called Venera 4 into the atmosphere (*Venera,* the Russian word for Venus, gave its name to a long series of Russian missions to this planet). This probe successfully entered the atmosphere of Venus and transmitted measurements to Earth as it descended by parachute. While the Venera 4 probe demonstrated that the atmosphere was thick and that CO_2 was the major gas, it failed to reach the surface. The last signals were received from the craft at an altitude of 23 km, with the atmospheric pressure already up to 20 bars and the temperature over 500 K. Many Soviet scientists thought that Venera 4 had reached the surface, but the real problem was spacecraft failure. The requirements for the early Venera probes had been set by a Russian scientist who did not believe the radio observations and had confidently assumed that the spacecraft would never have to withstand a pressure higher than 20 bars. The probe had been *designed* to fail at 20 bars of pressure.

Incidentally, if you are wondering why the first U.S. mission was numbered Mariner 2 and the first successful Soviet probe was Venera 4, it is because the first numbered craft in each series failed. In the early days of space exploration, more missions failed than succeeded.

In 1970 Venera 7, with a redesigned probe, successfully landed on the surface of Venus. As was by then expected, the pressure was 90 bars and the temperature a sizzling 740 K.

The discoveries of the late 1950s and 1960s changed forever our concept of our sister planet. No longer could the surface of Venus be imagined, even in science fiction, as a lush jungle populated by exotic creatures. Continuing optical and radio observations from Earth, in addition to direct measurements carried out by Soviet spacecraft, left no doubt that the entire surface of Venus was at temperatures hot enough to melt lead, tin, and zinc. We now know that Venus is the hottest planet in the solar system, despite the fact that Mercury is closer to the Sun. Indeed, the picture of Venus that has emerged is reminiscent of traditional concepts of hell.

10.2 THE ATMOSPHERE AND THE GREENHOUSE EFFECT

Why is Venus so hot? The explanation comes from the greenhouse effect, which we discussed in the last chapter. The challenge posed by Venus is that the greenhouse effect is so large that traditional theories were unable at first to explain it.

The Greenhouse Effect

When a planet's atmosphere contains gases that are opaque to infrared radiation, they inhibit the outward flow of heat and increase surface temperatures. Recall that the troposphere is defined as the part of an atmosphere where convective circulation takes place. Tropospheric convection establishes a temperature gradient, with higher temperatures near a planet's surface and cooler temperatures aloft. The equilibrium temperature for Venus, which is about 260 K, applies to the cloud layers high in the atmosphere, while below the clouds temperatures are much higher.

Although the clouds contribute to the greenhouse effect, it is primarily the infrared opacity of the gases that causes the lower atmosphere and surface to heat up. The infrared is the part of the spectrum where the bulk of the energy is radiated. The tiny amount of energy escaping directly from the hot surface of Venus in the form of radio emission is too small to affect the energy balance.

Not all gases are opaque to infrared radiation. For example, nitrogen and oxygen are virtually transparent at these wavelengths, so they cannot contribute significantly to an atmospheric greenhouse effect. Water vapor and carbon dioxide are very good infrared absorbers, and even the relatively small amounts of these two greenhouse gases in our own atmosphere are sufficient to raise Earth's average temperature some 25 C above the value our planet's surface would have if there were no atmosphere, or if only nitrogen and oxygen were present.

The greenhouse effect on Venus is obviously much more significant than the one on Earth. This is clear not only from the high value of the surface temperature, but also from the fact that the temperature is uniform to within a few degrees all over the surface. The entire planet acts as if it were encased in a heavy blanket.

We have already noted the presence of substantial amounts of carbon dioxide on Venus together with a relative lack of water vapor. It is clear that the CO_2 must be the major source of infrared opacity. The much greater greenhouse effect on Venus compared with that on Earth requires both that CO_2 be the major constituent of the atmosphere, and that the atmosphere itself be very massive.

Carl Sagan, the Cornell University astronomer who later became one of the best-known scientists of our time, began his research career working on theoretical calculations of the Venus greenhouse effect. At that time, about 1960, the large mass of the atmosphere was not suspected, and early calculations failed to explain the temperatures that were measured by the radio astronomers. During most of the decade of the 1960s, Sagan and his student and later colleague James Pollack produced theoretical models of increasing sophistication, using new measurements of the atmosphere and more and more powerful computers for their work. Only after a decade of this effort were they able to match the magnitude of the observed effect.

Mass of the Atmosphere

As Sagan, Pollack, and others struggled to interpret the evidence that our sister planet had an extremely high surface temperature, they were also learning how massive the atmosphere of Venus is. If you were standing on the surface of Venus at a pressure of 90 bars (in an air-conditioned, asbestos suit!) you would feel the kind of pressure a deep-sea diver deals with at a depth of 900 m. Venus has nearly the same surface gravity as the Earth, with one bar corresponding to the pressure exerted by a layer of water about 10 m deep. Heavy armor is required for a human to withstand such conditions. The high surface pressure adds to the infrared opacity of carbon dioxide, making it a much more efficient absorber of thermal radiation than the CO_2 in our atmosphere.

In order to maintain a greenhouse effect, at least some of the solar heat needs to be deposited at the surface. When scientists first realized the extent of the atmosphere and the thickness of the clouds, some questioned whether any light would reach all the way down to the ground. However, experiments on spacecraft that have reached the surface show that some sunlight does get through the cloud layer, providing an illumination about equivalent to a heavy overcast on Earth. This is sufficient to warm the ground and provide the necessary heating (Fig. 10.4).

Composition of the Atmosphere

The discovery that the atmosphere of Venus is 90 times more massive than that of Earth and consists of 97% CO_2 surprised many planetary scientists. Displaying a kind of Earth chauvinism, they had expected the atmosphere of Venus to be composed mainly of nitrogen. They also expected water vapor to be present, since water is so abundant on Earth. Even oxygen, a gas that might betray the presence of Earthlike life on Venus, was also anticipated.

Early searches for spectroscopic evidence of N_2, H_2O, and O_2 in the atmosphere of Venus were entirely negative. Nitrogen does not have any absorption bands in the part of the spectrum accessible to ground-based observers, and possible evidence of water and oxygen was blocked by the strong absorption by these gases in the atmosphere of the Earth. In the 1950s and 1960s efforts were made to put telescopes on

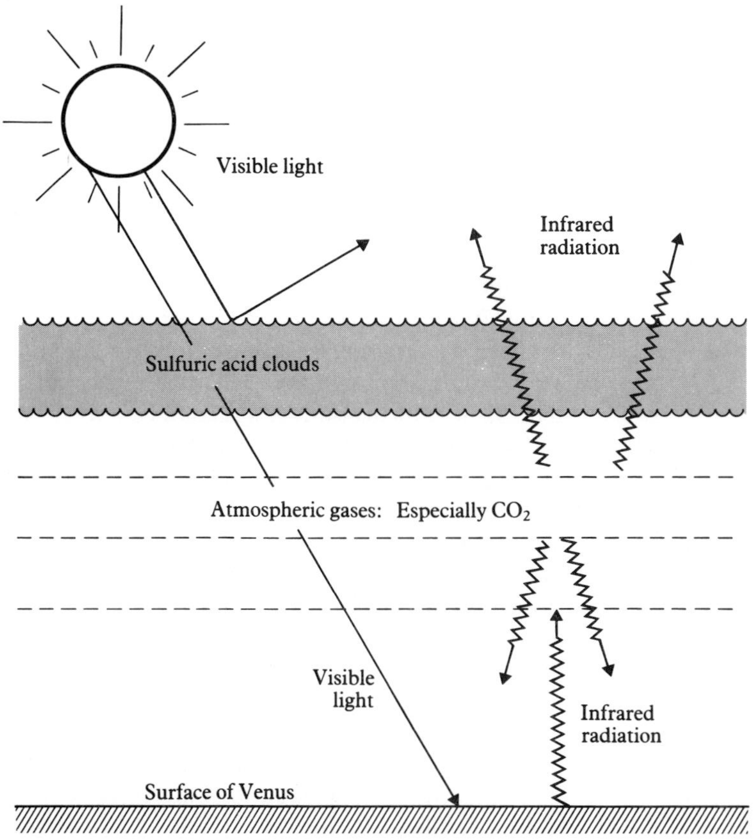

FIGURE 10.4 The atmospheric greenhouse effect on Venus. Although about 75% of the sunlight incident on Venus is reflected by the planet's brilliant clouds, some light still reaches the surface where it is absorbed. Infrared radiation from the warm surface tries to make its way out of the atmosphere but is strongly absorbed by carbon dioxide, causing additional surface warming.

high-altitude airplanes and balloons to get above most of the absorptions in our own atmosphere. The results were still negative. Not until the late 1960s were very weak H_2O absorption lines found, followed by carbon monoxide (CO), hydrochloric acid (HCl), and hydrofluoric acid (HF). A tiny amount of O_2 was discovered spectroscopically a few years later.

Evidently Venus is much, much drier than Earth. At its high temperature, any H_2O should be in the atmosphere, yet only traces were found, instead of the equivalent of oceans. The absence of H_2O on our sister planet is one of the major mysteries of the inner planets, and we will return to its solution later in Chapter 12.

Keep in mind that all of these observations were made from the Earth and refer only to the portion of the atmosphere of Venus that is above and just within the planet's ubiquitous clouds. These results were confirmed and extended by the space missions of the 1970s, when probes descended into the atmosphere and landed on the surface. A list of all the gases now known to be present in the atmosphere is given in Table 10.1.

TABLE 10.1 Composition of the atmosphere of Venus

Gas	Formula	Abundance
	Main Constituents	
Carbon dioxide	CO_2	96.5%
Nitrogen	N_2	3.5
	Trace Constituents	
Sulfur dioxide	SO_2	130[a] ppm[b]
Argon (40)	Ar-40	33
Argon (36)	Ar-36	30
Oxygen	O_2	30
Water vapor	H_2O	30[a]
Carbon monoxide	CO	20
Carbonyl sulfide	OCS	10[a]
Neon	Ne	9
Hydrochloric acid	HCl	0.6
Hydrofluoric acid	HF	0.005

[a]Abundances of these gases vary with altitude and latitude. They are not yet well defined.
[b]parts per million

Clouds of Venus

Ever since astronomers first realized that Venus was covered by a cloud layer, they have wondered about the composition of those clouds. At first the clouds were assumed to be made of H_2O (either liquid drops or ice), like those of Earth. Then the discovery of the very low concentrations of water vapor in the planet's atmosphere led some scientists to seek alternative explanations. Hydrocarbon droplets were in briefly in favor, but no proof surfaced to support this conjecture. The liquid or solid droplets of a cloud (like the surfaces of asteroids as discussed in Chapter 5) do not display the sharp, diagnostic spectral lines that would permit a definitive identification. Unfortunately for the astronomer, solids and liquids always present a more difficult problem for analysis than do the simpler gases.

The puzzle was finally solved by a series of observations in the 1970s. Improved data obtained from NASA's airborne telescope showed features in the infrared part of the spectrum corresponding to an unexpected material: concentrated sulfuric acid (H_2SO_4) (Fig. 10.5). The same conclusion had already been reached independently by observers studying the variations in brightness and polarization of sunlight reflected from the clouds in different directions. Results from both lines of evidence were announced at the same scientific meeting. At last, after decades of speculation and years of hard work, we knew the composition of the clouds of Venus, the brightest object in the sky after the Sun and Moon!

Sulfuric acid is produced by chemical reactions involving sulfur dioxide (SO_2) and H_2O. The most important cloud-forming process on Venus is **photochemistry**—chemical reactions

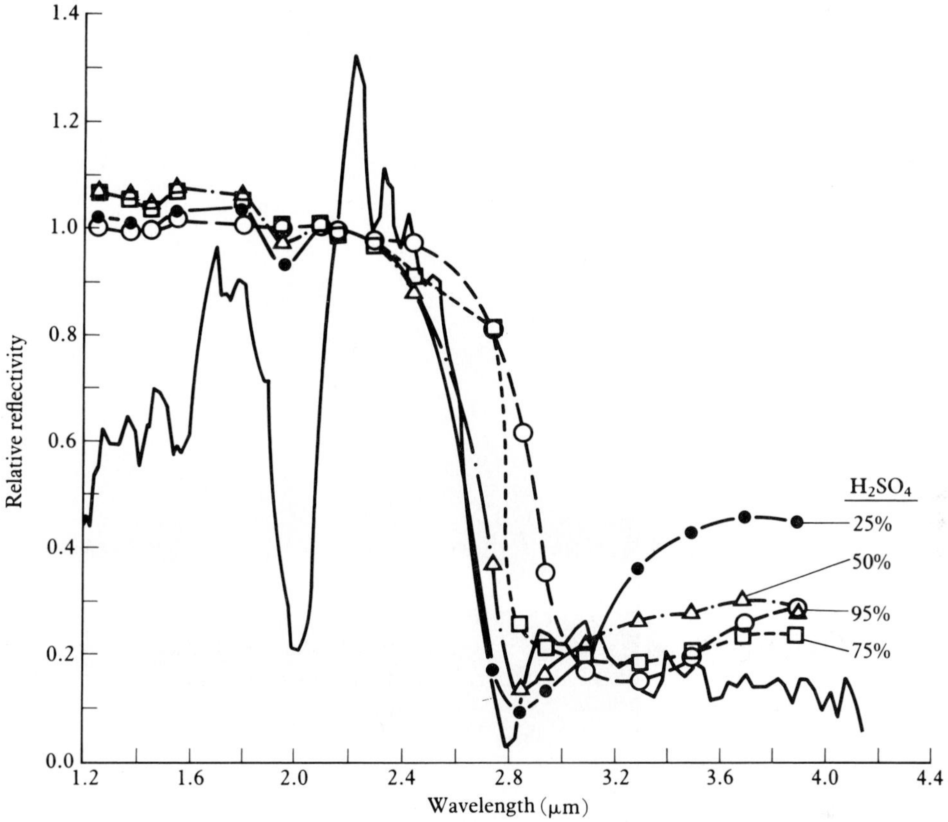

FIGURE 10.5 A comparison of the reflectivity of solutions of concentrated sulfuric acid and the clouds of Venus. The reflectivities are quite similar, indicating that the clouds of Venus could indeed be made of sulfuric acid.

driven by the energy of ultraviolet sunlight. Photochemical reactions are important in the upper atmospheres of planets, including the Earth, where the production of ozone from oxygen is an example of a photochemical process. On Venus, the H_2SO_4 and the unknown compound responsible for the ultraviolet-absorbing clouds (discussed later) are probably both produced and destroyed photochemically.

In addition to photochemical reactions, at least some of the basic cloud material may be supplied from below by active volcanism on the planet's surface. Observations over the past 25 years suggest that large fluctuations occur in the concentration of SO_2 in the atmosphere of Venus above the clouds. Sulfur dioxide is one of the products expected from volcanic eruptions; if there are occasional large eruptions of SO_2 on Venus, then we might understand the variations in the H_2SO_4 clouds.

Space probes that have passed through the clouds give us the picture shown in Fig. 10.6 of a series of discrete cloud layers. Clouds are seen extending from 30 to 60 km above the surface. But what are these various cloud layers made of? Are they all sulfuric acid, like the topmost layers? Only the Venera probes attempted compositional measurements of the clouds, and their

FIGURE 10.6 The vertical structure of the atmosphere of Venus. The clouds consist of several layers with different concentrations of particles. Below the clouds, the atmosphere is clear.

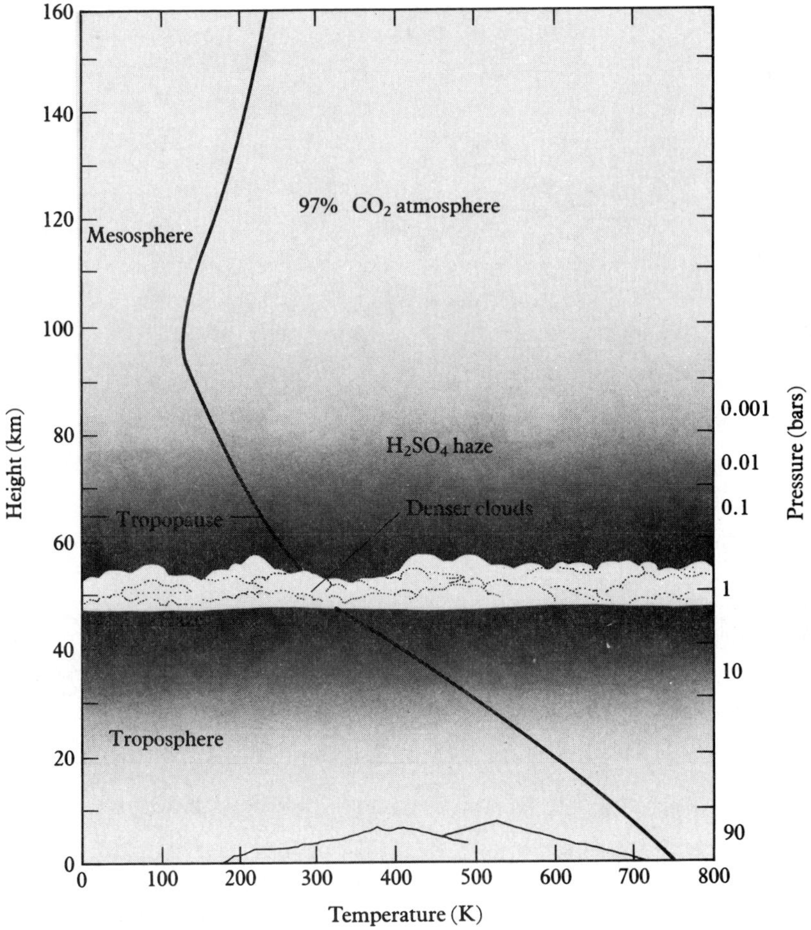

results have been contradictory. Sulfur or possibly chlorine compounds of some sort are indicated, but their exact identities are unknown.

Runaway Greenhouse Effect

In view of their many similarities, we might well ask ourselves why Venus has such a different atmosphere from that of the Earth. How did it acquire its massive quantities of CO_2, sulfuric acid clouds, and remarkable surface temperature? Or, alternatively, how did the Earth avoid this fate?

In the last chapter we discussed how the development and proliferation of life on Earth extracted CO_2 from our atmosphere and enriched it in O_2. Life, which in turn depends on the presence of liquid water, may be the key to the difference between the two planets. Perhaps Venus was just a little too close to the Sun, and therefore a little too hot to maintain liquid water and provide a suitable environment for life. We will discuss this possibility further in Chapter 12, after we have had a chance to look at Mars and see how it compares with both the Earth and Venus.

A second possibility is that Venus began with a more Earthlike climate and subsequently evolved the hellish conditions we see today. Suppose that Venus started with a surface temperature consistent with liquid water and a CO_2 atmosphere; is there a way it could have made the transition to its present state? Many scientists think it could, if its greenhouse effect got out of control. Such an atmospheric instability is called the **runaway greenhouse effect**.

Imagine what would happen if we could move the Earth into the orbit of Venus. Being closer to the Sun, our planet would absorb more solar energy and its temperature would rise. At the distance of Venus (0.72 AU), sunlight would be delivering about twice as much energy to every square meter of the Earth's surface. Higher ocean temperatures would lead to increased evaporation and more water vapor in the atmosphere. Water vapor is an effective greenhouse gas, so the result would be a stronger greenhouse effect and still higher surface temperatures, and hence more evaporation from the oceans. We have established a positive feedback loop, in which an initial disturbance—increasing the Earth's surface temperature—produces consequences that lead to an enhancement of that disturbance. The cycle would continue until the oceans literally boiled away and all the water was converted to vapor. This is the runaway greenhouse effect.

At this point the atmosphere is so hot that water vapor can easily rise to great heights where it is exposed to solar ultraviolet light. This is a crucial step. On Earth, water is protected from escape by a natural cold trap in the atmosphere. The air at the top of the troposphere is so cold that water cannot diffuse upward to levels where it could be attacked by ultraviolet light. A runaway greenhouse can raise the temperature throughout the lower atmosphere, giving water free access to high altitudes.

Just as in the case of evaporating water molecules from an icy comet nucleus, the H_2O in the upper atmosphere is broken apart into H and O atoms. Because of the large mass and high escape velocity of Venus, only the light hydrogen atoms escape into space. The oxygen remains behind to combine with rocks on the planet's surface. Thus the runaway greenhouse leads to the elimination of most water from a planet.

Note that the runaway greenhouse effect is not simply a large greenhouse effect. It is a *process*, whereby a planet's atmosphere can evolve from one composition and temperature to another different composition and much higher temperature.

Is this really what happened to Venus? Is there evidence that this planet ever had oceans of water that were subsequently lost? Might conditions on our sister planet once (perhaps briefly) have been suitable for the development of life? There are no firm answers to any of these fascinating questions. We will return to these issues in Chapter 12.

10.3 WEATHER ON VENUS

The massive atmosphere of Venus produces uniform surface temperatures and inhibits winds near the surface. These surface conditions, together with the very slow rotation period of Venus, result in a relatively simple tropospheric circulation pattern. In the stratosphere, however, wind speeds on Venus are remarkably high.

The Upper Atmosphere

Long ago, telescopic observation of the clouds of Venus showed that while they are featureless in visible light, dusky markings could be seen in pictures of the planet that were taken through filters that transmitted only ultraviolet light (Fig. 10.7). Studying such pictures in the 1960s, astronomers found that they could follow the motions of some of these dusky features long enough to see them move completely around the planet and back to their starting position. A complete circuit of Venus required only four days. The clouds moved in a retrograde direction, the same direction that was later found to hold for the planet's rotation. Thus the atmosphere and the surface move in the same direc-

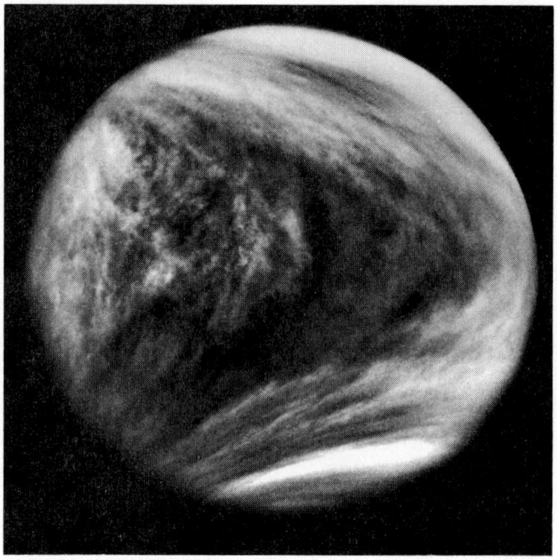

FIGURE 10.7 A picture of Venus obtained with the camera on the Pioneer Venus orbiter spacecraft, using an ultraviolet filter to bring out contrasts in the clouds (compare with Fig. 10.1). A large recurring Y-shaped cloud marking, sometimes dimly seen from Earth, is clearly visible here.

tion, although at very different speeds. This atmospheric motion corresponds to high-altitude winds blowing at a speed of 100 m/s (360 km/hr) from east to west.

The four-day stratospheric jet streams on Venus are the result of a complex interaction between the rotation of the planet and heating of the high clouds by absorption of ultraviolet sunlight. These winds are strongest at the planet's equator, tapering off toward either pole, unlike the pattern of tropospheric east and west winds found at temperate latitudes on Earth. More detailed imaging from spacecraft, also using ultraviolet filters, has permitted scientists to map these high-altitude jet streams on Venus. However, remember that we still do not know the chemistry of the thin, ultraviolet-absorbing clouds that make the dark patterns in these pictures.

Structure of the Atmosphere

We are fortunate in the case of Venus to have had two competing programs of exploration carried out independently by the United States and the Soviet Union. As a result, numerous probes carrying a variety of instruments were launched toward Venus in the years from 1962 to 1985, providing a great deal of information on the planet's atmosphere.

The atmospheric temperature profiles of Venus and Earth cross—Venus is colder than Earth at high altitudes and much warmer near the ground (Fig. 10.8). There is a region in the atmosphere of Venus where the pressure is near the sea-level pressure on Earth and the temperature a balmy 30 C. This would be a shirtsleeve environment for astronauts in the gondola of a balloon floating in the cloudy skies of Venus, were it not for the fact that those clouds are mainly sulfuric acid and the atmosphere unbreathable carbon dioxide.

In 1985, a joint Russian-French project deployed instrumented balloons at an altitude of

FIGURE 10.8 The atmosphere near the surface of Venus is much warmer than that of the Earth, but it is colder at higher altitudes.

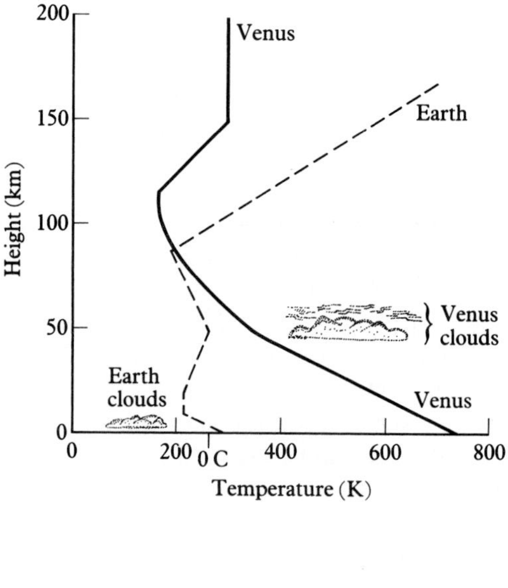

about 54 km. Instruments on the balloons measured pressure, temperature, and wind speed as part of a cooperative Soviet-French-U.S. experiment. The results indicated that winds at this altitude are more blustery than had been anticipated, with updrafts over the continent called Aphrodite. The balloons radioed back data for 48 hours, sufficient time for them to travel nearly halfway around the planet.

Near the surface of Venus, the wind speeds are low, with measured values ranging from zero to 2 m/s (approximately 0–6 km/hr). The pressure and density are so great at these levels that the atmosphere behaves in many ways more like an ocean than the air we are familiar with on Earth. Like the deep oceans of our planet, the surface of Venus has a nearly uniform temperature, from pole to pole and noon to midnight.

Atmospheric Circulation

Above the surface of Venus, there is a pattern of air rising near the equator and traveling north and south to descend near the poles (Fig. 10.9a). This simple type of atmospheric circulation is called a **Hadley cell**, after the British scientist who first proposed it as a model for the circulation of the Earth's atmosphere. While not describing our own planet very well, Hadley cell circulation turns out to be a very good model for the lower atmosphere of Venus.

FIGURE 10.9(a) In a simple Hadley cell circulation, warm air rises at the equator of a planet and travels toward the pole where it sinks and returns to the equator along the surface (see Fig. 9.16).

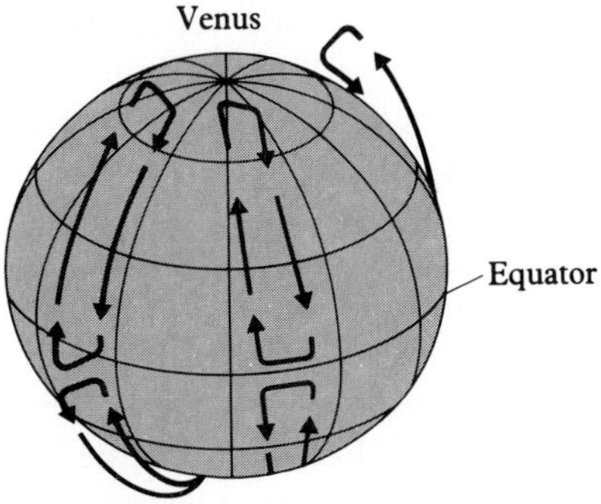

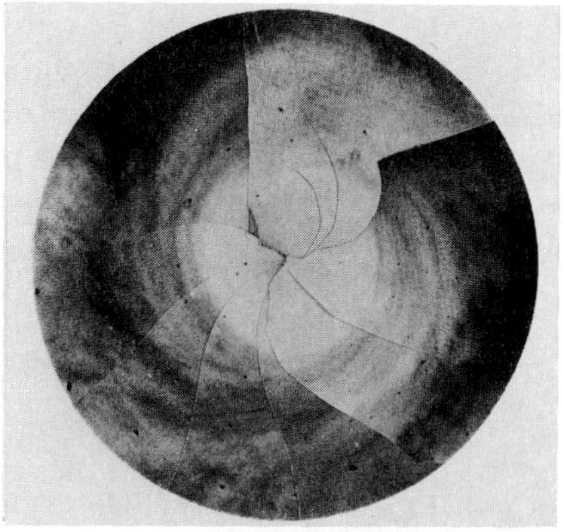

FIGURE 10.9(b) A mosaic composed of several pictures of the south pole of Venus, showing the spiral cloud structure associated with the polar vortex of descending air.

At higher elevations, near the main cloud layers about 50 km above the surface, the strong retrograde (westward) rotation described earlier becomes apparent. At this altitude, of course, the atmosphere is as thin as that of our own planet, so high wind speeds are possible. At the poles, these winds form a vortex of descending air, rather like the pattern of water swirling down a drain (Fig. 10.9b).

Calculations carried out with powerful computers suggest that the key to the difference between the circulation of Venus and of Earth lies primarily in the slower rotation of Venus. If the rotation of Earth were this slow, the high and low pressure systems that correspond to centers of fair and foul weather would fade away, the midlatitude stratospheric jets would disappear, the small Hadley cells now confined to the equatorial zone would expand all the way to the poles, and a globe-encircling wind would begin to blow at high altitudes.

The decrease in rotation does two things to produce these changes. First, it lengthens the duration of daylight, increasing the effects of heating in the daytime and cooling at night. Second, the Coriolis effect (discussed in Section 9.7) is less strong on a slowly rotating planet, and winds are less likely to be deflected into swirling, cyclonic motion. The combination of longer-term heating and a low Coriolis effect, aided by the larger total mass of the atmosphere, evidently produces a circulation pattern like the one on Venus.

The analysis given here is an example of the ways in which other planets can help us understand the Earth. In order to evaluate the true importance of various forces at work in natural systems, a scientist would like to be able to perform an experiment, to change the forces and then study the effect of these changes. Since we obviously can't slow down the Earth, the next best thing is to use a computer to model the effect. We are still left with some uncertainty, however, since a planetary atmosphere is very complex. Maybe we left out something important in our model. We can test the model by looking at another planet where basic conditions governing circulation are really different. Slowly turning Venus is one extreme; the giant planets with rotations more rapid than Earth's provide another. The best assurance we have that our models for the atmosphere of the Earth are accurate is their ability to deal correctly with these other worlds as well, where we find differences in rotation rates, distance from the Sun, and other basic parameters.

10.4 THE HIDDEN LANDSCAPE

Since we cannot photograph the surface of Venus through its clouds, we must use other techniques to map the topography. We have noted that the atmosphere is transparent to radio waves; thus radar can be used to beam microwaves through the clouds from the outside. These signals are reflected by the planet's solid surface and pass back through the clouds to our receivers. Radar mapping can be done with transmitters and receivers on the Earth, but is an especially powerful technique when used from a spacecraft in orbit about Venus. Being so much closer to its target, the radar on the spacecraft can afford to be much less powerful than its ground-based counterpart and still achieve higher resolution and more extensive surface coverage. In this and in the following two sections we examine the surface of Venus as revealed by radar studies.

Large-Scale Topography

The U.S. Pioneer Venus spacecraft was the first to map the surface of Venus, using a simple kind of radar that measured the altitude of the spacecraft. As the spacecraft orbited from north to south, the planet turned underneath from west to east, building up complete surface coverage over a period of about two years (1978–80).

The Pioneer Venus map of surface elevations has a horizontal resolution of 50 km, only slightly better than that of the Moon seen without a telescope. As with the lunar appearance at this resolution, this radar map cannot show features as small as individual craters or mountains. Unlike the view of the Moon as seen with the naked eye, however, the radar map measures topography, not color or reflectivity, yielding a global picture of large-scale features such as continents or mountain ranges.

Figure 10.10 shows the Pioneer Venus map and compares it with the topography of the Earth at the same resolution. In looking at the two maps, we are immediately struck by the absence of large continents or ocean basins on Venus. On Earth, most of the surface is either oceanic (typically several km below sea level) or continental (from sea level up to 11 km altitude). As we saw in the last chapter, these two terrain types reflect deep-seated differences in the composition and thickness of the crust, and they are formed by different processes. Venus, in contrast, has no deep basins and a smaller area of

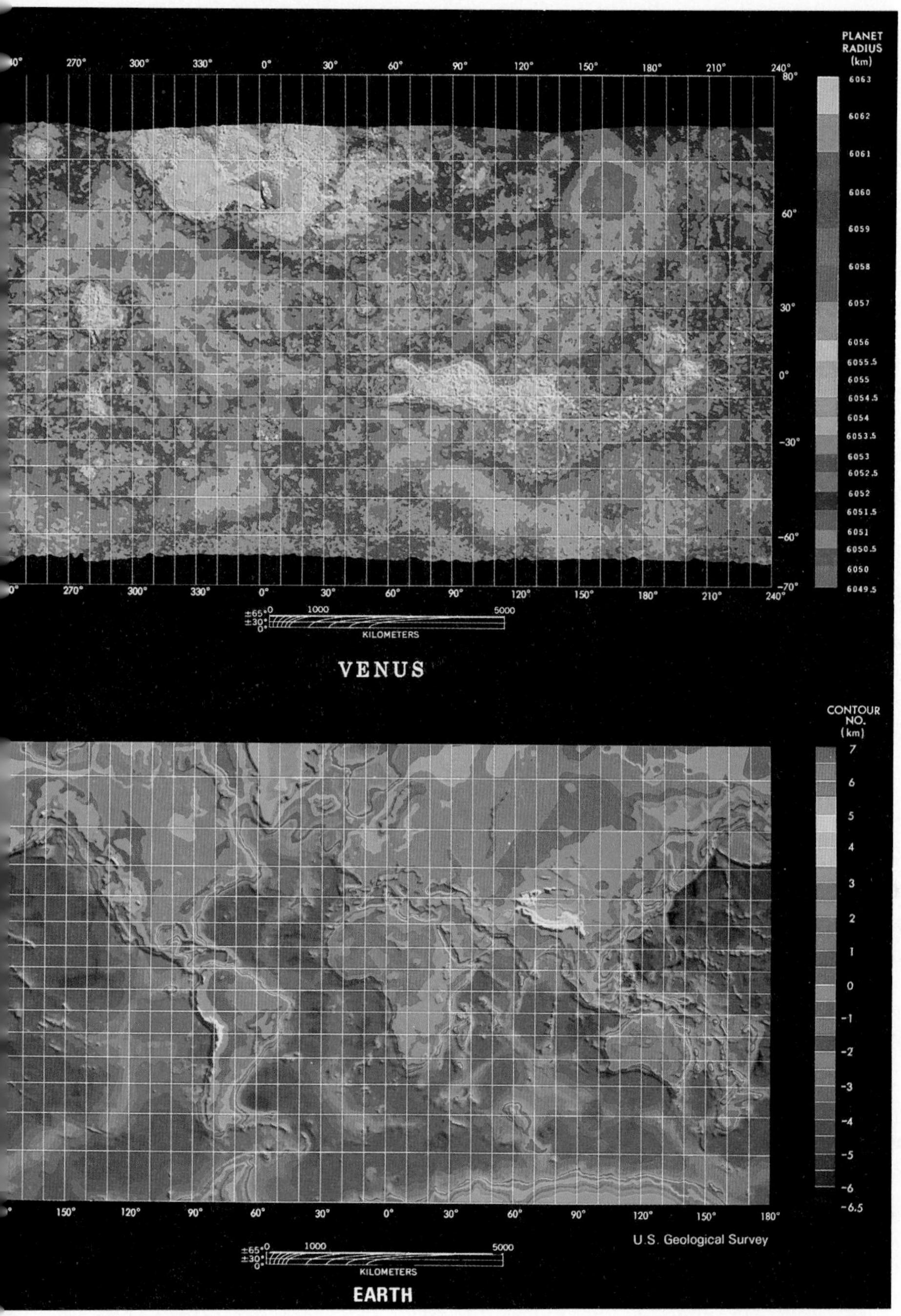

FIGURE 10.10 Comparison of the large-scale topography of Venus and Earth. The topographic map of Venus was obtained from radar altimetry carried out with the Pioneer Venus orbiter. The Earth's crust is divided about equally into continents and deep basins, while most of the surface of Venus consists of rather flat plains.

highlands. On Venus only 10% consists of highlands, compared with 45% continental surface on the Earth. As soon as the Pioneer Venus map was obtained, it became evident that Venus is not experiencing the same kind of plate tectonics as the Earth, or if it is, the continental masses are much smaller and the rifts and subduction zones less well defined.

Nomenclature

In order to discuss the processes that have been shaping the landscape of Venus, we must first gain some additional familiarity with the map in Fig. 10.10. With few exceptions, the topographic features of Venus have been named after real or mythical women, appropriate to the one planet in our solar system with a female name.

The largest upland or continental region is called Aphrodite, the Greek name for the goddess the Romans called Venus. Ishtar, named after the Babylonian goddess of love and beauty, is the prominent upland in the north, at about the same latitude as Greenland on our own planet. Ishtar is bigger than Greenland but only a small fraction of the size of North America. It contains the highest elevations on Venus, named the Maxwell Mountains after the nineteenth-century (male) Scottish scientist who first formulated the laws of electromagnetic radiation.

As improved radar images have been obtained from spacecraft, opportunities have multiplied to name more features. The large craters of Venus include, for example, Ariadne, Callas, Cleopatra, Dickinson, Joliot-Curie, Mead, Meitner, and Stuart. The circular tectonic features called coronae bear such names as Artemis, Gaia, Nefertiti, Nightingale, Sacajawea, and Sappho.

10.5 CRATERS AND TECTONICS

Many fundamental geological questions were left unanswered by Pioneer Venus. Without measuring individual impact craters, we could not determine the age of the surface or assess the level of geological activity. Without detecting individual volcanoes, we could not determine whether volcanic eruptions still take place on Venus. And without being able to see rift or subduction zones, we could not decide whether Venus shared with Earth any form of plate tectonics. To address these questions, we require higher-resolution radar images.

FIGURE 10.11 The Magellan spacecraft at the time of its launch.

Magellan

The Magellan radar mapper (Fig. 10.11) began orbiting Venus in 1990 and completed its mapping mission in late 1992. In addition to radar altimetry like that of Pioneer Venus, which simply measures the altitude of the spacecraft above the surface, Magellan carried an advanced **imaging radar**, also sometimes called a synthetic aperture radar, or SAR.

In radar imaging, the microwave signal is beamed to the surface at an oblique angle, and reflections are obtained from a relatively broad area. As the spacecraft moves, the angle at which it views the surface continually shifts, and consequently the details of the returned signal vary as well. Processing these data by powerful computers permits a radar *image* to be constructed of the planet's surface along the spacecraft track. This radar image looks very much like an ordinary photograph taken with oblique illumination—just the kind of information we need to reveal topographic detail, as we described for the Moon (Section 7.1). However, in this case brightness differences in the image correspond to variations in radar reflectivity, not color or albedo.

Magellan was not the first radar orbiter. The first radar imaging satellites at Venus were the Soviet Venera 15 and 16 spacecraft, which arrived in 1983 and mapped most of the north-

FIGURE 10.12 Global radar image of Venus, showing the bright, highly deformed bands near the equator that indicate prolonged periods of crustal compression.

ern hemisphere with a resolution of 2 km. Magellan, however, yielded 20 times the Venera resolution, mapping with a resolution of 100 m. Each day in orbit yielded a long, skinny image about 20 km wide and several thousand kilometers long. In its 2-year mapping program Magellan returned more data to Earth than all previous missions to all planets combined.

Radar is an ideal tool for the geologist. Radar waves reflect from the rock and are little affected by thin layers of soil or surface debris. In addition, of course, Venus has no vegetation or bodies of water covering the surface. Thus underlying geological patterns are more readily apparent than they would be on the Earth, making Venus a geologist's dream planet (Fig. 10.12).

Impact Craters

Magellan discovered almost 1000 impact craters on Venus, ranging in size from Mead, with a diameter of 280 km (larger than Chicxulub in Mexico), down to a few as small as 2 km (about twice the size of Meteor Crater in Arizona). The absence of larger craters (or of impact basins like Imbrium or Orientale on the Moon) tells us immediately that the surface of Venus does not date back to the heavy bombardment but must be at least as young as the lunar maria. The absence of smaller craters is a consequence of the thick atmosphere of Venus.

On Earth the atmosphere filters out most incoming projectiles with diameters smaller than about 50 m, which fragment and burn up before reaching the ground. With its much larger atmosphere, Venus is protected against most meteoroids smaller than about 500 m across, although the cutoff depends on the composition and speed of the meteoroid. An impact by a 500-m meteoroid typically makes a crater 5–10 km across. Because of this atmospheric filtering, Venus has relatively few craters less than 10 km in diameter, and none smaller than 2 km. Figure 10.13 shows the sort of complex, multiple crater that is formed when the meteoroid breaks apart just before striking the surface.

Three moderately large craters (30–50 km diameter) are shown in Fig. 10.14. The crater rims and ejecta are very bright, while the crater floors are dark. In a radar image, bright areas are rough and dark areas are smooth, because rough materials are better reflectors of radio energy. It makes sense that the ejecta, which consist of rocky blocks and fragments, are rough, but why is the floor smooth? This may be the result of impact melting of rock near the point of impact; since the surface temperature of Venus is so high, less additional impact energy is required to melt the rock than would be the case on the Earth or Moon.

FIGURE 10.13 Triple crater Stein, a complex feature formed when the impacting projectile broke apart in the thick atmosphere of Venus. The projectile had an initial diameter between 1 and 2 km.

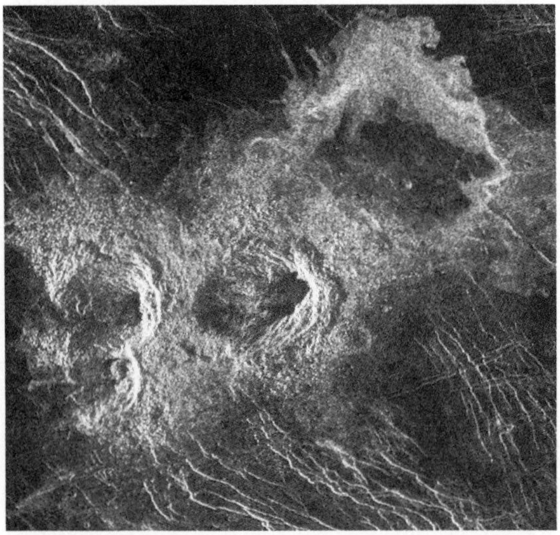

FIGURE 10.14 Three impact craters in the Lavinia region of Venus, the largest with a diameter of 50 km. The rough crater rims and ejecta appear bright because they are better reflectors of radar energy than are the surrounding smooth plains.

Long narrow flows of material are seen near some craters, extending beyond the ejecta blanket. These flows appear to be composed either of lava melted by the impact or of ejecta that flowed like a liquid along the surface. We will discuss evidence of similar fluidized ejecta around some martian craters in the next chapter. On Mars such features can be attributed to the presence of abundant water in the crust; however, Venus is dry, and so the presence of these flows is more difficult to understand.

One of the most striking and unexpected properties of the craters of Venus is their crisp, new appearance. Like the craters on the lunar maria, they look as if they formed just yesterday, yet we know that most of them must be many millions of years old. In spite of the presence of a thick atmosphere, erosion rates on Venus are very low, with craters neither worn down nor filled in by processes of erosion and sedimentation.

Crater Density and Surface Age

Venus has a total of approximately 1000 craters spread over a surface area nearly equal to that of the Earth. Our planet, in contrast, has only about 150 craters, and some of these are so old and degraded that we would not recognize their equivalents on Venus. Therefore we see that crater retention times on Venus are greater than on the Earth, and the level of geological activity is correspondingly less.

On the Earth and Moon the density of craters is strongly correlated with the type of surface terrain, with the most recently active areas having the fewest craters. Most lunar craters are found in the highlands, and most terrestrial craters are on the older continental cores. Is there a similar difference in crater density on Venus? Or, equivalently, are there regions of the venusian surface that differ significantly one from another in their age?

It is difficult to answer these questions because the craters are sparsely spread over the surface of Venus. The continental areas of Ishtar and Aphrodite appear to have fewer craters, not more, than the lowland rolling plains. Within the plains, crater densities seem to be about the same everywhere. The grouping of three craters close together in Fig. 10.14 seems to represent an anomalously high density, but this apparent clustering is probably just a statistical fluke.

The average crater density on Venus, in units of 10-km craters per million km^2, is 15% the value for the lunar mare, implying an age of about 500 million years. To within a factor of two, all crater retention ages on the plains of Venus are the same. Only Aphrodite and (perhaps) Ishtar are significantly younger. For comparison, remember that the age of the ocean basins on Earth is about 100 million years.

We know from the pristine appearance of the craters that erosion is not obliterating craters on a 500-million-year timescale. Some more fundamental geological process must be at work to renew the surface, either volcanic or tectonic or both. But whatever it is, this process does not erase craters gradually. For the most part, craters are either "fresh" or absent completely.

Tectonic Features

Tectonic geologic features are the result of either tension or compression in the crust of a planet. The cracks and ridges that result from tectonic forces show up readily in radar images. The many narrow bright lines in Fig. 10.14, for example, are all part of a tectonic grid of cracks.

A remarkably regular tectonic pattern is shown in Fig. 10.15, which shows a region of the Lakshmi plains about 40 km across. The fine straight lines, spaced about 1 km apart, are

FIGURE 10.15 Regularly spaced tectonic features in the Lakshmi plains. The surface has been fractured to produce a grid of parallel cracks and ridges with spacing of 1–2 km.

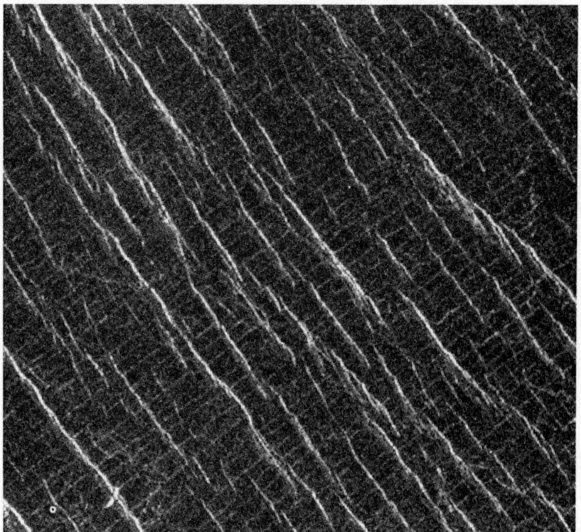

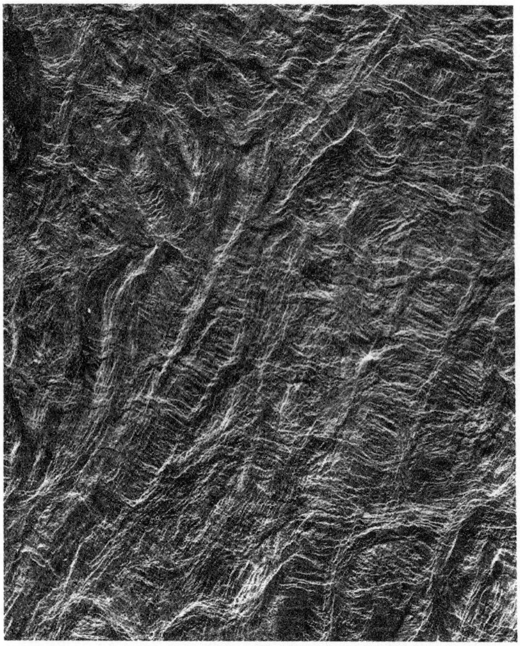

FIGURE 10.16 Detail of the Alpha region of Venus, part of the tectonic band that circles most of the planet's equator. The surface shows a dense network of ridges and mountains, resulting from compressional forces in the crust. The width of the figure is 475 km.

cracks resulting from stretching of the crust in one direction, while the less regular cross-pattern of ridges is due to compression of the crust in a direction at right angles to the stretching forces.

While there is evidence of tectonic modification of the surface everywhere, the crustal forces have been especially strong in certain areas. Figure 10.16 shows a large part of the equatorial region of Venus, stretching from the Aphrodite continent to the radar-bright region called Alpha. Many of the brightest regions of Venus are complex ridged terrain consisting of low mountain ridges spaced typically 10–20 km apart. These are areas of compression, where squeezing of the surface has caused it to wrinkle. Similar forces produce the mountains of our own planet, but it is rare on Earth to find belts of ridges or valleys on a scale approaching the ridged terrain of Venus.

Origin of Continents and Mountains

The highest and most dramatic folded mountains of Venus are in the northern Ishtar continent, which consists of a central elevated plain called Lakshmi surrounded by ranges of mountains (Fig. 10.17). The elevation of Lakshmi is 6 km, and the highest point in the adjacent Maxwell Mountains rises to an elevation of 11 km relative to the surrounding lowland plains. It is interesting to compare this area to the Tibetan Plateau (elevation 4 km above sea level) and the adjacent Himalayan Mountains (up to 10 km for the highest peaks) on Earth.

Recall that Tibet and the Himalayas are the result of the collision of the Indian subcontinent with Asia. This collision is happening today, with

FIGURE 10.17 Folded mountains surround the Lakshmi Plateau. In this region of the Danu Mountains, individual peaks are about 2 km high. The geology of this part of Venus may resemble that of the Himalayan Mountains and Tibetan Plateau of the Earth.

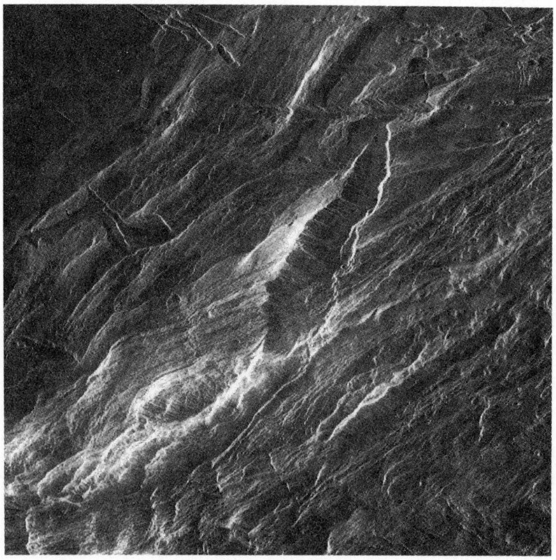

compressional forces maintained by the northward motion of the Indian plate. Without continuing application of force, the elevation of Tibet and the Himalayas would gradually decline. There is reason to think the same processes are at work on Venus. The mountains surrounding the Lakshmi Plateau are the steepest of Venus, with average slopes as great as 30°. Calculations show that such mountains would quickly collapse if they were not maintained by compression of the crust. The Ishtar continent thus seems to be a region where crustal forces converge, but is it the result of plate tectonics like that on the Earth?

One factor that distinguishes terrestrial plate tectonics is the relative ease with which the lithospheric plates can slide over the mantle beneath. As a result, each plate moves as a unit, and most of the forces are exerted along the edges in fault and subduction zones. On Venus there may be some small movement of plates, but in general it appears that the tectonic forces are distributed throughout the crust. There are no well-defined tectonic plates. In some places, like the folded mountains of Ishtar, the geological consequences are virtually the same on the two planets, but more commonly the tectonic activity on Venus is spread over the entire surface.

10.6 VOLCANOES ON VENUS

Volcanic activity is common on the surface of Venus. We don't know if any volcanoes are active today, but certainly such activity is recent on a geological timescale. Many of the volcanic landforms on Venus look very much like their terrestrial counterparts, but there are some surprises as well.

Volcanic Plains

The most common landscape on Venus is volcanic in origin. About 80% of the surface consists of lava plains roughly similar to the lunar maria, presumably resulting from high-volume eruptions of fluid lava that spread across large areas. It is these plains areas that are most reliably dated by crater counts as being roughly 500 million years old.

It is easy to estimate how much lava is produced on Venus and to compare this with the formation of new oceanic crust on the Earth. The total area of the plains is 400 million km^2, and we can estimate that the lava flows must be at least 2 km thick to have obliterated all preexisting craters. The total volume of plains lava is then about 800 million km^3. From the age of 500 million years, we estimate that about 1.6 km^3 of lava is deposited on the plains each year on the average. For comparison, the volume of new basaltic crust emplaced on Earth each year is about 10 km^3. Thus Venus is substantially less active volcanically than our own planet. In fact, the calculated eruption rate for the plains of Venus is fairly similar to the volcanic rate for the continents of Earth, neglecting oceanic eruptions.

Can we determine any more about the origin or timing of the plains volcanism? If this volcanic activity were a continuous process (as implied by the calculation of an annual eruption rate of 1–2 km^3 of lava), we would expect to see many craters modified by lava flows. As many as 20%, for instance, should have one wall or part of an ejecta blanket missing as a result of some nearby eruption that took place since the crater was formed. However, only about 5% of the craters on the plains show evidence of subsequent destruction or modification by lava flows, and the other 95% are pristine.

If this logic is correct, then we cannot be observing a steady state situation in the plains, with old craters removed by eruptions at the same rate new craters are formed from the impact of meteoroids. It appears that some time several hundred million years ago the activity level was very much higher than it is today. Indeed, we can imagine most of the volcanic

FIGURE 10.18 Computer-generated radar image of Venus showing the large shield volcanoes Sapas Mons (center) and Maat Mons (on the horizon). The vertical relief in this image is exaggerated ten times.

plains of Venus having been produced within a very short period about 500 million years ago in some sort of planetwide cataclysm, with a much lower level of activity continuing today.

Volcanoes and Lava Flows

Many thousands of individual volcanic mountains are seen in the Magellan images. The largest (as is also the case on the Earth and Mars) are shield volcanoes, characterized by shallow slopes and a summit crater or caldera. On Venus, shield volcanoes can be several hundred kilometers across and up to about 5 km high, generally similar to the size of Mauna Loa on Earth (Fig. 10.18). Individual lava flows mark their slopes, and judging from the radar images these volcanoes are indistinguishable from their terrestrial counterparts.

Venus also has volcanoes of a very different form produced by the eruption of thick, viscous lava. Many of these, called pancake domes, are remarkable circular domes as much as 45 km in diameter and 2–3 km high (Fig. 10.19). Terrestrial volcanic domes are never so symmetrical. Apparently all of the lava to make one of the venusian pancake domes is erupted at once from a single vent, rather like a large belch of material, and it then spreads out evenly to form a circular feature.

At the opposite extreme, Venus experiences eruptions of very low viscosity, fluid lava that can flow across the surface for great distances before congealing. The result is the formation of lava rivers of remarkable length (Fig. 10.20). About 40 such channels are longer than 100 km, and one, called Hildr, is 7000 km long—as long as the longest rivers on Earth, such as the Nile or

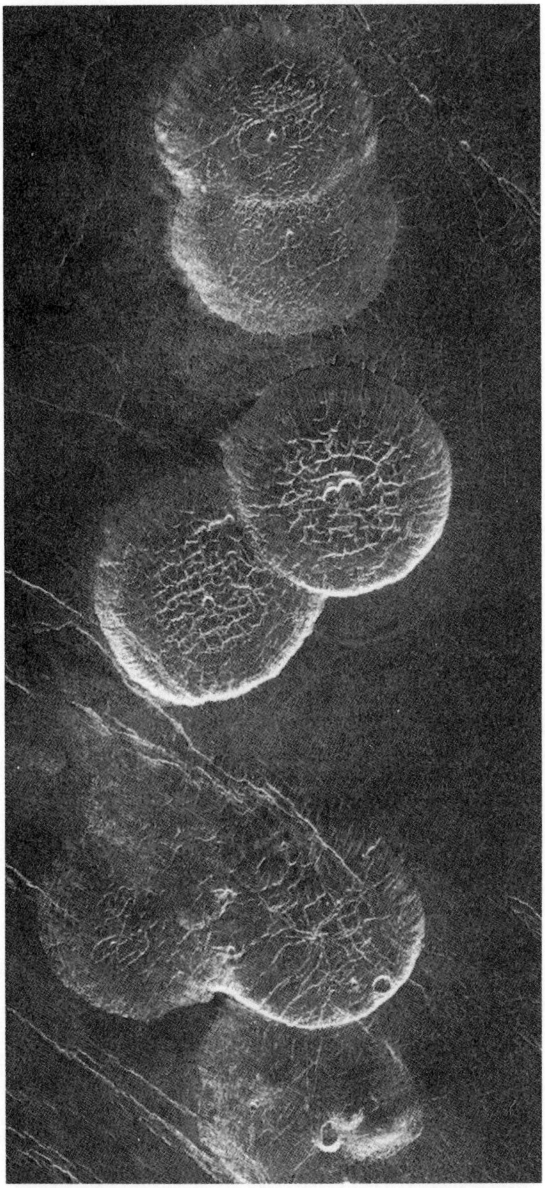

FIGURE 10.19 Pancake domes: circular volcanoes about 45 km in diameter and 2 km high.

the Mississippi. In contrast, the longest lava channels on Earth extend only a few tens of kilometers.

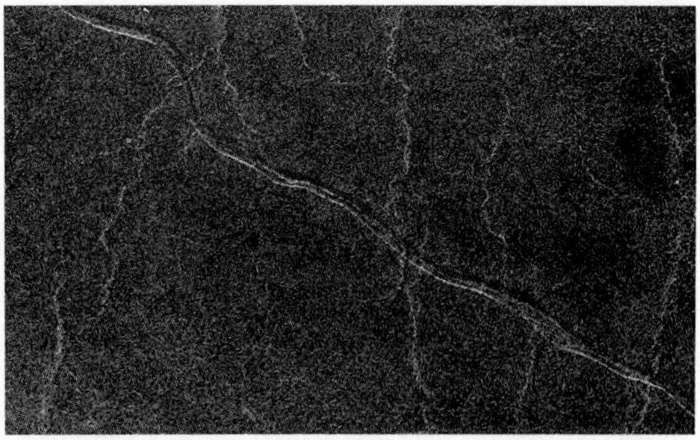

FIGURE 10.20 Small segment of the Hildr lava channel, which is about 7000 km in total length. Such lava rivers were formed by eruption of highly fluid lava.

Coronae

All of the tectonic and volcanic features we have discussed so far have their counterparts on Earth. Now we turn, however, to a unique class of features called **coronae**, which are found only on Venus. A corona is a circular or oval feature hundreds to thousands of kilometers across characterized by concentric and radial tectonic patterns and, often, by associated volcanic eruptions (Fig. 10.21). Approximately 400 coronae have been identified on Venus, with Artemis (diameter 2000 km) being the largest.

Each corona is characterized by a low central dome surrounded by a shallow trough and a great many concentric tectonic cracks. Associated volcanic activity may take the form of one or more shield volcanoes near the central dome. Apparently each corona is the result of a mantle hot spot—a plume of rising magma similar to the mantle plume that has created the Hawaiian islands on the Earth. Many of the coronae probably represent failed hot spots—plumes that turned off before the magma broke through to the surface. Alternatively, coronae may represent a stage in plume development that will later lead to surface eruptions and the construction of large shield volcanoes.

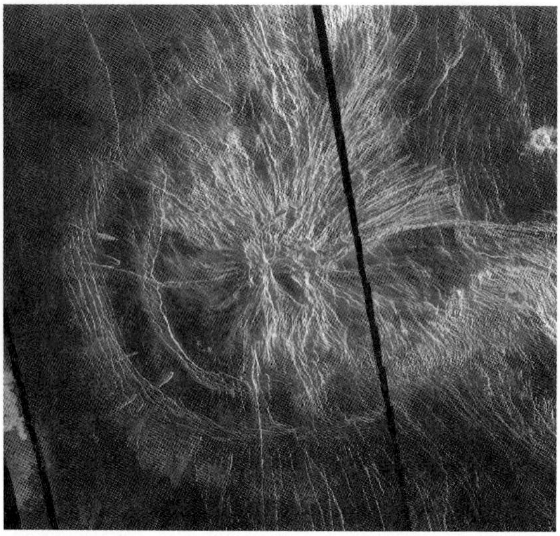

FIGURE 10.21 Pandora Corona, a circular feature about 350 km in diameter. Such features, which are unique to Venus, appear to have been produced over rising plumes of magma in the mantle of the planet.

Geological History

The Magellan data have revolutionized our knowledge of Venus. They clearly show that the planet is geologically active, with ongoing tectonic and volcanic modification of the surface. The energy for such activity is derived from convection currents in the mantle, as with the Earth. However, these currents have not led to planetwide plate tectonics, either because the plumes of rising magma are too small or else because the lithosphere of Venus is unable to slide over the mantle as it does on the Earth. The resulting pattern of coronae and regional deformation of the surface has sometimes been called blob tectonics, to distinguish it from terrestrial plate tectonics.

Major questions are also associated with the possibly episodic formation of the lowland lava plains. We are used to thinking of geological processes as slow and continuous. On Earth, for instance, we know the lithospheric plates move a few meters per century, and we confidently expect that they have been doing so for a long time in the past and will continue to move at this rate in the future. It is startling, therefore, to imagine that on Venus the entire surface of the planet may have been catastrophically resurfaced just 500 million years ago. Planets are not supposed to behave like that (or so we have been taught to think).

Whatever the outcome of these particular questions, we now know that Venus is lacking in ancient surface material. It is unlikely that any rock on Venus is as old as the oldest rocks composing the terrestrial continents. Venus probably has the youngest surface of any of the terrestrial planets, in the sense that its crust was most recently formed, with little if any memory of the way the planet was even a billion years in the past.

10.7 ON THE SEARING SURFACE

Among the most dramatic achievements of the space program of the former Soviet Union were the Venera spacecraft that successfully landed and operated on the hostile surface of Venus. These hardy robots allowed us to glimpse a place where no human will ever set foot.

The Venera Landers

To gain an appreciation for what conditions are like on the surface of Venus, you might begin by looking at the temperature control of a modern kitchen oven. You will find that the highest temperature the oven can achieve is about 500 F. The ground temperature all over Venus is within 15° of 860 F. And remember this temperature is coupled with an atmospheric pressure 90 times the value on Earth.

Despite these extremely inhospitable conditions, direct measurements were successfully carried out at seven locations on the surface of Venus by Soviet landers. These landers were suitably armored against the high pressure and

FIGURE 10.22 The VEGA spacecraft, launched by the Soviet Union in 1984, dropped probes at Venus which released balloons into the planet's atmosphere. The main spacecraft went on to fly past Comet Halley in 1986.

equipped with some internal cooling. The best of them survived nearly two hours, during which a variety of investigations were carried out, including pictures transmitted back to Earth, four of them in color. These are the only close-up pictures we have of the surface of another planet except for the Viking pictures of Mars and, of course, extensive surface photography of the Moon (Fig. 10.22).

The Venera cameras provided panoramic views of the surface extending from the soil directly in front of the spacecraft out to the horizon. The cameras were positioned about 1 m above the ground, providing a perspective comparable to that of an observer sitting on the surface of Venus. The cameras did not gaze straight out at the landscape as you would; instead, they were directed at mirrors set at a 45° angle, and the mirrors were rocked back and forth to produce the desired panoramas.

The immediate foreground of each picture shows the view toward the feet of the hypothetical observer, including a part of the spacecraft itself, with a scale reference provided by the triangular teeth spaced 5 cm apart. The white ladder-shaped boom extending out just left of center is about as long as your arm, while the distant white viewport covers (ejected after landing) are about the diameter of a human head (Fig. 10.23).

Images of the Surface

The first pictures of the surface of Venus, transmitted by Veneras 9 and 10 in 1975, showed rough, undistinguished landscapes dominated by loose rocks. These spacecraft both landed in regions of fractured lava plains on the lower slopes of the Beta volcanic complex; Venera 10 apparently photographed old basaltic lava flows, while Venera 9 landed near an area of more recent tectonic modification. Surface illumination, as we noted previously, was similar to that on a heavily overcast day on Earth.

The more detailed pictures transmitted by Veneras 13 and 14, which landed in March 1982, represent the best data available on the appearance of the surface of Venus. The Venera 13 picture shows many small rocks and some fine-grained soil, demonstrating that erosional processes on Venus break up large blocks into smaller material. Alternatively, the fine-grained material might be wind-borne lava fragments (so-called volcanic ash) from a large volcanic eruption. We know from the Magellan images that erosion is a very slow process on Venus. These pictures also show layering on rock surfaces, especially around Venera 14, perhaps indicative of lava flows (Fig. 10.24).

Magellan images of the Venera 13 and 14 landing areas are shown in Fig. 10.25. The uncertainties in the location of the landers amount to several hundred kilometers, so the actual landing sites could be anywhere in the central half of the images. The possibilities for Venera 13 thus include older lava plains, a belt of tectonic fractures, or even a pancake-type dome of viscous lava located 200 km southwest of the nominal landing site. There is nothing in the surface view to discriminate among these possibili-

FIGURE 10.23(a) A view of the surface of Venus obtained by Venera 13. The triangular teeth in the foreground are about 5 cm apart. Note the presence of fine-grained, soil-like material in the foreground. The bright crescent-shaped object in the center of the frame is the camera cover. The device to the left of the ejected camera cover measures the surface hardness. The banded boom to the right is for color calibration (see Fig. 10.24).

FIGURE 10.23(b) Venera 14 was surrounded by flat, platelike rocks resembling pieces of a dried lakebed. One of them was overturned by the shock of landing, revealing a bright surface (left foreground). In this case, the deployed camera cap unfortunately ended up under the surface-hardness measuring device, showing that Murphy's law also applies on Venus.

FIGURE 10.24 A close-up of the right-hand side of Fig. 10.23(a) in color. The rocks on Venus have not been bleached (by solar UV photons), like those on the Moon, but they do not show the red tint of iron oxides that dominates the martian landscape.

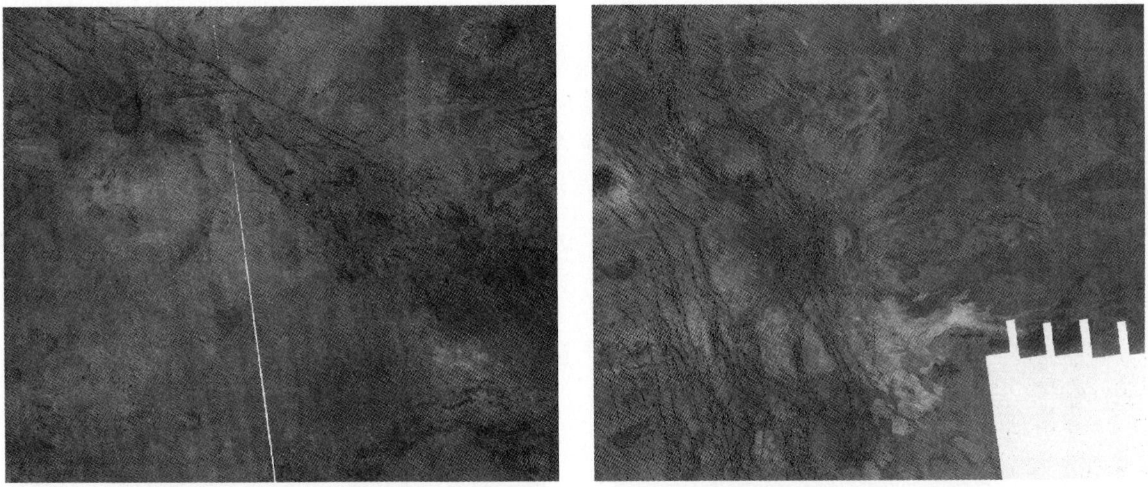

a b

FIGURE 10.25 Magellan images of the Venera 13 and 14 landing areas, as seen from orbit. Because of uncertainty in the landing location, it is unclear what sort of terrain was imaged by Venera 13(a), but Venera 14 very probably photographed the slopes of the low shield volcano that fills the central part of image (b).

ties. Venera 14, however, is probably located somewhere on the slopes of the low shield volcano that fills the central part of the orbital photo (Fig. 10.25b). The layered rock surfaces in the Venera image are likely to be basaltic flows from this volcano.

Surface Composition

Five of the successful Venera landers carried devices to detect gamma rays emitted from radioactive isotopes of uranium, thorium, and potassium in the surface rocks. By analogy with the levels of radioactivity of terrestrial rocks, their levels suggest basalt at all five sites, although the exact chemistry varies from one location to the next.

On Veneras 13, 14, and 18 a more sophisticated technique measured the surface composition (Figs. 10.26a, b). A drill was deployed to gather samples and bring them inside the spacecraft for analysis. Using an x-ray source to stimulate emission from the sample, this instrument was able to detect additional elements. The Venera 13 sample indicated a composition typical of oceanic basalts on Earth, while the result from Venera 14 resembled a much rarer kind of basalt with a high percentage of potassium. Venera 18, which landed near the Aphrodite continent, also revealed a basaltic composition, this time unusually rich in sulfur.

All five of the measured locations are in the common volcanic plains, where we expect basalt to be the primary rock type. Venera 13 very likely measured one of the young lava flows visible in the Magellan image of the site, but the other four landers appear to be on older and less distinctive surfaces.

In all of these investigations of Venus we lack much of the critical information that helps us to understand the geology of the Earth and Moon. We have no seismic probes of the interior and no measurements of subsurface temperature or heat flow. Since there are no samples from Venus in our meteorite collections, and of course no sample-collecting spacecraft has been sent to this hellish planet, we lack the chemical information and age measurements that have played such a

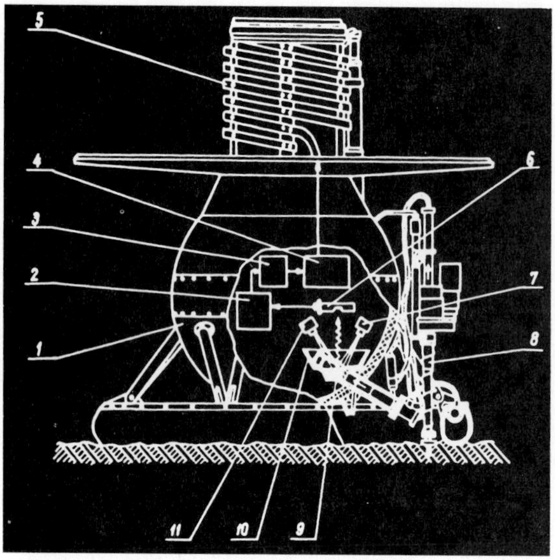

FIGURE 10.26(a) A line drawing of the landing module (Venera 18) from the VEGA spacecraft that reached the surface of Venus in June 1985. The large round pressure vessel (1) is surrounded by a coiled antenna (5) for communication with Earth. A drill (8) can obtain samples of surface materials which are then brought inside the module for analysis by x-ray fluorescence (7).

FIGURE 10.26(b) Yuri Surkov, of the Vernadsky Institute for Geochemistry and Analytical Chemistry in Moscow, holding a proof-test model of the x-ray fluorescence device he built for the VEGA lander (item 7 in Figure 10.26a).

large role in terrestrial and lunar studies. Since it is very difficult to envision spacecraft that could land on Venus and operate successfully for long periods on the surface, there are no prospects for obtaining samples or even roving across the surface. Thus, in spite of the fabulous success of the Magellan mission in providing detailed surface maps, Venus is likely to remain an enigma for a long time to come.

10.8 QUANTITATIVE SUPPLEMENT: THE GREENHOUSE EFFECT

Developing accurate models for the greenhouse effect is a challenging task that requires detailed understanding of how all of the gases in an atmosphere interact with visible and infrared radiation. However, we can look at a very idealized greenhouse model to get some sense of how the process works.

Consider a two-layer model as sketched in Fig. 10.27. The planet is illuminated by sunlight falling from straight above. The atmosphere is represented here by a single isothermal layer that is completely transparent to sunlight and completely opaque to the infrared radiation emitted by the surface. The incoming solar power is P_1, and the power radiated back into space (all of which comes from the atmosphere since radiation from the surface is blocked) is P_2. In equilibrium, the power emitted by the planet must equal incoming solar power, so $P_1 = P_2$.

The surface temperature is T_S, and by the Stefan-Boltzmann law it must be radiating a power P_S given by σT_S^4, all directed upward. The

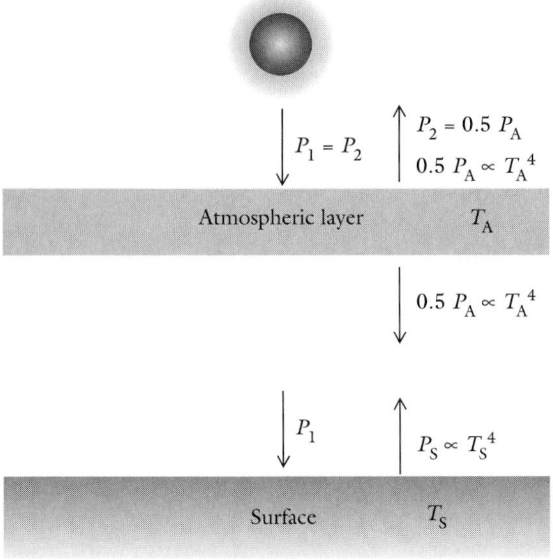

FIGURE 10.27 A planet with an atmosphere will have a surface that is heated both by direct radiation from the sun (P_1) and by radiation from the atmosphere (0.5 P_A).

atmospheric temperature is T_A, and in this model it radiates a power P_A, half upward to space and half downward to the surface, both powers given by 0.5 σT_A^4, where we have already noted that 0.5 σT_A^4 is also equal to P_2, the total power radiated by the planet.

We now write an algebraic expression for the energy balance at the surface. The power reaching the surface is P_1 (directly from the Sun) and 0.5 $\sigma T_A^4 = P_2 = P_1$ (the back radiation from the atmosphere). Thus the total power reaching the surface (and radiated from the surface) is twice the power from the Sun alone. If T_E is the equilibrium surface temperature without an atmosphere, you can see that

$$\sigma T_S^4 = 2\sigma T_E^4$$
$$T_S = \sqrt[4]{2}\ T_E = 1.2\ T_E$$

Thus if the equilibrium temperature (without a greenhouse effect) is 400 K, the surface temperature with the greenhouse effect is 480 K.

The next step in sophistication is to consider a multilayer atmosphere, with the energy balance at each layer calculated just as we have done here for the surface. Each layer of the atmosphere receives radiation from both above and below, and each layer radiates both up and down at its appropriate temperature. If you add up many such layers, you will find that the expression for the surface temperature is

$$T_S = \sqrt[4]{4}\ T_E = 1.4\ T_E$$

This is the limiting temperature for a perfectly effective, but very simplified, greenhouse model. In the case of Venus, however, the observed surface temperature is more nearly twice the expected value rather than being enhanced by a factor of only the square root of two. This is why the high temperature of Venus was such a shock to the early radio observers, and why it took many years of sophisticated modeling to understand what is really happening in this complex atmosphere.

SUMMARY

So similar to Earth in size and bulk composition, Venus is dramatically different in many other characteristics. It rotates slowly backward on an axis nearly perpendicular to its orbit; why, we do not know. Its surface is baking at a temperature of 740 K, at a pressure of 90 bars. The immense atmosphere that produces this high surface pressure is made mainly of carbon dioxide, which maintains the high surface temperature through a greenhouse effect. Some visible light from the Sun penetrates the thick sulfuric acid clouds and reaches the planet's surface.

The circulation of this atmosphere is considerably different from ours, owing to the planet's slower, retrograde rotation and greater proximity to the Sun. Hadley circulation, a simple equator-to-pole exchange, seems to dominate the flow. The thickness of the atmosphere maintains a nearly constant temperature over the entire surface of the planet.

Radar mapping has shown that the geology of Venus is substantially different from that of the Earth. Most of the planet (80%) consists of low, rolling volcanic plains produced by large eruptions of fluid basaltic lava. The crater densities on these plains are only 15% of the values for the lunar maria, indicating an age of about 500 million years. To our surprise, we find very few degraded or eroded craters, suggesting that the formation and destruction of craters are not in balance. It appears that a high level of volcanic eruptions, or possibly even a general overturning of the planet's crust, formed most of the plains surface about 500 million years ago, with lower levels of volcanism (roughly 1 km^3 of new lava per year) since that cataclysm.

The crust of Venus has been extensively modified by tectonic deformation to produce a wide variety of cracks and ridges, as well as major belts of folded mountains. The highest parts of the surface are found in the Ishtar continent, where the Lakshmi Plateau and the Maxwell Mountains resemble the Tibetan Plateau and Himalayan Mountains of the Earth, uplifted by compression of the crust. However, Venus is not experiencing terrestrial-type plate tectonics, perhaps because its lithosphere is too tightly attached to the mantle below. A variety of ongoing volcanic processes produce shield volcanoes, pancake domes, and lava rivers up to 7000 km long. Coronae are unique geological features that mark the top of subsurface plumes of magma rising through the mantle.

Both surface imaging and chemical analysis of the soil have been carried out by Venera landers. All of these spacecraft landed on the widespread lowland lava plains, and their analysis indicates the presence of basalt.

Many of the striking differences between Venus and Earth can be explained in terms of our sister planet's closeness to the Sun. At the distance of Venus, water cannot remain on a planetary surface. A runaway greenhouse effect will occur that ultimately leads to the breakdown of water molecules in the planet's upper atmosphere by solar ultraviolet light. At least we speculate that this may have happened on Venus. We will return to a comparative study of the evolution of the Earth, Venus, and Mars in Chapter 12.

KEY TERMS

corona
Hadley cell
imaging radar
photochemistry
runaway greenhouse effect

Review Questions

1. Compare the interior structures of the terrestrial bodies studied so far: Earth, Venus, Mercury, and the Moon. Consider also what we know about their thermal evolution. How do structure and evolution depend on the size and composition of a planet?

2. Compare the atmospheric structures of the Earth and Venus. Does the Venus atmosphere have the same regions as were identified on Earth: troposphere, stratosphere, ozone layer, ionosphere, magnetosphere? Why are there differences between the two planets?

3. Compare the atmospheric circulation patterns of Venus and the Earth. What are the roles of solar heating and planetary rotation? How does the presence of water on Earth give rise to kinds of weather that would be impossible on Venus? What is the effect on atmospheric circulation of the much larger mass of the Venus atmosphere?

4. Contrast the compositions of the atmosphere and clouds of Venus with those of the Earth. Do you understand why the clouds on both planets do not have the same composition as one of the primary gases in the atmosphere, but instead represent a trace constituent of the gaseous atmosphere?

5. Explain the runaway greenhouse effect. How does this differ from the ordinary greenhouse effect? If Venus once had oceans of water and later lost them through the runaway greenhouse effect, what evidence today might reveal this past history? Consider both chemical and geologic clues.

6. Describe the large-scale surface topography of Venus and compare it with those of the Earth and Moon. Does Venus have features analogous to the lunar highlands and maria? Or to the terrestrial continents and ocean basins? How does Venus differ from each?

7. Explain how craters are used to determine the ages of the surface features of Venus. How do these ages compare with those of the lunar highlands, the lunar maria, the terrestrial continents, and the terrestrial oceans? Does Venus have different surface units with very different ages, as do the Earth and Moon?

8. What features on Venus have a tectonic origin? What are the indications pro and con for plate tectonics on Venus? Can you think of reasons why the Earth and Venus should differ with respect to their plate tectonics?

9. How does the volcanic history of Venus compare with that of the Earth and Moon? Are any of the volcanic features on Venus distinctive or unique? Do scientists understand the reasons for these differences?

10. What would it be like to stand on the surface of Venus (with appropriate protection from the heat and pressure)? What would the ground and sky look like? How might you tell if you were located on volcanic plains, in a region of tectonic deformation like Alpha, or in the Maxwell Mountains? What would the Earth look like from the surface of Venus?

Quantitative Exercises

1. Venus requires 440 days to move from its greatest elongation (apparent distance) west of the Sun to its greatest elongation east, but only 144 days to move from greatest eastern elongation to greatest western elongation. Explain why.

2. Magellan scientists estimate that volcanism on Venus generates about 2 km^3 of new lava per year. At this rate of eruption, how long would it take to cover the planet uniformly to a depth of 1 km? Generally, if we wish to obliterate a large crater, we need to bury it to a depth of several kilometers. What time is required to cover Venus to a depth of 3 km?

3. The surface of Venus is nearly twice as hot as the surface of the Earth. At what wavelength is the maximum of its infrared radiation from the surface?

4. The terrestrial greenhouse effect produces an increase in the surface temperature of about 25 K, or a factor of about 1.1. Suppose that a huge impact generated a thick layer of dust at high altitude. In terms of the simple two-layer model discussed in Section 10.8, this would be equivalent to having the incident sunlight absorbed in the upper layer rather than at the surface. According to this model, what would happen to the greenhouse effect and the surface temperature?

Additional Reading

Chapman, C.R. 1982. *Planets of Rock and Ice: From Mercury to the Moons of Saturn* (Chapter 7). New York: Scribners. Good general discussion of the atmosphere of Venus and the difficult road to understanding its true nature.

Cooper, H.S.F., Jr. 1993. *The Evening Star—Venus Observed.* New York: Farrar Strauss Giroux. A popular account of the Magellan mission by the veteran space reporter of the *New Yorker.*

*Hunten, D.M., *et al.,* eds. 1983. *Venus.* Tucson: University of Arizona Press. Multiauthor technical overview of Venus, with emphasis on the atmosphere (has several chapters by Russian scientists).

**Journal of Geophysical Research* special issues (reprinted as book). 1992. *Magellan at Venus.* Washington, D.C.: American Geophysical Union. Highly technical summary of the initial results from Magellan (more than 3000 pages).

Morrison, D. 1993. *Exploring Planetary Worlds* (Chapter 3). New York: Freeman. Comparative discussion of Earth and Venus, using new Magellan observations.

*Indicates the more technical readings.

11

Mars: The Planet Most Like Earth

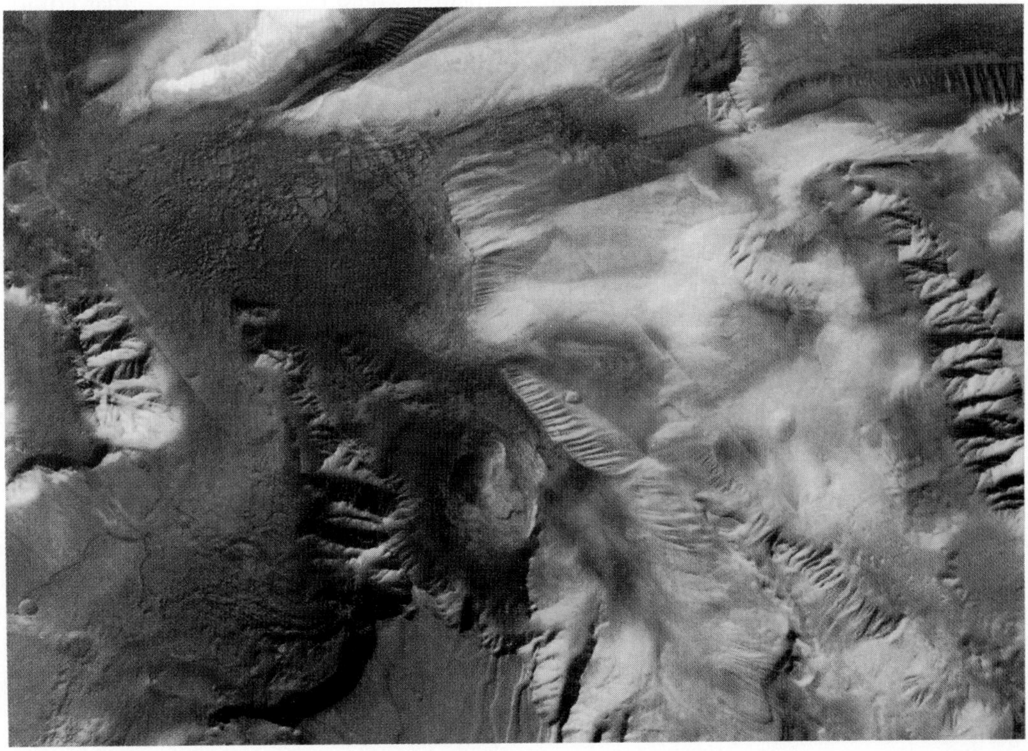

The heavily eroded martian canyonlands as imaged by Viking. The width of this image is about 50 km.

Of all the planets, Mars has the most romantic appeal. The Moon is closer, Venus brighter, Jupiter larger, and Saturn more beautiful, but Mars remains the one other place with which humans can identify. In the past, this planet was thought to harbor life, even intelligent creatures with civilizations in advance of our own. Now, even though we suspect that Mars may be biologically dead, we recognize its potential, both as an exciting target for exploration and as a possible site for human colonization in the twenty-first century.

Mars may have great potential, but today it is a foreboding world. With only a tenth the mass of the Earth or Venus, it has reduced levels of geological activity, with consequently less outgassing of volatiles from the interior. Being farther from the Sun, Mars is also noticeably cooler. It is a planet caught in a terminal ice age, with most of its water frozen in subsurface permafrost. The surface atmospheric pressure is so low that *liquid* water cannot exist, even if the temperatures should rise above the freezing point. Solar ultraviolet light bathes the surface in a lethal glow, unimpeded by the thin atmosphere.

What makes Mars so interesting in spite of these difficulties is its evidence of very different past climates. Volcanoes once erupted vast amounts of lava and, presumably, released water and other gases from the interior. Riverlike channels give testimony of a time when water flowed and, perhaps, rain fell from the martian skies. Layered deposits in the polar regions indicate cyclical climatic variations that appear to continue today. Mars is the one place besides the Earth where we see clear evidence of climatic cycles, in addition to long-term evolution of the surface and atmosphere. It is the only other planet where there is evidence that liquid water once flowed on the surface. And because Mars is in so many ways similar to the Earth, much of this knowledge can be compared directly with our understanding of our own planet.

11.1 A CENTURY OF CHANGING PERCEPTIONS

Mars is among the brightest objects in the sky, and its red color sets it apart from the other planets. As we saw in Chapter 1, observations of the movements of Mars played a critical role in the development of Kepler's laws of planetary motion. In the modern era, this planet has been a primary target for investigation by spacecraft. Although not as many spacecraft have been sent to Mars as to Venus and the Moon, it may be that the return from these missions has been higher. Mars is intrinsically more interesting than the Moon, without presenting the formidable environmental challenges of Venus. The two Viking landers of 1976, in particular, represent the most ambitious and sophisticated undertaking of two decades of robotic planetary exploration. As a result of Viking and its predecessors, we know a great deal about this planet.

Mars Through the Telescope

As seen from Earth, Mars follows a 26-month cycle from one opposition to the next. When the two planets are far apart, Mars looks like an inconspicuous red star, and even the best telescopes reveal no surface features. When they are close together, however, the distance from Earth to Mars can be as little as 55 million km, and telescopes can reveal features as small as 100 km across (Fig. 11.1). Note that this resolution is approximately the same as that of surface features on the Moon seen with the naked eye.

As we noted in Section 7.1, a resolution of 100 km is insufficient to reveal *topographic* features on a planet. In the case of Mars, the situation is made worse by the fact that we always see the planet at nearly full phase. Recall how little topography is visible at full moon, even through a telescope. Thus the features that are seen telescopically on Mars are markings that represent different colors and reflectivities of surface materials. Primarily, we map dark regions and light regions, corresponding to reflectivities near 15% and 30%, respectively.

In addition to these relatively permanent light and dark surface markings, Mars displays bright polar caps, which grow and shrink with the seasons, just as one would expect for deposits of ice or snow. Transient bright clouds of yellow or white color are also seen. From their presence, the existence of a martian atmosphere was correctly deduced by visual observers. Not until the 1940s, however, when CO_2 was identified spectroscopically, did astronomers begin to learn the composition of this atmosphere.

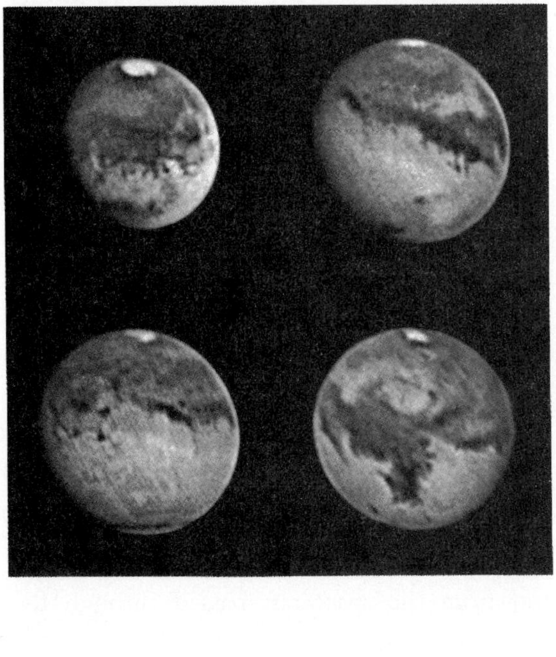

FIGURE 11.1 Some of the best Earth-based photos ever obtained of Mars, taken with a CCD electronic camera in 1988, when the planet was exceptionally close to the Earth.

The Canal Controversy

The history of martian studies notes well 1877, the year the Italian observer Giovanni Schiaparelli recorded the linear markings he called *canali,* or channels. These faint dark lines, glimpsed near the limit of detectability, seemed to stretch for thousands of kilometers across the surface. In English-speaking countries, the name was translated as canals, a term implying construction by intelligent beings.

By the early years of the twentieth century, the conviction was widespread that the canals of Mars proved the existence of intelligent life on that planet. The most vocal advocate of this position in the United States was Percival Lowell, who for two decades dominated the American public image of astronomy. Most astronomers, in fact, could not see the canals, but not until after World War I did their skeptical viewpoint prevail. Thereafter, the search for canals on Mars or other signs of intelligent life there was left largely to amateur observers and occasional cranks.

The idea that Mars was peopled by technologically advanced creatures faded with the canals. Mars as an abode of life, however, retained its appeal. The fact is, many of the phenomena seen on this planet could be interpreted in terms of widespread plant life. We will discuss the canal controversy further in Chapter 12, when we address the search for life on Mars.

The Seasonal Cycle

Since the tilt of its polar axis (25°) is nearly the same as that of the Earth, Mars experiences similar seasons, except they are about twice as long because the martian orbit is larger than ours. In each hemisphere, a polar cap forms during fall and winter under an obscuring layer of clouds, reaching its maximum extent (to latitude 65°) at the beginning of spring. It then recedes with the coming of warm weather, shrinking to a residual cap a few hundred kilometers across by the end of summer, when the cycle begins again. If the caps were composed of water ice, one would expect their retreat to release water, either as liquid or vapor; conceivably this water could enable the growth of plants in spring and summer, which is also the period of warmest weather.

Many twentieth-century observers saw changes in the dark surface markings that seemed to be seasonal. Often the regions darkened in summer, as might be expected if lifeforms were abundant and were growing in response to water released by the cap. Some observers claimed that the darkening was accompanied by a change of color to hues of green. Perhaps most persuasive of all, the dark areas reformed after major dust storms, as if new plants were pushing up through the layer of lighter material deposited by the storms. The circumstantial evidence for seasonal growth seemed strong, and nothing was known about conditions on Mars that precluded hardy plant life, perhaps similar to terrestrial lichens.

The First Flybys

Shortly after the successful Mariner 2 flyby of Venus, a similar spacecraft, designated Mariner 4, was launched toward Mars. The main difference between the two spacecraft was that among its instruments Mariner 4 carried a simple television camera. On July 15, 1965, 22 close-up pictures of Mars were transmitted to Earth. They showed impact craters, superficially similar to those on the Moon (Fig. 11.2). Today, we expect to find craters on planets, but in 1965 these pictures sent shock waves throughout the scientific world. The geologists had hoped for a more active planet with valleys, mountains, plains, and perhaps a volcano or two. In response to the new data, scientists speculated that Mars was *geologically* dead, but newspapers missed the adverb, and it was widely reported that Mars was a "dead world." The public image of Mars would never be the same again.

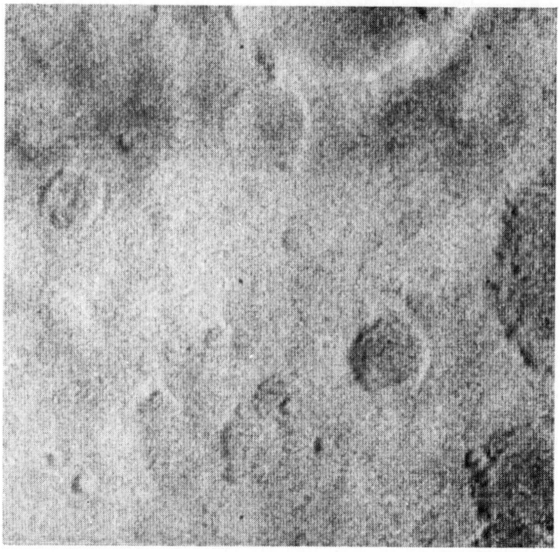

FIGURE 11.2 A small section of the surface of Mars as viewed by Mariner 4. The field of view is about 250 km on a side. Seeing the lunarlike craters in pictures such as this led several scientists to conclude that Mars was as geologically inactive as the Moon.

Mariner 4 also made an important discovery about the atmosphere of Mars. As it flew past the planet, the spacecraft was navigated to pass behind Mars as viewed from the Earth. Just before it disappeared, the radio signals to and from the spacecraft passed through the atmosphere of Mars, and the effects of the atmosphere on the signals could be measured. Analysis of these data showed that the surface pressure on Mars was slightly less than 0.01 bar, about a factor of ten lower than had been expected. The tenuous nature of its atmosphere further underscored that Mars was not very Earthlike after all.

In 1969 two more advanced spacecraft, Mariners 6 and 7, flew by the planet, photographing much of the surface at resolutions sometimes as high as half a kilometer. Again, they saw primarily cratered terrain, by pure bad luck missing the huge volcanoes, canyons, and channels we now know are there. Had we stopped after these three flybys, we would never have suspected the true geological complexity of Mars. Fortunately, plans for an orbiter were already well advanced, or the whole program would probably have been terminated after Mariner 7.

In addition to photographing the surface, these first flybys made many other measurements. The winter polar caps were determined to be made primarily of frozen carbon dioxide (dry ice) instead of water, CO_2 was established as the major constituent of the atmosphere, a tiny and locally variable quantity of O_3 was discovered, and the absence of any measurable planetary magnetic field was established.

The Mariner 9 Orbiter

While a flyby spacecraft has only a few hours to make its measurements, an orbiter can remain indefinitely to carry out detailed, long-term studies. On November 14, 1971, Mariner 9 became the first spacecraft to go into orbit about another planet. Initially the results were very disappointing. The spacecraft had arrived in the midst of a planetwide dust storm that obscured the surface from view; had the mission been a flyby, it would have been an embarrassing failure. This time, however, the controllers at the Jet Propulsion Laboratory (JPL) in Pasadena, California, could afford to wait. Over the next weeks the dust cleared, and soon a new world was being revealed to Mariner 9's cameras (Fig. 11.3).

In January 1972 a ten-month program began to map the entire martian surface at a resolution of about 1 km. One remarkable feature after another was discovered: the largest volcano in the solar system, the greatest canyon system, and

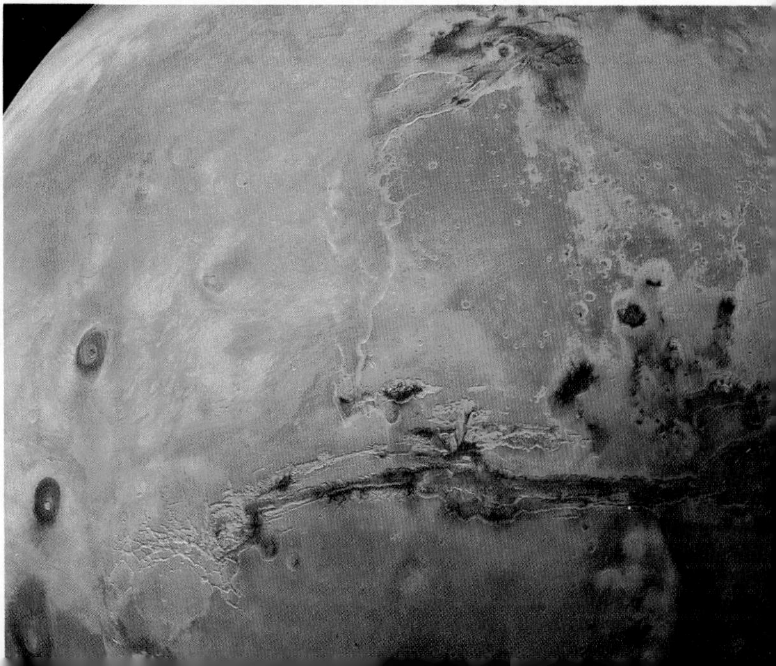

FIGURE 11.3 Global color view of Mars, obtained by combining more than 100 individual Viking frames.

even vast drainage channels that dwarfed terrestrial river systems. Repeated radio probes were also made of the structure of the martian atmosphere.

Scientists were ecstatic about the Mariner 9 results; however, it was not easy to communicate their findings to the public. The problem was the dust storm raging when the spacecraft arrived at Mars. The press, oriented toward instant news, reported only the problems encountered during the dust storm. Newspapers carried featureless pictures that showed only the pall of dust. A few months later, when spectacular pictures were being sent back, most of the U.S. press refused coverage on the grounds that Mariner was no longer news. The press from other countries, however, did not share this narrow perspective, and many of the Mariner discoveries reached audiences abroad while people in the United States remained ignorant of the advances being made.

Viking

The next step beyond orbital surveys is to land on the planetary surface. An ambitious Mars lander had been planned since the late 1960s, based on the huge Apollo Saturn V rocket. When the Nixon administration cut back the space budget, this program was scaled down to a more modest pair of landers and orbiters named Viking, to be launched by the Titan/Centaur rocket.

Arriving at Mars in June 1976, the first Viking was designed for touchdown on July 4, the two-hundredth anniversary of the Declaration of Independence, pending confirmation of a suitable landing site from orbital photographs. However, when the Viking orbiters photographed the surface at higher resolution (typically 100 m) than had been achieved by Mariner 9, they found unexpected hazards in the planned landing site. One day at a time the landing was postponed, while frustrated Viking scientists searched for a smoother place to land. Finally, on July 20 the Viking 1 lander successfully made its atmospheric entry and settled onto a rock-strewn surface in Chryse Planitia, the Plains of Gold. It was exactly seven years after Neil Armstrong took the first human step on the Moon. Two months later the second lander touched down in a region called Utopia (Fig. 11.4).

All four Viking spacecraft—two landers and

FIGURE 11.4 A close-up view of Utopia from the Viking 2 Lander. One of the footpads of the lander is visible in the lower right corner of the frame. Note the many holes in the nearby rocks, indicating that they were full of gases when they were molten.

two orbiters—were spectacularly successful. This time the public followed the exciting developments as the first pictures were returned from the surface and the search for life on another world began. The *New York Times*, for example, eventually wrote nine separate editorials in praise of Viking. Most of the material presented in this chapter and the next is derived from the Viking mission.

Mars Exploration Since Viking

After Viking, the United States suspended its Mars program. The reasons are complex, with the primary problem being a general decline in funds for planetary exploration. Disappointment that Viking did not find life was also a major negative influence, together with the fact that Viking was a hard act to follow. Several new missions were studied, including a rover that could drive for hundreds of kilometers across the martian surface and a scheme to return samples of martian material for analysis, but funding was not available. In 1986 a presidential commission recommended that the United States set the eventual human exploration of Mars as a major goal of its space program, and in 1989 U.S. President George Bush formally committed his administration to this goal, but it has since been abandoned, at least for the decade of the 1990s. A similar Soviet commitment to human exploration of Mars died with the end of the USSR.

Although these more ambitious proposals have so far led nowhere, the scientific study of Mars continues. The final planetary mission launched by the USSR was to Mars and its satellite Phobos; this 1988 mission obtained some data but failed a few weeks after its arrival at Mars. In 1992 the United States launched an orbiter called Mars Observer, but this spacecraft failed even before it arrived at Mars. Yet in spite of these discouraging events, both Russia and the United States are pursuing their efforts. In 1996, the United States is sending a small lander called Mars Pathfinder, which includes Rocky, a dog-sized rover on a tether, with a range of about 10 meters from the lander. A series of small Mars orbiters, as well as further landers, are also planned to compensate for the loss of Mars Observer. Russia has designed larger landers, and by 2001 hopes to operate a 150-kg rover on the martian surface which can travel for tens of kilometers (Fig. 11.5). If public interest can be maintained, we may be on the threshold of a new era of martian exploration, consisting of relatively modest missions but with more frequent launches and a steady flow of new data.

11.2 GLOBAL PERSPECTIVE

Mars is a middle-sized terrestrial planet. Its diameter of 6787 km is just over half that of the Earth, resulting in a surface area almost exactly equal to that of the continents on our planet. The mass of Mars is 11% that of the Earth, or about nine times that of the Moon. Because it is intermediate in size between the Earth and the Moon, we are not surprised that it has an intermediate level of geological activity.

Bulk Properties

Mars also has a density intermediate between the Earth and Moon, at 3.9 g/cm^3. When corrected for the weight of the planet, the uncompressed density of Mars is 3.8 g/cm^3. This planet must be deficient in metal relative to the Earth (uncompressed density 4.5 g/cm^3). Perhaps Mars represents more nearly the original ratio of iron to silicate material in the inner part of the solar nebula, with the Earth having lost some of its silicate mantle in one or more giant impacts. Alternatively, it may be that there was a lower proportion of iron available in the part of the solar system where Mars formed.

Detailed tracking of orbiting spacecraft has revealed evidence for a core on Mars. It is thought to consist primarily of iron sulfide (FeS) and to have a diameter of perhaps 2400 km, or

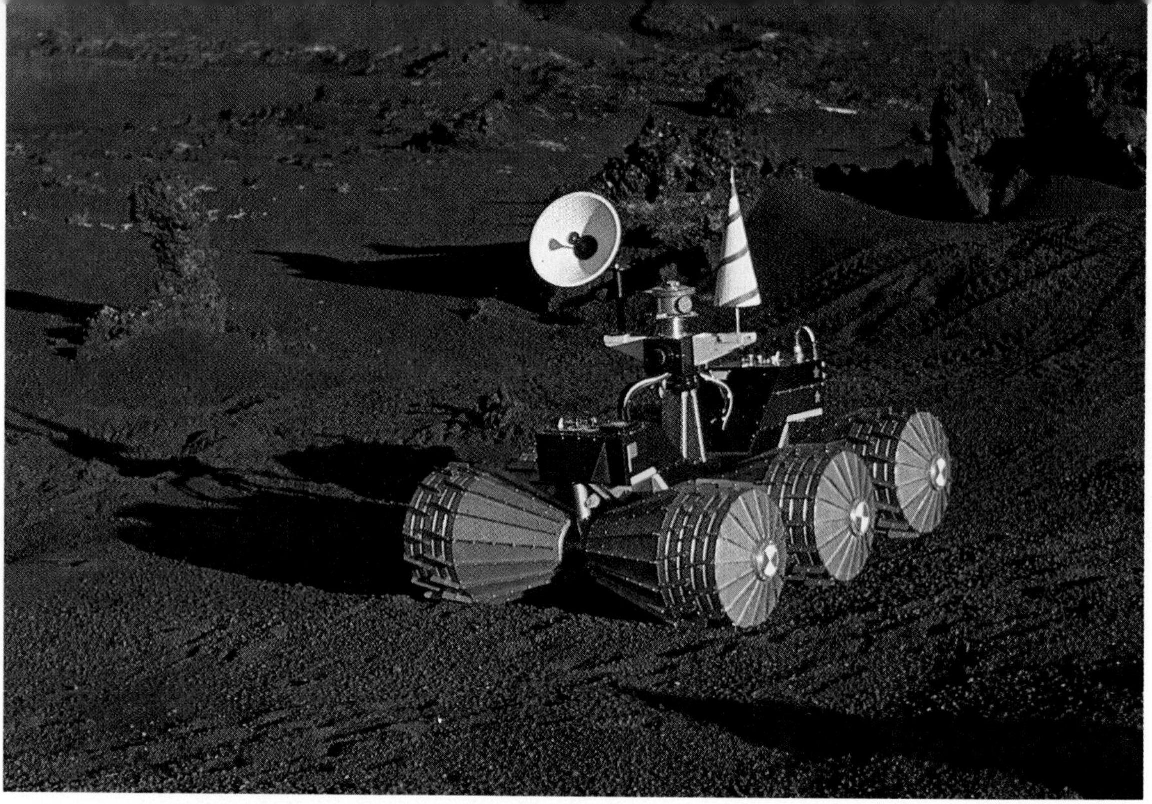

FIGURE 11.5 Test model of a Russian rover vehicle, planned for launch to Mars in 2001. The rover is shown here operating in volcanic terrain in Kamchatka.

40% of the diameter of the planet. If so, the cores on Mars and Earth take up a similar volume, relative to the size of the planet, with the main difference being that Mars's core contains a larger proportion of FeS, which is a lower-density material than metallic iron.

Calculations suggest that the core of Mars is solid, not liquid. If so, it should come as no surprise that Mars lacks the strong magnetic field that is generated on Earth by convective motions in its spinning, liquid iron core. Several Soviet orbiters, including the ill-fated Phobos spacecraft, have demonstrated that the magnetic field of Mars, if any, is exceedingly weak. There is no magnetosphere, except for the very small one created by the solar wind itself as it sweeps past the planet. Venus, Mars, and the comets are all examples of objects with atmospheres that interact directly with the solar wind.

As a differentiated planet, Mars has a mantle and a crust in addition to its iron sulfide core. Little is known about the properties of these layers, however. Although the two Viking landers each carried seismometers in hopes of probing the interior, no marsquakes were detected. One seismometer failed outright, while the other was incapable of distinguishing between seismic activity and trembling of the spacecraft due to wind.

Table 11.1 summarizes the bulk properties of Mars, in comparison with Earth and Moon.

Surface Nomenclature

A century ago, telescopic observers gave classical names (such as Hellas, Utopia, and Arabia) to the fixed light and dark features on Mars, which they thought might be deserts and seas, respectively. Before the first spacecraft visited Mars, scientists engaged in endless debates as to the nature of these features: for instance, whether they represented highlands and lowlands as do

TABLE 11.1 Comparison of Earth, Moon, and Mars

	Earth	Mars	Moon
Diameter (km)	12,756	6794	3476
Mass (Earth = 1)	1.0	0.107	0.0123
Density (g/cm^3)	5.5	3.9	3.3
Uncompressed density (g/cm^3)	4.5	3.8	3.3
Surface area (Earth = 1)	1.0	0.28	0.07
Atmospheric surface pressure	1.0	0.006	0.000

the lunar markings. To the surprise of many scientists, most of these markings turn out to have little to do with elevation or any other topographic property. Thus a new nomenclature was needed to deal with the topographic features photographed by Mariner 4 and its successors.

The spacecraft images revealed flat plains, mountain-ringed basins, canyons, valleys, volcanoes, and many impact craters. Following lunar convention, the craters were named for past scientists and others associated with the study of Mars. The great canyon system became Valles Marineris (the Mariner Valleys), named for the spacecraft (Mariner 9) that discovered it. (Actually, this feature had been seen from the Earth and called Coprates canal, the only canal that turned out to be real). The riverlike channels were given the names for the planet Mars in languages from all over the world: for example, Kasei Vallis (Japanese) and Nirgal Vallis (Assyrian).

Some large-scale topographic features on Mars correspond approximately to spots identified from Earth and given classical names. The largest martian volcano often collects bright clouds around it, and these clouds seen telescopically had been called Nix Olympica, the snows of Olympus. The volcano was therefore called Olympus Mons, Mount Olympus. Similarly, the largest martian impact basin also collects clouds in its interior and is one of the most prominent bright regions seen from Earth: Hellas. This name was applied to the basin itself, once its true nature had been established. A major lowland plain in the northern hemisphere had appeared on some old maps as Chryse Regio, or region of gold; now it became Chryse Planitia, plains of gold. In this way, much of the older nomenclature was preserved.

Surface Elevations

In an overview of a planetary surface, we begin with the largest-scale features, such as the continents of the Earth and Venus, or the highland and maria regions of the Moon. Thus we wish to know surface elevations. On Mars, these elevations are measured relative to the altitude at which the average atmospheric pressure is 0.0061 bar. This value was picked because it is the triple-point pressure of water, the pressure above which it is possible for water to be in liquid form. At its triple point, water can exist simultaneously as a solid, liquid, and gas provided the temperature is 0 C. Increasing the temperature even slightly at this low pressure causes the water to boil; lowering the temperature leads to freezing.

The range in elevations on Mars is very large, greater than that on any planet considered so far in this book. Four volcanic mountains rise to a height of 27 km above the reference level, while the lowest region, the Hellas basin, drops to 4 km below the reference level. Thus the total range in elevations on Mars is 31 km, compared with about 20 km on both the Earth and Venus

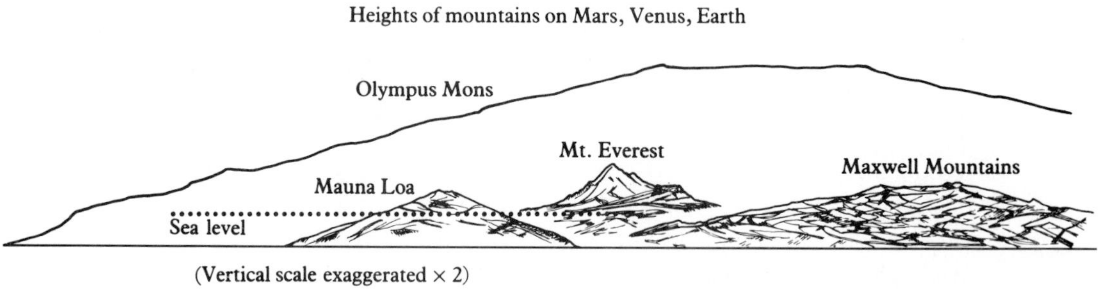

FIGURE 11.6 Comparison of highest elevations on Venus, Earth, and Mars.

(Fig. 11.6). The tallest individual mountain on Mars, Olympus Mons, rises 25 km above its surroundings, while the tallest mountains on Earth and Venus are only about 10 km high.

If you guessed that it is no coincidence that Venus and Earth have the same elevation range, in spite of the differences in their geology, you would be right. The explanation lies in the balance between the gravitational pull of a planet and the strength of its crustal rock. On planets the size of Earth and Venus, a mountain can rise only 10–15 km before its rock begins to deform under its own weight. A larger structure is simply unable to support itself, like a person who becomes too fat to walk.

If this self-limiting process is what determines the maximum elevations on Venus and Earth, we can understand why the martian mountains are higher. The surface gravity on Mars is two-fifths that of Earth or Venus, so its mountains can grow higher before they sag under their own weight. If there were similar volcanic mountains on the Moon they would probably be even taller, but the Moon never experienced mountain-building processes like those on Earth, Venus, and Mars.

Large-Scale Topography

Figure 11.7 is a map of Mars on which elevations have been indicated, measured in kilometers above (+) or below (−) the reference level. Two features are outstanding: a north-south hemispheric asymmetry; and the presence of a huge bulge near the equator at about longitude 100°.

Most of the southern hemisphere of Mars lies above the +3 km contour, while much of the northern hemisphere is below the 0 contour. In many places, the boundary between the two elevations is relatively sharp, with the surface dropping by 4 km in the span of a few hundred kilometers. A number of other properties of Mars also differ in the two hemispheres: in particular, the higher southern regions are more heavily cratered (hence older), and most of the darker regions are in the south.

The difference between the northern and southern hemispheres of Mars is one of the most fundamental and mysterious aspects of the planet. If the heavily cratered uplands of the south represent the older crust of the planet, then something peculiar has happened in the north, lowering the elevation by several kilometers and reworking the surface by volcanic or other processes. Whatever happened, it was more than a simple stripping away of the surface material, since there is no place to hide such a large quantity of excavated soil.

The second major feature of the large-scale topography of Mars is the Tharsis bulge, a volcanically active region the size of North America which rises about 10 km above the reference level. Tharsis straddles the boundary between

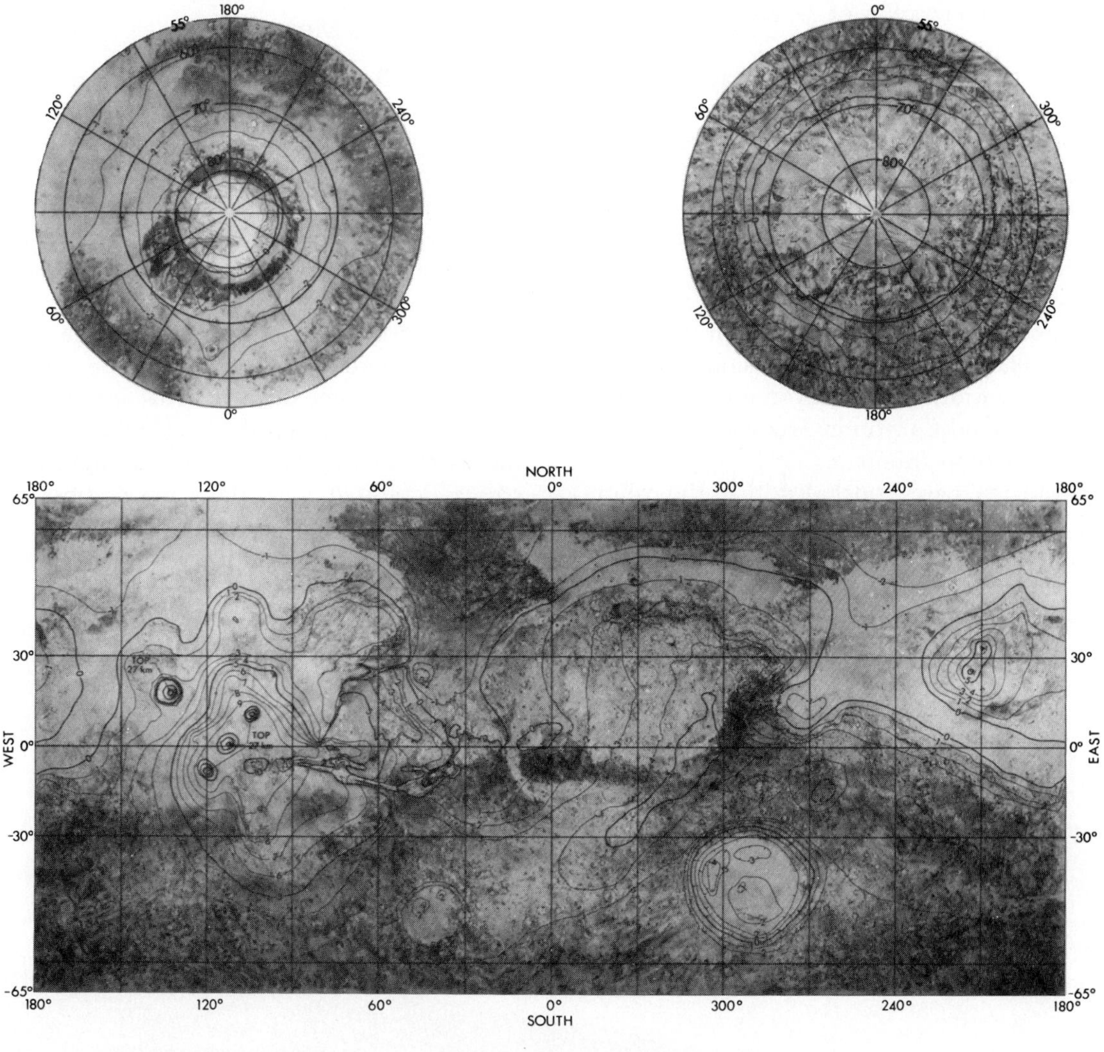

FIGURE 11.7 Map of the major topographic features of Mars as assembled from Mariner and Viking data. Note especially the Tharsis bulge near longitude 100° and the prominent boundary between the heavily cratered southern highlands and the smoother northern plains.

the southern uplands and the lower northern plains, and it is the least-cratered and therefore the youngest part of the planet's surface. To its east lies Chryse Planitia, which drops to an elevation of –3 km, while to the west is the shallower depression of Isidis Planitia. The deepest depression on Mars (6 km below its surrounding plains) is the Hellas basin, an ancient impact basin like Imbrium on the Moon.

Geologic Features

Near the middle of the Tharsis bulge are three great shield volcanoes, each rising 18 km to a summit at an elevation of 27 km. A still larger volcano, Olympus Mons, rises 25 km above the northwestern slope of Tharsis to reach the same height of 27 km above the reference level. Olympus Mons is nearly 700 km across at its base, about the size of France (Fig. 11.8).

Associated with the Tharsis bulge are the Valles Marineris, the central parts of an interconnected system of east-west-running canyons approximately 4000 km long. The individual canyons are about 3 km deep, but in the center of the Valles Marineris the depression is nearly 7 km and the width of the canyon more than 500 km. The Grand Canyon in Arizona could easily fit into one of its tributaries.

Geologists have concluded that the volcanoes, Valles Marineris, and the Tharsis bulge are all expressions of a major, long-lived center of volcanic activity that has dominated this one side of Mars. We will explore these ideas further in later sections of this chapter.

The Surface Material

Mars is the Red Planet, but what accounts for its color? The Viking landers confirmed what was suspected for decades: The color of Mars is due primarily to iron oxides in the surface soil. The soil itself is largely fine-grained sand or dust, which can be transported for great distances during major dust storms. Chemical analysis carried out by the Viking landers showed that the martian soil is similar in composition to some iron-rich clays found on Earth.

Most of the fine dusty material appears to be fairly light in color, like the sand in the deserts of the Earth. During dust storms, this light material is deposited over the surface, obscuring underlying rock, much of which is darker. Thus, after a major dust storm, the overall reflectivity of the martian surface is high. Subsequently, wind redistributes the lighter material, and part of the darker underlying terrain reappears.

FIGURE 11.8 Oblique Viking view of the Olympus Mons shield volcano, partly surrounded by high-altitude clouds.

This process of wind-blown dust, distributed over the surface by seasonal circulation patterns of the atmosphere, accounts for the seasonal changes in the surface markings recorded by generations of telescopic observers. Among the first scientists to advocate this nonbiological explanation for the surface changes were Carl Sagan and James Pollack, whom we encountered in the last chapter when discussing the Venus greenhouse effect. It is somewhat ironic that Sagan, who is so closely associated with the search for life on other planets, should have played this role in the history of martian studies.

Martian Samples

Although we have not sent astronauts to collect rocks from Mars or automatic sample collection spacecraft such as the Soviet Luna vehicles, we do have a handful of martian samples to study. These are the SNC meteorites, which we briefly discussed in Section 4.5. Most of the SNC meteorites are basalts with solidification ages of 1.3 billion years, presumably derived from one or more impacts in martian volcanic areas. One, however, is much older, dating back to the formation of the early martian crust. We are confident that they come from Mars in part because they preserve tiny bubbles of gas that match perfectly the analysis of the martian atmosphere carried out by Viking.

Like the lunar basalts, the SNC meteorites are derived from the upper mantle by partial melting. Detailed chemical analysis of these rocks, together with Viking data and remotely sensed information on the composition of the martian soil, permits some educated guesses concerning the chemistry of the crust of Mars. Ratios of isotopes of elements in the SNC meteorites can also be used to compare the bulk properties of Mars with those of the Earth and Moon.

One of the most interesting results from analysis of martian samples concerns the amount of water that may once have been present on Mars. On Earth, a major source of water is to be found in volcanic magmas, which carry water from the mantle and deposit it near the surface, where it is released to the atmosphere by the weathering of lavas. The martian magmas, as represented by the SNC meteorites, include about 1.4% H_2O (dryer than terrestrial magmas). Combined with the observed extent of volcanism on Mars over the past 3.9 billion years (the time since the heavy bombardment), this value indicates release of enough water from the interior to cover Mars to an average depth of 200 m. This is just a lower limit for the amount of water on Mars, since we presume that additional water was released earlier from volcanism as well as arriving in forms of impacting comets. With the discovery of one much older SNC meteorite, it may be possible eventually to trace the story of martian water to early times in the planet's history.

11.3 VIEW FROM THE SURFACE

Although no human has yet set foot on Mars, we have sent our surrogates in the form of the Viking landers. Through the eyes of these spacecraft we have looked more carefully at the surface of Mars than at that of any other planet beyond the Moon.

The Viking Landers

The Viking landers were each self-contained laboratories weighing about a ton and similar in size to a subcompact car. Powered by a radioactive electric generator, each spacecraft could communicate with Earth either directly or via one of the orbiter spacecraft. The scientific experiments on board included those aimed toward a general examination of the immediate environment (cameras, chemical analysis devices, a meteorological weather station, a seismometer, and a 3-m-long mechanical arm to probe and manipulate the soil), as well as a complex biology package to look for evidence of microscopic life on Mars. In addition, a special group of instruments investigated the atmosphere of Mars during spacecraft entry and descent.

Twin cameras on each Viking lander could record pictures in either color or monochrome, and operating together they provided stereoscopic views of the surface. The weather station measured atmospheric pressure, temperature, and wind speed and direction. Chemical analysis of soil was carried out with an x-ray spectrometer. An even more powerful general analysis device called the GCMS (discussed in more detail in Chapter 12) measured the composition of the atmosphere and of volatile components in the soil, such as organic molecules.

Each Viking lander was operated by controllers on Earth, who could command it to poke and dig in the soil, push rocks around, pick up and deliver samples to its various instruments, and take a wide variety of pictures of its surroundings. The Viking landers lacked mobility, however. Fixed at the spots where they landed, they could not show us the view over the next hill, or even behind a nearby rock. Thus the choice of the two places on Mars where the spacecraft would land was extremely important.

Selection of the Viking Landing Sites

The one imperative in selecting landing sites on Mars was safety. Just as had been the case for the first Apollo landings on the Moon, every effort was made to find dull, flat locations for Viking. This meant avoiding all the remarkable geologic features found by Mariner 9: No sites were considered involving volcanoes, canyons, channels, large impact craters, or anything else of geological interest. In addition, it was necessary to select sites at relatively low elevations, to make sure there was enough atmosphere to reduce speed of the descent. The only places that met these criteria were the lowland plains of the northern hemisphere.

The site selected for the Viking 1 landing was in Chryse Planitia, a depression east of Tharsis. In Mariner 9 images this region appeared flat and featureless, but the higher-resolution pictures from the Viking 1 orbiter revealed a much more complex topography, including features that appeared to be the result of water erosion. The Viking project manager rejected the preplanned site and began a search for a smoother

FIGURE 11.9 An orbital view of the site of the Viking 1 landing in Chryse, a low basin characterized by wrinkle ridges similar to those of the lunar maria.

area. It required nearly a month to identify a less rugged looking alternative, still in Chryse but several hundred kilometers northwest of the original site (22° N, 48° W).

Viking 2 was aimed for a more northerly latitude of 44°, where seasonal effects might be stronger. Again, the original site proved to be unexpectedly rough, and a search was made for a smoother area at the same latitude. A spot in Utopia, at longitude 226° W, was found. Utopia is a plains area northwest of Elysium.

The Plains of Gold

The Viking 1 site in Chryse is illustrated in Fig. 11.9. This orbiter view shows a marelike plain with a crater density suggestive of an age of about 3 billion years. Like the lunar maria, this area is thought to be volcanic, but with a surface possibly modified by later floods of water (see Section 11.6) as well as by wind erosion.

As seen from the lander (Fig. 11.10), Chryse is a desolate but strangely beautiful landscape not very different from some terrestrial deserts. The topography is gently rolling, and the surface is thickly strewn with rocks in sizes ranging from golfballs up to boulders. The absence of smaller rocks distinguishes this site from those on the lunar maria. The soil between the rocks is fine-grained and has a relatively hard, crusty surface.

The ubiquitous rocks in the Viking 1 site appear to be volcanic in origin, and they probably are ejecta from impact craters, just like their lunar counterparts. The scene has been modified by erosion, particularly by wind-blown dust. Erosion takes place slowly, however, and only a keen observer can see any differences between photos taken years apart at the Viking 1 site.

FIGURE 11.10 The view from the Viking 1 Lander showed a desolate plain with wind-sculpted hills and numerous rocks of all sizes.

Utopia

The landing site of the second Viking is shown in Fig. 11.11. Although the intent was to aim for an area with a smooth wind-deposited surface, it appears that the spacecraft landed instead on ejecta from the 90-km-diameter crater Mie, located about 200 km to the east. As seen from the lander (Fig. 11.12), this site is even rockier than Chryse. Indeed, we are fortunate that the lander did not wreck itself on a protruding boulder as it settled to the surface.

Viking 2 found itself in a very flat area, lacking the low hills that make the Chryse site more attractive. About all there is to see are the rocks. These are interesting enough to the geologist, however, with their distinctive angular shapes and a wide variety of colors and textures. Most are heavily pitted, perhaps an indication of their volcanic origin. The finer material that must once have been present in the ejecta from Mie Crater has been stripped away by wind, leaving the fragmented rocks.

While every effort was made to land both Vikings in flat regions with smooth surfaces, it is apparent from the pictures that this goal was not achieved. Both sites are rocky, with evidence of wind erosion. Presumably the fine material swept clear from these areas is deposited somewhere else, perhaps near the poles, where large fields of sand dunes have been photographed from orbit (Section 11.7). In any case, it may be that the two Viking sites are more typical than had been expected, and that a great deal of the surface of Mars consists of rock-strewn plains like these.

FIGURE 11.11 The Viking 2 landing site in Utopia, a more northerly plain that in this view from orbit appeared to be mantled in wind-driven dust.

The Thomas Mutch Memorial Station

Each Viking lander had a nominal lifetime of only 90 days, just sufficient to complete the life-detection experiments described in Chapter 12. However, scientists and engineers alike hoped for much longer operations on the martian surface, and they were not disappointed.

Viking Lander 2 operated in the Utopia plain until April 11, 1980, spanning two martian years. In spite of its far northern latitude, the vehicle survived two cold martian winters, providing unique meteorological information.

The first lander performed even better, and at the end of the regular Viking mission it was still in excellent condition. NASA then placed the lander into an autonomous operating mode, returning data to Earth only about once a month. This lander was designated the Thomas Mutch Memorial Station, named in honor of the Viking Imaging Team leader.

Thomas ("Tim") Mutch (Fig. 11.13), a geologist from Brown University, made the transition from studying the Earth to the Moon (during the Apollo era) and from there to Mars. In 1979 he accepted the highest-level science-man-

FIGURE 11.12 The Viking 2 Lander touched down in a plain heavily littered with rocks, probably derived from a nearby impact crater. There was no evidence of the mantle of dust that had been expected.

agement post at NASA, becoming Associate Administrator for Space Science. A little more than a year later, Tim Mutch died in a mountain-climbing accident in the Himalayan Mountains.

It was hoped that the Mutch Memorial Station could continue operations until the 1990s, but it fell victim to a human programming error that misdirected its antenna away from the Earth, breaking the vital communications link with controllers at JPL. The last data from Mars were received on November 5, 1982, a total of six Earth years and 106 Earth days after landing.

In 1984 NASA formally transferred ownership of the Viking 1 lander to the Air and Space Museum of the Smithsonian Institution. It is the only museum exhibit on another planet. Someday, perhaps, it will be brought back to a place of honor in Washington, next to the Wright brothers' first airplane and the Apollo 11 spacecraft. Or perhaps it will remain on Mars, for the benefit of the first human colonists and the generations that succeed them.

FIGURE 11.13 Thomas (Tim) Mutch was the highly popular leader of the Viking Lander Imaging Team.

It is appropriate to end this section with two quotes from Tim Mutch, one made at the time of the Viking 1 landing, the other after he became NASA associate administrator:

> *"Touchdown minus seven, six, five, four, three, two, one, zero." There was a long silence. It seemed like minutes. Over the loud-speaker came a muffled prayer, "Come on, baby." Nothing. I looked down at my shoes. I remember thinking, "You always wondered what a failure would feel like. Now you know." I mentally composed some remarks for friends standing nearby. Finally, the silence was broken. "We have touchdown."*

> *Even if the immediate future is uncertain, I have no doubts about the distant years. Some day man will roam the surface of Mars. Those wonderful Viking machines will be crated up, returned to Earth, and placed in a museum. Children in generations to come will stand before them and struggle to imagine the way it was on that first journey to Mars.*

11.4 CRATERS AND CHRONOLOGY

Like the Moon and Mercury, Mars has thousands of impact craters, concentrated particularly in its southern hemisphere. Generally speaking, these craters look much like their counterparts on airless planets, with the same sort of raised and terraced rims, flat depressed floors, and often central peaks. Presumably they were formed by the same impact processes. From the crater densities, which are substantially higher than those of Venus and the Earth, we conclude that the martian uplands date back to the first billion years of solar system history.

Martian Impact Craters

The martian impact craters themselves look like their lunar counterparts, but their patterns of ejecta are unique. On the Moon, a typical large crater is surrounded by a rough, hilly ejecta blanket close to the rim, with radial streaks and chains of secondary craters farther out. In contrast, many martian craters have smooth ejecta blankets with well-defined edges. There are no extended rays or streaks of material extending to great distances. These have been called **fluidized ejecta** craters or, more informally, splosh craters.

The martian crater ejecta are of three main types. Most common is a multilobed pattern, called a flower form. Several ejecta layers are present, with each lobe bounded by a steep, smooth edge (Fig. 11.14a). These flower craters are found primarily in the equatorial regions, within 40° of the equator. In the second type, illustrated in Fig. 11.14(b), the ejecta looks like a relatively smooth pancake of material with a slightly scalloped rim. We call these pancake craters, and they occur primarily at higher latitudes. In the third type, most of the ejecta is missing, leaving the crater perched on a relatively small elevated platform. These pedestal craters are probably the result of erosive action by wind, which has stripped away the outer layers of ejected material.

Evidently the martian ejecta blankets were formed by fluid debris that flowed along the ground rather than following explosive aerial trajectories. This hypothesis of fluid flow also explains a number of cases where the ejecta flowed over and around obstacles. The most likely explanation for the fluidized ejecta of martian craters lies in the presence of large quantities of water in the surface at the time the impacts occurred. If martian conditions then resembled those we see today, the water was in the form of ice until melted by the energy of the impact.

The martian craters are distinct from those of both the Moon and Venus. On the Moon there was no fluid flow because neither water nor an atmosphere was present. On Venus, in contrast, the high surface temperature and massive atmosphere leads to some fluidization, but the result looks quite different from the martian case. Only on Mars does ice vaporized by the impact domi-

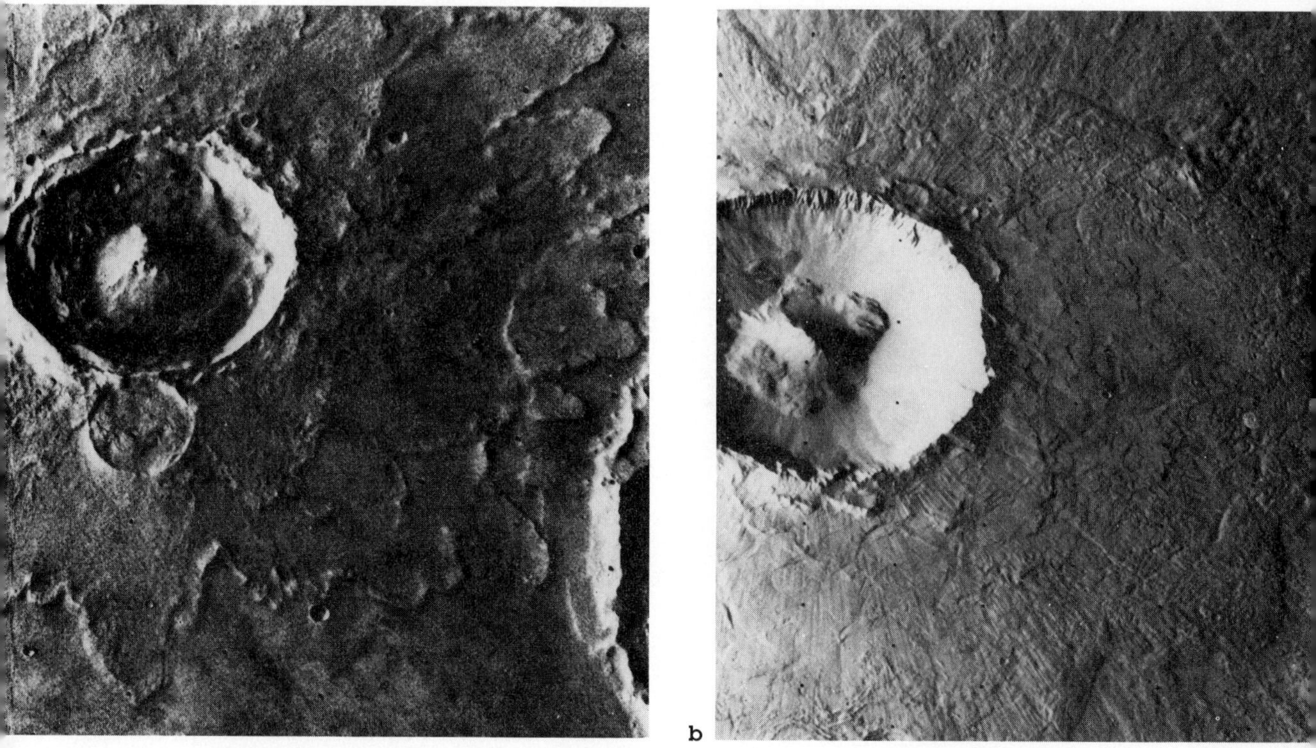

b

a

FIGURE 11.14 The unusual ejecta blankets of martian impact craters probably result from the presence of frozen water (permafrost) in the soil. (a) Yuty, an 18-km crater at 22° N, showing a central peak and several thin sheets of ejecta, with multiple lobes. (b) Arandas, a 28-km crater at 43° N, with a thick platform of ejecta terminating in a flow front.

nate in the emplacement of the ejecta blankets. Ice is also plentiful in polar terrains on the Earth, so possibly large impact craters near the poles of our planet would look more like the martian than the lunar or venusian craters, but the ejecta from such terrestrial craters have not survived to allow us to test this hypothesis.

Impact Basins

The largest martian basin is Hellas, about 1800 km in diameter and some 6 km deep. Unlike its lunar counterparts, Hellas has a simple form with a single rim of uplifted mountains. Its interior, which is frequently hidden by hazes of dust or water fog, appears to be relatively smooth, suggesting that the original floor has been buried by wind-blown dust. Parts of the rim mountains are missing. The second largest identified basin, Isidis, approximately 1200 km across, is eroded even more than Hellas.

The best-preserved large basin on Mars is Argyre (Fig. 11.15). Its diameter is roughly 700 km. The rim consists of a broad rugged area more than 200 km wide made up of mountains thrown up in the basin-forming impact. The next largest basins are only about half as large as Argyre, making them really no more than large craters. An example is the 210-km diameter crater Galle, shown in Fig. 11.15 lying on the rim of Argyre.

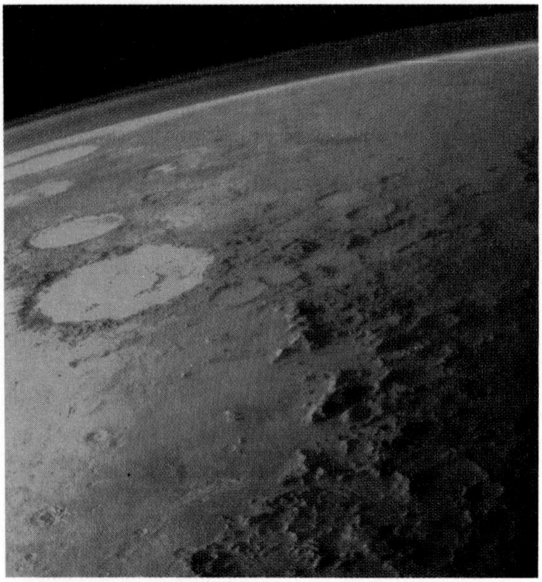

FIGURE 11.15 The Argyre impact basin, about 700 km in diameter. This Viking photo also shows haze layers in the atmosphere of Mars. The 210-km crater Galle is also visible superposed on the basin rim.

Mars has fewer large basins than the Moon, in spite of its much greater surface area. What does this mean? One possibility, reflecting a uniformitarian outlook, is that many of the basins that formed early in martian history have been degraded beyond recognition by subsequent geologic activity, including extensive erosion. Another way of stating this idea is that the martian crust stabilized later than that of the Moon, near the end of the terminal heavy bombardment. This would be consistent with the larger size of Mars, leading to slower cooling of the interior.

Age of the Martian Surface

When was the martian surface formed? As we have seen in our studies of the Moon, Mercury, and Venus, this question is fundamental to understanding the history of the planet. Unfortunately, we have no returned samples with which to measure the solidification age (except the SNC meteorites, but we don't know exactly where they originated). Thus any age estimate must depend on a hypothesis or model for the impact history of the planet. It is common to use the lunar history with a small modification in the flux to allow for the greater proximity of Mars to the asteroid belt.

A wide variety of crater densities exists on Mars. Some areas, such as the summits and slopes of the Tharsis volcanoes, have very few craters. The crater densities in these areas are substantially less than on the youngest lunar maria. In contrast, the densities on the older cratered uplands are higher than on the lunar maria, although still less than on the lunar highlands. This result dates the uplands to ages somewhere close to 4 billion years.

FIGURE 11.16 The cratered uplands that characterize most of the southern hemisphere of Mars. The craters are generally subdued by erosion, with a notable absence of small craters, relative to similar views of the Moon and Mercury. This frame is about 500 km across.

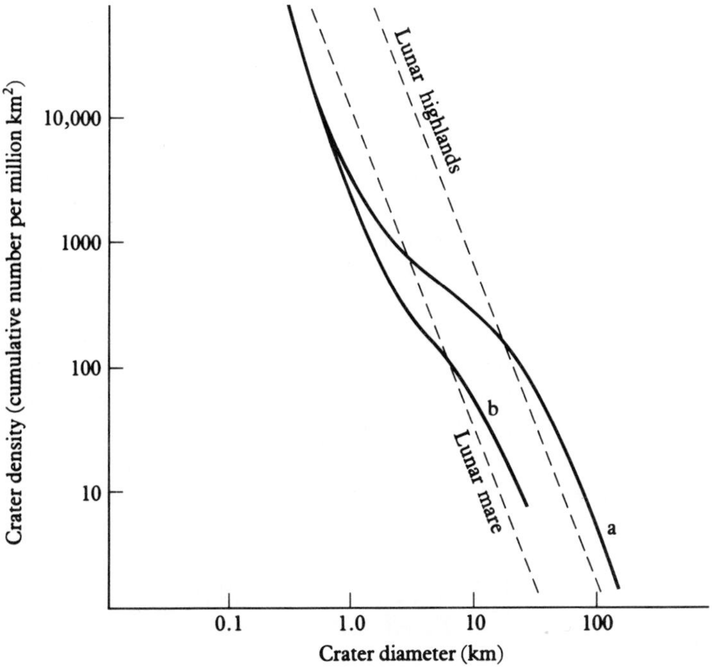

FIGURE 11.17 The distribution of crater sizes in the martian uplands and in the Chryse Basin, compared with crater densities on the Moon. Note the depletion of small craters on Mars, presumably the result of erosion. Curve *a* is for average martian uplands; curve *b* shows the data for Chryse.

One fundamental difference between the Moon and Mars must be considered when interpreting crater densities: erosion. On the Moon the only important agents that degrade old craters are further impacts. Mars, however, has an atmosphere. Even today, the erosive effects of martian windstorms are significant, and in the past it seems likely that the atmosphere was thicker. At that time, even running water and ice, with their major erosive effects, may have been present.

The effects of erosion are particularly evident in the most heavily cratered uplands (Fig. 11.16), where the large craters are shallow and subdued. In these areas, there are many fewer small craters than would be expected from the numbers of larger ones present. Figure 11.17 also displays this effect, plotting the density of martian craters of different sizes measured on both the uplands and the lower northern plains. The shortage of craters with diameters less than about 5 km is attributed to enhanced erosion early in the history of Mars. The fact that smaller craters are not in short supply relative to large ones in the younger terrains suggests that this era of high erosion rates terminated billions of years ago.

Venus also has few small craters, but in Chapter 10 we attributed this effect on Venus to the shielding effect of the atmosphere. Why is the explanation different on Mars? To produce a crater cutoff at 5 km requires an atmospheric pressure of about 1 bar (like the Earth), which is unlikely but certainly not impossible for Mars at the time the uplands were being cratered. It is indeed possible that the atmosphere played such a role on Mars. However, the presence on Mars of shallow degraded small craters strongly suggests that erosion is the primary cause of the missing craters. This situation is in dramatic contrast to that of the craters of Venus, which show no erosion at all at the resolution of the Magellan radar images.

In the northern plains the density of craters is much lower than in the uplands, as shown in the crater counts plotted in Fig. 11.17. The best-studied northern plains area is Chryse Planitia,

TABLE 11.2 Relative cratering rates for the terrestrial planets*

Mercury	Venus	Earth	Moon	Mars
2.0	1.0	1.5	1.0	2.0

*The higher value for Mars is due primarily to its proximity to the asteroid belt, while that of Mercury is a result of the Sun's gravity.

the place where Viking 1 landed (Fig. 11.9). Chryse is a flat basin of apparently volcanic origin. For craters with diameters between 1 and 10 km, the crater density on Chryse is virtually identical to that of the average lunar mare, while for larger crater sizes the density on Chryse is somewhat greater. Similar crater densities apply to the other northern lowlands.

Interpreting the Crater Densities

If the craters on Mars and the Moon are caused primarily by impacts with the same population of asteroids and comets, the cratering rates on these two bodies must have been roughly the same. To refine this result, actual impact rates must be calculated and allowance must be made for the differing impact speeds on individual planets. Shown in Table 11.2 are results of *one* such cratering model, calculated in 1981 by William K. Hartmann of the Planetary Science Institute in Tucson. Here cratering rates are expressed relative to the Moon.

Hartmann's model provides one way to date different martian terrains from their observed crater densities, if we assume the lunar cratering history, including a terminal heavy bombardment. We will use these age estimates throughout this chapter when discussing various features on Mars. However, note that cratering models developed by other scientists yield ages that differ by as much as a factor of two for the younger volcanic terrains. All investigators, however, agree that the oldest terrains date from about 4 billion years ago. Table 11.3 summarizes the results for some representative surface features.

11.5 VOLCANOES AND TECTONIC FEATURES

Like the other terrestrial planets, Mars has experienced extensive periods of volcanism. Most of the low-lying northern hemisphere appears to have been covered by mare-type flood basalts in the distant past. The most spectacular evidence of volcanism, however, is found in the shield volcanoes of Tharsis, which are the largest individual volcanic structures in the solar system.

TABLE 11.3 Crater ages for martian surface features

Feature	Crater Density Relative to Lunar Mare	Crater Retention Age (billion years)
Olympus Mons volcano	0.1	0.2
Arsia Mons volcano	0.1	0.2
Tharsis plains	0.5	1.6
Elysium plains	0.7	2.6
Chryse Planitia	1.1	3.2
Alba volcano	1.8	3.5
Hellas basin	1.8	3.5
Cratered uplands	10	4.0

Shield Volcanoes

The largest volcanoes of Mars are concentrated in the Tharsis and Elysium regions, with a few additional older examples in the southern hemisphere near the Hellas basin. These are all shield volcanoes, formed by the eruption of relatively fluid basaltic lavas, which build up massive domes with gentle slopes. Most shield volcanoes on both Earth and Mars are also distinguished by summit craters or calderas, produced by collapse of the surface when the underground pressure of the magma is withdrawn.

The four highest volcanoes, all in the Tharsis region, rise to 27 km above the martian reference level. The story of their discovery is an interesting one. As we noted in Section 11.1, Mars was shrouded by a planetwide dust storm when the Mariner 9 orbiter arrived in 1971. At that time, there was no reason to expect large volcanoes on Mars. As the dust began to clear from the upper atmosphere of Mars, the first surface features to appear were the tops of the four highest volcanoes, each with its large summit crater. Thus scientists faced an essentially featureless planet, with four faintly visible large craters.

On the basis of just this limited information, Harold Masursky of the U.S. Geological Survey reached the correct conclusion that Mars had volcanoes and that these volcanoes were much larger than their terrestrial counterparts. He thought that if this were not the case the coincidence of large craters being found at just the four spots that first emerged out of the dust was too unlikely. Note that this is just the reverse of the reasoning used by Gilbert to argue against volcanic craters on the Moon: Gilbert concluded that since the craters of the Moon were not on the tops of mountains they were not volcanic, while Masursky reasoned that if the craters were on the tops of mountains, they were volcanic. Subsequently, as the atmosphere cleared to reveal the great shield volcanoes, Masursky's insight was confirmed.

Tharsis and Olympus Mons

Each of the three Tharsis shield volcanoes is about 400 km in diameter, and they all rise to the same height. The southernmost volcano, Arsia Mons, has the largest caldera, 120 km in diameter (Fig. 11.18); the other two calderas are less than half as large. All three calderas are 3–4 km deep. Individual lava flows can be traced on the gently sloping flanks of these volcanoes; impact crater densities on the flanks and in the calderas suggest that these volcanoes have been active within the past few hundred million years.

FIGURE 11.18 The west flank and caldera of the Arsia Mons volcano as photographed by the Viking orbiters. This caldera is the largest on Mars, with a diameter of 120 km.

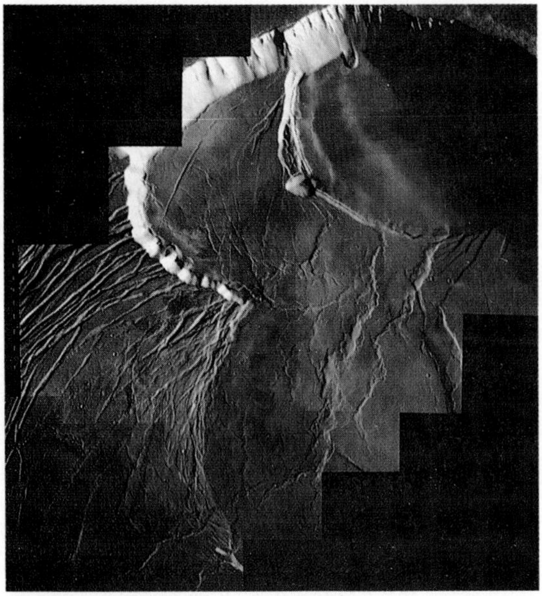

FIGURE 11.19 The caldera of Olympus Mons has a diameter of about 80 km. It consists of several interconnected, flat-floored depressions, similar to the calderas of terrestrial shield volcanoes.

Olympus Mons, which rises from a lower area on the side of the Tharsis bulge, is the largest volcano on Mars (Fig. 11.19), or, indeed, on any planet. Its diameter is 700 km, and the entire chain of Hawaiian volcanoes, from Kauai to the Big Island, would fit within it. Olympus Mons lies to the northwest of the main Tharsis bulge, and its summit is 25 km above the surrounding plains. The volume of this immense mountain is nearly 100 times greater than that of the largest volcano on Earth, Mauna Loa. The lava flows that make up its flanks slope downward at an angle of only about 4°. Individual channels about 100 m in width can be traced, together with many flow features that are virtually indistinguishable from their counterparts in the Hawaiian volcanoes.

Other Martian Volcanoes

The four volcanoes we have been discussing are the largest on Mars, but even modest martian volcanoes are huge by terrestrial standards. Figure 11.20 shows two shield volcanoes in the Tharsis region, each about 100 km in diameter. Hundreds of individual volcanoes have been identified in the high-resolution Viking surveys of the planet.

The older volcanoes on Mars tend not to be very high, perhaps as a result of subsidence of the surface. The best example is Alba, an immense ancient volcano north of Tharsis. Crater counts suggest that this feature may be 3–4 billion years old, nearly as old as the lunar maria. Alba is 1600 km across, with four times the area of Olympus Mons and nearly 200 times the area of Mauna Loa. However, it is only a few kilometers in height.

Shield volcanoes can be identified on Mars down to the resolution limit of the Viking surveys, about 5 km. At small sizes we also see many steep-sided cones apparently formed by more explosive eruptions. However, there are no viscous eruptions such as the pancake domes of Venus, nor any examples of larger composite volcanoes like those of Earth.

Volcanic Plains

Much of the northern hemisphere of Mars consists of plains, with about the same crater density as the lunar maria. Both of the Vikings landed on such plains, in the Chryse and Utopia basins. Three distinct types of plains can be identified, all apparently the result of volcanic processes.

The lower slopes of Tharsis and its surroundings are the youngest plains on Mars, judging from their low crater densities. They are clearly of volcanic origin, and they exhibit many individual lava flows, 20–30 km thick, some hundreds of kilometers in length. Most of the lava

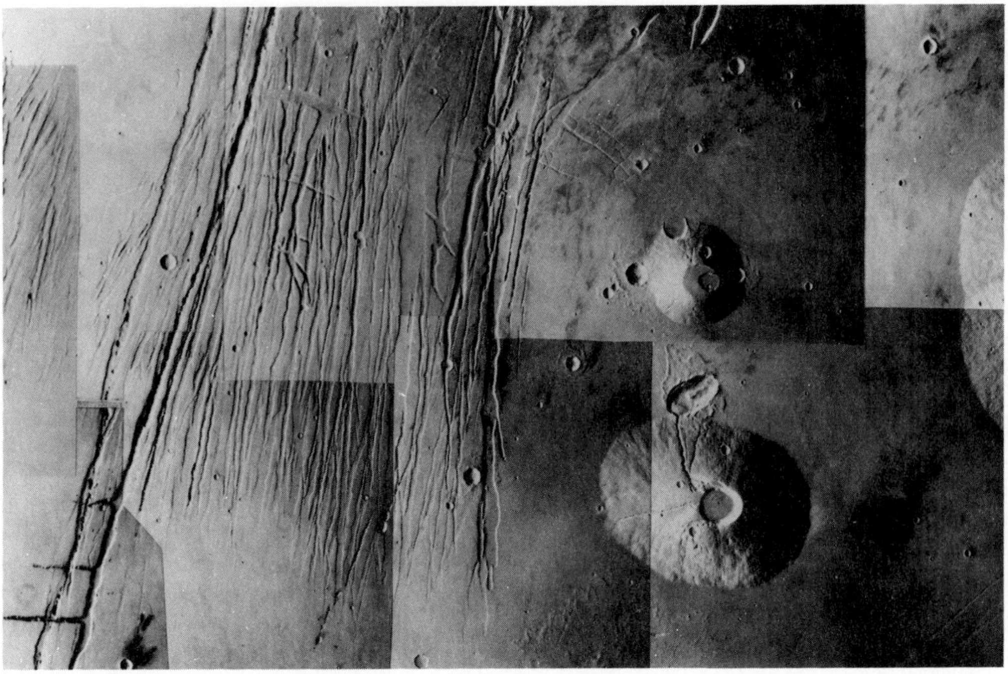

FIGURE 11.20 Two of the intermediate-sized Tharsis volcanoes are shown in this Viking orbiter mosaic of photos, together with a region of extensive tectonic fracturing associated with the formation of the Tharsis bulge. The width of this frame is about 600 km.

vents were in the plains themselves and not part of the shield volcanoes.

The second class of volcanic plains, which includes the Viking 1 site in Chryse, are called ridged plains. Their low domes and lava wrinkle ridges are similar to those of the lunar maria and some of the plains of Venus.

The plains at higher northern latitudes, above 30° N, are very different in appearance, partly as a result of wind erosion. The most striking of these northern plains show a variety of strange surface patterns: wrinkles, cracks, and low ridges. Viking 2 landed near some of these regions. Figure 11.21 shows a remarkable area with polygonal fractures, spaced a few kilometers apart. The origin of most of these features is poorly understood. Superficially similar polygonal fractures are found in arctic soils on Earth, where they are known as patterned ground, but it is very rare to find this terrain at the large scales exhibited on Mars.

Tectonics and the Tharsis Bulge

How does the Tharsis bulge compare with the continental masses of Earth and Venus? In part, the Tharsis bulge results from the accumulation of volcanic lavas over billions of years, but in addition it appears to be the product of tectonic forces that have bent the crust and uplifted an area as large as North America or Aphrodite.

Dramatic evidence of the tectonic forces that

FIGURE 11.21 Many of the northern plains areas of Mars have a strange patterned appearance when photographed from orbit. Individual ridges are separated by distances of a few kilometers. These fracture patterns have no analog on Earth and remain poorly understood even 25 years after Viking.

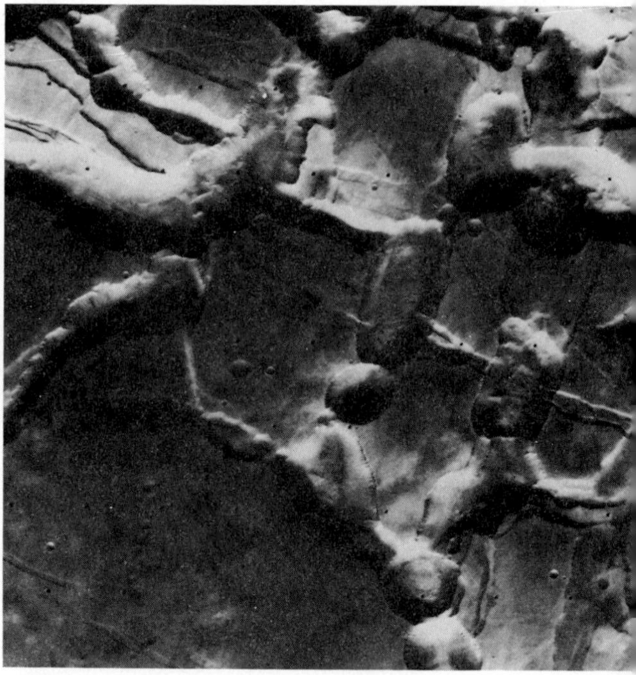

FIGURE 11.22 Noctis Labyrinthus, a region of intersecting closed depressions possibly caused by a combination of tectonic fracturing and collapse associated with the melting of permafrost. The frame is 120 km across.

shaped Tharsis is provided by the extensive fracturing of the surrounding crust (Fig. 11.20). Most of these fractures, which can be hundreds of kilometers long and a kilometer or more in width, point away from the center of the Tharsis bulge. Detailed studies of the apparent ages of these tectonic features based on crater densities indicate that the uplift began about 3 billion years ago and that the surface fracturing had tapered off by 2 billion years in the past. If these studies are correct, the uplift largely preceded the formation of the large Tharsis volcanoes.

The continents of Venus and the Earth are the product of compressional forces, resulting in folded and uplifted mountains (like Maxwell or the Himalayas) and plateaus (like Lakshmi or Tibet). The Tharsis continent on Mars more nearly resembles the top of a mantle plume, perhaps analogous to an upscaled corona structure on Venus. This close association of tectonic uplift with massive volcanism is unique to Mars. Thus Mars alone of the terrestrial planets has volcanoes, rather than tectonic mountains, as its tallest features.

Martian Canyons

The most spectacular tectonic features on Mars are the great equatorial canyons, especially the Valles Marineris. These canyons, each as much as 100 km wide and up to 7 km deep, stretch a quarter of the way around the planet. The upper parts of the canyon system, near the summit of Tharsis, are called Noctis Labyrinthus

(Labyrinth of Night) and consist of short intersecting segments that produce a labyrinth (Fig. 11.22). Farther down, the Valles Marineris descends the eastern flank of the bulge and extends into the old, cratered uplands east of Tharsis (Fig. 11.23 and chapter-opening photo).

The term *canyon* is really a misnomer, as the martian canyons were not carved by rivers. They contain no features characteristic of running water, nor do they terminate in lowland basins. They are also distinct from the channels, which are discussed in the next section. The canyons are tectonic in origin, basically huge cracks in the planet's surface, which subsequently have been widened and shaped by erosion. Figure 11.24 shows a detail in which several landslides tens of kilometers across can be seen. Note on the far wall that the upper part of the slide consists of blocky material, while the lower part of the slide shows evidence of a more fluid flow across the canyon floor, probably at speeds greater than 100 km/hr. Just as in the case of the fluidized impact crater ejecta, subsurface ice may have helped to lubricate these flows.

Landslides and possibly undercutting water springs have played a major role in widening the original fractures, but where has the eroded material gone? Perhaps some of the missing material was water ice, which is believed to make up a sizable fraction of the crust. Alternatively, the fine material may have been carried away by winds whistling down the canyons.

FIGURE 11.23 A small section of the heavily eroded martian canyonlands. A broader color view of this same area is shown on the chapter-opening page.

FIGURE 11.24 Landslides in Ganges Chasma in the martian canyonlands. Note the flow patterns in the large fan-shaped slides, and the indication of a harder caprock at the top of the canyon wall. Width of the frame is about 100 km.

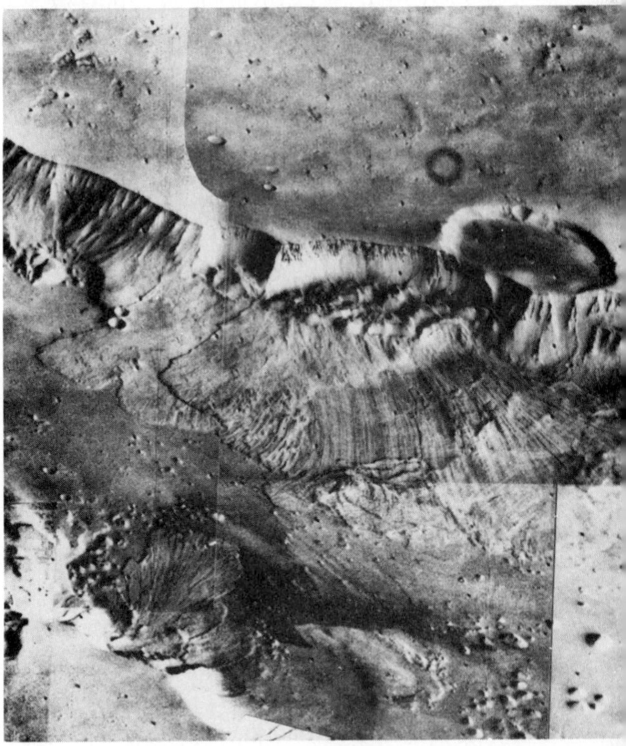

The Nature of Martian Volcanism

Detailed studies of the martian volcanoes have been carried out by geologists such as Michael J. Carr of the U.S. Geological Survey, who was leader of the Viking Orbiter Imaging Team. Even without surface measurements of chemical composition, Carr and his colleagues have concluded from the shapes of the volcanoes and the behavior of their flows that they were produced by basaltic volcanism similar to that which formed the Hawaiian volcanoes. Individual flows extend for great distances, indicating that the lavas had fairly low viscosity and that the rate of flow during eruptions was large. However, there are no equivalents to the remarkable lava rivers of Venus.

Estimates of ages of different flows suggest that the eruptions spread over vast time spans, with individual volcanoes active for hundreds of millions or even billions of years. Most of the SNC meteorites, with their solidification ages of 1.3 billion years, presumably are samples of some of these flows, probably from the Tharsis area. The scar from the impact that ejected these samples from Mars must be visible, but of course we do not know which crater it was.

The concentration of volcanoes in the Tharsis region is consistent with the idea that the Tharsis bulge is the result of a mantle hot spot. In this case, the same forces that raised the bulge provided the heat to generate the volcanoes. It is also easy to see why the martian volcanoes are so much larger than those on Earth. If there was no plate tectonic motion, the crust remained stationary over the hot spot. Thus each eruption was superimposed on the previous one, and the volcanoes just kept growing over billions of years.

One fascinating question pursued by Carr and other volcanologists is: Are any of the martian volcanoes still active? There is no evidence of eruptions taking place right now, but what of the youngest martian volcanoes, such as Olympus Mons? Carr and others have determined from crater counting that the youngest lava flows on Olympus Mons are less than 100 million years old. One hundred million years is a very short time on the scale of planetary history, and there is no reason to think a volcano that has been active for much of the past billion years or more will not continue to erupt periodically in the future. Therefore, scientists must conclude that Olympus Mons probably remains intermittently active, although the intervals between major eruptions could be many millions of years.

11.6 MYSTERIES OF THE MARTIAN CHANNELS

No geological discoveries on Mars have equaled the channels for either excitement or mystery. Impact craters, volcanoes, tectonic fractures, canyons—all of these might have been expected, and each has counterparts on other planets. The channels, however, speak to us of running water on Mars, a phenomenon previously thought to be unique to the Earth. From a study of the martian channels, we hope to illuminate the larger questions of the history of water on Mars and the possibilities that the planet could once have supported water-based forms of life.

There are two main types of channels on Mars, called runoff channels and outflow channels. Both were caused by running water, in contrast to the tectonic canyons discussed in Section 11.5. Both are, of course, dry today, since current conditions on Mars do not permit the existence of liquid water.

Runoff Channels

The martian features most like terrestrial dry river valleys are the **runoff channels**, found exclusively in the heavily cratered uplands of the southern hemisphere. Over much of this old terrain channels are extremely common, usually consisting of simple valleys tens to hundreds of meters wide and a few tens of kilometers long, often on steep slopes such as crater walls.

FIGURE 11.25 Drainage network of runoff channels in the old upland terrain near 48° S. These finely divided and branching erosional features were almost surely formed at a time when there was rainfall on the surface of Mars. Width of frame is 250 km.

More interesting than these simple valleys, because they are more clearly indicative of running water, are the networks of interconnected runoff channels. Dozens of major networks have been identified, typically several hundred kilometers in length. They have the characteristic form of terrestrial river systems, with small tributary channels connecting into larger channels to provide drainage for a wide area (Fig. 11.25). Compared with terrestrial drainage systems, however, even the best-developed martian runoff channels show relatively few tributaries.

In some cases, the smallest channels have been obliterated by subsequent erosion, and there are also instances where the water originated from underground springs rather than from direct surface runoff. Similar looking channels are formed on Earth by this process of sapping from springs. The process may begin with water seeping out at the base of a cliff from a subsurface reservoir or aquifer. As the water escapes, it carries some soil with it, and the cliff face above the flow begins to crumble. Thus the flow eats its way back into the surrounding terrain, forming a channel with a characteristically blunt appearance. The process continues until the groundwater reservoir is depleted or the spring is stopped by freezing.

Origin of the Runoff Channels

The shapes of the channel networks, and the way they clearly drain from higher areas into lower, point inescapably to erosion by liquid water. Many geologists have sought alternative explanations involving lava, wind, and ice, but none has survived detailed criticism. Unless we are badly misinterpreting the data, Mars once experienced real rivers flowing across its upland regions fed by large underground springs and, possibly, by drainage of rainfall.

When did the rivers flow? An important clue is given by the restriction of these channels to the cratered uplands. There are no such channels in the lower northern plains, which formed at about the time of the lunar maria, and none in any of the volcanic areas such as Tharsis. Apparently conditions have not been suitable on Mars for the formation of runoff channels during the past several billion years.

Detailed crater counts in the areas drained by the runoff channels confirm this timescale. The climatic conditions that permitted rivers to flow ceased before the formation of the northern plains or the emergence of the Tharsis bulge.

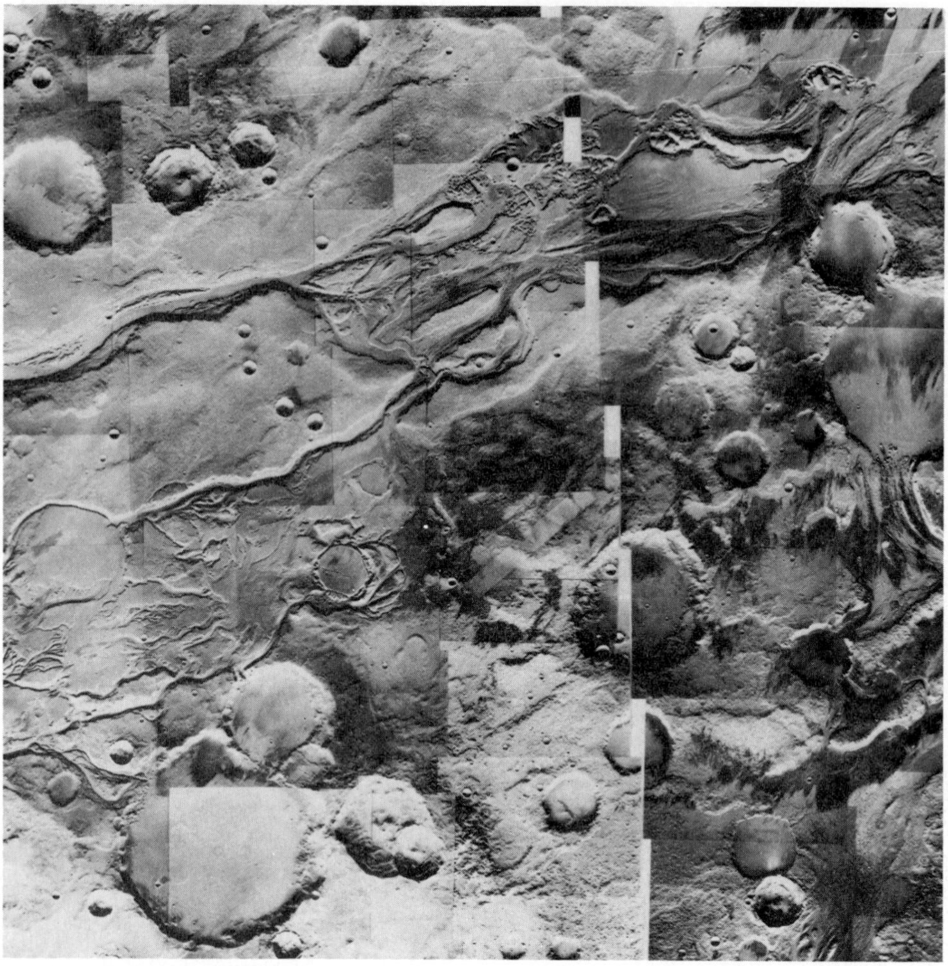

FIGURE 11.26 Several of the large outflow channels that drain from the southern uplands toward the northern basins. These appear to be the product of catastrophic water floods that took place about 3 billion years ago. Width of frame is 400 km.

According to the standard chronology, this puts the age of the rivers back about 4 billion years. Thus the channels we now see were active near the time of the terminal heavy bombardment, which is also the period suggested for enhanced levels of surface erosion. This association makes sense. Running water requires a thicker, warmer atmosphere than Mars has today, which would also have contributed to greater erosion than is taking place under present conditions.

Outflow Channels

Even larger than the runoff channels are the martian **outflow channels**. These tremendous valley systems are confined to the equatorial parts of Mars, primarily leading from the southern uplands into the northern plains. Like the runoff channels, they appear to have been carved by running water. However, in the case of the outflow channels the flow was probably intermittent and catastrophic, and the source of the water remains somewhat mysterious.

The largest and best-studied outflow channels lie north and east of the martian canyonlands and drain into the Chryse basin. These include Kasei Valley, Maja Valley, Simud Valley, Tiu Valley, and Ares Valley, together with smaller tributaries. Each major outflow channel is 10 km or more in width and hundreds of kilometers long, and each drops several kilometers in elevation along its length (Fig. 11.26).

Over most of their lengths, the outflow channels have cut multiple parallel channels that diverge and interconnect, each showing characteristic patterns of water erosion such as terraced walls, streamlined islands, and sandbars. In places, the flow seems to have broken out of these well-defined channels and spread over areas tens of kilometers wide.

The mouths of the outflow channels, where they descend from the uplands into the Chryse Basin, are characterized by teardrop-shaped islands and other sculpted landforms (Fig. 11.27). The islands are remnants of the plateau, shaped by the flood of water rushing out into the lowland plains. Within Chryse, evidence of water erosion extends for hundreds of kilometers from the channel mouths, but not as far as the location of the Viking 1 lander, so lander images cannot help us to understand the origin of these features.

FIGURE 11.27 Flow features, such as teardrop-shaped islands, mark the region where several outflow channels spread out into the lowland basin of Chryse, near the site originally targeted for the Viking 1 landing on Mars. Each island is approximately 40 km long.

Catastrophic Floods

The scale of the martian outflow channels and the evidence of massive flows of water dwarf terrestrial rivers. Estimates of the rates of flow in the larger channels are a hundred times greater than the flow from Earth's greatest river, the Amazon. To seek comparable rates of flow on Earth, we must look to catastrophic floods, caused by the sudden failure of natural dams holding back large lakes.

The best-studied example of such a flood on Earth is to be found in the states of Washington and Montana, where 18,000 years ago an ice dam broke releasing the water in a prehistoric lake called Lake Missoula. Within a few days the entire lake emptied westward across the Columbia Plateau, generating flows up to 120 m deep over a front tens of kilometers wide. The awesome force of this water erased previous surface features in western Washington State, cutting new channels up to 200 m deep through the bedrock. The resulting terrain, called the channeled scabland, is very different in appearance from that generated by slow, steady processes of water erosion. With its multiple parallel channels, teardrop islands, and huge bars, the channeled scabland is the closest analog we have to the martian outflow channels.

Presumably the martian features were formed in a similar way, by brief catastrophic floods arising in the uplands and dissipating in the broad northern basins. But where could such vast quantities of water have come from, and when did these remarkable events take place?

Chaos: The Sources of the Outflow Channels

To find the origin of the martian floods, we must look to the sources of the outflow channels. These are strange landscapes called **chaotic terrain**. They resemble regions where the surface of the old cratered uplands has collapsed into a jumble of irregular, rugged hills and valleys (Fig. 11.28). Both the chaotic terrain and the great floods could have a common origin in the sudden release of huge reservoirs of underground water, mud, or ice.

FIGURE 11.28 The chaotic terrain at the head of an outflow channel system. Evidently a local collapse of the surface was associated with the generation of a massive flood of water.

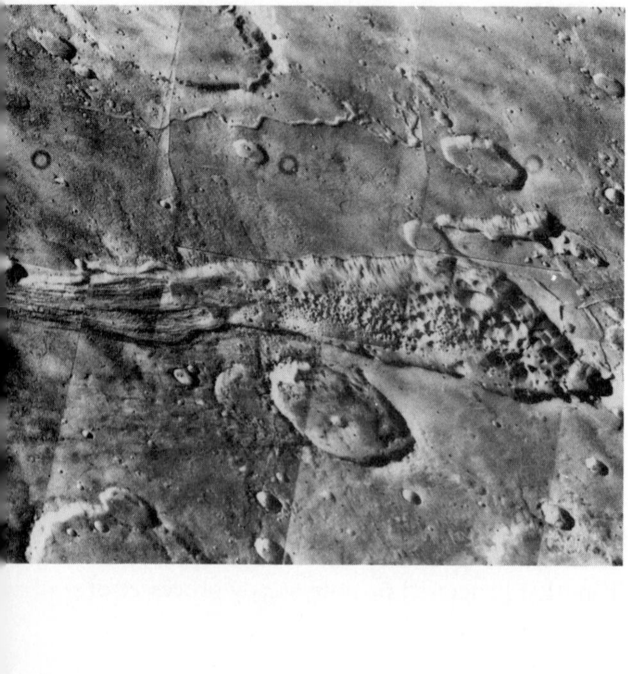

When did these events take place? From the fact that the outflow channels extend into the lightly cratered northern plains we can see that they represent a later era than that of the runoff channels. Crater counts in Chryse indicate that the deposits from channel outflow were formed there at just about the same time the volcanic plains were formed, which is probably about 3.5 billion years ago. This is also the time when internal forces were producing the Tharsis uplift.

While most geologists agree that the release of underground water generated the floods and produced the chaotic terrain, there is no consensus concerning the details of the mechanism. One possibility is the melting of subsurface ice by deep-seated volcanic activity. Another suggestion is the chemical release of water loosely bound to the clays of the martian soil. A third invokes the shifting of large quantities of underground water in response to the Tharsis uplift. Possibly all of these processes were involved.

Both the magnitude of the outflow channels and the huge depressions of chaotic terrain left behind indicate that water or ice must have made up a sizable fraction of the martian crust. Large quantities of ice are probably still present below the surface, but the conditions necessary to produce catastrophic floods ceased billions of years ago. We shall examine the fate of that water in Section 11.8.

11.7 THE POLAR REGIONS

The geological features of Mars described in the previous sections of this chapter are generally confined to equatorial and middle latitudes. Above about 70° north and south, the surface takes on a different character, strongly influenced by the polar caps.

Seasonal Polar Caps

The polar caps seen through a telescope—the white surface deposits that grow and shrink with the changing seasons—are called the seasonal caps. They are analogous to the seasonal snow cover on the Earth. We do not usually think of the snow that covers the ground in winter as a part of Earth's polar caps. If we looked at our planet from space, however, the seasonally changing snow cover would be obvious, and the area involved much larger than that covered by the permanent ice caps. It is the same for Mars.

The substantial eccentricity of Mars's orbit carries the planet farther from the Sun during southern winter, resulting in long cold winters and a large southern seasonal cap, extending down to about 55° S. The northern cap, half a martian year later, extends only to about 65°, although some surface frost was seen as far south as the Viking 2 landing site, at 48° N (Fig. 11.29). During spring, each cap retreats at a rate of about 20 km/day, reaching its minimum size in early summer.

FIGURE 11.29 Light dusting of surface frost photographed at the Viking 2 landing site in late winter.

Both seasonal caps are composed of CO_2 frost or dry ice. Since CO_2 is the main component of the atmosphere, it is always available to condense on the surface whenever the temperature falls below 150 K, the freezing point of CO_2. From changes in atmospheric pressure as the caps form and evaporate, the thickness of the seasonal deposit has been calculated to vary from a meter or so at high latitudes down to a few centimeters near its edge. This is less, but not much less, than the corresponding thickness of the seasonal snow cap on Earth.

Residual Polar Caps

As the seasonal polar cap retreats during southern spring, it reveals a brighter, underlying permanent cap, which persists through the summer months (Fig. 11.30). The diameter of this resid-

FIGURE 11.30 Residual south polar cap, 250 km in diameter and composed of both frozen CO_2 and H_2O.

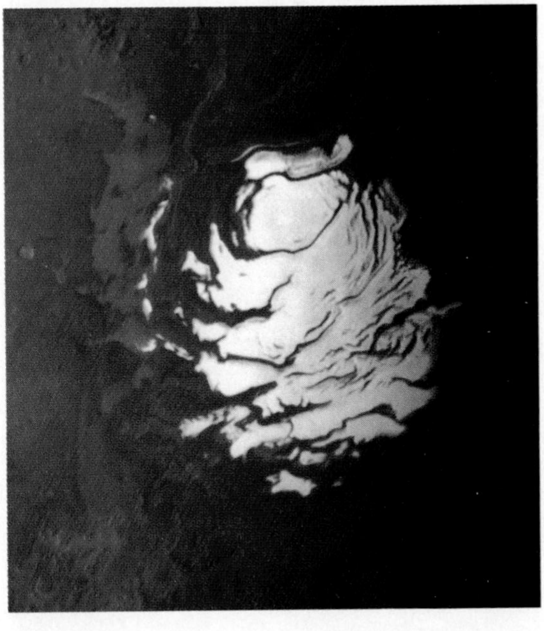

ual south polar cap is about 350 km, and it is located a little distance away from the true south pole, with its center at 30° W, 86° S. Measurements of its surface temperature made from orbit show that the residual cap remains at 150 K, the frost point of CO_2, throughout the summer, indicating that it is composed in part of CO_2. Water ice is undoubtedly present as well, perhaps in large quantities, but the temperature of the cap is controlled by the CO_2 ice, and this cold reservoir is sufficient to maintain itself in spite of the continuous summer sunshine.

The residual southern cap has a remarkable appearance, showing a spiral swirl pattern of dark lanes in the bright ice. These lanes are sunward-facing slopes and valley bottoms a few kilometers wide, which heat sufficiently to lose their ice cover. No one knows the thickness of the CO_2 ice deposit that makes up the bulk of the cap, but from the fact that these frost-free areas show through, we presume that the thickness is at most a few kilometers.

The permanent north polar cap (Fig. 11.31) is much larger than that in the south, never shrinking below a diameter of about 1000 km. It shows a similar swirl pattern of exposed sunward-facing slopes. Orbital temperature measurements show that in summer its temperature climbs above 200 K, far too warm for CO_2 ice to be stable. At the same time, the concentration of atmospheric water vapor rises sharply above the cap. Thus we see that the northern residual cap cannot contain frozen CO_2, but instead is composed of ordinary water ice, rapidly evaporating in the summer warmth. With its diameter of 1000 km and an unknown thickness, the residual northern cap may be one of the main storehouses for water on Mars.

FIGURE 11.31 Residual north polar cap, about 1000 km across and composed of frozen H_2O.

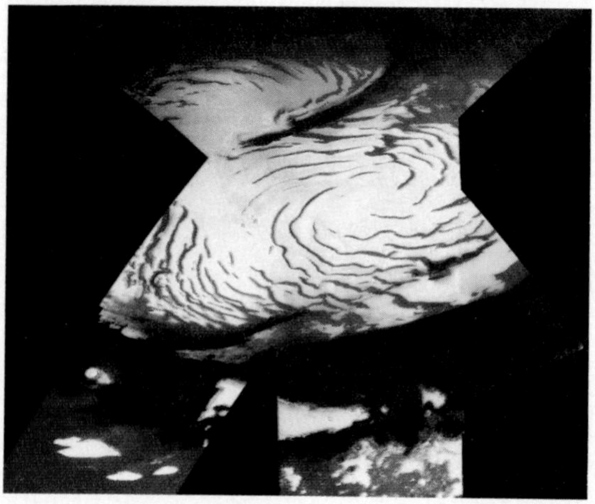

Why this major difference between the two martian polar caps? It is not simply a question of the summer temperatures, since the southern summers are hotter, yet it is the southern cap that stays cold enough to trap CO_2. The difference is probably associated with the global dust storms, which always take place during southern summer, when the northern cap is forming; therefore, the northern cap becomes dusty. Being dustier, it is also darker, and therefore it can absorb more sunlight. It may be just this difference in the albedo of the ice that allows the northern cap to heat up and lose its frozen CO_2.

Both of the martian polar caps contain large quantities of H_2O ice, but they may represent just the proverbial tip of the iceberg for martian water. Calculations suggest that ice may be stable as permafrost within the martian crust for all latitudes greater than about 40°. Nearer the equator, any ice originally present in the soil will eventually either melt or evaporate; the outflow channels may have originated as a byproduct of this loss of subsurface ice. At higher latitudes, the permafrost layer is progressively thicker and nearer the surface until it emerges into view near the poles in the form of the residual polar caps. Thus, unlike the terrestrial ice caps, the martian polar caps are directly connected with a much larger underground reservoir of frozen water.

Polar Terrain

So far we have concentrated on the polar caps of Mars, but the underlying polar terrain is equally interesting. At latitudes above 80° in both hemispheres, the ground is covered by deep layered deposits of sediment. Here the term *sediment* has its geological meaning of finely divided material; sediment does not have to be produced or deposited by water. These deposits lie on top of the old cratered uplands in the south, and overlie the volcanic plains in the north. In southern summer, the southern layered terrain is exposed by the retreating polar cap, while in the north the residual cap continues to cover most of these deposits throughout the year.

The polar layered deposits blanket all but the largest underlying topography, indicating a depth of several kilometers. They are sparsely cratered, and therefore relatively young. On the defrosted slopes within the residual polar caps, the layering is especially well shown (Fig. 11.32). The individual layers are between 10 and 50 m in thickness, distinguished by darker bands alternating with light, and sometimes by terraces between the bands.

In the north, there is another special polar terrain, consisting of extensive sand dunes. These dunes form a band about 500 km wide around the pole, between latitudes 70° N and 85° N, extending to the boundary of the layered deposits near the edge of the north residual cap. The martian dune field (Fig. 11.33) consists of evenly spaced, crescent-shaped dunes a little less than a kilometer long. In the other hemisphere there is no similar band of dunes, but smaller dune fields are common inside craters at high southern latitudes.

Formation of the Polar Deposits

The polar terrains we have been discussing differ from most other parts of the martian surface in that they are *depositional;* they represent wind-blown dust transported from elsewhere. These areas may therefore be the final resting place of much of the material scoured out of the canyons and other eroded areas of the planet.

FIGURE 11.32 A region near the north polar cap that is partially covered with frost. Many layers can be seen on partly defrosted slopes, presumably representing past climate cycles on Mars. Width of frame is 65 km.

FIGURE 11.33 A large dune field at latitude 81° N. The spacing between dune crests is a few hundred meters, and the total width of the frame is 62 km. Such dunes almost completely encircle the martian north pole just south of the layered terrain.

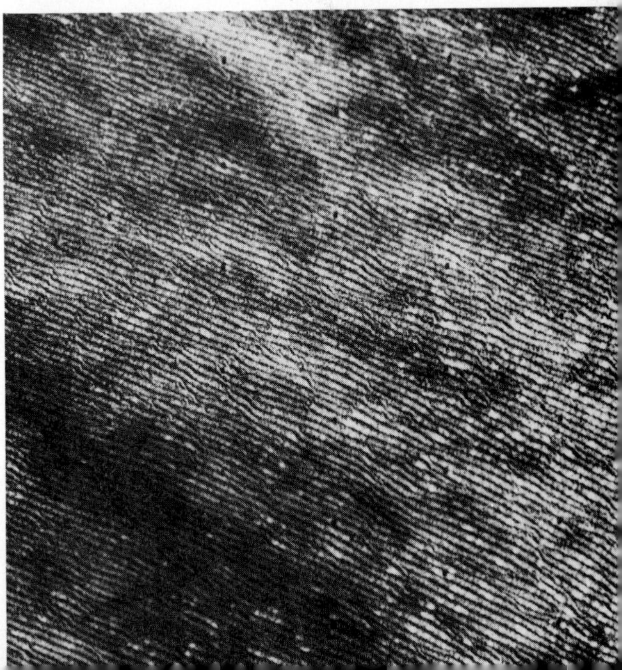

The distinct layering of the polar deposits is particularly intriguing. Calculations of the amount of dust raised on Mars during a major dust storm suggest that the maximum annual deposit from this source is less than 1 mm/yr, which of course we cannot see; the layers must represent much longer timescales. To build up a thickness of 10 meters would require about 10,000 years at this rate. Thus the polar layering must represent climatic changes over tens of thousands of years. What might these be?

The most likely causes of such climatic changes are to be found in variations in the orbit of Mars, caused by the combined gravitational attractions of the other planets. The Earth undergoes similar but smaller cyclical variations, which are probably a major cause of the ice ages that our planet has experienced at more or less regular intervals. One of these variations causes the role of the two martian poles to alternate with a period of 51,000 years, so that about 25,000 years in the future it will be the north pole of Mars, not the south, that will experience the hottest summers. Another effect is a periodic shift in the tilt of the martian axis of rotation. Calculations show that this tilt varies with a period of about 100,000 years. The eccentricity of the orbit also changes, this time with a period nearer to a million years.

As a result of all these orbital and rotational changes, conditions at the martian poles undergo regular cycles of change. At some times only the south residual cap will retain CO_2, at some times only the north, and perhaps at other times both or neither. Undoubtedly the polar layered deposits are the result of these changes, but the details of the processes of polar deposition and erosion remain unknown. The study of martian climatic change is a subject in its infancy.

11.8 THE MARTIAN ATMOSPHERE

Nineteenth-century astronomers recognized that Mars had an atmosphere. The existence of polar caps that waxed and waned with the seasons, the presence of white clouds over certain regions of Mars at certain seasons, and the occasional development of giant dust storms, all seemed to require the existence of an atmosphere. But it proved to be much more difficult to determine how thick the atmosphere is and what gases it contains, and definitive results were not obtained until the Viking landers carried out a direct analysis from the martian surface.

Viking Results

The Viking landers gave us atmospheric measurements along two descent trajectories as well as detailed measurements of atmospheric composition and weather at Chryse and Utopia. One of us (Owen) was a member of the team responsible for investigations of atmospheric composition using the GCMS (described in Chapter 12). After years of studying Mars with telescopes and spectrographs from Earth, it was thrilling actually to measure a sample of the atmosphere, finding gases such as nitrogen, argon, neon, krypton, and xenon which could not be detected by remote sensing.

The composition results are shown in Table 11.4, and the structure is summarized in Fig. 11.34. At first glance, the martian atmosphere looks remarkably like that of Venus, only a few thousand times thinner.

Like both Earth and Venus, Mars has a troposphere and a stratosphere. The thickness of the troposphere varies with location and season, however, and at night it practically disappears. Because the atmosphere is so thin and there are no moderating bodies of water, the temperatures of the surface and the lower few kilometers of the troposphere experience a wide daily fluctuation. Air temperatures at the two Viking sites reached summer daytime highs of about −30 C, while the ground itself occasionally exceeded the freezing temperature of water in strong afternoon sunlight. Under these conditions convection can take place, and the troposphere is typi-

TABLE 11.4 Composition of the martian atmosphere

Gas	Formula	Abundance (percent)
Carbon dioxide	CO_2	95.4
Nitrogen	N_2	2.7
Argon (40)	Ar-40	1.6
Oxygen	O_2	0.13
Carbon monoxide	CO	0.07
Water	H_2O	0.03*
Argon (36)	Ar-36	0.0005
Neon	Ne	0.0003
Krypton	Kr	0.00003
Ozone	O_3	0.00001*
Xenon	Xe	0.000008

*Abundance varies with season and location.

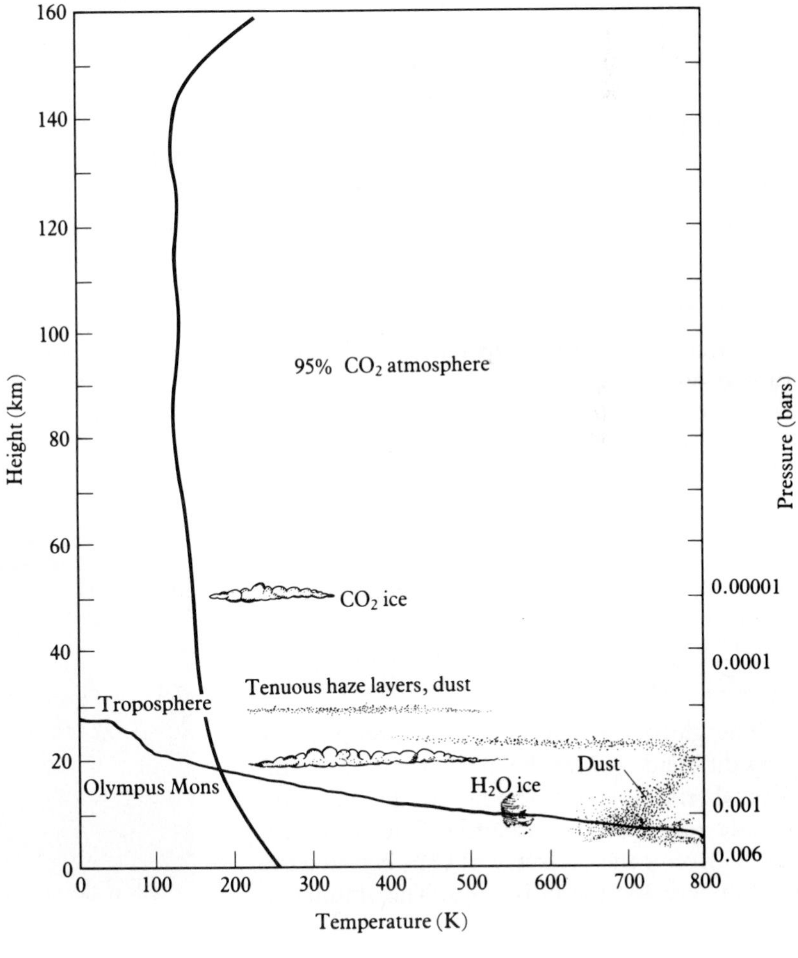

FIGURE 11.34 Vertical structure of the martian atmosphere. Both CO_2 clouds and H_2O clouds can form. Dust from the surface occasionally fills the atmosphere, warming it considerably.

cally 10–20 km thick. In contrast, nighttime temperatures of both ground and air dropped to nearly 100 C below freezing. Similar low surface temperatures are found over the polar caps. Convection ceases, and the atmosphere becomes highly stable. On Earth, with its much thicker atmosphere, the day/night variations are much less, and tropospheric convection rarely ceases entirely.

As shown in Table 11.4, the major constituents of the martian atmosphere are CO_2 (95.4%), N_2 (2.7%), and Ar (1.6%). These same gases exist in the atmosphere of Venus. However, the minor constituents of the martian atmosphere are different from those of Venus. No sulfur compounds or acids have been detected on Mars, and the distribution of noble gas abundances is very different. In particular, it seems that Mars has outgassed more Ar-40, the isotope produced by the decay of radioactive potassium, relative to Ar-36, the primordial isotope. Evidently both the chemistries and the histories of the two atmospheres are not the same, a topic we return to in the next chapter.

Martian Weather

The local weather at the two Viking landing sites turned out to be rather dull, repeating reliably from one day to the next. Here is the first weather report from the surface, as given by Viking meteorologist Seymore L. Hess:

> *Light winds from the East in the late afternoon, changing to light winds from the Southeast after midnight. Maximum winds were 15 miles per hour. Temperature ranged from −122 F just after dawn to a maximum of −22 F. Pressure steady at 7.70 millibars.*

The global circulation of the atmosphere is very different on Mars from what it is on Earth. The absence of large bodies of liquid water greatly simplifies the situation. The main seasonal driver is the exchange of CO_2 between the atmosphere and the polar caps. The amount of gas stored in the polar caps is so great that the average atmospheric surface pressure on the planet changes by 20% with the seasons. Such a change is unheard of on Earth, where a severe storm may cause a drop in pressure of no more than a few percent. Strong winds have been observed flowing off the sunlit caps, although no measurements of these winds have yet been made.

The clouds seen by early telescopic observers were confirmed by the Viking cameras. Morning fogs occur in low-lying areas (Fig. 11.35), and extensive clouds are regularly found around the high volcanoes, such as Olympus Mons. As it blows over the volcano, the rising air cools, and water vapor condenses to form the clouds. This association of mountains and clouds is well known on Earth.

Global Dust Storms

Only during the southern-hemisphere summer do conditions change dramatically. Since neither Viking lander was in the southern hemisphere,

FIGURE 11.35 Morning fog in the martian canyonlands, photographed by the Viking orbiter. Width of frame is about 1000 km.

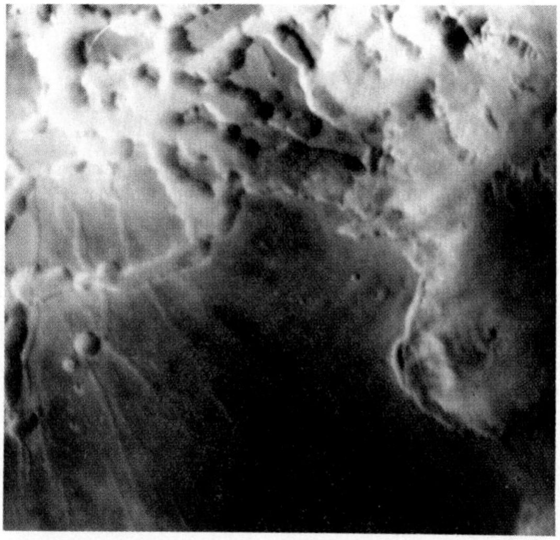

we are not sure of the details, but apparently the atmosphere becomes unstable, and conditions favorable to the formation of global dust storms develop. Once a large quantity of dust is raised, most of the sunlight is absorbed by the dusty atmosphere and blocked from reaching the surface. As a result, surface temperatures fall, but the atmosphere remains unstable and the dust clouds spread until nearly the entire planet is shrouded. It requires about a quarter of a martian year (six Earth months) for the dust to settle and the normal weather patterns to re-emerge.

How can such a thin mixture of gases produce and maintain the great dust storms of Mars? The answer must lie in high-speed local winds produced by sunlight heating the surface. Something like this occurs in terrestrial deserts during hot summer days. The air near the ground gets so hot that it suddenly soars upward, allowing cooler air to rush in. The result is a very local, circular wind that picks up surface dust and carries it aloft in a swirling column called a dust devil.

It is estimated that the local winds on Mars must achieve speeds of 150 km/h in order to raise the dust in such a thin atmosphere. The dust itself must include very fine grained material since larger grains would rapidly drop out, clearing the air. The orange skies revealed by the Viking landers give eloquent testimony to the continued presence of these fine dust particles in the atmosphere.

Water on Mars

Depending on your point of view, the atmosphere of Mars is either very dry or very wet. Because it is so thin and cold, the atmosphere can hold very little water vapor; in this sense it is exceedingly dry. But it is also true that the martian atmosphere normally holds nearly the maximum amount of water vapor possible for its pressure and temperature. Thus, relative to what it could contain, it is damp; we say that there is a large *relative humidity,* or that the atmosphere is close to *saturation*. As a result, it is common for H_2O clouds and fog to form at low elevation during the cold martian night.

In contrast to Mars, the atmosphere of Venus is truly dry, in both relative and absolute senses. This difference between Mars and Venus tells us something important about the water abundances of the two planets, a topic to be pursued in the next chapter.

SUMMARY

Mars has the most Earthlike surface environment of any planet. In spite of the smallness of Mars, the thinness of its atmosphere, and the low temperatures of its surface, we can recognize in this planet the closest we shall ever come to another Earth. It is not surprising that Mars has been the target of so many spacecraft, or that serious plans for human exploration are being discussed.

At the beginning of the twentieth century Mars was widely believed to have intelligent inhabitants, and even up to the 1960s the possibility of widespread plant life was assumed. Spacecraft exploration unfortunately failed to discover life, but the Mariner 9 orbiter and the four Viking craft have revealed a world equally fascinating in other ways.

The topography of Mars is dominated by a north-south asymmetry. Most of the southern hemisphere consists of ancient, cratered uplands, probably formed at about the same time as the lunar highlands but substantially modified by erosion, while the northern half of the planet is about 5 km lower and has a younger, largely volcanic surface, perhaps about as old as the lunar maria. A second, younger, east-west asymmetry is created by the Tharsis bulge, a volcanic area about the size of the continental United States with four huge shield volcanoes that rise to 30 km above the lowlands, creating elevation differences much greater than any on Earth, Venus, or the Moon. Because of its low surface gravity and persistent volcanic activity at a few sites, Mars has the highest mountains in the solar system.

The Viking landers provide detailed information for two surface sites, both rather dull lowland plains. Each is a rock-strewn volcanic basin, apparently stripped of fine-grained material by the powerful martian winds. The soil is a red clay consisting largely of iron oxides, while the rocks may be basalt fragments ejected from nearby impact craters.

Martian craters differ from those of the Moon primarily in their ejecta blankets, which show evidence of fluidization, presumably by the impact melting of permafrost. Counts of crater densities yield relative crater retention ages, but unfortunately we have no absolute chronology based on radioactive rock dating. Only with the aid of a calculated model for the impact history of Mars can we try to date the geologic events of the martian past.

The shield volcanoes of Mars look much like their terrestrial counterparts, except that they are vastly larger. Olympus Mons is the largest volcano in the solar system. In addition to volcanoes, the Tharsis uplift has produced numerous tectonic features, the most spectacular of which make up the 4000-km-long canyon system of the Valles Marineris. These canyons were not cut by running water, although water and wind erosion have both played a role in enlarging the original tectonic features.

The channels of Mars, which were cut by running water, should not be confused with the canyons, which were not. There are two main types of channels: the older runoff channels, which appear to have been produced by rain; and the larger and slightly younger outflow channels, which drain from the southern uplands into the northern basins. The outflow channels appear to have been created in short, catastrophic floods, presumably resulting from the sudden release of groundwater or permafrost.

The martian polar regions offer other fascinating enigmas. The two residual polar caps are quite different from each other; the southern is smaller and retains CO_2 as well as H_2O, while the northern is a thousand kilometers across and is composed exclusively of H_2O ice. Both caps are surface exposures of a much larger H_2O reservoir of permafrost. The underlying topography of both polar regions consists of a unique layered terrain, indicative of periodic changes in global climate on Mars.

The atmosphere of Mars has a surface pressure of only 0.006 bar; presumably it is a mere remnant of its former self. Just as the pressure on Venus is about 100 times greater than that on Earth, the martian surface pressure is 100 times less. The composition of the martian atmosphere, however, is very similar to that of Venus in its three main gases (CO_2, N_2, and Ar), but quite different in its minor constituents.

KEY TERMS

chaotic terrain
fluidized ejecta
outflow channel
runoff channel

Review Questions

1. Compare the capabilities of the Viking Mars landers with the exploration of the Moon carried out by Apollo astronauts. What did Apollo accomplish that Viking did not? Was the human presence critical? How important is it that we send humans to explore Mars?

2. Contrast the surface of Mars, as seen from the Viking landers, with that of Venus, as seen from the Venera landers. Compare both of these with the Earth and Moon.

3. Discuss how the age of the martian surface, and of major geological features such as the Tharsis volcanoes or the Valles Marineris, can be found. What assumptions must be made to apply the lunar chronology to Mars? How accurate do you believe the ages are in Table 11.3?

4. Compare and contrast impact craters on Mars, Moon, Venus, and Earth. What are the consequences of the different amounts of atmosphere on the three bodies? How much erosion takes place on each? Does atmospheric composition play a role?

5. Compare the martian volcanic features with those on the Earth and the Moon. Explain why the shield volcanoes on Mars are so much larger than their terrestrial counterparts. Why do you think there are no large shield volcanoes on the Moon?

6. Describe the various kinds of martian landforms that seem to be the result of flowing water. Are there similar features on the Earth? How old are the martian features? What do they indicate about the past climate on that planet?

7. Think about the history of water on Mars. How much might have been there once, where did it come from, and where has it gone? How much is present today in the atmosphere, in the polar caps, and in subsurface permafrost? What would be required to release some of this water and thereby perhaps create a more favorable environment on the planet?

8. What do the polar caps tell us about the climatic cycles on Mars? Describe what you would expect the polar caps to look like up close. What would you expect from a project to drill into the caps and extract a core? How would such a core compare with one obtained on Earth from the Greenland ice cap or from the Antarctic?

Quantitative Exercises

1. Calculate the flight time to Mars for a spacecraft launched on an orbit with Earth at its perihelion and Mars at its aphelion.

2. Estimate the amount of water ice in the residual north polar cap of Mars if the cap is 1 km thick. How does this compare with the estimate for release of enough water from the interior to cover the entire planet to an average depth of 200 m? Where might the missing water be hiding?

3. How much water would be released by the melting of permafrost from a chaotic area 100 km on a side and 3 km deep, if 33% of the permafrost volume is occupied by ice? For how long could this quantity supply an outflow channel 10 km wide and 100 m deep through which the water flowed at a speed of 100 km/hr?

Additional Reading

Anonymous. 1984. *Viking: The Exploration of Mars* (NASA EP-208). Washington, D.C.: U.S. Government Printing Office. Excellent short account of the Viking discoveries in the NASA educational publication series.

*Carr, M.H. 1981. *The Surface of Mars.* New Haven: Yale University Press. The definitive summary of our knowledge of Mars after Viking, written by the leader of the Viking orbiter imaging team in a highly readable format, with excellent illustrations.

Ezell, E.C., and L.N. Ezell. 1984. *On Mars* (NASA SP-4212). Washington, D.C.: U.S. Government Printing Office. Detailed history of Viking from the NASA history series.

*Greeley, R. 1985. *Planetary Landscapes* (Chapter 7). London: Allen & Unwin. Beautifully illustrated summary of the geology of Mars.

*Kieffer, H.H., B.M. Jakosky, C.W. Snyder, and M.S. Matthews, eds. 1992. *Mars.* Tucson: University of Arizona Press. Highly technical, detailed, multiauthor volume from the Arizona Space Science series.

Washburn, M.L. 1977. *Mars at Last.* New York: Putnam. Exciting journalist's account of the Viking missions.

*Indicates the more technical readings.

12

Life, Planets, and Atmospheres

Botticelli's famous painting *Primavera* ("spring") shows a profusion of the life we take for granted in our everyday experience on Earth. Does this remarkable property of matter, the quality of being alive, manifest itself on any other planet in our solar system? Mars seems the most likely candidate.

After four decades of space exploration, life on Earth is still the only life we know. This makes it very difficult for us to generalize about what kind of life (if any) we might expect to find elsewhere and how we should go about trying to find it. We must begin by considering terrestrial life and try our best to understand its fundamental properties. We can then see if the other inner planets ever offered environments in which these fundamental properties might have manifested themselves. The resulting perspective will give us a much better chance of realistically assessing the possibilities for finding life elsewhere in the universe.

In our own solar system, Mars is by far the best candidate for an Earthlike planet. The Viking mission described in the last chapter included a set of experiments specifically designed to search for evidence of life, and this search still provides an incentive for the new wave of Mars missions getting under way in the twentieth century's last decade. Why is this so? And why do the atmospheres and surfaces of Mars, Earth, and Venus exhibit so much variety, despite the common composition of the planets themselves? We have already provided some answers, and we will now use the study of habitable planetary environments to emphasize the critical differences among the three inner planets with atmospheres.

12.1 LIFE ON EARTH

Surely the existence of life on Earth is the most extraordinary characteristic of our planet. There must be some underlying attributes that make Earth uniquely suitable (in our planetary system) for the origin and continued existence of this remarkable property of matter. Having an appropriate and relatively stable surface temperature is certainly one essential attribute, as it enables the continued existence of open bodies of water, a phenomenon that is also unique to Earth. The availability of liquid water is essential for life as we know it, but to allow life to exist is not to ensure that it originates. Current ideas about life's origin stress the importance of the composition of the early atmosphere. The chemistry that occurred in that atmosphere got life started, and life has been closely coupled to the atmosphere ever since.

The Origin of Life

While no one has duplicated in a laboratory all the steps that originally led from inanimate matter to matter that is alive, it is clear that these events took place very early in the Earth's history, and that the necessary organic building blocks for life could have been produced in abundance only in an environment free of atmospheric oxygen. If this highly reactive gas had been present in the primitive atmosphere, the compounds necessary for the beginning of life would have rapidly been converted into oxides. Life has since evolved the capacity to protect itself from oxygen and even to use it to extract energy . . . but this is getting ahead of our story.

The primordial solar nebula, described in Section 4.1, must have been rich in simple compounds of carbon, hydrogen, oxygen, and nitrogen, the very elements that are the predominant constituents of life. We know this from studies of the molecular clouds that are the birthplaces of stars. We have learned from comets and primitive meteorites that substantial amounts of more complex organic compounds were also available in the early solar system (Sections 4.4, 6.2). Some of these materials must have been delivered to Mars and Venus as well as to Earth at the time these planets were forming.

Furthermore, laboratory experiments have demonstrated that a mixture of gases containing compounds of carbon, nitrogen, oxygen, and hydrogen (but no free oxygen gas) will form amino acids and a variety of other prebiological organic compounds when it is subjected to an electric discharge, ultraviolet light, or other sources of energy. Such experiments have been carried out to simulate conditions on the primitive Earth, starting with the fundamental work of Stanley Miller and Harold Urey at the University of Chicago in 1953 (Fig. 12.1).

The Miller-Urey experiment used a mixture of methane (CH_4), ammonia (NH_3), and water (H_2O) to imitate the early Earth. Most scientists now believe that the initial atmospheres of the inner planets were more likely composed of carbon monoxide (CO), carbon dioxide (CO_2), and molecular nitrogen (N_2), a mixture that also yields prebiological organic compounds when energy is supplied to drive the chemical reactions. The compounds generated from this atmospheric chemistry and those brought in by asteroids and comets formed the basis for the origin of life on Earth. Further reactions among these substances in ponds and tide pools eventually led to the development of complex molecules capable of self-replication. This was the beginning of life as we know it.

This model for the origin of life on Earth is still very general, since we do not yet know the

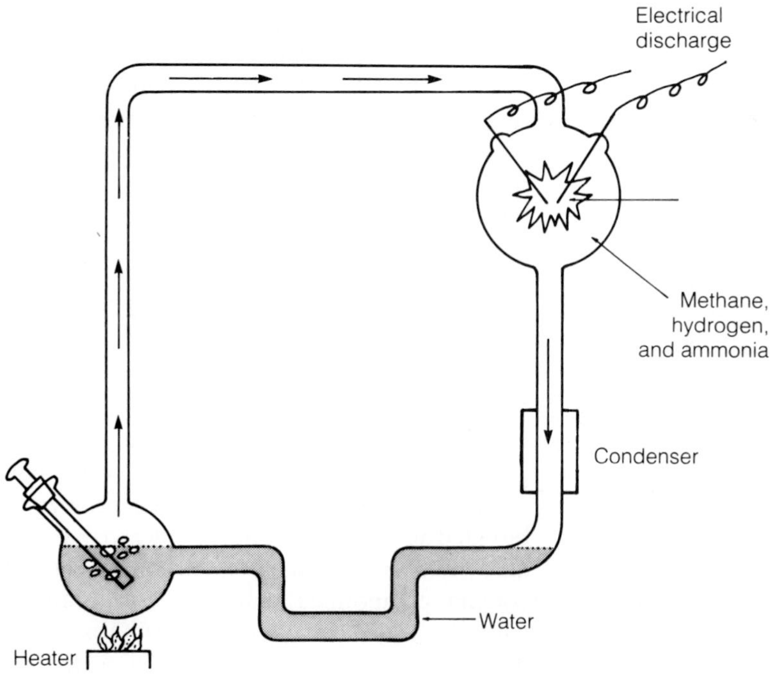

FIGURE 12.1 A schematic diagram of the experimental equipment used by Stanley Miller and Harold Urey shows the 5-liter flask that contained water vapor, methane, hydrogen, and ammonia to simulate their concept of the Earth's primitive atmosphere and oceans. The electrical discharge in the flask at the upper right reproduced the effects of lightning.

FIGURE 12.2 This is a cross section of a stromatolite found in Australia (cf. Fig. 9.14) which is 3.5 billion years old. It is also the oldest known example of life on Earth. The layered, domal structure is characteristic of these macroscopic fossils of microscopic organisms. The layers are produced as silt is trapped by successive colonies of photosynthetic bacteria struggling to receive as much unobstructed sunlight as possible. The rock is 20 cm across.

detailed reactions that led from inert organic matter to the ability to replicate. But there is nothing in this picture that singles out Earth. Both Venus and Mars should have begun with atmospheres similar to our planet's, and we know Mars had liquid water on its surface during the time the chemical reactions we just described were taking place on Earth. Hence at this stage of our knowledge, we cannot exclude the possibility that life began on Mars at the same epoch that it originated on Earth.

The Fossil Record

It would be immensely helpful if we could examine the rock record to see what was happening on Earth at the time life was first developing. As we saw in Section 9.3, however, this record has been irretrievably lost. The oldest rocks currently known solidified 700 million years after the Earth formed, and the oldest record of life preserved in the fossil record dates from still later, 3.5 billion years ago. By that time life had already progressed to the level of complex colonies of photosynthetic micro-organisms that form fossils called stromatolites (Fig. 12.2). The bacteria that made the stromatolites are quite sophisticated life-forms, compared with a self-replicating molecule. It may have taken hundreds of millions of years for life to evolve to this level of complexity.

We gain an idea of how much time was required by noting that the further evolution from the stromatolites to multicelled creatures required another 2.7 billion years. During this immense stretch of time, the most important developments were cells with nuclei and specialized subsystems for locomotion, which first appear in the fossil record about 1.5 billion years ago. After the development of multicellularity, the tempo of evolution increased. By the beginning of the Phanerozoic era, 590 million years ago, sexual reproduction had been invented, along with hard skeletons and other specialized tissues (Fig. 12.3). In another hundred million years, the backbone evolved, and shortly thereafter the colonization of the land began. From that point to the evolution of homo sapiens was but a short interval on the cosmic calendar.

12.2 LIFE AND THE EVOLUTION OF THE ATMOSPHERE

Our human presence on this planet is actually a gift from the green plants, since it is oxygen-producing photosynthesis that created the environment required for the development of warm-blooded animals. Rocks much older than 2.2 billion years are *suboxidized,* meaning that they contain compounds that have not combined with all the oxygen they could (CO is suboxidized compared with CO_2). Such rocks must

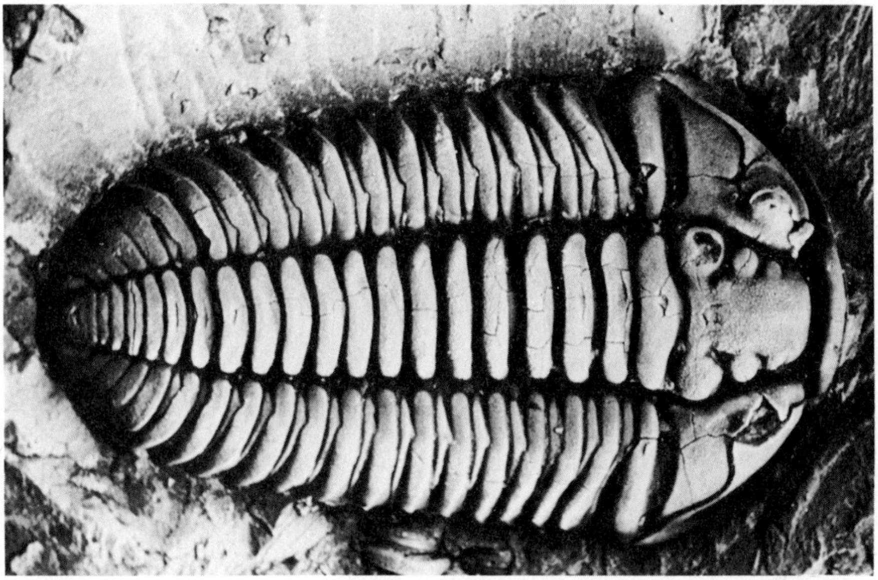

FIGURE 12.3 The trilobites that flourished in the seas at the start of the Phanerozoic eon had two eyes and a rather complex body structure. This photograph shows a trilobite fossil about 2 cm long.

have formed in the absence of free oxygen, which is certainly consistent with our model for the origin of life in an oxygen-free environment. The fossil record shows that stromatolites became common and widespread at about this time, indicating an increase in oxygen-producing photosynthesis by colonies of micro-organisms. Life not only contributes oxygen, it also removes carbon dioxide, but some of this exchange between water, rocks, and the atmosphere would also occur on planets with no life whatsoever.

Carbon Dioxide in Planetary Atmospheres

We have seen that CO_2 is the dominant gas in the atmospheres of both Mars and Venus. On our planet, it is now a minor constituent, primarily because of the activities of living organisms. Marine shell-forming creatures—most of them microscopic—manufacture their shells from calcium and carbon dioxide dissolved in the ocean. As these shells are incorporated into sediment, great deposits of carbonate limestone are created at the expense of the atmospheric carbon dioxide. The quantity of CO_2 represented by these carbonate deposits today is truly staggering. If all this gas were put back into the atmosphere at once, the pressure of CO_2 alone would be 70 bars, or 70 times the total atmospheric pressure today. As we saw in Chapter 10, this is close to the 90 bars of CO_2 that is presently in the atmosphere of Venus, where it causes an extremely large greenhouse effect.

Looking at the climate of Venus today, we can see that if all Earth's CO_2 was ever in the atmosphere at one time, most must have been deposited as carbonate rocks even before life arose. Otherwise, Earth would have been too hot for the chemistry leading to the origin of life to occur. The key to the removal of CO_2 from our planet's atmosphere was the presence of liquid water on Earth. Water dissolves CO_2, forming a weak acid that reacts with silicate rocks to make carbonates. As expressed in a highly simplified form by Harold Urey, who first described it, this reaction may be written as

$$CO_2 + CaSiO_3 \mathrm{T} CaCO_3 + SiO_2$$

It is therefore clear that with or without life, the early atmosphere of Earth would have evolved with time. Ultraviolet light from the Sun would continually break molecules apart in the upper atmosphere, a process called **photodissociation.** Heavy atoms like carbon and oxygen could simply recombine or form new compounds, a process called photochemistry, since it is driven by energy from light, primarily the high-energy photons at ultraviolet wavelengths. Hydrogen atoms, having very low mass, would steadily escape from the atmosphere into interplanetary space. Thus CH_4 (if present at all) would be converted to CO_2, NH_3 would inevitably form N_2, and CO would be oxidized to CO_2. The oxygen required for this process would be supplied by the photodissociation of H_2O. This transition to a fully oxidized environment must occur on any inner planet in any solar system, regardless of the presence of life.

Evolution on the Early Earth

We can see now that there is yet another threat to the survival of primitive forms of life on a young planet. Even the first, most simple organisms required food to supply the energy and raw materials for their continued existence. At first this food was free, since it was being produced spontaneously from the same chemical reactions that led to the origin of life itself. As we have seen, however, the character of the Earth's atmosphere was steadily changing, in ways that presaged the end of the free lunch.

To make the energy-rich organic compounds required for life, the gases in the atmosphere had to be suboxidized, on average. The microenvironments provided by volcanic activity and cometary impact were especially favorable in this respect. Free hydrogen generated from the reactions making the more complex molecules would help to maintain this suboxidized state, but Earth's gravity field is too weak to keep the hydrogen in the atmosphere. As hydrogen escaped from Earth and oxygen liberated from water molecules by photodissociation combined with other elements and molecules, the free food would soon have stopped forming and been replaced by carbon dioxide.

It was at some time just before this stage that life on Earth invented the remarkable process of photosynthesis, thereby connecting itself with the Sun, a source of energy that would last for a cosmic timescale. Photosynthesis produces both food and a secondary source of energy in the form of abundant free oxygen. This was an essential invention, and one can imagine that on some other planets, life could proceed up to this point and then die out, having failed to evolve a form that achieves this remarkable ability.

We conclude that the beginnings of life on a planet provide no assurance that life will survive there. Furthermore, if a cosmic accident should destroy all living organisms on a planet after the atmosphere has become oxidized, there will be no new origin of life. Once the early, suboxidized conditions are gone, they are gone forever.

The Gaia Hypothesis

Oxygen and carbon dioxide are two atmospheric gases that have been strongly influenced by the evolution of life. A careful examination of the interplay between life and the atmosphere reveals a variety of other less obvious relationships. The intricacy of these interactions has led two scientists, Lynn Margulis and James Lovelock, to suggest that life on Earth actually regulates the composition of the lower atmosphere. They have given the name Gaia (after the Greek goddess of Earth) to the total system of organisms, air, and oceans. Their **Gaia hypothesis** suggests that the system will evolve to maintain that particular equilibrium within which life can survive most successfully.

As an example of how the Gaia idea could work, consider a situation in which the average input of solar energy to Earth decreases, perhaps as a result of changes in the luminosity of the Sun. The Earth (Gaia) could respond by promoting the growth of organisms that liberate more CO_2 into the atmosphere. The additional CO_2 would result in a stronger greenhouse effect, trapping additional thermal radiation and warming the surface back to its original temperature.

The concept of Gaia is certainly intriguing and adds another dimension to our growing awareness of the interdependence of life on Earth. As we now realize, the growth of human population and its effects on the environment risk more than eliminating a few endangered species and producing aesthetically unpleasant effects such as smog and dirty rivers. We face the potential of disturbing a global balance, and the response of the Earth to this disturbance may have ramifications that are impossible to predict. Because of life, the responses of the Earth to change will likely be quite different from those on a sterile planet, as Mars and Venus both appear to be.

Why You Are Not Reading This on Venus

As we saw in Chapter 10, both Venus and Earth probably started out with approximately the same inventory of volatile compounds. Venus, being closer to the Sun, however, was subject to the runaway greenhouse effect described in Section 10.2. Liquid water simply can't remain on the surface of a planet at this distance from its star.

Support for this conclusion comes from a study of the isotopes of hydrogen on Venus. Recall that in the runaway greenhouse effect, all the water on Venus gets into the atmosphere and heats it up so much that water vapor rises to great heights. At these altitudes, solar ultraviolet radiation photodissociates the water molecules with great efficiency and a huge amount of hydrogen must then escape from the planet. Hydrogen has two stable isotopes. One, given the name deuterium, is twice as massive as the other. At a given temperature, less massive atoms or molecules in a gas move faster than heavier ones (also see Section 12.7, the Quantitative Supplement).

Since the light isotope of hydrogen will escape more readily from Venus, we expect to find evidence of this fractionation (loss) when we look at the isotope ratio in the tiny, residual amount of H_2O on Venus today. Indeed we do. Measurements of this isotope ratio by mass spectrometers on Pioneer Venus and by ground-based observations of water vapor in the atmosphere of Venus have demonstrated that deuterium has been enriched over 100 times its value in terrestrial oceans. A runaway greenhouse situation certainly occurred on Venus, leaving that planet in its present, highly desiccated state.

Without liquid water, life will not develop, and with no life and no liquid water, there is no way to remove carbon dioxide from the atmosphere. If Earth were as dry and lifeless as Venus, our planet would have a massive atmosphere dominated by CO_2 with a surface pressure of 70 bars, and a surface temperature of several hundred degrees celsius.

It is possible that during the first few hundred million years of its history, there were periods when Venus was cool enough for liquid water to exist on its surface. Such low-temperature epochs could have arisen because 4.5 billion years ago the young Sun was about 25% less luminous than it is today. As the luminosity of the Sun increased, however, the climate inevitably changed, turning Venus into the hellish oven that we see today. To find a planet that may once have resembled Earth more closely, we must turn our attention to Mars.

12.3 EXTRAPOLATING FROM EARTH TO MARS

Somehow, during the first 700 million years of our planet's history, the equilibrium among outgassing, chemical interactions with the rocks, and the escape of hydrogen left our planet with an atmosphere that had enough CO_2 in it to keep the Earth warm and enough surface pressure to allow liquid water to exist at that warm temperature. This was the environment in which life began. Did the same series of events occur on Mars? Even though the martian environment is obviously very hostile today, is it possible that life could have begun during that earlier epoch when the channels were cut and liquid water was present on the martian surface? If not, we at least have the opportunity to study the first billion years of planetary history on Mars, an opportunity we do not have on Earth. We may even be able to find a record of the composition and early evolution of the martian atmosphere, preserved in the ancient rocks.

Some scientists go further and hope for the existence of highly specialized forms of life on Mars *today*, organisms that have adapted to the extreme environment. The adaptations exhibited by life on Earth are certainly extraordinary, allowing habitation of Antarctic frozen lakes and geothermal hot springs, to mention just two extremes. The search for life remains one of the objectives for future Mars missions. We can put this search in perspective by considering how Martians might attempt to find life on Earth.

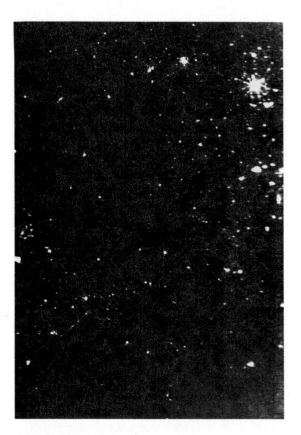

FIGURE 12.4 This satellite photograph shows Europe at night, with the bright, starlike pattern of Moscow in the upper right, connected by a line of lights to St. Petersburg. London is at the left center, with Paris below and to the right. The outlines of Spain and Italy are defined by the lights of their coastal cities.

The Search for Life on Earth

How can we determine whether there is life on another planet? The classic answer is, "Ask one of the inhabitants. Even a negative answer would be significant!" Certainly a wonderful idea, but not always possible. Suppose instead that we ask what kind of evidence would convince a martian astronomer that there was life on Earth.

There are three obvious signs of life detectable from a remote vantage point. First, a Martian might look for something in the physical appearance of the Earth that demonstrated the presence of life. Here one thinks of large-scale features that are clearly artificial in origin, or seasonal changes in surface markings that could be caused only by the growth and death of vegetation. The second clue is more subtle. We know that life on Earth has had (and is continuing to have) a profound effect on the composition of the atmosphere. Hence a spectroscopic search for gases that are not in chemical equilibrium with each other or with the surface of the planet (like O_2 on Earth) would be a clever way to look for evidence of life. Finally, a Martian could use the huge antennas and sensitive receivers normally employed by astronomers studying molecules to search for radio and TV broadcasts from a terrestrial civilization.

Until very recently, looking and listening would have failed to indicate the presence of life on Earth to a Martian who was using the same technology that we have. Clouds and the distorting effects of the Earth's atmosphere would obscure our planet's surface sufficiently that detection of large-scale human constructions or seasonal changes in vegetation would be very difficult. Although a martian astronomer would be able to view the night side of Earth (as we can observe the dark hemisphere of Venus), the nighttime glow of our cities would not have reached the level of easy detection before the widespread adoption of electric lighting less than a century ago (Fig. 12.4). Similarly, no radio transmissions would have been detectable until humans began to generate large amounts of radio power in the late 1920s. The second test, however—studying the composition of the atmosphere—would have yielded positive results, even if no humans had ever appeared on Earth.

Clues from Atmospheric Composition

Spectroscopic observations made from Mars would reveal the presence of a great number of oxygen molecules in the Earth's atmosphere, along with a small amount of methane (CH_4). This situation would provoke a Martian's curiosity, because ultraviolet light from the Sun rapidly stimulates the chemical combination of oxygen and methane to form carbon dioxide and water. The continued existence of methane on Earth therefore implies the presence of a source that replaces the methane in spite of the natural processes that destroy it. Terrestrial methane is produced primarily by bacteria that live in swampy marshes and in the forward stomachs of grass-eating animals.

In an equally striking way, the large amount of oxygen in the Earth's atmosphere would itself constitute a puzzle for our intelligent Martian, since oxygen is a highly reactive gas that rapidly combines with rocks in the crust and with magma that is continually being brought up from the interior. This process removes oxygen from our planet's atmosphere, so the continued existence of oxygen also requires a source—in this case, the presence of green plants that release oxygen through photosynthesis. Thus, a careful study of our planet's atmosphere could suggest to Martians that life exists on Earth, long before they thought about landing a spacecraft to sample the immediate environment.

Likewise, if astronomers on Earth had discovered large amounts of oxygen on Mars, they would have considered the probability of life to be very high. But after repeated efforts, they found that oxygen forms only 0.13% of the thin martian atmosphere, which has a total density less than 1% of our own. This tiny amount of oxygen can be easily explained as the result of the photodissociation of some of the water vapor in the atmosphere. Methane and other hydrocarbon gases remain undetected on Mars, so the planet shows no signs of the chemical clues that mark Earth as biologically active.

Reassessing Mars

When radio telescopes were developed in the 1950s, terrestrial astronomers used them to study Mars. These sensitive instruments detected the thermal radiation coming from the planet, allowing a determination of the average surface temperature. The results indicated that Mars at its best resembled a dry valley in Antarctica. These low temperatures agreed with earlier infrared measurements; unlike Venus, Mars held no surprises at radio wavelengths. And that was true for artificial broadcasts too. Our own planet radiates more energy than the Sun at the frequencies in the radio spectrum used for television transmission and military radars, but no artificial transmissions have ever been detected from Mars.

In short, Mars does not exhibit the three signs of life we discussed: no unexpected atmospheric gases; no "canals" or cities; no radio broadcasts. Seasonal changes were indeed observed on the surface, but as we saw in Chapter 11, these could be explained by dust blown about by martian winds. Yet the interest in Mars as a possible home for life remained high, partly in response to the detection of the dry river channels described in Section 11.6. Mars is the only planet besides our own on which we find evidence that liquid water was once present, and liquid water is essential for the existence of life. As a result, in the early 1970s, humans put nearly a billion dollars' worth of effort into a direct search for organisms on the martian surface: The United States built two Viking landers that each carried three separate experiments to test for the presence of martian microbes, plus an assortment of other instruments that could assess the habitability of the martian environment (Fig. 12.5).

12.4 THE SEARCH FOR LIFE ON MARS

Although the greatest excitement from a search for life on any planet would be the discovery of large, advanced creatures capable of communication, the history and present status of life on

FIGURE 12.5 The Viking landers packed an incredible ability to analyze the martian atmosphere and surface into a small volume. The extendable boom with a scoop that could bring soil samples to various experiments on board is at the lower left (cf. Fig. 12.8).

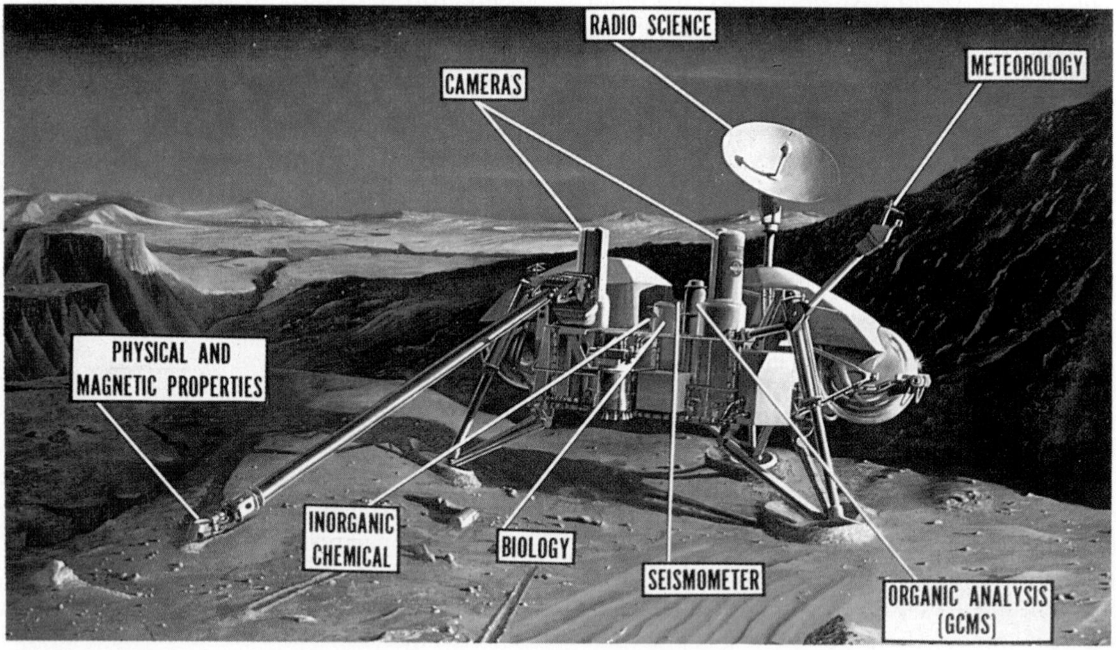

Earth demonstrate that microscopic organisms far outnumber large ones and are far more versatile in their adaptations to various environments. The fossil record shows that microbial life was the only kind of life on Earth for billions of years, far longer than larger creatures have existed. Nor have micro-organisms decreased in numbers or in adaptive ability. The dirt in your backyard contains more organisms than the number of stars in the Galaxy. There is no reason to expect another planet with life to differ from Earth in the overall development of living systems, and this is why the Viking project emphasized the search for martian microbes.

A major difficulty in any such search is trying to guess what the alien organisms would eat and drink and what effect they might have on their environment. The guiding principles have to come from life as we know it, which is carbon based and water dependent. We know water was present on Mars and is still there in the form of ice. What Viking needed to determine was the status of carbon on Mars. Is there some indigenous food for martian organisms to eat? If so, we expect to find energy-rich, carbon-containing molecules in the soil in addition to the CO_2 in the atmosphere.

Viking Results: The Search for Organic Compounds

The Viking lander made its soil analyses with a highly sophisticated instrument called a gas chromatograph–mass spectrometer, or GCMS. The GCMS baked soil in an oven to drive off volatile gases. These gases adhere to the gas chromatograph part of the instrument with different degrees of stickiness, and heating them makes the volatiles leave in sequence. The mass spectrometer analyzes the departing stream of gas to determine which compounds are present.

An example of the ability of the GCMS to detect organic matter is shown in Fig. 12.6, which illustrates the results of an analysis of Antarctic soil on Earth with a laboratory version of the Viking instrument. Although this soil contains barely enough living organisms to give weakly positive results in the three Viking life-detection experiments, the GCMS found lots of organic compounds. In other words, the fingerprint of life can be seen more easily than the living creatures themselves. For instance, the Antarctic soil sampled in Fig. 12.6 contains 10,000 times more carbon in dead organic molecules than in living micro-organisms.

FIGURE 12.6 A test model of the Viking GCMS was used to analyze Antarctic soil (a) and a piece of the Murchison carbonaceous meteorite (b). The graphs indicate that a rich variety of organic substances exists in each of these samples. Each peak in the graphs represents one or more organic substances. On this scale, the GCMS analysis of martian soil would be a straight horizontal line at zero, indistinguishable from the horizontal axis itself.

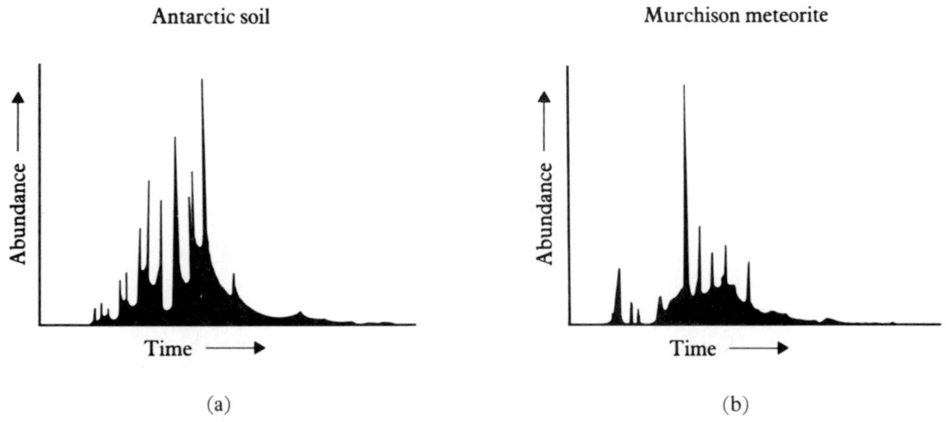

We must remember that not all organic compounds require living creatures to produce them. Figure 12.6 also shows a GCMS analysis of a carbonaceous meteorite that contains amino acids, the building blocks of proteins. Again carbon-containing molecules are present in abundance, but these organic molecules were not made by life. Thus, the detection of organic molecules in general would not prove that life exists on Mars, for the simpler kinds of organic compounds could have been brought to Mars by meteorites, or might even have arisen from photochemical reactions in the martian atmosphere.

In practice, these potential ambiguities did not plague the GCMS experiment, because it found *no organic compounds whatsoever* in the soil of Mars. On the scale of Fig. 12.6, the martian soil results would be indistinguishable from the zero horizontal line of the graphs. These results apply to both landing sites and to two different samples at each site, including one from underneath a rock that might (so it was thought) have sheltered organisms from deadly ultraviolet light. The upper limits on all likely organic compounds in the soil, such as benzene and other hydrocarbons, fall at a few parts per billion. This is the kind of sensitivity that would allow detection of the proverbial needle in a haystack.

These negative results set powerful constraints on any models of martian biology: How could life exist on Mars without leaving any trace of its presence? Could martian organisms be such efficient scavengers that no traces of their wastes, their food, or their corpses could be found, even with the great degree of sensitivity that the Viking landers brought to Mars? The biologists involved in the Viking project considered this an extremely unlikely possibility.

The Viking Biology Experiments

After considering many alternatives, the Viking project selected three experiments to search for micro-organisms on Mars. These experiments, called the gas exchange (GEX), the labeled release (LR), and the pyrolitic release (PR), all reflect scientific experience with life on Earth (Fig. 12.7). Thus, for example, all the organisms that we know derive their energy from two basic processes: oxidation (removal of hydrogen, combination with oxygen), and reduction (removal of oxygen, combination with hydrogen). The experiments were designed to measure these processes. The trick is to know what martian organisms would like to eat and drink, to supply these nutrients, and then to have a means to decide whether in fact the organisms consumed the nutrients or whether some nonbiological chemical reaction occurred.

The GEX and LR experiments both assumed that martian organisms would require water. The GEX supplied a mixture of several kinds of nutrients and then used a gas chromatograph to see what gases would be released by organisms that consumed this food. In effect, this was a search for evidence of metabolism. The LR experiment labeled its nutrient medium by using radioactive carbon in the organic compounds its nutrient mixture contained. CO_2 or CH_4 released by martian organisms could then be detected because of its radioactivity. This experiment was therefore looking for evidence of respiration.

The PR experiment was designed to imitate the martian environment as closely as possible. No nutrients were used; no water was supplied. The idea was simply to expose a sample of martian soil to a simulated atmosphere in which the CO_2 and CO were radioactive. The soil was illuminated by a lamp that matched martian sunlight as closely as possible. This experiment was thus a test for evidence of photosynthesis on Mars, by seeing whether organisms in the soil would incorporate the radioactive CO_2 and CO from the artificial atmosphere above them.

All three of these experiments were tested extensively on Earth and were found capable of detecting microbial life in extreme conditions, such as the nearly sterile dry valleys of Antarctica. The stage was set for the search for life on Mars.

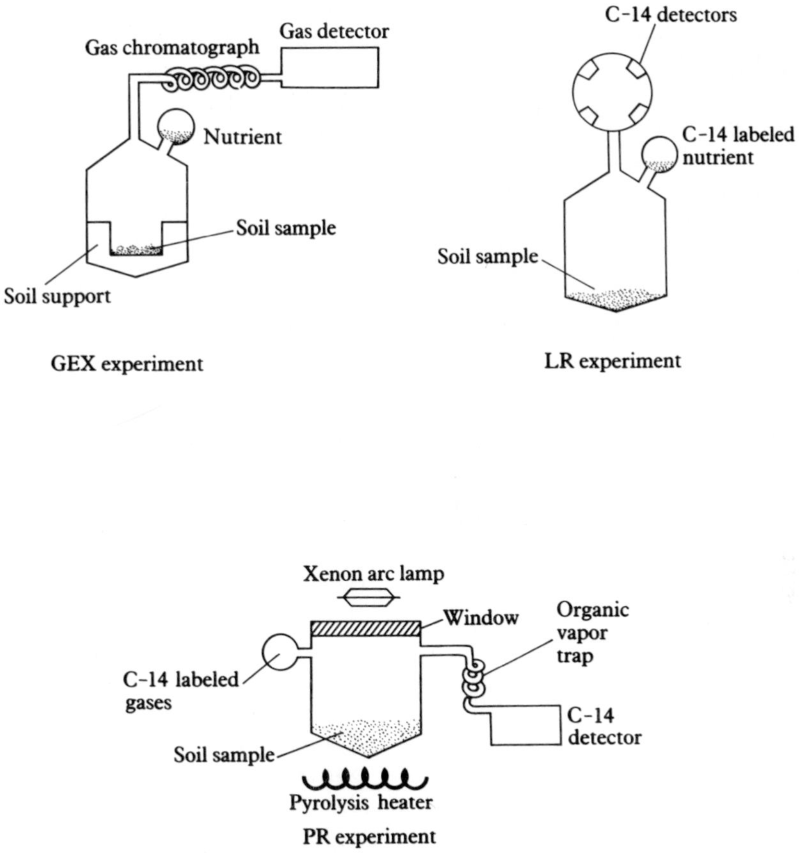

FIGURE 12.7 A schematic representation of the GEX, the LR, and the PR experiments shows that each of them analyzes the martian soil in a different way to test for the presence of life. The GEX experiment exposed a broth of nutrients to a few grams of soil and then looked for changes in the gas above the soil and nutrient mixture. The LR experiment tagged carbon-rich compounds with radioactive C-14 atoms in place of some of the usual C-12 atoms. These labeled compounds then dripped over the soil sample. Any biological processes should have caused some tagged compounds to appear in the gas above the sample. The PR experiment replaced the normal martian atmosphere with an equivalent set of gases labeled with radioactive carbon atoms. Any organisms that ingested some of these labeled molecules would produce a radioactive signal when the soil in which they lived had been roasted.

Results from GEX and LR

On July 28, 1976, the eighth day after the first landing on Mars, the Viking 1 lander scoop dug a trench in the soil and distributed samples to the various experiments (Fig. 12.8). The GEX placed about a gram of soil into a tiny, porous container positioned above the nutrient medium. Two days later, the first analysis of the gas in the container showed an exciting result: A large quantity of oxygen had appeared in the chamber, 15 times the proportion in the martian

FIGURE 12.8 In addition to taking samples directly from the surrounding soil, the Viking boom pushed aside a rock (left) to gather a sample from the ground beneath it (right). The purpose was to obtain soil that had been protected from ultraviolet light.

atmosphere. The simple exposure of martian soil to the humidity in the test chamber produced by the nutrient-laden fluid had apparently been sufficient to liberate oxygen from the soil.

Was this an indication of life on Mars? After months of testing, the biologists concluded that they were observing not biological activity but merely the chemical interaction of martian soil with a higher pressure of water vapor than had been present on Mars for millions of years. The day after the first GEX data, the LR experiment reported: again a positive result! After checking to be sure that the background radioactivity level was low, the LR added about two drops of the radioactive nutrient material to the soil that had been brought into its chamber. A sudden rise in the radioactivity of the gases above the soil sample appeared, a more dramatic reaction than biologists had found in their comparison experiments with many life-bearing soils on Earth.

Unfortunately for those who hoped for proof of life, the Viking scientists soon realized that the radioactive gas, almost certainly carbon dioxide, could arise from simple chemical reactions that involve oxygen compounds called peroxides. A second wetting of the soil showed no increase in the amount of radioactivity in the gas in the test chamber. Instead, the additional nutrient apparently absorbed some of the radioactive carbon dioxide that was originally released. Hence, the scientists concluded not that life exists on Mars, but rather that the martian soil may contain chemicals such as peroxides that release carbon dioxide when exposed to simple organic compounds.

PR Experiment Results

Since the first two experiments yielded information that seemed ambiguous, the Viking team eagerly awaited the results from the PR experiment, which did not use a water-based nutrient and thus avoided one of the primary agents (water) that the biologists suspected of causing purely chemical reactions in the soil. Analysis of the initial experiment revealed that radioactive carbon had indeed become part of compounds in the soil (Fig. 12.9). Weak as this signal was, it seemed clearly positive: To the PR experiment, martian soil behaved much like an Antarctic soil on Earth, nearly sterile but not entirely so. Yet even in this case the Viking scientists were skeptical that they had found life on Mars.

The reason for this skepticism comes from the fact that when the scientists arranged to heat the martian soil in separate experiments to 175 C and to 90 C for several hours before the radioactive gases were injected, they still found positive results from the PR experiment. The higher temperature reduced the reaction by 90%, but a positive signature still emerged; the lower temperature had no effect. Since Mars's surface temperature is never more than 30 C, the Viking scientists did not believe that any martian life could have adapted to survive three hours at

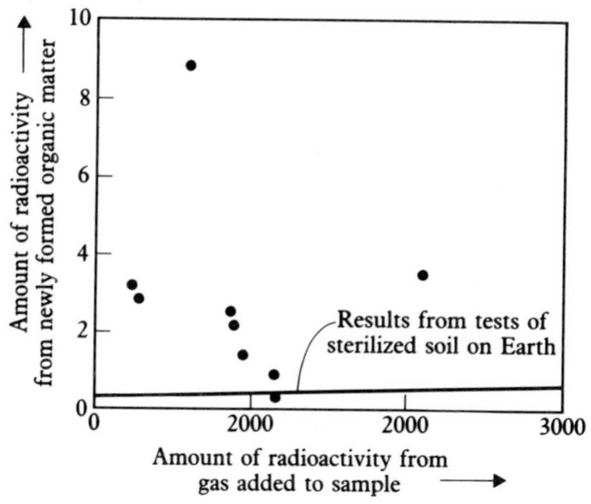

FIGURE 12.9 The PR experiment showed a rise in the amount of radioactivity in the gases above the soil being tested for living organisms once C-14-labeled CO_2 and CO were introduced. This suggested initially that living organisms had incorporated some of the radioactive carbon atoms from the simulated martian atmosphere.

175 C, which no terrestrial organisms can do. Further testing has suggested that a small amount of ammonia in the first PR soil sample—contributed by leakage from the descent engines on the Viking lander—may have been responsible for the soil chemistry that produced the first, weakly positive answer.

The Chemically Active Martian Soil

All three Viking biology experiments yielded positive results. How can we explain this, and how can we resolve this ambiguity between biology and chemistry?

There is only a tiny amount of ozone in the martian atmosphere, and it is present only in regions where water vapor is essentially absent. Thus high-energy solar ultraviolet radiation easily reaches most of the surface of Mars and apparently enhances the chemical activity of the iron-rich soil. Because the temperature on Mars is so low, and because liquid water is entirely absent from the planet's surface, compounds formed in this way can remain in chemical disequilibrium for long periods of time. If even a small amount of water is added to the soil, however, it produces chemical reactions that mimic the effects of biological activity.

The Viking missions have therefore demonstrated that martian soil probably contains no carbon-based living organisms, at least at the two landing sites. Perhaps the strongest arguments against a biological interpretation of the biology experiment results are provided by the "controls," in which heat-sterilized soil samples exhibited the same chemical reactions as the untreated samples. The absence of any organic compounds in the martian soil, to a level of parts per billion or less, is another compelling argument. These two findings effectively remove the ambiguity between chemistry and biology.

Could life on Mars be based on an element other than carbon? The GEX experiment, which showed that oxygen appears upon exposing martian soil to moisture, might be a characteristic reaction of some exotic alien biology. However, the rapid end of that reaction following the initial exposure to moisture speaks against a life-based interpretation and in favor of a chemical one.

Looking for Macro-Organisms

There was one device to detect life on Mars that was completely independent of chemistry—namely, the cameras on the two Viking landers. The pictures from Mars have been searched with extreme care for any evidence of movement, any burrow, footprint, trail, or artifact of any kind, and nothing has been seen, except the effect of winds on the dust in the empty landscape (Fig. 12.10). Since the landing sites were monitored over five years, even slow-moving organisms cleverly camouflaged to resemble rocks would have been detected.

FIGURE 12.10 These two photographs of the 2-m boulder near the Viking 1 lander (known familiarly to the Viking scientists as "Big Joe") were taken in September 1976 (left) and March 1977 (right). During the six-month interval, a small amount of sediment shifted from a rock in front of "Big Joe's" right flank (arrow).

Were the Vikings Simply in the Wrong Places?

What about life elsewhere on Mars? The similarity between the soil analyses at the two landing sites, separated by more than 2000 kilometers, argues against great variations from one place to another.

The soil particles on Mars are distributed over the entire planet during the periodic global dust storms. If every part of Mars eventually comes in contact with every other part, sampling the dusty material at one place should be equivalent to sampling at all locations. This argument gains strength from current models of the planet's surface, which suggest that the soil is turned over (gardened) by meteorite impacts, volcanic eruptions, and wind to depths of one to several meters, on timescales ranging from tens of thousands to tens of millions of years.

What about oases on Mars, regions where relatively high temperatures prevail, perhaps because of subsurface volcanism or some other cause of special microclimates? These exceptional environments are well known on Earth—in Yellowstone National Park in the United States, on the Kamchatka Peninsula in Russia, and near geothermal vents in the deep oceans. Such regions typically have very short lifetimes. As the sands shift to cover an oasis or dry spells exhaust it, as hot springs dry up in one place and appear nearby, seeds and micro-organisms must arrive from somewhere else to repopulate them. Without this outside reservoir of life, any oasis must soon become lifeless. The problem with the oasis model for life on Mars is that there doesn't seem to be any hospitable "somewhere else" on this dry, cold planet.

Comparing Mars as we see it now with our own planet, it seems clear that no known terrestrial organisms, including the toughest and most sophisticated microbes, could grow in the present martian environment. The main problem is the simple absence of liquid water. Some terrestrial organisms could survive on Mars in a dormant state, but even the algae that create their own temperate microclimate inside rocks in the dry valleys of Antarctica require more water vapor than exists in the thin atmosphere of Mars. Without a huge change in climate, even dormant life on Mars would eventually die.

12.5 ATMOSPHERES OF EARTH AND MARS

Dogmatic statements have fared poorly in the history of science, so we are somewhat reluctant to state categorically that no life exists on Mars. All the evidence accumulated so far points toward the absence of life, though some discovery may one day convince biologists that Mars does harbor some living organisms.

We have seen in the last chapter that there is ample evidence that liquid water existed on Mars during the first billion years of the planet's history. In Section 12.1, we found no reason why life couldn't have begun on Mars during this apparently favorable period. What happened to change this promising beginning? Why have the atmospheres of Mars and Earth evolved so differently, so that today the Earth is teeming with life and Mars appears totally barren?

Early Mars: What Went Wrong?

Mars is 1.52 times farther from the Sun than Earth, so the resulting decrease in solar illumination by a factor of 2.3 certainly decreases the solar heating. This means that the average surface temperature on Mars will be significantly lower than the terrestrial value. One could even imagine a "runaway refrigerator effect": as the temperature drops, less water vapor is available in the atmosphere so the greenhouse effect decreases, further decreasing the average surface temperature. The resulting increase in the latitudinal extent of the polar snow deposits will increase the planet's reflectivity, further decreasing the surface temperature, and so on. Unlike the runaway greenhouse effect on Venus, however, there is an easy way to counteract this cycle: simply increase the amount of CO_2 in the atmosphere. Adding the equivalent of only one or two Earth atmospheres of this gas (this is 1 or 2 bars, compared with the 90 bars of CO_2 on Venus) would create a greenhouse effect sufficient to warm the surface of Mars above the freezing point of water.

Is it reasonable to assume that Mars once had such an atmosphere? Based on the information currently at our disposal, the answer appears to be affirmative. Recall that our best model for the origin of the Earth's atmosphere invokes the delivery of volatiles by icy planetesimals and primitive meteorites. There is no reason to think that these messengers from the outer solar system would have missed Mars on their way to the Earth, especially when they seem to have endowed Venus with approximately the same volatile inventory as our own planet (Table 12.1). Thus the question is better phrased in

TABLE 12.1 Atmospheres of inner planets

Gas	Earth		Venus	Mars	
	Now	Reconstructed[a]	Now	Now	Reconstructed[b]
N_2	78.1%	1.9%	3.4%	2.7%	2%
O_2	20.9	trace	trace	trace	trace
Ar-40	0.93	190 ppm	40 ppm	1.6	20 ppm
CO_2	340 ppm	98%	96.5%	95.4%	98%
Water[c]	3 km	3 km	trace	trace	~0.9 km
Total pressure	1 bar	~70 bars	~90 bars	0.0065 bars	~7.5 bars

[a] For Earth, reconstruction removes the effects of water and life.

[b] For Mars, reconstruction removes the effects of impact erosion and escape.

[c] The abundance of water is given as the depth of a global layer that would contain all the water above the planet's surface.

another way: Why doesn't Mars have a thicker atmosphere today? What has happened to all those volatiles?

Reconstructing the Martian Atmosphere: Isotopes and Noble Gases

On Earth, we have recognized that the missing carbon dioxide is mainly bound up in the form of carbonate rocks. We can therefore reconstruct the total mass of outgassed carbon dioxide on Earth by surveying the rocks on our planet. Unfortunately, we cannot do the same thing on Mars. Brilliantly successful though they were, none of the Viking experiments was designed to look for the presence of carbonates, so we have no idea how much CO_2 might have been removed in this way. Instead, we must proceed indirectly in our efforts to understand whether indeed Mars ever had a truly dense atmosphere, and if so, when and for how long.

The approach we shall take is to look at abundances of the elements and their isotope ratios as we find them in the martian atmosphere today and to compare these values with similar data in the atmosphere of the Earth. We will pay special attention to the **noble gases**: neon, argon, krypton, and xenon. These gases are called "noble" because they don't associate with other elements. In fact, they don't even associate with each other! They are chemically inert and form gases consisting of single atoms, rather than molecules like N_2, O_2, H_2, and so forth. Thus once noble gases are introduced into a planet's atmosphere, they remain there; they don't combine with the rocks.

We met argon in our discussion of the Earth's atmosphere, where we discovered it constitutes 1% of our atmosphere. This argon is produced by the decay of radioactive potassium in rocks. There are two other isotopes of argon that are far less abundant. They represent primordial argon that was once part of the original solar nebula. Together with the other noble gases, the abundance of this primordial argon in the atmosphere provides a measure of the quantity of volatiles delivered to the planet and subsequently outgassed to the atmosphere.

The relative abundances of neon, argon, krypton, and xenon show the same pattern in the atmospheres of Mars and Earth. That suggests that other volatile elements, such as carbon and nitrogen, should also be present in similar relative abundances on the two planets. Knowing the ratio of total nitrogen and carbon to atmospheric argon on Earth and knowing the amounts of the noble gases on Mars, we can calculate how much carbon dioxide and nitrogen should be present there. This calculation yields a total surface pressure on Mars of 0.075 bars, about ten times the present value.

Another indicator of physical changes is afforded by the ratios of stable isotopes of common elements. For example, if a large amount of an element with more than one stable isotope has escaped from a planet's atmosphere, we expect fractionation of the isotopes to occur, just as in the case of the enrichment of deuterium on Venus. Hydrogen can certainly escape from Mars, but relatively little fractionation has occurred. Certainly no runaway greenhouse effect ever took place on this planet. A more interesting story is told by the isotopes of nitrogen.

Nitrogen atoms cannot escape from Earth, but Mars has a weaker gravitational field, and there is a kind of photodissociation of N_2 that gives the atoms enough energy to escape from the red planet. Viking mass spectrometers determined that the light isotope of nitrogen is strongly depleted on Mars. Knowing the escape process and the present value of the isotope ratio, it is possible to calculate the original nitrogen abundance. The result is that Mars must have started out with about ten times the nitrogen we now find in the atmosphere. To estimate the total atmospheric pressure corresponding to this much nitrogen, we use the ratios of CO_2 to N_2 on Venus and Earth, finding an average of $CO_2/N_2 \approx 40$. The total surface pressure on Mars corresponding to the original nitrogen abundance then works out to 0.065 bars.

Thus the two independent estimates of the total surface pressure—using the enrichment of the heavy isotope of nitrogen and the abundances of noble gases—overlap at a value of 0.07 bars, which we could use as the approximate total pressure of the early martian atmosphere. This is roughly a factor of ten more gas than we now find on Mars, but still 1000 times less than the volatile inventories on Earth and Venus. It seems that either Mars is fundamentally different from its two nearest neighbors in space, or we are still missing some crucial point about the early history of the red planet.

Water on Mars

The impression that Mars is different from Venus and Earth in some basic way is reinforced when we look again at the history of water on Mars. The geological evidence examined in Chapter 11 makes it clear that liquid water once flowed over the martian landscape. A surface pressure of 0.07 bars is only barely adequate to permit this; it provides a very narrow temperature range within which water can be in the liquid state. Furthermore, the amount of water corresponding to this reconstructed atmosphere is too small. Again using the Earth's volatile inventory for calibration, a 0.07 bar, CO_2-dominated atmosphere would be accompanied by an amount of water equivalent to a global layer of water only 9 m deep.

This 9-m layer of water derived from isotopic fractionation of the atmosphere may be compared with geologist Michael Carr's estimate of 500–1000 m for the amount of water required to cut the great flood channels on Mars, and the 200-m equivalent depth for water outgassed in the past 3 billion years deduced from the SNC meteorites. Apparently our attempt to reconstruct the martian atmosphere by determining the total inventory of volatiles produced by the planet since its formation has come up short. It is not simply that we are in sharp disagreement with what we find on Earth and Venus. Our reconstruction is internally inconsistent because we can't account for the amount of water necessary to cut the martian channels. Clearly we are still missing something that sets Mars apart from its two larger sisters.

The Role Played by Impacts

We have stated repeatedly that early impacts by comets and gas-rich meteorites are probably responsible for providing the volatile inventories on the inner planets. But impacts can take away volatiles as well as supply them. We must worry about the loss of volatiles by impacts, especially by impacts from relatively massive asteroids. Obviously, the giant impact on Earth that formed the Moon would have had a catastrophic effect on our planet's early atmosphere. Subsequent bombardment by comets could have replaced the missing gases. Similar events must have occurred on Venus, not forming a satellite in that case, but leading to approximately the same volatile inventories on these two, similar-size planets.

Mars, being both smaller in mass (about 0.1 of the Earth's mass) and closer to the asteroid belt, was much more vulnerable to atmosphere-stripping impacts. Even bombardment by smaller bodies would have had a devastating effect, and such impacts would have occurred more often. The consequent **impact erosion** of the atmosphere appears to be the missing process that accounts for the present low surface pressure on Mars.

The idea that Mars lost most of its atmosphere through successive impacts by asteroidal and cometary bodies is supported by calculations of the intensity of this bombardment during the first billion years of solar system history. These calculations suggest that Mars must have had an atmosphere with at least 100 times the present surface pressure to end up with the atmosphere we find today.

We can test this idea through further studies of the noble gases and their isotopes. The first thing we notice is that the isotopes of xenon and argon that are produced by the decay of radioactive parents are far more abundant relative to the primordial isotopes on Mars than on the Earth. Studies of martian geochemistry indicate that the radioactive parent elements were just as abundant on Mars as on Earth, so the difference must be the result of removal of the primordial isotopes rather than the enhancement of the isotopes produced by radioactive decay. Only impact erosion can accomplish this. Removal of the atmosphere through massive deposition of carbonates and nitrates won't do it, and neither will thermal escape. Thus there is observational support from the isotope abundances for the role played by bombardment in reducing the martian atmosphere to its present pitiful state. What was its original condition?

12.6 GOLDILOCKS AND THE THREE PLANETS

As we consider the three largest inner planets, we seem to find ourselves in the classic situation of Goldilocks confronting the three bowls of porridge: One is too hot, one is too cold, and one is just right. Our original objective in this chapter was to try to understand what basic properties make the Earth so different from its neighbors. Why only on our planet is the current climate "just right"?

The Trouble with Venus

Venus really is too hot. Being 41 million km closer to the Sun than Earth is has destroyed the habitability of this planet. The resulting increase in the intensity of sunlight at the surface of Venus is about a factor of two, compared with Earth. Under these conditions, liquid water cannot be stable and a runaway greenhouse effect is inevitable. The theoretical calculations that lead to this conclusion are brilliantly supported by the discovery of the huge enrichment of deuterium in the remaining water vapor in the atmosphere of Venus. An enormous amount of hydrogen must have left this planet to produce such an extreme fractionation of the hydrogen isotopes, and this is just what the runaway greenhouse scenario predicts.

When we correct for the losses, we find that the basic volatile inventory on Venus and the Earth appears to be the same. The amounts of carbon dioxide and nitrogen now in the atmosphere of Venus are very similar to the total amounts of these gases the Earth has produced over geologic time. Among the major volatiles, only the water is absent, and now we know why. There are some interesting discrepancies among the noble gases on Venus compared with Earth (and Mars), but these do not seem to affect the basic conclusion that all three of these planets should have started out with roughly the same proportions of carbon and nitrogen compounds, as well as water, relative to their individual masses. In other words, when a rocky inner planet of a given mass forms, we can expect a certain inventory of volatiles to accompany it, based on our experience with Venus and Earth. This realization will help us to reconstruct the early atmosphere of Mars.

Early Mars: Was There an Ancient Eden?

If the problem with Venus is being too close to the Sun, the problem with Mars is being too small. The 0.07-bar atmosphere we have reconstructed is nothing more than the volatile inventory left behind after the last major martian impact. That event has drawn a kind of screen over the early history of the atmosphere that we cannot penetrate with real confidence.

Once again we turn to the noble gases. If we assume that both Mars and Earth started out with atmospheres that were similarly dense, we should find the same amount of noble gases per gram of planetary rock on both planets today. We don't. We would need to add 165 times the amount of primordial argon that we find on Mars today to bring the ratio of gas to rock on Mars to the value we find on Earth. This is consistent with the factor of at least a 100 increase determined from the calculations of the effect of impact erosion. If we then add the carbon and nitrogen in their proportions on Earth, we would produce a CO_2-N_2 atmosphere on Mars with a surface pressure of about 7.5 bars (Fig. 12.11), over 1000 times the present value.

This high surface pressure may never have existed on Mars, just as the Earth's surface probably never felt the weight of the 70 bars of CO_2 produced by the planet over geologic time. However, this inventory certainly permits the existence of a 1- or 2-bar CO_2 atmosphere at a given time, which is sufficient to provide surface temperatures that would allow liquid water to exist on the martian surface. The inventory we have reconstructed also includes a ~0.9-km-deep global layer of H_2O, an ample amount to produce the channels and other erosional features we have discussed.

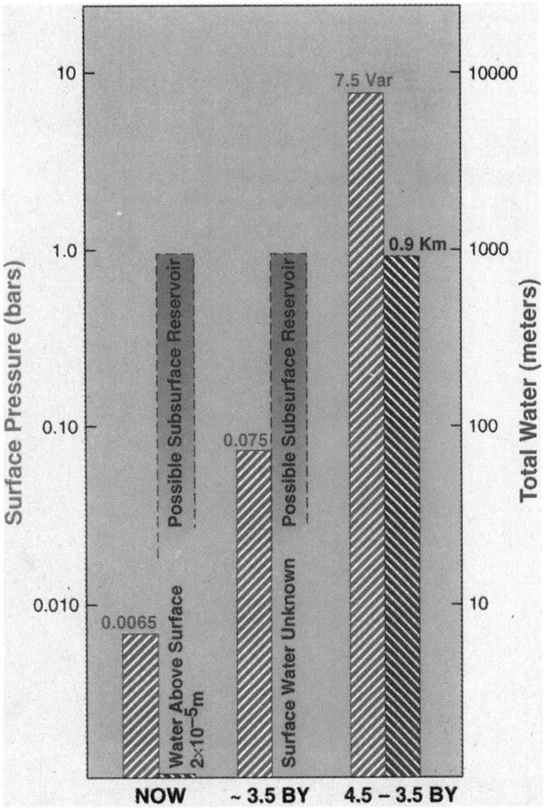

FIGURE 12.11 In this diagram, the original inventory of volatiles on Mars is shown by the bars marked 4.5–3.5 BY (right). During this 1-billion-year interval, impact erosion reduced the atmosphere to a pressure of about 0.07 bars (center). Much of this CO_2 may now be adsorbed on soil particles and frozen out in the polar caps, leaving the 0.0065-bar atmosphere we see today (left). (Note the logarithmic scale in which ordinates increase by factors of ten.)

Life on Mars?

We are now able to re-evaluate the problem of life on Mars. Perhaps no life exists on this intriguing red planet today, but some future mission might still find traces of life's early beginnings there. It is arresting to realize that the oldest evidence of life on Earth (Fig. 12.2) formed at an epoch when the availability of liquid water on the martian surface was just coming to an end. In addition to producing the spectacular channels and the smaller branching drainage systems, the water on Mars also seems to have collected in some places long enough to produce lakes and ponds, allowing deposition of layered sediments (Fig. 12.12). By 3.5 billion years ago, life on Earth had evolved to a level of complexity that enabled the production of macroscopic fossils. Might the same thing have happened on Mars?

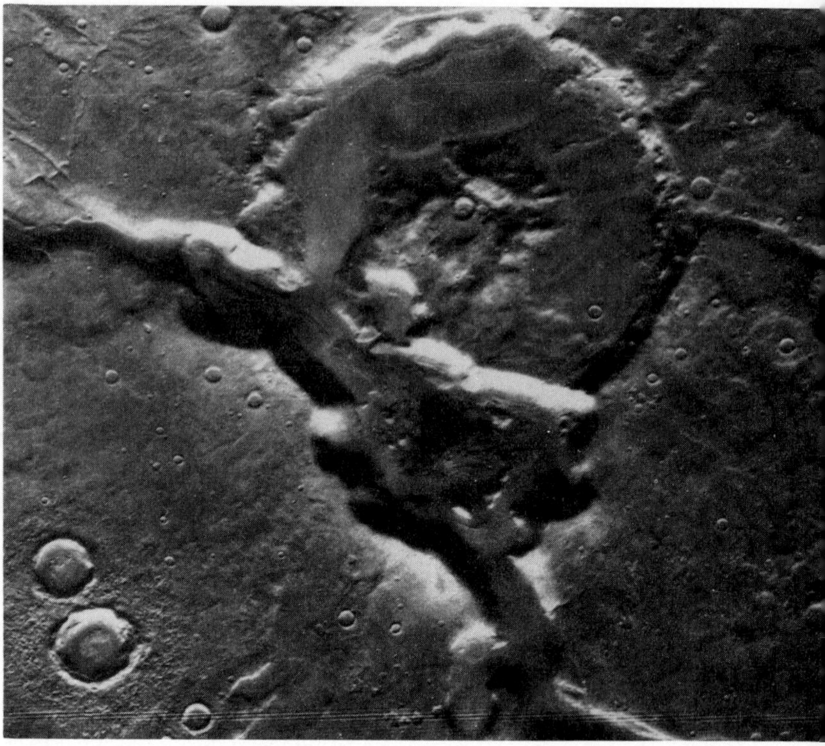

FIGURE 12.12 A flood channel intersecting martian craters. Evidently water from the channel has poured into the large crater at the left-hand side through the breach in the crater wall. Such craters may have been the martian equivalent of ponds or tide pools on the early Earth.

The answer to this question will not be easily won. After all, those ancient fossils on Earth were discovered only in the 1970s, although we know our planet rather well. The area of the surface of Mars is about the same as the area of all the land on Earth, so to find ancient fossils may require an extensive search. We will need roving vehicles that can move through the martian canyons, take samples as directed, and beam pictures back to Earth. Several possible landing sites for such missions have already been selected. In the best case, we should bring soil and rock samples back to our laboratories, where we can study them in detail. Missions that would involve this sort of sample return from Mars are now under consideration by the world's space-faring nations for possible implementation during the next two decades. It is a project worthy of the combined efforts of all interested peoples on our planet.

12.7 QUANTITATIVE SUPPLEMENT: ATMOSPHERIC ESCAPE

A molecule or atom can escape from a planetary atmosphere when its energy is sufficient to overcome the gravitational attraction of the planet. Consider a molecule (or atom) of mass μ in the atmosphere of a planet with radius R, located sufficiently high that if it begins moving outward with sufficient speed, it will not encounter any other molecules. Its kinetic energy is given by the formula

$$E = \tfrac{1}{2}\mu v^2$$

where v is the speed of the molecule. The condition necessary for escape is that this energy be greater than the molecule's potential (or gravitational) energy, given by

$$E = GM\mu/R$$

where M is the mass of the planet and G is the universal constant of gravitation ($G = 6.7 \times 10^{11}$ newton m^2 kg^{-2}).

If we equate the kinetic and gravitational energies, we can solve for the speed at which the molecule is just able to leave the planet, called the *escape velocity*, v_e:

$$v_e = \sqrt{2GM/R}$$

Note that the mass μ of the molecule is not a part of this equation, since it appears in the expressions for both kinetic and gravitational energy. Thus the escape velocity for a rocket, a baseball, or a molecule will be the same on the same planet. On the Earth, this escape velocity is 11 km/s.

In a planetary atmosphere, the speeds achieved by atoms and molecules depend on their masses and temperatures. From thermodynamics, we know that in a mixture of gases each species will have the same kinetic energy. If the energy is the same, then the less massive molecules in a mixture must be moving faster while the more massive ones must be moving more slowly. The average thermal velocity (v_t) of a molecule of mass μ is given, according to an expression derived by the nineteenth-century Scottish physicist James Clerk Maxwell, by

$$v_t = \sqrt{3R^*T/\mu}$$

where T is the local temperature and R^* is the universal gas constant ($R^* = 8.31$ joule deg^{-1} $mole^{-1}$).

If a planet is to retain its atmosphere over the lifetime of the solar system, we require that the average thermal velocity of the molecules be considerably less than the escape velocity, or

$$v_t << v_e$$

Massive molecules at low temperatures have the best prospect of remaining bound to their planet, since their thermal velocities will be lowest. We can write this using the expressions already given for the thermal and escape velocities:

$$\sqrt{3R^*T/\mu} << \sqrt{2GM/R}$$

Since there is a distribution of molecular velocities about this average value, at any given temperature some of the faster molecules will still escape. Calculations that allow for the range of speeds present indicate that a gas will be retained over billions of years if it meets the condition

$$\sqrt{3R^*T/\mu} < 0.2\sqrt{2GM/R}$$

SUMMARY

In spite of dedicated searches on other worlds, Earth remains the only inhabited planet we know. Current ideas for the origin of life indicate that our planet's early atmosphere must have contained no free oxygen. Instead, compounds such as N_2, CO, and H_2O would have permitted the chemical reactions necessary for the origin of life. Just how life began is not yet clear, but we find no reason at this stage to exclude the possibility that early Mars possessed the same conditions that led to the origin of life on Earth.

As the first primitive life-forms evolved, some of them developed the capability to perform photosynthesis, thereby connecting life on Earth with a source of energy that will last for billions of years. This connection eventually led to a change in the composition of Earth's atmosphere, as green planets produced abundant free oxygen. The discovery of several percent of oxygen in another planet's atmosphere would be a sure indication of the existence of life, even without observing evidence of plant growth or detecting signals from a technically advanced civilization. Life on Earth would have revealed its presence through its effects on the atmosphere for the last 2 billion years. No comparable evidence for life exists on Mars.

In fact, the current martian environment is extremely hostile to life as we know it, with short wavelength ultraviolet radiation reaching the planet's surface, temperatures that sink more than 100 C each night, and, most important, no possibility of liquid water. Nevertheless, knowing as little as we do about life as a general phenomenon, some scientists believe we should keep open minds and continue the search for life on Mars that began with the Viking landings in 1976.

The Viking spacecraft found no evidence for organic compounds in the martian soil at levels less than one part in a billion. Although each of three Viking biology experiments appeared to give a positive response, the same response was obtained from heat-sterilized martian soil, indicating that chemistry, rather than biology, was responsible. No indications of highly adapted large organisms were detected, and there seems little hope of finding environments on Mars that are radically different from the Viking landing sites, thereby allowing life to exist.

We can mentally reconstruct an early atmosphere on Mars that was ten times more massive than the present one by studying the abundances of the noble gases and the isotopes of nitrogen. This would not have been enough atmosphere to provide the water necessary to cut the giant outflow channels on the martian surface.

Investigating isotopes of noble gases on Mars and Earth produced by radioactive decay, we find that over 1000 times the present atmosphere is missing. This loss apparently occurred as the result of impact erosion, to which Mars, with its small size and its proximity to the asteroid belt, was especially vulnerable. Venus, in contrast, was large enough to have a stable atmosphere, but it is sufficiently close to the Sun that liquid water cannot remain on its surface, and a runaway greenhouse ensues. The enormous enrichment of deuterium in the tiny amount of water left on Venus today is a result of this runaway greenhouse effect.

Comparing Venus and Mars with Earth provides a sobering reminder of just how special our planet is. Earth is the right size and at the right distance from the Sun. It is now simply up to us, its human inhabitants, to avoid the destruction of its habitability.

KEY TERMS

Gaia hypothesis	noble gases
impact erosion	photodissociation

Review Questions

1. What conditions were necessary for the origin of life on Earth? And how did life begin?

2. Describe the interplay between life and the atmosphere. How does life affect the Earth's atmosphere today?

3. How would you go about finding life on another planet? Imagine yourself trying to find life on Earth. Which of your tests would work now? Which would work 1000, 1 million, or 1 billion years ago?

4. Describe the Viking GCMS experiment. What did it show? Why were these results important in the search for life on Mars?

5. Describe each of three Viking biology experiments, how they worked, and what results they obtained. Did they prove or disprove the existence of life on Mars? Explain.

6. What are the noble gases? How can we use these to understand the thinness of the present martian atmosphere?

7. Why was impact erosion of the atmosphere more significant on Mars than on Earth? When did it occur? What was happening on the Moon at that time?

8. Why is the atmosphere of Venus so much thicker than that of Earth? Cite some evidence from the isotopes of hydrogen in the present atmosphere of Venus which supports your argument.

9. Explain what is meant by the phrase "Goldilocks and the Three Planets." How does our experience with life on Earth and the evolution of the atmospheres of Mars and Venus make you feel about the prospects for finding life elsewhere in the universe?

Quantitative Exercises

1. We have repeatedly stated that Earth has the equivalent of 70 bars of CO_2 stored in carbonate rocks. Suppose you took a column of CO_2 that produced a surface pressure of 70 bars on Earth and moved it to Mars. What surface pressure would you measure on Mars?

2. Imagine that a comet with a diameter of 20 km suddenly struck Mars, with a velocity sufficiently low that everything stayed on the planet: the preimpact atmosphere plus all the volatiles contributed by the comet. The present water vapor content of the martian atmosphere is equivalent to a global layer of water 20 mm thick. Would the water from the comet make a noticeable contribution to this inventory?

3. If the average temperature in the atmosphere of the Earth is 260 K, find the mass μ of the lightest atom that can be retained over the age of the solar system. Repeat the calculation if the actual temperature of the upper atmosphere where escape takes place is 600 K.

4. Calculate the smallest molecular mass μ that would remain on Neptune's satellite Triton for 4.5 billion years, if its atmosphere has the same temperature as the surface, 37 K. What gases could you suggest that would satisfy this constraint?

Additional Reading

Cooper, H. 1980. *The Search for Life on Mars.* A description of the Viking mission by a science writer for *New Yorker* magazine.

Goldsmith, D., and T. Owen. 1993. *The Search for Life in the Universe, 2d ed.* Reading, Mass.: Addison-Wesley. A lively review of the origin of life on Earth and the search for life in our solar system and beyond.

Horowitz, N. 1986. *To Utopia and Back.* New York: Freeman. An exciting account of the Viking quest for signs of life on Mars and its significance, written by the principal investigator of the PR experiment.

Margulis, L. 1984. *Early Life.* Boston: Jones and Bartlett. Discussion of the origin of life by a noted biologist and proponent of the Gaia hypothesis.

13

Jupiter and Saturn: The Biggest Giants

This is an artist's rendition of the successive collisions between the many fragments of Comet S-L 9 and the planet Jupiter. We are looking toward the planet's southern hemisphere just as one fragment hits, while others are lined up, on their way in.

When ancient astronomers named the planet Jupiter for the king of the gods in the Greco-Roman pantheon, they had no idea of the planet's true dimensions. The name is entirely appropriate, however, since Jupiter is more massive than all the other planets combined. It has a faint system of rings and 16 known satellites. Of these, Ganymede and Callisto are larger than the planet Mercury, and Io is wracked by active volcanism. Jupiter itself has an internal source of heat; it is radiating about twice as much energy as it receives from the Sun. This giant also has the strongest magnetic field of any planet and a giant magnetosphere; at some radio frequencies, the magnetosphere can radiate more energy than the Sun itself.

Saturn is not far behind in these superlatives, even ahead in some. Most famous for its beautiful, intricate system of rings, this planet also has an internal source of energy, although its cause is not the same as Jupiter's. Saturn has 19 satellites; one of these, Titan, is only slightly smaller than Jupiter's giant Ganymede. Titan possesses a dense atmosphere of nitrogen and methane in which photochemical reactions are producing complex organic compounds, forming a dense smog that hides the surface from view.

Although these two huge planets are both made mostly of hydrogen and helium, their appearances are quite different. The visible surface of Jupiter is a deck of clouds that exhibits pastel shades of various colors in addition to the white condensation clouds one would expect. These clouds are organized into bands that run parallel to the planet's equator, producing a pattern of alternating dusky belts and lighter zones which is visible even with a small telescope. The planet Saturn is rather bland compared with Jupiter. The zonal structure in Saturn's cloud deck is much less pronounced, and the color of these clouds is more uniform. Still, storm systems are visible and occasionally a huge storm erupts.

13.1 THE PIONEER AND VOYAGER MISSIONS

Knowledge about the Jupiter and Saturn systems increased dramatically in the period 1974–81 as a result of two sets of spacecraft: Pioneers 10 and 11 launched in 1972, and Voyagers 1 and 2 launched in 1977 (Fig. 13.1). Most of what we know about these two worlds is derived from these four highly successful missions.

The Pioneer Encounters

It is considerably more difficult to travel to Jupiter and Saturn than to Mars or Venus. The greater distances the spacecraft must traverse mean longer trip times and thus a requirement for higher reliability of all the experiments and the spacecraft systems that make them work. In the inner solar system, electrical power can be obtained by means of solar panels that convert sunlight to electricity. But even at Jupiter's distance, the decrease in the intensity of sunlight prevents production of the power required to operate the spacecraft. Hence missions to the

FIGURE 13.1 The launch of Voyager 1 on Sept. 5, 1977, the first of two Voyager spacecraft to make the journey to Jupiter and Saturn. It has crossed the orbit of Pluto and is heading toward the Oort comet cloud.

outer solar system must include small electric generators powered by the heat released by the radioactive fuel they carry.

The large distances also create difficulties in communication. Just as sunlight decreases in intensity with the square of the distance, so do the radio waves carrying messages to and from the spacecraft. That means huge antennas and sensitive receivers must be used. During part of the Voyager 2 encounter with Neptune, for example, the signals from the spacecraft, traveling at the speed of light, took four hours to reach the Earth. At such large distances a spacecraft must be much more autonomous than if it were visiting Venus, where commands and responses can be sent in a matter of minutes.

Despite all these difficulties, the early missions to the outer solar system were very successful. The first of these were Pioneers 10 and 11, which encountered Jupiter in 1974 and 1975. Before these pathfinder missions, it was uncertain whether spacecraft could be sent safely to the outer planets. There was fear of destructive impacts with dust in the asteroid belt, which must be crossed to reach the giant planets. The Van Allen belts of the inner jovian magnetosphere also posed a hazard.

In addition to their role as scouts for the Voyagers, Pioneers 10 and 11 carried the first scientific instruments to the outer solar system. Well instrumented to map out the charged particle belts, the Pioneers made numerous discoveries about Jupiter's magnetosphere as they passed through the system. A few photographs and other measurements were made of the planet, but since this was a spinning spacecraft, the quality did not approach that achieved by the Voyagers a few years later.

The trajectory of Pioneer 11 took it close enough to Jupiter to allow a gravity-assisted deflection (Section 1.5) across the solar system for a close flyby of Saturn in 1979. This passage led to the discovery of a new ring (designated the F Ring), as well as to the first description of the planet's magnetic field and belts of trapped charged particles. Again, the Pioneer served as pathfinder, passing through the plane of the rings at the same distance from Saturn required by Voyager 2 for its gravity-assisted trajectory to Uranus and Neptune. Although an increase in the rate of the dust impacts was recorded, there were no adverse effects on the spacecraft.

The Voyager Encounters

The Voyagers (Fig. 13.2) were among the most sophisticated spacecraft ever used for planetary exploration, rivaling in size and complexity the two Viking landers. Launched in 1977, Voyagers 1 and 2 reached Jupiter in March and July of 1979. Each encounter actually stretched over many weeks, as the spacecraft approached the planet while making close flybys of the four large

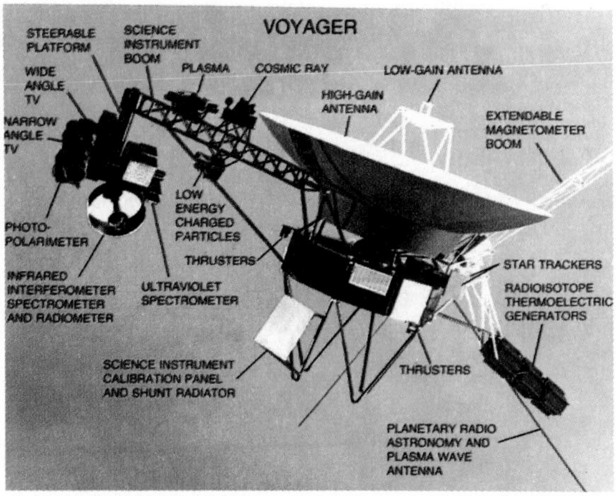

FIGURE 13.2 A diagram of the Voyager spacecraft showing the location of the various instruments. The big antenna at the top is about 4 m in diameter (cf. Fig. 3.14).

satellites. Gravity assists at Jupiter then sent both spacecraft on to Saturn, where Voyager 1 arrived in November 1980, followed by Voyager 2 in August 1981.

The Voyager 1 encounter with Saturn was targeted to provide a passage close to its largest satellite, Titan. This meant that the spacecraft could not pass close enough to Saturn to get the necessary boost required to reach Uranus or Neptune. However, the trajectory of Voyager 2 was chosen so that the spacecraft could fly on to the Uranus system in January 1986, where a third gravity assist directed it to Neptune, where it arrived in August 1989 (Fig. 13.3).

The Voyager spacecraft were controlled from the Jet Propulsion Laboratory (JPL) in Pasadena, as were the Vikings. During the planetary encounters, a staff of several hundred people operated these missions using complex computers and communication systems. In addition, more than 100 scientists associated with 13 spacecraft instruments were members of the Voyager team, which was led by Project Scientist Edward Stone, a physicist and magnetosphere expert from Caltech who headed the scientific operation and served as project spokesperson (Fig. 13.4). Stone was awarded the National Science Medal for his superlative direction of mission science. He subsequently became the director of JPL.

The two Voyager spacecraft returned so much information from the outer solar system that the data are still being analyzed some 15 years after the initial encounters. As the description of their trajectories indicates, these were flyby missions—no instruments were sent into the planets' atmospheres. A complement of sophisticated equipment on the spacecraft made measurements of the magnetic fields and the properties of charged particles in the immediate environment, and used remote-sensing techniques to study the radio, light, and infrared radiation reflected and emitted by the planets, satellites, and rings.

FIGURE 13.3 Because of a favorable lineup of the planets that occurs only once in 126 years, the Voyager spacecraft in 1977 could fly from Earth to Jupiter, Saturn, Uranus, and Neptune in less than 12 years. The decision to make a close flyby of Titan with Voyager 1 prevented this spacecraft from going on to Uranus and Neptune. Voyager 1 passed Saturn at too great a distance to obtain a sufficient "gravitational boost" to turn its course toward Uranus.

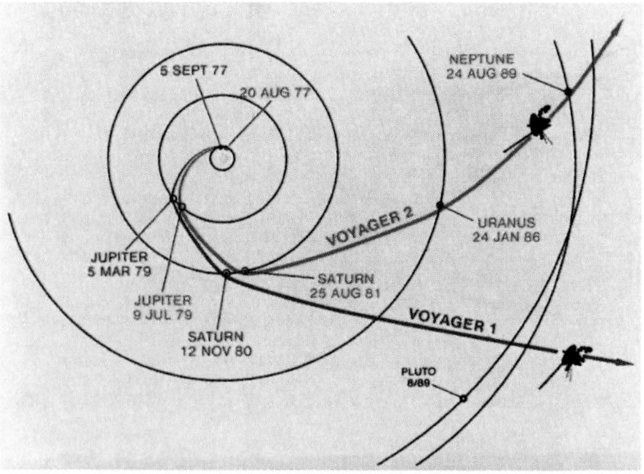

FIGURE 13.4 Voyager Project Scientist Edward Stone of Caltech, who supervised the Voyager science teams throughout the mission.

Successful as they were, the Voyagers by no means represent the end of our efforts to understand the outer solar system, since they simply flew past the planets. The next step in exploration must be orbiters and probes. The Galileo spacecraft that brought us our first picture of an asteroid went into orbit in December 1995 after deploying a probe in the planet's atmosphere (Fig. 13.5). A Saturn orbiter called Cassini with a probe named Huygens that will be delivered into the atmosphere of Saturn's satellite Titan is being prepared as a joint mission by NASA and the European Space Agency (ESA). It is scheduled for a launch in October 1997, to arrive in 2004.

13.2 INTERNAL STRUCTURE: JOURNEYS TO THE CENTERS OF GIANT PLANETS

We can see the basic difference between the outer and inner planets by looking at their respective densities. The density of Jupiter is only 1.3 g/cm^3, about the same as that of the Sun, while the density of Saturn is even lower: 0.7 g/cm^3. For comparison, recall that the densities of the inner planets tend to increase with the planets' masses (except for anomalous Mercury), because gravitational compression increases the densities of material in the interiors. Since Jupiter and Saturn would have densities far greater than 6 g/cm^3 if they were made of the same metal and rock as Earth, they must be made of something far less dense.

Constructing Theoretical Models

The explanation for the low densities of Jupiter and Saturn, as we noted back in Section 3.1, involves the two most common elements in the universe, which are also the two lightest and the two most difficult to condense: hydrogen and helium. Early studies of these two planets quickly demonstrated that they could not be composed of pure hydrogen. If they were, their densities would be even lower than they are. Adding helium to the models in about the same proportion that it is found in the Sun and other stars makes up the difference. There is also a

FIGURE 13.5 The Galileo spacecraft was forced to follow a rather complex trajectory en route to Jupiter, since it did not have the benefit of the powerful launch vehicles that sent out the Voyagers. Nevertheless, this excursion yielded additional scientific benefits in the form of encounters with Venus and the asteroids Ida and Gaspra.

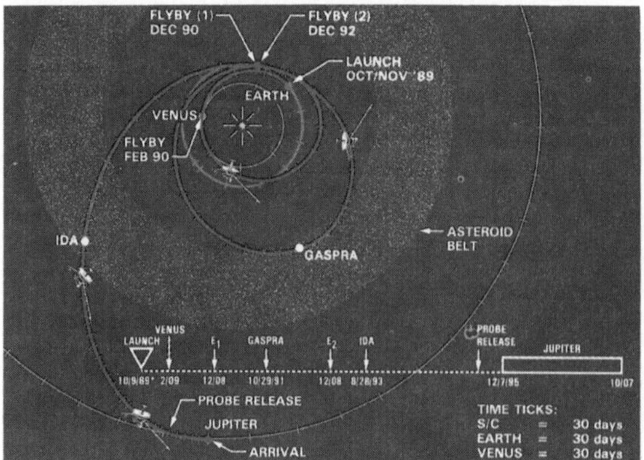

need for a dense core at the centers of these planets to account properly for the planets' gravitational fields.

It is difficult to go further in our description of the interiors with great confidence. We have no seismic data to tell us about properties of the interiors, nor can we expect to get any, since neither of these planets has a solid surface. Instead we must use the information we can obtain from remote investigations—mass, radius, rotation rate, heat balance, atmospheric composition, gravitational effects on satellite orbits—to construct models for the distribution and state of matter deep inside Jupiter and Saturn. This challenge is in many ways similar to that of modeling the interiors of the terrestrial planets (Section 8.5), without the advantages of seismology.

A basic problem in constructing such models of the giant planets is our limited knowledge of the behavior of the planets' principal constituents, hydrogen and helium, at appropriately high pressures and temperatures. The central temperature of Jupiter must be about 25,000 K to be consistent with the emitted thermal radiation, while the pressure there may be as much as 100 million bars, compared with just 4 million bars at the center of the Earth.

Even before these extreme conditions are reached, we know that hydrogen will first liquify and then assume a metallic state: The pressure squeezes the hydrogen atoms so much that the electrons are no longer bound to the nuclei. The liquid hydrogen then has the conductivity of a metal. This transition occurs at a depth of about 20,000 km, or about 75% of the distance out from Jupiter's center; the exact range of pressures and temperatures for this transformation is not well known. Above this zone, hydrogen is in the molecular form of two atoms linked together (H_2), but both the molecular and metallic states are liquids. At still greater distances from the center, the hydrogen assumes a gaseous state, but most of the mass of the planet is liquid. Looking at Jupiter from the outside, we simply see the upper layers of the deep gaseous atmosphere.

The Cores of the Giant Planets

Continuing our journey to the centers of these planets, we would find that beneath the deep layer of liquid metallic hydrogen there is a core of rock and ice. Here the terms ***rock*** and ***ice*** are used very generally to denote compounds of silicon, oxygen, metals, and the heavy volatile elements. We have no idea exactly what materials are present. We only know that there is a strong concentration of mass at the center of each planet, and that this mass must consist of elements heavier than hydrogen and helium.

Both Saturn and Jupiter have cores of about the same size: 10–15 times the mass of the Earth. Since Saturn is smaller than Jupiter, its core is ***relatively*** larger, compared with the planet as a whole. In other words, Saturn is depleted in hydrogen and helium, relative to Jupiter; if Saturn had its full quota of these two elements, it would be as large as Jupiter. We shall find that Uranus and Neptune also have cores of about this size: 10–15 Earth masses (Chapter 14). It appears that the formation of such a solid core may be a prerequisite for the formation of a giant planet.

Calculations based on theories for the behavior of matter at the temperatures found in the giant planets indicate that Jupiter is just about as large (in diameter) as a planet can be. If we perform the thought experiment of adding more mass to Jupiter, we would find that self-compression, due to the increased mass, would lead to a smaller diameter. Eventually, adding enough mass would cause the internal temperatures and pressures to increase to the point where nuclear fusion reactions would begin. At this point, corresponding to 70–80 times the current mass of Jupiter, a star is formed. Once

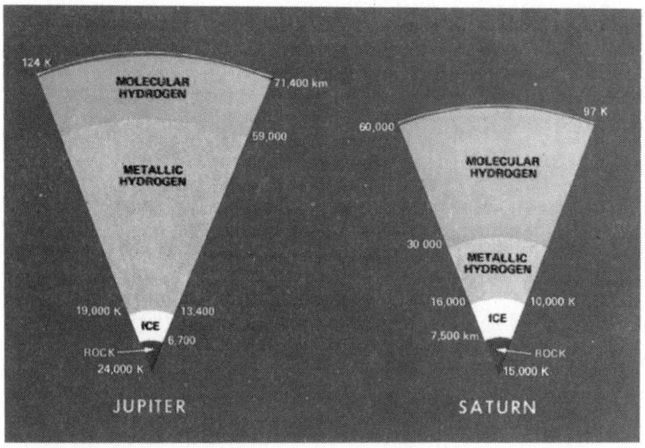

FIGURE 13.6 The interiors of Jupiter and Saturn probably consist of the same basic components, but they are distributed in different ways, as these simplified diagrams indicate. The interior of Saturn is not as hot as that of Jupiter, allowing helium to condense. Both have cores at their centers with masses approximately equivalent to 10–15 Earths.

internal energy is generated by nuclear reactions, a new adjustment of internal structure takes place, allowing the object to grow in size. Meanwhile, it is noteworthy that a super giant planet with 10 or even 50 times the mass of Jupiter would actually be smaller than Jupiter itself.

Figure 13.6 illustrates the conditions one would encounter on voyages to the interiors of these two planets. There are no well-defined solid surfaces or even gas-liquid boundaries. Such transitions are only sharp under relatively low pressures. Instead we find immensely deep atmospheres in which the gases gradually become compressed to the point where they liquify. In leaving Mars to venture into the outer solar system, we have left behind the planets on which it is possible to stand (except for Pluto—and don't forget the satellites!).

Two Overheated Giants

Both of these giant planets radiate more energy than they receive from the Sun: Jupiter twice as much, Saturn nearly three times. We can think of each planet as a giant lightbulb: Jupiter is radiating a total of 4×10^{17} watts, while Saturn is glowing with half this power, 2×10^{17} watts. These planets are therefore equivalent to 4 and 2 million billion 100-watt lightbulbs, respectively, but the energy they radiate is not emerging as visible light. The outer, visible layers of both planets are far too cold for that. Instead, the energy emerges as infrared radiation.

It is possible to calculate what the temperature of Jupiter should be from a knowledge of the local intensity of sunlight and the albedo of the planet. The answer is about 107 K (−166 C). Observations of Jupiter at infrared wavelengths show that the amount of radiation it produces corresponds to a temperature about 20° warmer than this calculation. Since the flux of radiation increases as the fourth power of the temperature, this means that Jupiter radiates twice as much energy as it gets from the Sun, as we stated earlier.

What is generating this extra energy? As we have seen, the answer does not lie in nuclear reactions, the ultimate source of sunshine and starlight. So instead of invoking nuclear physics, let's go back to Isaac Newton. If an apple falls from a tree, it acquires energy of motion from the Earth's gravitational field, which it gives up abruptly when it hits the ground. In the case of Jupiter, it is the contraction of the entire planet from an initial state when it was an extended cloud of gas and dust which produces most of the observed energy. This cloud was a condensation in the primordial solar nebula. Newton's apple may now be thought of as a piece of Jupiter itself, moving inward under the influence of the planet's gravitational field. The resulting energy of motion is transformed into faster motions of the gas molecules—an increase in the temperature of the gas. (This is the same process once invoked to power the Sun, before thermonuclear fusion was discovered; see Section 2.4.)

During the final stages of its formation process, when matter was streaming into the planet from considerable distances, Jupiter must have been very hot indeed, probably hot enough to produce a visible reddish glow. About 4.5 billion years ago, this hot young planet had a profound effect on its forming satellites, as we shall see in Chapter 15. Jupiter today is much quieter, slowly releasing primordial heat from that bygone era, possibly still contracting very slowly at a rate no larger than 1 mm/year.

Helium Rain: A Different Energy Source

The simple contraction scheme just described will not account for all of the energy radiated by Jupiter, and it doesn't work for Saturn at all. Saturn is sufficiently smaller than Jupiter that it never reached the red-hot stage in its early youth. Hence there has been ample time in the ensuing 4.5 billion years for this planet to cool down much further than its giant neighbor. Why then is it still emitting so much energy?

The answer, ironically, is found in the lower *internal* temperature that Saturn exhibits today. At sufficiently high pressures and temperatures, liquid helium dissolves in liquid hydrogen, in the same way that a cook can dissolve large amounts of sugar in hot water. But just as the cook has trouble stirring sugar into cold water, at the lower temperatures in Saturn's interior, helium does not dissolve. Droplets of helium form in the liquid hydrogen, and, being more dense, they move toward the center of the planet, like vinegar separating from the olive oil in a salad dressing. This very slow helium rainfall, deep in Saturn's interior, takes us back to Newton's apple: Once again gravitational energy is converted into energy of motion, which is transformed into heat and ultimately radiated into space.

This theoretical model for processes occurring far down inside Saturn can be tested by remote sensing. If helium is indeed raining out in the interior, it must be disappearing from the atmosphere. Therefore we expect to find the abundance of helium relative to that of hydrogen to be much smaller in Saturn's atmosphere than in the Sun.

The Voyager infrared spectrometer measurements show exactly this result. The solar mass of He/H is 26/100. On Saturn, He/H is about 3/100, an abundance distinctly lower relative to hydrogen than the amount observed in the atmospheres of Jupiter or the Sun. The amount of missing helium is consistent with a rate of helium precipitation that would produce the extra energy that Saturn and Jupiter radiate. Hence there is good consistency between the model for this internal energy source and its two observable consequences: the amount of helium in the atmosphere and the amount of energy radiated by the planet.

13.3 ATMOSPHERIC COMPOSITION AND STRUCTURE

The first gas to be identified in the atmospheres of Jupiter and Saturn was methane. Over 50 years ago, strong absorption bands in the spectra of both planets (Fig. 13.7) were found to coincide with absorptions in laboratory spectra of methane gas. Ammonia was discovered on Jupiter in the same way at about the same time, but whether it was present on Saturn remained a controversial problem until the late 1960s, when conclusive evidence for it was obtained. Hydrogen and helium, the most abundant gases in the atmospheres, were much more difficult to detect, but they were also identified in the 1960s.

At the pressures and temperatures in the regions of these atmospheres that we see, ammo-

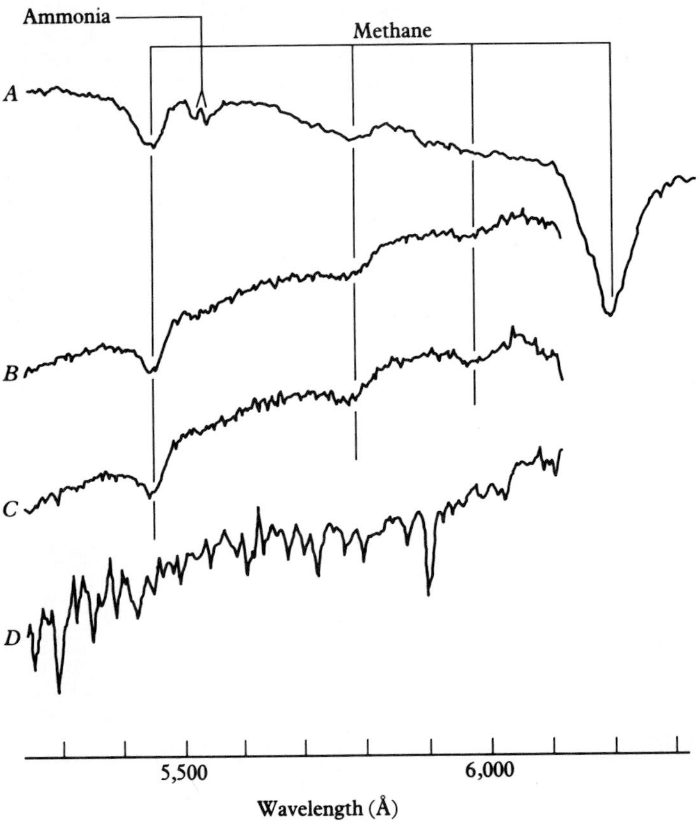

FIGURE 13.7 Spectra of Jupiter (*A*), Saturn (*B*), and Titan (*C*), showing methane and ammonia absorptions. These spectra have been divided by a spectrum of the Sun (*D*) to remove solar absorptions. If the planets had no atmospheres, their divided spectra would simply be gently curving lines. Note that there is more ammonia on Jupiter than on the other two objects.

nia can condense, just as water vapor condenses in the lower part of our own atmosphere. Depending on the circulation of the atmosphere, we expect to find cloudy regions and clear regions when we look at these planets. The deeper we can see in the cloud-free regions, the more gaseous ammonia vapor we find.

Variations in Composition with Altitude

Imagine looking down at the Earth from space. If you can only see the top of a thick white cirrus cloud, you are just looking through the stratosphere to the upper troposphere (Section 9.5). The air is very cold here, so only a little water vapor can be present. But now if your gaze wanders to a continent or an ocean peeking through the clouds, you are seeing down to the Earth's surface through a much warmer column of air that can hold more water vapor.

The same thing happens when we view the giant planets, but we are less aware of it, because these objects have no well-defined surfaces. Instead we are looking at superimposed layers of clouds, the highest of which is mainly ammonia cirrus. These are clouds of ammonia crystals instead of water crystals (Fig. 13.8). At deeper levels we expect clouds of liquid ammonia in combination with other substances.

FIGURE 13.8 Jupiter, as observed by Voyager 1, at a distance of 30 million km. We are looking through an atmosphere of hydrogen and helium to layers of clouds made up of ammonia and other compounds. The white oval to the lower left is almost as big as Earth. Everything we see is in motion (cf. Fig. 13.18). Colors are somewhat enhanced.

Saturn, being farther from the Sun than Jupiter, has a colder outer atmosphere; therefore ammonia condenses more readily, and less ammonia gas is present. That means that sunlight reflected by Saturn will suffer less absorption by ammonia than sunlight reflected by Jupiter, exactly what we see in the spectra of Fig. 13.7. This explains the early controversy about the presence of ammonia on Saturn: It was difficult for the astronomers to detect these weaker absorptions.

To put this difference between the two planets on a scale, the temperature in the atmosphere at an altitude where the pressure is equal to the sea-level pressure on Earth (1 bar) is about 175 K (−98 C) on Jupiter, but only 140 K (−133 C) on Saturn. This difference of 35° means that a balloon floating in Jupiter's atmosphere at an altitude where the pressure equals 1 bar would be below the white ammonia cirrus, but above the colorful clouds, while a balloon on Saturn at the same pressure level would be surrounded by clouds of ammonia crystals. The lower gravity of Saturn leads to a more extended atmosphere, hence a thicker cloud layer, which produces the much more uniform appearance of this planet as compared with Jupiter (Fig. 13.9).

Abundances of Atmospheric Gases

Methane and ammonia dominate the spectra of Jupiter and Saturn, but they are only minor atmospheric constituents, in smaller proportion on Jupiter than argon is in our own atmosphere.

TABLE 13.1 The compositions of the atmospheres of Jupiter and Saturn (by volume)

Gas	Formula	Jupiter	Saturn
Main Constituents (percent)			
Hydrogen	H_2	86.1	92.4
Helium	He	13.6	7.4
Methane	CH_4	0.1	0.2
Ammonia	NH_3	0.02	0.02
Water vapor	H_2O	0.2(?)	0.4(?)
Trace Constituents (parts per billion)			
Acetylene	C_2H_2	800	100
Ethane	C_2H_6	40,000	8000
Phosphine	PH_3	700	7000
Carbon monoxide	CO	3	2
Hydrogren cyanide	HCN	<1	?
Germane	GeH_4	0.6	0.4
Arsine	AsH_3	0.25	2.5
Methyl acetylene	C_3H_4	?	trace
Propane	C_3H_8	trace	trace

Although 500 times more abundant than methane, hydrogen has a much weaker absorption spectrum because it is a molecule of two identical atoms which interacts only weakly with light. (Nitrogen and oxygen are similar identical-atom molecules, and they are also transparent to visible and infrared light, hence their lack of importance for the terrestrial greenhouse effect.) Helium is even more difficult to detect, since its only absorptions at the temperature of Jupiter's atmosphere occur at very short wavelengths in the ultraviolet, which cannot be observed from Earth. Even from space, direct detection of helium is not easy, as these short wavelengths are screened by hydrogen in Jupiter's atmosphere.

Despite these difficulties, we now have rather good estimates for the relative abundances of hydrogen and helium, as well as many other gases that have been detected by spectroscopy. The instruments on the Voyager spacecraft succeeded in determining the helium abundances by measuring the effect of this gas on the observed spectrum of hydrogen. The hydrogen absorption bands are broader than if hydrogen were the only gas there. Some other, undetected gas must be present to supply the pressure that produces the observed broadening, and only helium is sufficiently abundant to serve this function. The atmospheric abundances for Jupiter and Saturn, based on Voyager and telescopic data, are summarized in Table 13.1.

Expected and Unexpected Gases

Table 13.1 is divided into two parts to distinguish the simple molecules that would be expected to form in these atmospheres from those whose presence is something of a surprise. Our expectation is that all the abundant elements should combine with hydrogen to make molecules. These should be the simplest possible molecules, given the huge excess of hydrogen and the relatively low pressures and temperatures in the upper atmospheres of these planets.

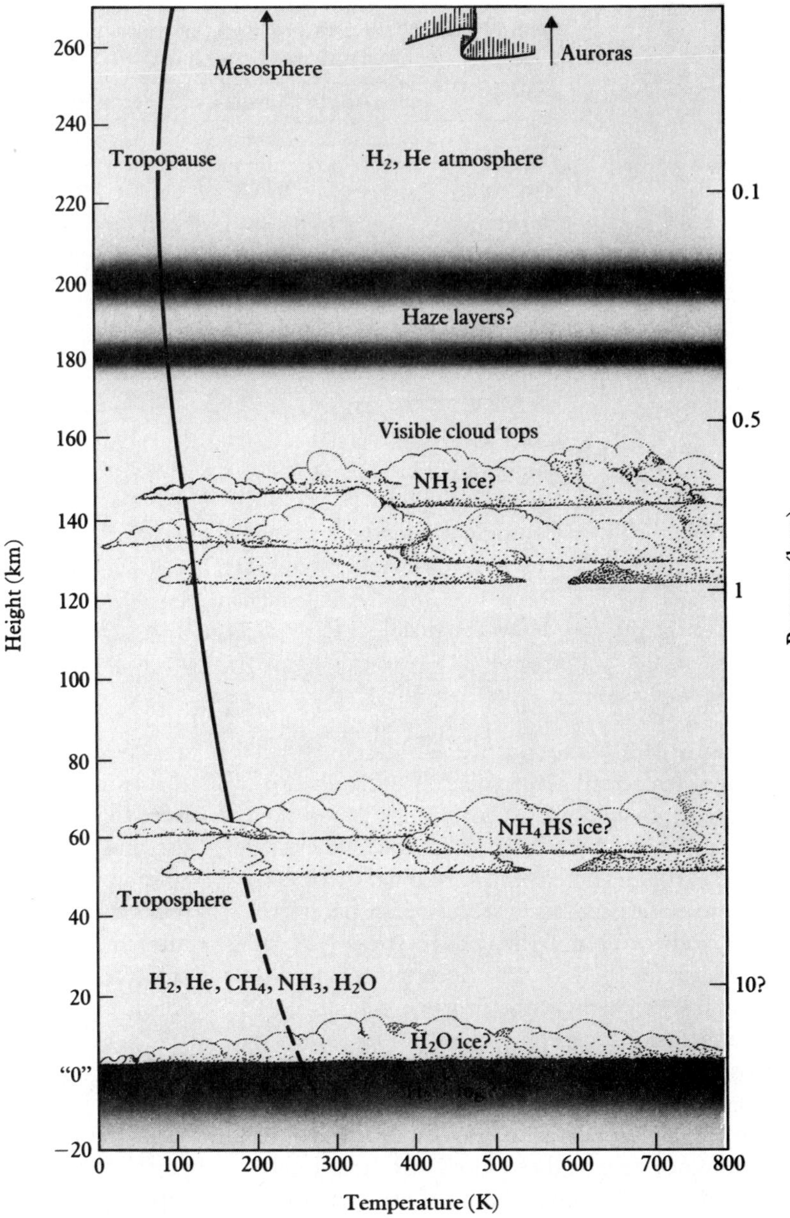

FIGURE 13.9 The atmospheres of both Saturn (a) and Jupiter (b) contain layers of clouds at different altitudes. Ammonia can condense over a greater altitude range on Saturn because of the planet's lower gravity. These diagrams show the approximate locations of major cloud systems on the two planets. The estimated position of the water clouds critically depends on the assumed abundance of H_2O in the atmosphere. (After W.K. Hartmann.)

The resulting compounds are indeed present and are shown in the upper part of the table.

This table lists a large number of trace constituents as well, such as carbon monoxide (CO), hydrogen cyanide (HCN), acetylene (C_2H_2), and other hydrocarbons (carbon-hydrogen compounds). Evidently, sources of energy other than the local heat are acting to produce these mole-

b

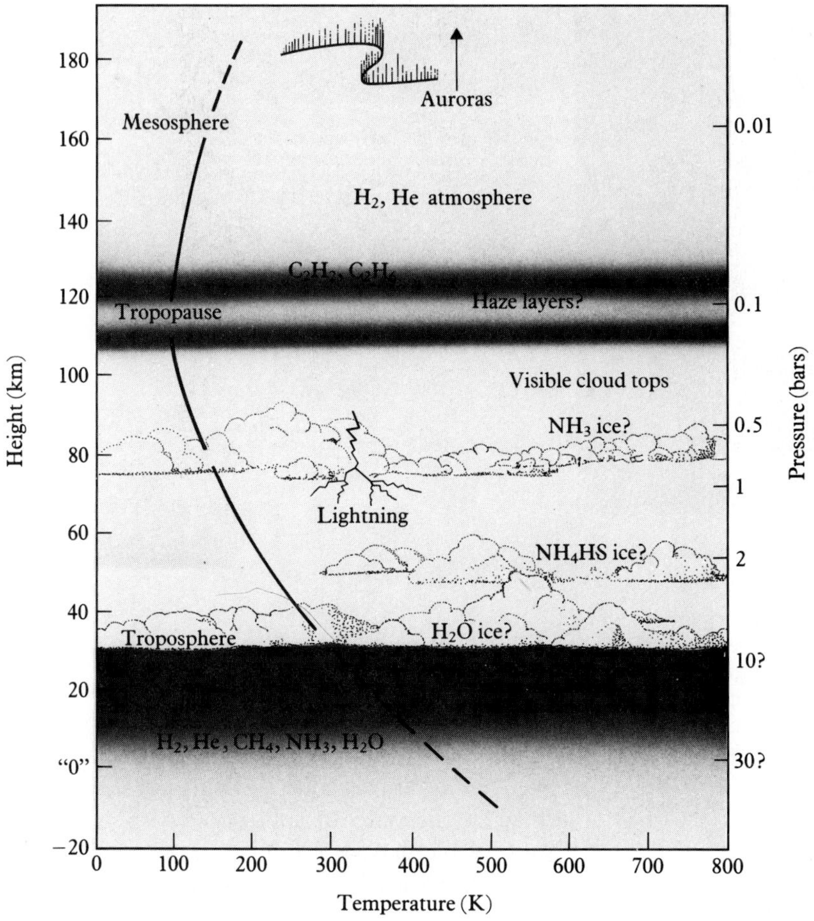

cules. In the upper atmosphere, there are photochemical reactions in which solar ultraviolet radiation breaks down methane, and its fragments form acetylene and other hydrocarbons. This is the same kind of photochemistry that produces ozone in the Earth's atmosphere and creates radicals like NH_2 and C_2 in the heads of comets.

In the convective region of the atmosphere, where we see the clouds, lightning discharges (detected by the Voyager spacecraft; Fig. 13.10) can contribute to these chemical processes and may contribute to the production of acetylene. Still deeper, at temperatures around 1200 K, carbon monoxide is made by a methane–water vapor reaction, which can be written as

$$CH_4 + H_2O \rightarrow CO + 3H_2$$

Since the CO produced in this way is observable, vertical currents must be sufficiently strong to bring it up to the visible layers of the atmosphere where it can be detected.

Table 13.1 makes Jupiter and Saturn look rather similar, but there are some significant differences. The smaller proportion of helium on Saturn can be explained by the condensation of this gas deep in the planet's interior, as discussed in the previous section. The higher proportion of methane also seems to be real and may reflect a difference in the mass of Saturn's core relative to the mass of the entire planet. A relatively

a

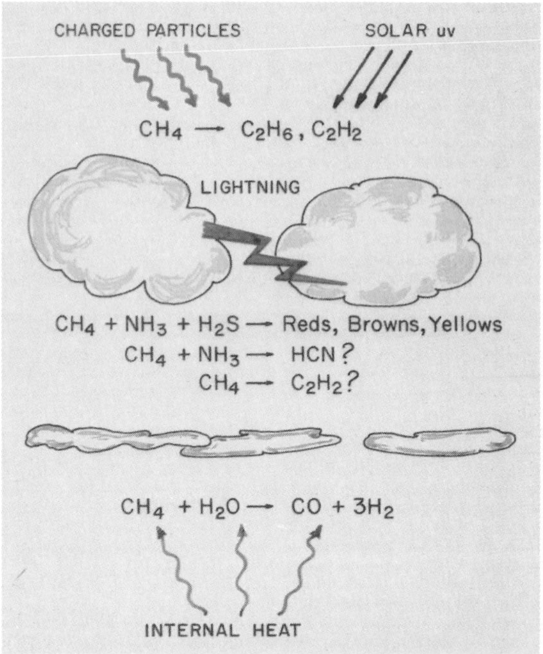

b

FIGURE 13.10 (a) This Voyager picture of the dark side of Jupiter was a time exposure lasting 3 min. 12 sec. A double auroral arc can be seen at north polar latitudes to the right, with some lightning flashes illuminating clouds in the lower center of the frame. (b) In addition to lightning and the bombardment of charged particles that cause the aurora, ultraviolet light from the Sun and internal heat from the planet are available to drive chemical reactions.

larger core would be expected to contribute more heavy elements. We see another manifestation of this enrichment in the abundances of ammonia and phosphine, but germane is actually about equally abundant on both planets. Evidently there are differences in the lower atmosphere chemistry that we still don't understand.

Gases other than those listed in the table are undoubtedly present in the atmospheres of both Jupiter and Saturn. In particular, neon should be about as abundant as ammonia, but like helium, it absorbs radiation only in the far ultraviolet, so it is very difficult to detect. The apparent absence of sulfur is more of a puzzle. Recall that sulfur is far more abundant than phosphorus or germanium in a cosmic or solar mixture of the elements (Table 2.2). One would expect hydrogen sulfide (H_2S) to form in such atmospheres. Yet careful efforts to detect this gas have failed, placing the ratio of sulfur to hydrogen in the observable part of Jupiter's atmosphere at least 1000 times lower than the solar value. Where is it? Perhaps the missing sulfur is bound up in one of the layers of clouds that form the lower boundary to the part of the atmosphere we can see.

The Change in Temperature with Altitude

All the abundances shown in Table 13.1 refer to the upper layers of the planets' atmospheres. We assume that the information obtained in this region will apply throughout the atmospheres for those gases that do not condense. We have already seen that the difference between the ammonia abundances above the clouds on Jupiter and Saturn is attributable to condensation. Observations made at radio wavelengths allow us to examine the atmospheres beneath the clouds, and here we find the amount of ammonia to be similar on both bodies. But there is no comparable information yet about water.

To understand this problem of condensation and mixing, we need to know how the temperature varies with altitude (or pressure) in these atmospheres. The best pre-Galileo information about the change of temperature with altitude has been derived from Voyager and Pioneer **occultation** measurements. As these spacecraft passed behind each planet, the radio signal they sent to the Earth had to pass through thicker and thicker layers of the atmospheres of the planets they were moving behind (Fig. 13.11). Occultations also have provided a powerful means to study the rings of Saturn and Uranus, as we will see in Chapter 18.

The temperature-pressure profiles shown in Fig. 13.9 represent averages of these various observations. As expected, Saturn is distinctly cooler than Jupiter at similar pressure levels, owing to its smaller size and greater distance from the Sun. Approximate locations of various cloud levels in the atmospheres are marked. It is interesting that on both planets, temperatures above the freezing point of water occur at pressures just a few times greater than the sea-level pressure on Earth. An astronaut in the gondola of a balloon could float quite happily here in

FIGURE 13.11 A schematic illustration of an occultation of a spacecraft by Jupiter. The way in which the radio signal is attenuated by the planet's atmosphere provides information about temperature, pressure, and composition as a function of altitude. Characteristics of the ionosphere are also obtained by this technique.

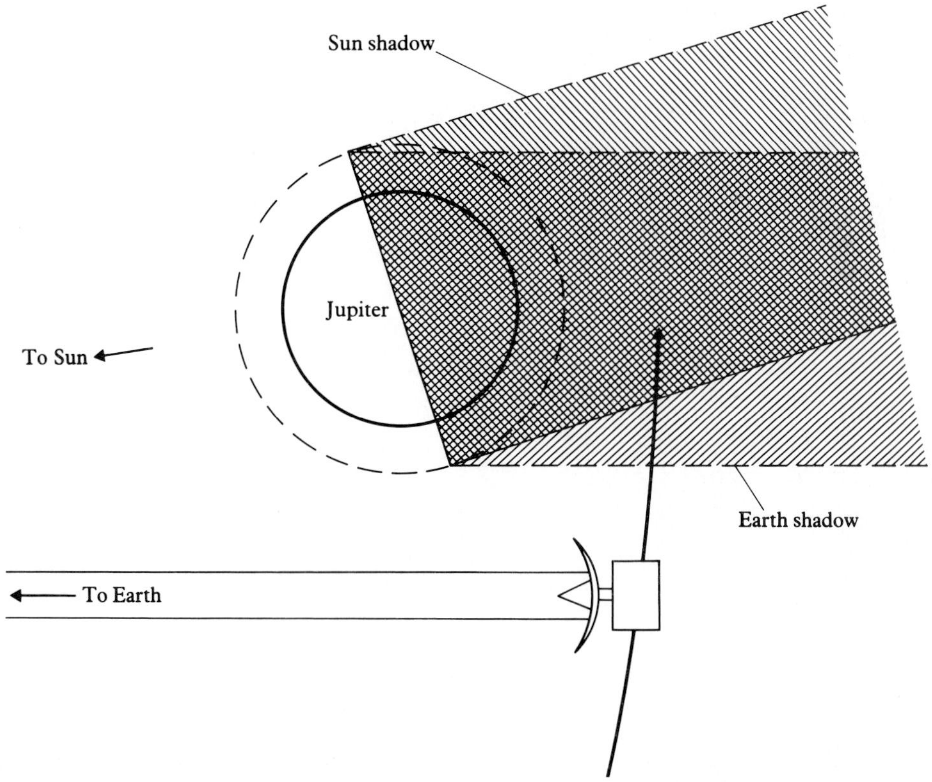

shirtsleeves and scuba gear, if one could just find a suitable gas to put in the balloon! These balmy conditions are a consequence of the planets' internal energy, although some warming would occur at these levels simply from an atmospheric greenhouse effect.

Thermal Inversions

The increase in temperature just above the tropopause is caused by the absorption of ultraviolet sunlight by gases and aerosol particles in this part of the atmosphere. Since temperature normally decreases with height in a planetary atmosphere, a region like this, in which the inverse occurs, is known as a **thermal inversion**. All four of the giant outer planets exhibit this effect; it is similar to the warming in the Earth's upper atmosphere produced by the ozone layer (Section 9.6). It is in this region of the atmospheres of Jupiter and Saturn that hydrocarbons form by photochemical reactions.

This solar ultraviolet radiation cannot penetrate to deeper layers, just as on Earth. Despite the dramatic differences between our oxygen-rich environment and these hydrogen atmos-

FIGURE 13.12 A high-resolution image of Jupiter taken by Voyager 1 at a range of 4 million km. The well-defined dusky band running diagonally through the white clouds defines the north temperate current with wind speeds of about 120 m/s (270 mph). The wavelike pattern at the bottom of the frame is similar to that in the "wake" of the Great Red Spot (cf. Fig. 13.17).

pheres, the physics of molecules dictates that most gases are excellent absorbers of ultraviolet radiation—hence the similarity in atmospheric structure between these very disparate planets. This same basic physics causes both Jupiter and Saturn to have ionospheres and exospheres. But the gravitational fields of these planets are so strong that even atomic hydrogen can escape only very slowly. So unlike the Earth, they have maintained hydrogen-rich environments throughout the lifetime of the solar system.

13.4 WEATHER AND CLIMATE

Even a modest telescope can show much detail on Jupiter. The region of the planet's atmosphere that we can see contains several different kinds of clouds. Individual pictures such as Fig. 13.12 provide snapshot views of these clouds at particular instants in time. While such pictures may suggest that all the clouds we see are at the same level in the atmosphere of the planet, Fig. 13.9 reminds us that this is not the case. Many of the color differences may be the result of seeing to different depths in different regions.

Even at telescopic resolution, which is about 2000 km for Jupiter, changes in the visible cloud systems can occur in a few hours. At Voyager resolution, the planet presents a constantly shifting pattern of clouds, but an underlying pattern of currents flowing parallel to lines of latitude has maintained its stability for decades. On Saturn, cloud changes are usually much more difficult to see (Fig. 13.13a), but again there is an underlying atmospheric circulation pattern. It has been traditional to describe the appearance of these planets (especially Jupiter) in terms of alternating bright zones and dark belts, but the currents seem to have a greater persistence than this cloud pattern. For example, a huge "storm" erupted in the equatorial region of Saturn in 1990, but the basic circulation system was maintained throughout this event (Fig. 13.13b).

FIGURE 13.13 (a) Saturn as viewed by Voyager 2. Compare with Jupiter in Fig. 13.8. (b) A rare and unpredictable outburst of giant storms on Saturn. This picture, obtained by the Hubble space telescope in Oct. 1990, shows a very different aspect of Saturn compared with its more normal appearance in 1981. Colors greatly exaggerated.

a

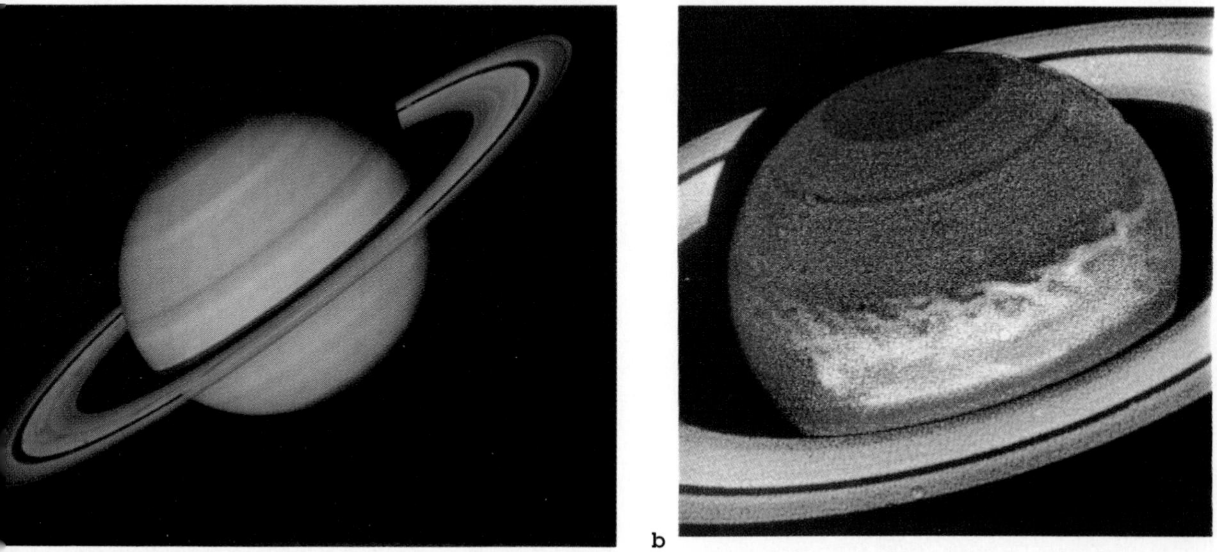

b

TABLE 13.2 Rotation periods for Jupiter and Saturn

	Jupiter	Saturn
Deep interior (System III)	$9^h55^m30^s$	$10^h39^m24^s$
Equatorial (System I)	$9^h50^m30^s$	10^h14^m
High latitudes (System II)	$9^h55^m41^s$	10^h40^m

Rotation Periods and Wind Speeds

Three rotational periods have been established on Jupiter. The two periods labeled Systems I and II (Table 13.2) are average values and refer to the apparent average speed of rotation at the equator and at higher latitudes, respectively. These periods are defined by observed motions of features in the planet's cloud layers which move with the local currents. Since there is no

FIGURE 13.14 (a) A map of wind currents on Jupiter. There is a strong equatorial jet, with currents at higher latitudes alternating between easterlies and westerlies, roughly following the pattern of boundaries between dark and light belts and zones. (b) The equatorial jet on Saturn is both stronger and wider than that on Jupiter. This planet does not have the same pattern of alternating currents as its larger neighbor, except at latitudes above 40°.

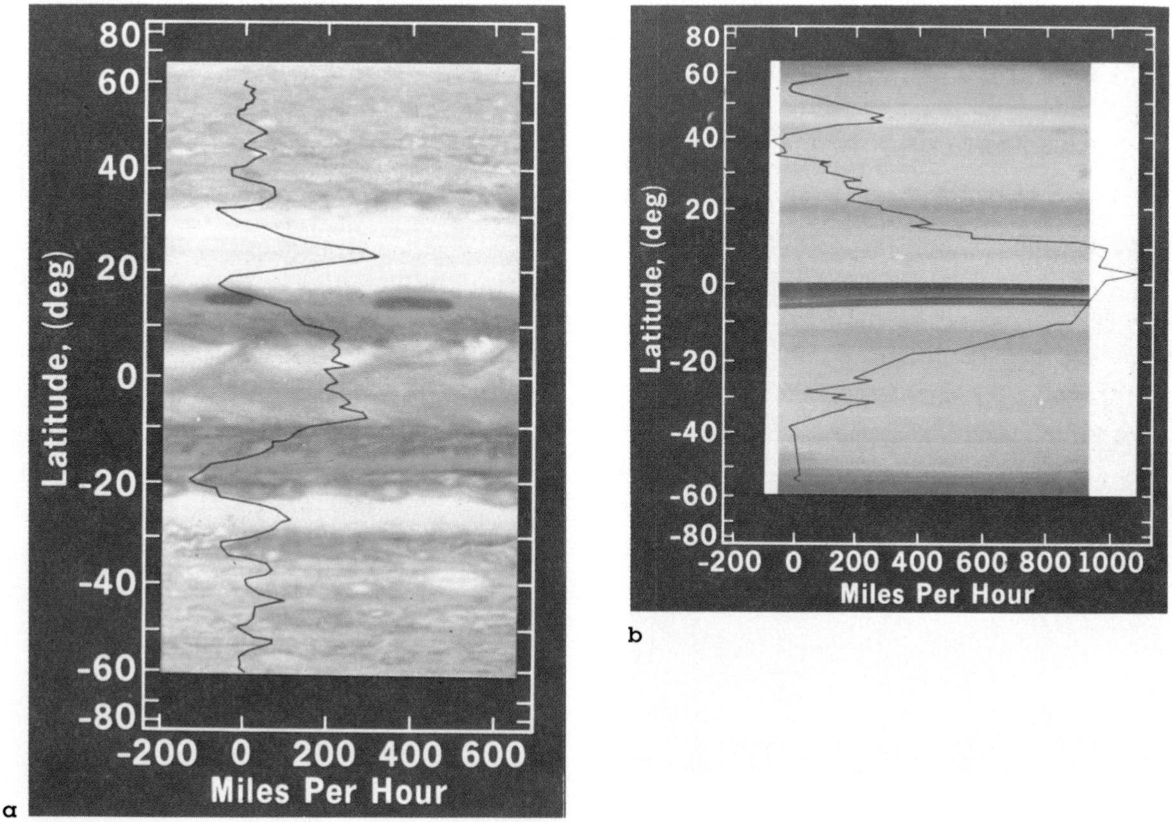

solid surface, we must look elsewhere for an absolute standard by which to judge these motions.

Studies of the jovian radio emissions made over the last 25 years have shown a definite, unchanging periodicity that must refer to the rotation of the planet's magnetic field. This deduction was verified by direct measurements of the field carried out by spacecraft. Since this field is generated deep in Jupiter's interior, the radio period, called System III, has become identified with the true rotation of the planet itself. Judged against this standard, we see that Jupiter exhibits an eastward-flowing equatorial jet stream (System I) with a relative velocity of about 300 km/hr, comparable to the velocities of jet streams on Earth. At higher latitudes in each hemisphere, an alternating pattern of easterly and westerly winds occurs (Fig. 13.14).

The visible clouds on Saturn appear much more uniform than those on Jupiter. Only rarely do spots appear that can be seen in telescopes from Earth. While a belt-zone pattern is evident, it is much more subdued (Fig. 13.13a). The Voyager cameras were able to define enough discrete features to map out Saturn's circulation, which is distinctly different from that of Jupiter (Fig. 13.14). Once again an equatorial jet is apparent, but now it extends to 40° on either side of the equator, and its peak velocity is a remarkable 1300 km/hr. At still higher latitudes, there is again an alternating pattern of eastward and westward currents, but they are not so closely tied to the faint belts and zones. Just as for Jupiter, the rotation of Saturn's magnetic field establishes the rotation period of the planet's core.

The most striking cloud features on Saturn recorded by the Voyager spacecraft were located near 40° N. This latitude marks a pronounced change in wind velocities: The great equatorial jet stream loses its strength and a high-latitude current takes over. Not surprisingly, we see a number of eddies and other atmospheric disturbances at this velocity boundary. One of the most striking of these is illustrated in Fig. 13.15.

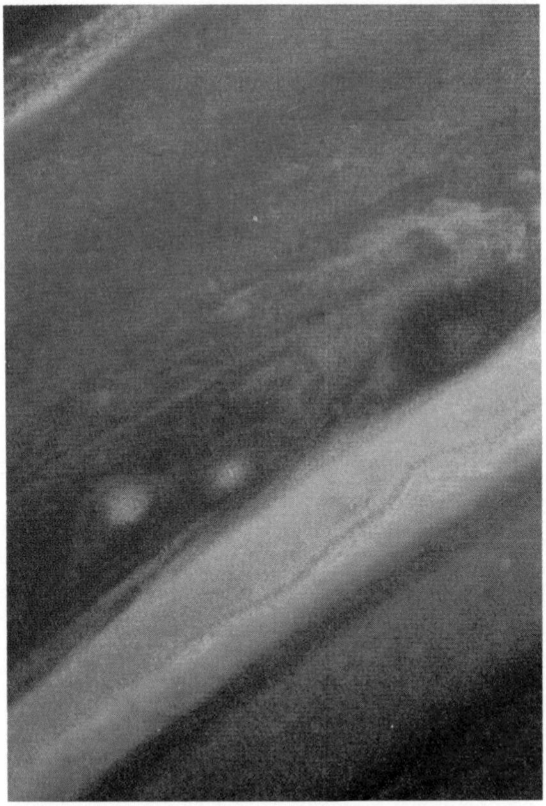

FIGURE 13.15 A close-up view of northern latitudes on Saturn. The contrasts have been greatly enhanced in this picture to reveal details that would otherwise be hard to see. The large dark oval, about 3000 km in diameter, is rotating in an anticyclonic direction, indicating that it is a high-pressure disturbance.

General Circulations of the Atmospheres

The reasons for the basic differences in atmospheric circulation between Jupiter and Saturn are not yet clear. One interesting possibility suggested by planetary meteorologist Andrew Ingersoll of Caltech is that the circulation patterns actually extend to very deep layers. The currents that we see may represent the outermost edges of concentric cylinders of gas, turning around the planet's rotational axis (Fig.

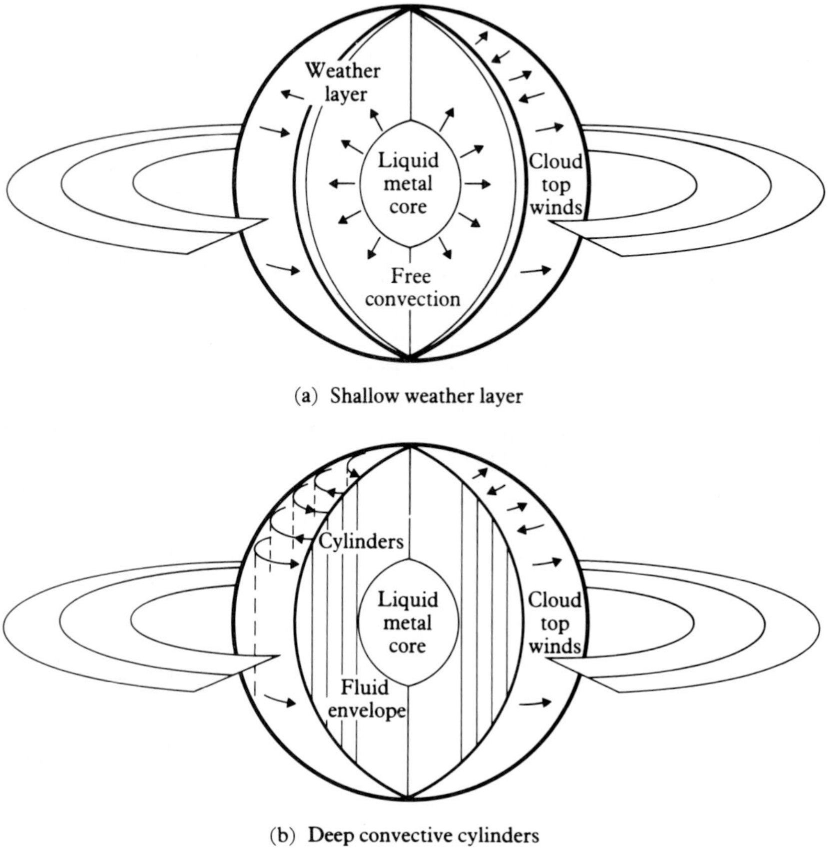

FIGURE 13.16 Two models for the global circulation on Saturn apply to Jupiter as well: (a) suggests that the circulation is driven in a rather shallow layer in the visible part of the atmosphere; (b) invokes deep convection following a pattern of concentric cylinders. (After M. Allison.)

13.16). The sizes and shapes of these cylinders would then be determined by the internal structures of the planets, so relative sizes of the cores of Jupiter and Saturn may play a role in determining the nature of the atmospheric currents. This is another illustration of just how different these hydrogen-helium giants are from the rocky inner planets.

We can compare the meteorology on these planets with the global circulation of the Earth's atmosphere (Section 9.7). On Earth, huge spiral cloud systems often stretch over many degrees of latitude and are associated with motion around high-pressure and low-pressure regions. These cloud systems are much less confined to specific zones of latitude than are the cloud systems on Jupiter and Saturn. Clouds on Earth can move in latitude as well as longitude, as the migration of hurricanes from the Caribbean along the East Coast of the United States dramatically demonstrates. This kind of motion simply doesn't happen on the giant planets.

Planetary meteorologists have simulated the general circulation of the giant planets by starting with a computer model of the Earth and simply increasing the rotation rate (in the model—not on the planet). The greatly enhanced Coriolis force (Section 9.7) appears to shrink disruptive eddies and produces alternating bands of strong winds that are extremely stable. Thus computer experiments play an important role in understanding the atmospheres of the outer

FIGURE 13.17 A close-up view of Jupiter's Great Red Spot. This giant storm system could swallow two planets the size of Earth, side by side. The white oval below it has existed for over 40 years; the smaller "doughnuts" at still lower latitudes have shorter lifetimes. Color contrasts are exaggerated in this and most other Voyager photos of Jupiter.

planets, just as they do in tracing the thermal evolution of the inner planets and the operation of the greenhouse effect.

High rotation rate is certainly one factor that makes the circulation on the giant planets very different from that on Earth. Another is the lack of a solid surface. Weather on Earth is often closely tied to the local environment, which in turn is determined by the varied nature of our planet's surface, especially the land-water dichotomy. On Jupiter and Saturn, the absence of physical boundaries contributes to the stability of the atmospheric circulation patterns.

The Great Red Spot

The permanence of underlying circulation patterns in the presence of constant change is most obvious on Jupiter, where one can see more features and variations in the cloud layers. The most famous of these features is the Great Red Spot (GRS) (Fig. 13.17), which has existed for at least 100 and perhaps more than 300 years. Its present dimensions are about 26,000 by 14,000 km, making it large enough to accommodate two planets the size of our Earth, side by side. These huge dimensions are probably responsible for the feature's longevity.

The true nature of this giant cloud system is still unknown, despite extensive observations. The reddish clouds within the spot exhibit a counterclockwise rotation with a period of about six days. (Remember that counterclockwise motion in the southern hemisphere of a planet is anticyclonic; see Section 9.7.) This cloud system thus appears to be an enormous anticyclone, a vortex or eddy whose center must be a region of locally high atmospheric pressure. Nevertheless, Voyager pictures failed to reveal any evidence of upwelling at the spot's center. The clouds here seem remarkably tranquil (Fig. 13.18). Cyclones and anticyclones are also found on Earth, but on our planet they are very short-lived. Furthermore, the clouds associated with them are white water clouds like all the others that form on Earth.

The huge lateral dimensions of the GRS are associated with a considerable vertical extent as well, allowing it to project well above and below the adjacent cloud deck. The upward projection has been verified by direct observation, but the lower reach of this enigma has not been established. We also don't know what chemicals are responsible for its color.

Ovals, Dark Brown Clouds, and Hot Spots

On Jupiter, cloud features much smaller than the GRS are also persistent and localized. The three white ovals found at a latitude just south of the Great Red Spot (Fig. 13.17) have existed for half a century; white ovals of this size (about as large as the planet Mars) are found nowhere else on the planet. Dark brown clouds, which are evidently deeper layers glimpsed through holes in

FIGURE 13.18 This is a mosaic of two pictures of the Great Red Spot, obtained by Voyager 2 at a distance of 2.6 million km. Note that the huge storm and its surroundings have changed in appearance since the image obtained three months earlier by Voyager 1 (Fig. 13.17).

the nearly ubiquitous tawny cloud layer are found almost exclusively at latitudes near 18° N. The blue-gray or purple areas (colors are somewhat subjective), from which the strongest thermal emission is detected, only occur in the equatorial region of the planet.

How does Jupiter "know" that it should have one type of cloud feature in only one location? These equatorial hot spots must be clear regions that allow us to look deep into the planet's atmosphere. We know that because of the intense thermal radiation we can detect there. Pictures of Jupiter made by recording just this escaping energy reveal these equatorial windows as the brightest areas on the planet (Fig. 13.19). Yet there is no brown cloud layer here, as there is at 18° N. And why is there only one Great Red Spot? We do not yet have answers to these questions.

One piece of this network of riddles seems fairly clear: The larger the disturbance, the longer it lasts. On Earth, our most powerful storms—the hurricanes and typhoons—die out (albeit with highly destructive effects) when they cross over land. The reason is that these storms get their energy from the condensation of water vapor that comes from the oceans. (Evaporation of water requires energy—that's why perspiration cools us—while condensation liberates energy, hence the destructive power of thunderstorms and hurricanes.)

When hurricanes and typhoons move off the ocean onto a continent, they no longer have an unlimited source of fuel, and they gradually dissipate. On Jupiter, there are no continents, and once these storms grow large enough to overcome the effects of encounters with smaller systems, they can evidently persist for a very long

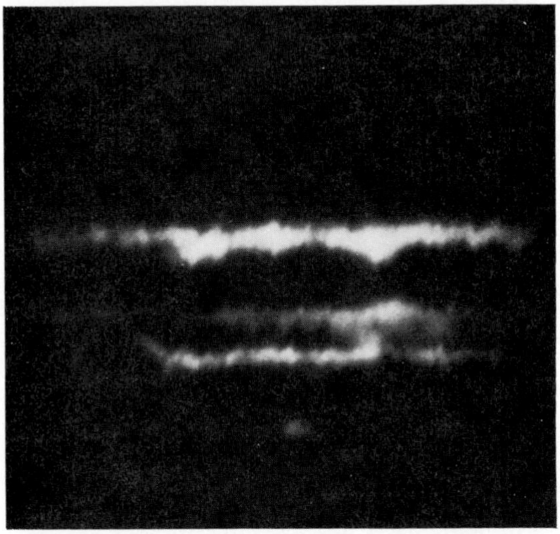

FIGURE 13.19 Images of the same hemisphere of Jupiter in visible light (top) and infrared light at a wavelength of 5 μm (bottom). The bright regions in the infrared picture are places where heat is escaping from the interior of the planet. The Great Red Spot appears dark in this image, indicating that it is opaque to this internal heat radiation.

time. Thus the largest such feature, the GRS, has existed for at least 100 years, while the smaller white ovals have been around for several decades. The still smaller white spots at more southerly latitudes have proportionately shorter lifetimes.

Seasons

Since both Jupiter and Saturn have major internal sources of energy, one might expect their weather to show no sign of seasons. This may be the case at low altitudes, where the internal heat is the most important source of energy for atmospheric circulation. But the upper atmospheres (including most of the regions we see) are strongly influenced by the solar radiation they absorb. We have already mentioned one indication of this in the discussion of the temperature inversions. Thus the key property for establishing seasons on Jupiter or Saturn remains the inclination of the planet's axis of rotation, just as it is for Earth.

Jupiter, with an axial inclination of only 3°, exhibits virtually no seasonal modulation of its weather pattern. But Saturn, inclined at nearly 27°, shows strong seasonal effects. One manifestation of Saturn's seasonal change is the increased abundance of ethane during the summer as a result of the irradiation of methane by ultraviolet light from the Sun. Since Saturn's satellites share the planet's inclination, we can anticipate that the atmosphere of Titan is also subject to seasonal effects.

13.5 CLOUDS, COLORS, AND CHEMISTRY

There are three basic kinds of clouds that can form in a planetary atmosphere. In our own skies, we are most familiar with white condensation clouds, formed when water vapor condenses into droplets or ice crystals. Dry areas of the Earth also produce clouds of dust, swept up by surface winds into the atmosphere. As we saw in Chapter 11, this is the most prominent form of cloud on Mars. Finally there is the photochemical haze or smog produced by chemical reactions that are mainly powered by the absorption of solar ultraviolet light. Having no solid surfaces, the giant planets can produce only chemical and condensation clouds.

Photochemical Smog

On Earth, we are accustomed to the formation of smog layers over our cities, since it is the cities that furnish the gases on which sunlight can act to produce the smog particles. But on the outer planets, the hydrogen-rich atmospheres themselves contain gases that are easily converted to compounds that form hazes. Because of these organic gases, photochemistry plays a very important role in the outer solar system, in spite of the large distances of the planets from the Sun.

Both Jupiter and Saturn are very poor reflectors of blue and ultraviolet light, indicating that the atmospheres of these planets contain layers of smog that absorb this radiation. There is so much excess hydrogen on these bodies that the hazes remain thin: As the particles drift downward, they are heated and converted back to simple substances. In the low-temperature, hydrogen-poor atmosphere of Saturn's satellite Titan, however, there is a smog layer so thick that we cannot see Titan's surface (Chapter 16).

These thin hazes of photochemical smog are found in the planets' stratospheres, since the high-energy ultraviolet light from the Sun does not reach lower altitudes. It is absorbed by methane and ammonia in the process of splitting these molecules apart. Below the tropopause, we encounter the thick clouds that appear in the photographs of these planets. These clouds, formed by condensation of atmospheric gases, are produced within certain specific ranges of temperature, as described in Section 13.3.

Condensation Clouds

The highest, coldest clouds are the white ammonia cirrus. If Jupiter and Saturn were simply weird varieties of terrestrial planets, we could reasonably expect that the next cloud layer to be encountered in a descent through the atmosphere would be composed of liquid ammonia droplets. These would be white, puffy clouds analogous to terrestrial cumulus, or perhaps layers of ammonia stratus clouds. But in these enormously deep atmospheres, we must anticipate the presence of other gases and ask what effect they will have on cloud composition. In fact, the next clouds we encounter are no longer white, but tawny in color. Evidently a change in chemistry has occurred. Some new compounds must be present to give color to the condensed ammonia.

There is no direct determination yet of the composition of these colored clouds, but a table of cosmic abundances of the elements gives us a clue. As discussed in Section 13.3, no compounds of sulfur have been detected yet on Jupiter and Saturn even though the relatively high abundance of this element leads us to expect such compounds to be present. Chemistry tells us that ammonia and hydrogen sulfide will combine to form ammonium hydrosulfide (NH_4SH). Since there is more nitrogen than sulfur in a cosmic (or jovian) mixture of the elements, this compound can use up all the H_2S and leave an excess of NH_3. Thus such clouds may account for the absence of detectable sulfur compounds above the tawny clouds. It is almost certainly the main constituent of this cloud layer. But ammonium hydrosulfide is white, so something else must be happening here.

The answer may lie in the ability of sulfur to combine with itself to form compounds that are indeed yellow or brown in color. However, this attractive hypothesis has been challenged by evidence that the ultraviolet sunlight required to drive these chemical reactions does not penetrate deeply enough into the atmosphere to do the job. Establishing the kind of chemistry that occurs in Jupiter's atmosphere is one of the tasks of the Galileo probe.

At much lower levels in the atmosphere, water clouds should form. Once again, these will not be pure water. They will certainly contain ammonia and may well resemble a very dilute solution of household ammonia, and other soluble gases will also be involved. The Galileo probe provides us with the first opportunity to study these deeper regions directly.

Chemical Clouds

With our present knowledge, it is already obvious that Jupiter and Saturn have remarkable atmospheres. Imagine floating on a world with colored clouds! The views would be magnificent, even if the aromas leaking into your spaceship might leave something to be desired. If you succeeded in ignoring the disagreeable odors, however, you would find that the nitrogen and sulfur compounds present in the atmosphere produce a complex and interesting chemistry. Some of the reactions taking place on Jupiter today may resemble those that occurred on the primitive Earth, producing organic compounds essential for the origin of life.

We must admit that we know only part of this interesting story. Sulfur compounds may also be responsible for the dark brown clouds on Jupiter, but they may not. Red phosphorus has been proposed as the substance responsible for the coloration of the GRS, but the theoretical calculations that led to this prediction have not been substantiated by observations of the planet itself. Yellow phosphorus is produced from phosphine (PH_3) in laboratory experiments, and it may contribute to colors seen on both Jupiter and Saturn. The various colored organic compounds produced by reactions of methane and ammonia provide additional possibilities.

These suggestions are based in part on theoretical studies as well as on a variety of laboratory experiments in which mixtures of gases are subjected to electric discharges, ultraviolet radiation, or bombardment by electrons and protons. Such experiments often produce colored materials; the problem is that there are usually no discrete absorption features in the spectra of these materials (the way there are for gases) that can lead to their identification on the planets. We may have to wait for a direct analysis by probes sent into these atmospheres to discover what the chemical agents are that produce the colors we observe.

Impact Scars in the Jovian Atmosphere

In July 1994 about 20 separate nuclei of Comet Shoemaker-Levy 9 smashed into Jupiter, as we described in Chapter 6. Each of these nuclei penetrated below the visible ammonia clouds before it disintegrated and exploded with an energy of millions of megatons. From the postimpact observations that showed only limited amounts of water injected into the jovian stratosphere, we can conclude that the explosions probably took place just above the dense water clouds, probably at a pressure level of about 5 bars.

Each of the cometary impacts generated a fireball that consisted of vaporized cometary material mixed with the gases of the jovian atmosphere. These materials were carried into the stratosphere and spread over thousands of kilometers (Fig. 13.20). These dark clouds included substantial quantities of carbon and sulfur, derived from both the comet itself and the planet's atmosphere. Twisted and dissipated by the strong jovian winds, the clouds gradually dispersed until they were no longer visible a few months after the impacts, although some evi-

FIGURE 13.20 A color image of Jupiter recorded by the Hubble space telescope in July 1994 reveals several regions of black debris left from the explosive impacts of fragments of Comet Shoemaker-Levy 9, all of which struck at the same longitude.

dence of enhanced haze at the latitude of impacts persisted for more than a year.

13.6 MAGNETOSPHERES AND RADIO BROADCASTS

In our discussion of Earth, we learned that our planet's magnetic field is strong enough to create a magnetosphere. This teardrop-shaped barrier to the flow of the solar wind contains the belts of trapped charged particles discovered by James Van Allen. We may reasonably expect to find similar environments around other planets with strong magnetic fields. The giant planets satisfy this criterion; all four have well-developed magnetospheres, with Jupiter's being the biggest and most complex.

Jupiter's Radio Broadcasts

Jupiter was the first planet found to be a source of radiation at radio wavelengths. This radiation was detected in 1955 at a frequency of about 20 megahertz, corresponding to a wavelength of 15 m. For comparison, a radio station on Earth that you tune in at 100 on your AM dial is broadcasting at 1 megahertz, or a wavelength of 300 m. If you prefer FM, a station at 100 on the dial of your radio is broadcasting at 100 megahertz, or 3 m.

These signals from Jupiter are not news reports or rock music; they are simply radio noise produced by the interaction of charged subatomic particles—primarily electrons—with the planet's magnetic field and its ionosphere. These radio emissions, caused by electrons and ions moving at very high speeds, are called **nonthermal radiation**, because they are not caused by the normal emission of energy associated with the heat of the source. The intensities of the jovian nonthermal noise bursts are occasionally great enough to make Jupiter the brightest object in the sky at these long wavelengths, except for the Sun during its most active periods.

Nonthermal radio bursts from Jupiter provided the first indication of a magnetic field on any planet other than Earth. Subsequent observations at wavelengths shorter than 1 m revealed that Jupiter is also a source of steady radio emission. It has become customary to refer to these two types of emission in terms of their characteristic wavelengths: *decameter* radiation (the erratic bursts) and *decimeter* radiation (the continuous source). Except at the very shortest radio wavelengths, the decimeter radiation is generated nonthermally, primarily by electrons spiraling at very high speeds around magnetic lines of force. In addition to these trapped electrons, Jupiter is also surrounded by a doughnut-shaped distribution of protons (Fig. 13.21). In other words, this giant planet has belts of trapped charged particles analogous to the Van Allen belts surrounding Earth (Section 9.6).

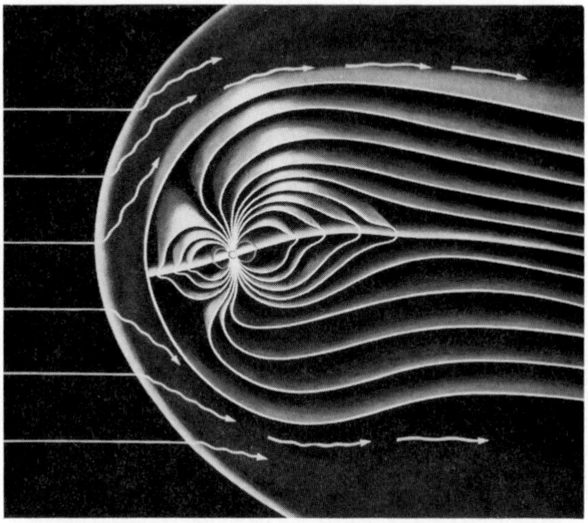

FIGURE 13.21 Jupiter's enormous magnetic field has trapped electrically charged particles, mostly electrons and protons, from the solar wind. When the electrons spiral around the magnetic field lines, they produce radio waves that can be detected from Earth or from spacecraft.

The Influence of Io

The noise storms at decameter wavelengths are strongly influenced by the position of the satellite Io in its orbit, as viewed from Earth. An example of this is an enhancement that occurs when Io and Jupiter form a 90° angle with the Earth. Evidently a cluster of magnetic field lines (called a **magnetic flux tube**) links Io to the planet (Fig. 13.22). As Io moves in its orbit, the foot of this flux tube sweeps across Jupiter, rather like the shadow of a satellite that is causing a solar eclipse. Electrons spiraling around the field lines and interacting with the jovian ionosphere cause the observed bursts of radio noise. Another way to think about this is to consider the flux tube as a wire along which an electric

FIGURE 13.22 As Io moves through Jupiter's magnetic field, it generates an electrical potential of 400,000 volts across its surface. At certain positions in its orbit, an electrical current of 3 million amperes will flow along Jupiter's electrically conductive ionosphere, where it triggers an auroral hot spot.

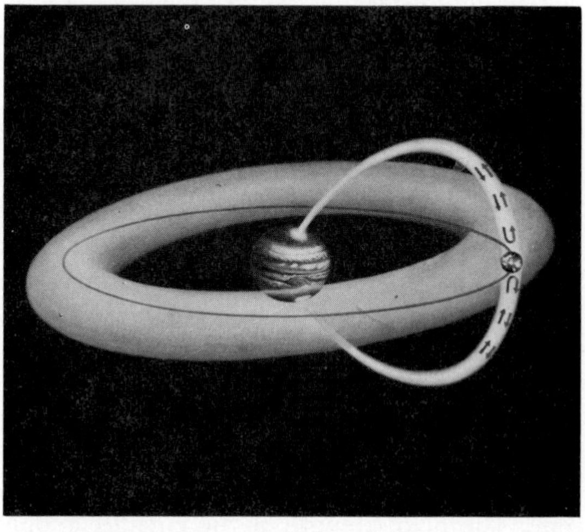

current (the spiraling electrons) passes. This 5-million-ampere current is generated by the motion of Io through the planet's magnetic field, just as a wire (armature) passing through a magnetic field generates a current in one of our power stations on Earth.

The flux tube and its influence on jovian radio bursts is not the only way Io interacts with the magnetosphere. As we will describe later, Io and its volcanoes are the primary source for heavy ions in the inner magnetosphere of Jupiter. These heavy, energetic particles are concentrated in a doughnut-shaped volume, the **Io plasma torus**, which surrounds Jupiter near the orbit of this satellite. This Io torus is the strongest part of what we might think of as the jovian Van Allen belts. It has been photographed from Earth in the glow from the various atoms and ions it contains, as shown for neutral sodium in Fig. 13.23.

Magnetic Fields

The deductions about the magnetic field of Jupiter from these Earth-based observations were refined and extended by the spacecraft that penetrated the jovian magnetosphere. The magnetic field of Jupiter is dipolar (like a bar magnet, the same type as Earth's) and is generated by a natural dynamo driven by convection within the electrically conducting layers of the planet's interior. The intrinsic strength of this field is 19,000 times greater than that of Earth, leading to a field strength at the equator of 4.3 gauss, compared with 0.3 gauss at the Earth's surface. (The surface field strength is reduced because Jupiter's volume is so much greater than Earth's.) The orientation of the jovian magnetic field is opposite to the present orientation of the Earth's field, so a terrestrial compass taken to Jupiter would point south, instead of north.

Saturn's magnetic field is somewhat weaker than Jupiter's, but it exhibits the same orientation. We have no idea whether these fields, like that of the Earth, reverse direction from time to time. The field strength at the equator is only 0.2 gauss. The most unusual characteristic of Saturn's field is that it is not inclined to the planet's axis of rotation, whereas the fields of Jupiter and Earth both have inclinations of about 10°. This lack of inclination is surprising, since most theories for the generation of planetary magnetic fields require such a tilt. Perhaps there is a conducting region around the field-generating core in Saturn's interior which modifies the external appearance of the field.

Saturn's Radio Broadcasts

Another difference between Jupiter and Saturn is the lack of strong radio emission from Saturn. It was not until the first Voyager spacecraft approached the planet in 1980 that long-wavelength radiation was detected. This radio noise occurs at frequencies from 3 kilohertz to 1.2 megahertz, overlapping our familiar AM radio broadcasting band, with a peak intensity at wavelengths near 2 km. Radio signals in this frequency range are reflected by the Earth's ionosphere. Thus the AM signals do not get out, and the Saturn radiation cannot get in—hence the need for a spacecraft to detect them.

The energy for this radio emission from Saturn is supplied by electrons from the impinging solar wind, so changes in solar wind pressure or speed produce large changes in the power radiated by Saturn. Since the interaction with the solar wind is related to the configuration of the planet's magnetic field, it has been possible to determine the rotation period of the field—and hence of Saturn's interior—exactly as in the case of the System III period for Jupiter. The resulting period of 10 hr 39.4 min is the standard against which the wind velocities are measured.

The Van Allen Belts of Jupiter and Saturn

Both Jupiter and Saturn are surrounded by huge seas of plasma. The protons, ions, electrons, and neutral atoms in these plasmas come from three

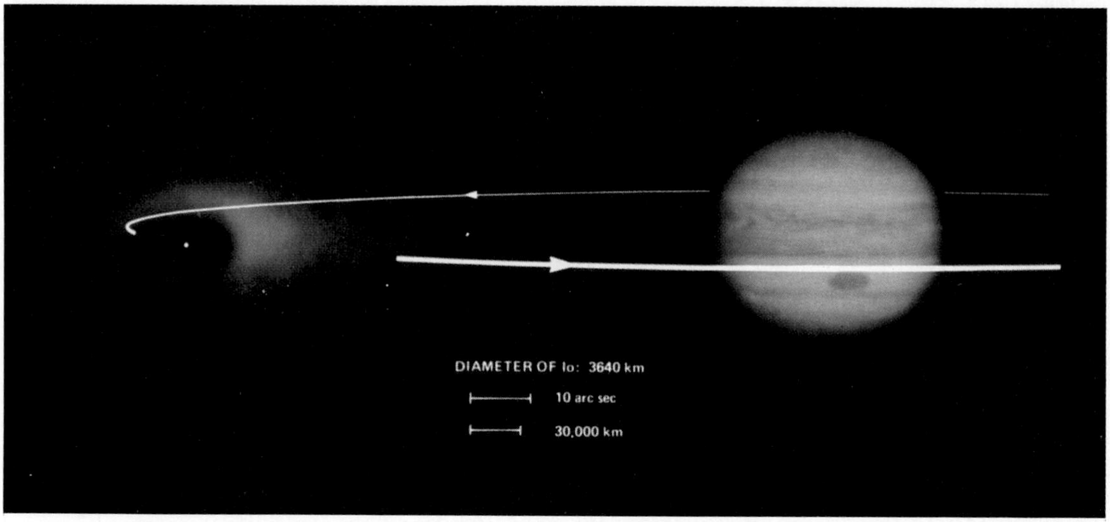

FIGURE 13.23 The orbit of Io is surrounded by a doughnut, or torus, of atoms and ions, as shown in a picture obtained from Earth in the light from the sodium atoms it contains. Images of Io and Jupiter have been added at their appropriate positions relative to the sodium cloud.

distinct sources: the solar wind, the atmospheres of the planets, and the surfaces of the satellites. The Earth's magnetosphere is not populated from this latter source, which is one of the major differences between our Van Allen belts and the magnetospheres of Jupiter and Saturn. This source contributes the heavy ions that are present in addition to the electrons and protons already mentioned. These ions are atoms stripped of one or more electrons. Being charged, they are under the control of the planetary magnetic fields.

Oxygen ions have been found around both Jupiter and Saturn. Sulfur and oxygen ions around Jupiter come from the volcanoes or surface of Io, which must also contribute the neutral sodium that is visible from Earth (Fig. 13.23). Some of these neutral atoms and ions escape directly from the volcanic eruptions on Io, which we will discuss in Chapter 15, while others are released from the surface by a process called **sputtering**. Sputtering results when an impacting ion from the plasma torus has sufficient energy to eject additional ions from a solid surface, almost like an atomic version of impact cratering. The process is important at Io because the more heavy ions that are injected into the plasma torus, the more "ammunition" there is to sputter other ions from the surface.

The heavy ions in the Io torus make the environment around this satellite lethal to human beings and very hostile to spacecraft. Pioneer 10 nearly died as it passed through this region, so considerable effort and expense were lavished on the Galileo spacecraft to protect its electronic components from this threat.

In the Saturn system, sputtering from the surfaces of the icy satellites and the rings furnishes oxygen ions, while nitrogen ions are supplied by the atmosphere of Titan. The details of the interactions of the electrons, protons, and heavy ions with the ices in the main rings are still not understood. Some charging of small solid particles in both Saturn's and Jupiter's rings may take place, leading to observable effects (Chapter 18).

Having seen where the ions and electrons in the magnetospheres originate, we now ask how they are lost. There are three major sinks for ions

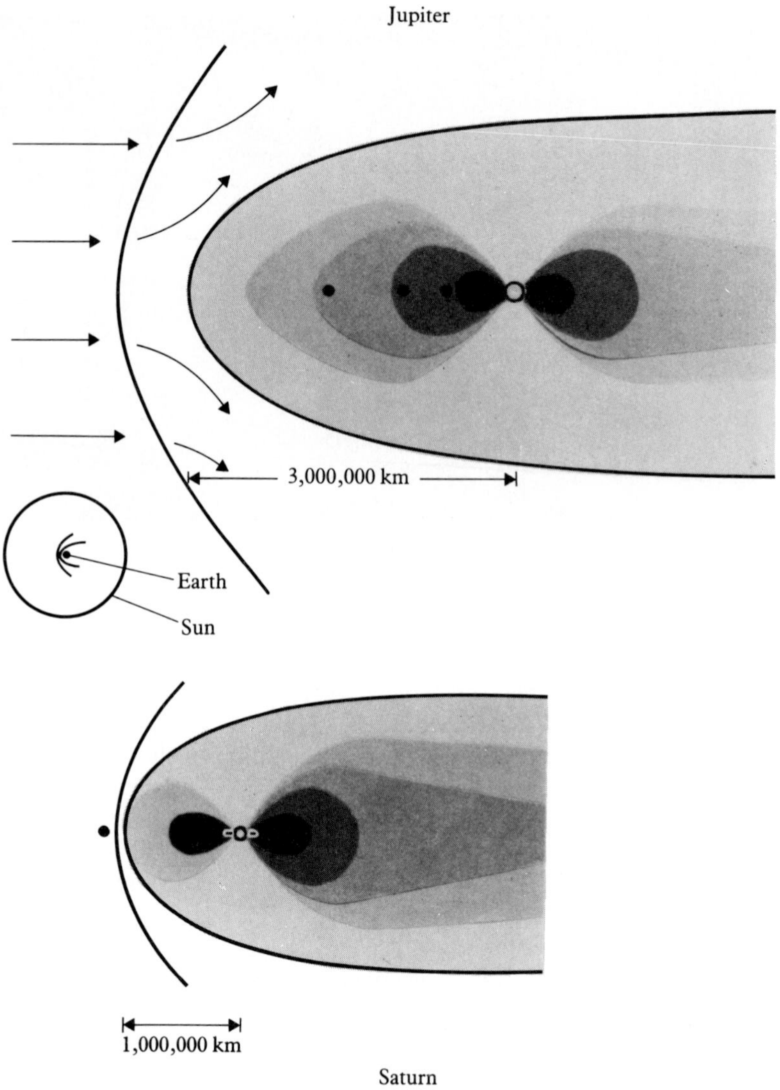

FIGURE 13.24 The small circle to the left in this diagram represents the Sun; inside it, to the same scale, is the Earth and its magnetosphere. The magnetospheres of Jupiter and Saturn are shown to scale for comparison.

and electrons in the Jupiter and Saturn magnetospheres, compared with two in the simpler magnetosphere of the Earth. Just as with the Earth, some escape from the outer part of the magnetosphere, and some are lost in collisions with the upper atmosphere of the planet. The third sink consists of the surfaces of the satellites and rings, which act as both a source (through sputtering) and a sink (through absorption of particles that strike their surfaces). The absorptions at Saturn are stronger than those in Jupiter's magnetosphere owing to the large surface area of the rings and the closely spaced inner satellites. These combined surfaces are more effective at absorbing charged particles than are the inner satellites and thin rings of Jupiter, leading to a major difference between the distributions of plasma in these two giant magnetospheres.

Dimensions of Magnetospheres

As we saw for the Earth, a magnetosphere is bounded on the upstream side (facing the Sun) by the equilibrium between the internal pressure exerted by the planet's magnetic field and the external pressure from the solar wind. The magnetic fields of Jupiter and Saturn are much stronger than Earth's, while the solar wind pressure is less at these greater distances from the Sun. The net result is that the two giant planets have magnetospheres so large that they exceed the size of the Sun (Fig. 13.24).

Since the intensity of the solar wind is not constant but changes with the activity of the Sun, the upstream boundaries of these magnetospheres also change. A strong solar wind pushes the boundary closer to the planet. At such times, Saturn's satellite Titan actually moves outside the magnetosphere when its orbit brings it around to a position between the planet and the Sun. At other times, Titan remains within the magnetosphere throughout the course of its orbital journey.

In the downstream direction, the magnetosphere stretches out into a so-called **magnetotail**, coaxed along in the anti-Sun direction by the solar wind (like the plasma tail of a comet; hence the name). Jupiter's magnetotail is so extended that it appears to reach all the way to Saturn's orbit, a distance almost equal to the distance of Jupiter from the Sun.

Energy to Power the Magnetospheres

Given the sources and sinks for the charged subatomic particles in the plasmas, what supplies their energy? Ultimately, it is the rotation of the parent planets. Rotation provides the energy to generate the planet's magnetic fields. Charged particles spiraling along magnetic field lines must revolve around the planet at the same rate as the field lines. This is the rotation period of the planet itself, as we discussed earlier.

As a result of this coupling of the plasma to the magnetic field, we have the curious situation that these spiraling electrons, protons, and ions are moving around the planet as if they were attached to it. They exhibit what we call *rigid body rotation*, rather than following Kepler's laws of planetary motion as they would if their motion were dominated by gravity instead of by magnetic forces. This means that a particle at Io's distance from Jupiter must follow the radio rotation period of System III: 9 hr 56 min. This is much faster than the period of revolution of Io, which is 42 hr 28 min. Hence the plasma flows past the satellite, leading to the unusual result that a wake *precedes* the satellite in its orbit about the planet (Fig. 13.25). The same thing happens with Titan in the magnetosphere of Saturn.

The charged particles become so energetic because of their rapid motion that when they strike the surface of a satellite they can knock out an ion that then joins the co-rotating plasma. Forcing this plasma to rotate with the field and to generate the radio noise that is equivalent to 10^{14} watts costs the planet energy. It means that its rotation rate must be steadily decreasing, but at an unmeasurably slow rate.

Auroras

By analogy with Earth, one would expect that charged particles from the magnetosphere will sometimes follow the magnetic field lines right into the planet's atmosphere. On Earth, this charged particle "precipitation" produces auroras in the north and south polar regions. Indeed, the same phenomenon has been detected on Jupiter, both by direct photography and by ultraviolet and infrared spectroscopy. Not only is light produced by these interactions, but new compounds are formed in the planet's polar atmosphere as molecules of methane break apart and the fragments recombine.

Ten years after the Voyager encounters, astronomers on Earth discovered that they could

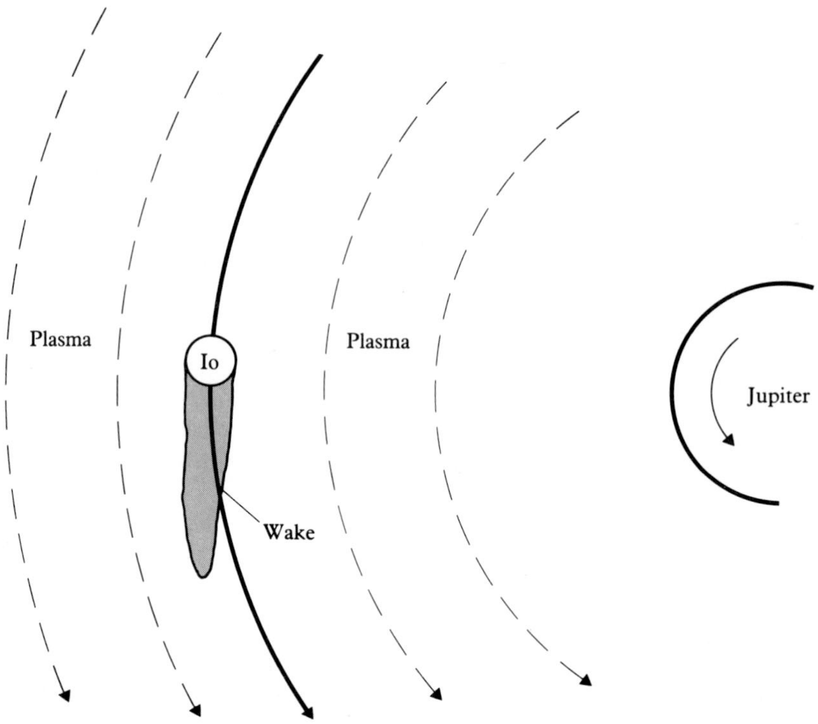

FIGURE 13.25 Jupiter's magnetic field is generated in the deep interior of the planet. Since it behaves as if it were rigidly connected to Jupiter, its period of rotation is constant with distance from the planet. Thus the magnetic field moves faster than the satellites move in their orbits, so they have magnetospheric wakes that *precede* them, instead of following them, as we might have expected.

detect evidence of the jovian auroras in the infrared spectrum of the planet. Molecules of H_3^+ (H_2 plus a proton) are formed in the planet's polar regions as protons spiraling in from the jovian Van Allen belts bombard molecules of H_2. The resulting emissions of radiation can be recorded by infrared detectors used with powerful Earth-based telescopes, producing images like the one shown in Fig. 13.26.

Ultraviolet auroras are also present on Saturn, but not the strong infrared emission from H_3^+ that is seen in Jupiter. Voyager's attempt to take pictures of Saturn's aurora was unsuccessful because the night sky on Saturn is never really dark. Sunlight scattered back into the dark hemisphere by the rings always leaves the planet in a twilight glow. This phenomenon is illustrated in Fig. 13.27, which shows a view of Saturn we cannot have from Earth. Looking back at the planet from Voyager 1, we see the shadow of Saturn crossing the magnificent system of rings. An

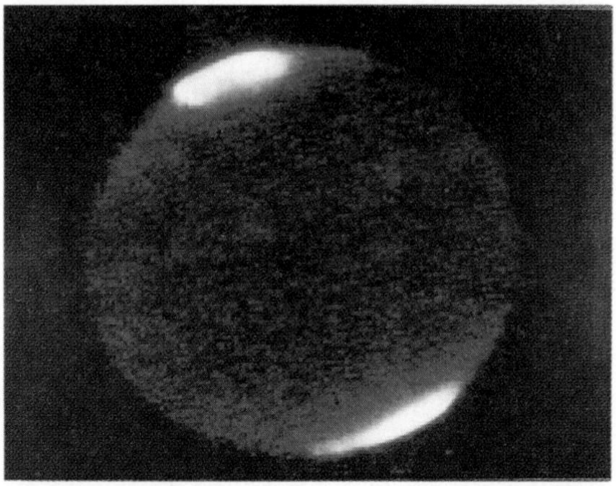

FIGURE 13.26 This picture of Jupiter was obtained with the 3-m NASA infrared telescope on Mauna Kea using a camera that recorded radiation at a wavelength of 3.4 μm. At this wavelength, methane in Jupiter's atmosphere strongly absorbs incident sunlight, so the planet appears very dark except at the poles, where emission of light from H_3^+ produces an aurora.

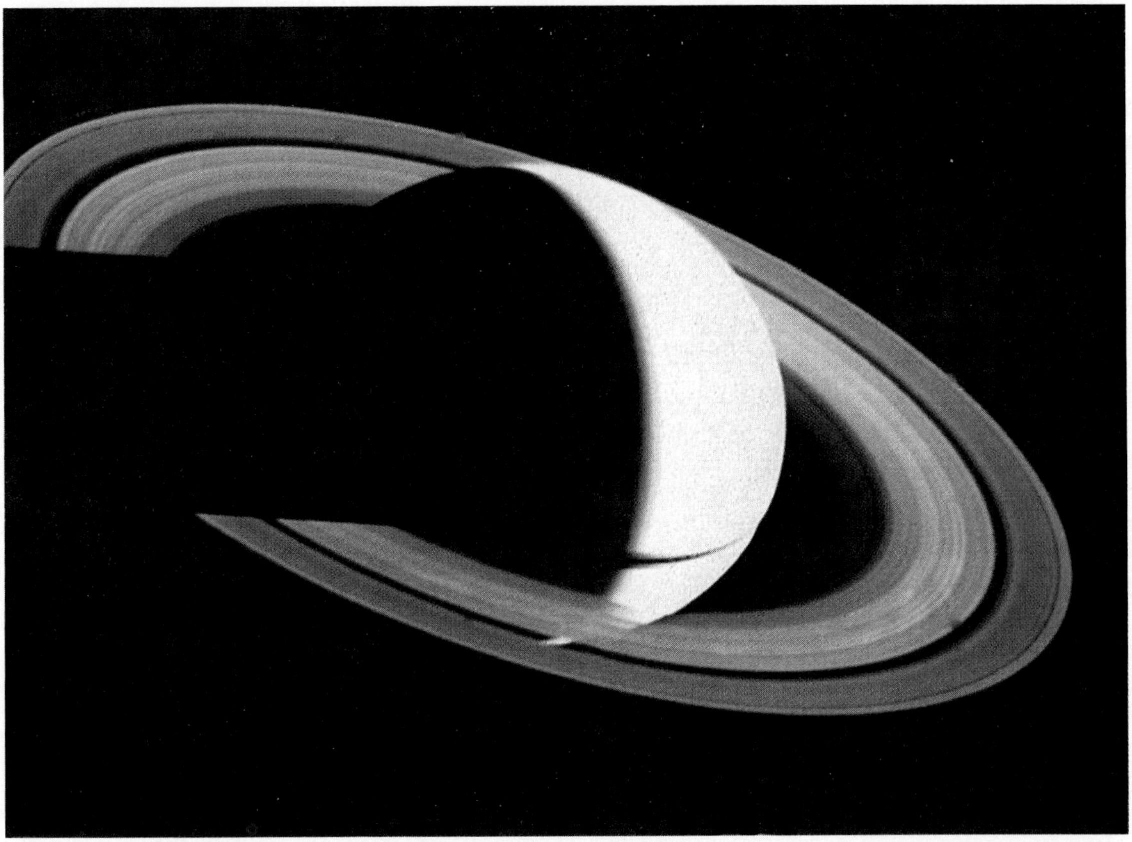

FIGURE 13.27 A dramatic picture looking back at Saturn, taken by Voyager 1 after its spectacular encounter with the ringed planet. In a view we can never see from Earth, Saturn appears as a crescent with its dark shadow crossing the magnificent ring system. (cf. Fig. 3.11).

observer floating in the atmosphere on the night side of the planet would see the bright rings arching up from the horizon and disappearing in shadow overhead. Ring light would illuminate the surroundings.

The illustration also reminds us that in leaving Saturn we have now visited all of the planets known to the ancients. As Voyager 2 ventured forward on its way to Uranus and Neptune, it was heading toward worlds unknown to Galileo, Shakespeare, Shah Jihan, Newton, Bashō, and Bach.

13.7 QUANTITATIVE SUPPLEMENT: BUILDING A GIANT PLANET

The main difference between Jupiter and Saturn on the one hand and the inner planets on the other is the ability of these giants to retain hydrogen and helium. Suppose we try to transform the Earth into a giant planet by adding those missing gases until the composition is the same as that on the Sun. How big would this new giant be?

We begin by referring to Tables 2.2 and 3.2. These emphasize that the greatest deficiencies of our planet are its lack of hydrogen and helium. Let us adopt silicon (Si) as our index element and add enough hydrogen and helium to yield a solar ratio of these elements to silicon. The solar ratio of H to Si (by the number of atoms) is about 2×10^4. However, we must convert this to a mass ratio by multiplying the numbers of H and Si atoms by their respective atomic masses. This gives us a ratio of H to Si in the Sun of about 850. A similar calculation for helium gives a mass ratio of He to Si of about 250.

We could continue this process for each element, but the fact is that the others add very little, relative to hydrogen and helium. Even though their individual masses are greater than those of atoms of H and He, their abundances are lower by an even larger factor.

To build our planet, we begin with the mass of silicon in the Earth, which is 7×10^{23} kg. We then add the appropriate ratios of H (850) and He (250) and multiply each by the mass of Si: $(1 + 850 + 250)7 \times 10^{23} = 8 \times 10^{26}$ kg. This is equivalent to 120 times the mass of the Earth. In other words, by restoring the missing light elements to the Earth, we have created a planet with a mass greater than that of Saturn.

We should notice, however, that this giant planet we have conjured up has only one Earth mass of heavy elements, whereas Jupiter and Saturn have cores of 10–15 Earth masses. Evidently, the heavy elements are enriched on those giants through the planet-forming process.

13.8 THE GALILEO JUPITER PROBE: PRELIMINARY RESULTS

On Dec. 7, 1995, the Galileo probe radioed back the first measurements made from within the atmosphere of a giant planet. For the previous 6 months the 339-kg probe had coasted alone toward Jupiter, after being ejected from the orbiter spacecraft that had brought it this far, rather like a bullet shot from a moving gun. The probe mission was a resounding success: All of the instruments returned useful data and the probe reached a deeper altitude than the nominal goal. As this book went to press, the following trajectory information and preliminary results were available.

During its 57-min parachute descent, the orbiter dropped 160 km to a pressure level of 22 bars (Fig. 13.28). It was also carried sideways another 600 km by strong winds. Measurements began in the region of the ammonia cirrus clouds; tenuous clouds were detected by the probe instruments at this altitude. At a pressure of about 1.6 bars a second, thin cloud layer was encountered, presumably the long-assumed but never established ammonium hydrosulfide layer. At 5 bars the probe was expected to encounter water clouds, but none were seen; the atmosphere remained clear down to the end of the mission at 22 bars. Measurements of the atmospheric water vapor abundance were also well below expectations. This surprising result probably does not apply to the planet as a whole, however. Instead, the probe investigators concluded that the probe went into one of the cloud-free areas on Jupiter, the regions that appear bright because of emitted thermal radiation at 5 micrometers (see Fig. 13.19). Caltech scientist Andy Ingersoll, who is a proponent of thick water clouds on Jupiter, characterized the entry point as one of the "deserts" of Jupiter and suggested that it was highly atypical. The probe also detected lightning from distant jovian thunderstorms.

One of the most important measurements was the ratio of helium to hydrogen in the atmosphere, which turns out to be 24% (mass fraction). This is just slightly lower than the solar helium abundance of 28%, suggesting that there is indeed some precipitation of helium in the interior of Jupiter, as there is on Saturn. The probe also confirmed the enhancement of carbon seen by Voyager and found that argon had an approximately solar abundance. These

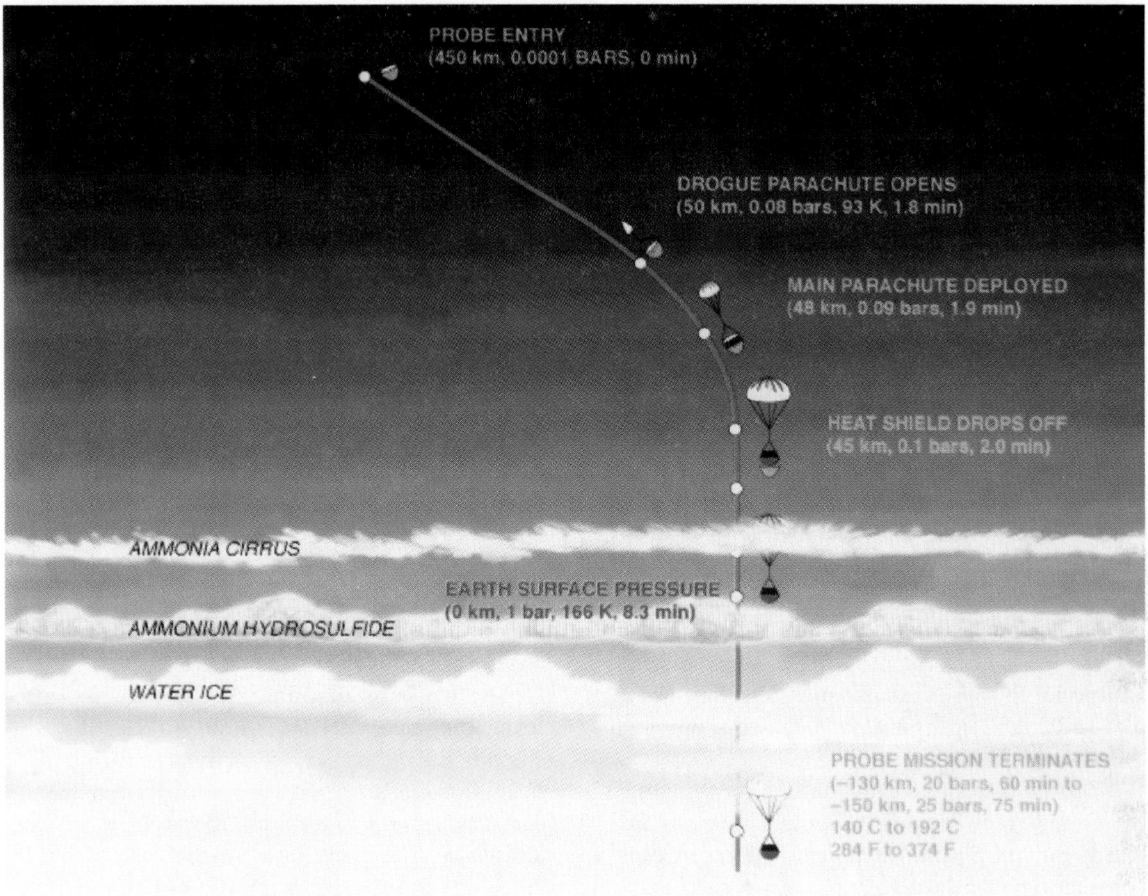

FIGURE 13.28 This artist's sketch shows the trajectory of the Galileo probe through the jovian atmosphere. The approximate locations of the various expected cloud layers are shown for reference, although the probe did not find these at the positions shown here (see text and Fig. 13.9).

results are consistent with a model for the formation of Jupiter's atmosphere that invokes the addition of solid material in the form of icy planetesimals to the infall of gas from the solar nebula.

The probe entered the atmosphere near Jupiter's high-speed equatorial jet, so it was no surprise that the measured winds were high near the level of the cloud tops. What had not been anticipated was the continued high wind speeds as the probe descended to a depth of more than 20 bars. On Earth the absorption of sunlight drives circulation in the troposphere, but this does not appear to be the case on Jupiter. Instead, the new results suggest that atmospheric winds are powered primarily by the internal heat of the planet.

About 9 hours after the end of its transmissions, the aluminum-titanium probe had sunk to a level in the jovian atmosphere where the temperature was so high that all components had first melted and then evaporated, becoming part of Jupiter's atmosphere.

SUMMARY

The two largest planets with their extensive retinues of satellites resemble miniature solar systems. Jupiter and Saturn are so massive that they have been able to retain thick atmospheres (and even more massive oceans) of hydrogen and helium throughout the 4.5 billion years they have existed. No solid surfaces exist, so there is no science of geology on Jupiter and Saturn.

Besides hydrogen and helium, these atmospheres contain other simple hydrogen-rich molecules. Some of these, such as methane and ammonia, are expected to form readily at the local temperatures and pressures. Others, like acetylene and carbon monoxide, require some additional source of energy such as solar ultraviolet radiation, lightning, or heat coming from the planetary interiors.

Both Jupiter and Saturn radiate more energy than they receive from the Sun. Jupiter is primarily radiating primordial heat, the result of its contraction from a giant protoplanetary phase. Smaller Saturn must have some other source of energy. Precipitation of helium deep in its interior provides the explanation, which is supported by the observed deficiency of helium in the planet's atmosphere. Both planets have layers of metallic hydrogen surrounding solid cores of rock and metal. These cores appear to be similar in size, about 10–15 times the mass of the Earth.

Despite the continuous motion of currents and clouds, some features in these atmospheres have remained remarkably permanent. Outstanding among these is Jupiter's Great Red Spot, which has lasted some 300 years. Its longevity is attributable to its size, which is large enough to swallow two Earths. The color of this giant high-pressure region, like the other colors found among the clouds in these atmospheres, is caused by traces of chemicals whose identities remain unknown. Hypotheses about the origin of the color of the GRS range from organic compounds to red phosphorus, and sulfur is regarded as a possible component of some of the brown and tawny-colored clouds.

Both Jupiter and Saturn are surrounded by giant magnetospheres. These magnetic fields, which are thousands of times stronger than Earth's, trap energetic plasmas. The atoms and fragments of atoms that make up the plasmas originate from the solar wind, from the atmospheres of the planets, and from the satellites and rings. Jupiter is an intense source of nonthermal radio noise at both long and short wavelengths, while Saturn radiates only at very long wavelengths. This difference is attributable to the presence of Saturn's extensive system of rings, which serves as an effective absorber of charged atomic particles near the planet.

The plasmas surrounding these giants differ from the environment found in Earth's Van Allen belts in that they contain large numbers of heavy ions. These ions are contributed by the surfaces of the satellites through a process known as sputtering and, in the case of Io, by volcanic eruptions. Precipitation of charged particles from the radiation belts causes auroras, which so far have been seen only on Jupiter.

KEY TERMS

Io plasma torus
magnetic flux tube
magnetotail
nonthermal radiation
occultation
sputtering
thermal inversion

Review Questions

1. Which spacecraft have visited Jupiter and Saturn so far? When were they launched and when did they arrive? What are the prospects for future exploration of these planets?

2. Why are Jupiter and Saturn called "giant" planets? What are they made of? How do their interiors differ from the interior of the Earth? How do we know?

3. Both Jupiter and Saturn radiate more energy than they receive from the Sun. What are the sources of that energy? Are they the same on both planets?

4. What gases are present in the atmospheres of these planets? What do you think would happen if an astronaut lit a match while visiting Jupiter? Why?

5. What sources of energy are available to form compounds in Jupiter's atmosphere? How does the chemistry going on there differ from what took place on the primitive Earth? What are the visible signs of this jovian chemistry?

6. How does the global circulation of Saturn differ from that of Jupiter? Why do scientists think this difference exists?

7. Both Jupiter and Saturn are broadcasting radio waves even though there are no advanced, technical civilizations with radio transmitters on either planet. Explain.

8. Compare the sizes of the magnetospheres of Jupiter and Saturn with each other and with the size of the Sun. If you could see Jupiter's magnetosphere with your unaided eyes, how big do you think it would appear in our nighttime skies?

Quantitative Exercises

1. Use Table 2.2 to show that the addition of missing carbon and nitrogen to the artificial planet described in 13.7 has only a small effect on its calculated mass.

2. Referring to Section 12.7, show that hydrogen and helium could not escape from this new planet once it formed, even though it is closer to the Sun and hence much warmer than Jupiter or Saturn. What does this result tell you about the origin of the differences in composition between the terrestrial and giant planets?

3. (a) How long would it take astronauts in a balloon floating in the atmosphere of Jupiter to circumnavigate the planet if the balloon moved at the same speed as the equatorial jet? (b) Repeat this calculation for Saturn.

4. Calculate the wind speed at the outer boundary of the GRS, given its size and six-day rotation period. Compare this speed with the east-west (zonal) wind speeds.

Additional Reading

Alexander, F.F. O'D. 1980. *The Planet Saturn.* New York: Dover. A history of naked-eye and telescopic observations of the planet, its rings, and satellites, oriented toward the dedicated amateur.

Beatty, J.K., and A. Chaikin, eds. 1990. *The New Solar System.* Cambridge, Mass.: Sky Publishing. An excellent compilation of articles by experts in the field, including fine reviews of the Voyager discoveries.

Beebe, R. 1995. *Jupiter: The Giant Planet.* Washington, D.C.: Smithsonian Institution Press. Excellent pre-Galileo overview of the planet and its satellites for the technically literate general audience, combining results from both astronomical and spacecraft investigations.

Burrows, W.E. 1990. *Exploring Space: Voyages in the Solar System and Beyond.* New York: Random House. History and politics, including extensive discussions of the tortured history of the Galileo mission to Jupiter.

Cooper, Henry S.F. 1982. *Imaging Saturn.* New York: Holt, Rinehart and Winston. A first-hand account of the Voyager Saturn encounter by a reporter from *The New Yorker.*

*Gehrels, T., and M.S. Matthews, eds. 1984. *Saturn.* Tucson: University of Arizona Press. Highly technical papers in another volume of the Arizona Space Science series.

Morrison, D. 1982. *Voyage to Saturn.* NASA Special Publication 451. Washington, D.C.: U.S. Government Printing Office. Popular account of the Voyager flybys of Saturn, including comparisons of the two planets and their satellites.

Morrison, D., and J. Samz. 1980. *Voyage to Jupiter.* NASA Special Publication 439. Washington, D.C.:

U.S. Government Printing Office. Popular account of the Voyager flybys of Jupiter.

Peek, B.M. 1958. *The Planet Jupiter*. London: Faber and Faber. The standard reference for observations of Jupiter with small telescopes.

Spencer, J. John, and J. Mitton, eds. 1995. *The Great Comet Crash*. Cambridge: Cambridge University Press. Multiauthor volume telling the story of the impact of S-L 9 with Jupiter and summarizing the initial scientific results.

*Indicates the more technical readings.

14

In Deep Freeze: Planets We Cannot See

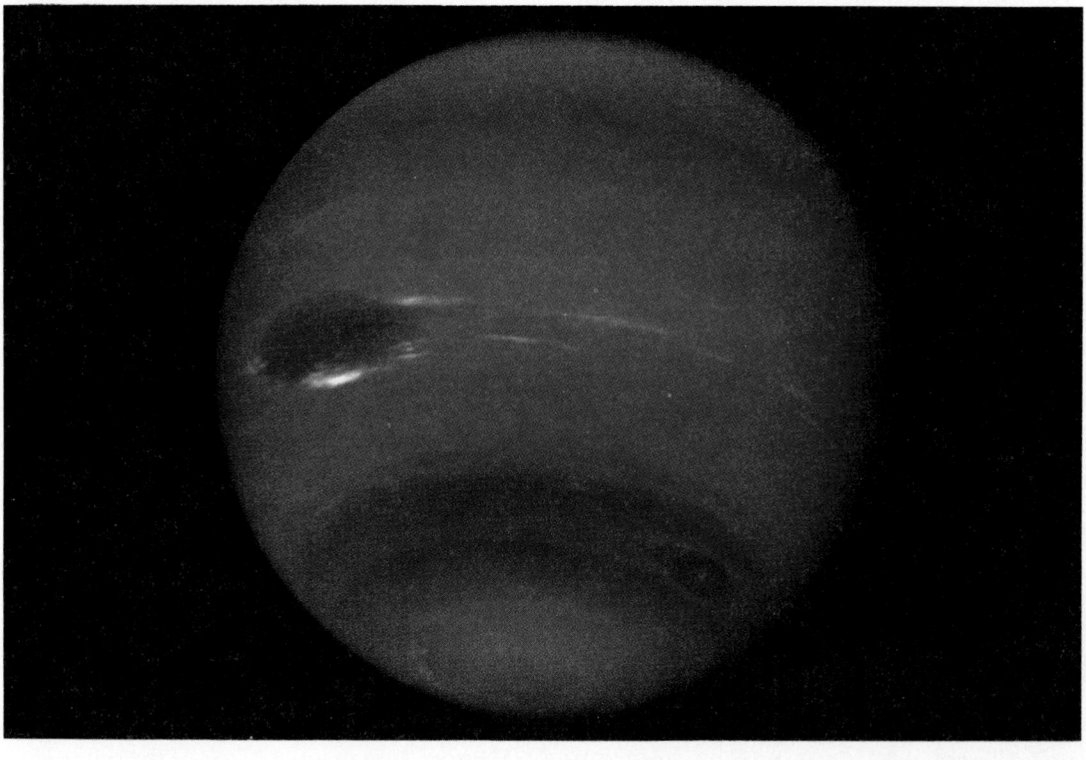

Neptune is the only other blue and white planet in our solar system besides the Earth. But here the white clouds are made of frozen methane instead of water ice, and the large, dark, oval feature we see is not a continent but a giant vortex in this planet's thick atmosphere.

Compared with Jupiter and Saturn, Uranus and Neptune are smaller, and have higher densities and a much higher proportion of methane in their atmospheres. Uranus and Neptune are not giants like Jupiter and Saturn; indeed, their masses are intermediate between the two giants discussed in the previous chapter and the terrestrial planets like Earth and Venus. Since their masses are so similar, it makes sense to discuss Uranus and Neptune together. When we look closely, however, we find differences between these two minigiants that are even more profound than those distinguishing the other planetary "twins," Earth and Venus.

The distance from Saturn to Uranus is about the same as the distance from Earth to Saturn. The next step, to reach Neptune, is only slightly larger. Crossing that gap, we find a planet that is more dynamic than Uranus and that possesses a strong internal heat source and an active meteorology in its visible atmosphere. The reasons for this difference between Uranus and Neptune are not yet clear, but the effect on the Voyager mission was to give the Neptune encounter a special edge of excitement that no one had anticipated based on the results from Uranus.

FIGURE 14.1 This perspective view of the Voyager trajectories clearly shows how far both spacecraft were from Pluto, even though Voyager 2 did reach Uranus and Neptune.

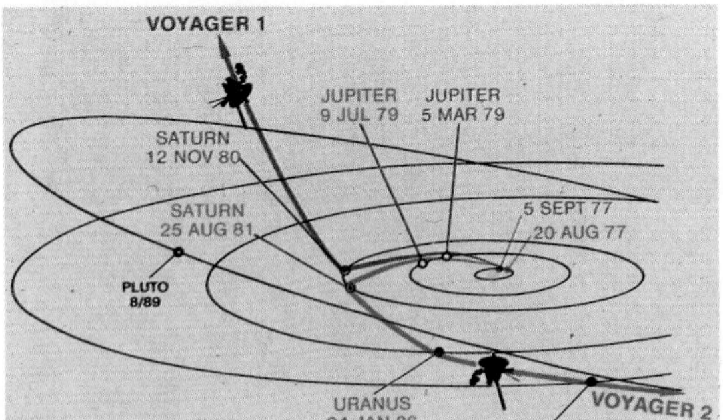

The last planet in our system is a completely different object altogether. Far from being a giant, Pluto is smaller than our Moon. It resembles one of the large icy satellites and will therefore be discussed in more detail in a later chapter, where this comparison can be elucidated.

14.1 DISCOVERIES OF THE OUTER PLANETS

Beyond Saturn, only the unchanging stars were known until two centuries ago. In extending our perspective to Uranus, Neptune, and Pluto, we enter new territory. These planets were *discovered*, in a literal sense: one by accident, one by calculation, and one as the result of a diligent search. In each case, the discovery generated at least as much public interest as when one of our spacecraft visits a planet for the first time.

Direct exploration of Uranus and Neptune by spacecraft did not occur until the late 1980s. Voyager 2, in its grand tour of the outer planets, reached Uranus in January 1986 and Neptune in August 1989. Unfortunately, Pluto could not be visited on this magnificent journey, since the alignment of the planets did not allow either Voyager spacecraft to reach it (Fig. 14.1). Thus

Pluto is now the last planet we perceive only in the ancient manner—as a moving point of light in the night sky rather than as a real, round world.

Discovery of Uranus

The night of March 13, 1781, was clear and dark over the town of Bath in southwest England. Pursuing his astronomical hobby that evening, a professional musician named William Herschel was continuing a project to chart all stars down to the eighth magnitude. He was using a 6-inch aperture telescope he had built himself, and on this particular evening he was examining the stars in a portion of the constellation Gemini. As he recorded in his notebook, one of the stars that came into view as he moved the telescope from one place to the next seemed unusual. Herschel described it as "a curious either nebulous star or perhaps a comet." Instead of a point of light, it appeared as a small disk.

Subsequent observations revealed that this object moved (unlike a star) but that the motion was too slow for a comet. Soon its path had been mapped well enough to establish that it was a new planet orbiting the Sun at a distance of 19 AU (nearly 3 billion kilometers). This is exactly where the Titius-Bode rule (Section 1.6) would predict a planet to be, and that realization greatly strengthened the belief of contemporary scientists that this rule must have some deep, causal meaning. A new world had been added to those known since antiquity, and the size of the solar system had suddenly doubled. In recognition of his accomplishment, Herschel received a knighthood and a royal pension, and went on to become one of the most productive astronomers of his era.

These days we have become rather blasé about new astronomical discoveries—volcanoes on Io, black holes, colliding galaxies—but five years after the United States declared its independence from Britain, the world was a very different place. Adding a new planet to the solar system was an extraordinary event. Perhaps the most famous comment on the effect of this discovery is in John Keats's beautiful sonnet "On First Looking into Chapman's Homer," where the poet uses Herschel's discovery as an image for his own feelings on being introduced to Homer's great epic:

> *Then felt I like some watcher of the skies*
> *When a new planet swims into his ken.*

Herschel named this new planet "Georgium Sidus" after his king and patron, George III of England. We have been spared the humor associated with having a planet named George's star by the suggestion of Herschel's contemporary Johann Bode (the same person responsible for the Titius-Bode rule) that the ancient tradition for planetary names be continued by using the name Uranus. In Greco-Roman mythology, Uranus was the father of Saturn, who was in turn the father of Jupiter. The name thus seemed appropriate for the most distant known planet.

Discovery of Neptune

Uranus is bright enough that even a small telescope can reveal it easily; in fact, it can just be glimpsed with the naked eye in a dark, clear sky. After Herschel's discovery, astronomers looked through records of previous observations and found some 20 cases where Uranus had been recorded. It had been mistaken for a star, since the earlier observers did not have telescopes as good as Herschel's and could not see the disk. With these additional observations (stretching back to 1690), it was possible to develop a very accurate path for the new planet.

Astronomers used these measurements to calculate the orbit of Uranus, but even when all of the gravitational effects from the known planets were accounted for, it was impossible to fit the observations with an elliptical orbit. By 1830 the discrepancy was 15 seconds of arc, which meant that the predicted position of Uranus was more than four times the planet's apparent diameter from its observed place in the sky. To understand what this discrepancy means, recall that any circle (such as the projection of a planet's orbit on the sky) contains 360°. A degree is divided into 60 minutes of arc, and a minute of arc is divided into 60 seconds of arc. In these units, the apparent sizes of the Moon and the Sun are ½°, or 30 minutes of arc. Jupiter at opposition is 45 seconds of arc, while Uranus appears roughly ten times smaller. Today, telescopes at excellent sites such as Mauna Kea routinely resolve an angular distance of 0.3 second of arc.

Even in Herschel's time, astronomers could easily measure a discrepancy of 15 seconds of arc. The validity of Newton's laws of planetary motion was universally accepted, so it seemed likely to astronomers of the time that this discrepancy was the result of a **perturbation** caused by the gravitational pull of some unknown body. A perturbation is a small deviation from normality, in this case a minor but cumulative change from a simple Keplerian orbit around the Sun.

Since scientists had now accepted one new planet, it was reasonable to suggest that perhaps another might be present, one orbiting the Sun at a still greater distance and perturbing the motion of Uranus by its gravitational attraction. How to find it among the millions of stars? The first clue astronomers used was the Titius-Bode rule. As Uranus very nearly obeyed this rule, perhaps the new planet would lie at the distance from the Sun predicted by this example of numerology, about 39 AU. With this starting point, it was then a question of solving some formidable mathematical equations that described various orbits for the hypothetical planet of unknown mass and position.

Two gifted mathematicians succeeded in solving this problem independently: John Couch Adams in England and Urbain Jean Joseph Leverrier in France. Although Adams obtained the first solution, in September 1845, he was unable to persuade any of his countrymen to initiate a telescopic search. After Leverrier published the preliminary results of his calculations in June 1846, the British astronomers realized that these agreed well with Adams's work. They then began to look for the object but, lacking adequate charts of this region of the sky, were unsuccessful.

Meanwhile, Leverrier finished his analysis in August 1846 and sent a letter to the German astronomer Johann Galle in Berlin suggesting that he try to find the new planet. The Berlin observatory had a newly completed set of star charts covering the region of Leverrier's prediction. Using these charts, Galle was able to find the planet on the evening of September 23, 1846, on his first attempt. It was just 1° (twice the apparent diameter of the Moon) from the position predicted by Leverrier and 2.5° from Adams's prediction. Galle could not discern the planet's tiny disk, but he identified it by its appearance as a wanderer among the fixed stars. Neptune is only about one-tenth as bright as Uranus, much too faint to be seen without a telescope.

The discovery of Neptune represented a stunning triumph for Kepler, Newton, and gravitational theory. Once again, a search was made for prediscovery observations, and once again several were identified, including one observation by the hapless English astronomer who first set out to find the planet two months earlier. But perhaps the most interesting early observation is one attributed to Galileo. In a sketch of observations of Jupiter made in December 1612 and January 1613, Galileo recorded a "star" that modern astronomers have been able to identify as Neptune by extrapolating the position of the planet backward in time along its orbit (Fig. 14.2). Even though Galileo's telescope was powerful enough to show Neptune, without charts he had no hope of identifying this tiny starlike object as a planet.

There was understandably some dispute over the proper credit for the discovery of Neptune. Matters were not helped by the intervention of the French astronomer François Arago, who suggested that the new planet be named Lever-

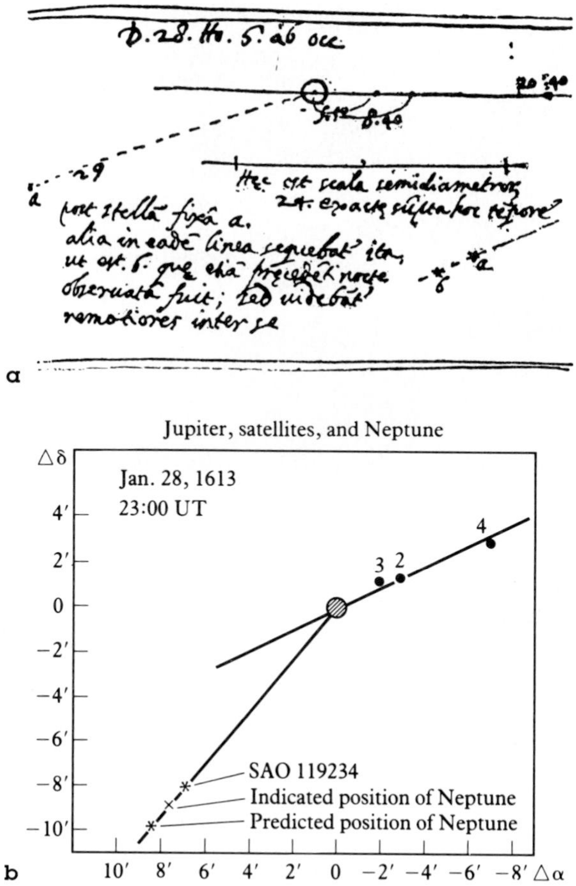

FIGURE 14.2 (a) An entry from the notebook in which Galileo recorded the relative positions of Jupiter, three of its moons, and two stars on Jan. 28, 1613. (b) A reconstruction of Galileo's observations by Charles Kowal. One of the "stars" was actually Neptune.

rier. Bowing to international disapproval, Arago later withdrew this idea and instead proposed Neptune, the Roman god of the sea. Happily, Adams and Leverrier managed to disassociate themselves from these disputes, and the two have been credited jointly with the discovery. Note that in this case the discoverer is considered to be the person who made the mathematical calculations, not the one who actually first saw Neptune through the telescope.

Discovery of Pluto

Continued tracking of both Uranus and Neptune during the rest of the nineteenth century suggested that yet another planet was required to account satisfactorily for their motions. Although the evidence was only marginally convincing, a number of astronomers undertook calculations similar to those of Adams and Leverrier. The most persistent of these individuals was Percival Lowell, already familiar to us from his work on Mars (Section 11.1). Lowell not only calculated an orbit for this hypothetical planet, he also initiated some systematic searches for it at his observatory. These were unsuccessful, but they provided valuable experience.

A special wide-field telescope was built at Lowell Observatory in 1929, 13 years after Lowell's death, and the search for a new planet was put in the hands of Clyde Tombaugh. On February 18, 1930, Tombaugh discovered the trans-Neptunian planet on photographs he had taken with the new telescope the previous month (Fig. 14.3). The ninth planet was announced to the world on March 13, Percival Lowell's birthday.

Although this discovery might seem to be another triumph for gravitational theory, it was actually the product of meticulous astronomical observations. Pluto's mass is far too small to have caused the apparent perturbations in the motions of Uranus and Neptune. Recent studies of this problem suggest that the difficulty lies in the observations of these two planets. The supposed perturbations do not, and never did, exist.

Pluto was the Roman god of the underworld, a fitting title for a planet at such a great distance from the Sun. The symbol for this planet, ♇, has the additional attribute of commemorating the initials of Percival Lowell, the main mover behind its ultimate discovery. This is as close as one might properly come to naming a planet for a mortal astronomer.

Pluto is so small and so far from the Sun that it is 10,000 times too faint to see with the unaided eye. It is only within the last 15 years

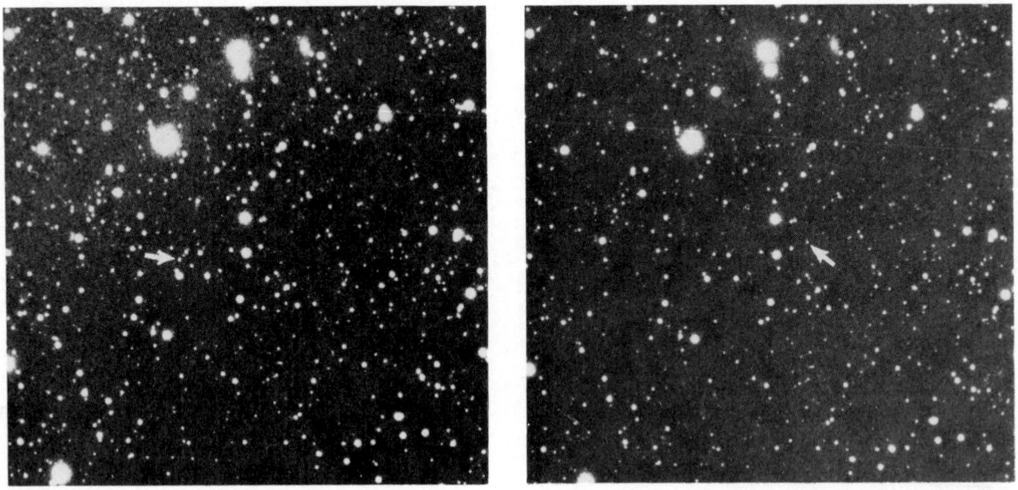

January 23, 1930 January 29, 1930

FIGURE 14.3 The two photographs from the Lowell Observatory that led to the discovery of Pluto. In comparing the two, taken six days apart, Clyde Tombaugh noticed on Feb. 18, 1930, that one of the "stars" (shown here with an arrow) had moved. It was the planet he had been seeking.

that we have learned anything more about it than Tombaugh knew 60 years ago.

14.2 THREE DISTANT WORLDS

Before discussing Uranus, Neptune, and Pluto in detail, it is useful to put them in perspective by comparing them with planets we already know. This comparison is summarized in Fig. 14.4 and Table 14.1, where we see that Uranus and Neptune are very nearly twins. While they can be classed as giant planets, with masses roughly 15 times that of the Earth, they are much smaller than either Jupiter or Saturn.

Density and Composition

In Section 13.2 we saw that Saturn was both less massive and less dense than Jupiter, which is consistent with the two bodies' having the same bulk composition. With a lower mass, Saturn suffers less internal compression; hence the overall density is lower, only 0.7 g/cm^3. If Uranus

TABLE 14.1 Basic properties of Uranus, Neptune, and Pluto

Planet	Distance from Sun (AU)	Diameter (Earth = 1)	Mass (Earth = 1)	Density g/cm^3
Jupiter	5.2	11.2	317	1.3
Uranus	19.2	4.11	14	1.3
Neptune	30.1	3.92	17	1.5
Pluto	39.4	0.19	0.002	2.1

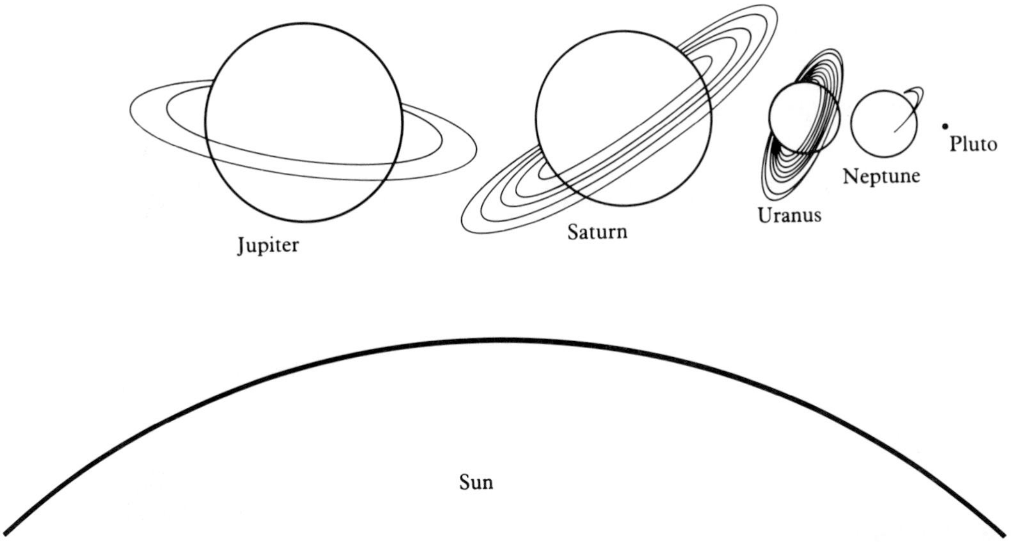

FIGURE 14.4 The relative sizes of the outer planets. Uranus and Neptune are distinctly smaller than Jupiter and Saturn, while Pluto is about the size of our Moon.

and Neptune had the same composition as Jupiter and Saturn, we would expect these bodies to have densities even less than that of Saturn, since they have only about 15% of its mass. Instead Uranus has about the same density as Jupiter, while Neptune has a greater density. This leads us to the conclusion that these planets each contain a higher proportion of heavier elements.

Ice and rock are the logical candidates for the dense material in Uranus and Neptune, as inspection of a table of cosmic abundances will convince you (Table 2.2). Oxygen is the most abundant reactive element after hydrogen, and these two combine readily to form water ice at low temperatures. What we call rocks are materials composed primarily of the common elements silicon and oxygen (Section 3.3). Scientists who have constructed models for the interiors of these two objects find that they must contain massive cores composed of the common heavy elements that form ice and rock. Surrounding these cores are relatively thin envelopes of liquid and gaseous hydrogen. Both Uranus and Neptune are too small to achieve the pressures and temperatures required for the formation of metallic hydrogen (Fig. 14.5).

Some scientists are currently suggesting that both of these planets may have thick layers of water clouds in their lower atmospheres. There is no direct evidence for the existence of these clouds, since water vapor has not been detected on either planet, but there are some chemical arguments in favor of this idea. In particular, the apparent deficiency of ammonia gas in both atmospheres (Section 14.3) could be explained if the ammonia were dissolved in these water clouds below the observable layers. As we shall see, however, there are alternative explanations for this apparent deficiency of ammonia.

Our knowledge of the interior structures of these two planets is still in a primitive state, so that even the existence of discrete cores is open to question. The only certainty is the marked concentration of heavy elements that gives both planets their relatively high densities.

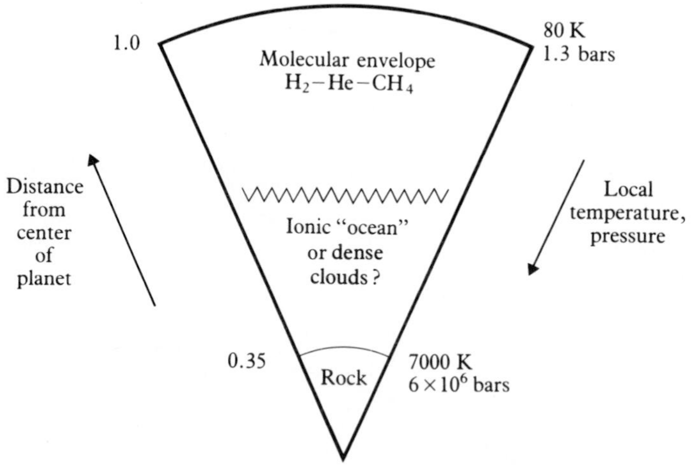

FIGURE 14.5 A schematic idea of what the interiors of Uranus and Neptune may be like. Pressures are not great enough for metallic hydrogen to occur (cf. Fig. 13.6). (After D. Stevenson.)

Differences Between Uranus and Neptune

Closer inspection of Table 14.1 indicates that the initial appearance of twinship may be incorrect, just as we found for Venus and Earth. The densities of Uranus and Neptune are distinctly different, indicating some internal structural or compositional differences. Probably even more significant is Uranus's lack of a significant internal heat source, while Neptune (like Jupiter and Saturn) derives much of its heat from internal sources. Models for its interior suggest that, like Jupiter, Neptune could still be radiating primordial heat. Despite its small mass, the high proportion of rock and ice to total mass leads to very slow cooling.

Perhaps the most puzzling enigma posed by Uranus is the bizarre orientation of its rotational axis, which is inclined by 98°, placing it practically in the plane of the planet's orbit (Fig. 14.6). The orbits of Uranus's satellites and rings are in the planet's equatorial plane, just as they are for Jupiter and Saturn, so the whole uranian

FIGURE 14.6 The axis of rotation of Uranus lies nearly in the plane of the planet's orbit. At the time of the Voyager 2 encounter, the axis pointed almost directly at the Sun (cf. Fig. 14.12).

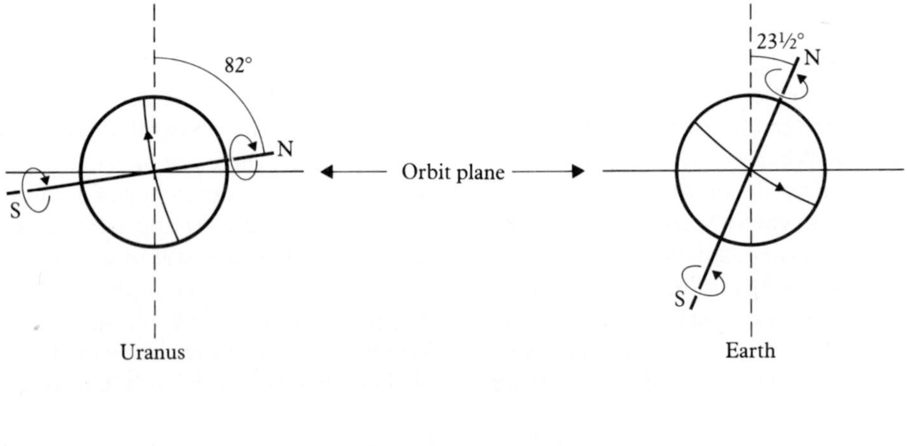

system is tipped on its side. It is often suggested that a glancing impact by another planet-sized object, which then added to the mass of Uranus, could explain this strange orientation, but there is no detailed theory yet that satisfactorily describes such an encounter and its effects.

Pluto: A Special Case

The above discussion has focused on Uranus and Neptune because these two planets are similar in many respects to Jupiter and Saturn. When we turn to Pluto, we are dealing with a different kind of object altogether. This planet is smaller than our own Moon and has a density of 2.1 g/cm^3. This combination of small size and low density suggests that Pluto is composed partly of water ice and probably resembles Triton, the large icy satellite of Neptune. In no way can it be considered one of the jovian planets, nor is it a displaced inner planet, roaming the cold and dark domains of space beyond Neptune. At best, it might be representative of the kinds of objects that collided with one another to form the icy cores of the giant planets, a fate it somehow avoided. We shall discuss it in Chapter 16, where we consider other icy objects with atmospheres.

14.3 ATMOSPHERES OF URANUS AND NEPTUNE

Through a telescope, the tiny disks of Uranus and Neptune appear greenish in color because the upper atmospheres of these planets are generally free of clouds. Although beautiful in their way, the pictures of Uranus obtained by Voyager 2 are singularly free of detail (Fig. 14.7). Neptune reveals some scattered white clouds, but the

FIGURE 14.7 The left-hand image shows Uranus as it would have appeared to someone riding along on the Voyager 2 spacecraft. The right-hand image has been specially processed to reveal the presence of a banded polar haze. (The rotational pole is nearly facing us.) The colors in this image are totally artificial.

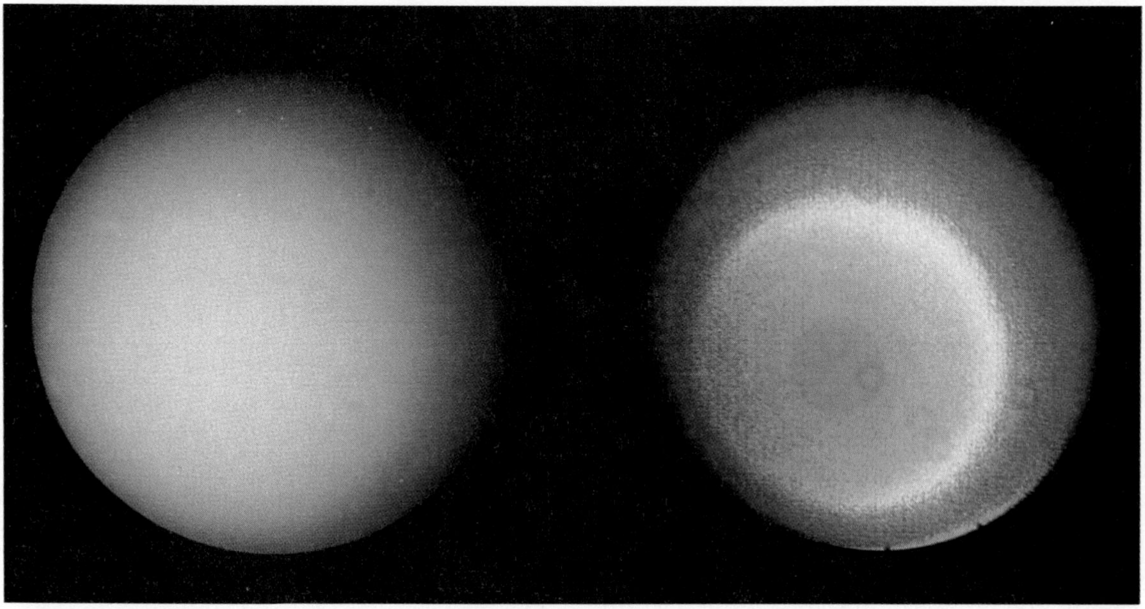

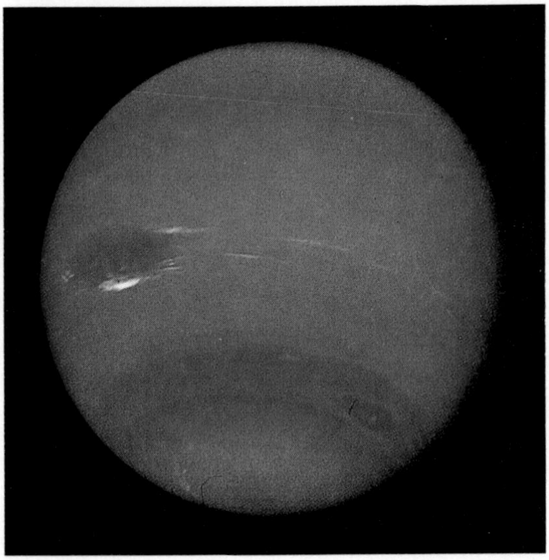

FIGURE 14.8 As Voyager 2 approached Neptune in the spring and summer of 1989, the images of the planet revealed a large, dark, oval storm system and several white clouds. Color is exaggerated.

underlying cloud deck in which storm systems appear is very low in the atmosphere (Fig. 14.8). Sunlight penetrates to great depths in the atmospheres of both planets before it is scattered back to space, and all along its path it is absorbed by methane gas (CH_4), giving both planets a faint greenish blue hue.

Composition

The same methane absorption bands that are present in spectra of Jupiter and Saturn appear in the spectra of Uranus and Neptune, but they are very much stronger, and many new absorptions are also evident (Fig. 14.9). The hydrogen (H_2) absorption lines on Uranus and Neptune are also stronger than in spectra of Jupiter and Saturn, but the increase is not as great as for methane. In other words, the proportion of methane in the atmospheres of Uranus and Neptune is much higher than in the atmospheres of Jupiter and Saturn (Table 14.2). This is consistent with the higher proportion of ices in the total masses of these smaller planets. Uranus and Neptune are clearly deficient in hydrogen and helium compared with Jupiter and Saturn, and this difference shows up in the composition of their atmospheres as well as in their average densities.

At a level in their atmospheres where the pressure is the same as the sea-level pressure on Earth, the temperatures on both Uranus and Neptune are about 73 K (− 200 C), or about the temperature of liquid nitrogen. This is much colder than conditions on Jupiter or Saturn at the same pressure level (Section 13.3). Ammonia will be frozen solid at this temperature but may be present as a gas at lower, warmer levels; at still greater depths there should be water vapor as well.

The Voyager results indicate that there is a tenuous haze in the upper atmosphere of Uranus centered over the rotational pole (Fig. 14.10). Evidently, this haze is formed as a result of the steady irradiation of the planet's upper atmosphere by solar ultraviolet light and thus is concentrated in the sunlit hemisphere. Even at this immense distance from the Sun, there is some atmospheric photochemistry forming a very thin smog layer that preferentially absorbs short-wavelength light.

The Voyager infrared spectrometer found that the ratio of the abundances of helium and hydrogen in the atmosphere of Uranus is even more similar to that in the Sun than is the case for Jupiter. This result is consistent with a two-step model for the formation of the giant planets. First, a huge core of ice and rock, 10–15 times the mass of the Earth, accumulates, and second, an envelope of gas (mostly hydrogen and helium) from the solar nebula collapses around the core.

In fact, the two steps are not so sharply separated. As the core forms, it will produce a secondary atmosphere of its own as a result of the outgassing of accreting material. This atmosphere will be composed primarily of nitrogen,

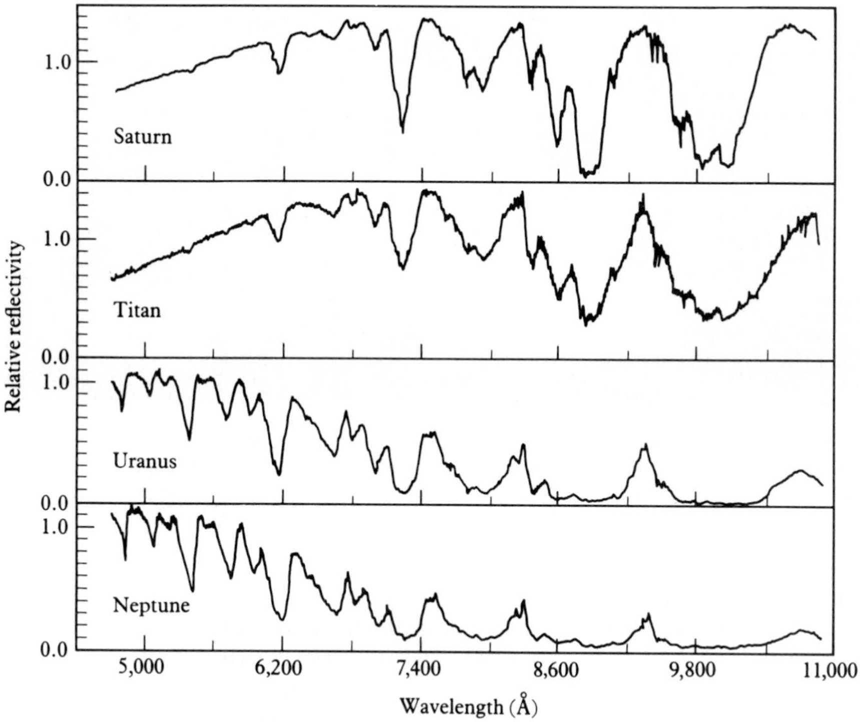

FIGURE 14.9 The spectra of Saturn, Titan, Uranus, and Neptune, each divided by the spectrum of the Sun. If these objects had no atmospheres, the spectra would be smooth, nearly flat curves. The deep absorptions are caused by methane, which is producing a much larger effect on Uranus and Neptune than on Titan and Saturn. (After R. Danhey.)

carbon monoxide, and methane. As the core grows, it will begin attracting hydrogen and helium from the solar nebula, and will finally develop an atmosphere that is dominated by these two gases. The core contributes the excess methane, since the CO will also be converted to methane while the bulk of the atmosphere is representative of the solar nebula. Apparently, Uranus and Neptune acquired less massive envelopes of solar nebula gas than Jupiter and

TABLE 14.2 Atmospheric compositions of Uranus and Neptune

A. Main Constituents (percent)			
Gas	Formula	Uranus	Neptune
Hydrogen	H_2	84	84?
Helium	He	14	14?
Methane	CH_4	2	2
B. Trace Constituents (parts per billion)			
Acetylene	C_2H_2	200	present
Ethane	C_2H_6	?	present
Hydrogen cyanide	HCN	<100	100
Carbon monoxide	CO	<40	1000

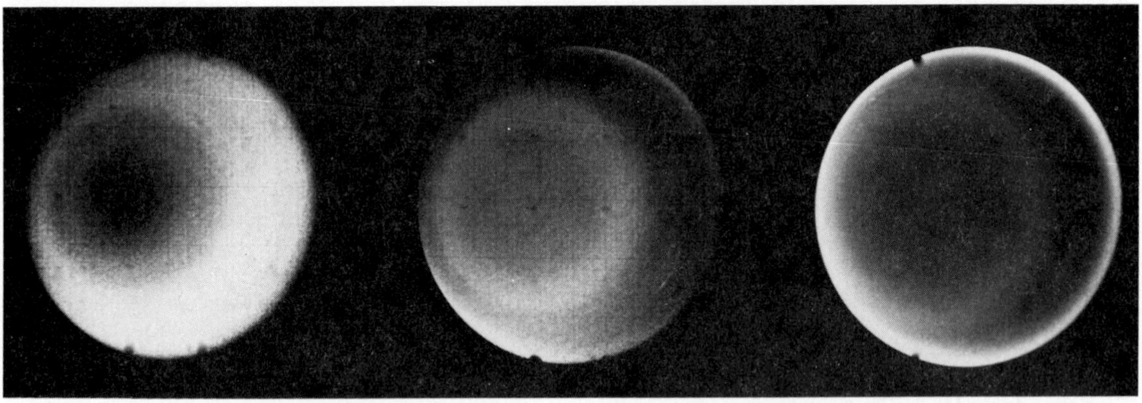

FIGURE 14.10 Voyager pictures of Uranus (left to right) taken through orange-, violet-, and methane-sensitive filters. The haze shown in Fig. 14.7 is easily seen in the violet image. The orange and methane images show two narrow cloud systems at temperate latitudes on Uranus. The bright edge of the image in the methane picture is caused by scattering from a high-altitude haze.

Saturn, and therefore they have a relatively higher proportion of heavy elements. We shall return to the problems of origins in Chapter 19.

Atmospheric Temperatures

At infrared wavelengths, where both Jupiter and Saturn exhibit strong emissions from methane and products of methane photochemistry, the spectrum of Uranus is essentially blank. In contrast, Neptune has emissions from both methane and ethane (Table 14.2). Why this difference? Evidently, there is a strong temperature inversion in the atmosphere of Neptune but only a weak one on Uranus.

When temperature increases with altitude in such an inversion, absorption bands produced by gases in a normal atmosphere become emission bands, as seen from the outside. This is because the local gas temperature in the inversion region is higher than that of the lower levels of the atmosphere. Emission bands produced in inversion regions can be detected with spectrometers sensitive to the infrared, typically at wavelengths near 10 micrometers, where there is little reflected sunlight to interfere.

As we saw in the discussion of Jupiter and Saturn, it is in the upper atmospheres of these planets that we expect trace gases to be produced by photochemistry and bombardment by charged particles trapped in the planet's magnetospheres. Furthermore, many gases have some of their strongest emission lines in the infrared, thereby making it easy to detect them at these wavelengths, even in tiny quantities. This combination of circumstances has led to the discovery of several trace constituents in these inversion regions above the tropopause, where the temperature is increasing with altitude.

It is useful to notice just how cold the tropopauses on Uranus and Neptune are (Fig. 14.11). At 55 K (− 218 C), our own atmosphere would condense to a mixture of ices! Hydrogen, helium, and neon can remain in the gaseous state, and nitrogen, methane, and carbon monoxide still have significant vapor pressures (like water vapor over ice on our planet). This is a small list of gases from which to make an atmosphere.

In the case of Neptune, we find evidence for all of these gases except neon, which is very difficult to detect. Hydrogen and methane were discovered by ground-based spectroscopy, and helium was detected by Voyager 2 using the same technique we described for Jupiter. Carbon monoxide emission lines from the region of the thermal inversion were detected in 1991 by astronomers using a radio telescope on Mauna Kea. The same set of observations revealed emissions from HCN, indicating that there must be a source of nitrogen in Neptune's upper atmosphere. N_2 seems the logical choice, since ammonia would be frozen out.

Presumably, the CO and N_2 on Neptune are convected up into the visible part of the atmosphere from the deep interior. Apparently, the reactions that efficiently convert CO to CH_4 and N_2 to NH_3 on Jupiter and Saturn are not working as well on Neptune. In fact, the amount of CO in Neptune's atmosphere is a thousand times greater than a **chemical equilibrium** model using these reactions would predict. Such a model uses the best current estimates for conditions in a planet's interior together with data for all relevant chemical reactions to predict what substances will form from the elements that are present. One such reaction is

$$N_2 + 3H_2 \rightarrow 2NH_3$$

in which nitrogen combines with hydrogen to produce ammonia. But this reaction requires a **catalyst** to proceed efficiently, and the problem on Neptune may be that an appropriate catalyst is not available. A catalyst is a substance that helps a reaction to proceed but does not take part in the reaction itself. In the case of nitrogen and hydrogen, iron compounds can serve as effective catalysts.

Temperatures Measured at Radio Wavelengths

Going to longer wavelengths, we find more puzzles in observations made with radio telescopes. No indications of nonthermal emissions were detected from either of these planets until Voyager reached them. Telescopes on Earth are not able to record noise bursts at decameter wavelengths, or steady signals from spiraling electrons in the decimeter region as they did for Jupiter. What the Earth-based radio observations do show, however, is very interesting indeed.

At short radio wavelengths, we can detect thermal radiation from deep within the atmospheres, so the corresponding temperatures should be higher than those detected elsewhere in the spectrum. In fact, one expects the measured temperatures to be higher as the wavelength of the observations increases because the atmospheric gases become increasingly transparent to radiation of longer wavelengths. This is how the high surface temperature of Venus was discovered (Chapter 10). The expected increase in temperature with increasing wavelength is certainly found on Jupiter, Saturn, and Neptune, but once again Uranus is the exception. Evidently, the temperature in this planet's atmosphere does not increase steadily with depth, as it does in the atmospheres of the other giants.

Unlike some of the other peculiarities of Uranus we have discussed, we can explain this unusual temperature structure. Recall that Uranus has no detectable source of internal energy. In other words, it doesn't have an excessively hot interior. That means that only a small amount of thermal energy will be trying to escape from deep within the planet. In that case, there will not be enough energy to maintain convection in the lower atmosphere, so the increase in temperature with depth at these deep levels will be very gradual. At higher altitudes, sunlight provides the energy to maintain convection. The situation resembles that in our oceans, where only the upper layers participate in the seasonal cycle driven by sunlight, while the murky depths remain at essentially the same temperature throughout the year. This combination of external sunlight and a weak internal heat source leads to the unusual temperature profile shown in Fig. 14.11.

The lack of a strong internal heat source probably also explains the absence of N_2 and CO in the upper atmosphere of Uranus. Without

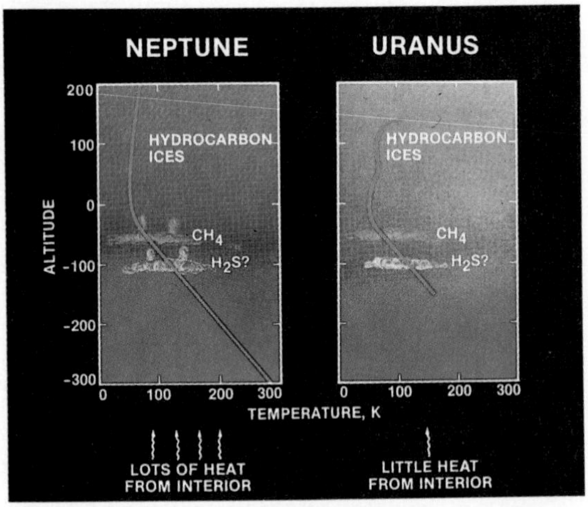

FIGURE 14.11 The variation of temperature with height on Uranus and Neptune. The solid lines show the Voyager results; the dashed lines are based on radio telescopic studies from Earth. Temperature increases more slowly with depth on Uranus than it would if the planet had a large internal heat source, as Neptune does.

strong, vertical convection from the planet's interior these gases will not rise to the part of the atmosphere that we can see.

The radio spectra of Uranus and Neptune reveal one other unusual characteristic. Radio brightness measurements for Jupiter and Saturn agree with calculations for the thermal emission expected from these planets, but the Uranus and Neptune data do not. The discrepancy is particularly marked in the 3–10 cm region of the spectrum, where the planets radiate more energy than the calculations would suggest. This means that the real atmospheres are more transparent at these wavelengths than the model atmospheres used in the calculations. This higher transparency indicates a lower amount of ammonia (NH_3), since this is the gas that absorbs at these wavelengths. Why is ammonia depleted? Perhaps the answer lies in those hypothetical water clouds described in the previous section. Since ammonia dissolves readily in water, it might be kept out of the atmosphere. Alternatively, perhaps the production of NH_3 from N_2 is inhibited, just as the production of CH_4 from CO seems to be. Or there might be a deficiency of nitrogen relative to sulfur, leading to clouds of NH_4SH that consume the NH_3. We need more information before we can expect an explanation for this puzzle.

14.4 CLIMATE, CLOUDS, AND WEATHER

We have seen that the upper atmospheres of both these distant worlds are remarkably clear. To study the local meteorology, we need to be able to observe clouds as tracers of global wind patterns. This proved to be especially difficult on Uranus, where we anticipated a climate radically different from that of every other planet owing to the extreme inclination of the planet's axis of rotation.

The General Circulation of the Atmospheres

Neptune experiences seasons similar to those we enjoy on Earth, since its axis is inclined by 27°, but the seasons are nearly 165 times longer than ours, since Neptune's period of revolution is 165 years. Now imagine the seasons on Uranus, whose axis is tipped over nearly in the plane of the planet's orbit. This planet requires 84 years to circle the Sun. This means that for 42 years, one pole receives some illumination from the Sun while the other pole is in total darkness. The pole-on configurations represent the winter and summer solstices, while the equinoxes occur when the equator faces the Sun (Fig. 14.12).

One might think that this orientation of the rotation axis should affect the circulation of the atmosphere. When one of the poles is facing the Sun, it receives continuous illumination that is independent of the planet's rotation. One would expect the circulation to consist of warm air ris-

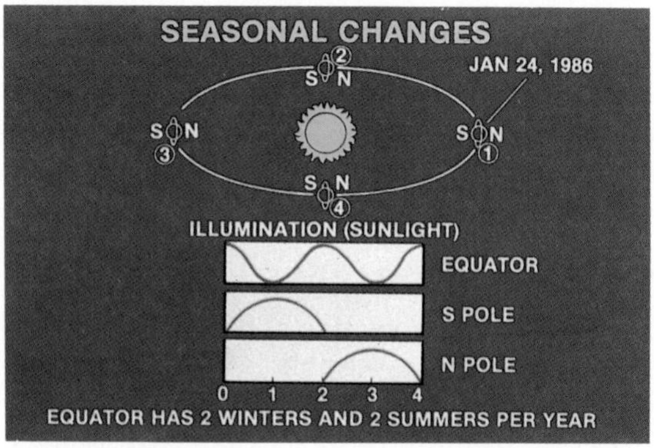

FIGURE 14.12 Seasons on Uranus last for 21 years; note that the equatorial region experiences two summers and two winters each year.

ing at the pole that points toward the Sun and then moving past the equator to the other pole, where it would sink. This is the simplest kind of Hadley cell (Section 10.3). However, Voyager found that the general circulation on Uranus is dominated by the planet's rotation, and the clouds therefore exhibit a banded pattern parallel to latitude lines, similar to that of Jupiter and Saturn.

This was not an easy conclusion to reach, since very few clouds were detected on Uranus (two are visible in Fig 14.10). The apparent rotation of the planet was revealed in sequences of pictures such as those shown in Fig. 14.13. These discrete clouds are probably condensed methane, rather than the ammonia or ammonium hydrosulfide clouds we encountered at Jupiter. They appear to be found at an altitude where the temperature is 80 K and the pressure is 1.3 bar.

Weather on Neptune

Unlike Uranus, Neptune has well-defined, high-contrast white clouds in its upper atmosphere (Fig. 14.14). These clouds are so prominent that their existence was deduced from Earth-based observations long before Voyager 2 arrived. The

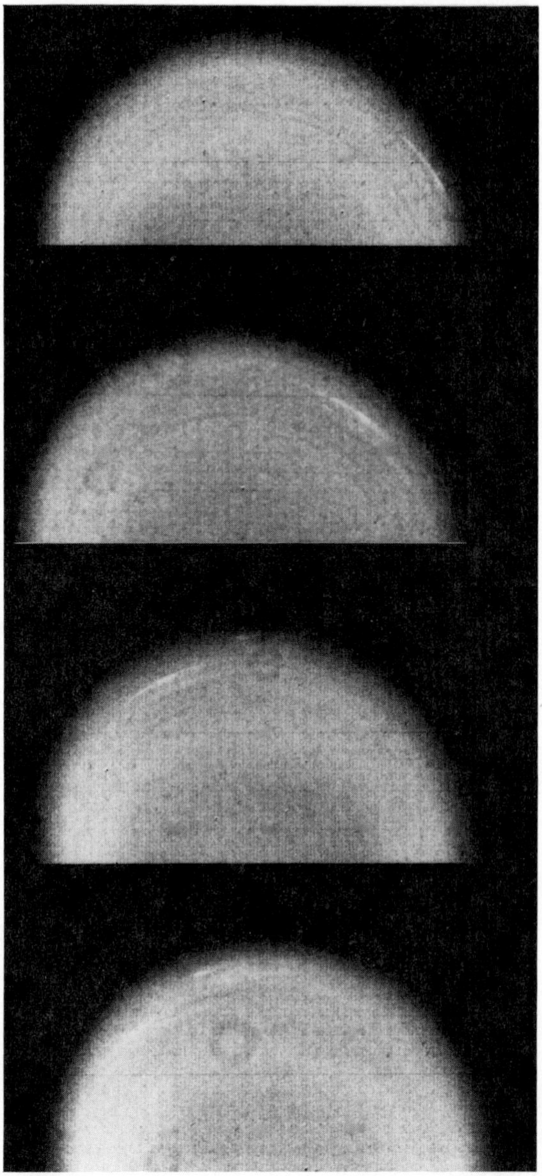

FIGURE 14.13 A sequence of four Voyager 2 images of Uranus taken through an orange filter indicates the motions of two narrow, bright clouds over 4.6 hours. The larger of the two clouds is at a latitude of 33° S; the smaller cloud, seen most clearly in the lower two images, is closer to the edge of the visible disk, at a latitude of 26° S.

FIGURE 14.14 This close-up view of Neptune's Great Dark Spot shows the associated white clouds that occur at higher altitudes.

spacecraft also found evidence for dark clouds in the lower cloud deck that provides a boundary to the visible atmosphere. The most prominent of these is the Great Dark Spot, a large elliptical vortex that is strongly reminiscent of Jupiter's Great Red Spot (GRS).

In the Great Dark Spot we are again dealing with a giant eddy, in this case about the size of Earth. As with the GRS, Neptune's Great Dark Spot occurs in the planet's southern hemisphere and represents an anticyclonic disturbance, with winds blowing counterclockwise around an area of high pressure. Smaller cyclonic/anticyclonic storms also occur at higher latitudes, again reminiscent of jovian weather patterns but on a smaller scale. While the white clouds continue to be seen in different shapes and locations, subsequent high-resolution Earth-based images reveal that the Great Dark Spot has faded.

The colors of the Neptune clouds give no hints that can be used to determine their composition. The white clouds certainly look like some form of cirrus—clouds of ice crystals—and the ice is probably methane, but we don't know for certain. The lower cloud deck is even less well defined, with suggestions for composition ranging from methane droplets to hydrogen sulfide ice crystals.

Rotation Periods and Wind Speeds

An accurate determination of the rotation period of Uranus was not possible before the Voyager encounter in 1986. As shown in Fig. 14.13, the spacecraft pictures revealed clouds that could be tracked as rotation carried them around the planet. These clouds indicate a period near 16 hours. As in the case of Jupiter or Saturn, however, what is actually measured by such observations is the sum of the underlying rotation and of the wind speeds at the location of the clouds. To disentangle these two effects, one must have an independent measurement of the rotation period of the deep interior, determined from observations of the magnetic field of Uranus to be 17.2 hours. Only then can the true wind speeds be determined. As shown in Fig. 14.15, the speed varies with latitude on Uranus, but in

FIGURE 14.15 When the speeds of planetary winds are plotted as a function of latitude, we find that the general circulations of the atmospheres of Uranus and Neptune are essentially identical. Neither planet exhibits the strong equatorial flow characteristic of both Jupiter and Saturn.

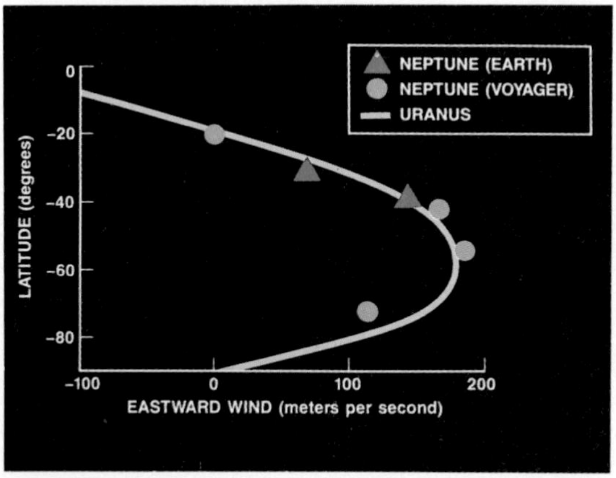

a markedly different way than on Jupiter and Saturn.

On Neptune it has been easier to determine a rotation period because the clouds are much more prominent. Again, the rotation period of the magnetic field provides the reference: Neptune's period of rotation is 16.8 hours, very similar to that of Uranus. This similarity extends to the pattern of wind currents on the planet as well. Uranus lies on its side, while Neptune's inclination is much more like Earth's, yet the change in wind speed with latitude is virtually identical on these two planets.

There is a curious paradox implicit in this result. The basic physics that governs moving air tells us that because the atmosphere of Uranus is getting most of its heat at the pole, wind speeds should increase toward the equator. However, the reverse is true: The fastest winds are at high latitudes. Furthermore, the Voyager measurements show that the temperature of the atmosphere of Uranus is the same at the north pole, which has just experienced 21 years of total darkness, as it is at the south pole, which has been heated for 21 years. Evidently, some dynamical processes—not yet understood—are redistributing the heat deposited from the Sun in ways that can account for both the unusual temperature distribution and the wind pattern. It is hard to escape the idea that something in the internal structures of these two planets is determining the wind speeds we see in the visible regions of their atmospheres.

14.5 MAGNETOSPHERES

As in the case of global circulation patterns in their atmospheres, the magnetospheres of Uranus and Neptune were assumed to be rather different from one another prior to the Voyager encounters. Experience with the Earth, Jupiter, and Saturn suggested that the axis of a planetary magnetic field should always be roughly aligned with the axis of the planet's rotation. Since the inclinations of the rotational axes of Uranus and Neptune are radically different, scientists expected the configurations of their magnetospheres and their interactions with the solar wind to differ as well. In fact, this turned out to be true, but in a different way from the original expectations.

A Tilted Field at Uranus

The first surprise came at Uranus, where the magnetic field turned out to be inclined 60° to the axis of rotation and to be significantly displaced from that axis as well. It is the close alignment of these two axes on Earth that leads inhabitants of our planet to say that a magnetic compass "points north." The magnetic pole is not too far from true north, as defined by the axis of rotation. But imagine the plight of a trekker on Uranus trying to establish directions on a cloudy day. There the magnetic axis is not only inclined by 60°, it is also offset from the rotational axis by one-third of the planetary radius (Fig. 14.16a). Hence the direction that a compass needle would point on Uranus (relative to the orientation of the rotational axis) would vary greatly with position on the planet. The strength of the field would also vary, from 0.1 to 1.1 gauss (compared with a nearly constant 0.3 gauss on Earth).

Uranus has a magnetosphere similar in size to that of Saturn. However, the composition of the atomic particles in the magnetosphere is simpler than in Saturn's. It consists almost entirely of protons and electrons derived primarily from hydrogen escaping from the planet's atmosphere. The magnetosphere also shares the high inclination of the planetary magnetic field to which it is bound. A magnetotail stretches out tens of planetary radii behind the planet, but because of this field inclination, it rotates like a corkscrew as the planet turns on its axis.

The entire sunlit side of Uranus is bathed in a glow of ultraviolet light emitted by escaping hydrogen atoms. This electroglow, as it is called,

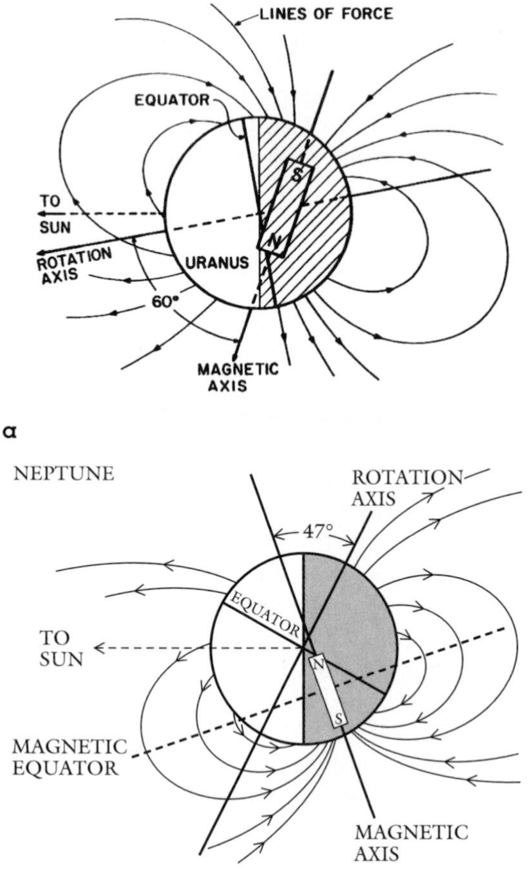

FIGURE 14.16 (a) The magnetic field of Uranus behaves as if there were a bar magnet inside the planet, tipped at an angle of 60° to the axis of rotation and offset as shown in the diagram. (b) To the surprise of most scientists, the magnetic field of Neptune exhibits a configuration very similar to that of Uranus.

was also seen on Jupiter, Saturn, and Titan. There is also auroral activity produced by the collision of magnetospheric ions and electrons with the planet's upper atmosphere. The aurora consists of ultraviolet emission lines detected by the spacecraft and infrared emission from H_3^+, which was discovered by Earth-based observations in 1992. On Uranus, of course, the aurora is not located near the north and the south poles. Instead, it occurs near the equator, since this is where the magnetic poles are found. The Uranians (if there were any) would speak of the beauties of the "tropical lights" instead of the "polar lights" that we extol.

Neptune Imitates Uranus

One might naively assume that the peculiar orientation of the magnetic field of Uranus has something to do with the high inclination of the planet's axis of rotation, but Neptune shows that this is not the case. Here the inclination of the rotational axis is only 27°, close to that of Earth, but the configuration of the magnetic field relative to this axis is similar to the situation on Uranus. Neptune's magnetic dipole is offset from the axis of rotation by a little over half of the planet's radius (for Uranus it is one-third) and is tilted relative to that axis by 47° (Uranus's tilt is 60°) (Fig. 14.16b). The strength of the magnetic field is about half that of Uranus, again varying over the globe in response to the offset of the dipole.

Once again we encounter a planetary magnetosphere, one that contains a rather tenuous plasma consisting of a mixture of light and heavy ions. Voyager instruments could not precisely identify these ions, but it seems likely that they consist of hydrogen escaping from the planet and nitrogen coming from the atmosphere of Triton, Neptune's large satellite (which we will discuss in Chapter 16).

How can we account for the strangely oriented magnetic fields of Uranus and Neptune, which are similar to each other but different from those of Jupiter, Saturn, or Earth? The best idea at present is that they are generated in mantles of pressure-ionized "ice" surrounding the dense, heavy-element, "rocky" cores of these two planets. (We use quotation marks here because the exact nature of these materials is not well defined.) What this means is that compounds of H, C, O, and N could become ionized at high pressure so they become conducting. They would then provide the conducting fluid in which the magnetic field is generated. The separation of "icy" and "rocky" compounds would be consistent with our present ideas about the internal structures of these planets, but these ideas are still in a rather preliminary state. The point is that we need a conducting fluid to generate these fields, and models for the interiors of Uranus and Neptune suggest that this configuration is reasonable. Much more research is needed before we can claim a complete understanding of exactly how these fields are generated or why they have such peculiar orientations.

14.6 QUANTITATIVE SUPPLEMENT: DISCOVERING A PLANET

When Galle first saw Neptune, it looked like a fuzzy, blue-green star, since his telescope was not clearly able to show the planet as a disk. Later observations with more powerful telescopes revealed that Neptune has an apparent angular diameter of 2.3 seconds of arc. (This may be compared with the apparent diameter of the Moon, which is 1865 arc seconds, or about 30 arc minutes.) Meanwhile, how could Galle decide whether this new object was really a planet and not simply an uncharted star? The answer, of course, is that he could see whether it moved, like a true wanderer among the background stars.

If this is a new, distant planet, it must be slowly traveling in a nearly circular orbit about the Sun. We can determine the period by watching Neptune for a few weeks. We then measure the length of its apparent path and calculate the angular speed of the planet in seconds of arc per day. If we divide this into 360°, the angular distance of a full traversal of the orbit, we obtain the period $P = 164.82$ years.

To determine the true size of this planet, we need to know how far away it is. We can use Kepler's third law to obtain this information. The relation between distance D and period P can be written in the form

$$D = P^{2/3}$$

We know P, so we obtain $D = 30.1$ AU. If the orbit is circular, the minimum distance of Neptune from Earth is then $30.1 - 1 = 29.1$ AU.

Now that we know how far away Neptune is and how big it appears to be (2.3 seconds of arc), we can determine its true size, using the small angle equation (Fig. 14.17):

$$\theta = 2r/(D - 1) \times 1.5 \times 10^8 \text{ km}$$

where $2r$ is the true diameter of Neptune and $(D - 1)$ is the distance of Neptune from Earth, which we convert from AUs to kilometers by multiplying by 1.5×10^8 km/AU. In this case the angle must be converted to radians—that is, fractions of an arc equal to the distance to that arc. (The circumference of any circle of radius r is $2\pi r = 360°$, so one radian $= 360°/2\pi$

FIGURE 14.17 The distance of Neptune from the Sun is D, so Neptune's distance from the Earth is $D - 1$, if D is expressed in astronomical units (AU). The diameter of Neptune is $2r$; seen from Earth, the planet subtends an angle $\theta = 2r/(D - 1)$ in the sky.

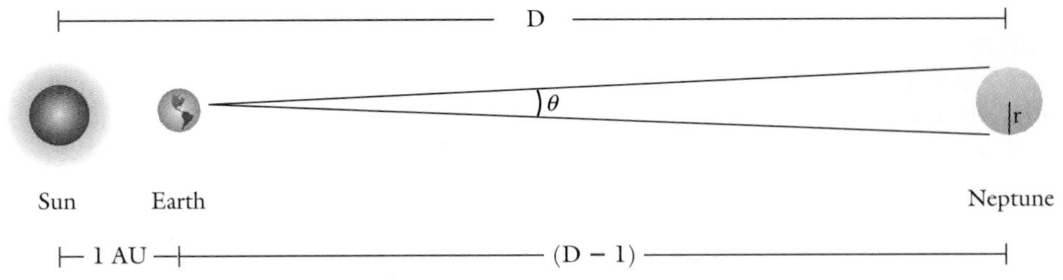

= 57.2958°.) Converting the apparent angular diameter of Neptune to these units, we find 2.3 arc seconds = 11.2×10^{-6} radians. This is the angle subtended by Neptune's disk. Using the small angle equation, we then find

$$2r = \theta(D - 1) \times 1.5 \times 10^8 \text{ km}$$

$$2r = 11.2 \times 10^{-6} \times 29.1 \times 1.5 \times 10^8 \text{ km, or}$$

$$2r = 49{,}000 \text{ km}$$

roughly four times the diameter of Earth.

SUMMARY

The three outermost planets were not known in antiquity; they had to be discovered. The stories of their discoveries illustrate an interesting variety of methods. Uranus was found accidentally, at a time when no one expected any planets beyond the five visible to the unaided eye. The search for Neptune involved theoretical predictions based on gravitational theory and the most sophisticated mathematical techniques of the nineteenth century; its discovery was a triumph for Newtonian celestial mechanics. The search for Pluto was similarly inspired by theoretical work, but ultimately the planet was found as the result of a very careful observational search. We now know Pluto is far too small to have perturbed the motions of other planets in a measurable way.

Uranus and Neptune form a pair of planets, substantially smaller than Jupiter or Saturn but much larger than the terrestrial planets. Their composition is also intermediate between that of the giants, with their nearly cosmic proportions of the elements, and that of the rocky, volatile-depleted terrestrial bodies. Uranus and Neptune are composed primarily of the common ices mixed with silicates and metals, with deep atmospheres dominated by hydrogen and helium.

Uranus is notable for its rotation, with its axis tipped on one side, and for its lack of a major internal heat source. Both planets have atmospheres composed of hydrogen, helium, and methane. The atmosphere of Uranus is largely free of clouds, but that of Neptune exhibits rapidly changing high-altitude clouds. Surprisingly, the circulation and temperatures in the Uranus atmosphere do not seem to be influenced by the peculiar orientation of its rotation axis, since Neptune exhibits a nearly identical circulation pattern.

Both of these planets have eccentric magnetic fields whose orientations are remarkably similar, inclined at 60° and 47°, respectively, to the axes of rotation. Charged particles trapped in their magnetospheres plunge into the atmospheres along magnetic field lines, creating auroras. The locations of these auroras are unusual from our terrestrial perspective, since the magnetic field axes on these planets are so displaced from the planets' rotational axes.

KEY TERMS

catalyst
chemical equilibrium
perturbation

Review Questions

1. How was Uranus discovered, and by whom? What gave scientists the idea that there might be another planet more distant than Uranus, and how did they try to find it? Who succeeded in seeing it?

2. How would you compare the discovery of Pluto with the discoveries of Uranus and Neptune? Consider the application of perturbation calculations versus luck.

3. Compare Jupiter and Saturn with Uranus and Neptune. Do you think all four planets have the same composition? Why or why not? How does Pluto fit into these two groups?

4. Methane and hydrogen have been detected spectroscopically in the atmospheres of Uranus and Neptune, but ammonia has not. Explain.

5. Describe the seasons on Uranus and compare them with the seasons on Earth. What is the reason for the difference?

6. Why was it so difficult for the Voyager spacecraft to determine the atmospheric general circulation of Uranus?

7. Describe the difference in the internal heat sources of Uranus and Neptune. What effect does this difference have on the structure (temperature versus pressure) of the atmospheres of these two bodies and on their respective weather patterns?

8. Describe the magnetic field orientations of Uranus and Neptune, and compare them with the fields of Earth and Jupiter. At what latitudes would you expect to find auroras on these two planets?

Quantitative Exercises

1. Calculate the acceleration of gravity on Neptune and on Earth. How do you explain the fact that the two values are so similar?

2. How many days would Herschel have needed to wait before Uranus moved a distance in the sky equal to the diameter of the full moon?

3. Calculate the speed of Neptune in its orbit in units of kilometers per second. Compare this with the orbital speed of Earth. Why are they so different?

4. Use the table of solar abundances (Table 2.2) to calculate the mean *molecular* weight of an atmosphere formed of H_2, He, CH_4, and NH_3, where the *elements* have solar abundances. Now compare this with a model for Neptune's atmosphere in which C/H is enriched 25 times relative to the solar value and all the nitrogen is present as N_2 instead of as NH_3. What change in He/H (compared with the solar value) would give the same mean molecular weight? Which possibility seems more likely to you, increased He/H or nitrogen as N_2? Explain how each possibility could occur.

Additional Reading

Beatty, J.K., and A. Chaikin, eds. 1990. *The New Solar System.* Cambridge, Mass.: Sky Publishing. Excellent chapters on the Voyager discoveries in the Uranus and Neptune systems.

*Bergstrahl, J.T., E. Miner, and M.S. Matthews, eds. 1990. *Uranus.* Tucson: University of Arizona Press. A collection of technical papers describing the results from the Voyager encounter and contemporary ground-based observations.

*Cruikshank, D.P., and M.S. Matthews, eds. 1995. *Neptune.* Tucson: University of Arizona Press. This is the companion volume to *Uranus*; it too highlights the Voyager discoveries.

Grosser, M. 1962. *The Discovery of Neptune.* Cambridge, Mass.: Harvard University Press. An engrossing account of the events leading up to the discovery of Neptune, including the discovery and early observations of Uranus.

*Hunt, G., ed. 1982. *Uranus and the Outer Planets.* Cambridge: Cambridge University Press. A series of historical and scientific papers presented at a meeting in Bath, England, on the two-hundredth anniversary of Herschel's discovery (before the Voyager encounters).

Littman, M. 1990 (revised printing). *Planets Beyond: Discovering the Outer Solar System.* New York: John Wiley and Sons. A popular account of the outer planets that includes the discoveries made during the Voyager encounters with both Uranus and Neptune.

*Indicates the more technical readings.

This is a view of the large volcanic structure on Io called Pele, after the supreme Hawaiian volcano goddess. We are looking through the plume toward the eruptive center on Io's surface, and we can see turbulent clouds in the plume projected against the dark sky at the middle left.

15

Worlds of Fire and Ice: The Large Satellites of Jupiter

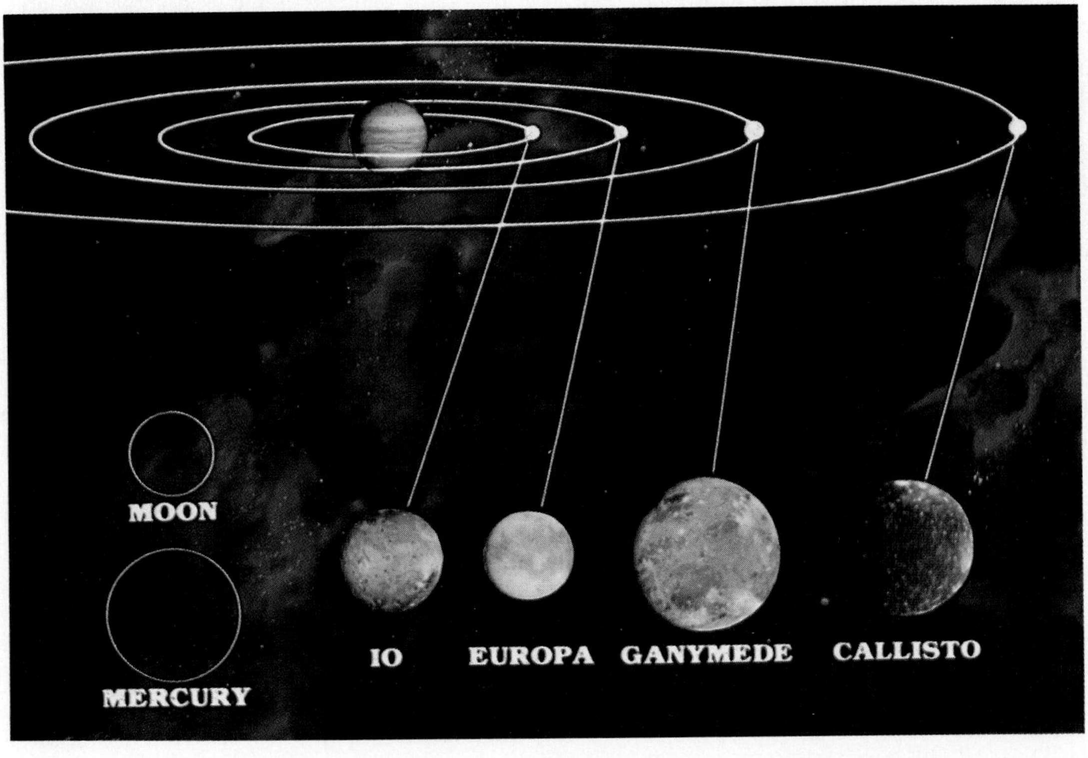

The four satellites discovered by Galileo form a miniature solar system as they orbit the planet. If they were in orbit about the Sun, we would regard them as interesting planets, two of which are the size of Mercury.

Except for Pluto, we have now described all the objects in the solar system that move in orbits around the Sun: the inner and outer planets, the asteroids, and the comets. We have seen that these objects can be roughly characterized as either rocky, icy, or gas-rich, with giant gas-rich planets incorporating huge amounts of rocky and icy material in addition to their deep hydrogen-dominated atmospheres.

Now we want to study the satellites that orbit the giant planets. We shall discover that some of these satellites are larger than Mercury and could easily be considered planets if they orbited the Sun. One even has an atmosphere denser than Earth's. Others are so tiny they resemble the dust in the asteroid belt or the debris left behind in the orbits of comets. These small particles are found in the spectacular ring systems that surround each giant planet. Again we will encounter a distinction between icy and rocky objects, but there will be nothing equivalent to the gas-rich planets, since none of the satellites has sufficient mass to attract these huge envelopes of hydrogen and helium.

We will begin with an overview of all the satellite systems and then describe the four large satellites of Jupiter, each one a unique world unto itself.

15.1 SATELLITE SYSTEMS

In the inner solar system, satellites are the exception. Neither Mercury nor Venus has one, and our own Moon is a peculiar object whose very existence has been difficult to understand. Only Mars has more than one satellite, and these are both tiny objects presumably captured from the early population of asteroids, as discussed in Chapter 5. In contrast, satellites are the norm for the giant planets, and it is quite reasonable to think of these planets and their retinues as miniature solar systems. Tiny Pluto boasts a satellite of its own, even though this planet is smaller than our Moon.

Discovery

The first satellites were discovered by Galileo in 1610, on the first night he turned his newly built telescope toward the sky (Fig. 15.1). These are known as the Galilean satellites of Jupiter. Titan, the largest satellite of Saturn, was found in 1655 by Christian Huygens of the Netherlands. Discoveries continued over the next two centuries at a rate of nearly one satellite per decade, with Cassini, Herschel, and Nicholson each equaling Galileo's record of four.

In the twentieth century, the most successful discoverer was Seth Nicholson of Mount Wilson Observatory, who spotted four faint outer satellites of Jupiter between 1914 and 1951. Another important contributor of the prespace era was Gerard P. Kuiper, who discovered one satellite each for Uranus and Neptune.

Recent satellite discoveries have been made primarily with the Voyager cameras, although ground-based work continues to be productive. From 1975 to 1990, three new satellites of Jupiter, eleven of Saturn, ten of Uranus, and six

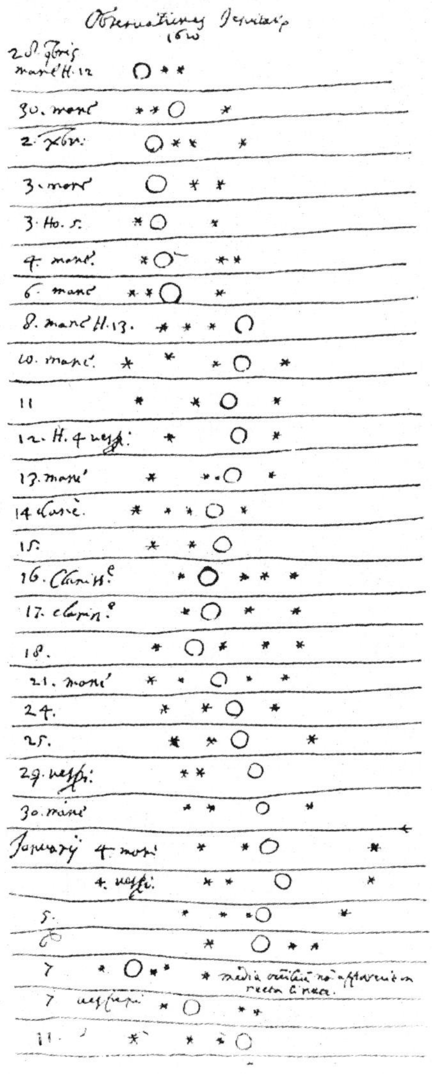

FIGURE 15.1 This reproduction of a page in one of Galileo's notebooks shows his sketches of the moons of Jupiter. What was absolutely astonishing to him was that these tiny points of light changed their positions from night to night. They were clearly moving in orbits about Jupiter; Earth was not the center of the universe!

of Neptune have been added to our catalog of outer solar system objects.

Table 15.1 summarizes the orbital and physical data for all the satellites in the solar system, including the Moon and the martian satellites for comparison. This table provides the background for discussions of individual objects in this and the next chapter.

Regular and Irregular Satellites

The satellites of the outer planets are conveniently divided into two groups on the basis of their orbits. All but one of the large moons and many of the small ones are **regular satellites**, meaning they revolve in orbits of low eccentricity near the equatorial plane of their planet. The regular satellites are the analogs of the planets in the solar system, which revolve in nearly circular orbits close to the Sun's equatorial plane.

The second group of moons are called **irregular satellites**. Their orbits are peculiar, having either large eccentricity or high inclination, or both. Several even revolve in a retrograde direction. The irregular satellites have orbits resembling those of comets or asteroids around the Sun.

Because they are so closely bound to their planets, each regular satellite system probably formed with its planet from a **subnebula**, a disk of dust and gas around the forming giant planet, just as the planets themselves formed from the primordial solar nebula. In contrast, the irregular satellites are probably interlopers captured from elsewhere. In the case of Neptune, an original, regular satellite system may have been disrupted by the capture process to produce the combination of regular and irregular orbits we see today.

The Jupiter System

The four most important members of the jovian satellite system are the Galilean satellites: Callisto, Ganymede, Europa, and Io. These are all large objects, ranging in size from a little smaller than our Moon to larger than the planet Mer-

TABLE 15.1 Satellites

Planet	Satellite Name	Discovery	Semimajor Axis (km × 1000)	Period (days)
Earth	Moon	—	384	27.32
Mars	Phobos	Hall (1877)	9.4	0.32
	Deimos	Hall (1877)	23.5	1.26
Jupiter	Metis	Voyager (1979)	128	0.29
	Adrastea	Voyager (1979)	129	0.30
	Amalthea	Barnard (1892)	181	0.50
	Thebe	Voyager (1979)	222	0.67
	Io	Galileo (1610)	422	1.77
	Europa	Galileo (1610)	671	3.55
	Ganymede	Galileo (1610)	1070	7.16
	Callisto	Galileo (1610)	1883	16.69
	Leda	Kowal (1974)	11090	239
	Himalia	Perrine (1904)	11480	251
	Lysithea	Nicholson (1938)	11720	259
	Elara	Perrine (1905)	11740	260
	Ananke	Nicholson (1951)	21200	631 (R)
	Carme	Nicholson (1938)	22600	692 (R)
	Pasiphae	Melotte (1908)	23500	735 (R)
	Sinope	Nicholson (1914)	23700	758 (R)
Saturn	Pan	Voyager (1985)	133.6	0.58
	Atlas	Voyager (1980)	137.7	0.60
	Prometheus	Voyager (1980)	139.4	0.61
	Pandora	Voyager (1980)	141.7	0.63
	Janus	Dollfus (1966)	151.4	0.69
	Epimetheus	Fountain, Larson (1980)	151.4	0.69
	Mimas	Herschel (1789)	186	0.94
	Enceladus	Herschel (1789)	238	1.37
	Tethys	Cassini (1684)	295	1.89
	Telesto	Reitsema *et al.* (1980)	295	1.89
	Calypso	Pascu *et al.* (1980)	295	1.89
	Dione	Cassini (1684)	377	2.74
	Helene	Lecacheux, Laques (1980)	377	2.74
	Rhea	Cassini (1672)	527	4.52
	Titan	Huygens (1655)	1222	15.95
	Hyperion	Bond, Lassell (1848)	1481	21.3
	Iapetus	Cassini (1671)	3561	79.3
	Phoebe	Pickering (1898)	12950	550 (R)
Uranus	Cordelia	Voyager (1986)	49.7	0.34
	Ophelia	Voyager (1986)	53.8	0.38
	Bianca	Voyager (1986)	59.2	0.44
	Cressida	Voyager (1986)	61.8	0.46

Diameter (km)	Mass (10^{23} kg)	Density (g/cm^3)	Albedo	Surface Material
3476	735	3.3	0.12	silicates
23	1×10^{-4}	2.2	0.05	carbonaceous
13	2×10^{-5}	1.7	0.06	carbonaceous
40	—	—	0.05	rock?
25	—	—	0.05	rock?
200	—	—	0.05	rock, sulfur
90	—	—	0.05	rock?
3630	894	3.6	0.6	sulfur, SO_2
3138	480	3.0	0.6	ice
5262	1482	1.9	0.4	dirty ice
4800	1077	1.8	0.2	dirty ice
15?	—	—	—	?
180	—	—	0.03	carbonaceous
40	—	—	—	carbonaceous
80	—	—	0.03	carbonaceous
30?	—	—	—	?
40?	—	—	—	?
40?	—	—	—	?
40?	—	—	—	?
15?	—	—	—	?
40	—	—	0.5	ice?
100	—	—	0.5	ice?
90	—	—	0.5	ice?
190	—	—	0.5	ice?
120	—	—	0.5	ice?
394	0.4	1.2	0.8	ice
502	0.8	1.2	1.0	pure ice
1048	7.5	1.3	0.8	ice
25	—	—	0.6	ice?
25	—	—	0.9	ice?
1120	11	1.4	0.6	ice
30	—	—	0.6	ice?
1530	25	1.3	0.6	ice
5150	1346	1.9	0.2	cloudy atmosphere
270	—	—	0.3	dirty ice
1435	19	1.2	0.4	ice/carbonaceous
220	—	—	0.06	carbonaceous?
25	—	—	<0.1	carbonaceous?
30	—	—	<0.1	carbonaceous?
45	—	—	<0.1	carbonaceous?
65	—	—	<0.1	carbonaceous?

(*Continued*)

TABLE 15.1 Satellites (continued)

Planet	Satellite Name	Discovery	Semimajor Axis (km × 1000)	Period (days)
Uranus	Desdemona	Voyager (1986)	62.7	0.48
	Juliet	Voyager (1986)	64.6	0.50
	Portia	Voyager (1986)	66.1	0.51
	Rosalind	Voyager (1986)	69.9	0.56
	Belinda	Voyager (1986)	75.3	0.63
	Puck	Voyager (1985)	86.0	0.76
	Miranda	Kuiper (1948)	130	1.41
	Ariel	Lassell (1851)	191	2.52
	Umbriel	Lassell (1851)	266	4.14
	Titania	Herschel (1787)	436	8.71
	Oberon	Herschel (1787)	583	13.5
Neptune	Naiad	Voyager (1989)	48	0.3
	Thalassa	Voyager (1989)	50	0.31
	Despina	Voyager (1989)	52.5	0.33
	Galatea	Voyager (1989)	62.0	0.43
	Larissa	Voyager (1989)	73.6	0.55
	Proteus	Voyager (1989)	117.6	1.12
	Triton	Lassell (1846)	354	5.88 (R)
	Nereid	Kuiper (1949)	5513	360
Pluto	Charon	Christy (1978)	19.7	6.39

cury. The orbits of the inner three are evenly spaced, so that the period of revolution for Europa is just twice that for Io, and Ganymede revolves in twice the period of Europa.

The even spacing of the Galilean satellites is more than a numerical coincidence, unlike the Titius-Bode rule for planetary distances. It represents a gravitational coupling, or *resonance*. If one of these orbits should change, the other two would change with it to restore the resonant relationship. The orbital positions of the three satellites are also coupled, so that all three can never be on the same side of Jupiter, for example. This relationship is called a Laplace resonance, named for the eighteenth-century French mathematician who first explained it.

The other 12 known members of the jovian system can be classified into three groups of four satellites each. Close to the planet are four small objects, three of which were discovered by Voyager. Far beyond the Galilean satellites are eight small irregular satellites, half in direct orbits and half retrograde.

The Saturn System

Saturn has the largest regular satellite system, just as it has the largest system of rings (and perhaps for the same reasons). Figure 15.2 compares the jovian and saturnian satellite systems. Of the sixteen satellites that are clearly regular, seven are co-orbital or nearly so; that is, they represent situations in which two or more satellites occupy nearly the same orbits. These cases, which are unique to the Saturn system, are discussed in Section 17.1.

Saturn's largest satellite, Titan, is about the same size as the two largest Galilean moons, Ganymede and Callisto. Next in size are six satellites with diameters of 400–1500 km, corresponding to surface areas ranging from that of France up to the whole of Europe. Jupiter has no satellites this size; in fact, the only objects of this size we have encountered previously are the largest asteroids, such as Ceres and Vesta.

The three outer satellites of the Saturn system are more irregular. Hyperion, although its orbit

Diameter (km)	Mass (10^{23} kg)	Density (g/cm^3)	Albedo	Surface Material
60	—	—	<0.1	carbonaceous?
85	—	—	<0.1	carbonaceous?
110	—	—	<0.1	carbonaceous?
60	—	—	<0.1	carbonaceous?
68	—	—	<0.1	carbonaceous?
155	—	—	0.07	carbonaceous?
485	0.8	1.3	0.3	dirty ice
1160	13	1.6	0.4	dirty ice
1190	13	1.4	0.3	dirty ice
1610	35	1.6	0.3	dirty ice
1550	29	1.5	0.2	dirty ice
60	—	—	~0.06	carbonaceous?
80	—	—	~0.06	carbonaceous?
150	—	—	0.06	carbonaceous?
160	—	—	~0.06	carbonaceous?
190	—	—	0.06	carbonaceous?
420	—	—	0.06	carbonaceous?
2700	214	2.1	0.8	ices of N_2, CO_2, CH_4, H_2O, CO
340	—	—	0.16	?
1200	—	—	0.4?	ice

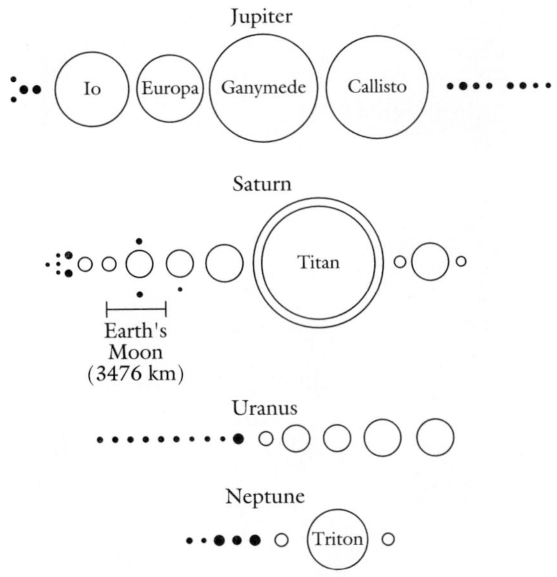

FIGURE 15.2 The satellite systems of Jupiter, Saturn, and Uranus are dramatically different, even in outline. Saturn has only one large satellite and only one distant irregular satellite. Ice is the dominant material composing all of Saturn's satellites, while it appears as a major component in only two of Jupiter's. Uranus has no satellites as big as our own Moon.

fits the definition of a regular satellite, has a resonance with Titan that introduces an irregular rotation. Iapetus has a moderate inclination and eccentricity, making it the Pluto (in orbital terms) of the Saturn system. Only outermost Phoebe, which is in a retrograde orbit at a large distance from the planet, is clearly an irregular satellite.

The Uranus and Neptune Systems

The 15 satellites of Uranus are all regular. Since they orbit in the equatorial plane of the planet, they share its high inclination. At the time of the Voyager 2 encounter, the polar axis of Uranus was directed nearly toward Earth, so that the satellite orbits made a bull's-eye pattern on the sky. The five larger uranian satellites are all in the same size range as Saturn's medium-size satellites. The ten objects closer to the planet that were discovered by Voyager are each less than 200 km in diameter.

Neptune is different. It has two systems of satellites, six in regular orbits and two that are highly irregular. Triton, its one large satellite (similar in many ways to the planet Pluto), is in a retrograde circular orbit. Triton is almost as close to Neptune as Io is to Jupiter. The second and much smaller irregular satellite, Nereid, has the highest eccentricity (0.75) of any satellite; it is the comet of the Neptune system. The regular satellites are all close to the planet, just like the inner satellites of Uranus (Fig. 15.2).

15.2 IMPACT CRATERS AND SURFACE AGES

Before we look at the satellites individually, we should ask ourselves (as did the Voyager scientists before the Jupiter and Saturn encounters) what cratering we should expect on solid bodies in the outer solar system. Will numerous impacts have taken place, and if so, will the resulting craters have been preserved for us to see?

Cratering in the Outer Solar System

Given the plethora of well-preserved craters on the Moon, Mercury, Mars, and even Venus, it may seem odd that we worry about the preservation of craters on outer planet satellites. The reason is that we are now dealing with a very different material. The surfaces of almost all the outer satellites are composed of ice or of ice-rock mixtures, like the comets. We know that on Earth ice is a plastic material, and that glaciers of ice can flow like a viscous fluid. If the same thing happens on these satellites, no craters would be preserved. However, surface temperatures are much lower on the satellites than on Earth, and at low temperatures ice becomes stiff and strong. How low is low enough?

At Saturn and beyond, ice should be as strong as rock, and craters should be preserved indefinitely unless there is local heating or erosion. For the jovian satellites, however, the ice may not be strong enough to preserve craters for billions of years. Thus the craters of Rhea (Fig. 15.3) are nearly indistinguishable from those on the

FIGURE 15.3 The surface of Rhea bears a superficial resemblance to the highland regions of our Moon (cf. Fig. 7.2), but there are no indications of marelike features on this mosaic of pictures of the north polar region obtained by Voyager. We are looking here at an icy surface, much more reflective than our Moon's dark rock. The smallest visible detail is only 1 km across.

FIGURE 15.4 Callisto revealed a heavily cratered surface to the Voyager 1 cameras in March 1979. This mosaic shows the huge ringed feature called Valhalla (center left), with a central spot about 60 km across. The concentric bright rings can be traced to a distance of 1500 km from the center, making this feature bigger than the somewhat similar ringed basin, Mare Orientale, on Earth's Moon (cf. Fig. 7.6).

Moon, while the craters on Callisto (Fig. 15.4) have been considerably modified.

What about the number of crater-producing impacts? We clearly know less about the meteoroid population in the outer solar system than in the space close to Earth. If, for example, most of the impacts on Earth and the Moon were from asteroids, we might expect a much smaller flux near the outer planets. So to answer the question of impacts in the outer solar system, we must first refine our understanding of the meteoroids closer to home.

The most complete census of meteoroids in the solar system has been carried out by Eugene Shoemaker and his colleagues. In Section 7.4 we discussed Shoemaker's contributions to the lunar cratering chronology and his survey of Earth-approaching asteroids. Following the Voyager discoveries that the outer satellites were heavily cratered, he generalized these studies to the entire solar system, with the following results.

In the inner solar system, there are two main sources of meteoroids large enough (at least 100 m) to produce visible impact craters. These are the asteroids and the comets. Shoemaker calculates that, for Earth and the Moon, the two sources have a ratio of 2:1—that is, about 30% of our craters are due to comets (both long-period and short-period) and about 70% to asteroids. These conclusions refer to the present, of course; different sources may have been important in the past, and certainly a different and much larger population of impacting bodies was present at the time of the terminal bombardment, 3.9 billion years ago.

Extending these results to the outer solar system, we see that the meteoroid population is reduced by about two-thirds without the asteroidal contribution. Impacts from short-period comets also are less important at Jupiter, and negligible at Saturn and beyond, further reducing the impact flux. Very roughly, the calculations of Shoemaker and others indicate a cratering rate about one-quarter that of the inner planets, primarily from impacts with long-period comets.

Variations in Cratering Rates

Once the meteoroid flux has been calculated, it should be possible to use crater densities on the satellites to calculate approximate crater retention ages, as described in Section 7.4. The uncertainties in the flux are substantial, probably amounting to a factor of two or three, but at least crater densities should tell us whether a surface age is measured in billions, or hundreds of millions, or perhaps tens of millions of years. First, however, two important corrections must be made to the calculated impact rates.

We saw when studying cratering on the Earth and Moon that the size of the crater excavated by a projectile depends on the impact speed. Earth, being larger, attracts meteoroids more strongly than does the Moon, and the resulting craters are larger. In comparing crater densities on Earth and the Moon, we must allow for the gravity field of Earth.

In the outer solar system, the satellite and its central planet are important, especially the latter. We can imagine each giant planet to be at the bottom of a gravity slope; meteoroids (mostly comets) are attracted toward the planet, and the closer they come, the faster they move as they fall down this slope. The result is that a satellite located far down the slope receives more impacts, and they occur at higher speeds. This effect can be very large: Mimas, the innermost large satellite of Saturn, can expect 20 times the cratering rate of Iapetus, far from the planet. Consequently, the inner satellites of Jupiter and Saturn should have higher cratering rates than Earth or the Moon.

The second correction must be made because almost all the outer planet satellites are in synchronous rotation; like our Moon, they always keep the same side facing their planet. Therefore they also always keep the same hemisphere oriented toward their direction of orbital motion, with the opposite hemisphere always facing backward. These are called the **leading side** and **trailing side**, respectively. The leading hemisphere thus receives more impacts, and they come at higher speed. Therefore, for a synchronous satellite, corrections in cratering rate must be made, depending on the position of the craters on the surface under study. These effects can also be substantial, amounting to a factor of ten between the leading and trailing hemispheres of the inner satellites of Saturn, which move at high speeds in their orbits.

Crater Retention Ages on the Satellites

Speaking very roughly, the effects of the local gravity field and the leading/trailing asymmetry increase the cratering rate and compensate for fewer meteoroids in the outer solar system. Consequently, crater densities on many of the satellites are related to age in a manner not too different from that derived for the inner solar system. For example, we can conclude that a crater density on one of these satellites similar to that of the lunar maria indicates an age of several billions of years. As we will see when we look at individual objects in this and the next chapter, many satellites include terrains (called lightly cratered plains) with crater densities of 100–200 10-km craters per million km^2, and these areas are probably of an age similar to that of the lunar maria and the northern plains of Mars.

Like the Moon, the satellites of the outer solar system also include surfaces with high crater densities. Such high densities cannot easily be explained from estimates of current cratering rates, which predict a maximum of a few hundred for the crater density even on a surface more than 4 billion years old. Thus we are drawn to the same conclusion described in Section 7.4 for the Moon: There was an early period when cratering rates were much higher than they are today.

This is an important conclusion, as has been emphasized by Laurence Soderblom of the U.S. Geological Survey. Soderblom was the deputy team leader and the senior geologist on the Voyager Imaging Team (Fig. 15.5). He played an important role in defining the standard cratering history for the inner solar system, with its late heavy bombardment ending about 3.8 billion years ago. Soderblom and his colleagues concluded that this unifying paradigm developed for the inner planets applies to the outer solar system as well. This bombardment is therefore associated with the entire solar system, rather than being limited to the terrestrial planets. With this perspective, we now turn to the individual satellites.

FIGURE 15.5 Larry Soderblom (right) shares a humorous moment with Bradford Smith (left) during the Voyager encounters with the Saturn system.

15.3 CALLISTO AND GANYMEDE: THE LARGE GALILEANS

Callisto and Ganymede are twins, in much the same way as Venus and Earth. They have nearly identical sizes and densities, and they occupy similar regions of space. They were the first solid objects to be studied that are composed about half of water ice, and it was with considerable anticipation that Voyager geologists awaited the initial high-resolution photos of these objects in March 1979.

Callisto: Basic Facts

Callisto has had the simplest geologic history, one dominated by impact cratering. Although its surface looks rather like the lunar highlands, we must remember that this satellite, like most of the others studied in this chapter, is composed in large part of ice (Fig. 15.4).

Callisto has a diameter of 4840 km, almost identical to that of the planet Mercury, yet its mass is less than one-third as great. From this simple comparison, we are immediately aware of a fundamental difference in composition. Detailed calculations for a body the size of Callisto and its density of about 1.9 g/cm^3 suggest that water ice is about equally as abundant as rocky materials in its interior.

The surface temperature of Callisto varies from 150 K near noon to about 100 K at night. Under such frigid conditions, water ice is stable even over very long intervals of time, and less than a meter of ice is expected to have evaporated from the surface during the past 4.5 billion years. A little closer to the Sun, in the main asteroid belt, an icy object could not survive. Thus abundant water ice is expected in the outer solar system, while ice and water can survive in the inner solar system only on the surfaces (and in the atmospheres) of planets with high gravity, such as Earth and Mars. This is why the discovery of icy polar caps on Mercury was so unexpected.

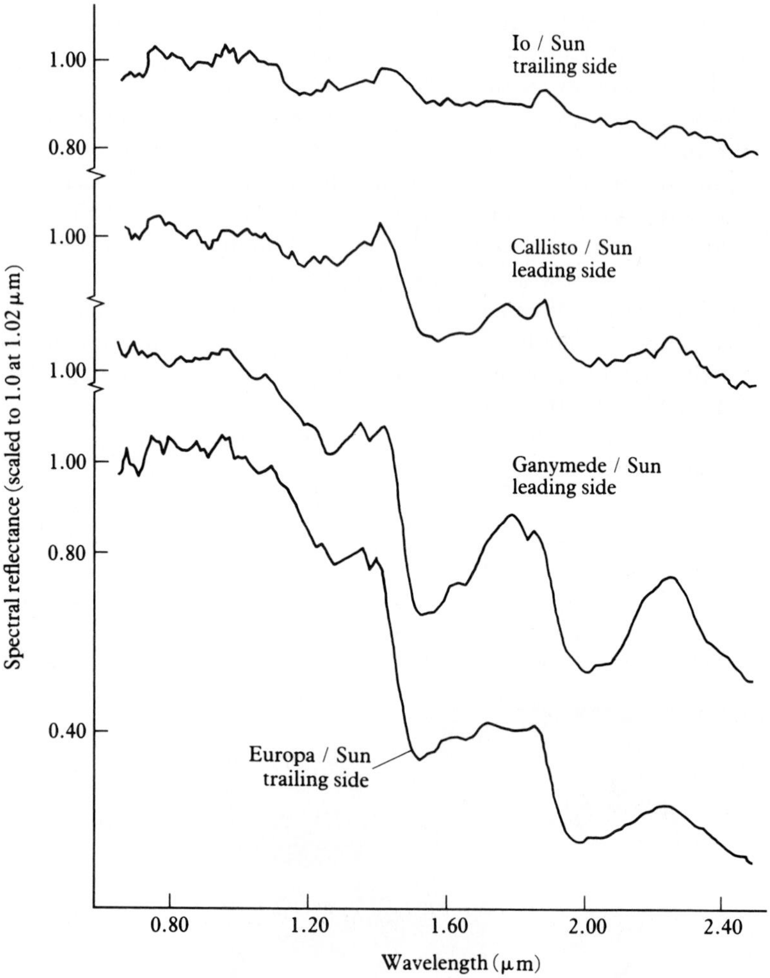

FIGURE 15.6 Long before the Pioneer and Voyager spacecraft reached the Jupiter system, we knew that Callisto, Ganymede, and Europa were covered with ice. This deduction is easily made from infrared spectra such as these, obtained with the Kitt Peak solar telescope.

Ice is also observed directly on the surface of Callisto, where its presence is revealed by prominent infrared absorption bands. This ice is by no means pure, however, and the reflectivity of Callisto is actually rather low, only 18%. If there has been no resurfacing with fresh ice for billions of years, the accumulation of meteoritic dust mixed with the original ice crust would be expected to be fairly dark. Figure 15.6 illustrates infrared spectra of all four Galilean satellites, showing the strength of the ice bands on each.

Geology of Callisto

Callisto is a good place to begin our detailed study of outer planet satellites because it is relatively simple to understand. The Voyager pictures reveal a heavily cratered world with a landscape that is almost exclusively the product of impacts (Figs. 15.4, 15.7). Over most of the surface, these impact craters are nearly as densely packed as those in the lunar highlands. Expressed in the same units that we have used

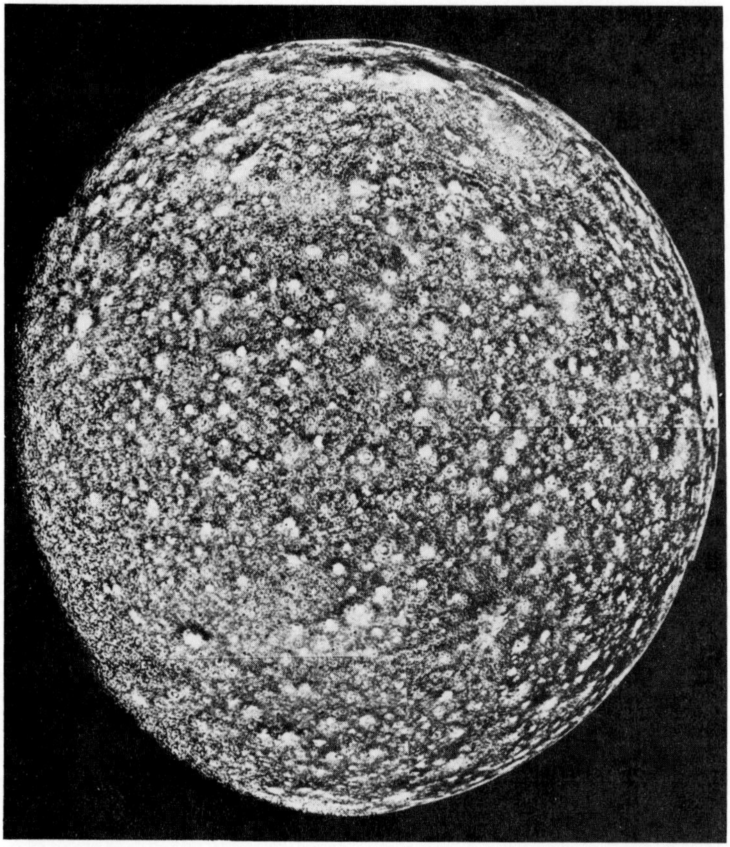

FIGURE 15.7 If you like craters, you'll love Callisto! This computer-enhanced mosaic of Voyager 2 pictures reveals that the satellite's surface is virtually covered with craters with diameters of up to 100 km. Yet larger craters are strangely absent, suggesting that the icy surface cannot support such large structures over geologic time.

before, the crater density on Callisto is 250 10-km craters per million km^2. Barring some improbable recent blizzard of impacts, it appears that the surface of Callisto is at least as old as the lunar maria, and that during this vast span of time little or no internally produced geological activity has been present to destroy the accumulating impact scars.

It also seems probable that many of these craters were formed at a time of higher cratering rates, the outer solar system equivalent of the late heavy bombardment on the Moon. It is tempting to assert that this bombardment ended on Callisto about 3.8 billion years ago, as it did on the Moon. Unfortunately, this attractive hypothesis cannot be tested at present.

A close inspection of Fig. 15.7 reveals some important differences, however, between the cratered surface of Callisto and the lunar highlands. The craters are not the same shape. Instead of having bowl-shaped profiles, the Callisto craters are subdued in topography, as though they had been flattened or the crustal material had undergone plastic deformation since the time when the craters were formed. Apparently, we see here one of the differences between a rocky and an icy object; at the temperature of Callisto's surface, ice, being less rigid, cannot preserve the sharp contours of an impact crater over hundreds of millions of years. Instead, it acts like a highly viscous fluid, which flows slowly, evening out topographic features.

Callisto also lacks the impact basins that we have encountered on the Moon, Mercury, and Mars. This apparent deficiency is probably another result of the icy composition of this satellite. Large topographic features are harder to preserve in a plastic material than small ones, and probably any really big craters or basins would simply have subsided into the surface.

This suggestion is supported by the presence of a few large circular white spots surrounded by low concentric ridges spaced about 10 km apart to make a bull's-eye pattern. These features resemble the pattern of waves formed by tossing a stone into a pool of water, and this analogy may be instructive. Apparently the bull's-eye basins are the ghostly remnants of great impacts nearly swallowed up by the icy surface of the satellite. Almost all topography has been destroyed by plastic deformation, and all that remain to mark the basin are the lighter surface and the concentric rings.

While we cannot probe the interior of Callisto and the other Galilean satellites, there are good reasons to believe that these are differentiated objects. Forming in the outer solar system where temperatures were low, they began their existence with a good deal of water ice mixed with rocky and metallic materials. Presumably the rocky component contained its fair share of radioactive materials as a heat source, while the water ice has a much lower melting temperature than rock. Inevitably, the interior temperature soon rose above the melting point of the ice, even if the satellite formed cold. The heavier materials then sank to form a rocky or muddy core, and the liquid water rose and eventually refroze to form a crust and mantle of ice. The resulting internal structure, illustrated in Fig. 15.8, should apply to both Callisto and Ganymede, and probably to Saturn's Titan as well.

Geology of Ganymede

Ganymede is the largest satellite in the solar system, and one that has experienced a unique geologic history. It occupies the next orbit inward from Callisto. Ganymede has a density nearly identical to that of Callisto, and we therefore believe its composition and interior structure are

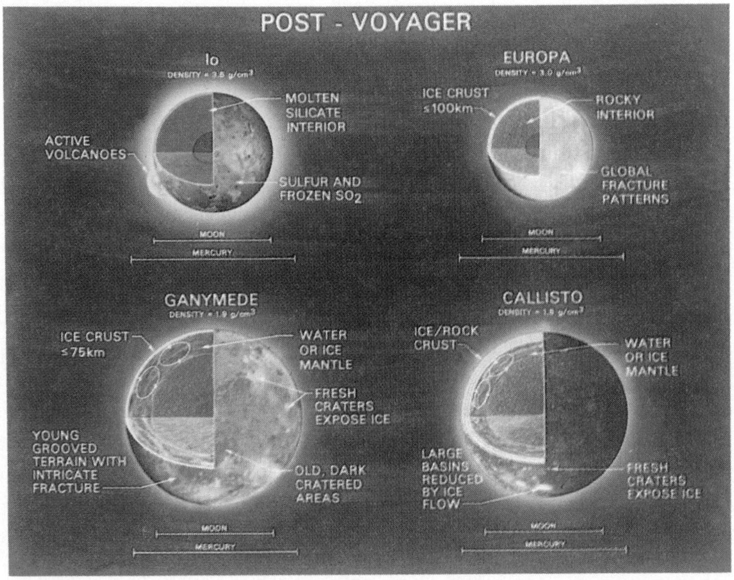

FIGURE 15.8 Conditions in the interiors of the satellites of Jupiter must be inferred from the observations of their surfaces, sizes, and densities. Here we see that Io, with the highest density, must have a molten interior to account for its volcanism. Callisto and Ganymede must be about 50% ice and 50% rock, while rocky Europa simply has a low-density crust consisting of ice or water and ice. The diameters of Mercury and the Moon are shown for comparison.

FIGURE 15.9 A distant view of Ganymede obtained by Voyager 2 shows that the surface has both dark and light terrains reminiscent of the appearance of our own Moon to the naked eye. Bright ray craters are visible everywhere.

similar. Both objects also have similar low temperatures, and both show the spectral signature of surface ice. Based on such overall comparisons, it would be reasonable to expect the two satellites to have experienced a similar geologic history. Yet even a glance at the images of Ganymede and Callisto reveals fundamental differences (Figs. 15.4 and 15.9).

About half the surface of Ganymede resembles the dark, ancient cratered surface of Callisto, including such features as crater ghosts and concentric bull's-eye ridge patterns. These old terrains, however, are interspersed with lighter, less cratered areas that have been modified by internal activity. The crater densities on these modified areas are typically 100–200 10-km craters per million km^2, significantly lower than those on the heavily cratered areas of either Callisto or Ganymede, although still indicating crater retention ages measured in the billions of years.

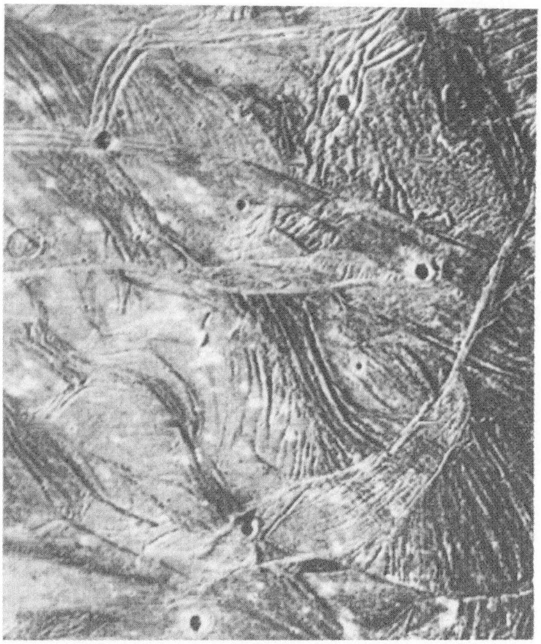

FIGURE 15.10 Surface features as small as 1 km across are visible in this high-resolution picture of Ganymede's grooved terrain. This kind of landscape is found only in the lighter regions of the satellite's surface. It is reminiscent of the Appalachian Mountains on Earth or the Maxwell Mountains on Venus. We will see it again on Uranus's satellite Miranda.

Indications of Internal Activity

Many of the younger, less cratered areas of Ganymede have systems of parallel mountains and valleys (Fig. 15.10). In horizontal scale these mountain systems on Ganymede resemble the low Appalachian Mountains of the eastern United States, with ridges separated by 10–15 km. These ridges are tectonic, but the mechanism of their formation is different from that of terrestrial mountains. While the mountains of Earth and Venus were created by wrinkling of crust subject to tectonic compression, the mountains of Ganymede appear to result from long cracks or faults separating strips of land that have been alternately lifted up or depressed (Fig.

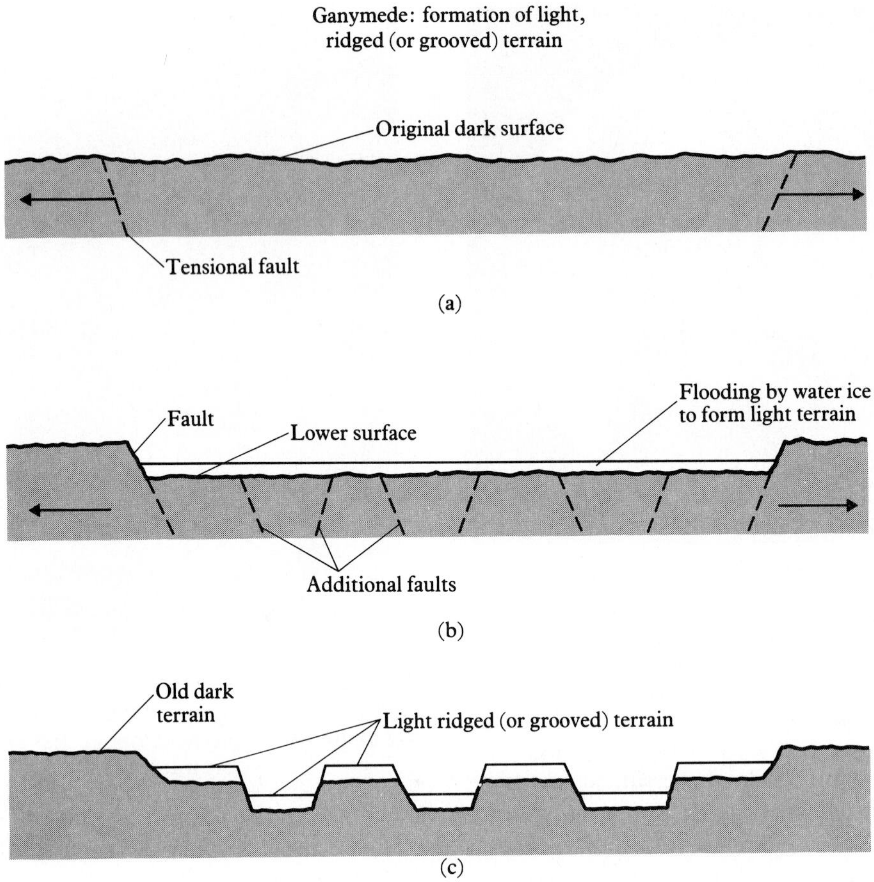

FIGURE 15.11 While we do not completely understand the processes that formed the grooved terrain on Ganymede, this diagram describes one possibility. (a) The original primitive crust is cracked by tensional stresses. (b) Subsidence, flooding, and freezing produce a new surface. (c) Additional faulting and subsidence produce the ridges and valleys seen today.

15.11). Many of the depressed areas in turn look as if they had been flooded with liquid—presumably liquid water—from the interior. We do not know exactly how long ago this activity took place, but it happened after the era of heaviest meteoroidal bombardment, and it appears to have stretched over hundreds of millions of years.

A compositional difference between the older and younger terrains on Ganymede can also be inferred. Both contain water ice mixed with dirty contaminants, but the geologically younger areas are lighter in color, suggesting less contamination of the ice. (Note that this is just the reverse of the case on our Moon, where the darker areas are younger.) The reflectivity of these younger terrains is about 40%, as compared with 25% for the heavily cratered areas. Ganymede is further distinguished by a few very bright craters and crater rays, which appear to have splashed nearly pure water across the surface (Fig. 15.12). No similar bright craters occur

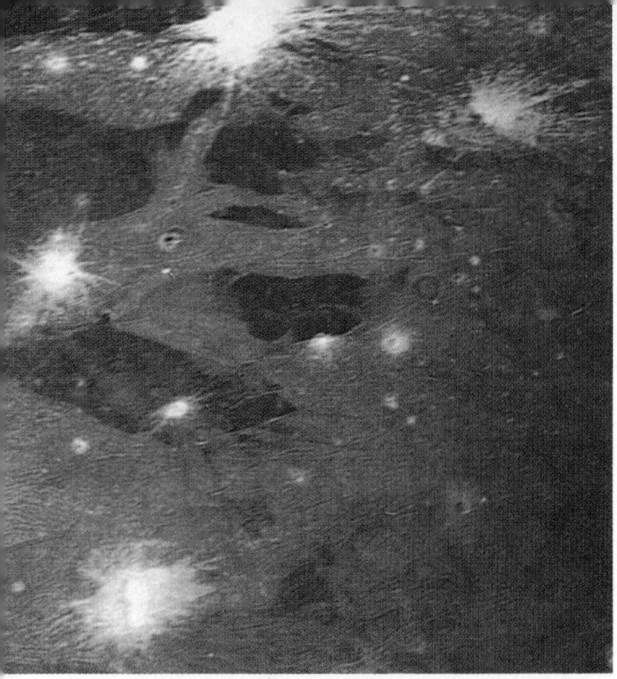

FIGURE 15.12 Impacts on Ganymede's icy surface have left a record of bright craters like these, surrounded by systems of rays. No dark craters have been found, indicating that the material that darkens some regions of Ganymede is probably confined to the thin surface layer.

on Callisto, suggesting the presence of fairly clean ice below the dirty crust of Ganymede but not of Callisto.

The evidence from the surface of Ganymede suggests that this satellite experienced a series of internal upheavals sometime during the first billion years of its existence. One suggestion is that these events were triggered by changes in the density and structure of ice as the ice in the interior altered its crystalline form with the slow cooling of the core. The resulting expansions and contractions could have cracked the surface and resulted in the submersion of low-lying areas by a "lava" of liquid water.

At one time the crust of ice may even have floated on a viscous mantle of slushy ice, resulting in an episode of icy plate tectonics analogous to that of Earth. The coupled subjects of internal evolution and surface geology on an icy world such as Ganymede are fascinating fields just beginning to be addressed by planetary geologists. The results from the Galileo mission are bound to advance our understanding of both Ganymede and Callisto.

15.4 EUROPA AND IO: THE ACTIVE GALILEANS

Even though their surfaces are made of ice, Ganymede and Callisto at least look reasonably familiar to us, with impact craters reminding us of our own Moon. Those craters tell us we are looking at very old surfaces. Nothing much has happened on these Mercury-size bodies for the last 3 billion years. When we turn to the smaller Galilean satellites, Europa and Io, we expect them to be even more quiescent, since the "rule of thumb" is that the smaller the body, the more quickly it cools off and assumes a dormant state. It was therefore surprising to find that these smaller satellites look very different from Callisto, Ganymede, or the Moon.

Geology of Europa

Of all the Galilean satellites, Europa remained the most enigmatic following the two Voyager encounters. This is a smaller satellite, similar in size to Io and our Moon. Its density of about 3 g/cm^3 indicates a primarily rocky composition with only about 10% water ice, much less than is present in Callisto or Ganymede, which are about half ice. Yet the surface of Europa is the brightest of any of the Galilean satellites (70% reflectivity), and its spectrum indicates a surface composition of relatively pure water ice. In this respect Europa is like Earth, which is made of dense materials on the inside but has a coating of water and ice over most of its surface. What is special about the icy surface of Europa, relative to those of Callisto and Ganymede, is its comparative purity, suggesting that some kind of continuing process periodically resurfaces the planet with fresh material from below.

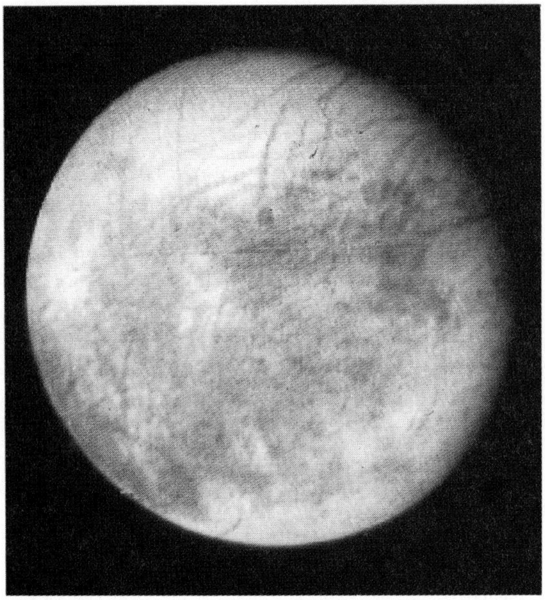

FIGURE 15.13 Europa is the smoothest satellite known. The absence of easily visible impact craters means that we are looking at a relatively young surface.

FIGURE 15.14 This view of Europa obtained by Voyager 2 shows streaks and ridges on the surface that may represent cracks in the icy crust of this satellite with extrusions of subsurface water. Perhaps the Galileo spacecraft will tell us whether there really is water below Europa's icy crust.

The visual appearance of Europa is also different from that of any other planetary body we have studied (Fig. 15.13). There are almost no impact craters visible at the Voyager resolution of a few kilometers, indicating that the record of early heavy bombardment by comets and asteroids has been erased. The crater densities on Europa are significantly less than on the lunar maria, roughly similar to those on the plains of Venus. Evidently, the formation of the icy crust was a relatively recent phenomenon on the geological timescale, which may have been caused by the tidal heating of Europa (see Section 15.5).

Linear Markings

There is little topographic relief of any sort on Europa; in fact, this satellite is the smoothest planetary object in the solar system. The markings that are visible on its surface (Fig. 15.14) are mysterious. Most of them take the form of long, narrow, light or dark lines stretching for hundreds or even thousands of kilometers across the landscape. Some of these lines are double or multiple, with one common form consisting of a light line running between two dark ones. If this all sounds familiar, it is because Europa actually has much of the appearance of the fanciful Mars globes drawn by Percival Lowell at the beginning of the twentieth century. The linear canals of Mars have found their true home on a satellite of Jupiter!

Topographically, the "canals" of Europa are not depressions but shallow ridges a few kilome-

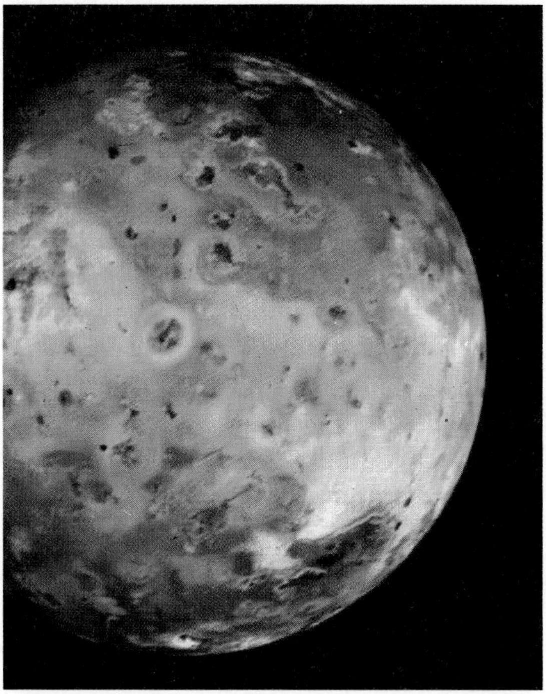

FIGURE 15.15 A partial mosaic of Io compiled from Voyager 1 pictures shows the astonishing surface features caused by this satellite's active volcanism. Again there are no impact craters, but now the surface we see is very young indeed, being continually reformed by volcanic eruptions. The colors in this photo are substantially exaggerated.

ters in width and a few hundred meters high. Perhaps these ridges are made of viscous material squeezed up through cracks in the icy surface. But squeezed up from where? Some scientists think that beneath the ice crust of this satellite is a global ocean of liquid water as much as a hundred kilometers deep. They suggest that the relatively thin crust occasionally cracks, resulting in a release of liquid from below. A few speculative individuals have even gone so far as to suggest that the oceans of Europa might conceivably support some exotic life-forms, maintaining themselves on the weak sunlight that could filter down through the thinner ice where recent cracking of the crust had taken place. One of the prime objectives of the Galileo orbiter is to obtain high-resolution pictures of this enigmatic satellite.

Composition of Io

The innermost of the large jovian satellites is by far the most spectacular, with its active volcanoes and constantly changing surface. Following Voyager, the name *Io* has almost become a household word (except that no one can seem to agree on how to pronounce it: EE-O or Eye-O). But before we look at the remarkable volcanoes of Io, let us discuss how it compares with the three other Galilean satellites.

The density of Io is 3.3 g/cm^3, slightly higher than that of Europa and indicative of a rocky object with little or no water. In fact, Io has nearly the same size and density as our Moon, and therefore probably a very similar bulk composition. Like Europa, Io has a high surface reflectivity. But here the similarities end. Instead of a bland white ice surface like that of Europa, or the gray dirty-ice surfaces of Ganymede and Callisto, Nature has given Io a multihued face of materials with yellow, red, brown, black, and pastel shades (Fig. 15.15). Note that the colors in Fig. 15.15 are exaggerated, as is usually the case in color pictures of Io.

The brightly colored materials on the surface of Io ought to be identifiable, but unfortunately the evidence from spectra is not entirely clear on this subject. The one material positively detected is sulfur dioxide (SO_2), which is an acrid volcanic gas on Earth but freezes on Io as a white frost (Fig. 15.16). A thin atmosphere of SO_2 gas has also been detected, while the breakdown products of this material, sulfur and oxygen, are major contributors to the charged particle population of the jovian magnetosphere, as described in Section 13.6.

It seems probable that sulfur dioxide is the primary component of the white areas on Io, but

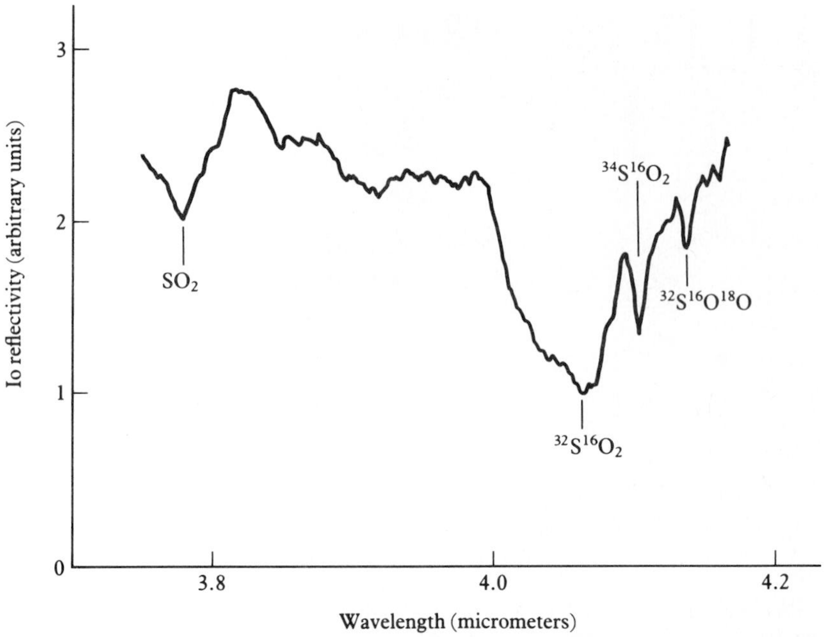

FIGURE 15.16 A spectrograph attached to an Earth-based telescope recorded this evidence of sulfur dioxide absorption on Io. Volcanic eruptions produce the gas. (After D.P. Cruikshank and R. Howell.)

what of the more colorful yellows, oranges, browns, and blacks? There is some spectral evidence for H_2S, but the most likely candidate is elemental sulfur, which has the property of taking on many colored forms depending on its temperature and the way it is cooled from the liquid state. Another possibility is provided by compounds of sulfur and sodium, which also display an appropriate diversity of hues. But even given the evidence for abundant sulfur and sulfur compounds, we do not know whether there is a thick crust of these materials or whether they make up only a thin coating, perhaps disguising more ordinary rocks.

Geology of Io

The underlying geology of Io is unique. There are no impact craters, indicating a geologically young surface. Note that this is true in spite of two factors that should increase the crater density, relative to the other Galilean satellites: the greater cratering rate so close to Jupiter and the absence of ice in its crust. Even more than Earth,

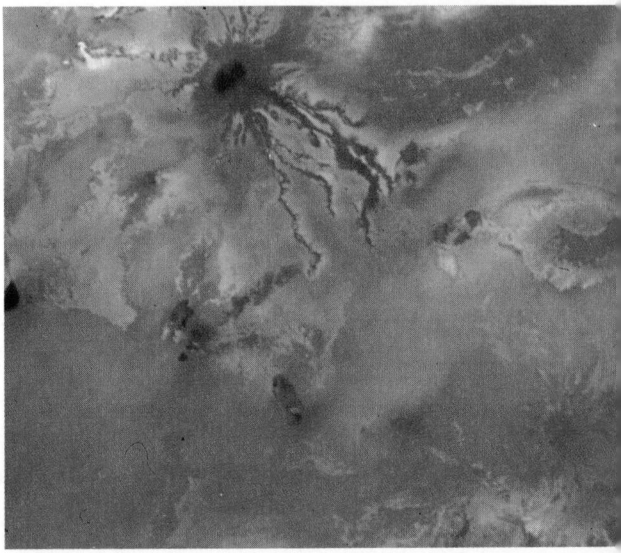

FIGURE 15.17 Seen in a close-up view by Voyager 1, the surface of Io reveals a rich variety of volcanic features. Here we see several large flows of "lava" (sulfur-dominated silicates?) emanating from volcanic calderas. Colors are exaggerated.

FIGURE 15.18 Near the north pole of Io, this mountain (Haemus Mons) soars to an altitude of 9000 m (30,000 ft). To reach such heights, it must be composed in large part of silicates, since simple sulfur compounds would not have sufficient strength to support such a tall structure.

Io is capable of erasing its craters as fast as they are formed. Calculations of the resurfacing rate required to destroy the craters on Io indicate that hundreds of meters of new crust must be formed each million years, a value similar to that of areas with high sedimentation on Earth.

Instead of impact craters, Io displays numerous volcanic features, including layered lava plains and flow fronts, volcanic mountains, calderas, and vents (Fig. 15.17). Long, twisting flows issue from some vents, while others are surrounded by white aprons, apparently deposits of sulfur dioxide frost. There are a few low shield volcanoes, and near the south pole there are also some isolated high mountains of unknown origin. With altitudes as great as 9 km, the mountains of Io are the highest features on the Galilean satellites (Fig. 15.18).

15.5 VOLCANOES OF IO

All this evidence of volcanic activity is dramatic enough to the geologist, but even more exciting was the Voyager discovery that ongoing eruptions were visible as the spacecraft flew past. For

FIGURE 15.19 A sequence of volcanic eruptions on Io. (a) Prometheus as it appears from straight above. The vent is surrounded by a ring of bright material deposited by the plume. In (b) we get a clearer view of the material raining down on the surface, while at (c) we see the plume in projection, with its characteristic mushroom shape.

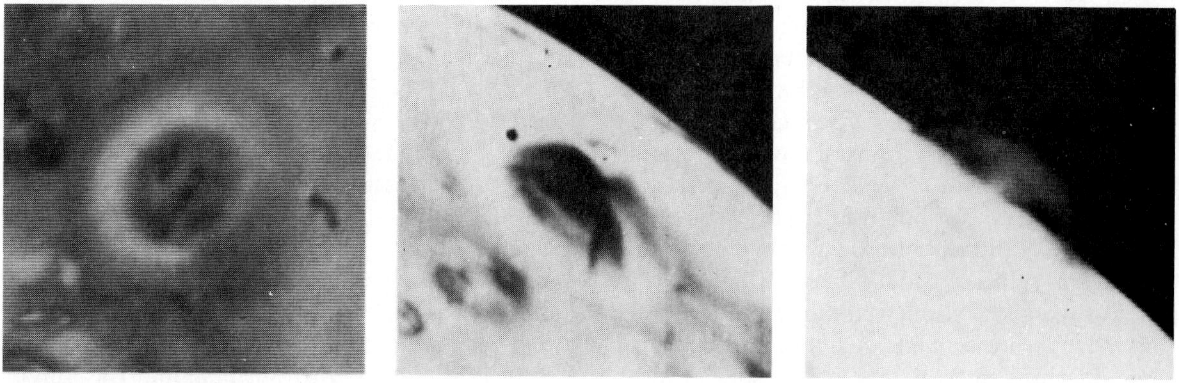

a b c

the first time outside Earth, it was possible to see geological changes as they took place, rather than being forced to infer them from evidence collected after the fact. The volcanoes of Io therefore became the focus for study of this remarkable world.

Plume Eruptions

The most spectacular volcanic events on Io are the plume eruptions, discovered by the Voyager cameras (Fig. 15.19). In March 1979, Voyager 1 photographed nine plume eruptions fountaining up to 300 km above the surface. Four months later, Voyager 2 found that one of these had ceased activity and one was unobservable, but the others were still going strong. The plumes arise from rifts or calderalike vents, and associated with them are characteristic patterns of white or dark red material deposited around the vents. By decision of the International Astronomical Union, the volcanoes of Io are named for volcano or fire gods from various cultures, such as Pele, the Hawaiian volcano goddess, and Loki, the Norse fire god.

Plume eruptions on Io apparently come in two varieties. The first kind is violent and short-lived, and produces a dark red surface deposit. The best example is Pele, the largest plume photographed by Voyager 1, which had ceased activity four months later when the Voyager 2 flyby took place. These plumes also apparently have the hottest vents, with temperatures up to 700 K, and the highest ejection velocities, 1000 m/s. Probably the primary material ejected is elemental sulfur. The second type consists of long-lived eruptions of a white material, very likely sulfur dioxide. These eruptions have lower temperatures and lower ejection velocities. Most of the plumes photographed by Voyager are of this sulfur dioxide class.

Both kinds of plume eruptions are produced by the high-pressure release of hot underground liquid, which turns to gas as it rushes upward and then condenses again as snow when released into the cold of space. Plume eruptions resemble steam geysers on Earth, such as the famous Old Faithful in Yellowstone National Park. However, the hot fluid that drives the eruptions on Io is probably sulfur or sulfur dioxide, rather than water. Analysis of the Voyager images indicates that about 100,000 tons of material is erupted each second in these plumes, enough to cover the entire surface of Io to a depth of tens of meters in a million years, or to alter the color of an area of thousands of square kilometers in a few weeks (Fig. 15.20).

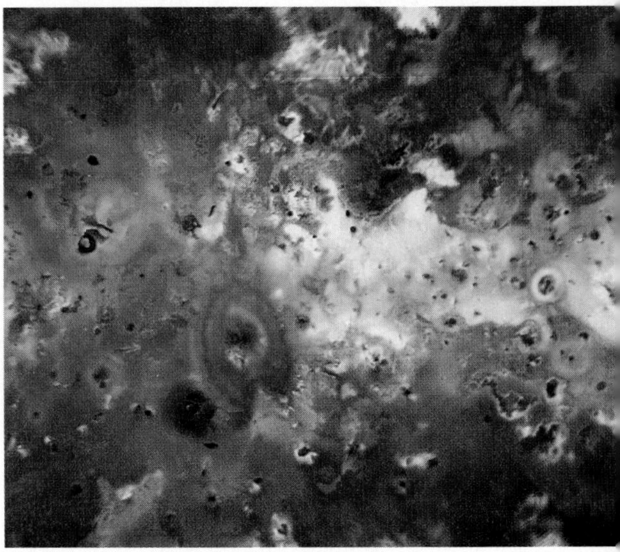

FIGURE 15.20 A map of the surface of Io in Mercator projection showing the remarkable variety of volcanic features produced by the incessant eruptions that wrack this satellite. Colors are exaggerated.

Most of the sulfur and sulfur dioxide in the eruption plumes rains back onto the surface, but a small fraction, probably amounting to 10 tons per second, escapes from Io. Quickly broken down by sunlight and the impact of energetic magnetospheric particles, this material provides most of the charged ions of sulfur and oxygen that are the primary constituent of the inner jovian magnetosphere. Thus the volcanoes of Io contribute directly to maintaining the huge magnetosphere of Jupiter.

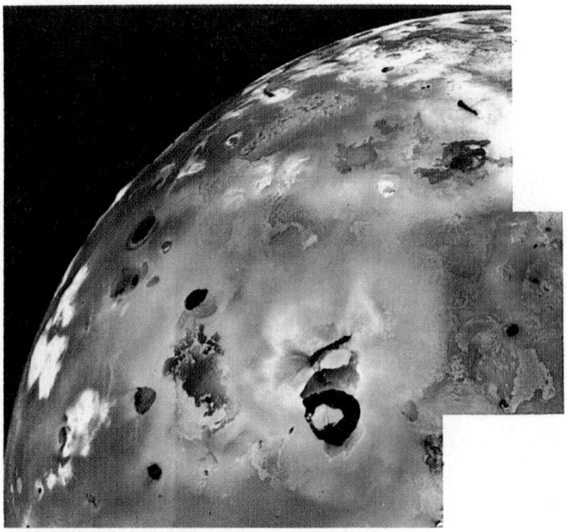

FIGURE 15.21 A close-up of the "lava lake" Loki (cf. Fig. 15.15). The lake itself is about 200 km across. At the time of the Voyager 1 encounter, this lake was about 150 C hotter than the surrounding plains.Color and contrast are enhanced.

The Hot Spots of Io

Io's remarkable level of volcanic activity produces a net flow of heat from the satellite that has been measured with infrared detectors attached to Earth-based telescopes as well as by the Voyager infrared detectors. Most of this heat is radiated from a few **volcanic hot spots** not necessarily associated with currently active plumes. In 1979, the largest hot spot was a "lava lake," 200 km in diameter, near the Loki eruption site (Fig. 15.21). The "lava" may be liquid sulfur, which has a melting temperature of 385 K.

Heat radiated from the Io hot spots can be monitored from Earth by measuring the infrared emission from the satellite when it is in eclipse. Once in each orbit, Io passes through Jupiter's shadow. With the reflected sunlight removed, the glow of the hot spots becomes more apparent, like city lights after sunset. Eclipse observations extending back to eight years before Voyager show these effects, although they were not recognized at the time. However, these early telescopic studies did not pinpoint the sources: All they told us is the total emission, and that it is coming from 1–2% of the surface, where the average temperature is about 300 K, similar to the surface temperature of Earth.

Beginning in 1983, ground-based infrared observations began to provide additional information on the nature of the hot spots. Observations carried out with sensitive infrared detectors at the NASA Infrared Telescope Facility on Mauna Kea confirmed that the Loki lava lake was still the single largest source of radiated heat. This was still true in 1995, even though other currently active hot spots have been identified. Io has proved to be highly variable, with eruptions coming and going from one month to the next.

We have learned three more important properties of the hot spots since Voyager: Most of the energy is coming from only one or two major sources, these sources have lifetimes of many years, and the intensity of the sources is highly variable. Other telescopic observations indicate that volcanic plume eruptions are also still taking place, but from the ground it has been impossible to locate specific eruption sites (Fig. 15.22). Just how different is the surface of Io after 16 years of continuous activity? A comparison of the Galileo and Voyager images promises to be very instructive.

Other Volcanic Activity

The large hot spots and the two kinds of plume eruptions do not exhaust the range of volcanism taking place on Io. The Voyager pictures show many surface flows similar to the long, twisting lava flows on the slopes of terrestrial volcanoes (Fig. 15.17); these are probably sulfur, but some of them could be basaltic rock similar to those from that produced by other volcanoes in the solar system. These quieter flows probably do more to create the surface landscape of Io than do the more spectacular plume eruptions.

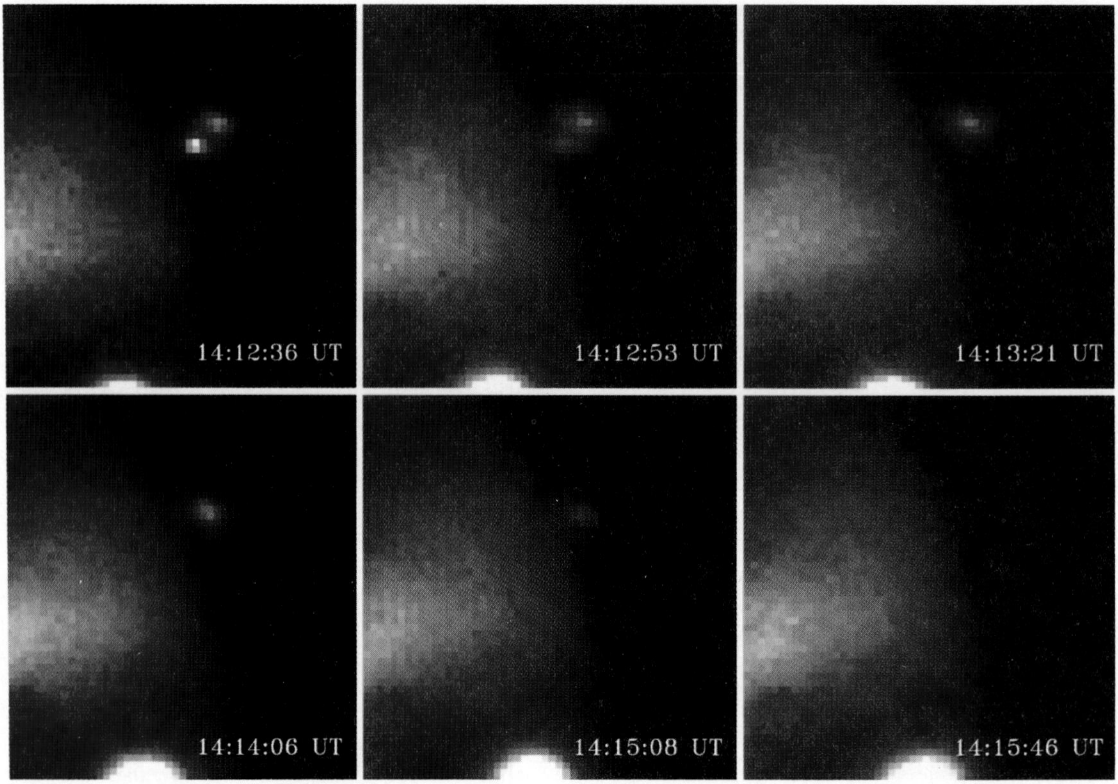

FIGURE 15.22 Io next to Jupiter, as recorded by John Spencer at a wavelength of 3.4 μm with a camera on the NASA infrared telescope. The satellite is in eclipse, so no sunlight is striking it. What we see is thermal radiation from Io's surface, coming primarily from "hot spots."

No astronaut is likely ever to stand on the surface of Io; the intense bombardment by energetic particles from the magnetosphere would kill a human in a matter of minutes. But perhaps someday we will be able to view the scene remotely, through the cameras of a lander spacecraft. From one hemisphere of Io, Jupiter itself would dominate the view, with 50 times the apparent diameter of the Moon in the skies of Earth. Even at night, the colored landscape of Io would be illuminated brightly by Jupiter. In the darkness of Jupiter's shadow, we might see the black sky above as glowing faintly with electrical discharges in the tenuous atmosphere of Io. The most spectacular sight would surely be one of the huge plume eruptions, sending a fountain high above the surface. Anywhere within hundreds of kilometers of such a vent, we would find ourselves in a gentle snowfall of crystals of sulfur or sulfur dioxide. After a few weeks, the surrounding landscape and the spacecraft itself would acquire a fresh coating of this falling material. Meanwhile, we would have to look over our shoulders, for who knows when a tongue of molten sulfur from some other vent might be headed in our direction.

Energy to Power the Volcanoes

Io's level of volcanic activity is unique. Yet Io is a small body, similar in size and composition to our Moon. Why is the face of Io constantly changing due to volcanic activity, while the Moon's volcanic fires cooled more than 3 billion years ago? What is supplying the heat that maintains such a high level of activity on just this one satellite? The answer is found in tidal heating, which we first discussed in Section 8.3 on Mercury.

Io owes its special character primarily to its unique location. It is about the same distance from Jupiter as is our Moon from Earth, yet Jupiter is more than 300 times more massive than Earth, causing tremendous tides on Io. These tides pull the satellite into an elongated shape, with a several-kilometer-high bulge extending toward Jupiter. This tidal bulge would not contribute to heating the interior if Io always kept exactly the same face turned toward Jupiter, which is the state it would eventually occupy. However, the gravitational pull of Europa and Ganymede will not permit Io to settle into an exactly circular orbit. Instead, the Laplace resonance forces Io's orbit to be slightly eccentric, with the result that it twists back and forth with respect to Jupiter on each orbital circuit, while at the same time moving nearer and farther from the planet. The twisting and flexing of the tidal bulge heats Io, much as repeated flexing of a wire coat hanger heats the wire. In this way, the complex interaction of orbit and tides pumps tremendous energy into Io, melting its interior and providing power to drive its volcanic eruptions.

The size of the tidal energy source of Io can be estimated from the heat radiated by the hot spots, which turns out to be much greater than the energy of the plume eruptions. The Io power plant generates about 100 million megawatts, more than ten times the total energy consumption of humans on Earth. Ultimately, the source of this energy is the spin of Jupiter itself, which is slowed very slightly by the interaction with Io. At the same time, the coupled orbits of Io, Europa, and Ganymede are being slowly pushed outward, just as tidal forces are pushing the Moon away from Earth.

After billions of years, this tidal heating has taken its toll on Io. Any water or other ices, as well as most carbon and nitrogen compounds, have long since been driven off, until now sulfur and sulfur compounds are the most volatile materials remaining. That is why sulfur and sulfur dioxide act as the working fluids for the volcanoes of Io, while a similar role on Earth is played by water and carbon dioxide. The interior of the satellite must be entirely liquid, with a solid crust probably no more than 25 km thick. The crust is constantly recycled by volcanic activity, and geologic features are probably short-lived.

15.6 COMPARING THE LARGE SATELLITES

Following our practice with the planets, it is useful to compare the satellites of Jupiter with one another to understand their individual peculiarities. We have two natural groupings: the outer icy "giants," Ganymede and Callisto, and the inner rocky objects, Europa and Io. For easy reference, Table 15.2 summarizes key data on these four satellites.

Ganymede and Callisto

The first problem is to determine why Ganymede maintained a substantial level of geological activity for hundreds of millions of years while Callisto did not. The distinctions between these two objects are in their distances from Jupiter and in small differences in their sizes and

TABLE 15.2 Galilean satellite summary

Satellite	Period (days)	Diameter (Moon = 1)	Mass (Moon = 1)	Density (g/cm^3)
Callisto	16.69	1.38	1.47	1.9
Ganymede	1.16	1.51	2.02	1.9
Europa	3.55	0.91	0.65	3.0
Io	1.77	1.04	1.22	3.6

densities. Could any of these factors be responsible for their different histories?

Because Ganymede is closer to Jupiter, it is subject to larger tidal stress, and it also experiences more impacts because of the attraction of Jupiter's gravity. Neither of these effects is large, but both could have resulted in slightly greater heating for Ganymede than Callisto early in their history.

Ganymede is larger than Callisto and also slightly more dense, indicating that it contains more radioactive material and therefore had (and still has) a greater internal heat source. In addition, Ganymede's larger size means that it can hold heat slightly more easily than Callisto. These effects also point toward a higher internal temperature and slower cooling for Ganymede.

Each effect just discussed could contribute to a more active geology on Ganymede than Callisto. The problem is that each individual factor is small, and even taken together they do not suggest that Ganymede should have been a great deal warmer than its twin moon. What must have happened is that these differences, even though small, managed to trigger a major change in the internal processes on Ganymede that was out of all proportion to the temperature differences. Such a major change can occur with ice, because it is a substance that can take on many different forms, each with its own density, depending on the temperature and pressure.

Calculations show that just a small difference in temperature early in the history of Ganymede may have led to convection in the upper mantle, perhaps sufficient to drive some kind of plate tectonic activity for a period of the satellite's history. Subsequently, about a billion years after the formation of Ganymede, a change of phase from liquid to solid in the interior may have caused a small expansion, sufficient to crack the crust and initiate the period of mountain building. According to these calculations, Callisto, by being just a little cooler, escaped these internal events.

Europa and Io

When we turn our attention to Europa and Io, we encounter other mysteries. Why are these two satellites so depleted in water (and other volatiles) relative to Callisto and Ganymede? Much of this difference may be a consequence of the high temperature of Jupiter early in its history, when it was contracting and forming its satellite system. Calculations of temperatures within the nebula surrounding this proto-Jupiter suggest that it may have been cool enough for ice to condense at the distances of Ganymede and Callisto, but not closer to the planet where Europa and Io formed. In that case, we should expect both Europa and Io to be dry, whereas Europa today probably contains about 10% water.

This simple interpretation fails to explain the existence of water on Europa because it neglects the role of impacting bodies from elsewhere. We know that the solar system was full of debris of various kinds early in its history and that a substantial part of that debris was cometary. The impacts of many icy comets must have supplied a great deal of water to all the Galilean satellites, probably enough to account for Europa's present icy surface. In the case of Io, however, this early infusion of water was doomed by the resonant lock that developed among Io, Europa, and Ganymede. Held in the gravitational grip of the other two satellites, Io could not circularize its orbit, with the result that tidal heating drove off all of the water and other volatile substances.

SUMMARY

Whereas the inner planets have few or no satellites, the outer planets resemble miniature solar systems. Jupiter, Saturn, Uranus, and Neptune each have more moons than the Sun has planets. Pluto, always the exception, has only one satellite. Almost half the known satellites in the solar system were discovered in the last 20 years, most of them by the cameras on the two Voyager spacecraft. There are surely more moons awaiting detection by future missions to these planets.

We begin our survey with Jupiter, whose large satellites are planet-size worlds of surprising variety and complexity. Ganymede and Callisto are larger than the planet Mercury. They are nearly the same size and density, inviting comparison between them. With densities less than 2.0 g/cm^3, they are of obviously different composition from the inner planets, which they resemble in size. Models indicate that they are nearly half composed of water ice and that their interiors are differentiated. Callisto's heavily cratered icy surface shows little sign of internal activity. Ganymede, in contrast, once experienced extensive geologic activity, with resurfacing and the widespread formation of long mountain ridges and valleys.

Io and Europa differ from their larger cousins in that they are not composed in equal parts of ice and rock. Both of these satellites are about the size and density of our Moon and presumably have a similar composition. Europa has a remarkably smooth, icy surface, possibly floating on a global ocean of water as much as 100 km deep. There are few craters, indicating that the surface we see was formed fairly recently, and there is a random pattern of linear markings. Much more spectacular, however, is brightly colored Io, the most volcanically active body in the solar system. Io has been extensively modified by internal heating and volcanic activity, so it shows no evidence of impact craters.

Io's energy source is tidal heating, a result of tides raised by Jupiter acting in conjunction with a noncircular orbit. This heating is sufficient to power widespread surface volcanism. Over its history, Io has outgassed and lost highly volatile materials, such as water and carbon dioxide, and today its surface is covered by sulfur and sulfur compounds, which are recycled through its volcanic eruptions.

Most of the internal heat of Io is radiated into space through hot spots, which appear to be large lakes of liquid sulfur heated from below. In addition, there are two kinds of plume eruptions, in which fountains of sulfur or sulfur dioxide rise tens or hundreds of kilometers into space and fall back to recoat the surface. Some of the sulfur dioxide from these eruptions also escapes, where it becomes ionized and contributes significantly to the inner magnetosphere of Jupiter.

When we compare these satellites with one another, the differences in the surfaces of Ganymede and Callisto are especially puzzling. There are good reasons to think that Ganymede has always been slightly warmer than Callisto, however, and that may have been crucial to the development of tectonic features. Both Io and Europa may have formed without water in the warm inner region of the jovian subnebula, but cometary impacts could have supplied the ice we find on Europa today. The tidal heating of Io prevented the accumulation of a similar icy crust.

KEY TERMS

irregular satellite
leading and trailing sides
regular satellite
subnebula
volcanic hot spot

Review Questions

1. How many satellites are known at the present time? How are they distributed between the inner and outer solar system? Do you expect some compositional differences to accompany this distribution? Explain.

2. Describe the general characteristics of Jupiter's four largest satellites. How do they resemble the characteristics of the planets in the solar system? How would you explain this resemblance?

3. What is the difference between a regular and an irregular satellite? Which planets have irregular satellites? Why don't all planets have some?

4. How would you compare the expected densities of impact craters on the satellites of outer planets with the observed values on the maria and highlands of the Moon? What accounts for the differences?

5. What is the evidence that the surfaces of Ganymede, Callisto, and Europa are covered with ice? How can impact craters be preserved in ice?

6. How does the surface of Ganymede differ from that of Callisto? What accounts for this difference?

7. What distinguishes the surface of Europa from the equally icy surfaces of Callisto and Ganymede? What makes this satellite interesting to scientists (and science fiction writers—see Arthur C. Clarke's *2010*) trying to understand the origin of life?

8. The appearance of Io has been compared to that of an anchovy pizza. What accounts for this remarkable moonscape? Why are there no impact craters? What is the source of Io's internal heat?

Quantitative Exercises

1. See if you can derive your own "rules" for the spacing of the regular satellites of the various giant planets, equivalent to the Titius-Bode rule for the spacing of the planets.

2. Compare the angular speed of Leda in its orbit around Jupiter with the angular speed of the entire Jupiter system as it moves in its orbit about the Sun. Will these two motions ever cancel one another so Leda appears stationary? Explain.

3. How fast must a molecule of SO_2 travel to escape from the gravitational field of Io? Compare this with the velocity required for a molecule of H_2O. How do you then explain the absence of H_2O and the presence of SO_2 on Io?

Additional Readings

Beebe, R. 1995. *Jupiter: The Giant Planet.* Washington, D.C.: Smithsonian Institution Press. Excellent pre-Galileo overview of the planet and its satellites for the technically literate general audience, combining results from both astronomical and spacecraft investigations.

*Morrison, D., ed. 1982. *Satellites of Jupiter.* Tucson: University of Arizona Press. Technical volume in the Arizona Space Science series dealing with all aspects of the jovian satellites.

Morrison, D., and J. Samz. 1980. *Voyage to Jupiter.* NASA Special Publication 439. Washington, D.C.: U.S. Government Printing Office. Popular description of the Voyager encounters with Jupiter, including an account of the discovery of volcanoes on Io.

Rothery, D.A. 1992. *Satellites of the Outer Planets.* Oxford: Clarendon Press. A well-illustrated compendium of information about all the satellites of the outer planets that were known as of 1992. Semitechnical.

*Indicates the more technical readings.

16

Titan, Triton, and Pluto: Icy Objects with Atmospheres

In December 2004, the Huygens probe will float down through the atmosphere of Titan to land on the satellite's mysterious surface. The artist has taken some liberty in showing Saturn in the background, since it is not possible to see through the thick layer of smog in Titan's upper atmosphere.

In the previous chapter, we described the four largest satellites of Jupiter in considerable detail. This is the customary approach to the study of small bodies in the outer solar system—namely, considering the satellites of each planet as a set, comparing the sets with one another. In this chapter we are adopting a radically different approach. We will study the largest satellites of Saturn and Neptune and the smallest planet in the solar system (with its satellite), considering their similarities and differences. We are collecting these three objects in a single chapter because they have many common properties, including atmospheres dominated by nitrogen gas. We begin with Titan, the closest, warmest, and best studied of the three.

16.1 TITAN: A RIDDLE WRAPPED IN AN ENIGMA

With a smog-filled nitrogen atmosphere that is denser than our own, Titan is one of the special places in the planetary system. Saturn's largest satellite, Titan is essentially the same size and the same density (1.9 g/cm^3) as Ganymede and Callisto. One might therefore expect it to look similar to these giant jovian moons, but sunlight reflected from Titan carries a message suggesting that this expectation will not be met.

First there is the color. As viewed with a telescope from Earth, Titan appears as a tiny, barely resolvable reddish disk, whereas Ganymede and Callisto are essentially colorless. But there is more. In the middle decades of this century, Gerard Kuiper carried out a systematic spectroscopic survey of the solar system using photographic plates that were specially sensitized to record short-wavelength infrared light. When he examined Titan with a spectrograph attached to the McDonald Observatory 82-inch (2.1-m) telescope in 1944, Kuiper found that Titan's spectrum showed absorption bands of methane gas in addition to the expected solar lines (Fig. 14.9). Unlike Ganymede and Callisto, Titan has an atmosphere (Fig. 16.1).

It may seem surprising that a satellite could have a substantial atmosphere; our Moon certainly doesn't. You will recall that the ability of a planet or satellite to retain an atmosphere is determined primarily by the body's mass and temperature. Titan is both more massive and colder than the Moon. Thus it has a stronger gravitational field, so molecules must move faster to have sufficient energy to escape into space. Yet the lower temperature means that molecular velocities will be slower, making it more difficult for them to escape. Thus both size and temperature favor retention of an atmosphere. But why Titan and why not Ganymede or Callisto? We shall return to this question after we learn more about Titan's atmosphere.

Temperature Measurements and Atmospheric Models

After Kuiper's discovery of methane, knowledge about Titan grew slowly, and it was not until nearly 25 years later that the next major discoveries were made. First, a measurement of Titan's temperature at the infrared wavelength of 10 μm showed it was surprisingly warm, although still a frigid −100 C (173 K). The second discovery came from polarization observations that suggested the surface was probably hidden below a haze of unknown composition. These observations were made with telescopes on Earth that could not magnify the image of Titan sufficiently to show the satellite as a disk. The deductions about conditions on this distant world had to be made by analyzing the thermal radiation or reflected sunlight from an object that appeared as a tiny point of light, indistinguishable in appearance from a reddish star.

The measured temperature was about twice as high as the expected value of 85 K, leading to speculation that Titan might have a major greenhouse effect. Recall from Section 10.2 that the first measured radio temperatures of Venus were also about twice as high as expected for its distance from the Sun, and that the greenhouse effect provided an explanation for this heating of the planet's surface. However, there was no assurance that the infrared radiation from Titan was coming from the surface, rather than from some warmer level in the upper atmosphere. Only in 1979, with the completion of the Very Large Array (VLA) radio telescope (Fig. 3.8), could the surface temperature of Titan be determined as it had been done for Venus some 20 years earlier. By using many telescopes linked by computer, this radio telescope provides high spatial resolution as well as good sensitivity. The resolution permits a clear distinction between Saturn and Titan, which had not been possible before. The answer from the VLA was that the surface temperature was between 78 K and 96 K, indicating that Titan was not supporting a Venuslike greenhouse effect. Evidently, the infrared temperature referred to a region of temperature inversion in Titan's upper atmosphere, similar to those found on the giant planets.

By the eve of the 1980 Voyager 1 encounter with Titan, many questions concerning the satel-

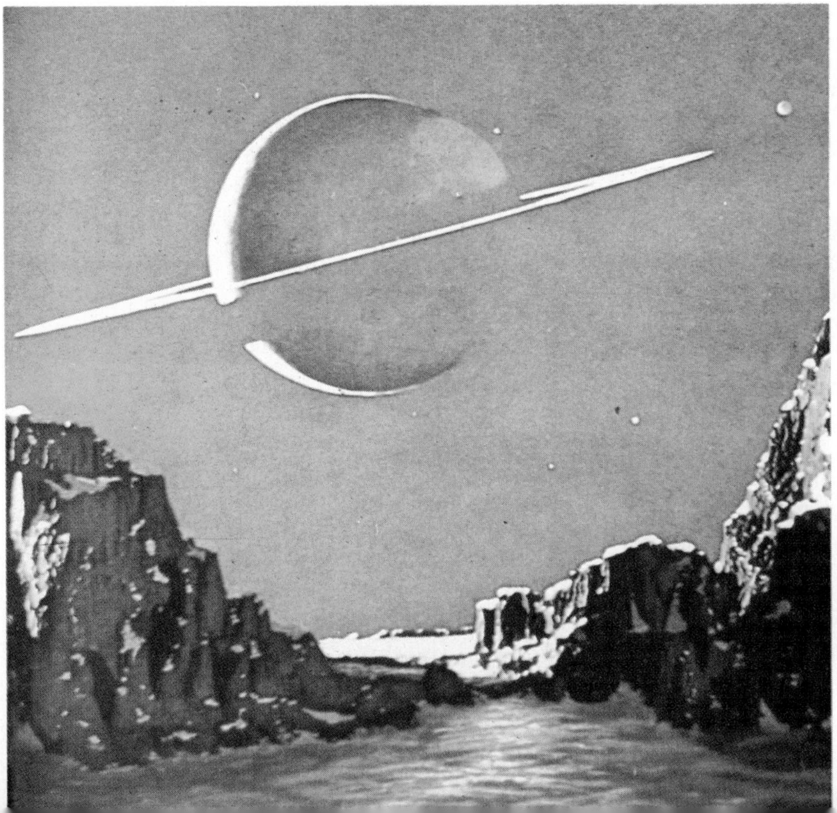

FIGURE 16.1 This is an early artist's conception of a view of Saturn from the surface of Titan. The blue sky is a result of scattering by Titan's atmosphere. However, we now know that views from Titan would be very different from this 1950s concept (see Fig. 16.6).

lite's atmosphere remained unanswered. Only methane and some other hydrocarbons had been detected spectroscopically, but the presence of nitrogen or argon was suspected. The composition of the clouds was unknown. A fundamental problem also remained in estimating the total amount of the atmosphere and therefore its surface pressure. Different scientists favored various possible models, in which the surface pressure ranged from a value almost as low as that on Mars up to several times the surface pressure on the Earth. Before Voyager the information available was insufficient to distinguish among these possibilities.

Voyager Results

The pictures of Titan sent back by the Voyager spacecraft were disappointing, since the surface of the satellite was completely obscured by a ubiquitous smog layer (Fig. 16.2). Most of the exciting results came from the other Voyager instruments.

FIGURE 16.2 From a distance, Titan shows a remarkably well defined contrast boundary between its northern and southern hemispheres, even though this equatorial division is simply set in a haze layer rather than on a solid surface. The north polar hood is also slightly darker than the adjacent haze. Contrast is exaggerated.

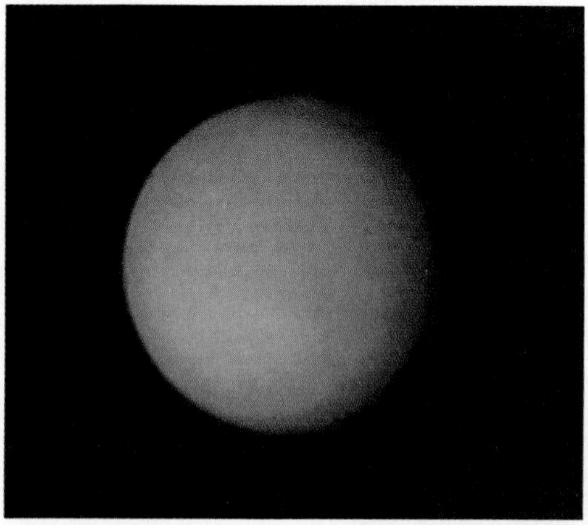

As the spacecraft passed behind Titan on November 11, 1980, its radio signal was attenuated by the atmosphere and ultimately blocked entirely by the satellite's solid surface. This occultation permitted Voyager scientists on Earth to determine how the density of the satellite's atmosphere varied with altitude above its surface. Assuming a nitrogen-rich composition based on results from the spectrometers on board the spacecraft, they deduced a surface pressure of 1.5 bars.

TABLE 16.1 Composition of the atmosphere of Titan

Gas	Formula	Amount
A. Main Constituents (percent)		
Nitrogen	N_2	82–99%
Argon	Ar	0–6[a]
Methane	CH_4	1–6[a]
B. Trace Constituents (parts per million)		
Hydrogen	H_2	2000
Hydrocarbons		
Ethane	C_2H_6	20
Propane	C_3H_8	20
Ethylene	C_2H_4	0.4
Diacetylene	C_4H_2	0.1–0.01
Methylacetylene	C_3H_4	0.03
Nitrogen compounds		
Hydrogen cyanide	HCN	0.2
Cyanogen	C_2N_2	0.1–0.01
Cyanoacetylene	HC_3N	0.1–0.01
Oxygen compounds		
Carbon monoxide	CO	50–150
Carbon dioxide	CO_2	0.015

[a]The presence of argon can be deduced only indirectly. There may be none at all, in which case the abundance of nitrogen would increase. The amount of methane varies with altitude and is still poorly determined.

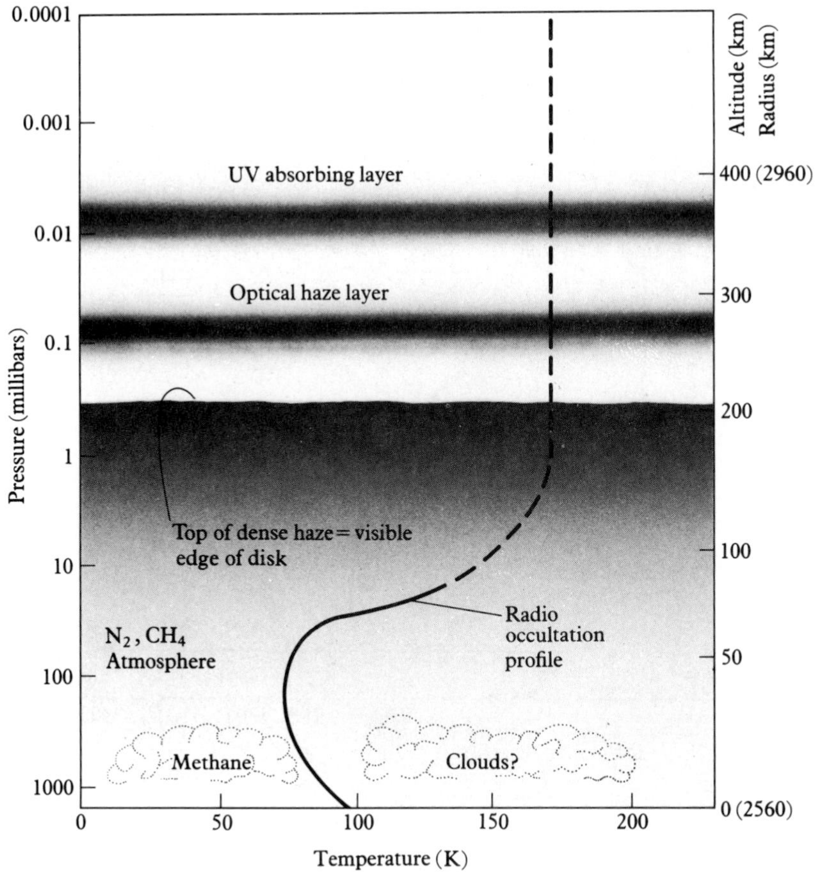

FIGURE 16.3 Unlike the atmospheres of Earth, Mars, and Venus, Titan's atmosphere is much warmer in the stratosphere than it is near the satellite's surface. There are several high-altitude haze layers.

As is often the case, the models had bracketed the true value. With this high surface pressure, Titan has a larger atmosphere than all of the terrestrial planets except Venus. Since its gravity is much less than that of the Earth, more gas is required on Titan to exert the same pressure as it would have here. Thus the amount of gas above the surface of Titan is approximately ten times greater than that above an equal area on Earth, and the atmosphere of Titan also stretches nearly ten times farther into space than does our own (Fig. 16.3).

Voyager made important discoveries about the composition of Titan's atmosphere as well as its structure. The methane detected by Kuiper turned out to be a minor constituent. Titan's atmosphere is mostly nitrogen, as is ours. Impacting electrons from Saturn's magnetosphere, cosmic rays, and solar ultraviolet light have sufficient energy to break molecules of methane and nitrogen apart. As the fragments recombine, new molecules are created, and a number of these were detected in Titan's atmosphere by the Voyager spectrometers (Fig. 16.4).

Table 16.1 shows us that some very interesting chemical reactions are taking place on Titan, especially when they are considered in the context of the **prebiotic chemistry** required for the origin of life on Earth. These reactions are producing hydrogen cyanide (HCN), a starting point for the formation of some of the components of DNA. The presence of both carbon

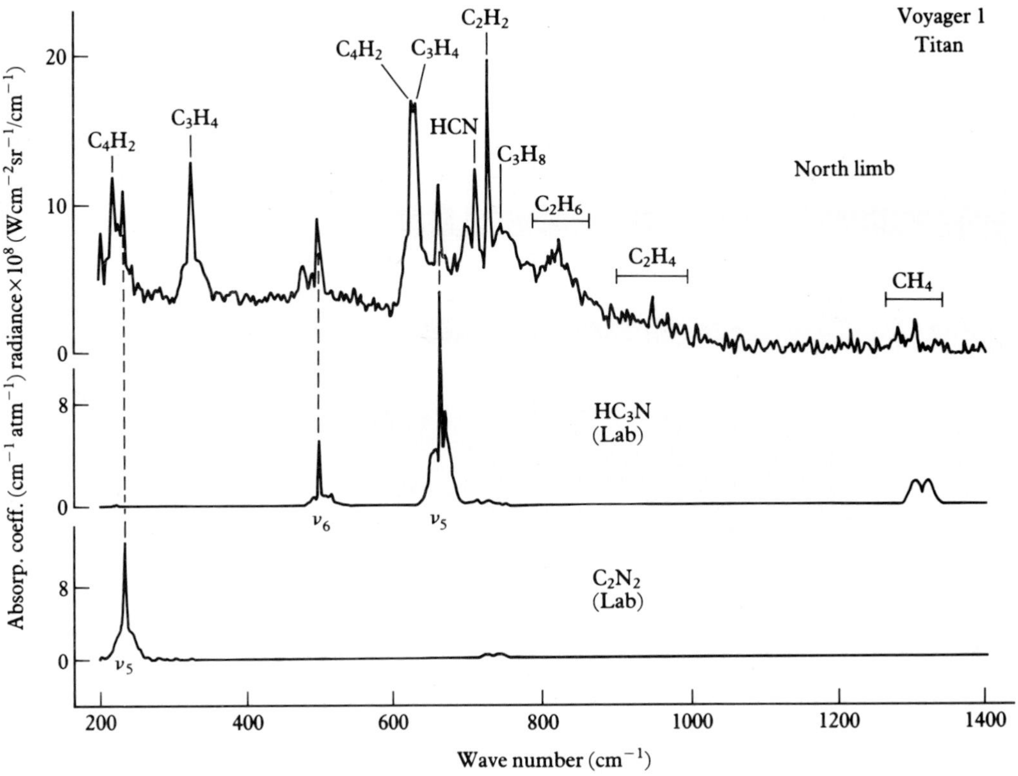

FIGURE 16.4 A section of the infrared spectrum of Titan recorded by Voyager 1 shows emission bands from several gases that are present on Titan in trace amounts (top). The two laboratory spectra of individual gases show how these substances can be identified in Titan's atmosphere.

monoxide (CO) and carbon dioxide (CO_2) in this mixture makes the formation of amino acids possible.

Photochemical Smog

In Titan we have discovered a world nearly frozen in time, where we can examine a primitive environment in which chemical reactions taking place today may resemble those that preceded the evolution of life on our own planet. It would be fascinating to study the end products of this chemistry, to use Titan as a natural laboratory for testing our ideas about chemical evolution. Are certain pathways toward complexity preferred? We know that there are more complex substances on this satellite than those already found because of the existence of the global haze layer, but what are these compounds?

The haze is all that we can see in the detailed pictures sent back by Voyager. Even pictures with much higher resolution than the one reproduced here, close-up views that show a small area at high magnification, reveal no detail at all: no gaps in the haze, no structure that could be used to map winds. The haze of Titan is even more uniform than the clouds of Venus. Only when viewed from the side does it reveal structure, in the form of a distinct layer hundreds of kilometers above the surface (Fig. 16.5).

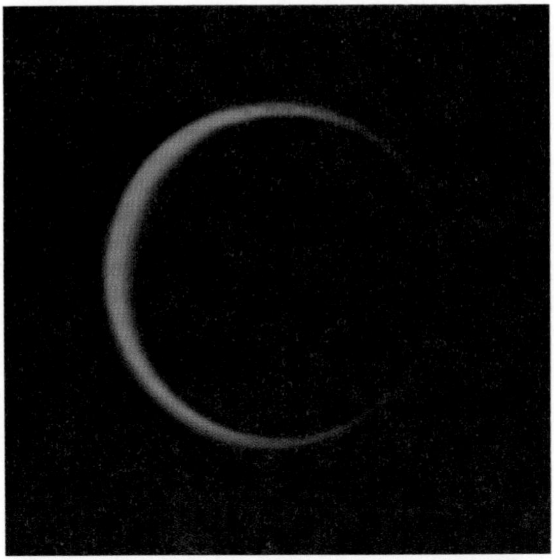

FIGURE 16.5 A detached haze layer can be seen in projection at the edge of Titan's visible disk. It lies above the thick haze that obscures the surface of the satellite from view. The image was recorded after the spacecraft had passed the satellite, hence the crescent phase.

Observations of a stellar occultation by the Voyager ultraviolet spectrometer revealed additional absorbing layers between 300 and 500 km above the surface. Some of the organic molecules formed in Titan's atmosphere, such as HCN, can form long chains called polymers. The reddish brown haze that pervades Titan's atmosphere is presumed to be a combination of these polymers and condensed organic compounds.

As the haze particles grow in size, they will gradually drop out of the atmosphere onto the satellite's surface, forming deposits in low-lying areas, drifts of organic material that by now may be several hundred meters thick. Since the lower atmosphere and surface of Titan are colder than the upper atmosphere, most of the products of this precipitation and photochemistry cannot evaporate and mix with the atmosphere again, the way water does on Earth. The only exceptions are gases like methane and ethane, which may exhibit something similar to the terrestrial water cycle.

Evolution of the Atmosphere

Although the surface of Titan is preserved in a deep freeze, the atmosphere is evolving. As the methane is broken apart by electrons and ultraviolet photons, the hydrogen that is produced escapes, since hydrogen atoms and molecules move fast enough even at Titan's low temperatures to reach escape velocity. A small amount of hydrogen is always present in Titan's atmosphere, since it is continuously being produced. Over the lifetime of the solar system, however, a large amount of hydrogen must have escaped into space. This suggests that deuterium should be enriched on Titan as it is on Venus, although not to the same extent. There is evidence that such enrichment has occurred, in the form of a high abundance of CH_3D—methane in which one of the hydrogen atoms is replaced by an atom of deuterium.

The loss of hydrogen must be balanced by a loss of carbon; this loss is represented by the precipitation of aerosol particles and hydrocarbons onto Titan's surface. Ethane (C_2H_6) is the most abundant end product of the reactions occurring in the atmosphere, and Titan's surface is sufficiently cold that this gas would condense on it. Thus we may imagine lakes, seas, even oceans of liquid ethane. Methane (CH_4) and propane (C_3H_8) would also form liquids, but CO_2, the other hydrocarbons, and the nitrogen compounds in Table 16.1 will solidify at these temperatures. If hydrocarbon seas are present, these solid compounds would sink rather than float, since their densities are higher than that of liquid ethane, methane, or propane.

Titan offers us the apparent paradox of an atmosphere that has undergone extensive evolution without losing its primitive characteristics. The chemistry occurring on Titan today must be very similar (if not identical) to that of 4 billion years ago. This situation, which is totally different from the more rapidly evolving conditions found among the inner planets, is due to Titan's very low temperature.

The average surface temperature on Titan is 94 K, a frigid −179 C. Titan is so cold that liquid water—so vital to life as we know it—is an impossibility. Even water *vapor* will be almost totally lacking from this atmosphere. That is the reason we still find methane and other hydrogen-rich compounds. If water vapor were plentiful, Titan's atmosphere would resemble that of Mars. The water molecules would be broken apart (just as the molecules of methane and nitrogen are dissociated on the real Titan), and the liberated oxygen would convert all the hydrocarbons to carbon dioxide. However, the only sources of oxygen on Titan are the ice particles the satellite sweeps up as it orbits around Saturn and the breakup of CO molecules in the atmosphere. These sources are sufficient to form the tiny amount of CO_2 detected on Titan but are hopelessly inadequate to oxidize all of the methane.

Surface of Titan

Now we have the components of a most unusual landscape! The basic structural element should be water ice, overlain in places by wind-blown drifts of aerosol particles and crossed by gullies formed by rivers of liquid hydrocarbons. Some residual topography from the cratering events that have marked the surfaces of other solar system bodies may be present, modified by these two processes. As on Earth, the atmosphere will screen out the smaller impacting bodies. Finally, there is the distinct possibility that the crust of Titan is subject to the icy equivalents of volcanism and plate tectonics caused by mantle convection, leading to mountain chains, ridges, scarps, volcanic peaks, and blocks of ice strewn over the surface.

What would it be like, sitting in a boat on an ethane ocean on Titan (Fig. 16.6)? Cold certainly, and probably gloomy. The level of illumination at Titan's surface at midday is estimated to be similar to that from a full moon at midnight on Earth. The entire sky would appear to be glowing faintly, as sunlight filters through the orange-brown smog layer. The Sun itself would be invisible, as it is when there is heavy smog over one of our cities. Yet the horizontal visibility would be very good (unless, of course, we were caught in an ethane fog bank), since the thick part of the haze is many kilometers above the satellite's surface.

We don't expect to be rocked by big waves, because global temperature differences on Titan appear to be small, and thus there should not be strong surface winds. Local weather may exist since the ocean is not global in extent, but the surface temperature is expected to be very uniform as a result of the moderating effect of the dense atmosphere. In principle, enough ethane is produced by the atmospheric photochemistry to form a global ocean of this compound; however, observations of Titan by Earth-based radar have indicated that the reflectivity of the satellite's surface at the wavelength of the radar signal is inconsistent with the presence of a global ocean deeper than 100 m. This conclusion agrees with the interpretation of Earth-based infrared observations that can detect sunlight returned from Titan's surface. (Infrared radiation can penetrate the haze on Titan, just as it penetrates the clouds of Venus.) A landscape made of dirty water ice provides the best fit to these infrared data. What we really need for exploration of Titan's surface is a spacecraft equipped with radar and an entry probe. This is exactly what the Cassini-Huygens mission is designed to provide when it arrives in 2004.

Origin of Titan's Atmosphere

Ganymede and Callisto are about the same size as Titan and are also composed of a mixture that is roughly half ice and half rock. Yet (as we saw in Chapter 15) these two satellites of Jupiter do

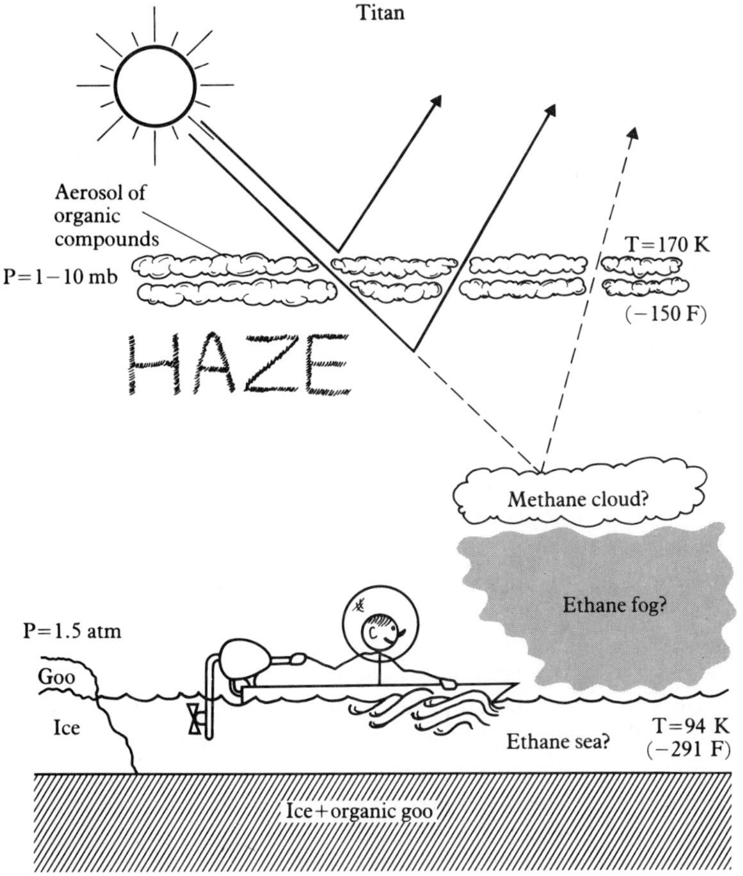

FIGURE 16.6 The surface of Titan may be predominantly ice, with a covering of organic matter and lakes and seas of hydrocarbons. Thus it could be an interesting place to explore by boat.

not have atmospheres, while Titan has an atmosphere denser than ours. What accounts for this difference? How did Titan acquire its atmosphere?

There is an easy test we can make to determine whether the mixture of gases we find on Titan today is a relic of a dense, primary atmosphere captured by the satellite as it formed from the solar nebula. If that were the case, we would expect neon to be as plentiful as nitrogen in the present atmosphere, since these elements are nearly equally abundant in the Sun and hence in the solar nebula (Table 2.2). In fact, the ultraviolet spectrometer on the Voyager 1 spacecraft detected no evidence of neon, setting an upper limit of 1% for the abundance of this gas in Titan's atmosphere. Thus we conclude that the satellite did not capture its gases directly from the nebula, but instead the atmosphere was formed by the release of gases from the solid materials making up the bulk of Titan, in much the same way as the inner planets developed their atmospheres.

Like the inner planets, Titan must also have accumulated some gases as a result of cometary impacts. The big difference in the case of Titan is that approximately 50% of its initial mass was water ice, which now forms a crust and mantle around the satellite's rocky core. Thus we can say that Titan is partially *made* of comets, so the external cometary contribution will add to a volatile mixture contributed by the icy planetesimals that accreted to form the satellite.

Differences Among Titan, Callisto, and Ganymede

Water ice formed at low temperatures (less than 75 K) is an excellent carrier of gases. There is room in the irregular structure of this low-temperature ice for molecules of gases to be accommodated. When the ice is warmed above 135 K, the lattice structure changes, becoming regular, and most of the trapped molecules are released. The ability of water ice to capture gases in this way depends on the size and electrical properties of the gas molecules. Hydrogen, helium, and neon are not trapped except at temperatures lower than 25 K, much lower than the temperature in the vicinity of Saturn during the time the planet and its satellites were forming, but argon, methane, nitrogen, and carbon monoxide can be captured in varying amounts, depending on the local temperature.

These properties of ice may explain why Titan has the atmosphere it does and why Ganymede and Callisto have no atmospheres. The jovian satellites were simply too warm for ice to trap large amounts of gas, warmer even than their present temperatures, since they were heated during formation by Jupiter itself. At the distance from the Sun where Titan formed, however, gas-rich ice was able to condense and serve as the carrier of nitrogen and the other volatiles now found on Titan.

There is a different possibility to consider, however. Another major difference between Jupiter and Saturn is the larger mass of Jupiter. Thus comets colliding with Callisto and Ganymede would crash with far higher energy than those hitting Titan, simply because of the greater gravitational attraction exerted by the more massive planet. (We discussed this effect in the last chapter in talking about cratering on outer planet satellites.) The result could have been that Titan was able to retain the gases delivered by comets that originated in the Uranus-Neptune region, while on Ganymede and Callisto, the gases escaped into space after the impacts. We will not know which (if either) of these possibilities is correct until the Huygens probe reaches Titan in 2004 and gives us the kind of data on isotope ratios and noble gas abundances that we are using to try to understand the origins of the atmospheres of the inner planets.

Argon may provide the key. Here we mean primordial argon, the isotopes Ar-36 and Ar-38, not the Ar-40 that is 1% of our own atmosphere and produced by the decay of K-40. Consulting our table of cosmic abundances (Table 2.2) we expect to find one atom of primordial argon for every 15 molecules of N_2, if the ices that formed Titan indeed captured a cosmic mixture of volatiles. At present, we cannot tell how much of this noble gas is present, since it is extremely difficult to detect spectroscopically.

An intriguing mystery in all this is the source of Titan's methane. We find almost no methane in the interstellar clouds that seem to represent the material from which the solar system formed. Instead we find CO, some CO_2, carbon itself, and various complex organic compounds. Perhaps the methane was made during the accretion of Titan, through chemical reactions driven by the heat released as impacting bodies gradually accumulated to form the satellite. (The hot interiors of the giant planets easily produce the methane in their atmospheres.) It is an important problem, because we find methane on both of the other subjects of this chapter as well.

16.2 TRITON: APPROACHING ABSOLUTE ZERO

Leaving the Saturn system behind, we take a big jump to Neptune to study Triton, Neptune's largest satellite. Neptune was the god of the sea in the Roman pantheon, and Triton was one of

FIGURE 16.7 This Triton fountain in Rome was made by the famous Renaissance sculptor Giovanni Bernini, whose name appears on the hotel in the background. Bernini might have been pleased to know that the satellite bearing the same name also produces "geysers" (cf. Fig. 16.13)

his children. This association has made Triton, the deity, a favorite subject for builders of fountains (Fig. 16.7).

Basic Data

Because Triton is relatively faint, we knew very little about it until the Voyager 2 flyby in 1989. In those days, Triton's most distinguishing characteristic was its unusual orbit. This satellite moves around Neptune in the direction opposite to the planet's rotation and opposite to the direction that all the planets follow in their orbits around the Sun. Nevertheless, its rotation is still synchronous with its period of revolution: Just as our own Moon keeps one hemisphere facing the Earth, the same side of Triton always faces Neptune. This retrograde motion means that Triton is an irregular satellite, and it is the largest such object in the solar system. The exact diameter measured by Voyager was 2705 km, making Triton a little more than half the diameter of Titan and three-fourths the size of our Moon. It is much larger than Phoebe or the irregular satellites of Jupiter.

Like these smaller satellites, however, Triton must be a captured object, a body that formed in the solar nebula rather than in the disk of material that surrounded Neptune at the time the planet itself was forming. This suggests that Triton may be more closely related to Pluto than to Titan, since Titan appears to be a product of the Saturn subnebula, the disk of material that must have surrounded Saturn at the time the planet formed. We should keep this distinction in mind as we compare these three objects with one another.

Spectroscopic observations from Earth in the early 1980s by astronomer Dale Cruikshank and his colleagues revealed that methane was frozen on Triton's surface. In addition to absorptions from methane ice, however, the spectrum revealed a feature that appeared to be caused by molecular nitrogen (N_2). It was clear that this could not be nitrogen gas because the amount of gas required would have been incompatible with the low temperature corresponding to the methane ice. Thus astronomers concluded that either liquid or solid nitrogen must be present on Triton's surface.

These observations also demonstrated that this satellite must have an atmosphere. The presence of solid methane and solid or liquid nitrogen at the likely temperatures existing on Triton would lead to the existence of a tenuous envelope of gases above the solid or liquid forms of these substances. The situation is analogous to that of water on the Earth. We have liquid water or ice on the Earth's surface and water vapor in the atmosphere, vapor that occasionally makes its presence known through the formation of clouds, rain, frost, and snow.

The View from Voyager 2

The pictures from the Voyager cameras (Fig. 16.8) revealed Triton to be yet another unique world, fundamentally different from the other satellites surveyed by the spacecraft as it toured the outer planets. From a distance, there was the unusual, pinkish color of the disk. Chemical reactions are evidently occurring in the icy methane and nitrogen, producing organic compounds with a reddish coloration. This could have been anticipated from our experience with the pastel shades in Jupiter's clouds, and the orange-brown smog on Titan, but we had never seen these effects on a solid surface.

Getting closer, the spacecraft failed to find any evidence of large impact craters. As in the case of Europa, Triton's surface is relatively young. Close inspection of the full set of pictures revealed that the amount of cratering varies over the surface, but even the most densely cratered areas only approach the crater populations of the lunar maria. Most of the surface is younger than this.

FIGURE 16.8 As Voyager 2 approached Neptune, the pinkish color of Triton and the lack of prominent craters were already apparent in this distant view.

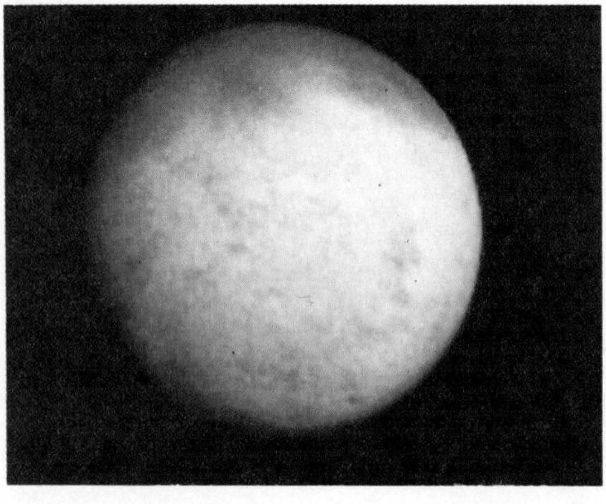

No evidence of nitrogen seas or lakes was found, but by the time these images were obtained, no liquid nitrogen was anticipated, because we knew the surface temperature of Triton was too low. Once the diameter of the satellite had been measured, it was possible to determine the reflectivity from the observed brightness.

Triton has one of the most reflective surfaces in the solar system, comparable to that of Europa or Enceladus. This high albedo means that very little of the radiant energy arriving from the distant Sun is absorbed, so the average surface temperature is just 37° above absolute zero, or 37 K. The melting point of nitrogen ice is 65 K, so only the solid form of nitrogen can exist on Triton. Subsequent ground-based spectroscopy of Triton has revealed the presence of ices of carbon monoxide and carbon dioxide on the surface in addition to those of nitrogen and methane.

For the same reason we expect it on Titan, we think argon should be present on Triton too, this time as ice. But argon is much more difficult to detect by remote observations than are the compounds that have been discovered so far. Water ice is also present, but it is mostly covered by the more volatile ices that dominate the spectrum.

Voyager images revealed several different types of terrain. The aspect of Triton at the time of the flyby was such that the spacecraft viewed mainly the southern hemisphere. Most of the visible surface (the south polar cap) consisted of a relatively flat plain, mottled with darker material (Fig. 16.9). Some of the dark material appeared to have been blown about by surface winds, forming triangular wind streaks reminiscent of those found on Mars (Fig. 16.10). Although these streaks appear dark in the Voyager pictures, measurements reveal that they have reflectivities in the range of 40–75%, far

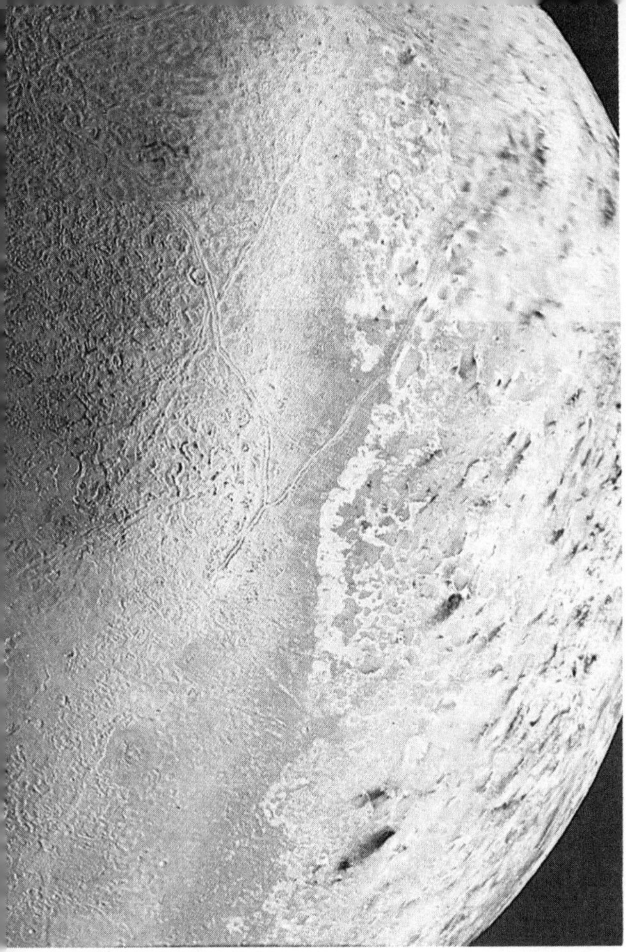

FIGURE 16.9 Closer to the satellite, the Voyager cameras revealed contrasting types of terrain and splotches of dark material on Triton's surface. This mosaic provides a view from the south pole (near right hand side) to latitudes about 30 degrees north of Triton's equator.

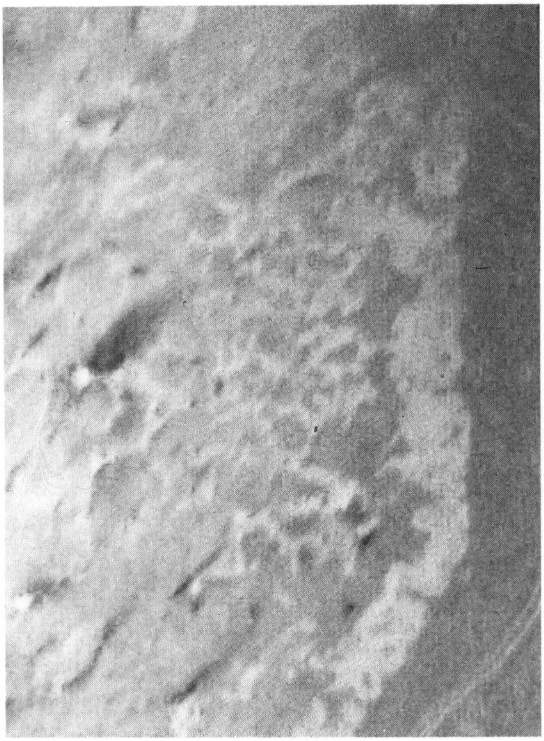

FIGURE 16.10 Seen at close range, the larger deposits of dark material on Triton's surface clearly showed signs of having been blown in well-defined directions by surface winds.

brighter than the surface of our own Moon. It is the contrast with the much brighter surroundings that makes them appear dark.

At higher latitudes above the equator, a smoother, more uniform band of material, neutral in tint, forms as a kind of collar around this polar cap. At still higher latitudes, a rougher terrain appears. Known as the cantaloupe terrain, for obvious reasons, this area is crossed by long groves, such as Slidr, 500 km long and 10 km wide, and is singularly free of splotches of dark material (Fig. 16.11).

Triton's Atmosphere

In addition to the indirect evidence of an atmosphere provided by the observations of wind streaks, the Voyager cameras also recorded the existence of haze and clouds (Fig. 16.12). The ultraviolet spectrometer on the spacecraft found nitrogen emission lines in the satellite's spectrum, confirming that this gas must be the major constituent of Triton's atmosphere. As the spacecraft was occulted by the satellite, attenuation of the radio signal by Triton's atmosphere allowed scientists to determine that the total atmospheric pressure is just 16 millionths of the sea-level pressure on Earth (16 microbars). This

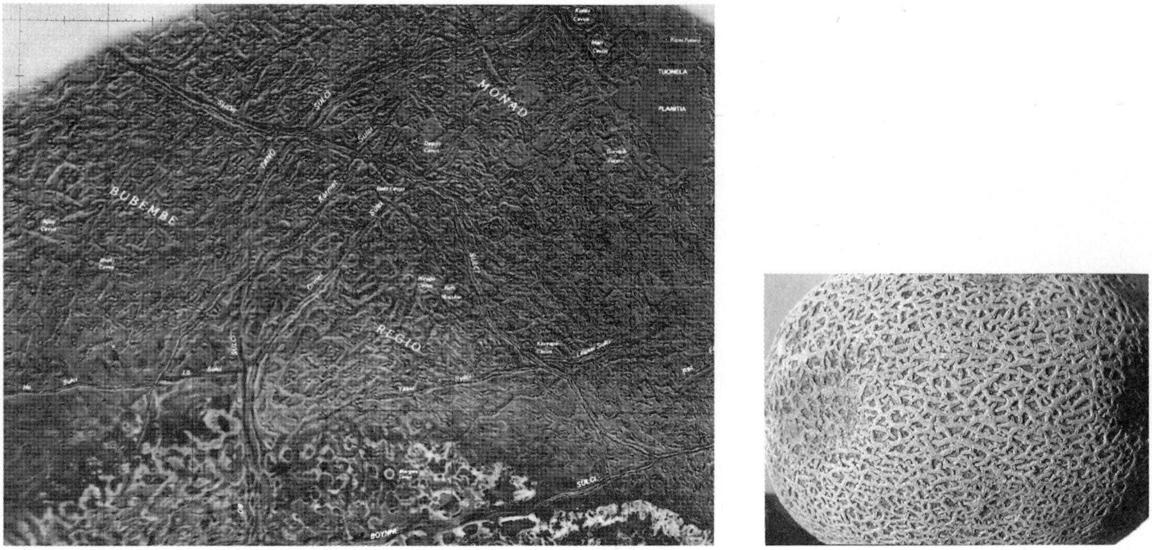

a b

FIGURE 16.11 The cantaloupe terrain on Triton (a) gets its name from its strange resemblance to the terrestrial fruit (b). Note the absence of dark material on this terrain.

FIGURE 16.12 In addition to the wind streaks (Fig. 16.10), the presence of an atmosphere on Triton is revealed by hazes, visible here at the horizon.

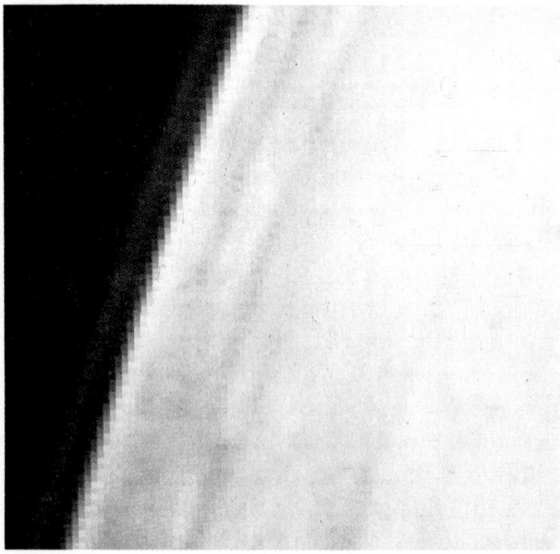

is the maximum pressure that an atmosphere of nitrogen could have with a surface temperature of 37 K. The spectrum revealed a trace of methane, but no evidence of neon, carbon monoxide, or argon.

Triton must have an evolving atmosphere, just as Titan does, and we can therefore anticipate the production of organic compounds from reactions between fragments of methane and nitrogen. Interestingly the composition of the *surface* of Triton is reminiscent of the *atmosphere* of Titan, except that we find larger abundances of CO and CO_2 on the colder satellite. The splotches of dark material on Triton's surface may therefore be the equivalent of the atmospheric smog particles found on Titan. Calculations based on the apparent escape of hydrogen suggest that a 6-m layer of hydrocarbons could have been produced on the surface of Triton over the age of the solar system. In fact, the dark splotches we see must be relatively recent, because they will subside into the ice with time.

Hence a large amount of material produced by the atmospheric chemistry on Triton may be buried under the visible surface.

Curiously, no one has yet found any evidence on Triton for those intriguing organic molecules that make the atmosphere of Titan so interesting. These compounds should be produced as part of the chemistry that leads from the methane and nitrogen that we know are present in the atmosphere to the dark, windblown material that we see on the surface. Or are we missing something here? Could it be that this dark material is actually contributed from an external source, like the dark coating we will find on the leading hemisphere of Saturn's Iapetus (Chapter 17)? The small satellites of Neptune also have dark surfaces. Perhaps the reason we see so little dark material on Triton is because its bright, nitrogen-coated surface was formed relatively recently.

Surely the most remarkable features of Triton discovered by Voyager 2 are geyser-like plumes that rise from the frigid surface into the tenuous atmosphere (Figure 16.13). These plumes appear as narrow columns of dark material that rise about 8 km from the surface, forming small dark clouds at their summits. It is as if these columns are rising smoothly through a very still layer of the atmosphere, eventually reaching a level where they are inhibited from rising any further, so the material they carry "piles up." At this level, there are apparently strong, smoothly flowing horizontal winds that carry the lofted material downwind for 100 to 150 km, while maintaining it in streamers that are about 8 km wide.

What is causing these geyserlike plumes to occur at a local temperature of only 37 K? The structure the plumes exhibit is consistent with a warm source that produces buoyant gas at the base of the plume. We must remember that a "warm source" in this context may still be colder than −200 C! The buoyant gas then rises through the lower atmosphere, where the occultation measurements indicate the presence of a thermal inversion. One can see this same effect on Earth in cold climates after a clear night. During the night, the ground cools rapidly by radi-

FIGURE 16.13 (Above) A plume of dark material rising 8 km above Triton's surface reaches a level where horizontal winds produce a streamer stretching over 150 km downwind. (Below) Arrows mark the positions of the plume and its streamer.

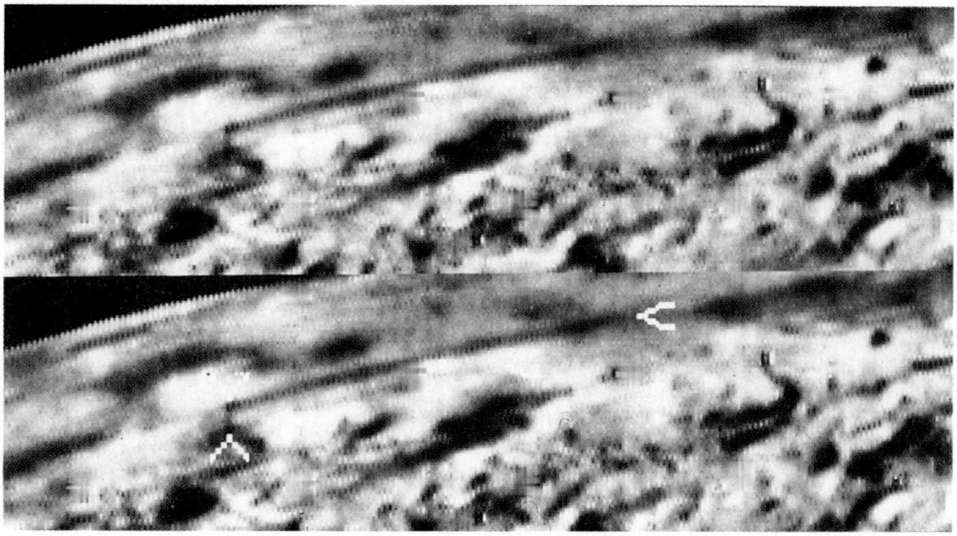

ating heat right through the atmosphere into deep space. The air near the ground therefore becomes colder than the air aloft, leading to a thermal inversion and very stable conditions near the ground—no winds. Smoke from a fire will rise vertically through the still air, piling up at the top of the inversion layer, where it may be blown away by a horizontal wind.

What is the tritonian equivalent of that fire? We don't have a clear answer yet. Suggestions have included dust devils and an unusual greenhouse effect, in which clear nitrogen ice plays the role of a dense atmosphere of gas. It is certainly interesting that the dark material and the geysers appear to occur together, as if some of that buried organic matter could be involved in the geyser phenomenon.

Even if we don't understand them, we can see that these geysers could have parallels on cold comet nuclei and may explain some of the activity that comets exhibit at large distances from the Sun. In any case, the production of a nitrogen atmosphere on Triton also explains why many comets appear to be deficient in this element: Nitrogen will start subliming (evaporating) from comet nuclei even at distances of 40 AU, so by the time the comets reach the inner solar system where we usually first detect them, they will have lost a significant fraction of this highly volatile gas.

Bulk Properties

Using the diameter of Triton and the mass determined from its gravitational effect on the trajectory of the Voyager spacecraft, we find the satellite's density to be 2.1 g/cm^3. This is about the same as the densities of Callisto, Ganymede, and Titan, but these satellites are much more massive than Triton. We would therefore expect Triton's density to be lower. Evidently, it is deficient in ice or it has acquired some additional high-density component that its more massive cousins do not possess. These conditions could occur together if Triton contains an interstellar mixture of organic compounds and oxidized carbon, which would provide components with densities greater than 1.0 and would incorporate some of the oxygen that would otherwise be available to make water ice.

16.3 PLUTO AND ITS MOON: A MOST UNUSUAL SYSTEM

Is Pluto really a planet? This is a question that scientists have been asking ever since the discovery of this unusual object. You might think that since it moves about the Sun in a roughly circular orbit it must be a planet by definition. Comets and asteroids also orbit the Sun, however, and Pluto's orbit is the most eccentric and highly inclined of any of the planets. Pluto is also very small, only two-thirds the size of our Moon. Its rotation period is unusually long; at 6.4 days, it is exceeded only by those of Venus and Mercury. Perhaps Pluto should be considered the largest known asteroid or comet rather than the smallest planet.

Pluto's orbit crosses that of Neptune, so there has been speculation that Pluto is actually an escaped satellite of this much larger planet: Perhaps a close encounter between Pluto and Triton when both were satellites of Neptune sent Pluto out into its own orbit about the Sun and gave Triton its puzzling retrograde orbit. Or perhaps a third body passed through the Neptune system, expelling Pluto and perturbing Triton. However, the orbital motions of Pluto and Neptune are in a resonance that prevents these planets from getting closer than 17 AU, making these intriguing ideas rather unlikely.

Pluto's Moon

The discovery that Pluto has a satellite of its own adds further weight to the current consensus that this planet had an independent origin. The satellite is named Charon (the *ch* is pronounced hard, like the *ch* in *chaos*) for the mythical boatman who carried the dead into the realm of Pluto. It was discovered in 1978 by James

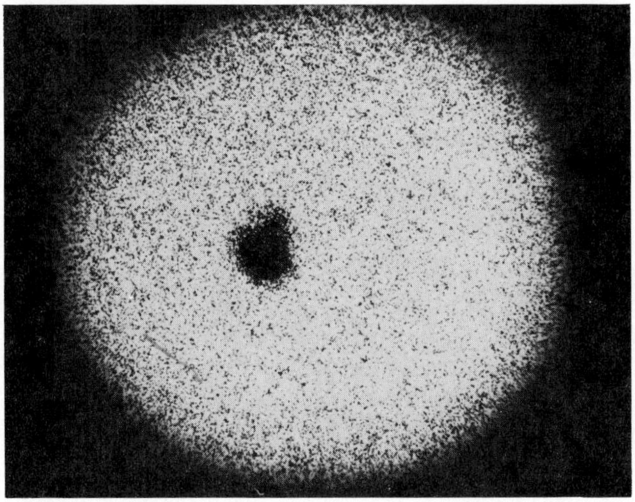

FIGURE 16.14 The discovery photograph of Charon, Pluto's only known satellite. Charon appears as a small bulge at the top of this highly magnified image of Pluto obtained by James Christy and Robert Harrington on July 2, 1978, with the 1.5-m reflector of the U.S. Naval Observatory.

Christy and Robert Harrington. The image of Charon appeared as a small bump on the otherwise circular photographic images of Pluto (Fig. 16.14) obtained at the Naval Observatory's Flagstaff station, less than 4 miles from the site of the discovery of Pluto itself.

Beginning in January 1985, the orientation of the orbit of Charon caused the satellite to pass alternately in front of and behind Pluto (Fig. 16.15). Careful observations of these eclipses and occultations, which continued through 1991, provided data on the relative size and brightness of the two objects, as well as on the orbit of Charon around Pluto and crude maps of Pluto itself. Improved infrared cameras and spectrometers are also coming into play, so we are steadily learning more and more about these distant objects even without a space mission.

FIGURE 16.15 As Pluto moves along its orbit around the Sun, the aspect of Charon's orbit changes as viewed from Earth. From 1985 through 1991, a series of eclipses and occultations occurred.

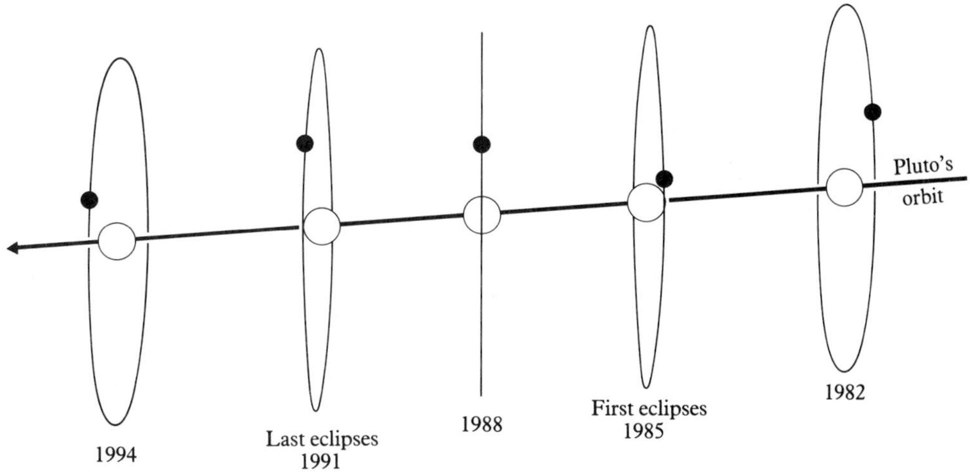

Charon orbits Pluto at a distance of just under 20,000 km, closer than any other natural satellite to its planet except Mars's Phobos. The period of Charon's revolution about Pluto is exactly equal to the period of the planet's rotation. Here is another case of tidal resonance in the solar system, and one that is different from all the others we know. It means that Charon is in a synchronous orbit that always keeps it above the same region on Pluto. Like two ballroom dancers, Pluto and Charon face each other as they waltz around the Sun. We give such geosynchronous orbits to some of our artificial satellites so they can function as television relays, beaming down programs to fixed antennas maintained by hotels and private individuals in areas remote from broadcasting stations. We don't expect a thriving television industry on Pluto, however. Instead, it appears that Charon was able to pull Pluto into a tidal lock because of its large relative mass, about one-tenth the mass of Pluto. This the largest satellite/planet mass ratio in the solar system, with Moon/Earth being second at 1/80.

Mass of Pluto

Until the discovery of Charon, the mass of Pluto could only be estimated indirectly by the apparent effect of Pluto on the motions of Uranus and Neptune (Section 14.1). These estimates were very uncertain, with values ranging all the way from 0.2 to 7 Earth masses. The corresponding uncertainties in the density of Pluto made it impossible to determine the composition of this planet.

By studying the motions of Pluto and Charon, however, it is possible to use Newton's laws to determine Pluto's mass. The result is a mass of 1/400 that of Earth, much smaller than the previous estimates. When this mass is combined with the occultation-derived diameter (2390 km), a density of 2.1 g/cm^3 is derived, essentially identical to the density of Triton. Evidently, both of these distant worlds are made of the same kinds of material.

The orbit of Charon defines the inclination of Pluto's axis of rotation, since it is reasonable to assume that a system that is tidally locked in this fashion will have the satellite's orbit in the equatorial plane of the planet. This was another surprise: The inclination of Charon's orbit to the orbit of Pluto around the Sun is 112°. This means that Pluto, like Uranus, has its rotational axis very nearly in the plane of its orbit (Uranus is at 98°). Three planets then rotate backward: Venus, Uranus, and Pluto.

Surface and Atmosphere

Long before Charon was discovered, the rotation period of Pluto was determined very accurately through observations of periodic changes in the planet's apparent brightness. The global reflectivity of Pluto varies from 0.3 to 0.5 as the planet rotates, compared with the constant values of 0.1 for our Moon, a rocky object, and 0.8 for Triton, Pluto's twin.

The first crude spectrophotometry of Pluto in 1975 revealed the presence of methane ice on the planet's surface. Observations made with improved instrumentation in 1992 by author Owen and colleagues show that ices of carbon monoxide and nitrogen are present as well (Fig. 16.16). Thus the composition of the surface of Pluto seems similar to that of Triton, with the difference that there is more dark material on Pluto and the methane to nitrogen ratio is higher than on Triton. Careful study of the eclipses and occultations of Pluto and Charon by one another has permitted the development of a crude map of the distribution of dark material on the planet's surface (Fig. 16.17).

The similarity in the surface ices on Pluto and Triton means that the atmospheres of the two objects must also be similar. Although we don't

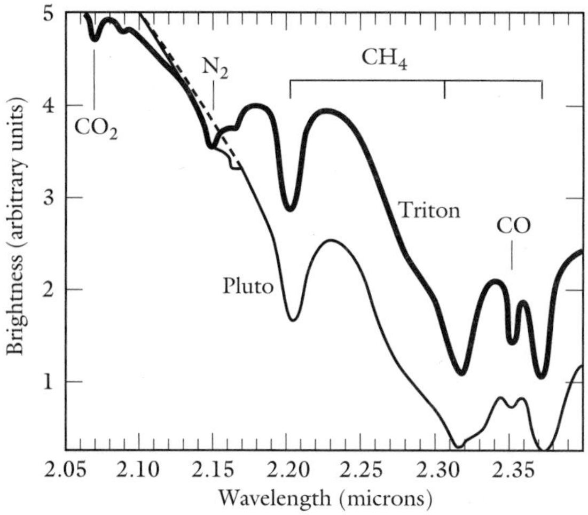

FIGURE 16.16 This spectrum of Pluto was recorded in 1992 with the United Kingdom infrared telescope on Mauna Kea. It shows the presence of ices of CH_4, CO, and N_2.

FIGURE 16.17 An artist's concept of a future flyby of Pluto and Charon (left foreground). The appearance of Pluto is based on observations of eclipses and occultations of Pluto and Charon plus Voyager images of Triton.

have any spacecraft data for Pluto that would allow us to establish this case, the fact that nitrogen ice is present means that this gas must dominate the atmosphere: Nitrogen has the highest vapor pressure, so it will be the most abundant gas. This conclusion is consistent with the observations of a stellar occultation that took place in 1988. The dimming of the star's light before it disappeared behind the planet proved that a thin, distended atmosphere was present. If nitrogen is indeed the major constituent, the structure of the atmosphere will be similar to that of Triton, leading to a surface pressure again in the 1–20 microbar range for a surface temperature of 35–40 K.

Observations of Pluto and Charon together and Pluto by itself (with Charon in eclipse or occultation) have allowed a determination of the surface composition of the satellite. The surprising result is that Charon is coated with water ice. This common substance is difficult to detect in spectra of Pluto, although it is clearly present on Triton. There is no evidence (yet) for solid methane on Charon. Evidently the process(es) responsible for the formation of this satellite heated it sufficiently to drive off the more volatile compounds. Alternatively, the methane on Pluto and Triton may have been made by impact heating (as we suggested for Titan), and the heat released in such events on smaller Charon was itself sufficient to drive off the methane. New observations of other small bodies in the outer solar system should provide the answer.

The eccentricity of Pluto's orbit must have a large effect on the atmosphere. At perihelion, Pluto's distance from the Sun is 29.7 AU, bringing it closer than Neptune, which occupies a nearly circular orbit at 30.0 AU. At the other extreme, aphelion, Pluto is at a distance of 49.7 AU. This means that the intensity of sunlight at the surface of the planet varies by a factor of about three around the orbit. Since we are dealing with an atmosphere that is nothing more than vapor in equilibrium with its ice, the resulting decrease in the temperature of the ice will essentially freeze out the atmosphere. We are very fortunate to be studying Pluto when it is so close to perihelion, which occurred in 1989.

At the Edge of the Planetary System

In Pluto we have found an outer planet that we could stand on, instead of being limited to exploration of a thick atmosphere from a balloon. The sky overhead would be black, even at midday, since the atmosphere is very thin and the Sun very far away. Seen from this vast distance, our star would be some 1600 times fainter than it appears from Earth, too small even to show a disk without optical magnification. Charon would be visible from only one hemisphere of Pluto, as a consequence of the synchronous rotation of the planet.

Pluto is the last large body in the solar system. Beyond this frigid world, there are only comet nuclei. The closest of these are relatively nearby in the Kuiper belt, still confined roughly to the plane of the planetary system. A thousand times farther out, the entire system is enveloped by the huge, spherical cloud of comets, first conceived of by Jan Oort.

It is as if we are on the beach of the outermost island of an archipelago, looking out to sea. There are a few rocks just offshore, their dark heads rising above the waves. Beyond them, the water stretches out unbroken to the horizon. From time to time, some pieces of driftwood wash up on our shore, but we know from examining them that these random wanderers began their journeys in the forests of the islands behind us. They do not signal the presence of new continents just over the horizon. The closest other land is very far away, and with the primitive means of transportation we have at our disposal, we will never reach it.

Meanwhile, we can try to find out more about Pluto. Improvements in ground-based and orbiting telescopes during the next ten years are bound to provide some new insights. The real increase in our knowledge will come from a spacecraft mission, currently planned for the early years of the next century. If this dream is realized, we will have completed a reconnaissance of the entire solar system by the year 2010.

16.4 QUANTITATIVE SUPPLEMENT: VAPOR ATMOSPHERES

The atmospheres of Pluto and Triton are like that of Mars because the gases that compose them are in equilibrium with the solid forms of the same substances. On Earth, the range of temperature and pressure in the atmosphere and on the surface keeps all of the major atmospheric constituents permanently in the gaseous state. Water, on the other hand, can be a solid, a liquid, or a gas, depending on the local conditions. When a gas is in equilibrium with its solid or liquid state, we commonly refer to it as a vapor. This is the situation for water on Earth and nitrogen on Pluto and Triton. Nitrogen is in the solid form on the surfaces of these bodies and is the dominant gas in their atmospheres.

One of the important consequences of this condition is the relationship it imposes on the local temperature and the pressure exerted by the vapor. As we increase the temperature, more evaporation occurs, and the pressure increases. At a given temperature, a limit to the vapor pressure is set by the point at which condensation begins. This limiting pressure is known as the saturation vapor pressure.

The relationship between the saturation vapor pressure (P_s) and the temperature (T) is a function of the latent heat of sublimation or evaporation (L_{12}) of the substance involved. (The latent heat is the amount of energy required to produce a change of state from solid or liquid to vapor.) This relationship is known as the Clausius-Clapeyron equation and has the form

$$(1) \qquad \ln P_s = \frac{-m_v L_{12}}{R^* T} + C \text{ or}$$

$$(2) \qquad P_s = Ce^{-(m_v L_{12}/R^* T)}$$

where m_v is the molecular weight of the vapor, R^* is the universal gas constant (8.317×10^{-8} cgs), and C is a constant determined by measuring the saturation vapor pressure at some standard temperature, such as 273 K.

Now you can see why we are fortunate to be observing Pluto near perihelion. The change in surface temperature has an exponential effect on the surface pressure: A small difference in T causes a large change in P_s. For the specific case of the solid-vapor equilibrium of molecular nitrogen, we can approximate equation (1) as

$$(3) \qquad \log P_s = \frac{-344}{T} + 4.53$$

where P_s is in bars. Thus for $T = 37$ K, $\log P_s \approx 9.30 + 4.53 = -4.77$ and $P_s = 17\,\mu$m bar. Reducing the temperature by 10% (2 K) reduces the pressure by a factor of 3.

SUMMARY

Titan, Triton, and Pluto seem somehow related to one another by virtue of their similar compositions. All three of these bodies have nitrogen-dominated atmospheres, despite their relatively small masses. The reason they can retain these gaseous envelopes is found in their low surface temperatures, ranging from 94 K for Titan to the frigid 37 K exhibited by Triton and Pluto.

Titan, the largest of the three, has a nitrogen-methane atmosphere with a surface pressure 1.5 times the sea-level pressure on Earth. This atmosphere contains a thick layer of photochemical smog, so that we have no pictures of Titan's surface. The nitrogen and methane are being continuously broken apart by solar ultraviolet radiation and electron bombardment from Saturn's magnetosphere. The fragments recombine to form a rich variety of organic compounds in the atmosphere, including the smog particles. This chemistry may resemble some of the reactions that took place on the primitive Earth, ultimately leading to the origin of life on our planet.

Neptune's largest satellite, Triton, is in a much colder environment than Titan. Its atmosphere is correspondingly thinner, because most of the nitrogen is now frozen out on the surface. Yet even at 37 K, chemical reactions are taking place that slowly destroy methane and nitrogen, building up complex organic compounds that we find in the form of dark, windblown material on Triton's surface. In addition to frozen nitrogen and methane, the surface includes ices of carbon monoxide and carbon dioxide. Most remarkable are the plumes of dark material rising into Triton's 16-microbar atmosphere and producing wind trails over 150 km long. We don't yet understand how these plumes are produced or whether we should expect to find something similar below the smog layer on Titan.

Pluto is the last planet in the solar system. It closely resembles Triton, which may be a Pluto clone captured by Neptune into its present, retrograde orbit at some early epoch. Pluto also has a surface coating dominated by nitrogen ice, with smaller amounts of carbon monoxide and methane. Solid carbon dioxide has not been found on Pluto, which also has a higher proportion of methane ice than the surface of Triton. The synchronous orbit of Pluto's satellite, Charon, is unique in the solar system, as is the large size of Charon relative to its planet.

KEY TERMS

photochemical smog prebiotic chemistry

Review Questions

1. Why don't we have any pictures of the surface of Titan?

2. What are the main constituents of Titan's atmosphere? What kind of chemistry takes place in that atmosphere and why?

3. Explain how Titan's atmosphere changes with time. What do you think will ultimately happen to it?

4. Compare the bulk composition of Titan, Triton, and Pluto. Do you regard these three objects as essentially identical except for their temperatures? Why or why not?

5. Describe the plumes on Triton. Have you ever seen anything like them on Earth? Explain.

6. Consider a "thought experiment" in which you move Titan to the position of Triton. What do you think would happen? How would Titan appear to passing spacecraft?

7. What is unusual about Pluto and Charon, considered simply as a dynamical system (planet and satellite)? What is unusual about the surface composition of these two bodies?

8. Why didn't scientists know the mass of Pluto before Charon was discovered?

Quantitative Exercises

1. Review Section 6.6 and calculate the surface temperatures of Pluto and Titan.

2. Using the equations of Section 12.7, derive an expression for the smallest molecular mass that Pluto could retain for 4.5 billion years. Repeat for Titan. How do these values compare with observations?

3. Calculate the extremes in surface temperature experienced by Pluto in its orbit about the Sun. Use the equations in Section 16.4 to calculate the change in atmospheric surface pressure corresponding to these variations in temperature.

4. Can you think of any cosmically common gases that would have higher vapor pressures on Pluto than nitrogen? Would any of these be stable on Pluto for 4.5 billion years? Justify your answer quantitatively and explain why these gases aren't present.

Additional Reading

*Burns, J.A., and M.S. Matthews, eds. 1986. *Satellites.* Tucson: University of Arizona Press. Articles by experts on all the satellites; an excellent reference for Titan, but the material on Triton and Pluto is somewhat dated.

*Cruikshank, D., and M.S. Matthews, eds. 1995. *Neptune.* Tucson: University of Arizona Press. A collection of technical papers that includes articles about Triton.

*Gehrels, T., and M.S. Matthews, eds. 1984. *Saturn.* Tucson: University of Arizona Press. Includes a series of technical articles about Titan.

Hoyt, W.G. 1980. *Planet X and Pluto.* Tucson: University of Arizona Press. This book discusses the general problem of searching for planets beyond the orbit of Neptune, including the discovery of Pluto.

Littmann, M. 1990 (revised printing). *Planets Beyond: Discovering the Outer Solar System.* New York: Wiley Science Editions. The revised printing of this semipopular account includes the Voyager results on Triton.

Tombaugh, C.W., and P. Moore. 1980. *Out of the Darkness: The Planet Pluto.* Harrisburg, Pa.: Stackpole Books. An account of the discovery of Pluto by the man who discovered it, assisted by a well-known popularizer.

*Indicates the more technical readings.

17

Small Satellites

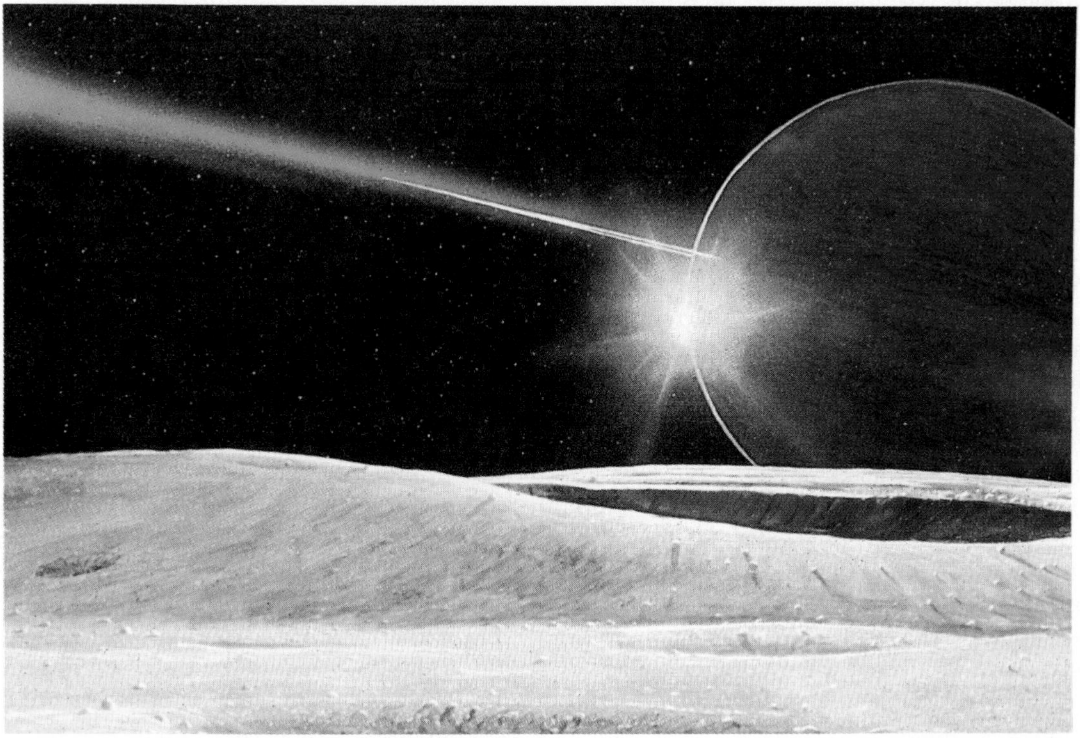

Astronomer-artist William Hartmann gives us an idea of the view of Saturn's bluish E Ring that we would enjoy from the surface of the satellite Tethys. Enceladus can be seen in the "half-moon" phase, buried inside the ring, while the classical ABC rings are visible close to the planet, just above the glare of the Sun.

In the previous two chapters, we have discussed the four large satellites of Jupiter, as well as Triton, Titan, and the Pluto-Charon system. This means that we have now considered all the objects in the solar system with diameters larger than 2000 km. To complete our survey of individual objects, we will now review the characteristics of the smaller satellites in the outer solar system. (We discussed Phobos and Deimos, the two moons of Mars, together with the asteroids in Chapter 6.) As most of these bodies lack individual personalities, we will consider them in groups.

These different satellite groups do have family traits, since we are dealing with several objects in each case. As we consider the different families in turn, you should recall the basic distinction we made in Chapter 15 between regular satellites, which move in direct, nearly circular orbits close to the plane of their planet's equator, and irregular satellites, whose orbits exhibit either high inclinations (including the possibility of retrograde motion) or high eccentricities, or both. It is also important to remember that the apparent ordinariness or normality of an object may simply be the result of inadequate observations, as we discovered when Voyager 2 made a close flyby of Miranda, a small satellite in the Uranus system.

Even from a distance, however, we can sometimes identify fundamental characteristics that test our abilities to provide satisfactory explanations for what we observe. The simple distinction between high and low reflectivity is one of these characteristics, since it carries implications for the composition and evolutionary history of these bodies. Picking the extreme example, we note that the small satellites of Saturn are surprisingly bright. We have seen that even comet nuclei, which we know are composed largely of water ice, are covered with a layer of dark, presumably carbon-rich material. Exposed rocky surfaces are always dark. Ice can be bright if it has a relatively recent origin, like the surface of Europa, but we will find bright, heavily cratered (and therefore old) surfaces on many of Saturn's satellites. Some kind of sorting or coating process has apparently occurred in this system that did not occur around Jupiter, Uranus, or Neptune.

17.1 THE SATELLITE SYSTEM OF SATURN

The five medium-size inner satellites of Saturn, together with the rings and about a dozen small satellites, form a regular and tightly knit group. This section focuses on four of these medium-size satellites: Rhea, Dione, Tethys, and Mimas. The special cases of two unusual Saturn satellites, Enceladus and Iapetus, are discussed in Section 17.2, and the small satellites of Saturn are covered in Section 17.4. We will return to the rings of Saturn in Chapter 18.

These unusual names were taken from Greek myths that describe the establishment of the Olympian gods: Zeus, Hera, Aphrodite, and so forth. To achieve their ascendancy, the Olympians had to fight with the Titans, a family of gods that included Kronos (Saturn), Rhea, Dione, Iapetus, Tethys, and Hyperion, who were assisted by giants, such as Enceladus and Mimas. This tradition has been continued with the satellites discovered in recent times, except for Janus, which was named after the two-faced Roman god who symbolizes the arrival of the new year.

When we consider the Saturn system, we are naturally inclined to make comparisons with Jupiter. In Section 15.1 we noted that Saturn has only a single giant satellite, while Jupiter has the four Galilean moons. It is also apparent that the satellites of Saturn (excluding Titan) show less variety than those of Jupiter. In particular, they do not exhibit the range of density, and hence of bulk composition, of the Galilean satellites (Table 17.1). These differences, as well as the existence of Saturn's extensive rings, clearly set this system apart from that of Jupiter. The reasons may have to do with the smaller mass of Saturn, the lower temperatures in the nebula that surrounded it as the system formed, or perhaps other processes not yet identified.

TABLE 17.1 Medium-size satellites of Saturn

Name	Period (days)	Diameter (Moon = 1)	Mass (Moon = 1)	Density (g/cm^3)
Mimas	0.94	0.11	.0005	1.2
Enceladus	1.37	0.14	.0011	1.2
Tethys	1.89	0.30	.010	1.3
Dione	2.74	0.32	.015	1.4
Rhea	4.52	0.44	.034	1.3
Iapetus	79.3	0.41	.026	1.2

Rhea

Rhea, as the largest Saturn satellite after Titan, provides a convenient point of reference for studying the medium-size icy satellites of the outer solar system. Its diameter is 1530 km, just half as big as Europa, but it is 60% larger than the largest asteroid, Ceres. Several other satellites of Saturn and Uranus have about the same size and

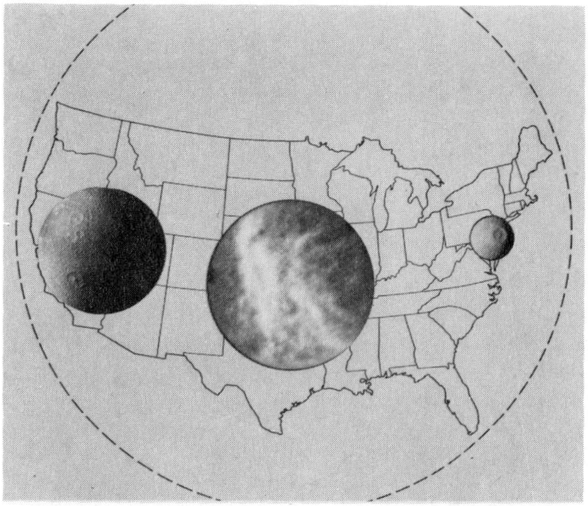

FIGURE 17.1 Three of Saturn's medium-size satellites superimposed on a map of the United States. From the left, they are Dione, Rhea, and Mimas. The dashed circle is the size of Earth's Moon.

mass. Figure 17.1 compares the size of Rhea and the other medium-size Saturn satellites with a map of the United States.

The density of Rhea is only 1.3 g/cm^3, substantially lower than that measured for any solid body discussed previously in this book (although we presume comet nuclei have similarly low densities). This does not mean, however, that the composition of Rhea is much different from that of Titan or the large icy satellites of Jupiter; rather, the density of Rhea is less primarily because of its smaller size. Ice is a fairly compressible material, resulting in a higher density for large satellites. Rhea, being smaller, has a less compressed interior. Taking this effect into account, we can estimate that Rhea, like Titan, has a composition that is roughly half water ice and half silicate minerals and metal.

Rhea is highly reflective (60%), and its infrared spectrum is dominated by the absorptions of water ice. That this satellite is nearly as bright as Europa might be taken as evidence that it too has occasionally resurfaced itself with fresh ice, but it is equally probable that the environment at Saturn is simply less dirty than at Jupiter.

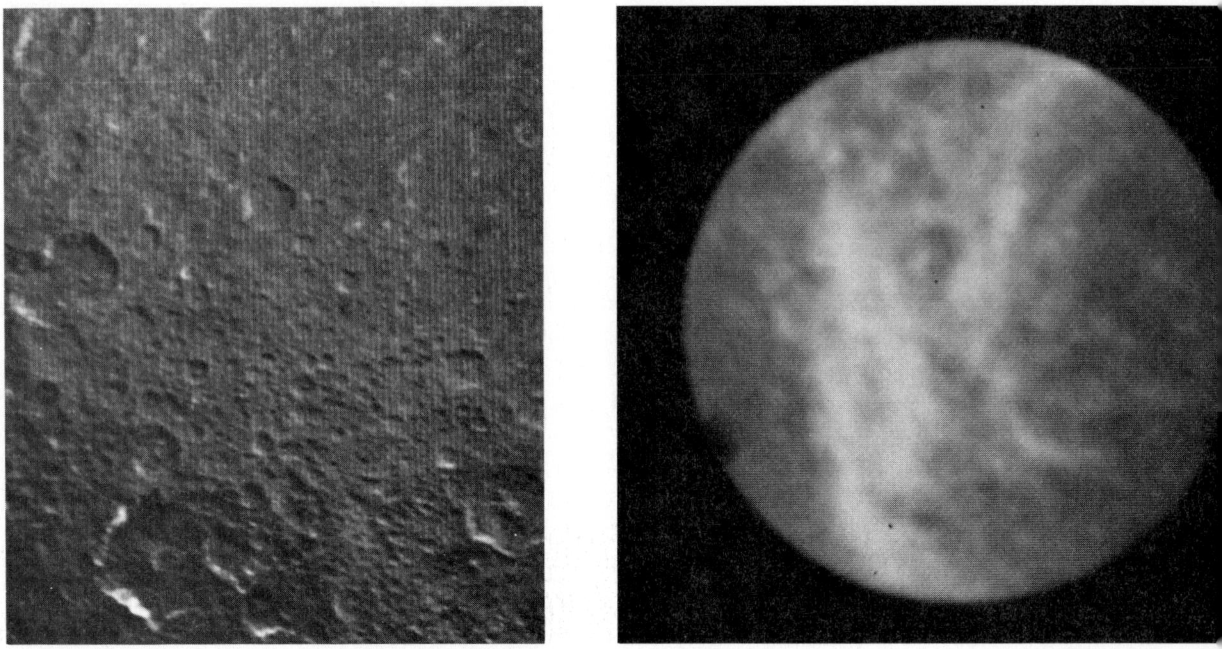

a b

FIGURE 17.2 (a) Craters are shoulder to shoulder in this close-up view of Saturn's satellite Rhea. The smallest features are 2.5 km across (cf. Fig. 15.3). (b) A distant view of Rhea, showing the bright streaks known as wispy terrain.

If most of the impacting debris is icy, a satellite surface can remain relatively clean even after billions of years of exposure.

The surface of Rhea is heavily cratered, as shown in Fig. 17.2(a). Unlike the subdued craters of Callisto and Ganymede, the craters of Rhea look remarkably lunarlike. In fact, most geologists, looking at this picture, would be hard put to distinguish Rhea from Mercury or the Moon, unless they were told that they were viewing brilliant white ice instead of a dark gray surface of rock. At the low temperatures (about 100 K) prevailing at the distance of Saturn, ice behaves very much like rock when a crater-forming impact takes place. The colder ice is, the less plastic and the more brittle it becomes.

The crater density on Rhea is about 1000 10-km craters per million km^2, as high as the value for the lunar highlands. Further, there is little if any indication of internal geological activity to erase or distort craters. Lack of geological activity should not surprise us; after all, what is there to heat a small icy world out at the distance of Saturn?

In the midst of all this evidence of a dull history, however, Rhea does display one peculiarity. While its leading side—the hemisphere that faces forward in its orbit as it moves around Saturn—is unmarked except by craters, its trailing side shows prominent streaks (Fig. 17.2b). Although the Voyager pictures are not good enough to reveal the exact nature of these markings, it seems probable that they are the remnants of some long-ago episode of activity in which water was released from the interior and condensed on the surface. Similar light streaks are observed on the trailing hemisphere of the next inner satellite, Dione.

Dione, Tethys, and Mimas

Dione is darker and smaller than Rhea (1120 km in diameter) and slightly denser (1.4 g/cm^3), but in other respects it is a very similar icy world. Its surface is heavily cratered, and on its trailing hemisphere the bright streaks (what geologists call "wispy terrain") are even more prominent, reflecting up to 70% of the incident sunlight and contrasting strongly with the underlying surface, which reflects only half as much (Fig. 17.3a).

The most significant new feature of Dione is its evidence of internal geologic activity. Over substantial parts of its surface, flooding or some other type of resurfacing has obliterated old craters, producing lower crater densities reminiscent of the lunar maria or the old lava plains of Mercury. Also visible on some Voyager pictures are valleys of unknown origin (Fig. 17.3b), apparently associated with the bright streaks. Perhaps the bright material represents frozen gases along the rims of these cracks in the crust.

Tethys, the next satellite in toward Saturn, is a close twin of Dione in almost every way. It too has a mixture of surface terrains, some very heavily cratered, others modified by geologic activity. The reflectivity of the icy surface is very high: 80%. Instead of several small valleys, however, Tethys has one giant valley system (Ithaca Chasma) that stretches three-quarters of the way around the satellite. The surface area of Ithaca Chasma, which is about 100 km wide over most of its length, is comparable to that of Valles Marineris on Mars. Such a tectonic feature could have been formed if the interior of Tethys expanded enough to increase the satellite's surface area by about 10%, not an impossibly large figure for some kinds of water ice, but we don't know why the surface cracked in only one place.

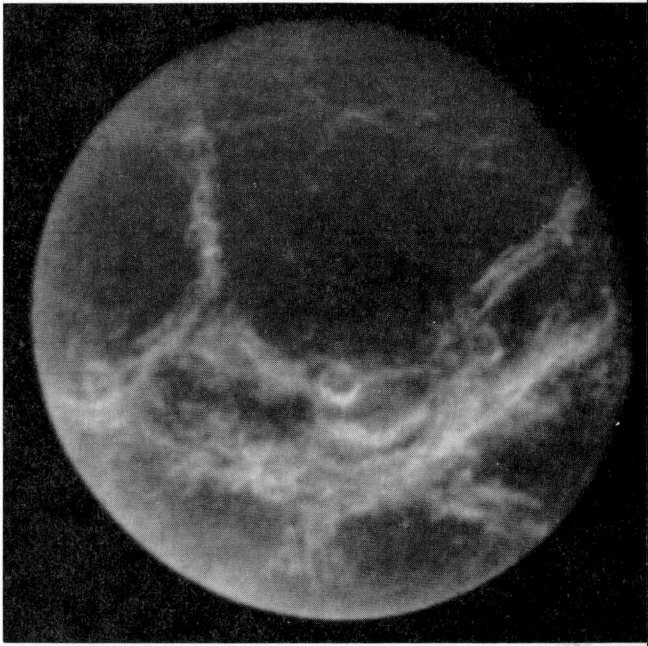

a

b

FIGURE 17.3 (a) A distant view of Dione shows the same type of wispy terrain seen on Rhea (Fig. 17.2b). (b) In addition to the expected craters, the surface of Dione shows sinuous valleys (lower right) sometimes bordered with bright deposits (upper left).

The innermost of the large satellites of Saturn, Mimas and Enceladus form a pair as do Tethys and Dione (Table 17.1). Both have diameters of about 500 km, and both have experienced the same environmental effects. Yet they are very different; so much so that we defer discussion of Enceladus to the next section.

Mimas is heavily cratered (1000 or more 10-km craters per million km^2) and shows little evidence of internal activity. Its most notable feature is a single very large crater on the forward-facing, or leading, hemisphere which has a diameter about one-third that of the satellite itself (Fig. 17.4). This is one of the largest craters in the solar system, not in absolute dimensions, but in relation to the size of the body struck. The energy released by such an impact does not fall far short of that required to shatter and disrupt Mimas itself.

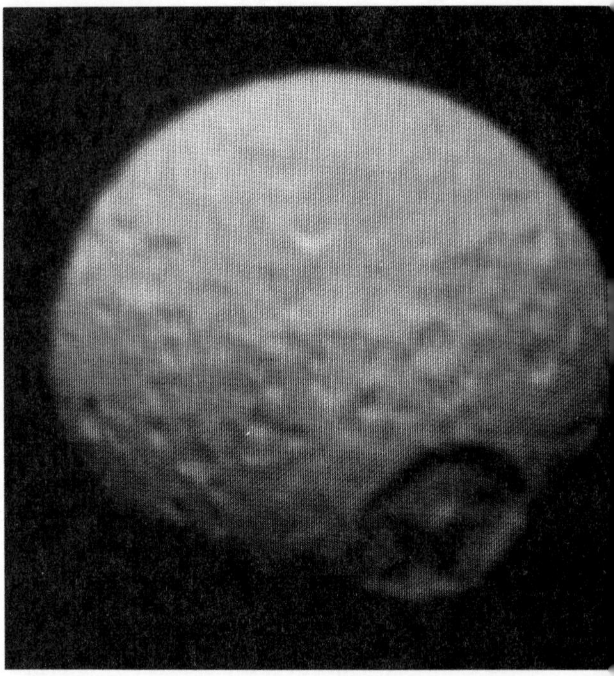

FIGURE 17.4 Mimas has one large crater, called Herschel in honor of the satellite's discoverer. If the object that made this crater had been slightly larger, it would have shattered Mimas entirely.

Comparing These Four Satellites

These four satellites of Saturn have many things in common. They all have surfaces of relatively pure water ice, and from their densities we can infer a bulk composition that is about one-half water ice as well. All are heavily cratered, testifying to a heavy meteoroidal bombardment. However, both Dione and Tethys show evidence of a surprising amount of past geological activity, which may have occurred only during the earliest period of planetary history.

Relative to the Galilean satellites of Jupiter, these four inner satellites of Saturn make up a compact group. It is interesting to ponder the effects of being so close to Saturn. Probably the most important influence of the giant planet is gravitational, as noted in Section 15.1. Passing meteoroids are pulled inward, converging toward the planet and increasing both the impact rate and the impact speeds for the inner satellites. The closer a satellite is to Saturn, the larger these effects. Thus the same flux of cometary impacts that will just build up a heavily cratered surface on Rhea will result in a much more severe pounding of Mimas, where Saturn's gravity produces a convergence of impacting bodies.

The presence of the very large crater on Mimas further reminds us of its violent history. It is easy to imagine that Mimas only barely escaped destruction during the early period of heavy bombardment, and that if there were once satellites of comparable size even closer to Saturn, they may well have been broken apart shortly after they formed.

17.2 ENCELADUS AND IAPETUS: TWO PUZZLING SATELLITES

Two of the medium-size Saturn satellites stand out sharply from their rather bland, icy, cratered siblings. Enceladus appears, in spite of its small size, to have remained highly active geologically. It also has a ring, the E Ring of Saturn, associated with it. Iapetus, in the outer part of the Saturn system, is a unique two-faced moon, with its leading hemisphere covered by a mysterious deposit of black material.

Enigmatic Enceladus

Because of their similar sizes, we would expect Mimas and Enceladus to resemble each other. Both are icy bodies with diameters of about 500 km, and both are close enough to Saturn to have experienced similar heavy bombardment by cometary projectiles. But they are not alike at all; in fact, Enceladus offers some major challenges to understanding the Saturn system.

Even from a distance, Enceladus looks strange. Its surface is blindingly white, reflecting nearly 100% of the incident sunlight. As far as is known, Enceladus has the highest reflectivity of any naturally occurring surface in the solar system, being 20% more reflective than Triton. The high albedo results in a low surface temperature, approximately 55 K. In addition, Enceladus seems to be the source of a ring around Saturn, appropriately called the E Ring (Fig. 17.5a). This faint, tenuous cloud of very small, spherical particles fills much of the space between the orbits of Mimas, Enceladus, and Tethys, with its maximum brightness at the orbit of Enceladus. Since the E-Ring particles are so small, they cannot survive for long in their present orbits; instead, radiation pressure would be expected to disperse them like the dust in the tail of a comet. We therefore conclude that either there is a continuing source of particles or the E Ring is young, having been formed by some recent event. Either way, Enceladus seems implicated as the most likely source of the E-Ring particles.

Seen at closer range, Enceladus lives up to its billing as a weird place (Fig. 17.5b). Over much of the surface, all impact craters have been erased, a sure sign of high levels of geological activity. Although the total cratering flux is not well known in the Saturn system, it appears that these smooth plains are no more than a few hundred million years old, about the same age as the early dinosaur fossils on Earth. Some of these smooth plains also show ridges and flow marks. Here, surely, we are seeing evidence of water volcanism. Even on the cratered terrain of Enceladus, many individual craters have been deformed by plastic flow in the crust, the result of higher temperatures below the surface.

It is tempting to compare Enceladus to Io, another outer planet satellite known to be currently active. Granted, the current activity rate on Enceladus is much lower, but then Enceladus is a much smaller body than Io. Both objects present essentially the same problem: to find a relatively large source of internal heating that is capable of maintaining geologic activity in spite of the rapid escape of heat from the interior.

In the case of Io, that mechanism has been identified as tidal heating. Could the same thing be happening to Enceladus? The difficulty is that nearby satellites do not force Enceladus to revolve in a noncircular orbit the way Io is constrained by Europa and Ganymede. Efforts have been made to construct scenarios in which Enceladus is occasionally forced into an eccentric orbit, resulting in episodic heating, but to date these ideas have failed to convince skeptics. The question of how Enceladus stayed hot therefore remains an open one.

The origin of the E Ring is also uncertain. Calculations of the lifetime of the particles suggest that the event or events that formed the ring took place within the past few tens of thousands of years, extremely recently on any astronomical or geological timescale. One suggestion is that a cratering event punched through to the liquid interior of Enceladus (if there really is a liquid interior), producing a fountain of fine water droplets that froze to produce the E Ring. The same mechanism might be responsible for the high reflectivity of the surface of Enceladus, which is coated with this fresh layer of tiny ice

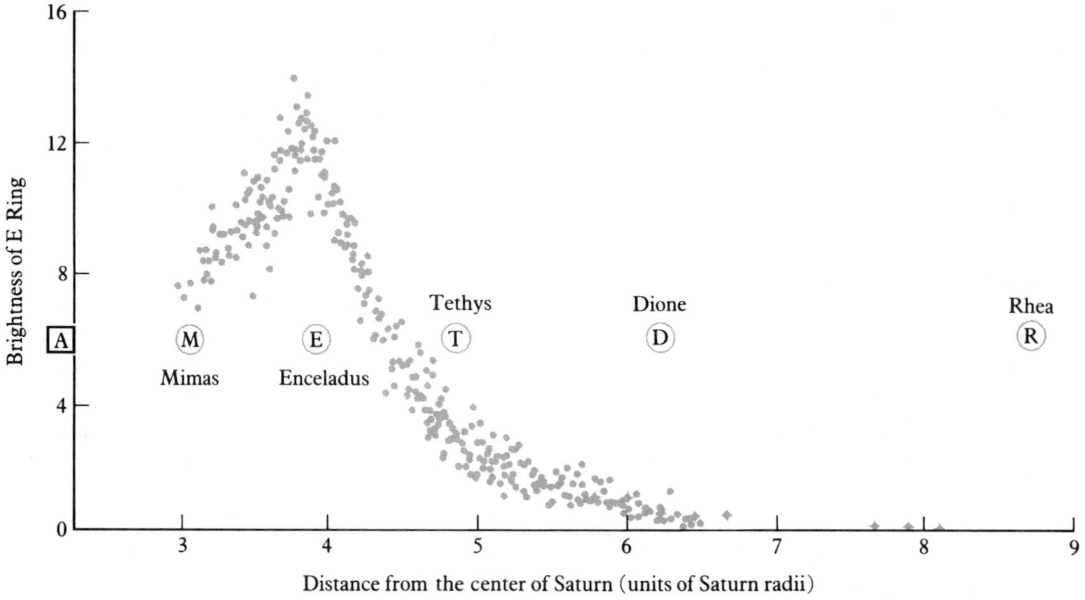

a

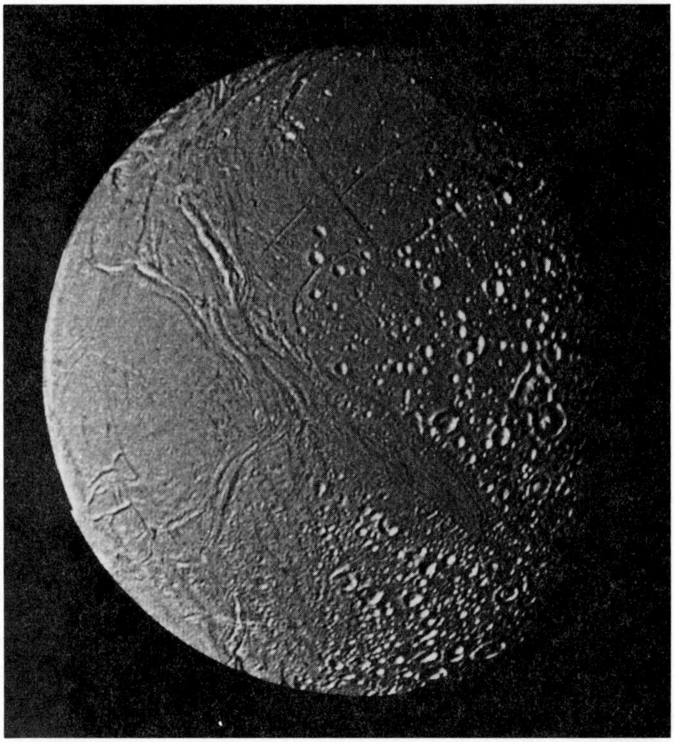

b

FIGURE 17.5 (a) Saturn's E Ring is closely associated with the satellite Enceladus, as revealed in this plot of ring brightness versus distance from the planet. Positions of the other satellites and A Ring (box on the *y*-axis) are shown for scale. (b) Enceladus is one of the most mysterious moons yet discovered. A large portion of the surface seems melted, with no evidence of impact craters, but the cause of this smoothing has not yet been established.

spheres. If this reasoning is correct, then the E Ring and the high reflectivity are only indirectly connected with the geological evidence of major internal activity.

Other hypotheses associate the expulsion of E-Ring particles from Enceladus with a cometary impact or with the activity sometimes observed on distant comets. Comet Schwassman-Wachmann I, in an orbit just outside that of Jupiter, is a remarkable object that exhibits explosive outbursts in which a cloud of particles is expelled from the nucleus. If this comet were in orbit around Saturn, such an outburst would produce a tenuous ring like the E Ring. Comet Halley produced such an outburst at a distance of 14.5 AU from the Sun. Unfortunately, no one knows the origin of these explosive eruptions of comets either. It seems most likely that they are associated with the sudden release of highly volatile gases like carbon monoxide and nitrogen (see the discussion of geysers on Triton in Chapter 16) that should have disappeared from Enceladus long ago. Although there may be a link between comets and Enceladus, there is not yet any real understanding.

Two-Faced Iapetus

Iapetus presents us with a different set of mysteries. Ever since its discovery in 1671 by the director of the Paris Observatory, Jean Dominique Cassini, the strangeness of Iapetus has been recognized by astronomers. It is a two-faced satellite with a dark leading hemisphere and a bright trailing hemisphere. Since Iapetus, like most satellites, always keeps the same face toward its planet, we see the brightness of the satellite vary dramatically as it moves around its orbit, presenting us first with its dark side, then its bright. The bright side is water ice with the typical reflectivity of about 50%. The dark side, however, is covered with a reddish-black material, probably organic in composition, that reflects only 3% of the incident sunlight.

Efforts to identify this dark coating have revealed that organic material extracted from the Murchison carbonaceous meteorite (Section 4.4) matches the spectrum of sunlight reflected from the dark side of Iapetus. But is it really the same material? Our ability to identify solids by remote measurements is not yet good enough to give us a definitive answer. Similar dark, reddish material seems to be common in the outer solar system, on some asteroids, and on some comet nuclei. Recall that the object called Pholus has the reddest color in the solar system (Section 5.4). Dark material, such as that making up the rings of Uranus and coating the inner satellites of both Uranus and Neptune, has equally low reflectivity but lacks the reddish color of Iapetus and the other objects just listed. Perhaps all of these surfaces represent different blends of primitive, carbon-rich substances from the solar nebula.

Although Voyager unfortunately did not come any closer to Iapetus than 1.1 million km, its cameras confirmed the two-faced nature of the satellite and provided additional information on the distribution of the dark material (Fig. 17.6). Voyager also provided the first measurement of the mass of Iapetus. In combination with the diameter of 1460 km, this mass yields a density similar to that of the other icy satellites of Saturn (Table 17.1). Thus Iapetus appears to be similar to the others in bulk properties, and the dark deposit seems likely to be of external origin. The zebra is revealed as a white horse with dark stripes, rather than the other way around.

Voyager pictures of the icy side of Iapetus show a heavily cratered surface with a crater density at least as great as that on Rhea, its twin in size. The surface of this side of Iapetus is therefore very old. Unfortunately, there is no way to establish the age of the other hemisphere, since the surface was too dark for the Voyager cameras to detect any features there.

The dark hemisphere is symmetric with respect to the orbital motion of Iapetus, so it

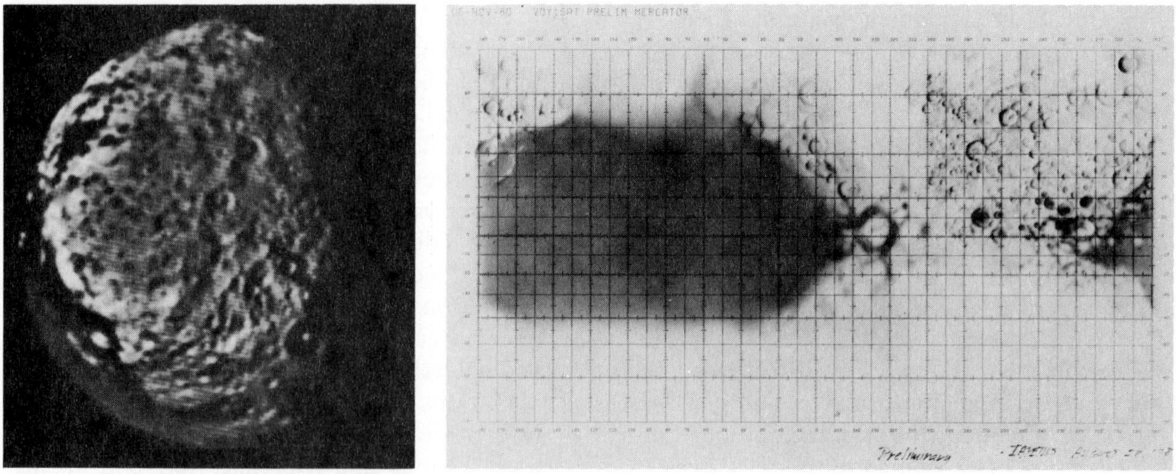

a b

FIGURE 17.6 (a) Iapetus exhibits a leading hemisphere that is roughly ten times darker than the trailing side. Here, special processing brings out the boundary of the dark side (lower left) but provides no detail. (b) A map of Iapetus based on the Voyager photographs. The large dark patch is the leading hemisphere, still totally unexplored.

must surely be the result of some external material striking the surface. But what might this material be? The next outer satellite, Phoebe, is dark and has been suggested as a source for the dark material. But the color of Phoebe does not match that of the dark side of Iapetus, so that theory is suspect. Perhaps the Phoebe-dust that is hitting Iapetus interacts with the surface material there to change its color. Alternatively, the dark material may even be indigenous to Iapetus after all, concentrated on the surface of the leading side by impacts that selectively evaporate away the water ice. Such a process would be similar to that by which a comet develops a dark surface by concentrating carbonaceous material as the ice evaporates.

These are intriguing ideas, but unfortunately they beg the question of the uniqueness of Iapetus. Why should this satellite alone have a dark leading hemisphere? A solution involving collisions—either to deposit dark material or to wear away bright material—seems to be the most reasonable possibility. But we do not know, and the Voyager data are probably not sufficient to provide a satisfactory answer. We hope to get a much closer look at the dark side of Iapetus in 2004 with the Cassini orbiter.

Enceladus and Iapetus are strange places, much stranger than had been anticipated before the Voyager missions. They are examples of a repeated phenomenon: When we send a spacecraft to a previously unexplored region of the solar system, we always seem to find more variety and more mysteries than we could previously have imagined.

17.3 THE SATELLITE SYSTEM OF URANUS

Unlike the other three giant planets, Uranus has no satellite with a diameter as great as 2000 km. Instead, it has five medium-size satellites with diameters and masses similar to those of the members of Saturn's regular satellite system. In order of their distance from Uranus, these five satellites are Miranda, Ariel, Umbriel, Titania, and Oberon—four named after Shakespearean figures, and the name Umbriel taken from Alexander Pope's "Rape of the Lock." An additional ten small satellites were discovered by Voyager during its 1986 encounter. Their orbits lie between the larger satellites and the rings, with two—Cordelia and Ophelia—on either side of the outermost ring. As in the case of Jupiter and Saturn, the existing tradition for nomenclature was honored, and these new small bodies were named for characters from Shakespeare's plays.

All 15 satellites of Uranus are regular in their orbits. They move in the equatorial plane of the planet, which is tilted nearly perpendicular to the orbital plane of the planetary system (Section 14.2). Also orbiting in this plane are a dozen narrow rings. In the 1980s, when the planet was nearly pole-on to the Sun, the ring and satellite orbits formed a bull's-eye pattern as seen from Earth.

Because the five main uranian satellites are so distant, they are difficult objects for astronomical investigation. Before the Voyager encounter, our meager knowledge of these frigid worlds had to be obtained from the light collected by large telescopes on the surface of Earth. As a measure of the challenge of studying this system, we note that it was not until 1977 that the rings were discovered, 1982 that the satellite sizes were first measured even approximately, and 1983 that the masses of two of the satellites were first determined by their mutual gravitational influence. Now that Voyager has left and headed for interstellar space, progress in studying this system depends on Earth-based astronomical observations, challenging the technology and ingenuity of planetary astronomers.

Titania, Oberon, Umbriel, and Ariel

In size, the satellites of Uranus resemble the inner satellites of Saturn (Table 17.2). Titania and Oberon are largest, each about the size of Rhea; Miranda is smallest, about the size of Mimas. The densities of the uranian satellites are also fairly similar to those in the saturnian system, although at 1.3–1.6 g/cm^3 they suggest a slightly larger proportion of silicate materials and metals, with correspondingly less ice. In general terms, however, it is fair to say that the satellites of both planets are composed of about half rock and half ice, with only minor variations from one object to another.

TABLE 17.2 Large satellites of Uranus

Name	Period (days)	Diameter (Moon = 1)	Mass (Moon = 1)	Density (g/cm^3)
Miranda	1.41	0.14	.0011	1.3
Ariel	2.52	0.33	.018	1.6
Umbriel	4.14	0.34	.018	1.4
Titania	8.71	0.46	.048	1.6
Oberon	13.5	0.45	.039	1.5

The surface compositions of the satellites of Uranus also resemble those of Saturn. Water ice is detected spectroscopically, but reflectivities are only 20–30%, suggesting that the ice is relatively dirty. Perhaps these objects are darker because their surfaces have been exposed longer to infalling debris, or perhaps there is more dark carbonaceous material present near Uranus to contaminate their icy surfaces.

The outer two uranian satellites, Titania and Oberon, have diameters of about 1600 km and are similar in appearance to Rhea, Dione, and Tethys. The main difference, aside from a lower surface reflectivity, is the absence of the wispy light streaks seen on the trailing hemispheres of the Saturn satellites. Both Titania and Oberon are heavily cratered at the Voyager resolution (about 10 km), with little indication of internal geologic activity. This cratering does show, however, that some sort of early heavy bombardment of satellite surfaces extended even out to the uranian system, more than 2 billion kilometers from the Sun.

Umbriel, the middle satellite of the five, is only about the size of Dione or Tethys (diameter 1190 km). It is the darkest of these uranian satellites and displays the least indication of any internal activity. Only one item of any interest shows up on the Voyager photos: a bright ring about 30 km in diameter near the edge of the illuminated hemisphere. (Since the satellites share the peculiar orientation of the planet, only one hemisphere of each could be seen; the other was experiencing a decades-long winter of darkness.) The nature and origin of this bright feature are unknown, and it will be very many years before we have another opportunity to look at it.

Ariel is about the same size as Umbriel but is much more interesting. Voyager was able to photograph it at 2 km resolution (Fig. 17.7), revealing a variety of geological structures. In addition to cratering, there are long tectonic valleys reminiscent of Ithaca Chasma on Tethys, and smoother areas that appear to have been resurfaced since the period of heaviest bombardment. Some of the valley floors have been filled in, and there is even evidence of flow patterns reminiscent of some of the valleys of Mars. Evidently, this satellite, in spite of its small size, experienced substantial geological activity early in its history, probably including a global expansion that stretched and cracked its crust. Interesting though this is, however, Ariel must take a back seat to the even more remarkable inner satellite, Miranda.

FIGURE 17.7 This mosaic of the four highest-resolution images of Ariel represents our most detailed view of this satellite of Uranus. Numerous valleys and fault scarps crisscross the cratered terrain.

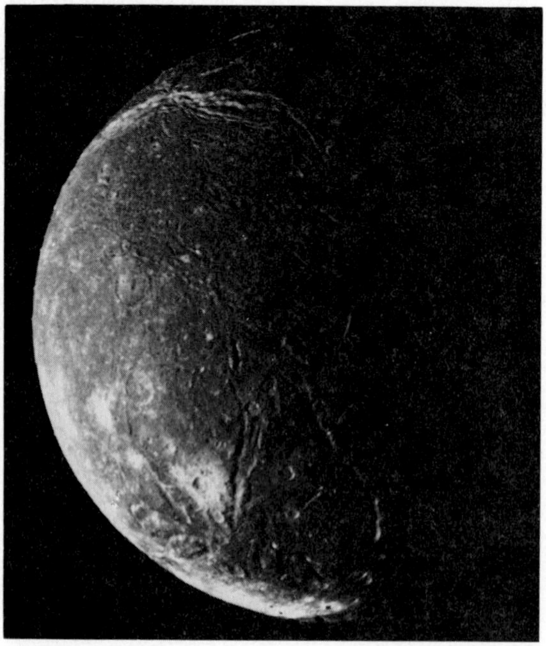

Bizarre Miranda

Miranda provided the big surprise of the 1986 Voyager encounter. The spacecraft passed only 28,260 km from this satellite, permitting excellent photography. The resolution, better than 1 km, equaled the best that had been obtained at Io or Rhea. However, the choice of Miranda to be the best-studied uranian satellite originally was not a happy one. Most geologists expected it to be a dull, cratered little world, perhaps resembling Mimas. The only reason Miranda was selected for close viewing was the requirement that Voyager pass through a point near Miranda's orbit to receive the gravitational boost necessary to reach Neptune in 1989. Hence Miranda was the only satellite for which a close approach was possible.

By good fortune, however, Miranda did not resemble Mimas at all, as is apparent in Fig. 17.8(a). Instead of being a dull, cratered satellite with little geological history of its own, Miranda turns out to have a surface extensively modified by internal forces. There are great valley systems up to 50 km across and 10 km deep, apparently produced by large-scale tectonic stresses. Peculiar oval or trapezoidal mountain ranges cover about half of the surface. In some areas the craters are apparently mantled and "softened" by overlying material, while elsewhere the craters are sharp and fresh looking. One huge cliff seen at the border of the illuminated area is between 10 and 15 km high (Fig. 17.8b).

Interpreting the images of Miranda is not easy. Many of the surface features of this satellite look similar to tectonic features encountered previously, such as the grooves of Phobos (Section 5.5), the mountain ranges of Ganymede (Section 15.3), or the valleys of Tethys (Section 17.1). But why are these phenomena all found together on this one small satellite? And what caused Miranda, which is less than 500 km in diameter, to be so much more active than its larger neighbors? One suggestion is that an impact shattered Miranda after it had differentiated, and the fragments fell back together randomly like a jumbled jigsaw puzzle. Another possibility is that the satellite, too small to complete its internal mixing, froze partway through the process of differentiation. We simply don't know exactly what happened here.

17.4 THE SMALLEST SATELLITES

Saturn has ten small satellites, most of them in the inner part of the system. Several of these inner satellites occupy unusual orbits: They share nearly the same path with each other or with the larger satellites. Some also interact strongly with the nearby rings, a topic we will return to in the next chapter. Jupiter has 12 small satellites, while Voyager discovered ten small uranian satellites in 1986 and six small satellites of Neptune in 1989.

Jupiter's Small Satellites

The 12 small satellites of Jupiter are not at all like those of Saturn. The outer eight were probably

FIGURE 17.8 (a) A close-up of part of the white "chevron" near the pole of Miranda (left) shows the white and dark grooved terrain with its sharp boundaries. To the upper right, a uniformly dark grooved terrain is visible. An older, heavily cratered section of the surface lies between these two features. (b) Looking toward the edge of Miranda below the bright chevron, an approximately 15-km-high scarp is visible. This would be a dramatic sight from the satellite's surface, or during the close-up view afforded during the 9 minutes (!) it would take for a falling object to reach the bottom.

a b

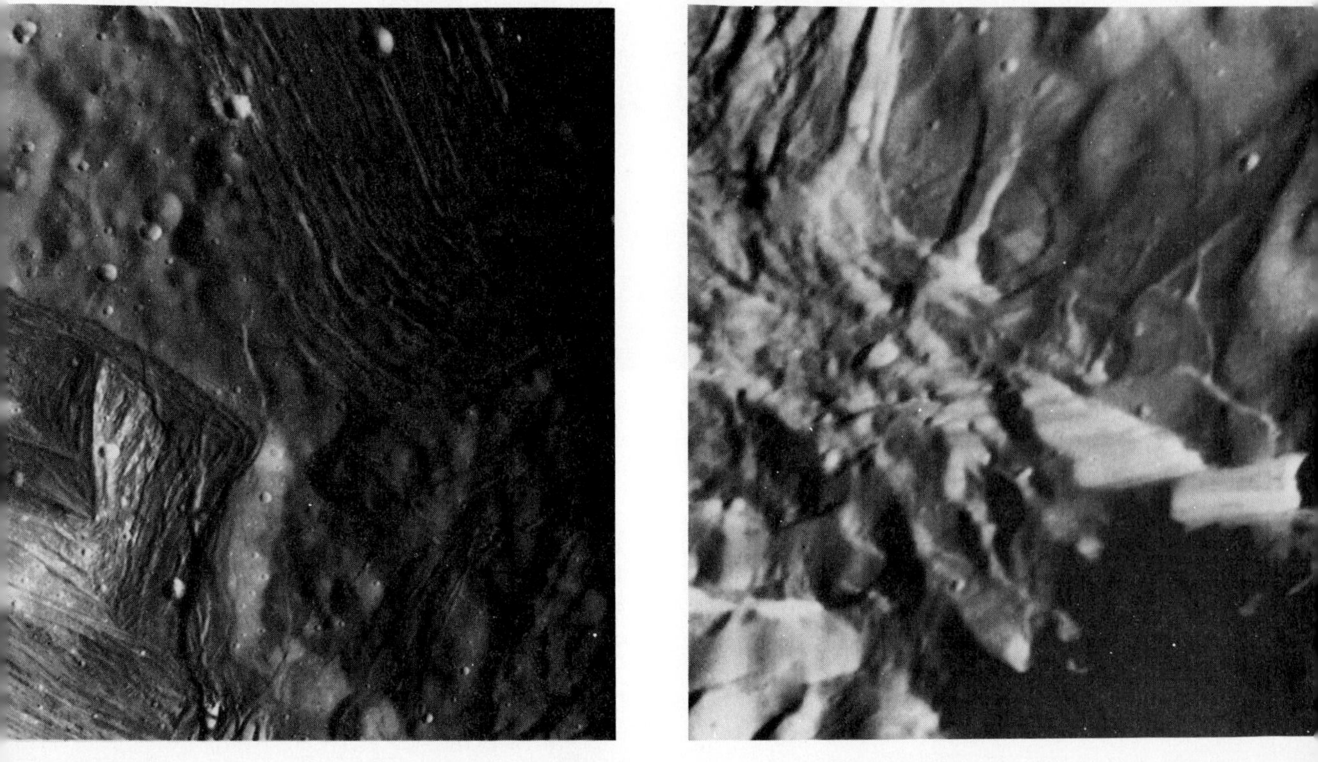

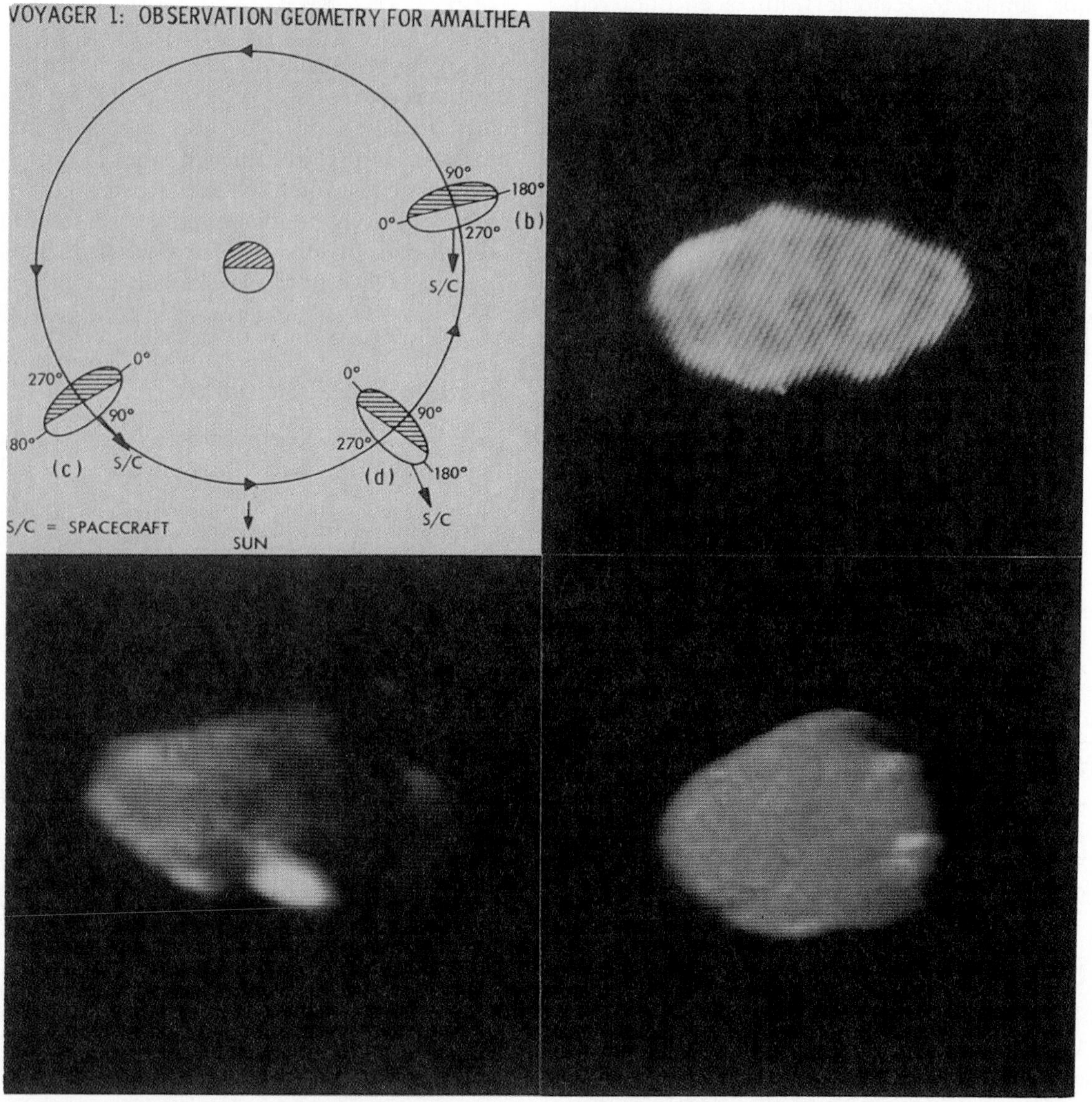

FIGURE 17.9 The sketch of Amalthea in its orbit on the left shows the different orientations of the satellite captured by the Voyager cameras. In the best color picture (right), the satellite appears reddish, with bright-rimmed craters.

captured early in solar system history, possibly from the same parentage as some of the primitive asteroids. Today they are found in two groups: the outer four in retrograde orbits at about 22 million km from Jupiter (nearly half the distance from Earth to Mars at closest approach), and the inner four in direct but highly inclined orbits at a distance of about 12 million km. Physically, these satellites are all smaller than about 200 km diameter. They have dark surfaces (with reflectivity of only about 3%), similar to that of the dark, primitive asteroids. Like the Trojan asteroids, they may be a remnant of the original population of planetesimals near the orbit of Jupiter.

Three of the four known inner satellites of Jupiter were discovered by Voyager. All three are less than 50 km in diameter, and the innermost two orbit very close to the planet, just at the outer edge of its tenuous ring. More substantial is Amalthea, which was discovered visually by the exceptionally capable observer E.E. Barnard in 1892. It is an irregular, cratered object about 250 km in its longest dimension, with a reddish color (Fig. 17.9).

The names of the so-called Galilean satellites were actually assigned by Galileo's contemporary and rival, an astronomer named Simon Marius. Encountering Kepler at a meeting, Marius suggested that it would be amusing to name these newly discovered objects after some of the amorous conquests of the king of the gods, who would be condemned to circle around him forever. This proved to be a rich resource, since there are well over 40 names of both genders on Jupiter's "list," providing a challenge to astronomers to find more satellites. (Amalthea, who was a nurse for the infant Jupiter, is an exception.)

Primitive Satellites

Three of the small satellites of Saturn and Uranus—Phoebe, Puck, and Hyperion—are examples of objects that probably retain a chemically primitive surface. We begin with two objects of 200 km diameter, Phoebe (the outermost Saturn satellite), and the similar-sized inner satellite of Uranus called Puck (Fig. 17.10). Unlike most other satellites we have discussed, both are dark objects, with 5–7% reflectivity. Their surface material is presumably composed of various types of carbonaceous matter left over from the formation of the outer solar system.

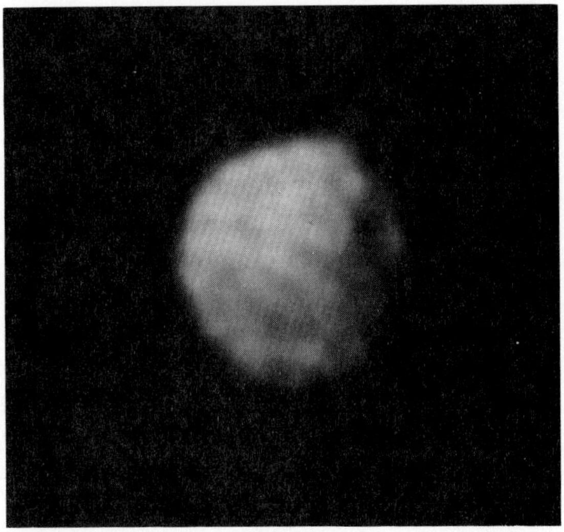

FIGURE 17.10 Puck, discovered on Dec. 30, 1985. Several craters are visible on the dark surface of this object, which is only 150 km in diameter.

Phoebe, an irregular satellite that must have been captured by Saturn during the planet's early history, is in a retrograde orbit. Phoebe is thus an icy planetesimal, rather like a giant comet nucleus, that somehow escaped incorporation by one of the giant planets or expulsion to the Oort cloud. As noted earlier, debris from this satellite may contribute to producing the dark deposit on the leading hemisphere of Iapetus.

Hyperion, which occupies a moderately eccentric orbit outside that of Titan, is apparently another primitive object. Voyager images (Fig. 17.11) show it to be irregular in shape, and spectroscopy has revealed that Hyperion has a surface of dirty ice; the dirt may be similar to the material on the dark side of Iapetus. Because of its small size, many scientists doubt whether Hyperion ever was heated or differentiated. Like a fresh comet, this satellite may provide a glimpse of the original, primitive dirty-ice material of the outer solar system.

One of the most remarkable aspects of Hyperion concerns not its composition or its orbit, but its rotation. As we have noted repeatedly,

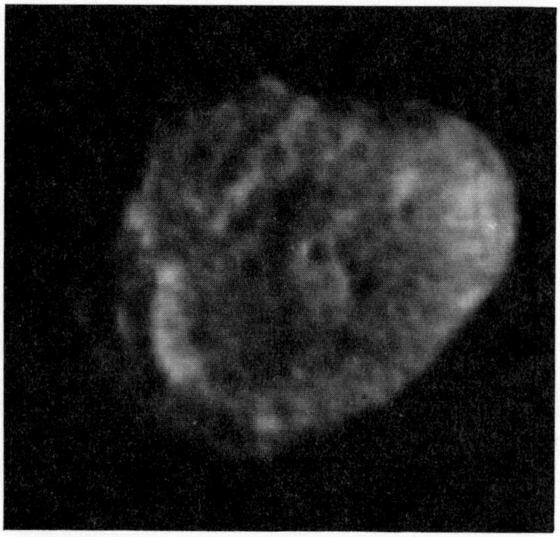

FIGURE 17.11 Hyperion is unique for its irregular shape and apparently chaotic rotation.

nearly all satellites are in synchronous rotation, always keeping the same side toward their planet. In Section 8.3 we saw how tidal forces lead to synchronous rotation. But Hyperion is different.

The problem first became apparent when members of the Voyager imaging team found that they could not match up their series of photos on the assumption that the satellite rotated synchronously. Different rotation periods were suggested, but none seemed satisfactory. Then several theorists calculated that it was possible for Hyperion not to have a fixed rotation period at all! This strange state, termed **chaotic rotation**, is possible because of the satellite's eccentric orbit and its close coupling with the much more massive Titan. In response to the many gravitational forces tugging it, Hyperion exchanges energy between orbital motion and rotation, causing it to tumble in a seemingly irregular way like a top that is running down. Careful telescopic observations confirmed this unique, and previously unknown, rotational state for Hyperion.

Small Inner Satellites of Saturn

The small inner satellites of Saturn are all icy objects with bright surfaces. Figure 17.12 is a composite group photo. Several of these satellites were discovered telescopically in 1966 and 1980, when the rings were turned edge-on to Earth, reducing the glare and allowing faint satellites in orbits close to the rings to be photographed. Others were discovered by the Voyager cameras in 1980 and 1981 (Table 15.1). As a matter of fact, Voyager photographed several more moons that are not listed in our tables of known satellites. There is little doubt as to the reality of these objects, but without a series of pictures, their orbits cannot be determined.

The most surprising aspect of these small inner satellites of Saturn is their orbits, each of which is special in some way. The three innermost satellites have orbits that are embedded within the rings, and the next three skirt the very edge of the rings and interact with them. Others share orbits with the larger satellites Dione and Tethys, providing the only known examples of Lagrangian orbits (see Section 5.4) within a satellite system. But perhaps most special are the co-orbital satellites, Janus and Epimetheus.

Co-orbital Satellites

The two **co-orbital satellites** were first spotted in 1966, when interference from the rings was at a minimum, but observers then failed to recognize that two objects might occupy essentially the same orbit. By treating the two objects as if they were one, astronomers initially calculated the wrong orbit, and for more than a decade textbooks listed a non-existent tenth satellite of Saturn. A later reanalysis of the observations suggested that two satellites might be present, and this possibility was quickly verified when the rings were again edge-on, in 1980. Meanwhile, the Voyager 1 spacecraft was approaching Sat-

FIGURE 17.12 A group portrait of Saturn's tiny satellites. At left is Atlas, guarding the outer edge of the A Ring, then Prometheus (top) and Pandora (bottom), which shepherd the F Ring, Janus (top) and Epimetheus (bottom), which occupy nearly the same orbit, two Lagrangian satellites of Dione, and the single Lagrangian of Tethys. Note the crater in the latter—again, a slightly larger impact would have shattered it.

urn, and its cameras soon confirmed the presence of two satellites in nearly the same orbit, about 13,000 km beyond the rings. The larger is about 200 km in its longest dimension, and the smaller is about 150 km.

If the two satellites had exactly the same orbital period, they might be able to avoid interfering, just as Lagrangian satellites stay clear of each other. But the so-called co-orbitals actually have orbits differing by about 50 km in radius. The orbital period of the inner one is 16.664 hours, and that of the outer one 16.672 hours. This difference causes the inner one to catch up with the outer at a relative speed of 9 m/s. About once every four years the inner satellite laps its slower-moving sibling, but since the space between orbits is smaller than the objects, there is no room to pass. How is a collision to be avoided? What happens is that short of a collision, the two satellites attract each other gravitationally and exchange orbits. They then slowly move apart, and the four-year cycle starts again. This strange orbital dance is unique, as far as we know, in the planetary system.

The co-orbital satellites are elongated and irregular in appearance, looking very much like fragments from some ancient catastrophe. Many scientists believe that they were once joined, and that some long-ago impact fractured their parent body to leave these two pieces in nearly the same orbit. All the small inner satellites of Saturn also look as if they might be remnants of past collisions. Recall the argument at the end of Section 15.2 indicating that the inner part of the Saturn system was subject to very heavy impact cratering early in its history. If we extend that argument to include these smaller bodies, we should not be surprised to see a few remnants of objects that were destroyed by such impacts. Indeed, the rings themselves might have been formed in this way. But we will return to the rings in Chapter 18.

The Smaller Satellites of Uranus and Neptune

In its 1986 flyby, Voyager discovered ten new satellites of Uranus. We have already discussed Puck, the only one of these objects photographed in any detail. Like it, the others appear to have dark surfaces, sharply distinguished from the larger satellites discussed in Section 17.3.

Prior to the Voyager 2 flyby, Triton and Nereid were the only two satellites of Neptune that were known. In addition to providing a large amount of new information about Triton, Voyager 2 obtained a few, distant images of Nereid and discovered six small satellites closer to the planet. Nereid is a generic name for oceanic nymphs. In keeping with this theme, the new satellites were named for mythological figures associated with oceans and rivers.

Nereid has the most eccentric orbit of any known satellite, suggesting that it, like Triton, was probably captured from the solar nebula. Voyager 2 passed too far from Nereid to obtain a good record of its appearance. We are not even sure whether it is irregular or spherical in shape. However, the observations were good enough to establish that Nereid's diameter is 340 km. Knowing the brightness of this satellite, scientists are then able to determine the reflectivity, which is 15%. This is much lower than the 80% reflectivity of Triton, but higher than the 5–6% reflectivities that characterize the small inner satellites of both Uranus and Neptune.

Of the six newly discovered satellites of Neptune, Despina, with a diameter of 400 km, is the largest. The others range from 50 to 100 km. Proteus and Larissa were seen well enough by the Voyager cameras to establish that they have cratered surfaces. The range of sizes, albedos, and locations of these small inner moons of Neptune is remarkably similar to that of the system of small, dark moons Voyager 2 found around Uranus. Both systems also have orbits with very low eccentricities and inclinations; these are highly regular satellites.

Irregular Shapes of Small Satellites

The small satellites of Saturn illustrate very well the point raised in Section 5.5 concerning the shapes of asteroids and other members of the planetary system. Small bodies can have irregular shapes, while larger bodies are more nearly spherical. If an object is large enough, its self-gravity overcomes the inherent resistance of its rock (or ice) and forces it into a spherical shape. In a similar way, the maximum height that a mountain can assume is determined by the point at which gravity overcomes the strength of rock.

For the icy satellites of the Saturn system, the limiting diameter below which objects still have the strength to resist gravity and remain nonspherical is about 400 km. Look at Mimas (Fig. 17.4) and Enceladus (Fig. 17.5b): Both have diameters of about 500 km, and both are clearly spherical. In contrast, Hyperion (Fig. 17.11), with a diameter of nearly 400 km, and the larger co-orbital satellite, Janus (Fig. 17.12), with a diameter of about 200 km, are quite irregular. Do you expect the same transition from irregular to spherical objects to apply to the satellites of Uranus? And what can you conclude about the shape expected for a comet nucleus?

The Limits of Completeness

Gerard Kuiper wrote a paper with this title in 1962, in which he summarized his efforts to find additional satellites. Kuiper took pains to point out that his searches for satellites close to their planets were limited by the glare from the planets themselves, whereas more distant satellites could not be found if they were too faint. The Voyager spacecraft circumvented the first problem by getting close enough to the planets to search for satellites against a black background. That is the main reason that the number of known satellites has grown from 36 at the time Kuiper wrote to the 61 listed in Table 15.1. But this list is also not complete. Kuiper's caveats are still applicable. In fact, we *know* there are more satellites than those listed, since both Voyager and ground-based observers of the outer planets have glimpsed satellites of Jupiter and Saturn that were not sufficiently well observed to yield definitive orbits. Future observations may find them, and may also reveal that Uranus has a distant, irregular satellite. It is presently the only giant planet without one.

17.5 QUANTITATIVE SUPPLEMENT: SEARCHING FOR NEW SATELLITES

Suppose you would like to try to see whether Pluto has a small, distant satellite in addition to Charon. To make such a search, you will need to use one of the biggest telescopes, since your target will be very faint. Just how faint will it be? And how can you convince the director of the observatory that some precious time on a major telescope should be used to add one more object to the list of satellites already known?

Because you will be talking to astronomers, you have to learn their language. This means mastering the magnitude scale. Since the days of Ptolemy, astronomers have used a scale for the brightness of stars that embodies the roughly logarithmic response of the human eye to variations in light. This scale was eventually set so that the brightest stars had a magnitude $m = 0$, and the faintest stars had a magnitude of 6. If the apparent brightness, or luminosity, of a star is L, then the difference in magnitudes of two stars of luminosity L_1 and L_2 is expressed as

TABLE 17.3 Magnitudes of selected planets and satellites

Object	Magnitude	Object	Magnitude
Venus	–4.4	Ganymede	4.6
Jupiter	–2.7	Titan	8.4
Uranus	5.5	Miranda	16.5
Neptune	7.8	Triton	13.6
Pluto	15.1	Naiad	24.6

$$m_1 - m_2 = 2.5 \log (L_2/L_1)$$

In other words, the larger the magnitude, the fainter the object. The difference in luminosity for a range of 5 magnitudes is thus a factor of 100. On this scale, Venus has a magnitude of –4.4 at its brightest, while dim and distant Pluto has a magnitude of 15.1. This represents a factor of 63 million in luminosity (Table 17.3). A 4-m telescope with a modern, high-sensitivity detector can record images of distant galaxies with magnitudes near 26. Observations over several nights can lower this limit further.

The "luminosity" (brightness) of a satellite will depend on its reflectivity p, its size (πr^2, where r is the satellite's radius), and the inverse square of its distance from the Sun, $1/D^2$:

$$L \propto p\pi r^2/D^2$$

Now consider a satellite the size of Phobos ($r \sim 10$ km) at the distance of Pluto. Its magnitude will be related to that of Pluto by the equation

$$m_{sat} - m_{Pl} = 2.5 \log L_{Pl}/L_{sat}, \text{ where } L_{Pl}/L_{sat} = r_{Pl}^2/r_{sat}^2 = 1.5 \times 10^4$$

assuming the reflectivities of Pluto and its hypothetical satellite are the same. Thus the satellite we are seeking is an object with about the same apparent brightness as an extremely distant galaxy at magnitude 25.5.

This would be a very tough observation indeed, one that requires the use of one of our biggest telescopes. You had better develop a strong argument for doing it!

SUMMARY

Each of the giant planets has a family of small satellites (with diameters less than 2000 km) in addition to the larger ones that are better known. Always the exception, Uranus has *only* small satellites. Each of these families has its own characteristics, determined largely by the distance from the Sun at which the system formed, but also influenced by the mass of the planet.

Saturn's family is notable for having so many bright satellites, including Enceladus, which has the most reflective surface in the solar system. Its surface includes large areas that are devoid of impact craters and look as if they had melted sometime in the past. Enceladus also occupies an orbit that is in the center of Saturn's E Ring, suggesting that the satellite may itself be a source of the ring particles. At present, we do not know how this could occur.

Iapetus poses a different kind of mystery, with one hemisphere over ten times darker than the other. No other satellite in the solar system shows a variation in surface brightness that even approaches this. Interpretations are divided between an internal and an external source of this dark material, whose composition is also undetermined at present.

The other small satellites of Saturn include two that share the same orbit, others in the Lagrangian points of the orbits of larger moons, and dark Phoebe, a captured satellite in a distant retrograde orbit. Hyperion exhibits chaotic rotation; Rhea, Dione, and Mimas each have unique characteristics, though they all share a common external appearance of being heavily cratered and icy.

In the Uranus system, we find five medium-size satellites that were discovered from Earth, plus ten small, dark bodies that were found by Voyager 2. These satellites are all remarkably regular. The most spectacular is Miranda, in part because it is the one best viewed by the passing spacecraft. This small moon appears to exhibit the results of a bewildering variety of geological processes, suggesting that it may have solidified before its differentiation went to completion.

The small outer satellites of Jupiter form two groups of four bodies each, one in a set of retrograde orbits, the other in direct orbits. They are apparently captured, dark, asteroidal bodies, possibly related to the Trojan asteroids. The small inner satellites include Amalthea, found visually in the last century, and three others discovered during the Voyager flybys.

Neptune's Nereid, with its high eccentricity and inclination, is the most irregular satellite in the solar system. Nevertheless, there is nothing especially unusual about the satellite itself. The six new satellites found by Voyager 2 occupy highly regular orbits close to the planet.

There may well be more satellites, as yet undiscovered, in each of these systems. The steady improvement in the sensitivity of Earth-based observations suggests that we may soon find out whether this is the case.

KEY TERMS

chaotic rotation
co-orbital satellite

REVIEW QUESTIONS

1. Make four separate charts showing satellite systems of the giant planets. Use the radius of each planet as the unit of distance and plot the positions of the satellites relative to their planets. What similarities and differences do you see in the different systems?

2. What is so strange about Enceladus? How would you try to explain this strangeness? Consider both Io and Europa as you develop your answer.

3. Describe the appearance of Iapetus. Our own Moon also exhibits a hemispherical asymmetry. How does the variation in the appearance of the surface of our own Moon differ from the asymmetry on Iapetus?

4. Does any of the other small satellites of Saturn show evidence of internal activity? Explain.

5. Describe the five large satellites of Uranus. Why is it strange that the surface of Ariel seems partially melted and that Miranda shows such varied surface features? Is there a correlation of activity (or reflectivity) with distance from Uranus? Explain why it is difficult to reach a conclusion.

6. Neptune has two highly irregular satellites. What are their names, and what makes them irregular? Contrast these objects with the regular satellites of Neptune, using size, reflectivity, distance from the planet, and orbital inclination.

7. Do you think the composition of Amalthea is closer to that of Ganymede or Io? Explain.

8. Which of the small satellites—in each system—is associated with its planet's rings? What is that association?

QUANTITATIVE EXERCISES

1. Check the albedos in Table 15.1 and calculate the effective temperatures of Nereid and Triton. From this calculation alone, would you expect to find nitrogen ice on either of these satellites? Explain.

2. Suppose you had a strong intuition that Uranus must have an irregular satellite (so far it's the only giant planet without one). If that satellite had a diameter of 20 km, how faint would it be relative to Miranda?

3. Cordelia and Ophelia are inside and outside the Epsilon Ring of Uranus. Compare the gravitational attraction of one of these satellites on a nearby ring particle with the attraction of Uranus on that same particle. Which is likely to have the larger effect on the ring particles, the planet or the satellite?

4. See if you can find some substance that is as dark as the leading hemisphere of Iapetus (reflectivity about 3%). Use the *Handbook of Chemistry and Physics*, think about coal, freshly cooled lava, and mud, and try your local paint store.

ADDITIONAL READING

*Burns, J., and M.S. Matthews, eds. 1986. *Satellites.* Tucson: University of Arizona Press. Standard technical reference on the satellites, from the Arizona Space Science series.

Rothery, D.A. 1992. *Satellites of the Outer Planets.* Oxford: Clarendon Press. A well-illustrated compendium of information about all the satellites of the outer planets that were known as of 1992.

*Indicates the more technical readings.

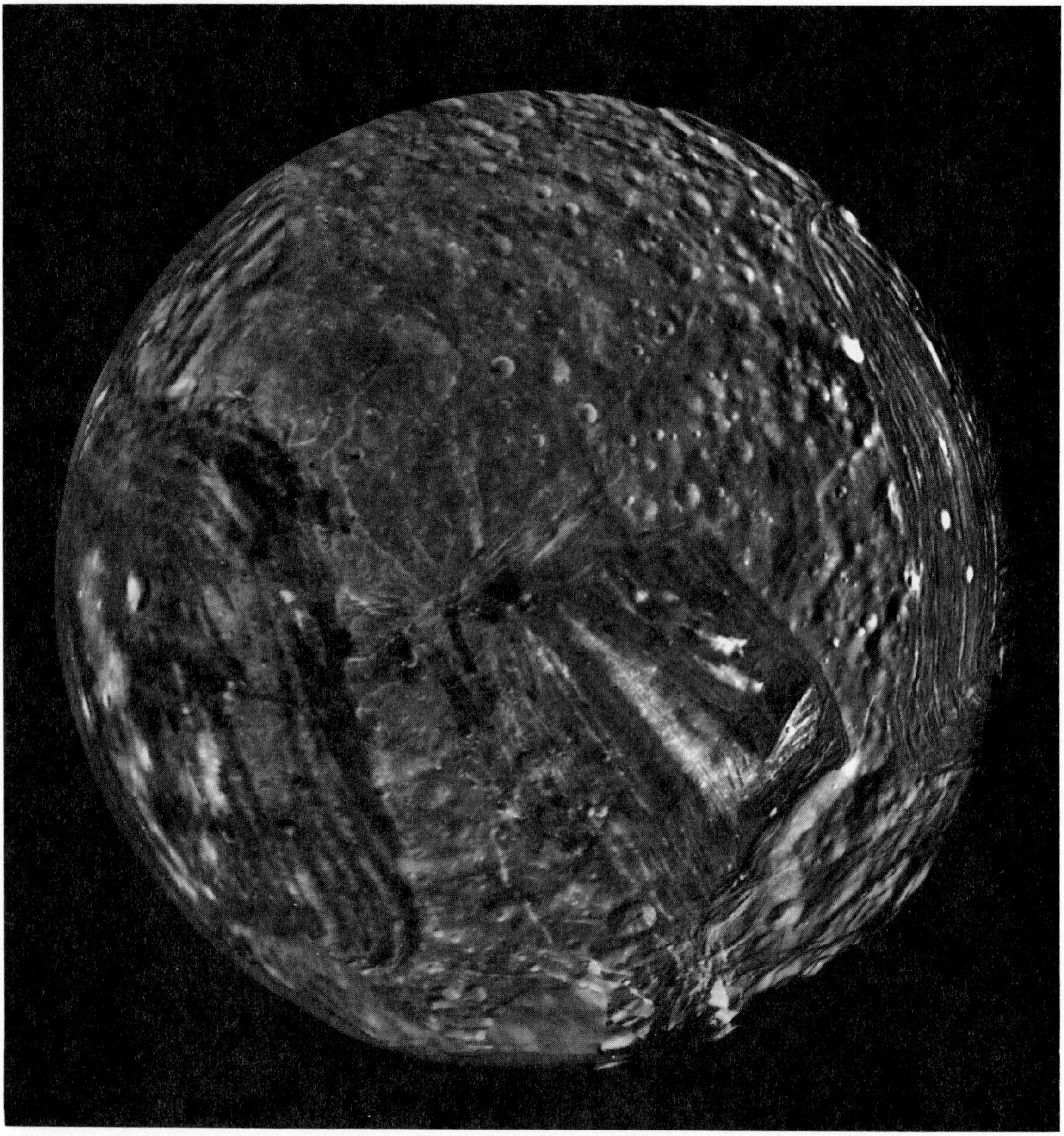

Miranda, the daughter of Prospero in Shakespeare's *The Tempest,* speaks the famous lines "Oh brave new world . . ." Voyager 2 discovered that the satellite bearing this character's name is itself a new world beyond the imaginings of the scientists who designed the mission. This picture is a mosaic of the best images obtained by the flyby in January 1986. Note the rich variety of geological features on this small (500 km) icy satellite.

18

Planetary Rings

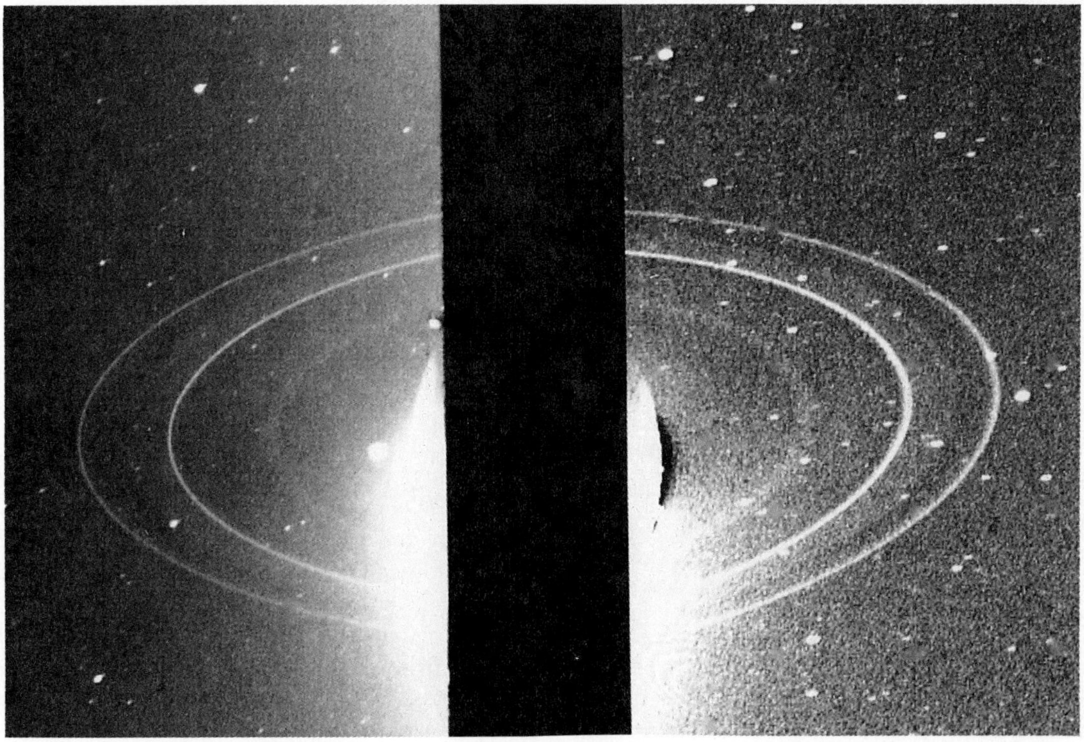

Voyager 2 took these two pictures of Neptune's rings looking back in the direction of the Sun when the spacecraft was hidden inside the planet's shadow. The complex structure of the ring system can be traced against the background of stars, which are visible in this deep exposure.

Why do some planets have ring systems while others do not? For 300 years, Saturn was the only planet known to be surrounded by rings. With their usual thoroughness, scientists wrote learned papers explaining why Saturn had rings and the other planets did not. We now know that all the giant planets have rings, but each system is unique. In Table 18.1, we give the dimensions of the rings in terms of the radii (R) of their respective planets. We will convert these dimensions to kilometers as we discuss each system. Note that despite their individual differences, all the rings are rather close to their planets, within a distance of 2.5 times the planet's radius. The ring systems are illustrated schematically in Fig. 18.1.

TABLE 18.1 Major ring systems

Planet	Outer Radius (R_{planet})	Mass (kg)	Reflectivity (%)
Jupiter	1.8	10^{10}	?
Saturn	2.3	10^{18}	60
Uranus	2.2	10^{14}	5
Neptune	2.5	10^{12}	5

We still have no complete theory for understanding why these systems are so different from one another, but we do know why the inner planets have no rings. The lifetime of a small particle in orbit around one of the inner planets is just too short. Gravitational perturbations by the Sun as well as the effects of solar radiation will cause such particles to spiral into the planets on relatively short timescales. Only in the outer solar system, with massive planets at large distances from the Sun, do we find the right conditions for stable ring systems. Yet it is still not clear just how these systems formed, or how stable they are.

In this chapter, we will discuss the various ring systems in order of increasing distance from the Sun, just as we have done with the planets. We follow this with a general discussion of the behavior of particles in the rings in response to the gravitational fields of the parent planets and nearby satellites, including small satellites (or large ring particles) embedded in the rings themselves. We conclude by discussing the possible origin of these systems.

18.1 RINGS OF JUPITER

The rings of Jupiter were discovered by Voyager in 1979 (Fig. 18.2) and confirmed within a few days by ground-based observations (it helps if an astronomer knows just what to look for!). One of us (Owen) led the Voyager effort to search for this ring, overcoming a certain amount of resistance from skeptics who thought it was impossible for Jupiter to have a ring. This negative expectation was based on the high temperatures of Jupiter early in its evolution. Ice, such as that in the rings of Saturn, could never have survived close to Jupiter.

The strategy selected by the Voyager sequence-planning team was to allow a single long-exposure picture to be taken by Voyager 1

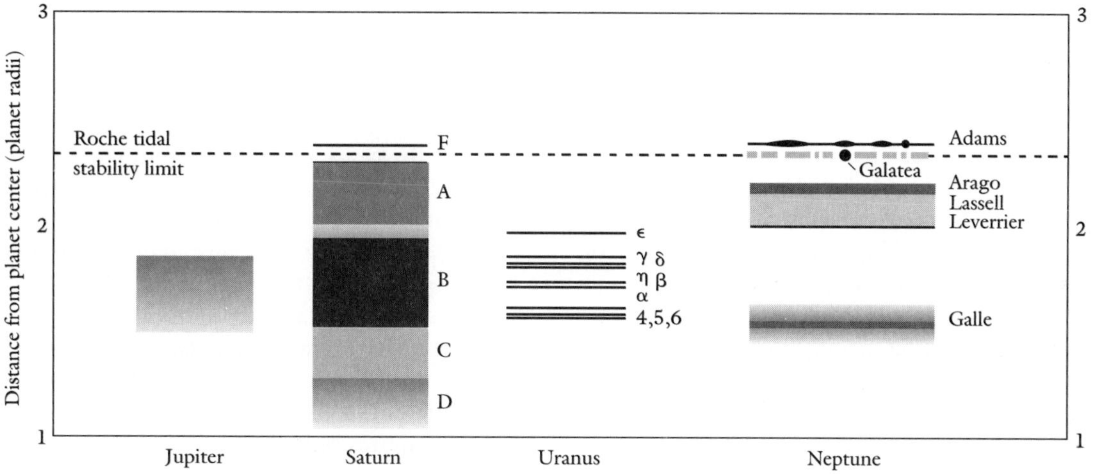

FIGURE 18.1 The four well-observed ring systems in the solar system are compared here in a diagram that presents each one in terms of its distance from its respective planet.

as the spacecraft passed through Jupiter's equatorial plane. The camera was aimed at a point in space that would correspond to Saturn's B Ring, halfway between the orbit of Amalthea and the jovian atmosphere. This single frame succeeded in catching the rings. Using the results from Voyager 1, the Voyager 2 cameras were programmed to take a variety of ring pictures, and most of what we know about the rings is derived from the Voyager 2 data.

FIGURE 18.2 (a) Owen (left) discussing the discovery picture of Jupiter's ring with Candy Hansen (right), who designed the Voyager 1 camera sequence that acquired it, while Jeff Plescia looks on. (b) Jupiter's ring, revealed in a mosaic of pictures obtained with the Voyager spacecraft in the planet's shadow so that small particles in the ring produce strong forward scattering of light.

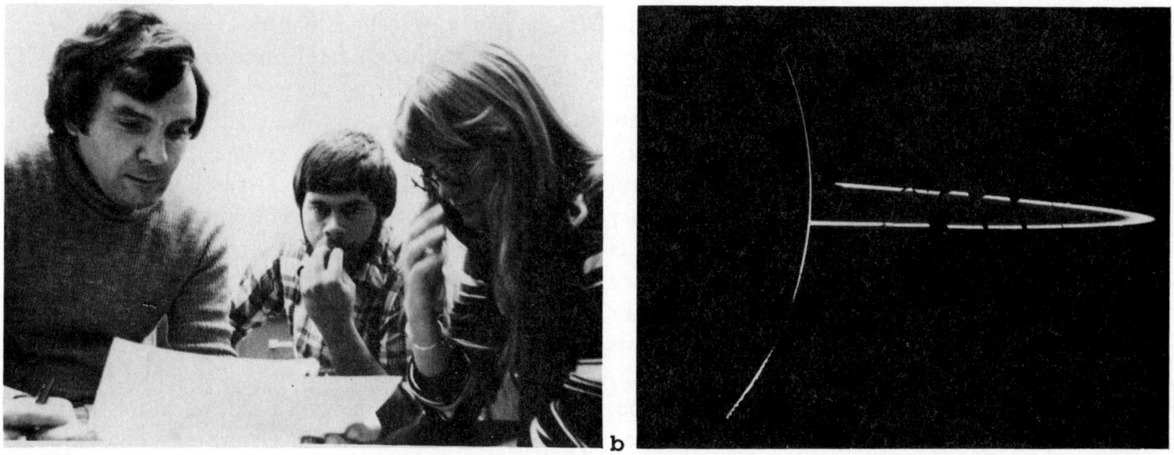
a b

The primary ring is 54,000 km from the planet and about 5000 km wide. It is much more tenuous than any of the other known rings, blocking only about 1 millionth of the light from a source seen through it. For comparison, a sheet of clear glass absorbs several percent of incident light, so one could say that the ring of Jupiter is 10,000 times more transparent than window glass! A still more insubstantial doughnut or torus of material surrounds the main ring.

Most of the particles in the jovian ring are very small, no larger than grains of dust. They are in fact made of dark silicate dust, since grains of water ice would evaporate at the temperatures found in the jovian system. Because of their small size, they easily acquire a small electrical charge from interactions with the jovian magnetosphere, and the mutual repulsive force of these charged particles lifts them up out of the ring plane. The processes involved in the formation of this extended cloud may be similar to those that generate the "spokes" in the B Ring of Saturn (see Section 18.2), but in candor neither phenomenon is very well understood.

18.2 RINGS OF SATURN

Many people believe that the rings of Saturn are the most beautiful sight that can be seen through a telescope. Even after the discovery of rings around Jupiter, Uranus, and Neptune, the Saturn rings remain *the* planetary ring system. The rings of Saturn were first seen by Galileo in 1610, but recognition of the nature of this new phenomenon did not come for another 50 years. Galileo thought he was glimpsing bumps on Saturn, or perhaps a triple planet. Some other seventeenth-century observers drew the rings as handles extending on either side of the planetary disk. Not until 1659 did the Dutch astronomer Christian Huygens recognize that Saturn was surrounded by a "thin flat ring, nowhere touching" the planet and lying in its equatorial plane. In 1675 Giovanni Cassini, who also discovered the two-faced nature of Iapetus, found that there were at least two concentric rings, not a single, continuous one. The dark lane that separates the two parts is still called the Cassini Division in his honor.

That the rings are not solid but composed of billions of tiny moons orbiting the planet was not demonstrated until the second half of the nineteenth century, although many astronomers had suspected this for some time. According to Kepler's laws of planetary motion, the particles closest to Saturn must travel faster in their orbits around the planet than the more distant ones. At the inner edge of the main rings, one circuit of Saturn requires only 5.6 hours, while the period at the outer edge is 14.2 hours.

Characterizing the Ring Particles

Further understanding of the nature of the ring particles awaited the development of modern astronomical instrumentation. Infrared spectroscopy revealed in 1970 that the particles were composed primarily of water ice. Interpretation of the first radar signals bounced from the rings in 1973 led to information on the sizes of the particles, which were found to be typically tens of centimeters in diameter. The ability of the rings to serve as strong reflectors of radar further indicated that the particles must be rather loosely distributed, not collapsed into a single layer just one particle in thickness.

Some additional information was provided by the 1979 flyby of Pioneer Saturn, but most of what we now know about the rings is the direct result of the Voyager encounters in 1980 and 1981. Popular interest in these encounters was so great that Saturn and its rings appeared on the covers of both *Time* and *Newsweek*, with long articles in the magazines describing the Voyager discoveries.

The rings of Saturn consist of a thin sheet of small icy particles, ranging in size from grains of sand up to house-size boulders. Just as with the asteroids and other populations of debris, there are many more small particles than large ones. Most of the individual particles are bright, reflecting 50–60% of the incident sunlight and exhibiting the strong spectral signature of water ice. Some particles are darker, perhaps indicating the presence of organic material or silicate. An insider's view of the rings would probably resemble a bright cloud of floating hailstones with just a few snowballs and larger objects, many of them loose aggregates of smaller particles.

With the few exceptions noted later, each ring particle follows an almost perfectly circular orbit around Saturn in the equatorial plane of the planet. It is easy to see why this must be. Imagine a particle with an eccentric orbit. During each circuit of Saturn, it would move in and out, crossing the orbits of other particles. Low-speed collisions would result, and the particle would lose energy, almost as if it were rubbing against its neighbors. The same sort of "fender-benders" would be encountered by a particle in an inclined orbit, swinging back and forth across the plane of the rings. Either way, the effect of the friction is to circularize the orbit and bring the particle into the same plane with the rest of the ring particles. This is generally what we see at Saturn, and yet there remain many surprises when we look closely: narrow gaps, waves, eccentric rings, kinky rings, and even braided rings!

Overview of the Rings of Saturn

The rings of Saturn are very broad and very thin. Their dimensions are summarized in Table 18.2. The main rings, those visible from Earth, stretch from about 7000 km above the atmosphere of the planet outward for a total span of more than 70,000 km (Fig. 18.3). The distance from one edge of the rings through the planet to the opposite edge is almost as great as the distance from the Earth to the Moon. Yet the thickness of this vast expanse is only about 20 m, the width of a typical house lot. If we made a scale model of the rings out of paper the thickness of the sheets in this book, we would have to make the rings more than 1 km across—about eight city blocks. On this scale, the planet would loom as high as a 100-story skyscraper.

TABLE 18.2 Dimensions of the main Saturn rings

Name	Outer Edge (*R*)	Outer Edge (km)
D	1.233	74,400
C	1.524	91,900
B	1.946	117,400
Cassini Division	2.212	133,400
A	2.265	136,600
F	2.324	140,180

Three distinct rings, called the A, B, and C rings, can be seen from Earth; an additional narrow F Ring was discovered in 1979 by Pioneer. Voyager, however, revealed much greater complexity. The ring material is organized into tens of thousands of ringlets visible in Voyager photographs. These ringlets are not generally separated from each other by gaps, but rather represent local enhancements or depletions of the concentration of ring particles. Only a few empty gaps exist, providing natural boundaries between parts of the rings. We will discuss the origin of these gaps in Section 18.5, but we first take a tour of the rings, noting the major features as we proceed outward from the planet (see Fig. 18.3).

The Inner Rings

Between the upper atmosphere of the planet and the inner edge of the C Ring, Voyager discovered several thin rings invisible from Earth, collectively called the D Ring. However, the substantial part of the ring—the C Ring—starts 7000 km out, where the particles are packed densely enough to reflect a fair amount of sunlight. The planet is easily visible through the C

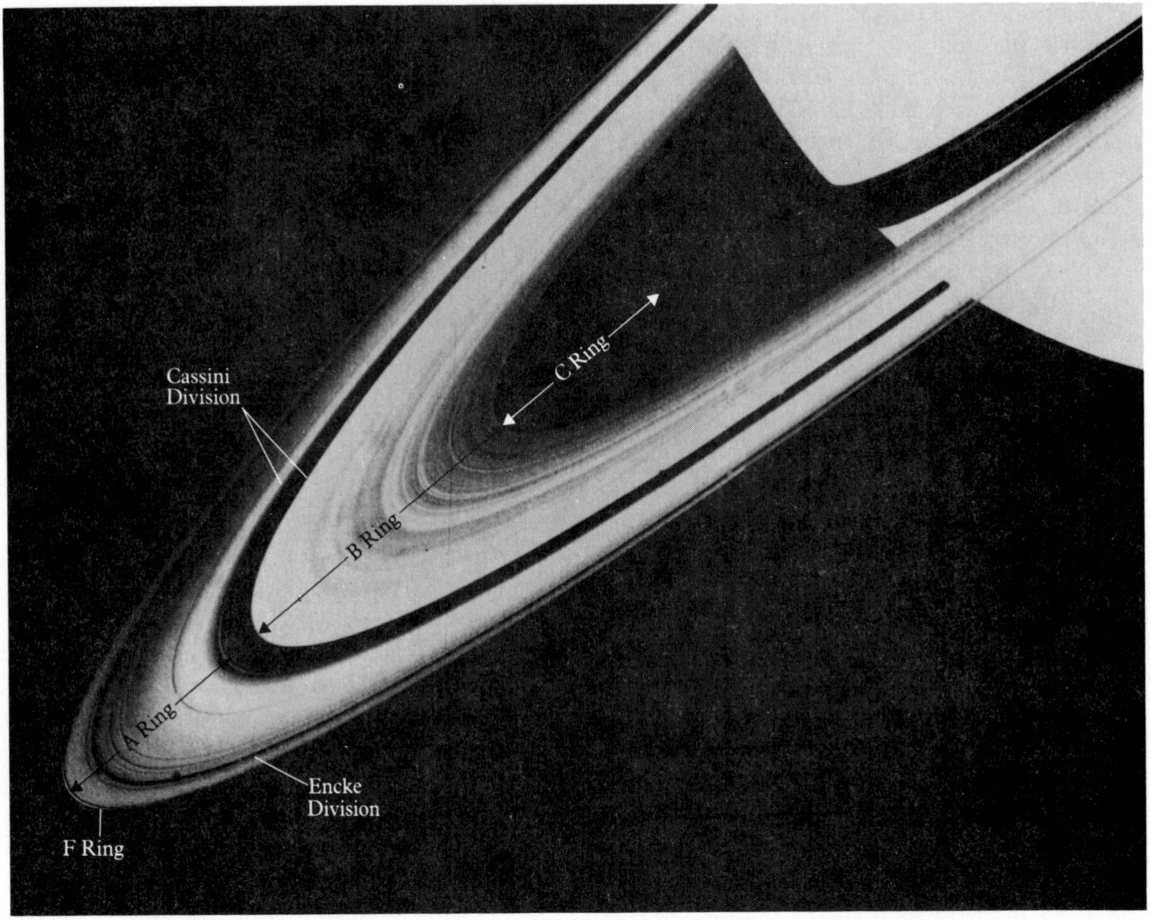

FIGURE 18.3 A Voyager picture of Saturn labeled to show the various sections of the ring system. (The C Ring is not visible in this high-contrast picture.) Note that the shadow of the rings on the planet appears as a dark band with a white line in it. The white line is caused by sunlight passing through the Cassini Division (cf. Fig. 18.5).

Ring, which is so thin that it used to be called the "crepe ring." There are two major gaps in this ring, each several hundred km wide.

Inside one of the C Ring gaps is a remarkable narrow **eccentric ring**. This entire ribbon behaves as if each of its particles orbited Saturn with the same eccentricity, and with all the individual orbits exactly aligned. The color of this ring is subtly but definitely different from that of its neighbors, suggesting that the particles that make up the eccentric ring have a different composition. There are only three of these eccentric ringlets in the Saturn system, each very narrow and each lying in an empty gap. As we will see later, the rings of Uranus and Neptune are also narrow and eccentric.

The A and B Rings

At a distance of 32,000 km from Saturn the ring particles suddenly become more densely concentrated, and the structure of the concentric

FIGURE 18.4 (a) A picture centered on Saturn's B Ring. The mottled appearance of the ring to the left center is caused by the superposition of so-called spokes. (b) Here we are looking back at the rings after the spacecraft passed them. The spokes appear bright (the diagonal streak to the left), indicating that we are looking at small particles that are preferentially scattering the light forward, along the same direction it was traveling.

ringlets becomes more complex. This is the edge of the B Ring, the brightest part of the ring system and the part that contains most of the mass. Throughout much of the B Ring, which stretches out to 57,000 km, the particles are so closely spaced that the ring is nearly opaque. Particles are typically from tens of centimeters up to meters in diameter. There are no empty gaps in the entire 25,000 km span of this ring.

Perhaps the most enigmatic features of the B Ring seen by Voyager are the dark radial shadings that turn as the ring rotates like the spokes on a wheel (Fig. 18.4). These transient features consist of very fine particles hovering in a cloud above the plane of the rings. Each long cloud is formed within a few tens of minutes and then survives for as long as one revolution of the rings, about ten hours. No one knows how these mysterious clouds are formed, but the mechanism probably involves electrically charging fine particles in the ring so that they can rise out of the ring plane.

At the outer edge of the B Ring lies the 3500-km-wide Cassini Division, the one break in the rings that can easily be seen from the Earth. As shown in Fig. 18.5, however, the Cassini Division is by no means an empty gap. Within the division there are a number of discrete ringlets, separated by several true gaps, including one eccentric ring, and a great deal of fine structure visible in the spacecraft photos. At least one small satellite also orbits inside this division. At a time when it was still thought that the Cassini Division was empty, a proposal was considered to target the Pioneer Saturn spacecraft, which reconnoitered the planet in 1979, to pass through the rings within this division. Had this been attempted, the Pioneer Saturn mission would have come to a very sudden end!

Outside the Cassini Division, beginning at 61,000 km above the planet, lies the last major ring, the A Ring. The A Ring is intermediate in brightness and transparency between the opaque B Ring and the translucent C Ring. Its most outstanding feature is in one of its gaps, the 360-km-wide Encke Division, which contains two

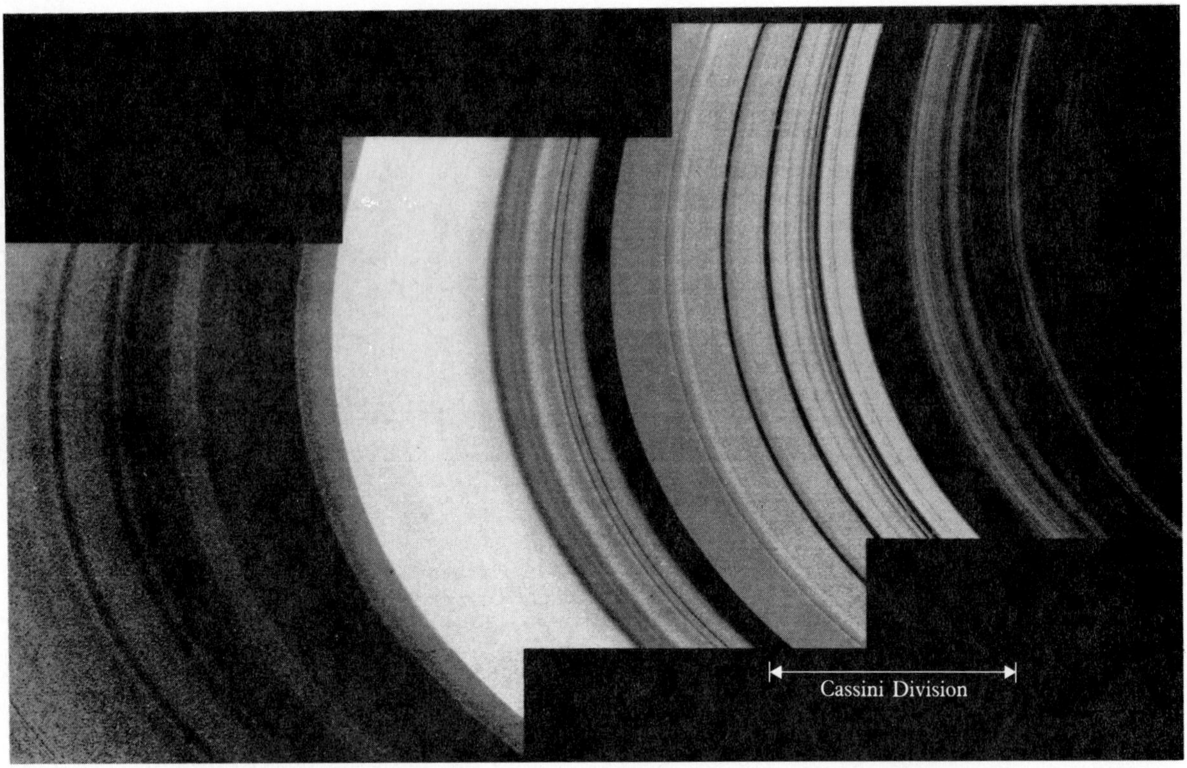

FIGURE 18.5 Seen in this greatly magnified view, the Cassini Division in Saturn's rings, which looks dark and empty when viewed from Earth, is actually found to contain several ring structures of its own. The classical Cassini Division, just to the right of the center, contains five bands of material bounded by dark gaps.

discontinuous, kinky ringlets (Fig. 18.6) and one known small satellite. These peculiar ribbons of material are only about 20 km wide and were observable from the Voyagers only when the spacecraft passed very close to the rings. The problem of the origin of kinky rings is discussed in Section 18.5, but don't expect any answers—kinky rings remain a complex problem in celestial mechanics.

Beyond the Main Rings

The A Ring ends abruptly 96,000 km from the planet. There is no tapering off of ring particles,

FIGURE 18.6 A ring that appears to defy Kepler's laws, this discontinuous kinky segment lies in the Encke Division in the A Ring.

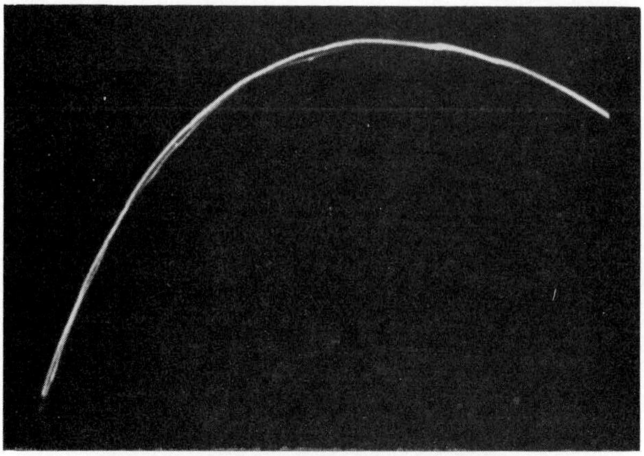

FIGURE 18.7 Saturn's F Ring consists of at least five separate strands, of which two are easily seen here. They exhibit a braided appearance, with knots of greater density (moonlets?) embedded within them.

no stray ringlets drifting off into space. There is, however, the fascinating F Ring 4000 km farther out. Unlike the rings discussed so far, the F Ring is an isolated bright ribbon whose width varies from 30 to 500 km. It is the third eccentric ring in the Saturn system. It also has the most complex and peculiar structure of any Saturn ring, including areas where it divides into multiple strands that appear to be intertwined or braided (Fig. 18.7).

The discovery of braiding in the F Ring was one of the sensations of the Voyager encounters, leading to newspaper headlines asserting that these rings disobeyed the laws of physics. They don't, but they do make it clear that scientists are sometimes unable to interpret the laws of nature to explain very complex situations. What makes the F Ring both complicated and fascinating is the presence of two small satellites of Saturn that orbit the planet on either side of the ring. The boundaries of the ring are clearly set by the presence of these satellites, and it is their gravitational influence that generates its fine structure. We will return to this subject when we discuss satellite-ring interactions.

Continuing outward from Saturn, we discover the G Ring, a tenuous band some 1000 km wide, composed of particles whose average size is about half a millimeter. This is a mysterious ring indeed, since it is far enough from Saturn (at a distance of 2.8 times the planet's radius) to be outside the boundary normally associated with ring formation, as we will see in Section 18.6. Furthermore, the particles composing this ring are so small that they should disappear in a few thousand years as a result of interactions with the plasma in Saturn's magnetosphere and erosion by random micrometeorite impact. Perhaps there are small boulders or icebergs orbiting Saturn at this distance, and a recent impact on one of these objects has produced the particles we see. There is no known nearby satellite to hold the ring in place or to supply the particles of which it is composed.

At least these latter two problems are solved in the case of the gossamer E Ring, which we met in our discussion of Enceladus (Section 17.2). Spreading across a radial distance of three to eight times the radius of Saturn, the particles of the E Ring are most concentrated at the orbit of Enceladus. The average diameter of these tiny bits of matter is only μm, comparable to the size of particles in cigarette smoke. Like the smoke, the ring is blue! It must also be a relatively recent phenomenon, and attention is certainly focused on Enceladus as the likely source of the ring material. As we discussed in the last chapter, however, there is no rigorous explanation of the way in which this mysterious moon could produce this remarkable ring.

The general problem of the origin and structure of Saturn's rings is a deep one, generating scores of papers by scientists intrigued by these issues. We will discuss ideas about the origins of ring systems in Section 18.6. To address properly the structure of these rings, we need to look at the fine-scale organization of the ringlets revealed by Voyager and to compare the structure of the Saturn's rings with the very different rings of Uranus.

18.3 RINGS OF URANUS

Occultations of stars by planets first attracted the attention of most planetary scientists in 1977, when they yielded the unexpected discovery of the rings of Uranus. This discovery, however, was the accidental byproduct of another experiment aimed at studying the atmosphere of this planet.

FIGURE 18.8 James Elliot making an adjustment in the 0.9-m telescope of the Kuiper Airborne Observatory (KAO), with which he discovered the rings of Uranus.

Discovery of the Rings of Uranus

On March 10, 1977, an occultation of a star by Uranus occurred which was particularly favorable, except for one thing: The event could not be seen from any of the more populated parts of the Earth. Only from Antarctica and the Indian Ocean was it a "sure thing," with marginal conditions in Australia, southern India, and South Africa. Nevertheless, observers set up equipment at all the observatories that might be in the path to await this exceptional event. To supplement the ground-based observations, James Elliot and his colleagues from MIT proposed to use NASA's Kuiper Airborne Observatory (Fig. 18.8) to fly above the southern Indian Ocean, well within the predicted zone for the occultation. To accommodate this project, the airplane had to fly to the Southern Hemisphere and operate out of an Australian airbase.

As expected, Elliot's team succeeded in measuring the occultation of the star by Uranus, while the ground-based observers saw the planet skim by the star without blocking its light. But the most important results that night were not from the planet at all, as Elliot has recounted in detail in his book about rings (see the Additional Reading list). As measured from both the airplane and the ground, the occulted star began to wink out about 40 minutes before it should first have been affected by the upper atmosphere of the planet. Several times the star dimmed dramatically for intervals that lasted from 2 to 8 seconds, then returned just as suddenly to full brightness (Fig. 18.9).

Clearly something was briefly blocking the starlight, and at first Elliot and his colleagues on the plane thought they might be seeing a swarm of small uranian satellites. The occultation events were not randomly spaced, however. As Uranus moved beyond the star, exactly the same sequence of occultations occurred, but in reverse order. Further, the occultation pattern was the same at the different observing sites, even though some of them did not detect an occultation of the star by the planet itself. What had

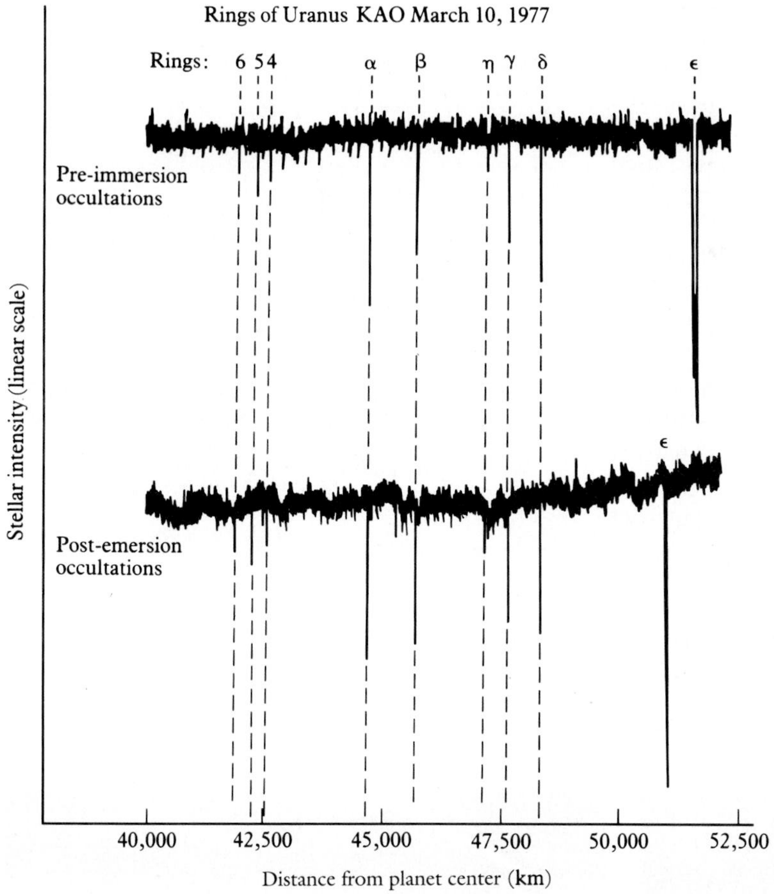

FIGURE 18.9 The two intensity tracings that revealed the presence of Uranus's rings. Top: As Uranus passed in front of a star, the intensity of starlight recorded by a photometer on the KAO telescope suddenly dimmed, then recovered, and repeated this behavior nine times. Bottom: The same thing happened on the other side of the planet, except that the eccentric Epsilon Ring was closer to Uranus and narrower on this side of the planet.

been discovered was a series of narrow, opaque rings, invisible in reflected sunlight but detected by their ability to block the light from a distant star. The situation resembles that of counting the number of cars in a train on a dark night by recording the visibility of a streetlight viewed through the passing cars.

The Advantage of Occultations

The uranian rings had not been detected before because they are narrow and composed of dark particles. But even if they had been seen in reflected sunlight, the occultation technique reveals much more than could be photographed with any telescope. The reason is that the occultation resolution is not limited by the size of the telescope or the shimmering of the Earth's atmosphere. The level of detail probed by an occultation depends on how rapidly the rings appear to move across the sky and on how often we can sample the changing brightness of the star. In the case of the Uranus occultation, structure as small as a few kilometers could be measured.

Since their discovery, the rings of Uranus have been repeatedly studied using occultations, with the results described in Section 18.5. This technique is very well suited to the task for three reasons: (1) Since the rings are dark, they produce almost no reflected light to interfere; (2) at near-infrared wavelengths, the planet itself is also very dark, further reducing interference; and (3) because of Uranus's high tilt we now can see the rings nearly face-on, presenting a wide target. In contrast, stellar occultations cannot be used from Earth to study the saturnian rings, which are very bright and are viewed obliquely. Ironically, therefore, the *fine structure* of the faint and distant uranian rings can be studied better from the Earth than can that of the much brighter saturnian rings.

The rings of Uranus are just the opposite of the Saturn rings in several respects. At Saturn we have broad rings interrupted by a few narrow gaps; at Uranus we have very narrow rings separated by broad gaps. The Saturn ring particles

TABLE 18.3 Rings of Uranus

Ring Name	Distance (km)	Width (km)	Eccentricity
6 Ring	41,850	1–3	0.0010
5 Ring	42,240	2–3	0.0019
4 Ring	42,580	2	0.0011
Alpha	44,730	8–11	0.0008
Beta	45,670	7–11	0.0004
Eta	47,180	55	0
Gamma	47,630	1–4	0
Delta	48,310	3–9	0
Lambda	50,040	1–2	?
Epsilon	51,160	22–93	0.0079

FIGURE 18.10 The Uranus rings as revealed by the Voyager 2 cameras corroborated and extended the occultation observations. (a) A distant view of the planet shows only the Epsilon Ring, revealing its variation in thickness (compare upper right with lower left). (b) Here we see all of the pre-Voyager rings. Starting from the bottom, they are 6, 5, 4, α, β, η, γ, δ, ϵ.

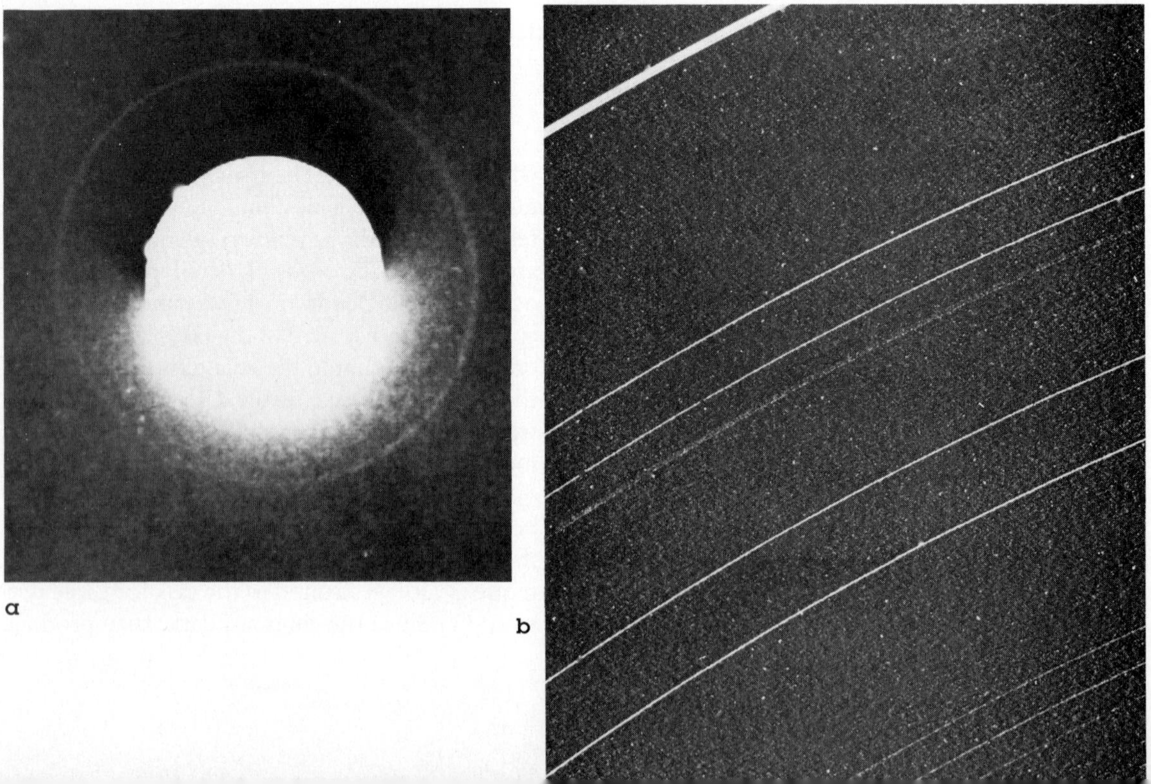

a b

are bright and composed of ice; those at Uranus are dark and apparently made of (or at least coated by) some sort of carbonaceous material.

The dimensions of the rings of Uranus are shown in Table 18.3. There are five main rings, named the Alpha, Beta, Gamma, Delta, and Epsilon rings in order outward from the planet (Fig. 18.10). Their distances from the atmosphere of Uranus range from 19,000 to 26,000 km. An additional complex ring called the Eta Ring lies between the Beta and Gamma rings at 22,000 km from Uranus. All six of these rings, as well as three less prominent rings inside the Alpha Ring, were known from occultation observations before Voyager 2 reached Uranus.

Voyager Observations of the Rings of Uranus

In 1986, the Voyager cameras found one more narrow ring, bringing the total to one dozen. Two occultations of stars by the rings were also observed by the spacecraft, yielding ring profiles of substantially higher resolution than that obtained either by the Voyager cameras or by Earth-based occultation observations.

Most of the 12 rings of Uranus are nearly circular and exceedingly narrow, no more than 10 km in width. In other words, their lengths are nearly 100,000 times their breadth, like a piece of spaghetti several city blocks long. Yet particles within each ring are very close to one another, blocking most of the starlight when the ring passes between us and an occulted star.

Two of the rings have a special structure, nearly as peculiar as that of the F Ring of Saturn. The Epsilon Ring, which probably contains as much mass as all the other rings combined, is both eccentric and variable in width. Where it comes closest to Uranus, its width is 22 km; at the opposite side, where the ring is 700 km farther from the planet, it has a much greater width of 93 km. Like the eccentric rings of Saturn, the Epsilon Ring must be held together by some gravitational force to maintain its shape. The Eta Ring is equally misshapen. It consists of a relatively broad, low-density ring 55 km wide, with a narrow, denser component at its inner edge.

Because the rings are so dark and difficult to photograph, even the best Voyager images are rather disappointing. However, the nonimaging data from the spacecraft are more exciting. The best stellar occultation profiles resolve features as small as 10 m in size, revealing intricate structure and demonstrating that the thickness of these rings, like those of Saturn, is no more than a few tens of meters (Fig. 18.11). In addition, the results of the occultation of the spacecraft by the rings, during which the effects of the rings on the transmitted radio signal were measured from

FIGURE 18.11 A profile of the Epsilon Ring of Uranus obtained by the Voyager 2 spacecraft as the ring occulted a star (cf. Fig. 18.9).

Earth, show that the Epsilon Ring is composed primarily of larger particles than those characteristic of Saturn's rings. Very little of the mass in the Uranus ring system is to be found in particles smaller than a few centimeters in diameter. Apparently such fine dust is either not formed, or else it has long since been swept out of the rings, leaving only the larger particles behind. This is another, unexpected difference between the rings of Uranus and those of Saturn.

18.4 RINGS OF NEPTUNE

Prior to the Voyager 2 flyby, information about possible rings around Neptune was contradictory. Following the success of the stellar occultation technique for detecting the rings of Uranus, several such occultations were observed for Neptune. In some cases, no evidence for rings was found. In others, the star's light disappeared briefly on one side of the planet but not on the other. These results gave rise to the idea that Neptune might be surrounded by discontinuous rings, or arcs, as they came to be called, and small satellites. An arc or a satellite would block the light of a star on only one side of the planet, and only if it happened to be in the proper place in its orbit at the time.

The View from Voyager 2

The eagerly awaited images from Voyager 2 revealed a situation even more complex. The arcs turned out to be discrete features in a very narrow, faint, but continuous ring, which is one of three prominent, narrow rings surrounding the planet (Fig. 18.12). These rings are named Galle, Leverrier, and Adams, in order of increasing distance from the planet. (Recall that Adams and Leverrier independently calculated the position of Neptune, while Galle found it with the telescope.) The four discrete, sausage-shaped concentrations of material known as arcs are in the Adams Ring. There is still no accepted theory that explains how the material in these arcs remains concentrated in this way, instead of spreading around the orbit with the rest of the ring particles. The satellite Larissa indeed turns out to have been responsible for one of the other occultation events observed from Earth.

The Leverrier and Galle rings, as well as the material in the orbit of the satellite Galatea and the wide ring stretching outward from Leverrier (Fig. 18.13), consist of particles that are too widely dispersed to be detected by the occultation technique. The reason they show up so well in the Voyager pictures is a result of an optical effect called **forward scattering**. When particles

FIGURE 18.12 The ring system of Neptune recorded by Voyager 2 as the spacecraft passed behind the planet. The faintness of the rings required a long exposure, which reveals the background stars as well as intense, scattered light from the planet itself. The overexposed image of the planet is omitted from this mosaic.

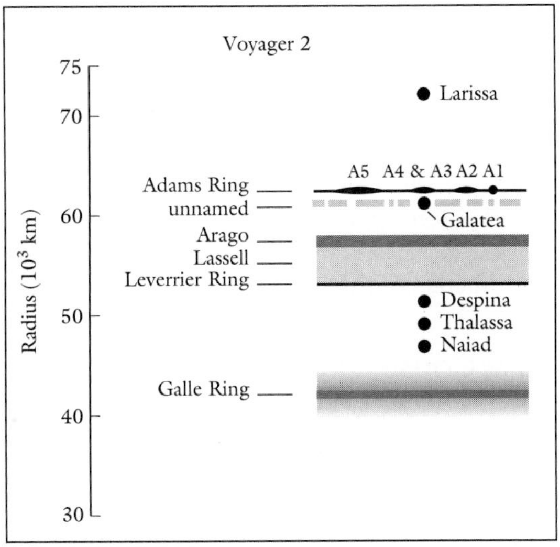

FIGURE 18.13 The structure of the Neptune ring system is illustrated in this diagram, which also shows the radial position of the inner satellite orbits (cf. Fig. 18.12).

in a certain range of sizes are struck by a beam of light, most of that light is scattered forward in the same direction as the incident beam. This is why the windshield of your car looks so much dirtier when you are driving toward the Sun. As Voyager 2 approached Neptune (or one of the other giant planets), the light scattered back to the cameras by the small ring particles was very faint. But after the spacecraft passed the planet, got into the planet's shadow, and was able to look back in a sunward direction, the rings just "lit up."

18.5 RING DYNAMICS AND SATELLITE-RING INTERACTIONS

As the resolution of ring images improves, we find more and more structure. What appeared to Huygens as a single sheet of material surrounding Saturn was found by Cassini to be divided in two parts. Johann Encke found a second gap that was established beyond any doubt by James Keeler, late in the nineteenth century. The Voyager pictures demonstrated that the particles in the rings are organized into thousands of ringlets, including some within the Cassini Division, and the occultation results have shown that still more complex structures are present.

In the Voyager occultation experiments, the brightness of the star was recorded 100 times per second as it winked on and off through the wide expanse of the rings. At Saturn, the result was approximately a million separate measures of the transparency of the ring along a line 82,000 km long. Thus features could be resolved down to 100 m in size, much better than the best Voyager imaging resolution of a few kilometers. The narrow F Ring, for instance, was resolved by the spacecraft cameras into a series of ringlets each a few kilometers across, while in the occultation data the brightest ringlet itself showed even more structure at the 100-m level. The many regular patterns revealed by both imaging and the occultation data in the A and B Rings were equally interesting.

Waves in the Rings

Remarkably, the fine structure of Saturn's rings varies with time and location. Figure 18.14 illustrates the structure in the outer part of the B Ring, at the border of the Cassini Division. Four different photos, taken at different times, are shown side by side. While the major features line up well, it is clear that the smaller ringlets do not. These ring structures are not fixed, but shift from hour to hour. We should think of them as transient waves, flowing back and forth over the rings' spinning surfaces like waves on the ocean.

Tens of thousands of ringlets can be identified, many of them grouped into patterns of regularly spaced bright and dark lines. Many of these patterns are examples of **spiral density waves** (Fig. 18.15). As the name implies, these are waves that follow a spiral pattern, like the grooves on an old-fashioned phonograph

FIGURE 18.14 These four radial segments of the outer edge of the B Ring of Saturn show that the fine structure in the ring is not uniform around the planet. The outer edge of the B Ring itself is eccentric, so it occurs at different distances from the planet at different places along its circumference.

record. Density waves had been predicted theoretically but never observed before Voyager arrived at Saturn. They are a phenomenon peculiar to a flat spinning disk in which individual particles can interact gravitationally. Even though the individual particles rarely touch, the ensemble of billions of particles behaves in many ways like a thin sheet of rubber, because of mutual gravitational attractions. This is especially true in the B Ring, where the particles are most closely packed together.

Another similar spiral wave pattern can occur when the thin sheet of spinning particles is bent upward or downward by the gravitational pull of one of the inner satellites. These **bending waves** (also shown in Fig. 18.15) are wrinkles in the rings, and they help to give it some thickness instead of letting the particles collapse down into a single very thin layer, as they would do without the influence of the satellites.

What Causes the Structure in the Rings?

Dozens of individual spiral wave patterns have been identified, primarily in Saturn's A Ring, but a great deal of the structure, particularly that seen in the B Ring, cannot be accounted for by density waves or bending waves. Some of this structure is probably the result of the superposition of unidentified wave patterns, but much of it may have quite different causes. We could

FIGURE 18.15 The same 5:3 gravitational resonance with the satellite Mimas generates both these systems of waves. The one on the left is a bending wave with ripples 1–2 km high, sufficient to cast shadows. On the right is a spiral density wave consisting of local concentrations of material within the ring plane.

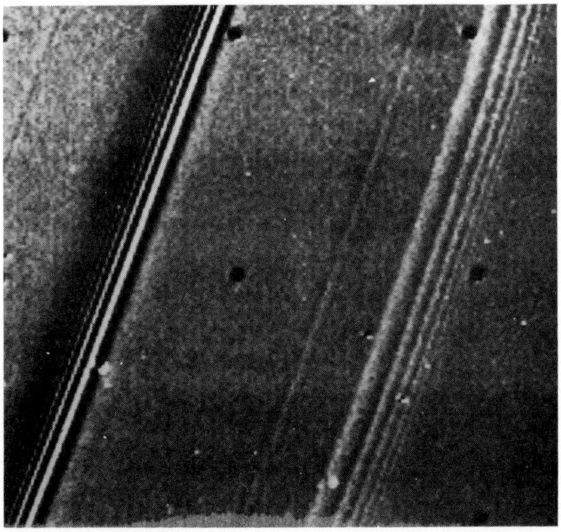

understand the behavior of the rings better if we had moving pictures of the changing ring patterns. Unfortunately, however, there was only a single Voyager occultation, and the high-resolution images are only snapshots. A Saturn orbiter spacecraft is required to obtain the data we need, and this is one of the tasks of the Cassini-Huygens mission.

While we wait for these new data, we can try our best to understand what we have seen so far. Why are the edges of the rings so sharply defined? What causes the Cassini Division and the smaller gaps in the rings of Saturn? Why are three saturnian ringlets, as well as most of the uranian rings, eccentric? Why are the rings of Uranus so narrow? And what generates the spiral wave patterns? It turns out that the answers to all of these questions involve interactions between the rings and the satellites.

Sharp Edges of Rings

Left to itself, a planetary ring will slowly spread out. Over hundreds of millions of years, the interactions between closely spaced ring particles will force them apart, with some spiraling toward the planet to disintegrate like meteors in its atmosphere and others expanding out to ever greater distances from the planet. To be stable, a ring must be bounded.

The binding force that limits the outward spreading of the rings of Saturn is gravitational. The outer edge of the A Ring is in a six-sevenths resonant orbit with Janus and Epimetheus, and the gravitation of these two co-orbital satellites keeps the ring particles firmly trapped at this particular distance from Saturn.

Shepherd Satellites

A different kind of gravitational influence holds the thin F Ring of Saturn in place and even explains its slightly eccentric shape. In this case there are two satellites, Prometheus and Pandora, one on each side of the ring (Fig. 18.16).

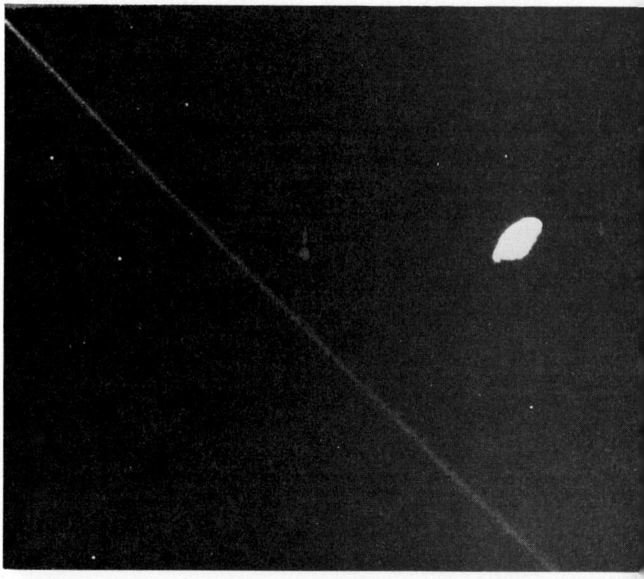

FIGURE 18.16 The F Ring of Saturn appears to be held in place by the two small shepherding satellites Prometheus and Pandora. This view shows a segment of the F Ring with Prometheus in the same field of view.

These are called the **shepherd satellites** for their role in keeping the particles of the F Ring narrowly confined. Probably it is the influence of the shepherd satellites that also generates the braids in the F Ring, but a complete theory for this process has not yet been worked out.

The rings of Uranus were discovered at a time when the Saturn rings were still thought to be broad and relatively unstructured, and at first the existence of such narrow rings utterly confounded astronomers. It was in response to the challenge posed by the narrow rings of Uranus that theorists Peter Goldreich and Scott Tremaine (both then at Caltech) first developed the idea that a ring could be confined by shepherding satellites.

These shepherds, predicted for Uranus, were confirmed in concept by the discovery of Pandora and Prometheus on either side of the F Ring of Saturn. It was therefore with great interest that the Voyager cameras searched for similar shepherds at Uranus. Unfortunately, however, the search was only partially successful.

Two uranian shepherds (Cordelia and Ophelia) were discovered in association with the Epsilon Ring. Each is a small dark satellite less than 50 km in diameter. These shepherds orbit about 2000 km on either side of the ring, very much like the F Ring shepherds at Saturn. However, no other shepherd satellites were seen near the other uranian rings. Since the cameras were sensitive only to dark objects 10 km or more in diameter, smaller shepherds may be present. Thus we still have no confirmation of the postulated 22 satellites needed to explain all 11 of the other known rings of Uranus.

Embedded Satellites

If there are satellites inside the rings, they can have gravitational effects also. Within the broad main rings of Saturn, the presence of unseen satellites is thought to produce the few genuine gaps. By its gravitational herding effect, a small satellite could sweep clear a lane much wider than its own diameter. To confirm this explanation, very careful searches were made of the Voyager images for these **embedded satellites**.

The first evidence that embedded satellites are indeed present was the discovery of scalloped edges in the Cassini Division and the Encke Gap. These scallops are the result of waves generated in the bounding ring material by the passages of the satellites, much like the wake that stretches far behind a ship moving across the oceans of Earth. From the properties of their wakes, the positions and masses of these embedded objects can be calculated, and both turn out to be less than 15 km in diameter. So far, only one of them, Pan, has been found in a Voyager picture.

Embedded satellites have also been suggested as an important agent for forming narrow, eccentric rings. Contradictory though it may seem, calculations have shown that under some circumstances—generally involving closely packed ring particles—an embedded satellite will produce a narrow ring rather than clearing a gap. If the orbit of the embedded satellite is eccentric, so will be the path of the thin ring it controls. This is an especially attractive theory for trying to understand the kinky rings in the Encke Division of Saturn's A Ring, and it has also been suggested as an alternative way of forming some of the rings of Uranus and Neptune.

Resonant Effects of the Larger Satellites

The larger external satellites also influence the ring structure by their gravitational effects. The Cassini Division, for example, is caused by Mimas. At the inner edge of the division a ring particle orbits Saturn exactly twice for each orbit of Mimas, and this strong resonance is apparently responsible for the resonant gap located here, just as the resonant gaps in the asteroid belt are caused by Jupiter (Section 5.2).

Weaker resonances with other, smaller satellites do not produce gaps, but they do disturb the ring particles having orbital periods that are simple fractions of those of the satellites. The two co-orbital satellites generate a large number of these resonances. The disturbances in these cases produce the spiral density waves. Each spiral represents a wave pattern that originates at the resonant position and then winds its way outward.

Mimas is the main cause of the spiral bending waves, which result from a gravitational tug up out of the ring plane. Since Mimas does not orbit Saturn in exactly the same plane as the ring, it exerts such forces each time it circles the planet. These bending waves are seen to extend inward, rather than outward from their place of origin.

By invoking the presence of satellites both

seen and unseen, we can explain a great deal of complex structure found in the rings of Saturn. Not everything makes sense, of course, but most of the prominent features seem understandable. The main problem with this approach is that it depends in part on the existence of a variety of unseen satellites embedded within the rings. These satellites have been searched for but not found. If they are present, we wonder how they have eluded detection; if they are absent, then our understanding of ring structure is much poorer than we would like to think. Similarly, the presence of two shepherds for the Epsilon Ring of Uranus seems to provide an adequate explanation for the existence of this ring, but the satellites to control the other 11 rings of Uranus remain at present only speculation.

18.6 ORIGIN OF RING SYSTEMS

The planetary ring systems have several features in common: They are composed of small particles of a wide variety of sizes; they are close to their planets; and much of their structure appears to be controlled by a few larger objects, including the external satellites. The interactions with satellites in particular seem to be necessary to bound a ring and keep it from gradually spreading out and fading away. But what could have produced the original swarm of billions of particles in a disk surrounding the planets?

There are two basic theories of ring origin. First is the breakup theory, which suggests that the rings are the remains of a shattered satellite. The second theory, which takes the reverse perspective, suggests that the rings are made of particles that were never able to come together to form a satellite in the first place.

The Tidal Stability Limit

In either theory of formation, an important role is played by tidal forces. When tides were discussed in Section 8.3, we noted that they are very sensitive to distance, varying as the inverse cube of the separation between two bodies. The effect of tides is to distort a satellite, raising bulges on it that face toward and away from the planet it orbits. The closer a satellite comes to its primary, the larger the tidal distortion. Ultimately, the satellite can be torn apart, if its internal strength is not great enough to withstand the tidal stress.

Around each planet there exists a **tidal stability limit**, the distance within which tidal forces can destroy an intruding satellite. This limit was first calculated by the nineteenth-century French mathematician Edouard Roche, and it is often called the Roche limit. Its exact value depends on the density and internal strength of the satellite. If we simply consider two particles that are just touching each other, the tidal stability limit is at about 2.5 planetary radii from the center of the planet. In this case, the limit defines the distance from the planet at which the difference in the gravitational force exerted by the planet on each of the two particles is larger than their mutual gravitational attraction. Thus they will not be able to coalesce and form a larger body. As shown in Fig. 18.1, the major ring systems of each of the four giant planets lie within the respective Roche limits. The G and E rings of Saturn are notable exceptions, their distant locations emphasizing their relatively recent origins.

We stress that the tidal stability limit is calculated for an intruding satellite with no intrinsic strength of its own. It tells us where a liquid satellite, or one made up of disconnected bits of gravel, would be pulled apart by tidal forces. The stronger a satellite, the less it is affected by tides. As an extreme example, consider the Space Shuttle or any other low-orbit artificial Earth satellite. These operate well inside the stability limit, yet they are not pulled apart by tides. On the other hand, a loose tool left floating in the Shuttle bay will drift away under tidal forces and be lost in space. This is what it means to be inside the Roche stability limit.

Another way of looking at the stability limit is to think of it as the distance within which individual particles will not come together under their own gravity to form a larger body. If a very large disk of particles once surrounded Saturn, for instance, it is easy to imagine them coalescing to form individual satellites everywhere except inside the stability limit, where they might remain to this day as the ring particles. In such a case, the ring particles would probably all be very small, and they would retain their primitive composition.

Ring Formation by Satellite Breakup

In the breakup theory of ring formation, we might imagine a satellite or even a passing comet coming too close and being torn apart by tidal forces. There is an even more likely scenario, however, in view of the very heavy meteoroidal bombardment that took place early in the history of the Saturn system. We have already suggested that a number of small inner satellites might have been broken up during this early bombardment, leaving fragments such as the co-orbital and shepherd satellites. If a satellite were disrupted inside the tidal stability limit, it would be unable to reform itself, and the fragments would spread out into a ring. In this case we would expect a wide range of particle sizes, including a few large fragments several kilometers across. The embedded satellites found in the rings of Saturn may be examples of such fragments. If the presence of many such bodies in the size range from 1 to 10 km could be confirmed, this would also argue for the breakup theory.

How much material is actually present in the four ring systems we are discussing? This is difficult to determine, since what we see is primarily the smaller particles, while most of the mass may be contained in just a few particles (or moonlets) near the upper end of the size distribution. This caution about unseen large particles aside, however, it has proved possible to measure the mass of the main B Ring of Saturn from Voyager data. Since the spiral density waves depend on the gravitational interactions between ring particles, their behavior is sensitive to the mass of material present. Within the B Ring, there is nearly 1 ton of material per square meter, spread through the 20-meter thickness of the ring. Summing over the entire ring system, we obtain a total mass of about 10^{18} kg. The mass of the Uranus rings is guessed to be at least a thousand times less, and that of the Jupiter rings more than a million times less, than the Saturn rings. Neptune's system may have even less mass than Jupiter's. Estimates for the masses of the rings are given in Table 18.1.

The total mass of the saturnian rings of 10^{18} kg is equal to that of an icy satellite about 250 km in diameter—about the same size as the larger co-orbital satellite, Janus. This mass is therefore consistent with the idea that the rings might have resulted from the breakup of a typical inner satellite near the tidal stability limit. Alternatively, it is easy to imagine there being this much left-over material that never formed a satellite.

Although the presence of small moons appears to provide a gravitational explanation for much of the ring structure, it does not answer intriguing questions about the lifetime of rings. Calculations show that if the rings were made only of small particles—the kind of particles that we deduce from Voyager data—they would not survive for more than a few hundred million years. In that time the ring particles would be eroded away by meteoritic impacts and sputtering by charged particles in the magnetosphere. Thus either the rings are relatively young, or else they are renewed from within by the continuing breakup of kilometer-sized particles, just as the near-Earth asteroids are renewed by collisions among larger objects in the main asteroid belt. So far, these kilometer-size ring particles remain conjectural, as they are too small for the Voyager cameras to detect.

We are left with the mystery of the different chemical compositions of the four known ring systems. Saturn seems to make the most sense, since both its rings and its satellites are made of the same icy material. Jupiter can be understood also when we remember that icy bodies, either satellites or ring particles, could not have survived near the planet early in its history, when it radiated more heat than at present. The dusty rings of Jupiter represent the results of erosion from rocky satellites. In fact, even the bright rings of Saturn must include some non-icy material. Voyager observations show subtle color differences between the C and B rings as well as between discrete ringlets in the B system and their immediate neighbors. Ground-based observations that record spectra of the entire ring system demonstrate that the rings are absorbing more ultraviolet light than pure ice would. Perhaps some silicate or organic dust is mixed in with the ice, as one would indeed expect if these rings represent the disruption of an icy object like a small satellite.

Uranus and Neptune remain perhaps the most strange. Their large satellites are icy, but both the rings and the small inner satellites are made of (or at least coated by) very dark, presumably carbonaceous material. The association of the dark satellites with the dark rings suggests that the rings might have been formed by the breakup of one or more inner satellites. This possibility is supported by some models for the evolution of these systems: The inner satellites may once have been even closer to their parent planets, but may have moved out to their present positions during the lifetime of the solar system in response to tidal forces.

Formation of Distant Rings

The ideas we have just discussed relate to rings formed inside the tidal stability limit. What about the E and G rings of Saturn, which are well outside this boundary? We have already noted that these rings must be very young because the tiny particles composing them will diffuse away into space very rapidly in relation to the 4.5-billion-year lifetime of the solar system. What process could produce such small particles? We see dust like this in the zodiacal cloud, which originates from collisions within the asteroid belt and from debris left behind by expiring comets. The latter source may have some relevance here. If a comet broke up because of an impact while it was passing through the Saturn system, one might expect the debris to form a temporary ring. Objects 10–15 km in diameter orbiting the planet at the distance of the G Ring would have escaped detection by Voyager cameras and would furnish targets for such impacts.

In the case of Enceladus, there is the alternative speculation that the E Ring particles are somehow generated by the satellite itself, as a spray of liquid water from the satellite's interior freezes in the vacuum of space. This is an interesting approach, but no one has produced a theory explaining how this tiny satellite could have a molten interior, despite the evidence for melting of its surface (Fig. 17.5).

18.7 QUANTITATIVE SUPPLEMENT: THE TIDAL STABILITY LIMIT

The tidal stability limit, also called the Roche limit, is the distance from a planet at which an unconsolidated satellite (one made of liquid, for example) will be disrupted by tidal forces. We can calculate this distance by finding the balance between the self-gravitation of an object that holds it together and the differential gravitational forces from the planet that tend to pull it apart.

For the sake of simplicity, consider the situation for two identical spherical particles, each with mass m and radius r, that are just in contact with each other. The gravitational force between them is

$$F_g = G\frac{m^2}{4r^2}$$

These two particles will be at the tidal stability limit when the differential gravitational force on them is equal to F_g. Suppose they are in orbit about a planet of mass M at a distance d from the center of the planet. If they remain aligned as they orbit so that one is at a distance from the planet of $d + r$ and the other is at $d - r$, the differential force on them is simply

$$F_d = G\frac{Mm}{(d - r)^2} - G\frac{Mm}{(d + r)^2}$$

Equating the differential force to the self-gravitational force we obtain

$$G\frac{m^2}{4r^2} = GMm\left(\frac{1}{(d - r)^2} - \frac{1}{(d + r)^2}\right)$$

Suppose that these are two particles of a planetary ring. Farther from the planet than this limit, the gravitational attraction between them can allow them to merge into a larger particle that can in turn continue to grow. But inside this tidal stability limit the particles will not stick together because of their mutual gravitational attraction.

The precise distance at which these forces balance depends on the densities of the particles and, to some extent, on their exact shapes. The term *Roche limit* is usually applied to the case considered by Roche, in which the orbiting particles have the same density as the planet itself. Algebraically, we can derive it from the above equation by expanding the right side and simplifying, to obtain

$$G\frac{m^2}{4r^2} = GMm(4rd_R^{-3})$$

Solving for d_R, the Roche limit, for the case where the densities of planet and particle are the same, we have

$$d_R = 2.5R$$

where R is the radius of the planet.

SUMMARY

Each giant planet has a ring system made up of small bodies orbiting within the tidal stability limit. The particles forming the bright rings of Saturn vary in size from 10 m to dust. The rings of Neptune and Uranus are made of very dark material, and most of the particles span a more limited size range, from a few centimeters up to tens of meters, while the tenuous ring of Jupiter, the E and G rings of Saturn, and the wide ring of Neptune are little more than bands or ribbons of smoke-size particles. In the case of Jupiter, these particles must consist of dust eroded from the inner satellites. For Saturn, water ice should be the dominant constituent of the small particles as it is for the larger ones. The dark ring particles surrounding Uranus and Neptune probably derive their low reflectivity from organic matter.

Structurally, the rings display a remarkable diversity. Within the rings of Saturn, by far the best studied, we see examples of eccentric rings, kinky rings, and braided rings. In some places the boundaries of the rings are established by shepherding or embedded satellites, and in some cases resonances with more distant satellites fill this role. Other resonances stimulate spiral density waves and bending waves that propagate across the rings, generating a complex structure of thousands of ringlets. All the rings of Uranus and all but one of Neptune's rings are confined to narrow ribbons. The outer Epsilon Ring of Uranus has two shepherding satellites, as does the narrow F Ring of Saturn, but no such shepherds have been found for the other narrow rings of Uranus or Neptune. The outermost, or Adams, ring of Neptune displays four concentrations of material, called arcs, that resemble sausages on a string.

The rings of Uranus were discovered by observations of a stellar occultation, and this technique has proved to be very powerful for studies of the other ring systems as well. Occultation observations from and with spacecraft have permitted resolution of detail within the rings to scales of a few tens of meters.

We are still not sure of the ways in which the various rings originated, but tidal forces and impacts must have played important roles. At least two possibilities exist. Perhaps the rings within the tidal stability limit represent material that was prevented from forming a large satellite by the presence of strong tidal forces. The alternative option, which is currently the most favored, suggests that a satellite near the tidal stability limit was broken apart by one or more impacts and the fragments were unable to reaccrete owing to tidal forces. In the case of distant rings such as Saturn's E and G rings, one must invoke a local source that produces very fine particles, as these rings must be relatively young.

KEY TERMS

bending wave
eccentric ring
embedded satellite
forward scattering
shepherd satellites
spiral density wave
tidal stability (Roche) limit

Review Questions

1. Compare the four outer planet ring systems. What do they have in common? What are the most striking differences?

2. Describe the individual particles in the rings of Saturn. Do they all have the same composition? How do these particles compare with those found in the other three systems? Why are they different?

3. Compare the discoveries of the rings of Uranus and Neptune. Why was the interpretation of the Neptune results so difficult? What are "arcs" and where are they found?

4. Explain how the occultation technique is used to probe ring structure. What exactly is measured? What determines the resolution? How do ground-based and spacecraft data compare?

5. Ring structure is determined in part by satellite interactions. Explain the roles of shepherd satellites, embedded satellites, and satellite resonances. Which of these produces what structures in the known ring systems?

6. Explain the difference between bending waves and spiral density waves. Which type produces vertical structure in the rings? Why?

7. What is the tidal stability limit? How is it possible for artificial satellites to circle the Earth inside this limit?

8. Describe alternative ways in which rings can form. Would you expect frequent formation of short-lived ring systems early in solar system history? Explain.

Quantitative Exercises

1. Find the distance in kilometers above the surfaces (or cloud tops) corresponding to the tidal stability limits for Mars, Earth, Saturn, and Neptune.

2. Make a chart showing the positions of the A, F, G, and E rings of Saturn and the orbits of Atlas, Pan, Pandora, and Prometheus with respect to the Saturn tidal stability limit. What happens if you use a density of 0.7 g/cm^3 for Saturn and 1.4 g/cm^3 for the satellites?

3. Repeat exercise 2 for the Neptune system, plotting rings and inner satellites. Do you think there may be some significance to the fact that arcs are found only in the Adams Ring and not in Leverrier or Galle? Explain.

4. The NASA Space Shuttle typically orbits the Earth at an altitude of about 200 km. Is this inside or outside the Earth's Roche limit? How does the orbit of Phobos relate to the Roche limit of Mars?

Additional Reading

*Bergstralh, J.T., E. Miner, and M.S. Matthews, eds. 1991. *Uranus.* Tucson: University of Arizona Press. A collection of technical articles dominated by results from the Voyager 2 encounters, including two chapters specifically devoted to rings of Uranus.

*Cruikshank, D.P., and M.S. Matthews, eds. 1995. *Neptune.* Tucson: University of Arizona Press. A companion volume to *Uranus* from the Arizona Space Science series, again emphasizing the Voyager 2 results, including a chapter on the rings of Neptune.

Elliot, J., and R. Kerr. 1984. *Rings: Discoveries from Galileo to Voyager.* Cambridge, Mass.: MIT Press. A well-written survey of planetary rings prior to the Voyager 2 encounters with Uranus and Neptune; the product of a collaboration between a scientist and a science journalist.

*Greenberg, R., and A. Brahic, eds. 1984. *Planetary Rings.* Tucson: University of Arizona Press. A collection of technical articles by experts in the field, this is the standard reference for the rings of Jupiter, Saturn, and Uranus.

*Indicates the more technical readings.

19

The Origin of Planets

Shiva, the Hindu god of destruction and rebirth, is here shown in his form as the cosmic dancer, Nataraja. The Hindu concept of the recurring cycle of life, death, and rebirth is a good analogy for the formation of the planetary system from the gas and dust produced in the interiors of multiple generations of stars.

In this book we have discussed more than 50 worlds, some in considerable detail. These planets, satellites, asteroids, and comets display an incredible diversity of composition and history. Yet they were all presumably formed at about the same time, condensing from the same primordial solar nebula that gave birth to the Sun. Exactly how this happened is not yet clear, but there is nothing in our current models that would make the origin of planets a rare event. Instead, it appears that the formation of planetary systems is a natural part of star formation, opening the possibility that there may be many such systems around other stars, some similar to our own. If we can understand the processes that formed the planets we know, we can then try to predict how probable it is that these same processes produced other planetary systems. Conversely, if we could find planets orbiting other stars, their existence might help us to understand the origin and evolution of our own system.

Having stated what we would like to do, we must admit right away that it is not yet possible to do it. We are unable to work backward from the wealth of data on the present state of the solar system to derive a unique, detailed description of how the system began. Neither can we work forward from a theory of star formation to the production of a solar system with all the properties we find in ours today. Even the most basic part of the process, the formation of the Sun itself, is only poorly understood. Instead of a unique and all-encompassing theory, we must work with a collection of reasonable explanations for those properties of the solar system that seem especially basic.

This unsatisfactory state of affairs may improve dramatically within the next decade or two. Infrared and radio observations are constantly revealing more about the formation and early evolution of stars. A number of techniques have been developed recently to enable planets as large as Jupiter to be detected. Planet-size bodies have already been found in orbits around neutron stars and in 1995 the first "Jupiter" was discovered orbiting the nearby solar-type star 51 Peg. The direct detection of other planetary systems, together with a deeper understanding of the star formation process, will give us the great leap forward that we need to know how the Sun and its planets, satellites, asteroids, and comets came into existence from a formless interstellar cloud.

19.1 BASIC PROPERTIES OF THE PLANETARY SYSTEM

No theory of planetary formation can deal with the wealth of detail about individual objects presented in this book. To begin, therefore, we should try to identify the really basic properties of the system that may be understandable in terms of its origin. We have made a list of 13 of these properties in Table 19.1. In the subsequent discussion, we shall refer to them as Fact 1, Fact 2, and so forth.

The "facts" from Table 19.1 will be used in the following sections to constrain theories of the origin and early evolution of the planetary system. Of course, the concept that emerges must be consistent with other data as well, particularly the detailed chemical and isotopic composition of the planets and the primitive bodies of the system, such as the comets and meteorites. In addition, any theory of origins will make use of the increasingly detailed models for star formation that astronomers are developing from improved observations and applications of theoretical astrophysics.

There are other items that we could add to the list. We may wonder about the exceptionally large mass of Jupiter, for example, or the fact

TABLE 19.1 Facts that any theory of origins should explain

1. The most ancient age recorded in the solar system is just over 4.5 billion years, even though the Galaxy that contains the solar system is much older than this. Further, many meteorites share this common age.
2. The planets all move around the Sun in the same direction that the Sun rotates and nearly in the plane that passes through the Sun's equator.
3. Although the Sun has 99.9% of the mass in the solar system, the planets have 99.7% of the system's angular momentum.
4. The inner planets, which are composed primarily of the cosmically rare silicates and metals, are smaller and more dense than the outer planets. The giant planets, in contrast, have a more nearly cosmic (or solar) composition, and their satellites have lower densities, indicative of water ice and other volatiles.
5. The asteroids, which represent a composition intermediate between the metal-rich inner planets and the volatile-rich outer solar system, are located primarily between the orbits of Mars and Jupiter.
6. The primitive meteorites are composed of compounds representative of solid grains that are expected to have formed in a cooling gas cloud of cosmic (solar) abundance at temperatures of a few hundred K.
7. Comets, like the surfaces of some outer planet satellites, appear to be composed primarily of water ice, with significant quantities of trapped or frozen gases like carbon dioxide and nitrogen, plus silicate dust and dark carbonaceous material.
8. Variations in isotopic ratios established prior to solar system formation are preserved in objects formed at different distances from the Sun.
9. Volatile compounds (such as water) must have reached the inner planets even though the bulk composition of these bodies suggests formation at temperatures too high for these volatiles to form solid grains.
10. Despite the general regularity of planetary rotation, Venus, Uranus, and Pluto all rotate in a retrograde direction.
11. All the giant planets have systems of regular satellites orbiting in their equatorial planes. They resemble miniature versions of the solar system itself, with the addition of rings, which the Sun does not possess, unless you consider the zodiacal dust and the asteroid belt.
12. Three of the giant planets, Jupiter, Saturn, and Neptune, have one or more highly irregular satellites—either in retrograde orbits or with high eccentricities.
13. All of the giant planets have ice-rock cores equivalent to 10–15 Earth masses, all have atmospheres rich in hydrogen and helium, and all except Uranus are radiating substantial quantities of heat from their interiors.

that the distances between pairs of outer planets are so large. You should think about whether the model we develop is capable of addressing other phenomena that may have seemed especially puzzling when they were presented in earlier chapters.

19.2 THE LIFE OF A STAR

At the beginning of the twentieth century, many astronomers thought that the planets originated as the result of a remarkable accident: the near-collision of the Sun with another star. Since then, we have come to realize that stars form in a spinning cloud of dust and gas that we have called the solar nebula in the case of our own system. These same disk-shaped nebulae can produce smaller objects as well. Further, we have located many kinds of matter orbiting nearby stars, and their existence strengthens the idea that even single stars do not form alone. Finally, we now know that the ages of the Sun and of the planetary system are approximately the same. As we have seen, the Moon, the meteorites, and the Earth all formed 4.5 billion years ago. Astrophysicists who study stellar evolution give this same value for the age of the Sun. From all of these arguments, we conclude that the Sun and the planets probably formed together from a common source of material (Fact 1).

Stellar Nurseries

On a galactic timescale, the Sun is a relative newcomer. We are not among the latest arrivals, however. Stars that are more massive than the Sun have much shorter lifetimes. Their internal temperatures are hotter and their nuclear fires burn with greater intensity. The bright blue-

FIGURE 19.1 The Horsehead Nebula in Orion, as photographed with the 4-m telescope of the Kitt Peak National Observatory. Deep within the clouds of gas and dust that lie behind this obscuring veil, new stars are forming now, some perhaps with planets (cf. Fig. 1.1).

white stars that dazzle us at night are all younger than the Sun. Some of them are only a few *million* years old. For example, the stars in the Pleiades, a familiar cluster of stars in winter skies, came into existence after the dinosaurs had disappeared from Earth. The most brilliant and massive of these stars are destined to explode as supernovas, generating a very special group of elements that can be created only under the unique conditions that briefly occur during these cosmic cataclysms.

Even younger stars exist, since they are being formed today. Star formation takes place in clouds of interstellar gas and dust such as those in the constellation of Orion (Fig. 19.1). A typical interstellar cloud in which star formation is occurring has a mass hundreds of times greater than that of the Sun. It is composed primarily of hydrogen and helium, the predominant elements in the stars that it will spawn. The other elements that can be studied appear to be present in roughly the same proportions as they are found in the Sun and other young stars, exactly as one would expect (Table 2.2).

What may seem surprising, however, is the richness of the molecular chemistry that takes place in these clouds. Instead of just simple compounds like methane and ammonia, a large array of molecular species is continually being formed. A list of those known at the time of this writing would include more than 60 entries. Among the more interesting interstellar compounds, we call attention to ethyl alcohol (CH_3CH_2OH), the essential ingredient of liquor, beer, and wine; formaldehyde (CH_2O), good for preserving corpses; and hydrogen cyanide (HCN), a deadly poison that is also a vital compound in experiments designed to simulate chemical evolution on the primitive Earth. New molecules are constantly being discovered as astronomers use more sensitive radio and submillimeter telescopes and study new segments of the radio spectrum.

It is interesting to compare a list of interstellar molecules with the molecules found in comets (Tables 6.2, 6.3). There are many species in common: H_2O, CO_2, HCN, CH_3OH, OH, etc. As we noted in Chapter 6, some scientists think that the icy nuclei of comets contain unaltered interstellar material, trapped during the earliest stages of the formation of the solar system. Alternatively, it may be that the similarities simply reflect the universality of the processes that produce these compounds, whether they take place in the solar nebula or in interstellar clouds.

A Star Is Born

One of the first things to notice about stars is that many of them are members of multiple systems. Doubles are almost as common as singles. In principle, a double star could have planets in stable orbits if those orbits either are close to one of the components or are at a large distance from both of them, circling the center of mass of the system. In practice, we don't know whether planets will form under such conditions, so we will concentrate our attention on the much simpler case of single stars, like the Sun.

We start with a slowly rotating cloud of interstellar gas and dust that may itself be part of a much larger complex such as one of the giant clouds in Orion. At some point the cloud begins to collapse. Perhaps some gravitational instability has been created in its interior by a random coming together of some of the material; or possibly a nearby star has exploded as a supernova, seeding the cloud with short-lived radioactive elements and sending out shock waves that begin to compress the cloud. The collapse is possible as long as the energy of motion (internal pressure) of the gas in the cloud is less than the gravitational energy represented by the mass of the cloud and the distance through which it collapses.

As the collapsing cloud becomes smaller, three things happen: (1) its rate of rotation increases; (2) it flattens into a disk; and (3) it heats up, especially near the center. The heating

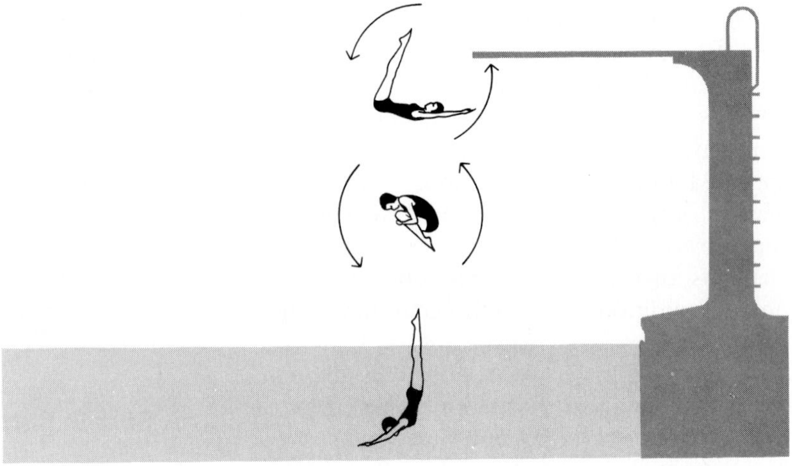

FIGURE 19.2 A high-diver who contracts her body in the direction perpendicular to her spin axis will increase her rate of spin. Her angular momentum remains constant as she falls, and the angular momentum depends on the product of her rate of spin and the square of her size in the direction perpendicular to her axis of spin.

is simply the conversion of gravitational energy to thermal energy (like the main source of internal energy for Jupiter). The increase in rotation rate results from conservation of angular momentum: As the mass of the cloud comes closer to the center, the angular velocity must increase to keep the momentum constant (Fig. 19.2). The more rapid spin in turn causes the material to flatten into a disk.

In the disk, the gas and dust can radiate energy to space much more easily than in the center, where a spherical condensation develops. Here the temperature continues to rise until it finally reaches the point where nuclear fusion of hydrogen to helium can occur. At this stage, this spherical assemblage of matter "turns on" and begins its life as a star.

Mass and Dimensions of the Disk

There is still a great deal of dispute about the mass and the dimensions of the original solar nebula and of the disk itself. The central condensation must have had a mass approximately equal to that of the present Sun, but what about the disk?

We can gain an idea of the minimum amount of mass that must have been present if we simply ask how much material of cosmic composition would be required to make all of the present planets. The idea behind this calculation is that the solar nebula started with cosmic composition and then the individual planets formed from it with compositions reflecting the local temperature. If it had been cooler close to the Sun, massive planets like Jupiter and Saturn might have formed there.

To discover what these hypothetical planets would have been like, we can perform the thought experiment of adding hydrogen and helium to the existing inner planets until the ratio of these light elements to a key heavy element like silicon or iron is the same as it is in the Sun. The masses of the resulting planets are indeed similar to those of Jupiter and Saturn. Thus we might conclude that the initial disk must have had a mass roughly equal to at least ten times the present mass of Jupiter, or about 1% of the solar mass. In this case, the minimum mass for the entire nebula is about 1.01 solar masses.

More likely, planet formation is not so efficient, and there was probably much more material available that has since been lost by the blowing away of the nebula or by gravitational ejection of larger bodies after the planets formed. Hence these days scientists generally adopt a value for the entire nebula of about 1.1

times the mass of the Sun. Recent observations of similar disks of matter around young stellar objects indicate masses of this magnitude (Fig. 19.3).

19.3 THE PROBLEM OF ANGULAR MOMENTUM

Stars are born at the centers of spinning disks of gas and dust. This configuration can still be thought of as the primordial solar nebula, even though deep in the interior of the central condensation, nuclear reactions are beginning to convert hydrogen to helium. This is probably a good place to stress again that the exact sequence in which events occurred in the early nebula is not known. Thus some of the processes we will describe in the next section must have started *before* the angular momentum transfer that we discuss here was completed.

Angular Motion of the Solar Nebula

The disk of material revolves around the forming star in the same sense that the star itself spins on its axis. The disk must also be very symmetric about a plane passing through the star's equator. This configuration provides a ready explanation for Fact 2: All the planets revolve around the Sun in the same sense in which the Sun rotates, and they move approximately in the Sun's equatorial plane. Planets forming out of this disk will have exactly that configuration and motion.

It was considerations such as these that led Immanuel Kant and Pierre Simon Laplace to propose (independently) nebular theories for the origin of the solar system as early as the eighteenth century. The main reason these theories were challenged was the attention given to Fact 3: Why do the planets have more angular momentum than the Sun? Or, to pose this question in the framework of the nebula theory, how could the Sun slow down by transferring angular momentum to the disk and hence the planets?

It is enlightening to put this problem in context. It turns out that all stars with masses at least 15% greater than the Sun's rotate much more rapidly than our star, whereas all stars with comparable or smaller masses exhibit low rotation rates. Furthermore, if one calculates the rotation rate the Sun should have, given conservation of angular momentum as the original cloud collapsed, this rate turns out to be similar to that of the more massive stars. It is also the rate that would result if the present angular momentum of the planets were put back into the Sun. We conclude that the planets have the correct angular momentum for the system as it formed; some process slowed down the Sun.

Magnetic Braking

One solution to this angular momentum problem invokes **magnetic braking.** Here one imag-

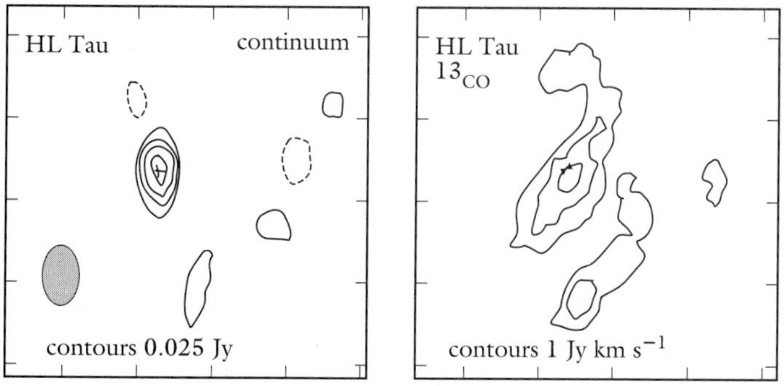

FIGURE 19.3 The gaseous disk surrounding the star HL Tau is visible in this plot of the intensity of radiation emitted by carbon monoxide molecules at radio frequencies. The star is at the position of the cross, and the disk extends diagonally on either side of it (compare Fig. 19.7). This disk is estimated to have a mass of about 10% of the stellar mass.

ines the magnetic field of the Sun moving through the disk of material around it. Since the material in the disk is following orbits defined by Kepler's laws, it is orbiting the Sun more slowly than the Sun is rotating. The situation is analogous to the interaction of the plasma in the Io torus with the rapidly spinning magnetic field of Jupiter. The material in the inner part of the disk is ionized. The Sun's magnetic field encounters stiff resistance from this plasma as the Sun rotates, slowing down the spin. When the material in the disk is later dissipated, it takes the excess angular momentum with it.

This idea has several problems. It does not explain why only stars with low masses exhibit slow rotation, since some stars with large masses have strong magnetic fields, far stronger than the Sun's. A way out of this dilemma would be to postulate that only low-mass stars form with disks. But we now know that disks are not uncommon around young stars with masses greater than the Sun's. For example, the star Beta Pictoris, which we will discuss in section 19.7, has twice the mass of the Sun and yet is surrounded by a clearly visible disk of material.

The Solar Wind: Blowing the Problem Away

A second solution involves the solar wind. Recall that the gases in the outer fringes of the solar atmosphere have enough energy to escape into space, flowing steadily outward through the solar system at speeds of about 400 km/s. The amount of matter lost in this way is a tiny fraction of the Sun's total mass, yet it is carrying angular momentum with it, and this momentum is lost by the Sun. Furthermore, observations of very young stars with masses like the Sun's indicate that they generate particularly intense stellar winds shortly after they form. While it is difficult to make an accurate estimate of this effect, it appears sufficient to account for the present slow rotation of the Sun.

An especially appealing aspect of this theory of solar wind braking is its natural explanation of the mass dependence of stellar rotation. Only stars with masses comparable to or less than that of the Sun have the proper atmospheric structures to produce steady stellar winds. Hence the same strong wind that ultimately clears residual gas and dust from the disk can slow down the rapidly rotating star that generates it. More massive stars will not produce such winds, so they will continue to exhibit rapid rotation.

Attractive as this idea is, it is still just a theory. We need more observations of real disks to see what actually takes place in them. For example, in a real disk of gas and dust, there will be complex gravitational interactions among different components as soon as any clumping of material destroys the disk's homogeneity. These interactions themselves can transfer angular momentum, and a complete model for the nebula must take all these effects into account.

19.4 EVOLUTION OF THE DISK: CONDENSATION, AGGREGATION, ACCRETION, AND DISSIPATION

We return now to the evolution of the material in the disk. How is it possible to get from this flattened mixture of dust and gas to a system of planets, satellites, and comets moving through nearly empty space? Of course the space is not completely empty, since we have the solar wind and the myriad dust grains produced by collisions in the asteroid belt and the dissipation of comets. Compared with the solar nebula, however, the present interplanetary space is a pretty good vacuum, better in fact than the best vacuum we can produce in our laboratories. We must try to reconstruct the processes of agglomeration and dispersal that led to this condition.

Early History

The first thing to realize is that temperatures in the disk were not uniform. The initial condensation would generate heat as the gravitational energy of the extended cloud was converted to thermal energy. Most of the mass in the collapsing cloud became concentrated toward the center, converting enough potential energy to raise the internal temperature of this central condensation above the ignition point for thermonuclear reactions. The material in the outer part of the disk was collapsing only in the vertical direction, as particles in the cloud sank to the **midplane** (the plane that ran through the middle of the disk), forming a dense sheet of material. Local sources of heat from short-lived radioactive elements and electromagnetic discharges were available in the disk.

As the central condensation grew and the dust in the nebula gravitated to the midplane, the entire nebula was heating up, especially near the center. Temperatures above 2000 K were reached in the disk near the Sun, according to present estimates. This means that the dust in this region was vaporized and the nebula was totally gaseous. Subsequent cooling of this hot gas allowed the condensation of molecules and the formation of grains again, but with some sorting of condensates according to distance from the Sun (Fig. 19.4). This temperature-dependent condensation was one of the primary causes of the chemical *fractionation* of material in the planetary system. Calculations indicate that the collapse of the cloud to a disk took about 10 million years. At this point, the young Sun provided a source of energy to keep the temperatures high in the inner parts of the disk, inhibiting the condensation of most volatiles in the region now occupied by the inner planets.

This scenario provides a partial explanation for Facts 4 through 7. The inner planets are deficient in light elements compared with the cosmic

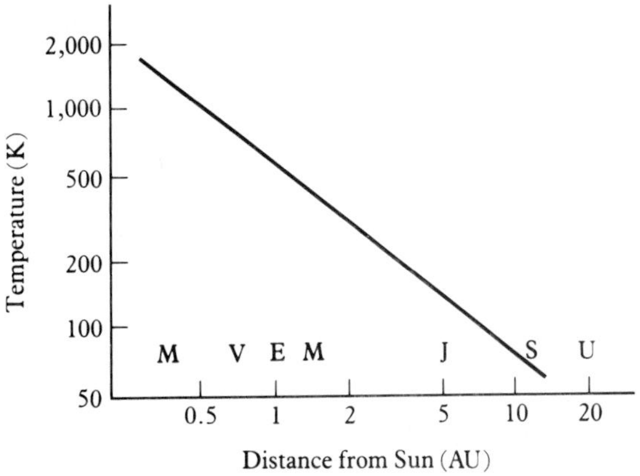

FIGURE 19.4 This graph illustrates current ideas about the decrease in temperature with increasing distance from the Sun. The positions at which the various planets formed are marked.

distribution. Our model suggests that the inner planets formed at relatively high temperatures, at which volatile compounds of light elements (H_2O, CO_2, etc.) could not condense. The increasing fraction of volatiles in the solid objects we find today as we move outward through the asteroids to the outer solar system and the comets roughly follows the temperature gradient in the original solar nebula.

When we reach Uranus, a further change in the composition of solid bodies occurs. The larger satellites of Uranus, Neptune's Triton, and the Pluto-Charon system all exhibit densities significantly higher than those of the icy satellites of Saturn. It is not yet clear what change in composition is responsible for this. Perhaps the more distant objects contain a higher proportion of oxidized carbon, especially CO_2, than the Saturn satellites, which formed at relatively higher temperatures. Solid CO_2 has a higher density than H_2O, and it would also use up some of the available oxygen that might otherwise form water. The net result would be a higher ratio of rock to ice than we find in the Saturn system, and hence satellites with higher mean densities.

There is thus a reasonably good agreement between the decrease in temperature with increasing distance from the Sun in the original nebula and the chemical fractionation we observe today, as we move from the rocky inner planets to the ice- and gas-rich outer planets. This agreement is attractive, but it implies a segregation of material in the early nebula that was not rigorously maintained. There must have been some mixing, as solid, condensed material from the outer part of the disk (large orbital radius) reached the inner part (small orbital radius) because it was moving in eccentric orbits. This allowed material formed under one set of pressure and temperature conditions to be mixed with material from another environment. A prime example is the acquisition of water by Earth and Venus, even though these planets formed in a part of the nebula too hot for the condensation of water or ice from the vapor state.

This **radial mixing** explains Fact 9: Despite the temperature gradient in the solar nebula, some volatile constituents found their way to the surfaces of the inner planets, yet this is consistent with Fact 8. Although solids were mixed to some extent, the high-temperature chemistry that took place in the inner nebula did not wipe out isotope anomalies throughout the disk. Condensed matter from the outer parts of the nebula has preserved its original isotopic signatures.

The early vaporization of the interstellar grains in the inner part of the nebula helps us to understand the age cutoff in Fact 1: Despite the comparatively young age of the Sun, nothing older than 4.5 billion years has yet been found in the solar system. In other words, we have not yet detected solid interstellar material in sufficient quantity to determine its age, although there are indications of the presence of interstellar carbon and silicon in some meteorites and the grains in the Allende meteorite are dated at 4.57 billion years, suggesting an early origin. As mentioned earlier, interstellar grains may also be present in comets, which formed (and remained) at low temperatures in the outer nebula. If we could obtain large enough samples from comets for study in the laboratory, we should find ages older than the 4.5-billion-year limit.

Condensation in the Solar Nebula

It is possible to calculate what compounds and minerals would form from this cooling cloud by investigating various possible chemical reactions among the elements that were present. Such calculations indicate that there is a definite temperature sequence in which various compounds form (Fig. 19.5).

As the hot gas cools, one of the first things to condense is spinel, a magnesium-aluminum compound with the formula $MgOAl_2O_3$. This is the same substance found in the white inclusions of the Allende meteorite. Further cooling leads to the condensation of a wide variety of silicates, which must have appeared as dust grains in the inner nebula. Since silicon and oxygen are both abundant elements, silicates should have become the dominant minerals in the inner disk, where temperatures were too high for the condensation of volatiles.

The inner boundary for the condensation of H_2O ice in the solar nebula was located at a distance of about 4 AU from the Sun. This is a major threshold, since in a cosmic mixture, ice is potentially far more abundant than rock, even though the reverse is true on Earth and the other inner planets. Recall that in the Sun and stars oxygen is approximately 18 times as abundant as silicon. This means that in the solar nebula, H_2O should be approximately nine times as abundant as SiO_2. After forming all possible oxides and silicates, there is still plenty of oxygen left to make ice. As we saw in Chapter 13, crossing the threshold for the formation of ice enables the growth of giant planets. But how are planets formed from the tiny grains of ice and dust that condense in the cooling nebula?

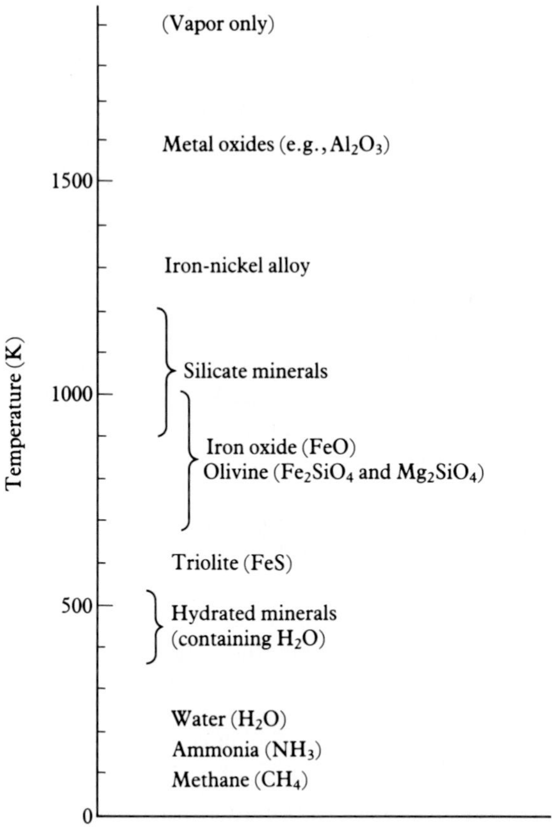

FIGURE 19.5 The chemical condensation sequence in the solar nebula, showing the primary chemical species that would be expected to form in a cooling gas cloud of cosmic (solar) composition. As the temperature drops, each material forms and condenses into liquid droplets or solid grains. The calculations upon which this diagram is based, originally carried out by John Lewis of the University of Arizona, presume that the grains remain in chemical equilibrium with the remaining gas. A slightly different condensation sequence results if the grains are immediately swept up into planetesimals and taken out of contact with the gas, but the general results are the same. These solid grains then become the building blocks of the planetesimals, and ultimately of the planets themselves.

Aggregation of Grains

The next step toward making planets after condensing grains from the gases in the solar nebula is to get these grains to accumulate into larger aggregates. This process is aided by the fact that at any given point in the nebula, most of the material should be in nearly identical orbits, so the relative velocities between nearby grains will be small. This enhances the chances that two colliding grains will be able to stick together, although it tends to reduce the total number of impacts. Sticking is especially likely if the grains are fluffy.

The evidence we have for the appearance of grains in the solar nebula suggests that they indeed had this desirable fluffy characteristic. Figure 4.11 shows some of the grains, probably of cometary origin, collected in high-altitude flights by a project designed to collect interplanetary particles. These should be the most unmodified grains remaining in the solar system, since they have never been heated. These fluffy particles form clumps that will gradually grow through additional collisions. While this is occurring, the solid material continues to settle toward the midplane of the nebula, the plane that cuts through the middle of the entire assemblage.

Within the disk, the fluffy clumps of grains stick together to form loosely bound objects as large as 10 km across, and we speak of these objects as **planetesimals**, the building blocks of the planets (see Section 4.1). The planetesimals are large enough to exert a significant gravitational attraction on each other, leading to further rapid growth.

Timescales for the Formation of Planetesimals

The idea that the planets formed from intermediate-size bodies called planetesimals was originally the work of Russian theorist V.A. Safronov, who worked on this problem in the 1950s and 1960s. Previously, many scientists had thought that the formation of planets probably had taken place directly from the solar nebula and its tiny dust grains, but we now recognize in the asteroids and comets surviving examples of planetesimals, and increasingly sophisticated calculations support Safronov's hypothesis.

It is possible to estimate the length of time required for the various stages in the formation of planetesimals. The theoretical models suggest that it may have taken as long as 100,000 years for 1-μm particles to reach the midplane of the disk, whereas 1-cm particles would settle out in only 10 years. The initial loose condensations of dust should evolve into solid bodies with diameters of about 10 km in roughly 1000 years.

Confirmation of these amazingly short timescales is provided by investigations of the abundances of isotopes formed from radioactive parents with very short half-lives. Such studies indicate that solid material formed from the solar nebula within a few million years. Most of the meteorites that scientists have examined were formed from these grains within the first 20 million years after the nebula collapsed to a disk. The Earth and the other solid planets were essentially complete in less than 100 million years from the time the solar nebula formed.

From Planetesimals to Planets

The inner planets and the cores of the giant outer planets formed from the continued accretion of planetesimals. Rapid gravitational growth quickly used up much of the solid material in the disk to form thousands of objects with diameters of a few hundred to a few thousand kilometers. These accumulations, spanning the size range from the largest asteroids up to objects larger than the Moon, are called **planetary embryos**.

Once most of the mass was in the form of planetary embryos, collisions were much rarer. But they were also more violent, with impact speeds of several kilometers per second. As the gravitational reach of the embryos extended farther and farther, the violence of the impacts increased. When a small embryo struck a large one, the material was accreted and the large embryo grew, but when two embryos of comparable size hit, they were likely to be fragmented into thousands of smaller pieces. A competition developed between the processes of accretion and the processes that break planets apart.

Several scenarios have been calculated for the evolution of the embryos in the inner solar system. The models of George Wetherill, in particular, suggest that a great many objects of lunar size or larger were formed, in addition to the growing protoplanets themselves. According to this picture, the late stages of the accretionary process were characterized by impacts of incredible violence as these lunar-size objects bombarded the planets (or, in some cases, were gravitationally ejected from the solar system). The possibility of a few discrete, random impacts of this magnitude can explain some of the peculiarities of the planets, as we shall see in the next section. These models also indicate that so much mixing took place in the inner solar system that all of the terrestrial planets should have about the same composition.

Gradually, over a period of tens of millions of years, the protoplanets emerged from this chaotic situation. It seems inevitable that they were heated by the energy of these collisions once again above the melting point of silicates. Thus the planets probably differentiated as they formed, and no primitive material survived the violence of the accretionary process.

In the outer part of the solar system, more mass was present in the form of condensed water ice, and the velocities of the planetary embryos were lower. As a result, the level of violence was less and the protoplanets grew larger. Calculations indicate that within less than a million years, cores as large as ten Earth masses could form and begin to attract the gas as well as the solid matter in the disk. At first these protoplanets were too hot to hold on to much gas, but after several millions of years they cooled to the point where large quantities of hydrogen and helium were acquired, at least by Jupiter and Saturn. In the case of Uranus and Neptune, it appears that the nebular gas dissipated before much of it could be captured.

Dissipation of the Nebula

At some point after the planets formed, a short-lived but highly intense blast from the solar wind cleaned out the remaining gas and small dust. Observations of young stars have revealed that such intense winds are a natural part of the star formation process. In some cases they take the form of bipolar outflows, in which two streams of gas oriented in opposite directions are seen to emanate from young stars. Most of the observations revealing these winds are indirect, involving the interpretation of Doppler-shifted lines in stellar spectra that indicate the presence of moving clouds of gas. Occasionally, however, it is possible to record the effects of jets from a young star as they clear away material from the larger cloud of dust and gas from which the star (and associated planets, if they are present) has formed (Fig. 19.6).

19.5 SMALL BODIES AND THE IMPORTANCE OF IMPACTS

We know from our detailed study of the impact history of the lunar surface that all of the inner planets must have continued to accumulate mass for the next 600 million years. Although this late bombardment effectively created the surfaces we see today, it contributed only a small percentage of the total mass of an individual planet. It is just the icing on the cake. On the other hand, massive impacts that took place in the first 100 million years had a profound effect on the ultimate structure of the solar system.

A Return to Catastrophism

The science of terrestrial geology took a great leap forward in the nineteenth century by adopting the principle of uniformitarianism. As stated by one of its foremost promoters, Charles Lyell, this principle holds that the landscapes we find around us were formed by processes we see operating today. This idea, which was so important for understanding the great age of the Earth and the slowness with which it changes, failed utterly to account for the landscape we find on the Moon, as discussed in Chapter 7. The processes acting today could not supply the

FIGURE 19.6 An image of the region around a young star (not seen here) showing how symmetrical jets of gas from the star have cleared an elliptical space whose boundaries include gas that has been shocked and compressed by the force of the jets.

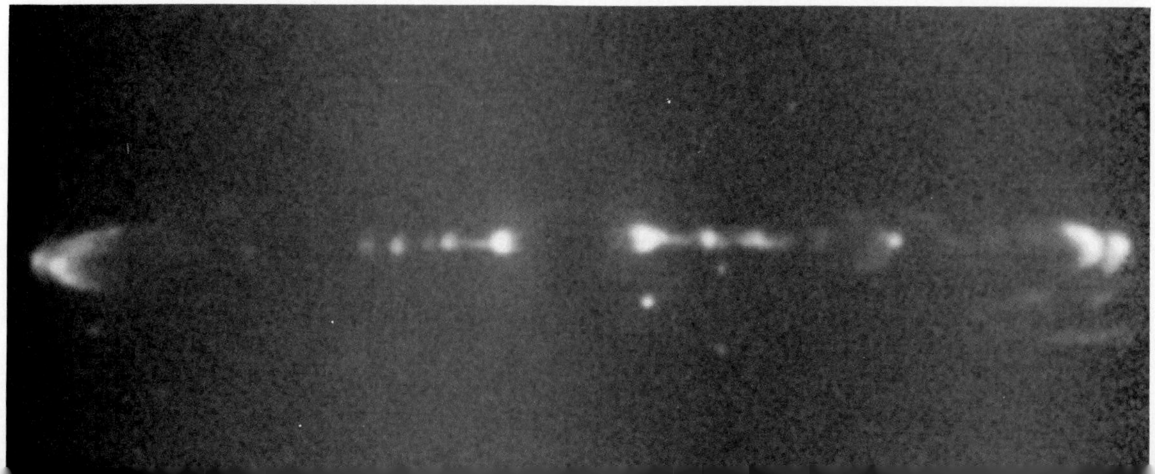

observed number of craters in the total history of the solar system. It is necessary to invoke bombardment by the fragments left over from the formation of the planets to account for the craters we see on every ancient solid surface in the solar system, from Mercury out to Triton, a distance of nearly 6 billion km.

A careful reader of this book will have realized that impacts have had even larger effects than the formation of craters and basins on planetary and satellite surfaces. A giant impact between the Earth and a Mars-size planetary embryo was probably the reason we have the Moon. A massive impact with Mercury apparently stripped away a large fraction of that planet's mantle, leaving Mercury with a disproportionately massive metallic core. Giant impacts on Venus, Uranus, and Pluto are the most likely causes of the strange rotations of these bodies (Fact 10). At least some of the planetary rings must have resulted from impacts with objects already in orbit about the planets. The dark face of Iapetus, the chaotic rotation of Hyperion, and the strange difference in composition between Pluto and Charon are other examples of conditions we see today that may have been caused by unique events probably associated with impacts.

Contrary to normal human intuition, which likes to think of nature as smoothly following fundamental laws, it is becoming clear that the early solar system was a chaotic place, with bodies randomly smashing into each other, sometimes destructively, sometimes creatively, often with long-lasting and fundamental effects. We regard the recognition of the importance of impacts for the origin and evolution of planets and satellites and the origin and evolution of life on Earth as a major contribution of solar system exploration to the history of ideas.

This bombardment is not yet over, as the K/T extinction, the lunar crater Tycho, and Saturn's G Ring make abundantly clear. To this extent, impacts do not flagrantly violate Lyell's hypothesis—they too are processes that are ongoing today, but fortunately they just don't happen very frequently. They also no longer occur on the large scales that formed the Moon or tipped Uranus on its side. Asteroids and comets can still alarm us, however, which is one reason we should try to understand how they formed.

Origin of Asteroids

The small bodies now in independent orbits about the Sun represent material that was left over from the planet formation process. The asteroids appear to have several sources. Some may simply represent a case of planetesimal growth that was halted at an early stage, when planetesimals had reached only a maximum size of 100–1000 km. The gravitational effect of nearby Jupiter removed most of the planet-forming material from the region of the asteroid belt before it could accumulate into a full-size planet (Fact 5). Perhaps the gravitational influence of Jupiter was also responsible for the circumstance that Mars is so much smaller than Earth and Venus.

Some of the dark C-type asteroids may actually be the carbon-rich remains of huge comet nuclei, whose volatiles were gradually exhausted once they were captured into their present orbits. There are indications that some of these bodies may still have water ice in them. Other objects in the asteroid belt may be the remains of some of the giant collisions that seem to have been an important part of the history of the inner planets. In other words, the main asteroid belt may be a kind of quasi-stable dumping ground, rather like the little pile of debris that sometimes accumulates at the exact center of the intersection of two busy highways.

Was There a Late Heavy Bombardment?

The lunar cratering record suggests to many geologists that there was a discrete burst of impacts—the late heavy bombardment—about 600 million years after the formation of the Moon. Where did these impacting bodies come from, at a time when the inner solar system had been largely cleared of debris left over from the accretionary process? Many scientists think they may have originated near the orbits of Uranus and Neptune.

Because the distances between the outer planets are so great, it required a much longer time to clear the debris from this region. Quite possibly, the late heavy bombardment consisted largely of icy bodies gravitationally scattered inward by Uranus and Neptune. Perhaps this was the time when most of the volatiles that make up the terrestrial planet atmospheres arrived (Fact 9 again). At the same time, an even larger number of such objects should have been scattered outward, where they became the comets of the Oort comet cloud.

In addition to contributing volatiles to the atmospheres of the terrestrial planets, the cometary ices would also have been added to the "primordial soup" of organic chemicals accumulating in the oceans of Earth, ultimately to become part of organisms as life began. You may be pleased to realize that some of the carbon, nitrogen, and oxygen atoms in your cells were formed in stellar interiors, shot into interstellar space by supernova explosions, trapped in the ices of comets and delivered to the surface of the primitive Earth. Some of the water you drank today may also have come from comets, but the alcohol in your beer was produced by local yeast.

This apparently benign contribution of comets to the course of life on our planet must be contrasted with a darker side of the bombardment of the early Earth by small bodies. During the first 700 million years of our planet's history, impacts by objects of 10 km diameter or larger were far more frequent than they are today. Recall that one 10 km object striking our planet 65 million years ago caused widespread extinctions of life on Earth. Now imagine the planet being bashed by still larger bodies. Some scientists believe that the results of such impacts would have been sufficiently severe that they speak of the **impact frustration** of the origin of life during this period.

The last of the large lunar basins, such as Orientale and Imbrium, were formed about 4 billion years ago by the impact of planetesimals approximately 100 km in diameter. If the Moon was hit several times by such objects between 4.1 and 3.9 billion years ago, the much larger Earth must have been struck at least a hundred times. Each of these impacts would have blasted away a substantial fraction of the Earth's atmosphere and blanketed the entire planet with molten rock, effectively baking the surface and even bringing the upper layers of the ocean to the boiling point. Impacts by objects 250 km in diameter, which may have taken place before 4.0 billion years ago, would have provided enough energy to boil the oceans away entirely, effectively sterilizing the planet. If life originated at still earlier epochs, it would not have survived such impacts. This may be what happened, with life forming and then being snuffed out several times. In any case, it may be more than a coincidence that the first chemical evidence of life on Earth dates from about 3.8 billion years ago, just after the end of the heavy bombardment.

In the outer solar system, some planetesimals (such as Saturn's Phoebe) were captured into stable orbits when they passed close to a giant planet. These captures could occur as a result of frictional drag when these bodies penetrated the extended atmospheres (or subnebular disks) of the forming planets. The friction between the moving planetesimal and the gas that still existed in the planetary subnebula would cause free-roaming planetesimals to lose energy and enter orbits about the planets, providing a natural explanation for Fact 12: All the outer planets except Uranus have one or more satellites in retrograde orbits. Satellites forming with the planets would not assume such orbits, but captured satellites could.

Formation of the Comets

A large percentage of the icy bodies that were not incorporated into the giant planets were forced into orbits with very high eccentricities that either expelled them from the solar system or took them out to a radial distance of 50,000 AU from the Sun to form the Oort cloud of comets, which we discussed in Chapter 6. The long-period comets we discover from time to time are coming from this distant region, where the gravitational perturbations of passing stars occasionally cause these icy nuclei to change their orbits, bringing them close enough to the Sun for us to see.

A second group of icy bodies exists just beyond the orbit of Pluto in a disk-shaped distribution called the Kuiper Belt, which is the main source of the short-period comets. Some of these bodies presumably formed in the same region of space in which we find them today. They represent the outer fringe of planetesimal formation in the solar nebula, where objects could no longer accrete to the size of Pluto. The Kuiper Belt probably contains some comet nuclei scattered into it from the Uranus-Neptune region as well.

19.6 GIANT PLANETS AND THEIR SATELLITE AND RING SYSTEMS

From the model we are presenting for the formation of the solar system, it seems natural to expect that the giant planets with their systems of satellites and rings must have formed in a manner similar to that of the Sun and its retinue of planets. In the outer reaches of the solar nebula, enough matter was present to form giant protoplanets surrounded by disks that we have called subnebulae. The satellites evidently developed from these disks, followed or accompanied by the formation of rings.

Growth of the Giant Planets

As we saw in Chapters 13 and 14, the giant planets consist of cores of rocky and icy material surrounded by huge envelopes of gas. All four of these planets have cores amounting to 10–15 Earth masses. These massive cores are apparently the key to the further attraction of gas from the solar nebula to make up the hydrogen and helium atmospheres observed today.

As the rock and ice cores of the outer planets approached their maximum masses (Fact 13), they caused the surrounding gas in the nebula to collapse, simply from the strength of their gravitational attractions. Thus the cores are now surrounded by atmospheres that are a mixture from two sources: the gases released by in-falling solid material and a primordial mix of gases collected from the solar nebula.

It is this mixture that explains the inference that all four giant planets started out with atmospheres that had a solar abundance of hydrogen and helium, while methane is enriched (or hydrogen and helium depleted) to various degrees. The hydrogen and helium were contributed by the solar nebula, while the extra methane was produced by the growing core as one of the components of its outgassed atmosphere. On Saturn, we find a slightly greater enrichment of methane than on Jupiter, where it is about twice the value expected from a solar mixture of the elements. On Uranus and Neptune, methane is 25 times more abundant than a solar mixture would predict, in keeping with the much higher ratio of core mass to total mass exhibited by the planets. For reasons not yet understood, Uranus and Neptune were not able to capture as much hydrogen and helium from the solar nebula as their more massive cousins closer to the Sun.

Formation of Satellite Systems

Just as a disk of gas was left around the forming Sun, the matter that formed the giant planets produced both central condensations and disks. The disks shared the sense of rotation of the forming planets, so satellites forming in these disks orbited their planets in the same direction that the planets rotated. These became the regular satellites that we find today (Fact 11).

Within the protoplanetary disk, one could expect to find a temperature and pressure gradient similar to the one we just discussed for the solar nebula. Near the planet, the density of gas would be greater and the temperature higher than at large distances. Obviously these protoplanetary disks never got as hot as the solar nebula near the Sun. Nevertheless, this change in physical properties with distance from the planet should leave some signature on the forming satellites.

In the case of Jupiter, the hottest of these planets because it was the largest, we can see the effects of this heat on the forming satellites when we examine them today. Recall that at Jupiter's distance from the Sun, water ice is stable, even over the lifetime of the solar system. Thus we expect to find satellites that reflect the cosmic proportions of the elements, roughly half ice and half rock by mass. This is indeed the case with the two outer satellites, Callisto and Ganymede.

Io and Europa, close to Jupiter, have higher densities (3.1–3.4 g/cm^3), exactly as this scenario would predict. The corresponding proportion of water is no more than 10% on Europa, while Io has been baked out by tidal heating and volcanic activity. This low proportion of water on Europa can be understood in terms of higher subnebular temperatures than those in the regions where Callisto and Ganymede formed.

The satellite systems of Saturn and Uranus do not show a similar variation of satellite density with distance from the planet. Several of the inner satellites of Saturn actually appear even less dense than the half-ice, half-rock composition of Callisto and Ganymede. These satellites may be fragments of the icy mantles of disrupted larger bodies. There may also be variations in other volatiles trapped in the ice. For example, such variations provide the most probable explanations for the absence of methane-nitrogen atmospheres on Ganymede and Callisto and the presence of such an atmosphere on Titan.

Formation of Rings

Why do some planets have rings while others don't? What determines the kinds of rings a given planet may have? These are questions that are now under active investigation. In Chapter 18 we discussed several scenarios for the origin and evolution of ring systems.

The general picture of rings as debris that did not accrete into large bodies provides a possible explanation for the existence of the rings of Saturn. However, the close association of this ring system and all of the others with a number of inner satellites has led scientists to argue in favor of the impact-induced breakup of one or more satellites as a more likely cause. Since it is clear that the bombardment of inner satellites by infalling debris must have been very heavy in the first few hundred million years of planetary history, it seems inevitable that satellites close to the planet, if they existed, were at least heavily eroded if not completely disrupted by large impacts. The debris of such events, if it were inside the tidal stability limit, could not have pulled itself together gravitationally to reform the satellite.

If these ideas are correct, the rings did not form directly from a subnebula surrounding the giant planets. Instead, the ring systems are the natural result of the formation of satellites, combined with the high intensity of impacts produced close to the planet by its massive gravitational field. This process may still be taking place. For example, if a passing comet broke up as it made a close approach to a giant planet, it might lose enough energy to be captured into an orbit about the planet (as happened in 1992 for Comet Shoemaker-Levy 9). Under some circumstances, this cloud of orbiting debris might evolve into a short-lived ring. The non-gravitational forces produced as gas was suddenly liberated from freshly exposed surfaces could slow the comet sufficiently to allow the capture.

19.7 THE SEARCH FOR OTHER PLANETARY SYSTEMS

One of the most powerful methods available to scientists is that of comparison. By examining different manifestations of the same kind of phenomenon, it is possible to derive the general laws and principles that are at work. For example, finding that craters on the Moon had the same ratios of depth to diameter as bomb craters on Earth helped scientists to deduce that explosive impacts rather than volcanism were responsible for these features of the lunar landscape. Newton discovered the law of gravity by noticing the effects of this force on the motions of apples and the Moon. The comparative approach to planetary science has been a central theme of this book.

Planetary Systems

It would be wonderful if we could apply this same approach in our efforts to understand the origin of the solar system. Many scientists have concluded that we will never be able to gain a full understanding of our own system without having some additional examples. We would begin by examining several different planetary systems to see what common properties they exhibited. These properties would help us determine which aspects of our own system are truly fundamental, requiring a common theory for their explanation, and which are random, the result of chance events. Do all systems consist of small planets close to their star and massive planets farther out? Do the planets always revolve around the star in the same direction? Are comets and asteroids always "left behind" when solar systems form? There are certainly many questions we can ask.

Unfortunately, we don't yet have the answers. At the present time, ours is the only planetary system we know well, despite the existence of billions upon billions of other stars in our Galaxy. Since there is a higher probability for stars to form in pairs or multiples than alone, we might speculate that when a single star is observed, there was not enough material in the collapsing cloud to form a stellar companion, so a planetary system formed instead. Nature's preference for forming systems of small bodies in orbit around more massive ones is well illustrated by our own system's giant planets with their satellites and rings. On the other hand, the excess material may have been lost during the early stages of the star's formation and the star may truly be a single object in space.

These ideas are now only speculation. We need real data on other planetary systems to move beyond this stage. Fortunately, just such data are beginning to become available, and the prospects for learning more about other planetary systems during the next few years are bright.

Clouds and Disks Around Young Stars

Because individual planets are so small and faint, it turns out that the easiest observational searches for other planetary systems are those that look for clouds of dust and gas equivalent to the primordial solar nebula that gave birth to our own planetary system. The temperature of such a nebula—a few hundred K—makes it a strong source of infrared radiation. Using various observational techniques, several astronomers have recently succeeded in identifying disks of gas and/or dust around relatively nearby young stars. Some of these disks appear to have the dimensions (radii of about 100 AU) expected to make them capable of forming planets.

The identification of dust disks analogous to the solar nebula is certainly important, but it does not provide direct evidence of the processes of aggregation and accretion that we think led to planetesimals and eventually to the planets themselves in our own system. Hence the significance of the 1983 discovery by IRAS (the Infrared Astronomical Satellite) of excess thermal radiation from a number of bright, relatively nearby stars, including Vega and Fomalhaut. The thermal radiation detected by IRAS was considerably greater than the amount that the star itself could emit. Analyses of these objects soon showed that their excess radiation was coming from disks or rings of orbiting solid particles. While the sizes of the emitting particles are not known, they are definitely larger than the sizes of dust particles associated with forming stars. Evidently, some aggregation of particles has taken place, and the solid material survived the transition of the star to a mature state. The dimensions of these disks are a few hundred AU, and estimates of the amount of material present have ranged up to several times the mass of Earth.

The next step was to image some of this newly discovered material. Bradford Smith of the University of Hawaii (who was also the Voyager Imaging Team leader; see Fig. 15.5) and Richard Terrile of JPL had previously designed a special viewing system to permit the faint inner satellites of Saturn to be imaged in spite of the scattered light from the much brighter planet and rings. They modified this camera and with it obtained the image shown in Fig. 19.7 of Beta Pictoris, one of the stars IRAS had discovered to be emitting excess infrared radiation. The picture reveals a disk of solid particles, seen nearly edge-on in orbit around the star.

The Beta Pic disk has a diameter of nearly 1000 AU. To understand what that means, we have to realize that Beta Pictoris is a hotter, more massive star than the Sun. Thus a radial distance of 500 AU in the Beta Pic disk is equivalent to a distance of 250 AU from the Sun. In Fig. 19.7, we are therefore seeing a band of material on each side of the star which corresponds to the region in our solar system from somewhere in

FIGURE 19.7 A composite of a positive image of the star Beta Pictoris divided by a negative image of the star Alpha Pictoris. The stray light from the two stars is nearly canceled out, leaving dark rings around the resulting composite image. Dark and bright star images from the two original pictures appear in the surrounding field. The disk of the material surrounding Beta Pic is visible as a bright diagonal band, which actually stretches some 500 AU on either side of the star.

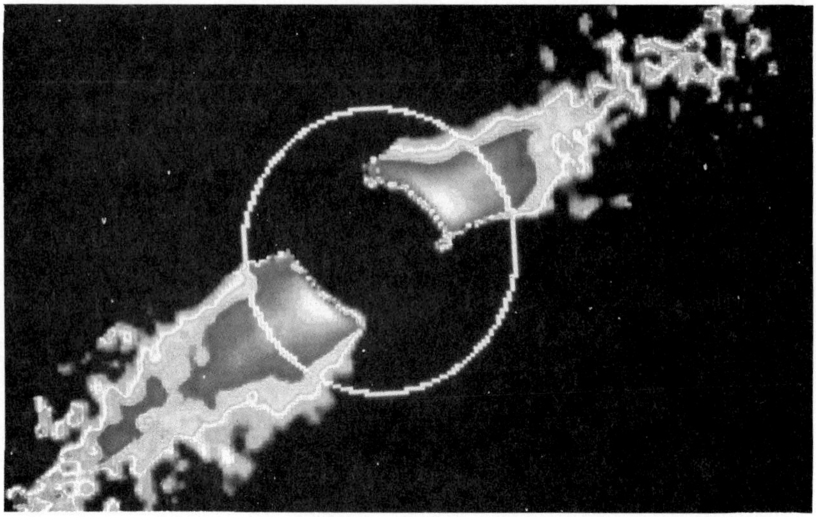

FIGURE 19.8 The Beta Pictoris disk recorded with a special, antiblooming detector and a small coronagraphic mask. The circle shows the size of the mask used in Fig. 19.7. Here we can see to a distance of 30 AU from the star.

the inner Kuiper Belt (at 50 AU) out to a distance of 250 AU. Clearly there is nothing with the density of these Beta Pic particles at this position in our solar system today.

It would obviously be more interesting to investigate the inner region of the disk where we might expect planets to form, but this is very difficult owing to the overwhelming brightness of the star. A group of French scientists led by Alfred Vidal-Madjar has achieved a significant improvement by using a special detector that suppressed the scattered light. They have recorded the disk to within 30 AU of the star (Fig. 19.8), equivalent to 15 AU in our system—that is, just inside the orbit of Uranus (19 AU). The French found no indication that the particle density in the disk was lower closer to the star, as we might expect to be the case if giant planets were forming in this region.

Although several other stars are known to be surrounded by disks of dust from their anomalous brightness at infrared wavelengths, Beta Pic remains the only star whose disk has been successfully recorded in visible light. It is therefore understandably the subject of intensive, continued investigations. The most exciting recent development has been the observation of Doppler-shifted absorption lines in the spectrum of Beta Pic which indicate that material from the disk is being vaporized and is falling into the star. The amount of mass that is being consumed in this process is equivalent to the loss of 50 comets per year from the disk, assuming an average diameter of 20 km for the comet nuclei. Are these "particles" really comet nuclei? We do not yet have spectra of the disk that can tell us the composition of the material we see in Figs. 19.7 and 19.8. Still closer to the star, the signature of silicate grains has been detected in spectra—just as silicates are found in spectra of comets. We do not yet have signs of ice.

Whatever the particles turn out to be, the Beta Pic disk and other similar features discovered by IRAS are not planetary systems. They may be analogous to the Kuiper Belt, or to some other form of solid material left over from the formation of the star. Certainly there is no way to tell from these data whether there might also be planets present. But one thing these discoveries have shown us is that other single stars have formed with solid material in orbit around them. In this sense, at least, our Sun is not alone. The next step is to look for evidence of planets, a much more challenging problem.

Gravitational Evidence for Planets Orbiting Other Stars

As a planet moves around its star, Newton's third law tells us that both the planet and the star must be exerting gravitational forces on one another. In fact, both planet and star are moving in orbits around the **center of mass** of the system (Fig. 19.9), the planet's orbit being far larger than the star's. The challenge to the astronomer is to find a way of measuring the orbital motion of the star, which will then allow a deduction of the existence of one or more unseen planetary companions. Two different approaches to this problem have been pursued. The first, *spectroscopy,* is the detection of the motion of the star through measurements of Doppler shifts in the stellar absorption lines as the star periodically approaches or recedes from the observer. The second is **astrometry**, the direct measurement of the star's apparent motion, compared with a reference frame of background stars. Note that geometrically, spectroscopy is most successful if we view the system edge-on (in the plane of the orbits), so that the star's velocity is directly in the line of sight.

FIGURE 19.9 A planet and its star are both moving around the center of mass of the system: (a) face-on view, (b) edge-on view. Note that the size of the star relative to the size of the planetary orbit has been greatly exaggerated.

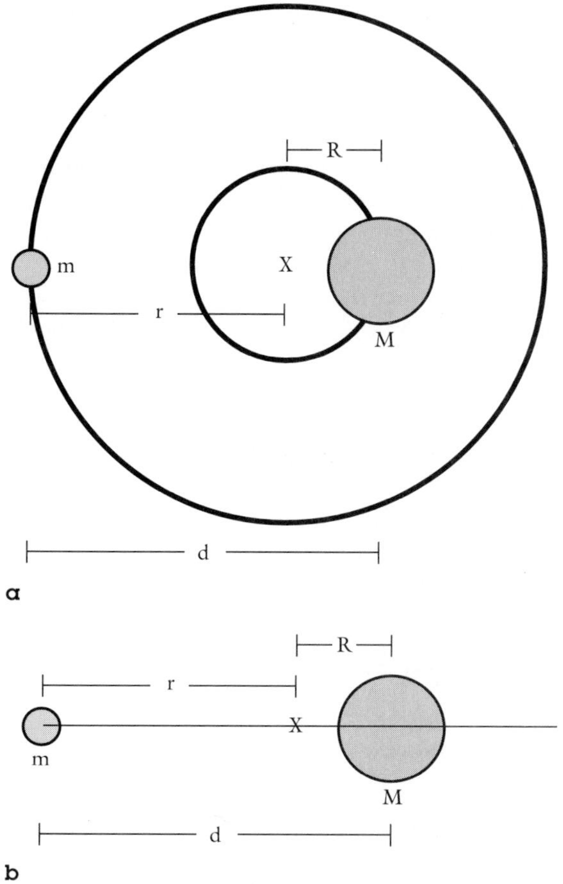

The main difficulty for the spectroscopic method is in measuring the spectral lines of the star with sufficient precision to detect the very small Doppler shifts corresponding to the gravitational attraction of the orbiting planet. A Jupiter-size planet will produce a Doppler shift corresponding to a velocity of only 5 m/s, or 11 mph. In effect, the astronomer is becoming a galactic traffic cop, looking for a small deviation from the "speed limit," which in this case is the normal, unperturbed velocity of the star through space. Whereas the police officer uses the Doppler shift from a radar pulse to check a driver's speed, the astronomer is using the Doppler shift in the light from the star itself. The underlying physics is identical.

The velocities to be measured are so small that one has to worry about phenomena intrinsic to the star that could mimic perturbations by a planet. However, any such phenomena should also affect the shapes of the spectral lines, not just their displacement, and therefore can be eliminated as a cause of error. Consequently, this approach has been adopted by several investigators, who have been able to achieve the necessary precision in measuring Doppler displacements.

The astrometric technique has been in use for many years as a by-product of the measurement of stellar distances and motions. In this approach, the astronomer observes the motion of a star through space, watching for any deviation from a simple orbital path. Stars move around the center of the Galaxy in elliptical orbits, just the way planets move around the Sun. But if there is also a companion moving around the star, the star's path through space is wavy rather than straight. The reason is the same one we encountered before: The center of the star moves around the system's center of mass with the same period as the companion. If the mass of the companion is a large enough fraction of the mass of the star, and if the orbit of the companion is not too small, a distant observer will be able to notice the star moving around the center of mass of the system. It is this orbital motion that causes the star to exhibit a wavy path as it follows its apparent track through the sky (Fig. 19.10).

By examining the motions of nearby stars with very high precision over many years, it has been possible for astronomers to identify a few that indeed follow slightly wavy paths. Unfortunately, the most reliable examples lead to masses for the companions indicating that they are more like small, faint stars rather than planets such as those we have in our own system.

The minimum mass for a star to sustain itself by nuclear reactions in its core is 0.08 times the mass of the Sun, or approximately 80 times greater than the mass of Jupiter. Thus there is a substantial range of masses—up to 80 times the mass of Jupiter—that corresponds to neither stars nor planets, as we are familiar with the terms. Such objects are called **brown dwarfs**. Invisible companions in this mass range are the objects most likely to be discovered by ground-based astrometric searches, if such brown dwarfs actually exist in substantial numbers in the Galaxy. So far the evidence is not definitive; a few such objects may be present, but they are not as abundant as many astronomers had expected.

The First Extrasolar Planets

A very special case of planet-size bodies in orbit around stars was discovered at the beginning of this decade. Late in the lifetime of a star, its nuclear fuel becomes exhausted and the star collapses as the internal temperature and pressure drop. If the star is at least 1.4 times the mass of the Sun, the collapse is catastrophic, a supernova explosion occurs, and the star achieves a new equilibrium only when its internal pressure is great enough that electrons and protons are forced together to form neutrons. The result is a tiny star composed of neutrons with a diameter

FIGURE 19.10 If a star has one or more sufficiently massive planets around it, the star itself will move in a small orbit around the center of mass of this planetary system. The center of mass of the system will move along an elliptical orbit about the center of the Galaxy, while the star (which is what we observe) will execute a periodic wiggle across this orbit as it pursues its path.

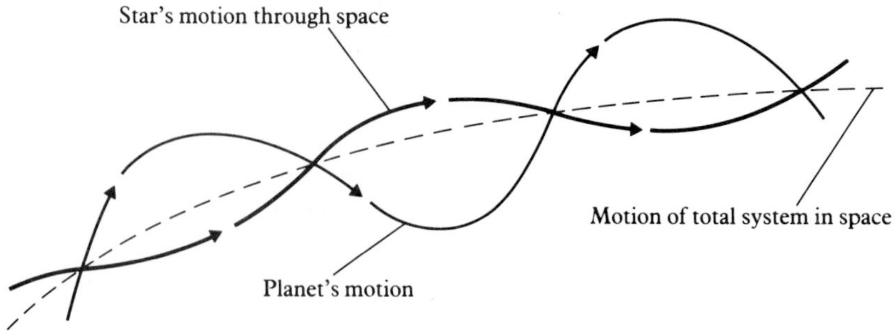

of only 10–15 km but a mass greater than that of our Sun!

Such **neutron stars**, remnants of supernova explosions, have many interesting properties. Of special relevance to us is that they are very stable, high-speed rotators. We know this because synchrotron emission from electrons traveling along the star's magnetic field lines creates beacons of radio waves that we can observe from Earth. Timing the arrival of these beacons as they sweep across our radio telescopes has allowed astronomers to find that some of these exotic objects rotate with periods of just a few thousandths of a second. A rapidly rotating neutron star that broadcasts radio waves in this way is called a **pulsar**.

If a pulsar has planet-size objects orbiting around it, the motions of the planets could be detected by their effect on the star's radio broadcast, again because of the star's motion around the center of mass of the system. Now, instead of simply rotating on its axis, the pulsar is also moving in a tiny orbit around the center of mass (Fig. 19.9). That means that the pulses of radiation broadcast by the spinning star will be periodically Doppler-shifted as the star first moves toward us, then away from us in its orbit.

It is possible for astronomers to measure the frequency of pulsar broadcasts with extremely high precision, and in most cases, no such periodic effects are observed. In one well-documented instance, however, a pulsar with a period of only 6.2 milliseconds has been shown to have two bodies in orbits around it, with masses of 2.8 and 3.4 times the mass of the Earth. These objects are located at 0.47 and 0.36 AU, respectively, from the pulsar, distances that may be compared to the 0.39 AU separation of Mercury and the Sun. A third planet about the size of our Moon appears to orbit still closer to the star.

What are these pulsar planets? It is highly unlikely that planets as we know them could have survived the supernova explosion that formed the pulsar. Thus astronomers believe that these three objects must have accreted *after* the explosion, presumably from material left behind from a now-vanished companion star. Thus these bodies don't really qualify as the planets we are seeking, but they do demonstrate that planet-size objects can form under a variety of conditions, which adds some support to our idea that planetary systems may be common.

Nevertheless, recent results suggest that our understanding of solar system formation may not be as good as we have thought. In 1995 the patient measurement of stellar radial velocities finally paid off with the discovery of the first "real" planet outside our own solar system. This planet orbits a faint nearby star called 51 Peg, which is very similar to the Sun in mass, luminosity, temperature, and rotation rate—quite different from the pulsars. According to Geoff Marcy, who studied this system in the autumn of 1995, the planet has an orbital period of 4.22 days, placing it at a distance from the star of only $\frac{1}{20}$ of an AU, about five times closer than the perihelion distance of Mercury from the Sun. The mass of the planet, inferred from the observed radial velocity variations of the star, is 0.45 times the mass of Jupiter. The planet is therefore what we would call a giant planet, intermediate in mass between Jupiter and Saturn, but located in a very unjovian orbit. (It is the closeness of the planet to its star and the short period of revolution that made it relatively easy to discover; the same techniques would not have detected this planet in a jovian orbit.) In January 1996, Marcy found two larger planets (3 and 6 times the mass of Jupiter). One is in a circular Mars-size orbit around the star 47 UMa, the other in an eccentric Mercury-size orbit around 70 Vir. Based on these discoveries, Marcy suggests that giant planets will be found in such short-period orbits for something like 5% of the nearby solar-type stars. These results demonstrate that other planetary systems may be common, but they are not necessarily constructed like our own. We are on the threshold of a new era in planetary astronomy, and undoubtedly there will be many more surprises as additional planets are found.

Improved Searches

A much more powerful search than any carried out to date could be accomplished with an astrometric instrument in Earth orbit. Above the blurring effects of the atmosphere, a rather small telescope (aperture 1.5 m) with its optical alignment continuously monitored and held constant would be able to detect changes in star positions as small as the apparent width of a dime seen at the distance of the Moon.

With such a space instrument, it should be possible to detect planets the size of Uranus or Neptune in orbits as small as Jupiter's out to a distance of about 30 light-years. (A light-year is the distance light travels in one year at its velocity of 3×10^{10} cm/sec. This distance is 9.5×10^{12} km, or 63,240 astronomical units.) In other words, any star within this limit would suffer a sufficient perturbation by a Uranus-size planet that we could detect the effect on the star's motion, provided the planet were as close to the star as Jupiter is to the Sun. We choose a planet with the mass of Uranus because all four of the outer planets in our system have cores of about this mass. Since Jupiter and Saturn are actually much more massive than this, we are making a conservative assumption about the sizes of the planets we hope to detect. If they are actually the size of Jupiter or Saturn, it will just be that much easier to discover them.

Even with the most optimistic assumptions, it is clear that the detection of planets as small as the Earth from their gravitational perturbations is far beyond current technology. Similarly, we are probably decades away from a space telescope that could image a terrestrial planet near another star. Yet for many persons, the ultimate goal of planetary searches is to find another world like our own that might be potentially habitable. Discovering a Jupiter or a Neptune, however satisfying scientifically, would still leave the issue of the frequency of terrestrial planets unresolved.

Fortunately, there is another technique for detecting planets like the Earth in orbit about stars like the Sun. Imagine how our solar system would appear to an observer on a planet circling some distant star. If this vantage point were located in the plane of our planetary orbits, then once every year the Earth would pass between the Sun and the observer, blocking a small fraction of the sunlight for a period of about ten hours. The light blocked would amount to only $\frac{1}{10{,}000}$ of the Sun's brightness, but if the solar brightness were monitored with sufficient accuracy, this "photometric signature" of the presence of the Earth could be detected. From the time between these events the orbital period of the Earth could be derived, and from the amount of light lost the size of our planet could be estimated.

Bill Borucki of NASA's Ames Research Center has shown that current technology is capable of measuring stellar brightnesses with the requisite precision, and he has proposed building a small telescope that could monitor the light of more than 1000 solar-type stars continuously from Earth-orbit. The telescope must be in orbit in order to escape the variable transmission of the atmosphere. If Earth-size planets are common around solar-type stars, Borucki should quickly discover dozens of them with this approach. At present, this photometric technique is the only way we know to leapfrog giant planet detection and move directly to a search for terrestrial planets, in the quest for what Carl Sagan has called the "Pale Blue Dot" of another Earth, blessed with oceans and possibly even an atmosphere like our own.

Signals from Extraterrestrial Civilizations

An obvious way to accelerate the process is to search for a very special kind of planetary system: one in which *intelligent life* has evolved. If such a place exists, and if its inhabitants are transmitting high-powered radio signals in our general direction, then it is well within the capabilities of our current radio telescopes to detect these signals. The discovery of such signals would not only demonstrate the existence of other planets, but would also of course tell us a great deal more. And if we could ever manage to decipher the signal, or even to establish a two-way communication, we could learn all we wanted about their system simply by asking.

The search for extraterrestrial intelligence (abbreviated **SETI**) is a matter of profound significance for the future of humanity. In the mid-1980s, The Planetary Society, a public interest group, began the first serious SETI program aimed at surveying the entire sky with a radio receiver sensitive simultaneously to a million different frequency bands. The society's president, Carl Sagan, has long been a proponent of the search for such signals, and major funding for the million-channel receiver was provided by Steven Spielberg, maker of such science fiction films as *E.T.*, *Close Encounters of the Third Kind*, and *Jurassic Park*.

The next step for SETI was to be a program funded by NASA that would have used a spectrum analyzer capable of receiving 16- to 32-million frequencies simultaneously. This receiver was to be employed for both an All Sky Survey and a Targeted Search. The Survey was planned to cover the entire sky with a sensitivity far superior to that of any previous efforts. The Targeted Search would have examined in greater detail the 700 nearest Sun-like stars, those regarded to be most likely to have planets. Both modes would have covered thousands of frequencies that had not been previously explored.

The NASA SETI program began on Oct. 12, 1992, commemorating the 500th anniversary of Columbus' landing in the New World. The program was terminated one year later by the U.S. Congress. The Senate vote was 77 to 23 to end what astronomer Frank Drake has called "the last great adventure left to mankind." The leading congressional antagonist, Senator Richard Bryan of Nevada, referred to this project as "The Great Martian Chase," derisively pointing out that "Not a single Martian has said 'take me to your leader,' and not a single flying saucer has applied for FAA approval."

Fortunately, the search for intelligent life on Earth is not compelled to stop with a study of the U.S. Congress. In the aftermath of this cancellation, the SETI program has risen like a phoenix from the ashes of senatorial ridicule. The NASA Targeted Search has been taken over by a nonprofit corporation known as The SETI Institute, which has raised nearly $10 million in private and corporate donations. Renamed Project Phoenix, the next stage of the targeted search was carried out during 1995 with the Parkes Radio Telescope in Australia (Fig. 19.11). If this fundraising effort continues to be successful, the SETI program still has a chance to detect an extraterrestrial civilization before the end of the millennium.

FIGURE 19.11 Astronomer Jill Tarter, chief scientist of Project Phoenix, studying a computer screen in the control room of the 64-m radio telescope at Parkes, Australia. Equipped with a highly sophisticated multichannel spectrum analyzer, this huge antenna will systematically survey the southern sky, searching for signals from civilizations far more advanced than any that have ever existed on Earth.

19.8 QUANTITATIVE SUPPLEMENT: ANGULAR MOMENTUM IN THE SOLAR NEBULA

The distribution of angular momentum in the solar system has always challenged astronomers investigating the origin of the planets. To see why, it is instructive to calculate a table of the angular momentum associated with the Sun, Jupiter, Saturn, and the Earth, relative to the center of the solar system. For the Sun, this momentum is rotational, and it is given approximately by $0.1\ mR^2/p$, where m is the mass of the Sun, R is its radius, and p is its rotation period. For the individual planets, almost all of the angular momentum is contained in their orbital motion, given approximately by mr^2/p, where r is the radius of the orbit and p is the period required for one orbital revolution. With these formulas and the data given in the endpaper of this book, we calculate the numbers shown in Table 19.2. Examining these values, we see that there is much more angular momentum in the planets, especially Jupiter, than in the Sun. The disparity is even more dramatic if we consider the angular momentum per kilogram, obtained by dividing the numbers in the last column by the mass: For the Sun this is 2×10^{10} while for Jupiter it is 2×10^{15}, a hundred thousand times greater.

For comparison, we can calculate the angular momentum for a condensing solar nebula of about one solar mass (2×10^{30} kg). Let us assume that the original nebula has dimensions typical of interstellar distances, say 1 light-year, or about 10^{16} m. Its minimum rotation period corresponds to one orbit around the center of the Galaxy in 200 million years, or 6×10^{15} s. Then its angular momentum is that given by the formula above for the Sun, or about 3×10^{45} kg m^2 s^{-1}, which is a hundred thousand times greater than the angular momentum of the Sun today. The angular momentum of the solar nebula per kilogram is, however, about 10^{15}, similar to that of Jupiter. Therefore we can conclude that the planets have about the correct quantity of angular momentum, while the Sun is highly deficient. Hence some process in the past must have robbed the Sun of its angular momentum, but most of this momentum was not transferred to the planets; it was simply lost from the system.

SUMMARY

This final chapter reviews many ideas presented throughout the book, considering them here in the context of the formation and early evolution of the planetary system. Planetary scientists would like to regard the solar system as but one example of a more general phenomenon, but unfortunately at this writing there is no concrete evidence for the existence—or absence—of other planetary systems like ours. Thus we must first look to our own system to see whether we

TABLE 19.2 Angular momenta of the Sun, Jupiter, Saturn, and Earth

	m (kg)	r (m)	p (s)	Momentum (kg m^2 s^{-1})
Sun	2×10^{30}	7.0×10^{8}	2×10^{6}	5×10^{40}
Jupiter	2×10^{27}	7.8×10^{11}	4×10^{8}	3×10^{42}
Saturn	6×10^{26}	1.4×10^{12}	9×10^{8}	1×10^{42}
Earth	6×10^{24}	1.5×10^{11}	3×10^{7}	5×10^{39}

can generalize from the knowledge we have gained concerning its members. Table 19.1 summarizes some basic properties of our planetary system related to its origin.

There is a consensus among astronomers that the solar system was born about 4.5 billion years ago from a condensation in the interstellar medium called the solar nebula. Similar dense interstellar clouds of gas and dust are observed widely in the Galaxy as the birthplaces of stars. One approach to understanding the origin of the solar system is to study these other, similar processes that are taking place around us today.

The solar nebula was the nursery of both the Sun and the planetary system. The Sun, forming in the center, attracted most of the mass and evolved into a self-sustaining star. The rest of the gas and dust collapsed into a disk, where it was heated by the protosun. As part of this process, the protosun lost most of its angular momentum, leaving most of the rotation in the spinning nebular disk.

As temperatures dropped in the disk, solid grains condensed, sorted compositionally to some extent by the temperature structure in the disk. These grains rather quickly aggregated into larger bodies, building up what are called planetesimals, according to the scenario originally proposed by Safronov. These planetesimals, tens to hundreds of kilometers in diameter, became the building blocks of the planets. The planets, growing by accretion and heated by impacts, differentiated as they formed.

For the first few hundred million years of solar system history, a great deal of solid debris remained between the planets, and impacts were much more frequent than today. The heavily cratered surfaces of the Moon and other solid bodies throughout the system date from this time. A small fraction of the leftover icy planetesimals from the outer solar system contributed to the atmospheres of the terrestrial planets, while others were gravitationally ejected from the solar system or into the Oort comet cloud. A second contemporary reservoir of comets is the Kuiper Belt, just beyond the orbit of Pluto. Both reservoirs still contribute objects that occasionally impact the inner planets today.

In the outer solar system, solid planetary cores formed that were large enough (10–15 Earth masses) to attract and hold part of the remaining gases of the solar nebula, thus creating the giant planets. These planets formed within their own subnebulas, which gave rise to their regular satellite systems. The rings may have originated in the same way, but it seems more likely that they are the product of later impact fragmentation of inner satellites, possibly augmented by cometary debris. Ultimately the remnants of the original nebula were blown away by strong solar winds from the young Sun.

This picture of the origin of our system is consistent with the basic properties listed in Table 19.1. It also suggests that the formation of other systems like our own should be a common aspect of the origin of single stars. In this case, the search for other planetary systems is especially compelling. Unfortunately, existing techniques fall just short of being able to detect other planetary systems like our own, although they have already revealed such interesting phenomena as substellar brown dwarfs, companions to neutron stars, disks of small bodies orbiting some other stars, and a handful of giant planets.

Finally, there is a question that goes far beyond the existence of other planetary systems. Might there not also be other planets with life, even intelligent life, orbiting nearby stars? We do not know, but the search—SETI—has begun for radio signals from such other technical civilizations. The discovery of such signals, and even more the possibility of establishing communication with extraterrestrial intelligent creatures, would be one of the most significant events in human history.

KEY TERMS

astrometry
brown dwarf
center of mass
impact frustration
magnetic braking
midplane
neutron star
planetary embryos
planetesimals
pulsar
radial mixing
SETI

REVIEW QUESTIONS

1. Review Table 19.1. Do you agree that all of the "facts" listed there are fundamental in the sense that a theory of solar system origin must explain them before we can believe it? What facts would you add or delete?

2. Describe the current model for star formation. Do you expect stars to form near the Sun—just outside the Oort cloud, for example? Why or why not?

3. Explain how angular momentum is conserved as an interstellar cloud collapses to form the solar nebula. Why is there a problem in accounting for the difference in the angular momentum carried by the Sun and by the planets? What solutions could you suggest for this problem?

4. Make a chart depicting the various steps in the evolution of matter in the solar nebula disk, from its initial dispersed state to the formation of planets. Describe what happens in each step and give the length of time required.

5. What role (if any) did icy bodies play in the formation of the inner planets?

6. How did comets form? What are the two major reservoirs of comets in the solar system and how do we distinguish between them? Or, to put it differently, when a comet is discovered, how do we know from which reservoir it originated?

7. Explain the steps in the formation of satellites and rings. Why don't Venus or Earth have rings or extensive satellite systems?

8. What kind of observational evidence exists to support the "nebula hypothesis" for star and planet formation? Why don't we have any direct observations of other planetary systems? What kind of indirect evidence is likely to accrue in the next decade?

9. You have just completed a survey of our planetary system and the current theories for how it formed. There are approximately 20 billion stars very similar to our Sun in the galaxy. How many of them do you think have planets? Why? How would you try to make contact with civilizations living on these planets?

QUANTITATIVE EXERCISES

1. The outer planets with their regular satellite systems offer us many opportunities for comparison with the solar system as a whole. Calculate the rotational angular momentum of Jupiter and the orbital angular momentum of each of the four Galilean satellites. How do these values compare with the situation for the Sun and planets?

2. Jupiter and its Galilean satellites are thought to have formed from the collapse of a protojovian nebula. Suppose this nebula had an initial radius of 2 AU and shared the orbital motion of Jupiter around the Sun. Use the same methods described for the solar nebula to calculate the angular momentum of this protojovian nebula, and compare your results with the angular momenta found in Exercise 1. What can you conclude about the loss of angular momentum in the jovian system? Why might the situation here differ from that of the Sun and planets?

3. There is a famous equation developed by astronomer Frank Drake for calculating N, the number of advanced civilizations in the Galaxy:

$$N = N_* \times f_s \times N_p \times f_e \times f_l \times f_i \times L/L_{MW}$$

where N_* = the number of stars in the galaxy, f_s = the fraction of stars that could sustain planets with life, N_p = the average number of planets per star, f_e = the fraction of (Earthlike) planets suitable for life, f_l = the fraction of those planets on which life develops, f_i = the fraction of planets with life on which intelligence develops, L = the average lifetime of an advanced civilization, and L_{MW} = the lifetime of the Milky Way Galaxy. Given that N_* = 300 billion, f_s = 0.3, and L_{MW} = 10 billion years, solve this equation for N, justifying the value you give for each term in the equation.

ADDITIONAL READING

*Burke, B.F., et al., eds. 1994. *Planetary Systems: Formation, Evolution, and Detection.* Dordrecht: Kluwer Academic Publishers. A collection of technical papers including discussions of the technology to detect other planetary systems.

Drake, F., and D. Sobel. 1992. *Is Anyone Out There?: The Scientific Search for Extraterrestrial Intelligence.* New York: Delacorte Press. Highly readable historical account by the premier SETI pioneer.

Goldsmith, D., and T. Owen. 1992. *The Search for Life in the Universe,* 2d ed. Reading, Mass.: Addison-Wesley. A general discussion of those aspects of astronomy, biology, and planetary science that relate to the question of the cosmic prevalence of life, and strategies to detect and communicate with advanced civilizations.

Hutchison, R. 1983. *The Search for Our Beginnings.* New York: Oxford University Press. Highly readable discussion of solar system origin with emphasis on cosmochemical issues.

*Lunine, J., E.H. Levy, and M.S. Matthews, eds. 1993. *Protostars and Planets III.* Tucson: University of Arizona Press. A collection of technical articles by scientists whose specialties range from studies of meteorites to the interstellar medium, all focused on the problems of star and planet formation.

*Weaver, H.A., and L. Danly, eds. 1989. *The Formation and Evolution of Planetary Systems.* Cambridge: The University Press. A collection of articles by specialists on various aspects of the problem, including both theories and observations.

*Indicates the more technical readings.

APPENDIX

Units and Dimensions

Throughout this book, we have tried to emphasize the metric system, since it is the system of measurements commonly used in science at the present time. It is also the most widely used system in the world, with only the stubborn Americans, Burmese, and South Africans adhering to the foot-pound system. That adherence makes metric values unfamiliar to many of us, however, so we present tables below with some equivalences.

Note that in the metric system, units of measurement increase by factors of 10, 100, 1000, and so on:

1 centimeter = 10 millimeters

1 meter = 100 centimeters = 1000 millimeters

1 kilometer = 1000 meters = 100,000 centimeters = 1,000,000 millimeters

English Unit	Metric Equivalent
1 inch	25.4 millimeters 2.54 centimeters
1 foot	30.48 centimeters
1 yard	91.44 centimeters 0.9144 meter
1 mile	1609.3 meters 1.6093 kilometers
1 ounce	28.4 grams
1 pound	453.6 grams 0.454 kilogram
1 quart	1136.5 cubic centimeters 1.137 liters
1 mile/hour	1.609 kilometers/hour 1609 meters/hour 0.45 meter/second

Metric Unit	English Equivalent
1 millimeter	0.04 inch
1 centimeter	0.39 inch
1 meter	39.37 inches
1 kilometer	0.62 mile
1 gram	0.04 ounce
1 kilogram	35.27 ounces 2.2 pounds
1 liter	0.88 quart
1 kilometer/hour	0.62 mile/hour
1 meter/second	2.2 miles/hour

In discussing the dimensions and masses of atoms, planets, and satellites, as well as distances between objects in the solar system and between waves of electromagnetic radiation, we encounter some very large and very small numbers. Instead of coping with a lot of zeros, it is easier to resort to what is called exponential notation, or powers of 10. This way we can write one million as 10^6 instead of 1,000,000. If this is unfamiliar, you can think of the exponent as giving you the number of zeros following the 1. Thus $10^0 = 1$.

Similarly for small numbers: 1/10 is simply 10^{-1}; 1/10,000 is 10^{-4}. Now the negative exponent tells you the number of places to the right of the decimal point. The diameter of a hydrogen atom is $0.00000001 = 10^{-8}$ cm.

Still further simplifications are used. The distance from the Earth to the Sun is 93 million miles or 93×10^6 miles or 149.6 million kilometers or 149.6×10^6 kilometers or 1.5×10^{13} centimeters or 1 astronomical unit. This last convenient unit is often used to give other distances in the solar system. For example, Jupiter is 5.203 astronomical units from the Sun.

At the short end of the scale, 10^{-8} centimeters is defined to be 1 angstrom (abbreviated Å, or simply A), in honor of Anders Ångstrom, a Swedish physicist who was a pioneer in spectroscopy. (Hence the diameter of a hydrogen atom is 1 angstrom.) This unit is convenient for describing wavelengths of visible and ultraviolet light. The longer wavelengths of infrared light are often given in terms of micrometers (sometimes microns) where 1 micrometer (abbreviated μm) equals 10^4 angstroms, or 10^{-4} centimeters or 10^{-6} meters.

Glossary

absorption line. Removal of energy from a narrow range of wavelengths or frequencies in the electromagnetic spectrum due to atomic or molecular absorption. (Ch. 2)

accretion. Gravitational accumulation of mass in a planet or protoplanet. (Ch. 4)

albedo. The reflectance or reflectivity of an object; specifically the ratio of reflected energy to incident energy (also called the Bond albedo). (Ch. 3)

angular momentum. A measure of the momentum associated with rotational motion about an axis. (Ch. 4)

anorthosite. Primary igneous rocks of the lunar highlands, composed almost entirely of the mineral plagioclase feldspar. (Ch. 7)

anticyclonic. Rotation induced in a high-pressure mass of atmosphere by Coriolis forces; clockwise in the northern hemisphere of a directly rotating planet. (Ch. 9)

aperture. The diameter of the primary lens or mirror of a telescope; hence the best single measure of the light-gathering power of a telescope. (Ch. 3)

aphelion. For an object orbiting the Sun, the point in the orbit that is farthest from the Sun. (Ch. 1)

asteroid. Any small body (less than 1000 km in diameter) orbiting the Sun that does not display the atmosphere or tail associated with a comet; also called minor planets. (Ch. 4)

astrometry. Precise measurement of positions in astronomy; hence an approach to searching for external planets by measuring the tiny periodic shifts in a star's position caused by the gravitational pull of invisible objects in orbit about it. (Ch. 19)

astronomical unit (AU). The semimajor axis of the Earth's orbit, or equivalently the average distance of the Earth from the Sun; approximately 150 million kilometers. (Ch. 1)

atmospheric window. The part of the electromagnetic spectrum within which a planetary atmosphere is more-or-less transparent; for Earth, the wavelength regions where astronomical observations can be carried out from the ground. (Ch. 3)

aurora. Light radiated by atoms and ions in a planetary ionosphere, mostly in the magnetic polar regions; also called polar lights. (Ch. 9)

bar. Unit of pressure equal to 10 dynes/cm^2, or approximately the atmospheric pressure at the surface of the Earth. (Ch. 9)

basalt. Common igneous rock, composed primarily of silicon, oxygen, iron, aluminum, and magnesium, produced by the rapid cooling of lava. Basalts make up most of Earth's oceanic crust and are also found on other planets that have experienced volcanic activity. (Ch. 3)

bending wave. Spiral wave structure in a ring or disk of material caused by the gravitational perturbations of a satellite with an orbit inclined to the ring plane. (Ch. 18)

Bode's law. See *Titius-Bode rule.*

breccia. Any rock made up of recemented fragments of material, usually the result of extensive impact cratering. (Ch. 4)

brown dwarf. An object intermediate between a planet and a star; roughly, with mass less than 0.07 solar masses and greater than a few times the mass of Jupiter. (Ch. 19)

C-type asteroid. One of the most populous group of main belt asteroids, characterized by dark, spectrally neutral or slightly red surfaces; thought to be primitive, similar in composition to the carbonaceous meteorites. (Ch. 5)

caldera. Volcanic crater resulting from the collapse following withdrawal of magma, often found at the summit of a shield volcano. (Ch. 9)

carbonaceous meteorite. A primitive meteorite made primarily of silicates but including chemically bound water, free carbon, and complex organic compounds. Also called carbonaceous chondrites. (Ch. 4)

carbonate. A chemical compound that contains CO_2, such as calcium carbonate ($CaCO_2$), the primary constituent of the shells of marine organisms. When decomposed by heating, carbonates release carbon dioxide. (Ch. 9)

catalyst. Atom or molecule that enables or accelerates a chemical reaction without itself being altered or consumed. (Ch. 14)

catastrophism. The concept that the geology of the Earth and planets has been greatly influenced by rare events of large magnitude, such as impacts by asteroids. In the eighteenth and nineteenth centuries, catastrophism was associated specifically with attempts to explain most geological features as results of the biblical flood, but today the term is used in the much broader sense given above. (Ch. 7)

center of mass. The average position of the various mass elements of a body or system, weighted according to their distances from that center of mass; also called center of gravity. (Ch. 19)

CFC. Chlorofluorocarbons, nontoxic industrial chemicals used in refrigerators and air conditioners, as propellants in spray cans, and for cleaning electronics—but with the unintended consequence of destroying stratospheric ozone. (Ch. 9)

chaotic rotation. Rotation that does not have a fixed period or axis of rotation, such as that of Saturn's satellite Hyperion. (Ch. 17)

chaotic terrain. The regions of the martian uplands consisting of jumbled depressions and isolated hills, thought to have been produced by collapse induced by the withdrawal of subsurface water or ice. (Ch. 11)

chemical equilibrium. Composition that reflects a chemical balance of different atoms and molecules, without any active sources or sinks of new material. (Ch. 14)

chondrule. A small silicate spherule (typically a few millimeters in diameter) commonly found in primitive meteorites. (Ch. 4)

coma. Atmosphere or head of a comet, forming a visible halo around the nucleus. (Ch. 6)

comet. Any of the most primitive Sun-orbiting members of the solar system, consisting of a small nucleus composed of ices, silicates, and carbonaceous material, which when heated generates a tenuous temporary atmosphere as its volatiles evaporate. (Ch. 4)

comet nucleus. The solid part of a comet, typically a few kilometers (up to hundreds of kilometers) in diameter. The nucleus consists primarily of a mixture of ices and solid grains of silicate and carbonaceous composition. (Ch. 6)

composite volcano. Common type of terrestrial volcano, with a cone built up by repeated (and sometimes explosive) fountains of lava mixed with hot gas. (Ch. 9)

compound. A substance composed of two or more chemical elements, such as H_2O (formed from hydrogen and oxygen). (Ch. 2)

conduction. One of the basic ways of transferring energy, caused by the motion of atoms and electrons in a solid. Although it is not as intrinsically efficient as convection or radiation, conduction is usually the dominant means to transfer energy in a solid. (Ch. 2)

constellation. Originally a configuration of stars; now one of 88 specific areas of the sky with internationally agreed-upon boundaries. (Ch. 1)

continental drift. A gradual motion of the continents over the surface of the Earth due to the motion of lithospheric plates, as described by the theory of plate tectonics. (Ch. 9)

convection. One of the basic ways of transferring energy, caused by the large-scale (macroscopic) motion of material, such as the rising of pockets of hot gas and sinking of cooler gas. When convection occurs, it is usually more effective at transferring energy than either radiation or conduction. (Ch. 2)

co-orbital satellite. Informal term for the two Saturn satellites Janus and Epimetheus, which share almost the same orbit, or for any other satellites found in similar dynamical situations. (Ch. 17)

Coriolis effect. The deflection of material moving across the surface of a rotating planet, producing in the case of the Earth's atmosphere the familiar cyclonic and anticyclonic patterns that characterize our mid-latitude weather. (Ch. 9)

corona (Sun). The tenuous outer atmosphere of the Sun, consisting of gas at temperatures of more than a million degrees. (Ch. 2)

corona (Venus). Large, circular tectonic features unique to Venus, apparently caused by a rising plume of mantle material, often with associated volcanic activity. (Ch. 10)

cosmic rays. Atomic nuclei (mostly protons) that strike the Earth's atmosphere with exceedingly high energies. Some originate in the Sun, but most cosmic rays have a galactic origin. (Ch. 7)

crater. A circular depression (from the Greek word for bowl or cup), generally of impact origin. (Ch. 7)

crater density. The degree to which impact craters are packed together on a planetary surface, measured in units of the number of craters of a given size per unit area (for example, the number of craters larger than 10 km in diameter per million square kilometers of surface). (Ch. 7)

crater retention age. The time over which a planetary surface has accumulated impact craters; hence the time since the surface was mobile or extensive erosional or tectonic forces were acting to destroy craters. (Ch. 7)

crust. The outer solid layer of a planet; on Earth, the upper 10 to 60 kilometers. (Ch. 7)

cyclonic. The rotation induced in a low-pressure mass of atmosphere by Coriolis forces; counterclockwise in the northern hemisphere of a directly rotating planet. (Ch. 9)

differentiated meteorite. A meteorite from a differentiated parent body, as contrasted with a primitive meteorite. (Ch. 4)

differentiation. The gravitational separation or segregation of different densities of material into different layers in the interior of a planet, as a result of heating. (Ch. 3)

Doppler effect. Apparent change in wavelength (and frequency) of the radiation from a source due to its relative motion toward or away from the observer. (Ch. 8)

dust tail. A cometary tail, usually broad, somewhat curved, and yellow-white in color, made of dust grains released from the nucleus of the comet. (Ch. 6)

Earth-approaching asteroid. See *near-Earth asteroid.*

eccentric ring. A planetary ring that has the form of an ellipse, as a result of the alignment of the individual elliptical orbits of the particles that comprise it. (Ch. 18)

eccentricity (of ellipse). The degree to which an orbit is noncircular; specifically, the ratio of the distance between the foci to the major axis of the ellipse. (Ch. 1)

eclipse of the Moon. The phenomenon visible when the Moon passes wholly or in part through the shadow of the Earth. (Ch. 1)

eclipse of the Sun. The blocking of all or part of the light of the Sun by the Moon. (Ch. 1)

ecliptic. The apparent annual path of the Sun on the celestial sphere. (Ch. 1)

ejecta blanket. Rough, hilly region surrounding an impact crater made up of ejecta that has fallen back to the surface—usually extending one to three crater radii from the rim. (Ch. 7)

electromagnetic radiation. Radiation consisting of electric and magnetic waves; they include radio, infrared, visible light, ultraviolet, x rays, and gamma rays. (Ch. 2)

electromagnetic spectrum. The whole array or family of electromagnetic waves, usually ordered by wavelength (or equivalently by frequency or energy). (Ch. 2)

electron. Basic subatomic particle with a negative electric charge. The number and configuration of electrons in an atom or molecule are critical to both its chemical properties and its spectrum. (Ch. 2)

element. Basic form of matter; the smallest unit (atom) that retains the chemical properties of an elemental substance. There are 92 naturally occurring elements, with numbers of protons (and electrons) from 1 (hydrogen) to 92 (uranium). (Ch. 2)

ellipse. A closed curve (one of the conic sections) that describes the orbit of one object about another subject only to their mutual gravitational attraction. (Ch. 1)

embedded satellite. A small (perhaps invisible) satellite orbiting within a ring system and gravitationally influencing the structure of the ring. (Ch. 18)

equinox. Either of the two intersections of the ecliptic and the celestial equator; occupied by the Sun on about March 21 and September 21, when the day and night are of equal length all over the planet. (Ch. 1)

escape velocity. The minimum upward speed required to escape entirely from the gravitational attraction of a body. (Ch. 1)

eucrite. One of a class of basaltic meteorites believed to have originated on the asteroid Vesta. (Ch. 4)

exosphere. The part of the upper atmosphere of a planet from which atoms or molecules of gas can escape into space. (Ch. 3)

family of asteroids. A group of asteroids with similar orbital elements, indicating a probable common origin in a past collision. (Ch. 5)

fault. In geology, a tectonic crack or break in the crust of a planet along which slippage or movement can take place, accompanied by seismic activity. (Ch. 9)

fluidized ejecta. Ejecta from an impact crater that flows along the surface like a liquid rather than arching freely through space, apparently a common phenomenon in the past on Mars. (Ch. 11)

forward scattering. The tendency of small particles to reflect or scatter light primarily in a forward direction, that is, in nearly the same direction from which light is incident on the particle. (Ch. 18)

fractionation. A process that changes the relative abundances of elements or isotopes in a planetary body (or atmosphere), such as by the selective loss of one component. (Ch. 4)

fragmentation. A process that breaks up objects, usually as the result of high-speed collisions. (Ch. 4)

Gaia hypothesis. The suggestion that the Earth responds to changing external circumstances (such as variations in the luminosity of the Sun) in ways that preserve conditions suitable for life. (Ch. 12)

gas-retention age. A measure of the age of a rock, defined in terms of its ability to retain radioactive argon. (Ch. 4)

geocentric. Earth-centered, as in the pre-Copernican idea that the Sun, Moon, and planets all circled the Earth, which was thought to be located at the center of the universe. (Ch. 1)

geologic timescale. The history of the Earth over the past 4.5 billion years, as determined from the rocks deposited in its crust. (Ch. 9)

granite. Igneous rock associated primarily with the Earth's continental crust, composed chiefly of the minerals quartz and alkali feldspar. (Ch. 9)

greenhouse effect. The blanketing of infrared radiation near the surface of a planet by infrared-opaque gases in the atmosphere (for example, carbon dioxide), producing an elevated surface temperature. (Ch. 9)

Hadley cell. A theoretical mode for the circulation of a planetary atmosphere, in which air heated near the equator rises and moves toward the poles, where it descends and flows back toward the equator. The circulation of the atmosphere of Venus approximates this situation. (Ch. 10)

half-life. The time required for half of the radioactive atoms in a sample to disintegrate. (Ch. 4)

heliocentric. Centered on the Sun; specifically, the Copernican theory that the planets are in orbit around the Sun rather than the Earth. (Ch. 1)

highlands (lunar). The older, heavily cratered crust of the Moon, covering 83% of its surface and composed in large part of anorthositic breccias. (Ch. 7)

ice age. One of the periods in the Earth's climatic history when global cooling led to the formation of extensive ice sheets over polar and even temperate land masses. (Ch. 9)

igneous rock. Any rock produced by cooling from a molten magma. (Ch. 3)

imaging radar. Radar carried on a moving platform (airplane or spacecraft) that can, with suitable data processing, produce images of the ground beneath; also called synthetic aperture radar. (Ch. 10)

impact basin. A large impact feature, usually 300 km or more in diameter. (Ch. 7)

impact erosion. Loss of atmosphere produced by repeated, large impacts. (Ch. 12)

impact frustration (of the origin of life). Environmental conditions unfavorable to the survival of life on a planet and caused by large impacts in its early history. (Ch. 19)

inclination (of an orbit). The angle between the orbital plane of a revolving body and some fundamental plane—usually the plane of the celestial equator or of the ecliptic. (Ch. 1)

Io plasma torus. Prominent feature of the jovian magnetosphere located near the orbit of Io, which consists of relatively dense concentrations of sulfur and oxygen ions trapped in the planet's rapidly rotating magnetic field. (Ch. 13)

ion. An atom that has gained or (more usually) lost one or more electrons and thus has a net electric charge. (Ch. 2)

ionize. To add or subtract one or more electrons from an atom, making it an ion. (Ch. 2)

ion tail. See *plasma tail.*

ionosphere. The upper region of the Earth's atmosphere in which many of the atoms are ionized, or any similar feature of the atmosphere of another planet. (Ch. 9)

irregular satellite. A planetary satellite with an orbit that either is retrograde or has high inclination or eccentricity. (Ch. 15)

iron meteorite. A meteorite composed primarily of metallic iron and nickel and thought to represent material from the core of a differentiated parent body. (Ch. 4)

isotope. Any of two or more forms of the same element, whose atoms all have the same number of protons but different numbers of neutrons. (Ch. 2)

K/T event. Major break in the history of life on Earth (a mass extinction) that occurred 65 million years ago, between the Cretaceous and Tertiary periods, due to the impact of a comet or asteroid in the Yucatan region of Mexico. (Ch. 9)

Kuiper Belt. Disk-shaped region beyond the orbit of Neptune that is thought to contain many icy objects and to be the main source of the short-period comets. (Ch. 6)

leading side. The hemisphere of a synchronously rotating satellite that always faces forward in the direction of its orbital motion. (Ch. 15)

lithosphere. The upper layer of the Earth, to a depth of 50 to 100 km, involved in plate tectonics. (Ch. 9)

long-period comet. A comet with a period of revolution of 200 years or more. (Ch. 6)

luminosity. Intrinsic brightness; specifically the total energy output of the Sun or a star. (Ch. 2)

M-type asteroid. Asteroid composed primarily of metal, presumably related to the iron meteorites. (Ch. 5)

magma. Melted rock in the interior of a planet; called lava when it is erupted on the surface. (Ch. 7)

magnetic breaking. Process proposed to account for the slow rotation of some stars, in which angular momentum is transferred from the star to the surrounding plasma through its magnetic field. (Ch. 19)

magnetic flux tube. Feature of the jovian magnetosphere consisting of a loop of electrical current connecting Io and the planet. (Ch. 13)

magnetosphere. The region around the Earth (or other planet) occupied by its magnetic field, and within which the planetary field dominates over the interplanetary field associated with the solar wind. (Ch. 9)

magnetotail. The region of a planetary or cometary magnetosphere in which the magnetic field lines stream away from the object as they are carried "downstream" by the solar wind. (Ch. 13)

main belt asteroids. Asteroids that occupy the main asteroid belt between Mars and Jupiter, sometimes limited specifically to the most populous parts of the belt, from 2.2 to 3.3 AU from the Sun. (Ch. 5)

mantle. The part of a planet between its crust and core; on Earth, the mantle is the greatest part of the planet, with about 65% of the mass. (Ch. 7)

mare (plural: **maria**). Latin for "sea"; name applied to the dark, relatively smooth features consisting of basaltic lava flows that cover 17% of the Moon. (Ch. 7)

mass extinction. The sudden disappearance in the fossil record of a large number of species of life, to be replaced by new species in subsequent layers. Mass extinctions are indications of catastrophic changes in the environment, such as might be produced by a large impact on the Earth. (Ch. 9)

metamorphic rock. Any rock produced by the physical and chemical alteration (without melting) of another rock that has been subjected to high temperature and pressure. (Ch. 3)

meteor. The luminous phenomenon observed when a bit of material (cosmic dust) enters the Earth's atmosphere and burns up; popularly called a falling star or shooting star. (Ch. 4)

meteor shower. Many meteors appearing to radiate from a common point in the sky caused by the intersection of the Earth with a swarm of meteoric particles. (Ch. 6)

meteorite. A portion of a meteoroid that survives its fiery passage through the atmosphere and strikes the ground. (Ch. 4)

meteoroid. A meteoritic particle in space before any encounter with the Earth. Often used to describe objects larger than cosmic dust but smaller than an asteroid (e.g., from a few centimeters up to 50 m in diameter). (Ch. 4)

midplane. The central plane of rotation of a disk, such as the early solar nebula. (Ch. 19)

mineral. A solid compound (often primarily silicon and oxygen) that forms rocks; the term could also be applied to condensed volatiles such as ice. (Ch. 3)

model. See *scientific model.*

molecule. A combination of two or more atoms bound together; the smallest particle of a chemical compound or substance that exhibits the chemical properties of that substance. (Ch. 3)

near-Earth asteroid. An asteroid with an orbit that crosses the Earth's orbit or that will at some time cross the Earth's orbit as it evolves. (Ch. 5)

neutron. Basic subatomic particle with zero electric charge. The neutrons in an atom are located in the nucleus, and their number determines the isotope of the element. (Ch. 2)

neutron star. Extremely dense, collapsed star composed of neutrons, which can be formed in the aftermath of a supernova explosion. Rapidly rotating neutron stars can be observed as pulsars. (Ch. 19)

noble gases. The chemically inactive or inert elements: helium, neon, argon, krypton, xenon, radon. (Ch. 12)

nonthermal radiation. Any radiation from an astronomical body that is not thermal in origin, such as radio emission (synchrotron radiation) from electrons spiraling in the magnetosphere of Jupiter. (Ch. 13)

occultation. The passage of an object of large angular size in front of a smaller object, such as the Moon in front of a distant star, or the rings of Saturn in front of the Voyager spacecraft. (Ch. 13)

Oort comet cloud. The spherical region around the Sun from which most "new" comets come, representing objects with aphelia at about 50,000 AU, or extending about a third of the way to the nearest other stars. (Ch. 6)

opposition. The position of a planet when it is opposite the Sun in the sky, rising at sunset and setting at sunrise. (Ch. 1)

outflow channel. Martian channel, typically several kilometers wide and hundreds of kilometers long, that once drained large floods of water from the northern uplands to the southern lowlands. (Ch. 11)

oxidizing. Chemically dominated by oxygen; tending to form compounds of oxygen rather than hydrogen; opposite of reducing. (Ch. 3)

ozone. Molecule of oxygen with three atoms rather than the more common two (symbol O_3). (Ch. 3)

parent body. In planetary science, any larger original object that is the source of other objects, usually through breakup or ejection by impact cratering—for example, asteroid Vesta is thought to be the parent body of the eucrite meteorites. (Ch. 4)

perihelion. For an object orbiting the Sun, the point in the orbit that is closest to the Sun. (Ch. 1)

perturbation. The small gravitational effect of one object on the orbit of another. (Ch. 14)

Phanerozoic period. The most recent eon in the Earth's history, covering the past 590 million years. It is divided into the Paleozoic, Mesozoic, and Cenozoic eras. (Ch. 9)

photochemical smog. The aerosol of complex organic materials produced in a reducing atmosphere by various photochemical reactions. (Ch. 16)

photochemistry. Chemical reactions that are caused or promoted by the action of light—usually ultraviolet light that excites or dissociates some compounds and leads to the formation of new compounds. (Ch. 10)

photodissociation. Breakup of molecules by ultraviolet light. (Ch. 12)

photon. A discrete unit of electromagnetic energy. (Ch. 2)

photosphere. The part of the Sun from which the visible light originates; hence the apparent surface of the Sun. (Ch. 2)

planetary embryos. Hypothetical objects of roughly lunar mass proposed as an intermediate step between planetesimals and the final formation of the terrestrial planets. (Ch. 19)

planetesimals. Hypothetical objects, from tens to hundreds of kilometers in diameter, formed in the solar nebula as an intermediate step between tiny grains and the larger planetary objects (or planetary embryos). The comets and primitive asteroids may be leftover planetesimals. (Ch. 19)

plasma. A hot gas consisting in whole or in part of ions (charged atoms). (Ch. 2)

plasma tail. A cometary tail, usually narrow and bluish in color, extending straight away from the Sun and consisting of plasma streaming away from the comet's head under the influence of the solar wind. Also called an ion tail. (Ch. 6)

plate tectonics. The motion of segments or plates of the outer layer of the Earth (the lithosphere), driven by slow convection in the underlying mantle. (Ch. 9)

polar cap. A permanent or periodic deposit of ice or other volatiles near the polar region of a planet. (Ch. 8)

prebiotic chemistry. Organic chemical reactions that take place in the absence of life and are thought to have been essential to produce the chemical building blocks for life itself. (Ch. 16)

Precambrian period. The period of Earth history from the formation of the planet until the beginning of the Phanerozoic, 590 million years ago. The Precambrian is divided into the Priscoan, Archean, and Proterozoic eons. (Ch. 9)

primitive. In planetary science, an object or rock that is little changed, chemically, since its formation—hence is representative of the conditions in the solar nebula at the time of formation of the solar system. Also used to refer to the chemical composition of an atmosphere that has not undergone extensive chemical evolution. (Ch. 3)

primitive meteorite. A meteorite that has not been greatly altered chemically since its condensation from the solar nebula—also called a chondrite (either ordinary chondrite or carbonaceous chondrite). (Ch. 4)

primitive rock. Any rock that has not experienced great heat or pressure and therefore remains representative of the original condensates from the solar nebula. (Ch. 3)

proton. Basic subatomic particle with positive electric charge. The protons in an atom are located in the nucleus, and their number specifies the atomic number of the element. (Ch. 2)

proton-proton chain reaction. The most important process for generating the energy (luminosity) of the Sun through the fusion of hydrogen nuclei to form helium. (Ch. 2)

pulsar. Rapidly rotating neutron star, which emits periodic pulses of radio energy. (Ch. 19)

radial mixing. Condition in the early solar nebula that allowed material (e.g., planetesimals) formed at different distances from the Sun to intermix. (Ch. 19)

radiation. Electromagnetic radiation; also one of the three basic ways of transferring energy from one location to another. Since radiation travels through a vacuum, it is the primary means of transferring energy from one object in space to another, such as from the Sun to the Earth. (Ch. 2)

radioactive half-life. See *half-life.*

radioactivity. The process by which certain kinds of atomic nuclei naturally decompose with the spontaneous emission of subatomic particles and gamma rays. (Ch. 4)

reducing. Chemically dominated by hydrogen; tending to form compounds of hydrogen rather than oxygen. Opposite of oxidizing. (Ch. 3)

regolith. The broken or pulverized upper layers of a planetary surface, fragmented by impacts; specifically the lunar soil. (Ch. 7)

regular satellite. A planetary satellite that has an orbit of low or moderate eccentricity lying approximately in the plane of the planet's equator. (Ch. 15)

remote sensing. Any technique for measuring properties of an object from a distance; used particularly to refer to measurements (imaging, spectrometry, radar, etc.) of a planet carried out from terrestrial observatories or an orbiting spacecraft. (Ch. 3)

resolution. The degree to which fine details in an image are separated or visible. Resolution can be specified in either angular or linear units, but is used here usually in the sense of the linear dimensions (in km) of the smallest features that can be studied on a planet. (Ch. 7)

resonance. An orbital condition in which one object is subject to periodic gravitational perturbations by another—most commonly arising when two objects orbiting a third have periods of revolution that are simple multiples or fractions of each other. (Ch. 5)

resonance gap. A location in an ensemble of orbiting particles that is empty because it corresponds to periods that are simple fractions of those of a perturbing external body, such as the Kirkwood gaps in the main asteroid belt. (Ch. 5)

retrograde motion. An apparent westward motion of a planet with respect to the stars, caused by the motion of the Earth. (Ch. 1)

retrograde rotation (or **revolution**). Movement that is backward with respect to the common direction of motion in the solar system; counterclockwise as viewed from the north; going from east to west rather than from west to east. (Ch. 1)

rift. In geology, a place where the crust is being torn apart by tectonic forces; generally associated with injection of magma from the mantle and with the slow separation of lithospheric plates. (Ch. 9)

runaway greenhouse effect. A process whereby the heating of a planet leads to an increase in its atmospheric greenhouse effect and thus to further heating, quickly altering the composition of its atmosphere and the temperature of its surface. (Ch. 10)

runoff channel. A branching river channel with many tributaries, found in the old martian uplands and presumably formed at a time when higher temperatures and a more massive atmosphere permitted rain to fall on Mars. (Ch. 11)

S-type asteroid. The second most common class of asteroids, located primarily in the inner part of the main belt and characterized by moderate reflectivities (20%) and spectra that indicate the presence of silicate minerals similar to those in many meteorites. (Ch. 5)

scientific model. A description (usually mathematical) of processes in nature that should be self-consistent, provide a framework for understanding observations and experiments already carried out, and predict additional observations or experiments with which the model can be tested against alternatives. (Ch. 8)

secondary crater. An impact crater produced by ejecta from a primary impact. (Ch. 7)

sedimentary rock. Any rock formed by the deposition and cementing of fine grains of material. On Earth, sedimentary rocks are usually the result of erosion and weathering, followed by deposition in lakes or oceans, but breccias formed on the Moon by impact processes should also be considered as sedimentary rocks. (Ch. 3)

sedimentation. The process on Earth in which eroded material is transported (by water or wind) to lower areas, often the sea floors, where it slowly accumulates to form new sedimentary rock. (Ch. 9)

seismic waves. Waves in the solid Earth (or other planet) caused by earthquakes or impacts, which can be used to probe the structure of the interior. (Ch. 9)

SETI. The Search for Extraterrestrial Intelligence, carried out so far by looking for radio signals from other civilizations in the Galaxy. (Ch. 19)

shepherd satellite. Informal term for a satellite that maintains the structure of a planetary ring through its close gravitational influence—specifically, the two Saturn satellites, Prometheus and Pandora, that orbit just inside and outside of the F Ring, and the two small satellites of Uranus that orbit on either side of its Epsilon Ring. (Ch. 18)

shield volcano. A broad volcano built up through the repeated nonexplosive eruption of fluid basalts to form a low dome of shield shape, typically with slopes of only 4–6 degrees, often with a large caldera at the summit. Examples include the Hawaiian volcanoes on Earth and the Tharsis volcanoes on Mars. (Ch. 9)

short-period comet. A comet with a period of revolution of less than 200 years. Most known short-period comets have periods less than 15 years. (Ch. 6)

silicate. Minerals containing silicon and oxygen; the most common constituents of ordinary terrestrial rock. (Ch. 3)

SNC meteorite. One of a class of basaltic meteorites that are impact-ejected fragments from Mars. (Ch. 4)

solar activity cycle. The 22-year cycle in which the solar magnetic field reverses direction, consisting of two 11-year sunspot cycles. (Ch. 2)

solar day. The "day" as we usually think of it, as the average interval (24 hours on the Earth) between the times when the Sun rises or crosses the meridian. (Ch. 8)

solar nebula. The disk-shaped cloud of gas and dust from which the solar system formed. (Ch. 3)

solar wind. A radial flow of plasma (mostly protons and electrons) leaving the Sun at an average speed of about 400 km/s and spreading through the solar system. (Ch. 2)

solidification age. The most common age determined by radioactive dating techniques—the time since the rock or mineral grain being tested solidified from the molten state, thus isolating itself from further chemical changes. (Ch. 4)

solstice. The position in the Sun's apparent path when it is farthest south (around December 21) and farthest north (around June 21). (Ch. 1)

spectrometer. An instrument for forming and recording a part of the electromagnetic spectrum; in astronomy, a spectrometer is usually attached to a telescope to record the spectrum of an individual star or planet. (Ch. 2)

spiral density wave. Waves produced in a self-gravitating rotating sheet of material. In the rings of Saturn such waves, generated by resonant gravitational effects of the inner satellites, produce spiral patterns that wind around the rings like grooves in a phonograph record. (Ch. 18)

sputtering. The process by which energetic atomic particles striking a solid alter its chemistry and eject additional atoms or molecular fragments from the surface. (Ch. 13)

stony meteorite. A meteorite composed mostly of stony (silicate) minerals; the term can be applied to either primitive or differentiated meteorites if they are made of silicates. (Ch. 4)

stony-iron meteorite. A fairly rare kind of differentiated meteorite composed of a mixture of silicates with metallic iron-nickel, thought to have originated near the core-mantle boundary of a differentiated parent body. (Ch. 4)

stratigraphy. The study of rock strata, particularly the sequence of layers and the information this provides on the geologic history of a region. (Ch. 7)

stratosphere. The cold, stable (nonconvective) layer of the Earth's atmosphere above the troposphere and below the ionosphere; also, the similar layer above the troposphere on any planet. (Ch. 9)

subduction. In terrestrial geology, the tectonic process whereby one lithospheric plate is forced under another; generally associated with earthquakes, volcanic activity, and formation of deep ocean trenches. (Ch. 9)

subnebula. A miniature version of the disk-shaped solar nebula that is hypothesized to have formed around each of the giant planets at the time of their formation, giving rise to their regular satellite systems. (Ch. 15)

sunspot. A region of the solar photosphere that is cooler than its surroundings and therefore appears dark. (Ch. 2)

synchronous rotation. Rotation of a body so that it always keeps the same face toward another object; the situation where the periods of rotation and revolution of an orbiting body are equal. (Ch. 8)

tectonic. In geology, associated with stresses in the crust of a planet, often leading to the formation of faults (cracks) and folded ridges; in the case of the Earth, associated with even the large-scale motion of lithospheric plates. (Ch. 9)

thermal inversion. Situation in a planetary atmosphere where local heating causes the temperature to increase with altitude, rather than decreasing as it normally does. (Ch. 13)

thermonuclear fusion. The joining or fusion of atomic nuclei at high temperatures to create a new, more massive atom with the simultaneous release of energy; the main source of energy in the Sun and stars. (Ch. 2)

tidal force. A differential gravitational force that tends to deform a body or to pull it apart due to the tidal effect of its neighbor. (Ch. 8)

tidal heating. Heating of one body by tidal friction or repeated stressing resulting from its motion within the strong tidal field of its neighbor, as in the tidal heating of Io. (Ch. 8)

tidal stability limit. The distance—approximately 2.5 planetary radii from the center—within which differential gravitational forces (or tides) are stronger than the mutual gravitational attraction between two adjacent orbiting objects. Within this limit, fragments are not likely to accrete or assemble themselves into a larger object. Also called the Roche limit. (Ch. 18)

Titius-Bode rule. A numerical scheme by which a sequence of numbers can be obtained that give the approximate distances of the planets from the Sun in astronomical units. Also called Bode's law. (Ch. 1)

trailing side. The hemisphere of a synchronously rotating satellite that always faces backward, away from the direction of its orbital motion. (Ch. 15)

troposphere. Lowest level of the Earth's atmosphere, where most weather takes place; any region in a planetary atmosphere where convection normally takes place. (Ch. 9)

uncompressed density. The density that a planetary object would have if it were not subject to self-compression from its own gravity—hence the density that is characteristic of its bulk material independent of the size of the object. (Ch. 8)

uniformitarianism. The principle in geology that the landforms we see were formed through the action of the same processes that are acting today, continuing over very long spans of time. In the nineteenth century this approach, as opposed to the biblical catastrophism of previous generations, laid the foundations of modern geology. (Ch. 7)

volatile. A substance that has a relatively low boiling temperature. Although usage of this term depends on the context, *volatile* in this text usually refers to substances that are gaseous at temperatures above about 400 to 500 K—such as water or carbon dioxide. (Ch. 3)

volcanic hot spot. Location of a particularly large flow of heat from the interior of a planet; on Io, the name given to the areas covering about 1% of the surface from which most of the internal heat escapes to space. (Ch. 15)

zodiac. A belt around the sky 18 degrees wide and centered on the ecliptic, and within which are found the Moon and planets. (Ch. 1)

zodiacal dust cloud. A tenuous, flat cloud of small silicate dust particles, derived from comets and collisions in the asteroid belt, that pervades the inner solar system. (Ch. 6)

Figure Credits

Chapter One *Opener*, courtesy Hale Observatories; 1.1, Harvard College Observatory; 1.2, 1.3, Dale Smith; 1.4, Donald Goldsmith; 1.7, NASA; 1.9–1.11, Yerkes Observatory; 1.14–1.16, Donald Goldsmith; 1.17, Yerkes Observatory; 1.18, *Le Baiser* (1886), marble sculpture by A. Rodin, S. 1002 in Musée Rodin, Paris, photographed by Bruno Jarret (183.6 cm × 110.5 cm × 118.3 cm). Copyright ARS NY/ADAGP 1987. Courtesy ARS and Musée Rodin; 1.20, 1.21, Donald Goldsmith.

Chapter Two *Opener*, Isabella Gardiner Museum; 2.3, Donald Goldsmith; 2.6, Burndy Library/AIP Neils Bohr Library; 2.9, 2.10, NOAO Solar; 2.11, Yokoh/ISAS; 2.12(a), University of Hawaii photo by M.W. Buie, D.P. Cruikshank, and A. Storrs; 2.12(b), Ann Ronan Picture Library; 2.13, NOAO Solar.

Chapter Three *Opener*, University of Hawaii, Richard Wainscoat; 3.2, David Morrison; 3.4, Donald Goldsmith; 3.7, National Optical Astronomy Observatories; 3.8, Courtesy NRAO/AUI; 3.9, University of Hawaii, Richard Wainscoat; 3.10–3.14, NASA.

Chapter Four *Opener*, courtesy Department Library Services, American Museum of Natural History; 4.1, Mt. Wilson and Las Campanas Observatories, Carnegie Institution of Washington; 4.2, Donald Goldsmith; 4.5, NASA; 4.7, John E. Bortle, W.R. BROOKS OBSERVATORY; 4.8, *New Solar System,* Sky Publishing Co./Cambridge University Press; 4.10, NASA; 4.11–4.13, Smithsonian Institution photos; 4.14, NASA.

Chapter Five *Opener*, NASA; 5.2, Discovered/photographed by Eleanor F. Helin, Palomar Observatory; 5.4, Adapted from data assembled by Benjamin Zellner; 5.5, courtesy Robert Millis (published in the *Astronomical Journal*); 5.7, Reprinted with permission from *Geochim. et Cosmochim. Acta 40*, 701–719, C.R. Chapman (1976) Pergamon Journals, Ltd.; 5.8, University of Arizona photo by Gill McLaughlin; 5.13, courtesy of Karen Meech; 5.16–5.23, NASA.

Chapter Six *Opener*, H. Reitsema and A. Delamere, Ball Aerospace; 6.1, Avec autorisation speciale de la Ville de Bayeux; 6.2, Bill Golisch; 6.3, courtesy of Vern Suomi; 6.4, courtesy of D.K. Yeomans; 6.5, Mt. Wilson and Las Campanas Observatories, Carnegie Institute of Washington; 6.6, Arlene Ammar, CNES; 6.7, Mt. Wilson and Las Campanas Observatories, Carnegie Institute of Washington; 6.8, Palomar Observatory photograph; 6.9, David Morrison; 6.10, H. Reitsema and A. Delamere, Ball Aerospace; 6.12, NASA; 6.13–6.14, Hubble space telescope photo, courtesy of NASA; 6.15, NASA, courtesy of Mark V. Sykes; 6.16, University of Arizona; 6.17, Photos from the SOLWIND satellite coronagraph, Naval Research Laboratory, Washington, D.C., courtesy D.J. Michaels.

Chapter Seven *Opener*, NASA; 7.1, Mt. Wilson and Las Campanas Observatories, Carnegie Institute of Washington; 7.2, Lick Observatory; 7.3, NASA; 7.4, SOVPHOTO; 7.6–7.13, NASA; 7.16, NASA; 7.19, David Morrison; 7.20–7.28, NASA.

Chapter Eight *Opener* (foreground), National Gallery of Art; *Opener* (background), NASA; 8.4, Courtesy of the National Astronomy and Ionosphere Center; 8.5, 8.9, Donald Goldsmith; 8.10–8.13, NASA; 8.14, Yosio Nakamura, from Nakamura *et al., Proc. Lunar Sci. Conf.*, 5th, 2883–2890 (1974); 8.15, Foster, *Earth Science*, © 1982 The Benjamin/Cummings Publishing Co. Reprinted with permission of Addison-Wesley, Reading, MA.

Chapter Nine *Opener*, Johnson Space Center; 9.1, NASA; 7.2–7.3, Adapted from W.K. Hartmann; 9.4 (left), G.W. Stose; 9.4 (right), U.S. Geological Survey; 9.5, David Morrison; 9.6, Dale P. Cruikshank; 9.8, Foster, *Earth Science*, © 1982 The Benjamin/Cummings Publishing Co. Reprinted with permission of Addison-Wesley, Reading, MA.; 9.9, Nafi Toksoz; 9.10–9.12, Robert J. Foster; 9.13, 9.14, NASA; 9.16, Foster, *Earth Science*, © 1982 The Benjamin/Cummings Publishing Co. Reprinted with permission of Addison-Wesley, Reading, MA.; 9.18, Data courtesy of P. Tans, NOAA/CMDL; 9.19, SOVFOTO; 9.20(a), Meteor Crater Enterprises, Northern Arizona, U.S.A.; 9.20(b), NASA; 9.21, C.J. Orth *et al.*, in *Science 214*, 1341–1343 (1981).

Chapter Ten *Opener*, Galleria degli Uffizi, Florence; 10.1, Palomar Observatory photograph; 10.4, Tobias Owen; 10.5, James Pollack, NASA Kuiper Airborne Observatory; 10.6, Based on *Moons and Planets*, Second Edition, by William K. Hartmann, © 1983 by Wadsworth, Inc. Adapted by permission; 10.7, NASA, courtesy of Larry Travis; 10.9(b), S.J. Limaye, Ph.D. thesis, Department of Meteorology, University of Wisconsin, 1977; 10.10, NASA/U.S. Geological Survey, Courtesy Harold Masursky, NASA; 10.11–10.21, NASA; 10.22, courtesy Roald Sagdeyev; 10.23–10.24, Valery Barsukov; 10.25, NASA; 10.26(a), Valery Barsukov; 10.26(b), David Morrison.

Chapter Eleven *Opener*, NASA; 11.1, University of Arizona; 11.2–11.5, NASA; 11.7, U.S. Geological Survey, Courtesy Michael Carr; 11.8, NASA; 11.9, NASA, Courtesy Michael Carr; 11.10, NASA; 11.11, NASA, Courtesy Michael Carr; 11.12, NASA; 11.13, Brown University, Courtesy James W. Head; 11.14–11.16, NASA; 11.17, Adapted from data given by Michael H. Carr in *The Surface of Mars* (Yale University Press, 1981); 11.18–11.30, NASA; 11.31, NASA/U.S. Geological Survey; 11.32–11.33, NASA; 11.34, Based on *Moons and Planets*, Second Edition, by William K. Hartmann, © 1983 by Wadsworth, Inc. Adapted by permission; 11.35, NASA.

Chapter Twelve *Opener*, Galleria degli Uffizi, Florence; 12.1, Goldsmith, D. and Tobias Owen, *The Search for Life in the Universe*, © 1992 Addison-Wesley Publishing Company, Reprinted with permission; 12.2, David Usher, Cornell University; 12.3, Goldsmith, D. and Tobias Owen, *The Search for Life in the Universe*, © 1992 Addison-Wesley Publishing Company, Reprinted with permission; 12.4, courtesy Vern Suomi; 12.5, 12.6, NASA; 12.7, Goldsmith, D. and Tobias Owen, *The Search for Life in the Universe*, © 1992 Addison-Wesley Publishing Company, Reprinted with permission; 12.8, NASA; 12.10, Goldsmith, D. and Tobias Owen, *The Search for Life in the Universe*, © 1992 Addison-Wesley Publishing Company, Reprinted with permission; 12.12, NASA.

Chapter Thirteen *Opener*, NASA; 13.1, 13.2, NASA; 13.4–13.6, NASA; 13.7, Tobias Owen; 13.8, NASA; 13.9, Based on *Moons and Planets*, Second Edition, by William K. Hartmann, © 1983 by Wadsworth, Inc. Adapted by permission; 13.10, NASA; 13.12–13.20, NASA; 13.21, 13.22, Paintings by Davis Meltzer © National Geographic Society; 13.23,13.24, NASA; 13.26, 13.27, NASA; 13.29, NASA.

Chapter Fourteen *Opener*, NASA; 14.2, Charles T. Kowal; 14.3, Lowell Observatory photograph; 14.7, NASA; 14.9, Robert Danehy and Tobias Owen; 14.10, NASA; 14.12–14.16, NASA.

Chapter Fifteen *Opener*, NASA; 15.3–15.5, NASA; 15.6, Roger Clark; 15.7–15.10, NASA; 15.12, 15.13, NASA; 15.14, NASA/U.S. Geological Survey; 15.15, NASA; 15.16, R. Howell, D.P. Cruikshank, and T. Gebelle; 15.17–15.19, NASA; 15.20, NASA/U.S. Geological Survey, Courtesy Alfred McEwen; 15.21, 15.22, NASA.

Chapter Sixteen *Opener*, NASA; 16.1, courtesy Charles Bonestell; 16.2, NASA; 16.4, Virgil Kunde, from Kunde *et al., Nature 292*, 686 (1981); 16.5, NASA; 16.7, © Rosenthal; 16.8–16.10, NASA; 16.12–16.14, NASA; 16.17, NASA.

Chapter Seventeen *Opener*, © W.K. Hartmann, courtesy W. A. Feibelman; 17.1–17.4, NASA; 17.5(a), From W.A. Baum *et al., Icarus 47*, 84–96 (1981); 17.5(b), NASA; 17.6–17.8, NASA; 17.9–17.12, NASA.

Chapter Eighteen *Opener*, NASA; 18.2(a), David Morrison; 18.2(b), NASA; 18.3–18.7, NASA; 18.8, Reproduced with permission from the *Annual Review of Astronomy and Astrophysics*, Vol. 17, © 1979 by Annual Reviews, Inc.; 18.10–18.16, NASA.

Chapter Nineteen *Opener*, Rijksmuseum Amsterdam; 19.1, National Optical Astronomy Observatories; 19.2, Donald Goldsmith; 19.6, Calar Alto Observatory, Hans Zinneckes and Mark McCaughrean; 19.7, R.J. Terrile, Jet Propulsion Laboratory, and B.A. Smith, University of Arizona, Las Campanas Observatory; 19.8, Courtesy A. Vidal-Radjou, IAP, CNRS, INSU, France; 19.11, Seth Shostak/SETI Institute.

Front Cover Background image ©1996 PhotoDisc, Inc. Central Jupiter image courtesy Mt. Stromlo and Siding Spring Observatories, Australian National University.

Index

A Ring of Saturn, 493–496, 503–505
Absorption lines, 38, 39
Accretion, 83
Adams, John Couch, 398, 502
Adams Ring, 502
Albedo
    description of, 67–68, 156–157
    of Enceladus, 68, 473
Allende meteorite, 94, 95, 522
Alpha Centauri, 26, 33, 74–75
Alpha Pictoris, 531
ALSEP (Apollo Lunar Surface Experiments Package), 170, 171
Amalthea, 480, 481
Ancient Greeks, 11
Angstrom (A), 544
Angstrom, Anders, 544
Angular momentum, 81, 519–520, 538
Anighito, 79
Annular eclipse, 8
Anorthosites, 183
Antarctic meteorites, 87, 96
Anticyclonic, 245–246
Aperture, 68
Aphelion, 24
Apogee, 24
Apollo Moon program, 167, 169–173
Arago, François, 398–399
Archean period, 228
Arcturus, 48
Argon, 348, 350, 454
Argyre basin, 307, 308
Ariel, 476–478
Aristarchus Plateau, 187
Armstrong, Neil A., 169, 294
Asteroids
    basic statistics regarding, 107
    beyond Jupiter, 118–120
Asteroids *(continued)*
    collisions of, 121, 129
    composition and appearance of, 126
    C-type, 113–115, 118, 127, 526
    densities of, 59
    differentiated, 64
    discovery of, 104, 106, 124–127
    distinction between comets and, 104, 118
    families of, 113
    Gaspra, 121–123, 128
    Ida, 103, 123–124, 128
    main belt, 105, 106, 111–115
    near Earth, 120–121
    orbits of, 25, 111–112
    origin of, 526
    as parent bodies of meteorites, 88, 99, 105, 115
    physical studies of, 109–110
    radar images of, 124
    size and reflectivity of, 108–109
    size-frequency distribution of, 108
    surface processes of, 127–128
    Trojan, 117–118
    Vesta, 97, 104, 105, 110, 113, 115–116
Astrometry, 533, 534
Astronomical units (AU), 17, 544
Astronomy, 11–18
Aten, 106
Atlas, 483
Atmosphere. *See also specific planets.*
    circulation of, 244–245
    definition of, 225
    differences in, 65
    of Earth, 237–238, 244–245
    Earth's, 237–238
    escape of molecule or atom from, 352–353
Atmosphere *(continued)*
    getting and holding, 64–65
    life and, 238, 332–337
    relationship between ocean and, 235
    structure of, 238–240
    vapor, 464–465
Atmospheric windows, 67
Atoms, 35
Aurora
    description of, 241
    production of, 387–388, 411–412

B Ring of Saturn, 491, 493, 495, 503, 504
Bar, 227
Barnard, E. E., 481
Basaltic meteorites, 97–99
Basalts, 62
Bending waves, 504, 506
Beta Pictoris, 520, 531–532
Big Dipper, 3, 4
Binary stars, 33
Binoculars, 69
Binzel, Richard, 116
Blackbody radiation, 51
Bode, Johann, 397
Bode's law, 27
Bond albedo, 156–157
Borucki, Bill, 536
Botticelli, Sandro, 257, 331
Breccias
    description of, 86
    example of, 87
    parent bodies of, 97, 99
    rocks of lunar highlands as, 182–183
Brown dwarfs, 534
Brownlee, Donald, 95
Brownlee particles, 95

Bruno, Giordano, 19, 32
Bryan, Richard, 537
Bush, George, 295

C Ring of Saturn, 493–494
Calculus, 19
Calderas, 230
Callisto
  basic facts of, 358, 427–428, 454
  compared to Ganymede, 441–442
  composition of, 61, 452, 529
  craters on, 425, 428–429
  density of, 59
  geology of, 428–430
Caloris Basin, 208
Capture theory, 215, 216
Carbon dioxide
  global warming and, 247–248
  as greenhouse gas, 244, 248
  identification spectroscopically of, 291
  life and, 335–336
Carbonaceous meteorites, 93–96
Carr, Michael J., 316, 349
Cassini, 361, 418
Cassini, Giovanni, 492, 503
Cassini Division, 492, 493, 495, 496, 503, 505, 506
Cassiopeia, 15
Catastrophism, 181, 525–526
Catholic Church, 14, 19
Cenozoic period, 228–229
Center of mass, 533
Ceres
  compositional class of, 113
  discovery of, 104
  size of, 107, 113, 114
Cerro Tololo, Chile, 71
CFCs (chlorofluorocarbons), 240, 244
Chaotic rotation, 482
Chaotic terrain, 320
Chapman, Clark R., 110, 111
Charon, 460–464, 521
Chassignites, 97
Chiron, 118, 120, 146
Chondrites, 93. *See also* Primitive meteorites.
Chondrules, 93, 97
Christy, James, 460–461
Chryse Planitia, 297, 299, 302, 303, 309–310
Climate, 243
Cochran, Anita, 154
Coma
  composition of, 141
  description of, 135, 139–140
  tail of, 141–143
Comet Biela, 151
Comet Encke, 147, 152
Comet Hale-Bopp, 136, 138, 144
Comet Halley
  explosive outbursts from, 475
  head of, 140
  nucleus of, 133, 143–147
  orbit of, 28, 118, 119
  plasma tail of, 48
Comet Honda-Mrkos-Pajdusakova, 138
Comet IRAS-Araki-Alcock, 138, 144, 146
Comet Mrkos, 142
Comet Schwassman-Wachmann I, 475
Comet Shoemaker-Levy 9 (S-L 9), 147–150, 155, 252, 381, 382
Comet West, 155
Comets
  aging of, 155–156
  albedo and temperatures of, 156–157
  appearance of, 135–136
  atmosphere of, 139–143
  collisions of, 147–150
  densities of, 59
  description of, 104
  distinction between asteroid and, 104, 118
  dust released by, 149, 151–153
  early study of, 134–135
  formation of, 528
  list of famous, 139
  nineteenth and twentieth century discoveries regarding, 137–138
  nucleus of, 133, 135, 143–149
  orbits of, 25, 138–139
  origin and evolution of, 153–156
  as parent bodies of meteorites or interplanetary dust, 99, 104
  short-period, 153
  sixteenth and seventeenth century discoveries regarding, 136–137
  tails of, 36, 141–143
Composite volcanoes, 230
Compounds, chemical, 36, 94
Comte, August, 36
Conduction, 44
Conservation of angular momentum, 81
Conservation of Energy Law, 41
Constellations
  appearance of, 3–4
  movement of Sun through, 4–5
  in zodiac, 7
Continental crust, 234–235
Continental drift, 231
Convection, 44, 46
Co-orbital satellites, 482–484
Copernican system. *See* Heliocentric system (Sun-centered).
Copernicus, Nicholaus, 12–14, 16, 32, 196, 257
Copernicus (crater), 176, 214–215
Coprates canal, 297
Cordelia, 477, 506
Core, 225, 227, 376
Coriolis effect
  description of, 245
  planetary rotation and, 245–246
  Venus and, 270
Corona, 46, 47, 50, 279, 280
Cosmic rays, 190
Crater retention age, 177, 426–427
Craters/cratering (lunar). *See also* Impact craters.
  absolute and relative ages of, 179–180
  densities and surface ages of, 177, 179
  description of, 165–167
  ejected material from, 175–176
  evolution of, 176–177
  impact origin of, 173–174, 525, 527
  on inner planets, 180–181
  process of impact, 174, 175, 426
  size distribution of, 177–179
Cruikshank, Dale P., 63, 146
Crust, 225–226
C-type asteroids, 113–115, 118, 127, 526
Cyclonic, 245
Cygnus, 4

D Ring of Saturn, 493
Dactyl, 103, 123
Darwin, Charles, 215
Daughter theory, 215–216
Davida, 113, 114
Debris, 104
Decameter radiation, 383
Decimeter radiation, 383
Deimos
  composition and appearance of, 127–129
  discovery of, 125
  orbit of, 126
  origin of, 121–122
Density
  composition and, 61
  metric calculation of, 56–58
  of planets, 58. *See also specific planets.*
Despina, 484
Deuterium, 35, 337
Differentiated meteorites. *See also* Meteorites.
  abundances of, 87

Differentiated meteorites. *(continued)*
basaltic, 97–99
characteristics of, 88
description of, 86
irons and stony-irons as, 96–97
sources of, 98–99
Differentiation, 64
Dione
basic facts regarding, 469, 471, 472
Lagrangian orbit and, 482, 483
naming of, 468–469
Dirty-snowball model, 143, 147, 149, 151
Diurnal motion, 4
Donati's Comet, 48
Doppler effect
applied to Mercury, 201
description of, 199–200
radar and, 218
Double stars, 33
Drake, Frank, 537
Dust tails, 142–143

E Ring of Saturn, 473–475, 497, 507, 509
Earth
absence of craters on, 248–249
circulation of atmosphere on, 244–245
comparison between Venus and, 258
composition of atmosphere of, 237–238
core of, 227
crust of, 225–226
cyclic processes of, 223–224
death of dinosaurs on, 250–252
geologic activity on, 230
geologic timescale of, 227–229
global warming on, 247–248
greenhouse effect on, 243–244, 248
Ice Age on, 246–247
impacting bodies on, 249–250, 252–253
interior of, 224–225
ionosphere of, 240–241
life and atmosphere of, 238, 332–339
magnetosphere of, 241–243, 385
main crustal plates of, 231, 232
mantle of, 226–227
movement of, 7, 33
ocean composition and, 236
ocean temperatures and, 236–237
origin of, 215
plate tectonics and, 230–235
primitive rock and origin of,

Earth *(continued)*
62–64
rotation and Coriolis effect on, 245–246
size and density of, 57–58, 209, 258
stratosphere of, 240
structure of atmosphere of, 238–240
Sun as source of energy for, 243
terrestrial geology of, 223
as viewed from space, 222–223
volcanism on, 229–230
Eccentric ring, 494
Eccentricities
description of, 16
of Nereid, 424, 484
of orbits, 24, 25, 28, 111–112
Eclipses, 7, 8
Ecliptic, 4, 5
Einstein, Albert, 42
Ejecta blankets
description of, 175–176
martian, 306, 307
Electroglow, 411–412
Electromagnetic radiation, 36, 39
Electromagnetic spectrum, 36–37
Electrons, 35
Elements
description of, 34, 35
solar abundances of, 40
states of, 36
Elliot, James, 498
Ellipse
formula for, 27
Kepler's law involving, 16
Embedded satellites, 506
Enceladus
albedo of, 68, 473
basic facts regarding, 469, 472, 473, 484
E Ring and, 473–475, 497, 509
volcanism on, 473
Encke, Johann, 503
Encke Division, 495–496, 506
Energy
early ideas regarding sources of, 41–42
equivalence of mass and, 42
generation and transport of, 44–45
of sun, 40–41, 44–45
through thermonuclear fusion, 42–44
Eos family, 113
Epicycles, 11, 13, 14
Epimetheus, 482, 483
Epsilon Ring, 499, 501, 502, 506, 507
Equinox, 4, 5

Eros, 120–121
Erosion
by ice, 230
impact, 349, 350
of martian craters, 309
Escape velocity, 24, 74
Eta Ring, 501
Eucrites, 97, 115–116
Europa
compared to Io, 442–443
composition of, 61
density of, 59, 433, 529
geology of, 433–434
linear markings on, 434–435
Evolution. *See also* Planetary system origin.
of atmosphere, 334–337
on early Earth, 336
impacts and, 248–252
Exosphere, 64
Exploded planets, 99
Explorer I mission, 241, 242
Exponential notation, 544

F Ring of Saturn, 493, 497, 503, 505
Falls, 85–87
Fault, 234
Finds, 86–87
Fission theory, 215–216
Fluidized ejecta, 306
Fomalhaut, 531
Forward scattering, 502–503
Fra Mauro formation, 167, 184
Fractionation, 81, 337, 521, 522
Fragmentation, 84, 108

G Ring of Saturn, 497, 507, 509
Gaia hypothesis, 336–337
Galilean satellites. *See also* Callisto; Europa; Ganymede; Io
discovery of, 418
naming of, 481
orbital and physical data for, 420–422
Galileo Galilei, 12, 17–18, 22, 32, 163, 398, 399, 418, 419, 492
Galileo spacecraft, 361, 381, 390–392
Galle, Johann, 398, 413, 502
Galle Ring, 502
Ganymede
atmosphere of, 64–65, 454
compared to Callisto, 441–442
composition of, 61, 452, 529
density of, 59
geology of, 430–431, 441–442
internal activity on, 431–433
magnitude of, 485
orbit of, 27–28
size of, 358

Gas, 60, 64
Gas exchange (GEX) experiments, 342–344
Gas retention age, 92
Gaspra
  description of, 122–123, 128
  observations of, 121
GCMS (gas chromatograph–mass spectrometer), 341–342
Geocentric system (Earth-centered), 11, 13, 14
  description of, 11, 13, 14, 197–198
  failures of, 18–19
Geologic timescale, 227–228
Geometric albedo, 157
Geostationary orbit, 23
Gilbert, G. K., 173–174, 311
Granites, 225–226
Gravitational theory, 20–21, 363, 398, 533
Great Comet of 1577, 136, 137
Great Comets of 1531, 1601, and 1682, 136
Great Nebula (Orion), 82
Great Red Spot (GRS), 377, 378, 410
Greenhouse effect
  on Earth, 243–244, 337
  models for, 284–285
  prediction of changes in, 248
  runaway, 266–267, 337
  on Venus, 261–263, 266–267, 285

Hadley cell, 269, 409
Hadley Rille, 188
Halley, Edmund, 136
Halley's Comet, 134
  discovery of, 136
  illustration of, 137
  twentieth century viewings of, 138
Hansen, Candy, 491
Harrington, Robert, 461
Hartmann, William K., 310, 467
Hawaiian Islands, 233, 234, 312
Hektor, 117
Heliocentric system (Sun-centered)
  description of, 13, 14, 197, 198
  theories based on, 13–19
Helium
  discovery of, 40
  thermonuclear fusion and, 41–42
Hellis basin, 297, 299, 311
Helmholtz, Hermann von, 41–42
Henbury iron meteorite, 96
Herschel, 418
Herschel, William, 66, 397
Hess, Seymore L., 326
Hidalgo, 118
Highlands, 164–165
Hodges, Annie, 89
Hubble, Edwin, 72
Hubble space telescope (HST), 72
Huygens, 361, 445
Huygens, Christian, 418, 492, 503
Hydrazine, 22
Hydrogen
  properties of, 59
  as simplest atom, 35
Hydrologic cycle, 238–240
Hydrosphere, 225, 236. *See also* Oceans.
Hygeia, 114
Hyperion
  orbit of, 422–423
  rotation of, 481–482
  shape of, 484
  surface of, 481

Iapetus
  basic facts regarding, 469, 475
  dark coating on, 475–476
  orbit of, 423
Icarus, 121
Ice, 60
Ice Age, 246–247
Ida, 103
  appearance of, 128
  description of, 123–124, 128
Igneous meteorites. *See* Differentiated meteorites.
Igneous rock, 62
Imaging radar, 272–273
Imbrium Basin, 167, 176, 184, 185, 527
Impact basins
  description of, 166
  on Mars, 307–308
Impact craters
  on Mars, 306–307, 349
  on Moon, 173–177, 525. *See also* Craters/cratering.
  on outer planet satellites, 424–427
  on Venus, 273–275
Impact energies, 191–192
Impact erosion, 349, 350
Inclinations, orbital, 24, 25
Inclusions, meteorite, 95
Infrared Astronomy Satellite (IRAS), 72, 121, 531, 532
Ingersoll, Andy, 390
Interamnia, 113, 114
International Astronomical Union Minor Planet Center (Cambridge, Mass.), 106
International Ultraviolet Explorer (IUE), 72
Interplanetary dust particles (IDPs)
  comets as parent bodies of, 99
  description of, 95–96
Io
  compared to Europa, 442–443
  composition of, 61, 435–436
  density of, 59, 435, 529
  orbit of, 383
  shadow on Jupiter of, 8
  tidal heating and, 205
  volcanism on, 75, 358, 437–441, 473
Io plasma torus, 384, 385
Ionization, 35, 240–241
Ionosphere, 240–241
Ions, 35
Iridium, 251–252
Iron meteorites, 85, 87, 96–97
Irregular satellites, 419
Isidis basin, 307
Isochron, 100
Isotopes, 35

Janus, 482–484
Jefferson, Thomas, 84
Jet engine, 22
Jewett, David, 154
Johnson Spaceflight Center (NASA), 87
Juno, 104, 105
Jupiter
  atmosphere of, 364–374, 390–392
  auroras on, 387–388
  calculation of mass of, 27–28
  circulation of atmosphere on, 375–377
  clouds of, 377–381
  composition of, 60, 61, 81
  constructing theoretical models for, 361–362
  core of, 362–363
  density of, 361
  eclipse of Sun on, 8
  Galileo's discoveries regarding, 18–19
  Great Red Spot on, 377, 378, 410
  magnetosphere of, 382–389
  magnitude of, 485
  orbits of, 112, 117–118
  origin of, 83
  overview of, 10–11, 358
  Pioneer missions to, 358, 359
  radio emission from, 382–383
  rings of, 490–492, 509
  rotational periods for, 374–375
  satellites of, 418–422, 479–481. *See also* Galilean satellites; *specific satellites.*
  seasons on, 379
  size and density of, 58, 59
  temperature of, 363–364
  Van Allen belts of, 384–385
  Voyager missions to, 358, 359

Jupiter *(continued)*
 weather and climate of, 373–379

Keats, John, 397
Keeler, James, 503
Kelvin scale, 34
Kepler, Johannes, 12, 15–17, 20, 32, 124, 136
Kepler's laws
 description of, 16–17, 19, 205
 in mathematical form, 27–28, 413
 observation of Mars and development of, 290
Kesei Vallis, 297
Kimberlite, 225
Kirkwood, Daniel, 112
Kirkwood gaps, 112
Koronos family, 113, 124
Krypton, 348
K/T event, 251–252
Kuiper, Gerald P., 154, 418, 446, 484–485
Kuiper Airborne Observatory, 71
Kuiper Belt, 118, 154, 156, 528, 532

Labeled release (LR) experiments, 342–344
Lagrange, Joseph Louis, Comte, 117
Lagrangian orbits, 117–118, 482, 483
Lakshmi Plateau, 276, 277
Landers, 76
Larissa, 484
Law of Conservation of Energy, 41
Leading side, 426
Leverrier, Jean Joseph, 398, 502
Leverrier Ring, 502
Life
 evolution of atmosphere and, 334–337
 on Mars, 340–346, 351–352
 origin of, 332–334
 searches for, 536–537
 searching for, 338–340
Liquid fuel rockets, 22
Lithosphere, 226
Long-period comets, 138–139
Lovelock, James, 336
Lowell, Percival, 291, 399, 434
Luminosity
 description of, 32, 485
 of Sun, 34, 41, 43, 45, 46
Luu, Jane, 154
Lyell, Charles, 525
Lyra, 4

Magellan radar mapper, 272–273, 278, 280, 283
Magma, 188
Magnetic breaking, 519–520
Magnetic field
Magnetic field *(continued)*
 in core of Earth, 227
 in magnetosphere, 242–243. *See also* Magnetosphere.
Magnetic flux tube, 383
Magnetopause, 242–243
Magnetosphere
 description of, 225, 241–243, 382
 dimensions of, 387
 of Earth, 241–243, 385, 387
 energy to power, 387
 of Jupiter, 382–389
 of Neptune, 411–413
 of Saturn, 384–387, 441
 of Uranus, 411–413
Magnetotail, 387
Magnitude scale, 485
Main belt asteroids
 collisions among, 121
 compositional classes of, 113–115
 description of, 106, 111
 families of, 113
 illustration of, 105
 large-size, 113, 114
 orbits and resonances of, 111–112
Manicougan Lakes Crater, 250
Mantle
 description of, 225
 of Earth, 226–227
Mare Crisium, 166
Mare Imbrium, 166, 167, 186
Mare Orientale, 166
Mare Serenitatis, 166
Mare Tranquillitatis, 180
Margulis, Lynn, 336
Maria, 165, 186–187
Mariner 2, 260, 292
Mariner 4, 292–293
Mariner 6, 293
Mariner 7, 293
Mariner 9, 293–294, 302, 311
Mariner 10, 205–207
Marius, Simon, 481
Marius Hills, 187
Mars
 atmosphere of, 65, 292–293, 324–327, 346–351
 bulk properties of, 58, 295–296
 canal controversy and, 291–292
 canyons on, 314–315
 channels on, 316–320
 composition of, 61
 crater densities on, 310
 early studies of, 291–292
 eclipse of Sun on, 8, 9
 effect of impacts on, 349–350
 first flybys for, 292–293
 geological features of, 299–300
 impact basins on, 307–308
 impact craters on, 306–307, 349
Mars *(continued)*
 large-scale topography of, 298–299
 movement of, 10, 16
 overview of, 9–10, 290
 polar regions on, 320–324
 recent studies of, 295
 samples from, 301
 satellites of, 420, 421. *See also* Deimos; Phobos
 search for life on, 340–346, 351–352
 seasonal cycle of, 292
 soil of, 345
 surface age of, 308–310
 surface elevations on, 297–298
 surface material of, 300–301
 surface nomenclature for, 296–297
 surface studies of, 293–294, 301–306
 tectonic forces on, 313–314
 volcanism on, 299–300, 310–313, 316
 water on, 301, 319–320, 327, 349
Mars Pathfinder, 295
Martin, Jim, 75
Mass, center of, 533
Mass extinction, 251–252
Masursky, 311
Matter, categories of, 59–61
Mauna Kea, Hawaii, 55, 71
Mauna Loa, Hawaii, 229–230
McCord, Thomas B., 110
Mercury
 appearance of, 196
 atmosphere of, 65, 212–213
 composition of, 60–61, 196, 206
 craters on, 207–208
 days and seasons on, 202
 density of, 58, 196, 209–210
 geocentric and heliocentric perspectives on, 197–198
 geological history of, 213–215
 geology of, 207–208
 heat flow on, 211
 magnetic field of, 211–212
 movement of, 10
 orbit of, 24, 196
 origin of, 217–218
 overview of, 8
 polar caps on, 207
 rotation of, 200–202, 218
 surface of, 206–207
 surface temperature on, 207
 synodic period of, 219
 tides and spin of, 202–205
 as viewed from Mariner 10, 205–207
Mesozoic period, 228
Metal, 60–61

Metamorphic rock, 62
Meteor Crater, 249–250
Meteor showers, 151
Meteorites
- age of, 91–92, 97
- Antarctic, 87
- as asteroids or exploded planets, 88, 99, 115
- classification of, 84–90
- description of, 80, 84
- differentiated, 86, 96–99. *See also* Differentiated meteorites.
- effect of humans of, 88–90
- falls of, 85–87
- finds of, 86–87
- lunar, 98
- orbits and origins of, 88
- parent bodies of, 85, 88, 97–99, 104
- primitive, 86, 88, 92–96, 98. *See also* Primitive meteorites.
- radioactive age dating and, 99–100
- SNC, 97–98, 301, 316, 349
- striking Earth, 249–250

Meteoroids
- description of, 80
- lunar craters and, 173–174
- outer solar system and, 425–426

Meteors, 80, 151, 152
Metric system, 543–544
Midplane, 521
Miller, Stanley, 333
Mimas
- basic facts regarding, 469, 472, 484
- Cassini Division and, 506
- naming of, 468–469

Minerals, 62. *See also* Rock.
Miranda, 476–479, 485
Molecules, 36
Month, 6, 7
Moon. *See also* Craters/cratering.
- albedo of, 68
- apparent motion of, 5–6, 10
- appearance of craters on, 165–167
- atmosphere of, 65, 212–213
- basins of, 184–185
- calculation of impact energies on, 191–192
- catastrophism and, 181–186
- comparison of Mercury to, 196
- composition of, 61, 196
- dating craters on, 177–181
- density of, 209–210
- direct observation of, 73
- early history of, 214
- eclipse of, 7, 8
- expeditions to, 167–173
- highlands of, 164–165, 182–184, 275
- impact craters on, 173–177
- interior of, 210–212

Moon *(continued)*
- nomenclature for surface features on, 207
- origin of, 215–217
- phases of, 6–7
- possibility of ice deposit near north pole of, 207
- regolith of, 189–190
- resolution of surface of, 162–164
- rotation periods linked to orbital motion of, 202–205
- size of, 162
- soil of, 190
- stratigraphic studies of, 167
- surface of, 162–164, 188–191
- valleys and mountains of, 187
- volcanism on, 186–188, 214

Moonquakes, 210
Morabito, Linda, 75
Morrison, David, 111
Morrison asteroid, 106
Motion, Newton's laws of, 20, 33
Mount Fuji, Japan, 230
Mr. Spock, 106
M-type asteroids, 113–115
Murchison meteorite, 94–95, 475
Mutch, Thomas, 304–306

Naiad, 485
Nakhlites, 97, 98
Near-Earth asteroids, 120–121
Neon, 348
Neptune
- atmosphere of, 403–411
- basic properties of, 400–403
- clouds of, 410
- composition of, 60, 61, 395, 404–406
- core of, 362
- direct exploration of, 396
- discovery of, 397–399, 413
- Great Dark Spot on, 410
- magnetosphere of, 411–413
- magnitude of, 485
- rings of, 502–503, 506, 509
- rotation of, 411
- satellites of, 422–424, 479, 484. *See also* Nereid; Triton.
- seasons on, 408
- size and density of, 39, 58
- temperature of, 406–408
- weather on, 409–410

Nereid
- basic facts regarding, 484
- eccentricity of, 424, 484

Neutron stars, 535
Neutrons, 35
Newton, Sir Isaac, 19–22, 27, 33, 137, 363
Newton's laws, 19–21, 33, 533
Nicholson, Seth, 418
Nirgal Vallis, 297
Nix Olympica, 297
Noble gases, 348, 351
Noctis Labyrinthus, 314–315
Nonthermal radiation, 383
Normal reflectivity, 156
North, 24–25
Nuclear energy, 42

Oberon, 476, 477
Observatory sites
- locations for favorable, 70–71
- orbiting, 72

Occultations, 371, 498–501, 503
Oceans
- composition of, 236
- formation and destruction of crust of, 233–234
- relationship between atmosphere and, 235
- temperature of, 236–237

Oceanus Procellarum, 186, 187
Olympus Mons, 297, 298, 300, 311–312, 316
Oort, Jan, 153–154
Oort comet cloud, 153–154, 156, 528
Ophelia, 477, 506
Opposition, 10
Orbit
- of asteroids, 25, 111–112
- description of, 23, 24
- eccentricities of, 16, 24, 25, 28, 111–112, 424, 484
- geostationary, 23
- Lagrangian, 117–118, 482, 483

Orbital velocity, 23, 24
Orbiters, 76
Organic compounds, 94
Orientale Basin, 166, 184–185, 214, 527
Orion, 3, 4, 516, 517
Orion Nebula, 3, 526
Osmium, 251–252
Ostro, Steven, 124
Outflow channels, 318–320
Oxidizing, 59
Oxygen, 59
Ozone, 66, 240
Ozonosphere, 240

Pallas
- compositional class of, 113
- discovery of, 104, 105
- size of, 113, 114

Pallasites, 97
Pandora, 483, 505

Pangaea, 234
Parallax, 33
Parent bodies
  asteroids or exploded planets as, 98–99, 104
  description of, 85
  of differentiated meteorites, 97–99, 115
  of primitive meteorites, 98, 115
Patientia, 113, 114
Payne, Cecilia, 40
Peary, Robert, 79
Perigee, 24
Perihelion
  calculation of, 28
  description of, 24
  Sun at, 202
Perturbations, 27, 398, 399, 490
Phaethon, 121
Phanerozoic period, 228, 252, 334
Phase integral, 157
Phobos
  composition and appearance of, 127–129
  discovery of, 125
  escape velocity from, 24
  gravitational field of, 23
  orbit of, 126
  origin of, 121–122
  shadow on Mars by, 8, 9
Phoebe, 423, 476, 481, 527
Pholus, orbit of, 118
Photochemical smog, 380, 450–451
Photochemistry, 264–265
Photodissociation, 336
Photons, 36
Photosphere, 34
Piazzi, Guiseppe, 104
Piazzia, 106
Pic du Midi, France, 71
Pickering, William, 242
Pioneer 10, 74, 358, 359
Pioneer 11, 74, 358, 359
Pioneer Venus, 270, 271
Planck Radiation Law, 51
Planetary embryos, 524
Planetary motion
  ancient interpretations of, 11
  direction of, 24–25
  Kepler's Laws of, 16–17, 290
  retrograde, 10
The Planetary Society, 537
Planetary system origin
  angular momentum and, 81, 519–520, 538
  evolution of disk and, 520–525
  impacts and, 525–528
  life of stars and, 516–519
  overview of, 514
  properties of planetary system and, 514–516
  satellite and ring systems and,
Planetary system origin *(continued)*
  528–530
Planetary systems
  basic properties of, 514–516
  dimensions of, 26
  direct observation of, 72–76
  distant observation of, 65–72
  search of other, 530–537
Planetesimals, 83, 523–524, 526
Planets. *See also specific planets.*
  atmosphere of, 64–65
  basic properties of, 56–59
  building giant, 389–390
  chemistry of, 59–61
  classification of rocks and, 61–64
  concept of cratering on inner, 180–181
  discoveries of outer, 396–400
  distances between, 26, 27
  extrasolar, 534–535
  growth of giant, 528
  interiors of, 208
  orbits of, 25
  overview of, 8–11
  thermal evolution of, 213–214
Plasma, 47, 240
Plasma tail, 141–142
Plate tectonics
  description of, 231
  on Earth, 230–235, 258
  Venus and, 258, 277
Plescia, Jeff, 491
Pluto
  basic properties of, 400, 401, 403
  composition of, 61
  discovery of, 399–400
  exploration of, 396, 397
  magnitude of, 485
  mass of, 462
  orbit of, 24
  as planet, 460
  satellite of, 460–462
  size and density of, 58, 396
  surface and atmosphere of, 462–464
Pollack, James, 262, 301
Pope, Alexander, 20
Prebiotic chemistry, 449
Precambrian period, 228
Primitive meteorites. *See also* Meteorites.
  carbonaceous, 93–94
  characteristics of, 88
  composition and history of, 93
  description of, 86
  inclusions in, 95
  interplanetary dust and, 95–96
  organic matter in, 94–95
  solidification ages of, 92
  sources of, 98
Primitive rock, 62–64
Probes, 76
Project Phoenix, 537
Prometheus, 483, 505
Proterozoic period, 228
Proteus, 484
Proton-proton chain reaction, 42
Protons, 35
Protoplanets, 83, 84, 529
Psyche, 113
Ptolemaic system. *See* Geocentric system (Earth-centered)
Ptolemy, Claudius, 7, 11, 13
Puck, 481, 484
Pulsar, 535
Pyrolitic release (PR) experiments, 342–345

Radar
  Doppler effect and, 218
  imaging, 272–273
  used for asteroids, 124
  used to view Mercury, 199–200
  used to view Venus, 272–273
Radial mixing, 522
Radiation. *See also* Electromagnetic radiation; Magnetosphere; Thermal radiation.
  analysis of, 65
  description of, 44
  laws of, 50–51
  nonthermal, 383
  thermal, 50–51
Radio telescopes
  application of, 259–260, 407–408
  description of, 69–70
Radioactive age dating, 99–100
Radioactive decay rate, 91, 92
Radioactive half-life, 91
Radioactivity, 91
Reducing, chemically, 59
Reflecting telescopes, 69
Reflectivity, 156
Regolith, 189–190, 207
Regular satellites, 419
Relativity theory, 42
Remote sensing, 65
Resolution, 163
Resonance gaps, 112–113
Resonances, 112
Retrograde motion, 10, 14
Retrograde rotation, 25
Rhea
  basic facts regarding, 469–470, 472
  craters of, 424, 470
  naming of, 468–469
Rifts, ocean, 233

Righetti, Francesco, 195
Rigid body rotation, 387
Rings
embedded satellites and, 506
of Jupiter, 490–492
of Neptune, 502–503
origin of, 507–509, 529–530
particles of, 492–493, 497, 502–504
of Saturn, 492–498
sharp edges of, 505
shepherd satellites and, 505–506
structure of, 504–507
tidal stability limit and, 507–510
of Uranus, 498–502
waves in, 503–504
Roche, Edouard, 507
Roche limit, 507–510
Rock
dating of, 91–92, 99–100
description of, 60
granitic, 225–226
origin and classification of, 61–64
suboxidized, 334–335
Rockets
description of, 22
escape velocity and, 24
orbit of, 23
Ropy pahoehoe lava, 62, 63
Rotation
chaotic, 482
Coriolis effect and, 245–246
retrograde, 25
rigid body, 387
synchronous, 205
Runaway greenhouse effect, 266–267, 337
Runoff channels
description of, 316–317
origin of, 317–318

Safronov, V. A., 523–524
Sagan, Carl, 262, 301, 536, 537
Sagdeyev, Roald, 143
San Andreas fault, 234–235
Satellites. *See also specific satellites.*
co-orbital, 482–484
densities of, 58–59
discovery of, 418–419
embedded, 506
formation of, 528–529
list of, 420–422
regular and irregular, 419
revolution of, 25
searching for new, 485–486
shepherd, 505–506
Saturn
atmosphere of, 364–374
circulation of atmosphere on, 375–377
composition of, 60, 61, 81
constructing theoretical models for, 361–362
core of, 362–363
density of, 361
direct observation of, 73
helium rain on, 364
magnetosphere of, 384–387, 411
motion of, 25
origin of, 83
overview of, 10, 11, 358
Pioneer missions to, 360
radio emission from, 384
rings of, 473–475, 492–498, 503–509
rotational periods for, 374, 375
satellites of, 420–423, 468–476, 481–484, 521, 529. *See also specific satellites.*
seasons on, 379
size and density of, 58
temperature of, 363, 364
ultraviolet auroras on, 388–389
Van Allen belts of, 384–385
Voyager missions to, 360
weather and climate of, 373–379
Schiaparelli, Giovanni, 291
Schmitt, Harrison H., 161, 169
Schwabe, Heinrich, 49
Scientific model, of planetary history, 213
Search for extraterrestrial intelligence (SETI), 537
Seasonal cycles, 4, 245
Secondary craters, 176
Sedimentary rock, 62
Sedimentation, 230, 323
Seismic waves, 224
SETI Institute, 537
SETI (search for extraterrestrial intelligence), 537
Shakespeare, William, 134
Shepherd satellites, 505–506
Shergorrites, 97
Shield volcanoes, 229–230
Shoemaker, Eugene, 181, 425–426
Short-period comets, 153
Silicates, 60
Sinous rilles, 187
Sister theory, 215, 216
61 Cygni, 33
Smith, Bradford, 531
SNC meteorites
composition of samples of, 301, 316, 349
description of, 97–98
Soderblom, Larry, 427
Solar cycle, 49–50
Solar day, 201
Solar nebula. *See also* Planetary system origin.
accretion and fragmentation and, 83–84
aggregation of grains in, 523
angular momentum and, 519, 538
condensation in, 81–83, 522, 523
description of, 60, 61, 81, 332–333
dissipation of, 525
final stages of, 84, 95
Solar system
fundamental properties of, 81
history of, 60
scale model of, 25–26
Solar wind, 47–48, 520
Solid fuel rockets, 22
Solidification age, 92
Solstice, 4, 5
Space Shuttle, 22
Spacecraft. *See also specific spacecraft.*
basic operation of, 75–76
direct observation by, 73–75
Spectral analysis, 39–40
Spectral lines, 37–39
Spectrometer, 37–39
Spectrometry, 109–111
Spectroscopy
application of, 39, 40, 492
description of, 37, 39, 65–67, 533
Spencer, John, 440
Spicules, 46
Spielberg, Steven, 537
Spiral density waves, 503–504
Sputtering, 385
Stefan-Boltzmann Law, 51
Stone, Edward, 360
Stony meteorites, 85–86, 114. *See also* Primitive meteorites.
Stony-iron meteorites, 86, 87, 96–97
Stratigraphy, 167
Stratosphere, 240
Stromatolites, 334
S-type asteroids, 113–115
Subduction, 233
Subnebula, 419
Sun
apparent motion of, 4–5, 11
basic properties of, 33–34
chemistry of, 34–36
composition of, 36–40, 81
eclipse of, 7, 8
energy of, 40–43
heating of Earth by, 243
interior structure of, 45
Kepler's laws regarding, 16, 17
life history of, 43–45
radiation laws and, 50–51
solar wind and, 47–48
as star, 32–34
sunspots and, 48–50

surface activity of, 46, 47
Sunlight, reflected, 67–68
Sunspot cycle, 49–50
Sunspots
description of, 48–49
illustration of, 46
solar cycle and, 49–50
Supernovas, 517, 535
Swift, Jonathan, 124–125
Synchronous rotation, 205
Synodic period, 218–219

Tarter, Jill, 537
Taylor, Stuart Ross, 183
Tectonics. *See also* Plate tectonics.
description of, 231
martian, 313–314
Venus and, 275–277
Telescopes
invention of, 18
types of, 68–70
viewing Mars with, 291
viewing Moon with, 163
Terrile, Richard, 531
Tethys
basic facts regarding, 469, 471, 472
Lagrangian orbit and, 482, 483
naming of, 468–469
Tharsis region, 297–298, 300, 311–313, 316
Themis family, 113
Theory of Relativity, 42
Thermal inversion, 372–373
Thermal radiation
description of, 50–51
detection of asteroids from, 107
measurement of, 259–260
Thermonuclear fusion, 42–44
Thomas Mutch Memorial Station, 304–305
Thomson, William, 41–42
Thorium, 99
Tidal forces, 203, 507–508
Tidal heating, 205, 473
Tidal stability limit, 507–510
Tides
friction due to, 203–205
Mercury and effect of solar, 205
nature of, 203
Titan
atmosphere of, 64, 65, 358, 446, 448, 451–454, 529
composition of, 61
discovery of, 418
magnitude of, 485
size of, 422, 446
smog layer on, 380, 450–451
surface of, 452
temperature measurements of, 446–448
Voyager discoveries regarding, 448–450
Titania, 476, 477
Titius, Johann Daniel, 27
Titius-Bode Rule, 27, 104, 398
Tombaugh, Clyde, 399, 400
Toutatis, 124
Trailing side, 426
Triangulation, 32–33
Triton
atmosphere of, 457–460, 462–463
basic facts of, 455–456, 460
composition of, 61, 521
magnitude of, 485
orbit of, 424
Voyager discoveries regarding, 456–457, 484
Trojan asteroids, 117–118
Troposphere, 238
Tunguska (Russian Siberia), 249
Tycho Brahe, 12, 14, 136–137, 167
Tycho (lunar crater), 167, 169, 177, 252

U.S. Viking, 8
Umbriel, 476–478
Uncompressed density, 208
Uniformitarianism, 181
Universe, relevance of, 2–3
Uranium, 99
Uranus
atmosphere of, 403–411
basic properties of, 400–403
circulation of atmosphere on, 408, 409
composition of, 60, 61, 404–406
core of, 362
direct exploration of, 396
discovery of, 27, 66, 397
magnetosphere of, 411–413
magnitude of, 485
orbit of, 397–398
rings of, 498–502, 505–507, 509
rotation of, 410–411
satellites of, 420–424, 476–479, 484, 521, 527, 529
size and density of, 58, 396
temperature of, 406–408, 411
Urey, Harold, 215, 333

Valles Marineris, 289, 297, 300, 314, 315
Van Allen, James, 241–242, 382
Van Allen belts, 242, 384–385
Vapor atmospheres, 464–465
Vega, 33, 531
Venera missions, 261, 265–266, 272, 280–283
Venus
Venus *(continued)*
atmospheric circulation of, 269–270
basic properties of, 61, 258–259
clouds of, 259, 264–266
comparison between Earth and, 258
composition of atmosphere of, 65, 244, 263–264, 350
continents and mountains on, 276–277
coronae on, 279, 280
density and surface age of craters on, 275
direct observation of, 76
early radar observations of, 260–261
geological history of, 280
greenhouse effect on, 244, 261–262, 266–267, 284–285, 337
impact craters on, 273–275
large-scale topography on, 270–272
life and, 337
Magellan radar imaging of, 272–273, 278, 280, 283
magnitude of, 485
mass of atmosphere on, 262–263
movement of, 10
orbit of, 196
overview of, 8–9
phases of, 18, 19
rotation of, 200, 260, 261, 269–280
size and density of, 58, 258
structure of atmosphere on, 268–269
surface composition of, 283–284
surface images of, 281–283
surface temperature of, 259–260, 280, 285, 350
tectonic features on, 275–276
upper atmosphere on, 267–268
Venera missions to, 261, 265–266, 272, 280–283
volcanic plains on, 277–278
volcanoes and lava flows on, 278–279
weather on, 267–270
Vesta, 97
compositional class of, 113
discovery of, 104, 105
eucrite as parent body of, 115–116
remote sensing studies of, 110
size of, 113
volcanic activity on, 116
Vidal-Madjar, Alfred, 532
Viking 1, 289, 302, 303, 312, 313
Viking 2, 303, 304, 312, 313, 321

Viking landers, 290, 294–295, 301–302, 324, 340–346
Volatiles, 60, 351
Volcanic crater hypothesis, 173–174
Volcanic hot spots, 439
Volcanism
  on Earth, 229–230
  on Io, 75, 358, 437–441, 473
  on Mars, 299–300, 310–313, 316
  meteorite formation due to, 99
  on Moon, 186–188, 214
  on Venus, 265, 272, 277–279
Volcanoes
  along rift and subduction zones, 233
  composite, 230
  shield, 229–230, 311
Von Braun, Werner, 242
Voyager missions, 358–361, 372, 373, 388, 389, 396, 403, 404, 409, 424, 438, 448–450, 453, 456–457, 475, 478, 481–485, 490–498, 500, 501–506, 508, 509

Water cycle, 238–240
Wavelengths, 544
Weather
  circulation of atmosphere and, 244–245
  Coriolis effect and, 245–246
  description of, 243
  stemming from troposphere, 238
Wetherill, George, 217–218, 524
Whipple, Fred L., 143, 146
Wien Radiation Law, 51
Wien's Law, 260

Xenon, 348, 350

Year, 5

Zellner, Benjamin, 111
Zodiac
  constellations in, 7
  description of, 5–6
Zodiacal dust cloud, 152–153, 509
Zodiacal light, 153

PLANETARY EXPLORATION SPACECRAFT: MOST IMPORTANT HISTORICAL MISSIONS

Spacecraft	Country of Origin	Launch Date
Sputnik 1	USSR	4 October 1957
Sputnik 2	USSR	3 November 1957
Explorer 1	U.S.	31 January 1958
Luna 2	USSR	12 September 1959
Luna 3	USSR	4 October 1959
Mariner 2	U.S.	27 August 1962
Mars 1	USSR	1 November 1962
Ranger 7	U.S.	18 July 1964
Mariner 4	U.S.	28 November 1964
Mariner 5	U.S.	14 June 1965
Venera 3	USSR	16 November 1965
Luna 9	USSR	31 January 1966
Luna 10	USSR	31 March 1966
Surveyor 1	U.S.	30 May 1966
Lunar Orbiter 1	U.S.	10 August 1966
Venera 4	USSR	12 June 1967
Mariner 6	U.S.	24 February 1969
Venera 7	USSR	17 August 1970
Luna 16	USSR	12 September 1970
Mars 2	USSR	19 May 1971
Mariner 9	U.S.	30 May 1971
Pioneer 10	U.S.	2 March 1972
Venera 8	USSR	26 March 1972
Pioneer 11	U.S.	5 April 1973
Mariner 10	U.S.	3 November 1973
Venera 9	USSR	8 June 1975
Venera 10	USSR	14 June 1975
Viking 1	U.S.	20 August 1975
Viking 2	U.S.	9 September 1975
Voyager 1	U.S.	5 September 1977
Voyager 2	U.S.	20 August 1977
Pioneer Venus Orbiter	U.S.	20 May 1978
Pioneer Venus Probe Carrier	U.S.	8 August 1978
Venera 11	USSR	9 September 1978
Venera 12	USSR	12 September 1978
Venera 13	USSR	30 October 1981
Venera 14	USSR	4 November 1981
Venera 15	USSR	2 June 1983
Venera 16	USSR	7 June 1983
VEGA 1	USSR	15 December 1984
VEGA 2	USSR	21 December 1984
Giotto	ESA*	2 July 1985
Magellan	U.S.	4 May 1989
Galileo	U.S.	18 October 1989
NEAR	U.S.	16 February 1996(?)
Mars Global Surveyor	U.S.	1996 (?)
Mars Pathfinder	U.S.	1996 (?)
Mars 96	Russia	1996 (?)

*ESA = European Space Agency, a consortium of European countries.